MW00997328

The Electric Power Engineering Handbook

ELECTRIC POWER GENERATION, TRANSMISSION, AND DISTRIBUTION

THIRD EDITION

The Electric Power Engineering Handbook

Third Edition

Edited by

Leonard L. Grigsby

Electric Power Generation, Transmission, and Distribution, Third Edition

Edited by Leonard L. Grigsby

Electric Power Transformer Engineering, Third Edition

Edited by James H. Harlow

Electric Power Substations Engineering, Third Edition

Edited by John D. McDonald

Power Systems, Third Edition

Edited by Leonard L. Grigsby

Power System Stability and Control, Third Edition

Edited by Leonard L. Grigsby

The Electric Power Engineering Handbook

ELECTRIC POWER GENERATION, TRANSMISSION, AND DISTRIBUTION

THIRD EDITION

EDITED BY

LEONARD L. GRIGSBY

CRC Press is an imprint of the
Taylor & Francis Group, an **informa** business

CRC Press
Taylor & Francis Group
6000 Broken Sound Parkway NW, Suite 300
Boca Raton, FL 33487-2742

Printed in the United States of America on acid-free paper
Version Date: 20111104

International Standard Book Number: 978-1-4398-5628-4 (Hardback)

Library of Congress Cataloging-in-Publication Data

Electric power generation, transmission, and distribution / editor, Leonard L. Grigsby. -- 3rd ed.
p. cm. -- (Electric power engineering series)
Includes bibliographical references and index.
ISBN 978-1-4398-5628-4 (hardback)
1. Electric power production. 2. Electric power distribution. 3. Electric power transmission. I. Grigsby, Leonard L.

TK1001.E25 2013
621.31--dc23 2011044128

Visit the Taylor & Francis Web site at
http://www.taylorandfrancis.com

and the CRC Press Web site at
http://www.crcpress.com

Contents

PART III Transmission System

PART IV Distribution Systems

PART V Electric Power Utilization

PART VI Power Quality

Preface

The generation, delivery, and utilization of electric power and energy remain one of the most challenging and exciting fields of electrical engineering. The astounding technological developments of our age are highly dependent upon a safe, reliable, and economic supply of electric power. The objective of the Electric Power Engineering Handbook is to provide a contemporary overview of this far-reaching field as well as a useful guide and educational resource for its study. It is intended to define electric power engineering by bringing together the core of knowledge from all of the many topics encompassed by the field. The chapters are written primarily for the electric power engineering professional who seeks factual information, and secondarily for the professional from other engineering disciplines who wants an overview of the entire field or specific information on one aspect of it.

The first and second editions of this handbook were well received by readers worldwide. Based upon this reception and the many recent advances in electric power engineering technology and applications, it was decided that the time was right to produce a third edition. Because of the efforts of many individuals, the result is a major revision. There are completely new chapters covering such topics as FACTS, smart grid, energy harvesting, distribution system protection, electricity pricing, linear machines. In addition, the majority of the existing chapters have been revised and updated. Many of these are major revisions.

The handbook consists of a set of five books. Each is organized into topical parts and chapters in an attempt to provide comprehensive coverage of the generation, transformation, transmission, distribution, and utilization of electric power and energy as well as the modeling, analysis, planning, design, monitoring, and control of electric power systems. The individual chapters are different from most technical publications. They are not journal-type articles nor are they textbooks in nature. They are intended to be tutorials or overviews providing ready access to needed information while at the same time providing sufficient references for more in-depth coverage of the topic.

This book is devoted to the subjects of power system protection, power system dynamics and stability, and power system operation and control. If your particular topic of interest is not included in this list, please refer to the list of companion books referred to at the beginning.

In reading the individual chapters of this handbook, I have been most favorably impressed by how well the authors have accomplished the goals that were set. Their contributions are, of course, key to the success of the book. I gratefully acknowledge their outstanding efforts. Likewise, the expertise and dedication of the editorial board and section editors have been critical in making this handbook possible. To all of them I express my profound thanks.

They are as follows:

- Nonconventional Power Generation — Saifur Rahman
- Conventional Power Generation — Rama Ramakumar
- Transmission Systems — George G. Karady
- Distribution Systems — William H. Kersting

- Electric Power Utilization — Andrew P. Hanson
- Power Quality — S. Mark Halpin
- *Transformer Engineering* (a complete book) — James H. Harlow
- *Substations Engineering* (a complete book) — John D. McDonald
- Power System Analysis and Simulation — Andrew P. Hanson
- Power System Transients — Pritindra Chowdhuri
- Power System Planning (Reliability) — Gerry Sheblé
- Power Electronics — R. Mark Nelms
- Power System Protection — Miroslav M. Begovic*
- Power System Dynamics and Stability — Prabha S. Kundur†
- Power System Operation and Control — Bruce Wollenberg

I wish to say a special thank-you to Nora Konopka, engineering publisher for CRC Press/Taylor & Francis, whose dedication and diligence literally gave this edition life. I also express my gratitude to the other personnel at Taylor & Francis who have been involved in the production of this book, with a special word of thanks to Jessica Vakili. Their patience and perseverance have made this task most pleasant.

Finally, I thank my longtime friend and colleague—Mel Olken, editor, the *Power and Energy Magazine*—for graciously providing the picture for the cover of this book.

* Arun Phadke for the first and second editions.

† Richard Farmer for the first and second editions.

Editor

Leonard L. ("Leo") Grigsby received his BS and MS in electrical engineering from Texas Tech University, Lubbock, Texas and his PhD from Oklahoma State University, Stillwater, Oklahoma. He has taught electrical engineering at Texas Tech University, Oklahoma State University, and Virginia Polytechnic Institute and University. He has been at Auburn University since 1984, first as the Georgia power distinguished professor, later as the Alabama power distinguished professor, and currently as professor emeritus of electrical engineering. He also spent nine months during 1990 at the University of Tokyo as the Tokyo Electric Power Company endowed chair of electrical engineering. His teaching interests are in network analysis, control systems, and power engineering.

During his teaching career, Professor Grigsby received 13 awards for teaching excellence. These include his selection for the university-wide William E. Wine Award for Teaching Excellence at Virginia Polytechnic Institute and University in 1980, the ASEE AT&T Award for Teaching Excellence in 1986, the 1988 Edison Electric Institute Power Engineering Educator Award, the 1990–1991 Distinguished Graduate Lectureship at Auburn University, the 1995 IEEE Region 3 Joseph M. Beidenbach Outstanding Engineering Educator Award, the 1996 Birdsong Superior Teaching Award at Auburn University, and the IEEE Power Engineering Society Outstanding Power Engineering Educator Award in 2003.

Professor Grigsby is a fellow of the Institute of Electrical and Electronics Engineers (IEEE). During 1998–1999, he was a member of the board of directors of IEEE as the director of Division VII for power and energy. He has served the institute in 30 different offices at the chapter, section, regional, and international levels. For this service, he has received seven distinguished service awards, such as the IEEE Centennial Medal in 1984, the Power Engineering Society Meritorious Service Award in 1994, and the IEEE Millennium Medal in 2000.

During his academic career, Professor Grigsby has conducted research in a variety of projects related to the application of network and control theory to modeling, simulation, optimization, and control of electric power systems. He has been the major advisor for 35 MS and 21 PhD graduates. With his students and colleagues, he has published over 120 technical papers and a textbook on introductory network theory. He is currently the series editor for the Electrical Engineering Handbook Series published by CRC Press. In 1993, he was inducted into the Electrical Engineering Academy at Texas Tech University for distinguished contributions to electrical engineering.

Editor

Contributors

Math H.J. Bollen
Swedish Transmission Research Institute
Ludvika, Sweden

Simon W. Bowen
Alabama Power Company
Birmingham, Alabama

Robert P. Broadwater
Department of Electrical Engineering
Virginia Polytechnic Institute and State University
Blacksburg, Virginia

Steven R. Brockschink (retired)
Stantec Consulting
Portland, Oregon

Jim Burke
Quanta Technology
Raleigh, North Carolina

Daryl Chipman
Salt River Project
Phoenix, Arizona

William A. Chisholm
Kinectrics/Université du Québec à Chicoutimi
Toronto, Ontario, Canada

George L. Clark
Alabama Power Company
Birmingham, Alabama

Patrick Coleman
Alabama Power Company
Birmingham, Alabama

Anthony M. DiGioia, Jr.
DiGioia, Gray and Associates, LLC
Monroeville, Pennsylvania

Murat Dilek
Electrical Distribution Design, Inc.
Blacksburg, Virginia

Dale A. Douglass
Power Delivery Consultants, Inc.
Niskayuna, New York

Michael L. Dyer
Salt River Project
Phoenix, Arizona

Richard G. Farmer
School of Electrical, Computer and Energy Engineering
Arizona State University
Tempe, Arizona

Jacek F. Gieras
Department of Electrical Engineering
University of Technology and Life Sciences
Bydgoszcz, Poland

Yancy Gill
Salt River Project
Phoenix, Arizona

Jim Green
Salt River Project
Phoenix, Arizona

Charles A. Gross
Department of Electrical and Computer Engineering
Auburn University
Auburn, Alabama

John V. Grubbs
Alabama Power Company
Birmingham, Alabama

James H. Gurney (retired)
BC Hydro
Vancouver, British Columbia, Canada

S. Mark Halpin
Department of Electrical and Computer Engineering
Auburn University
Auburn, Alabama

Andrew P. Hanson
The Structure Group
Houston, Texas

James R. Hunt
Salt River Project
Phoenix, Arizona

Géza Joós
Department of Electrical and Computer Engineering
McGill University
Montreal, Quebec, Canada

Peter M. Kandaris
Salt River Project
Phoenix, Arizona

John G. Kappenman
Metatech Corporation
Duluth, Minnesota

George G. Karady
School of Electrical, Computer and Energy Engineering
Arizona State University
Tempe, Arizona

Brian Keel
Salt River Project
Phoenix, Arizona

John R. Kennedy
Georgia Power Company
Atlanta, Georgia

William H. Kersting
Department of Electrical and Computer Engineering
New Mexico State University
Las Cruces, New Mexico

Christopher J. Melhorn
EPRI PEAC Corporation
Knoxville, Tennessee

Roger A. Messenger
Florida Atlantic University
Boca Raton, Florida

Vaughn Nelson
Alternative Energy Institute
West Texas A&M University
Canyon, Texas

Robert E. Nickerson
Consulting Engineer
Fort Worth, Texas

Paul I. Nippes
Magnetic Products and Services, Inc.
Holmdel, New Jersey

Hugh Stewart Nunn II
Salt River Project
Phoenix, Arizona

Vishal C. Patel
Southern California Edison Company
Rosemead, California

Joe C. Pohlman
Consultant
Pittsburgh, Pennsylvania

Saifur Rahman
Department of Electrical and Computer Engineering
Virginia Tech
Arlington, Virginia

Rama Ramakumar
School of Electrical and Computer Engineering
Oklahoma State University
Stillwater, Oklahoma

Manuel Reta-Hernández
Electrical Engineering Academic Unit
Universidad Autónoma de Zacatecas
Zacatecas, Mexico

Kenneth H. Sebra
Baltimore Gas and Electric Company
Dameron, Maryland

Douglas B. Seely
Stantec Consulting
Portland, Oregon

Tom A. Short
Electric Power Research Institute
Burnt Hills, New York

Raymond R. Shoults
Department of Electrical Engineering
University of Texas at Arlington
Arlington, Texas

Larry D. Swift
Department of Electrical Engineering
University of Texas at Arlington
Arlington, Texas

Rao S. Thallam
Salt River Project
Phoenix, Arizona

F. Ridley Thrash
Southwire Company
Carrollton, Georgia

Giao N. Trinh (retired)
Hydro-Québec Institute of Research
Boucherville, Quebec, Canada

I

Electric Power Generation: Nonconventional Methods

Saifur Rahman

Saifur Rahman is the founding director of the Advanced Research Institute (www.ari.vt.edu) at Virginia Tech, where he is the Joseph R. Loring professor of electrical and computer engineering. He also directs the Center for Energy and the Global Environment (www.ceage.vt.edu). He is a fellow of the IEEE and the editor in chief of the *IEEE Transactions on Sustainable Energy*. He is also the vice president of the IEEE Power and Energy Society (PES) and a member-at-large of the IEEE-USA Energy Policy Committee. He currently serves as the chair of the U.S. National Science Foundation Advisory Committee for International Science and Engineering. He is a distinguished lecturer for the IEEE PES and has lectured on smart grid, energy-efficient lighting solutions, renewable energy, demand response, distributed generation, and critical infrastructure protection topics in over 30 countries on all 6 continents.

1 Wind Power

Vaughn Nelson
West Texas A&M University

Before the industrial revolution, wind was a major source of power for pumping water, grinding grain, and long-distance transportation (sailing ships). Farm windmills for pumping water are still being manufactured and used around the world [1], with the peak use in the United States in the 1930s and 1940s when there were over 6 million.

During the 1930s, small wind systems (100 W to 1 kW) with battery storage were installed in rural areas; however, these units were displaced with power from the electric grid through rural electric cooperatives. There were a number of attempts to build wind turbines for the utility grid; however, most operated only a short time due to technical and economic problems. The only exception was the Gedser turbine (35 m diameter, 200 kW) in Denmark, which operated from 1958 to 1967. More information on history of wind turbines is available from Refs. [2, Ch. 1; 3,4].

After the first oil crisis in 1973, there was a resurgence interest in small systems, with the sale of refurbished units and manufacturer of new units. Also as a response to the oil crisis, governments and utilities were interested in the development of large wind turbines as power plants for the grid. In the 1980s the market was driven by installation of distributed wind turbines in Denmark and the wind farm market in California, which led to today's significant wind industry.

The major advantages of wind energy are similar to most other renewable energy resources; renewable (nondepleting), ubiquitous (located in many regions of the world), and in addition wind energy and photovoltaics do not require water for the generation of electricity. The disadvantages are that wind is variable and comes from a low-density source, which then translates into high initial costs. In general, windy areas are distant from load centers, which means that transmission is a problem for large-scale installation of wind farms.

The rapid growth of wind power (Table 1.1) has been due to wind farms with 194,400 MW installed by the end of 2010 and in addition there are around 1,200 MW from other applications. There will be an overlap between large and small (≤100 kW) wind turbines in the diverse applications of distributed and community wind, wind diesel, and village power (primarily hybrid systems). Wind turbines for producing electricity for stand-alone applications or grid connected for households and small businesses, telecommunications are primarily small wind turbines. Numbers installed and capacity are estimates with better data for wind farms and rougher estimates for the other applications.

TABLE 1.1 Wind Energy Installed in the World, Estimated Numbers and Capacity, End of 2010

Application	#	Capacity, MW
Wind turbines (primarily in wind farms)	161,000	194,400
Distributed[a]–community	1,300	400
Wind diesel	270	28
Village power	2,000	50
Small wind turbines	700,000	250–300
Telecommunication	500	2–5
Farm windmill	310,000	Equivalent, 155

[a] The overlap between distributed wind turbines and wind farm installations is difficult to distinguish. For example in Denmark, the large number of distributed units is counted as part of national capacity.

As of 2010, over 70 countries have installed wind power as most countries are seeking renewable energy sources and have wind power as part of their national planning. Therefore, countries have wind resource maps and others are in the process of determining their wind power potential, which also includes offshore areas.

1.1 Wind Resource

The primary difference between wind and solar power is that power in the wind increases as the cube of the wind speed,

$$\frac{P}{A} = 0.5 * \rho * v^3 \quad (\mathrm{W/m^2}) \tag{1.1}$$

where

ρ is the air density
v is the wind speed

The power/area is also referred to as wind power density. The air density depends on the temperature and barometric pressure, so wind power will decrease with elevation, around 10% per 1000 m. The average wind speed is only an indication of wind power potential and the use of the average wind speed will underestimate the actual wind power. A wind speed histogram or frequency distribution is needed to estimate the wind power/area. For siting of wind farms, data are needed at heights of 40–50 m and generally at hub heights. Since wind speeds vary by hour, day, season, and even years, 2–3 years of data are needed to have a decent estimate of the wind power potential at a specific site. Wind speed data for wind resource assessment are generally sampled at 1 Hz and averaged over 10 min (sometimes 1 h). From these wind speed histograms (bin width of 1 m) the wind power/area is determined.

1.1.1 Wind Shear

Wind shear is the change in wind speed with height and the wind speed at higher heights can be estimated from a known wind speed. Different formulas are available [2, Ch. 3.4], but most use a power law:

$$\frac{v}{v_0} = \left(\frac{H}{H_0}\right)^{\alpha} \tag{1.2}$$

where

v is the estimated wind speed at height H
v_0 is the known wind speed at height H_0
α is the wind shear exponent

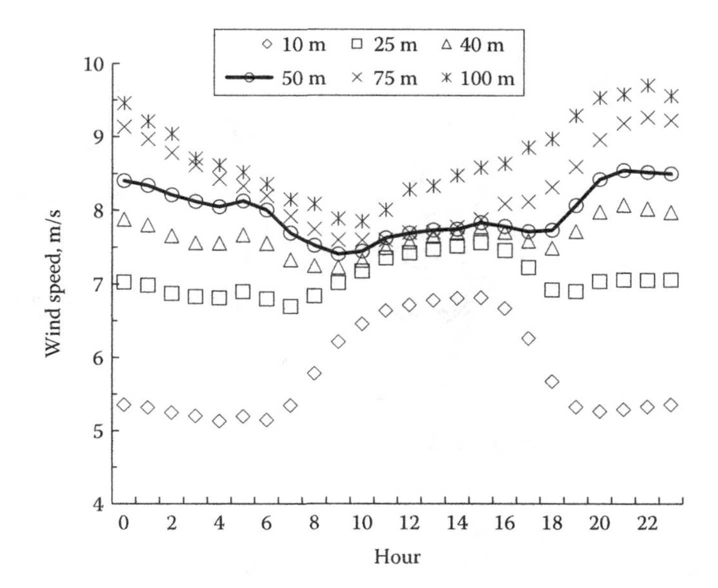

FIGURE 1.1 Average annual wind speed by time of day for White Deer, Texas (10, 25, 40, 50 m), and Washburn, Texas (75, 100 m).

The wind shear exponent is determined from measurements and in the past, a value of 1/7 (0.14) was used for stable atmospheric conditions. Also this value meant that the power/area doubled from 10 to 50 m, a convenient value since the world meteorological standard for measurement of wind speed was 10 m height.

In many continental areas the wind shear exponent is larger than 0.14, and the wind shear also depends on the time of day with a change in the pattern from day to night at a height around 40 m (Figure 1.1). This means that wind farms will produce more energy at night when the load of the utility is lower, a problem for the value of the energy sold by the wind farm. The pattern of the data at 50 m for Washburn (not shown on graph) was similar to 50 m data at White Deer; however, there even was some difference between the two sites. Both sites were in the plains around 40 km apart. This shows that wind power is fairly site specific, even in the plains. Wind data for wind farms have to be taken at heights of at least 40–50 m, as at these heights and above the wind pattern will be same and the wind speeds at higher heights at the same site can be estimated using Equation 1.2. There are some locations, such as mountain passes, where there is little wind shear, so taller towers for wind turbines would not be needed. Note that with large MW wind turbines, hub heights are 60 m to over 120 m.

1.1.2 Wind Maps

Wind power maps, W/m^2, are available for many countries, regions, and states/provinces within countries. Early maps were for 10 m height with an estimate for 50 m using the power law for wind shear Equation 1.2 and a wind shear exponent of 0.14. The wind power map [5] for the United States (Figure 1.2) shows large areas with wind class 3 and above. More detailed state maps are available. In addition, wind power potential is estimated using geographic information systems (GIS) with land excluded due to urban areas, highways, lakes and rivers, wildlife refugees, and land that is at a distance from high voltage transmission lines. The wind power potential is very large, so it is not a question of the wind energy resource, but a question of locations of good to excellent wind resource, national and state policies, economics, and amount of penetration of wind power into the grid. For example, the catchable wind power potential of Texas is estimated at 223,000 MW, which is much larger than the 110,000 MW generating capacity of the state [6].

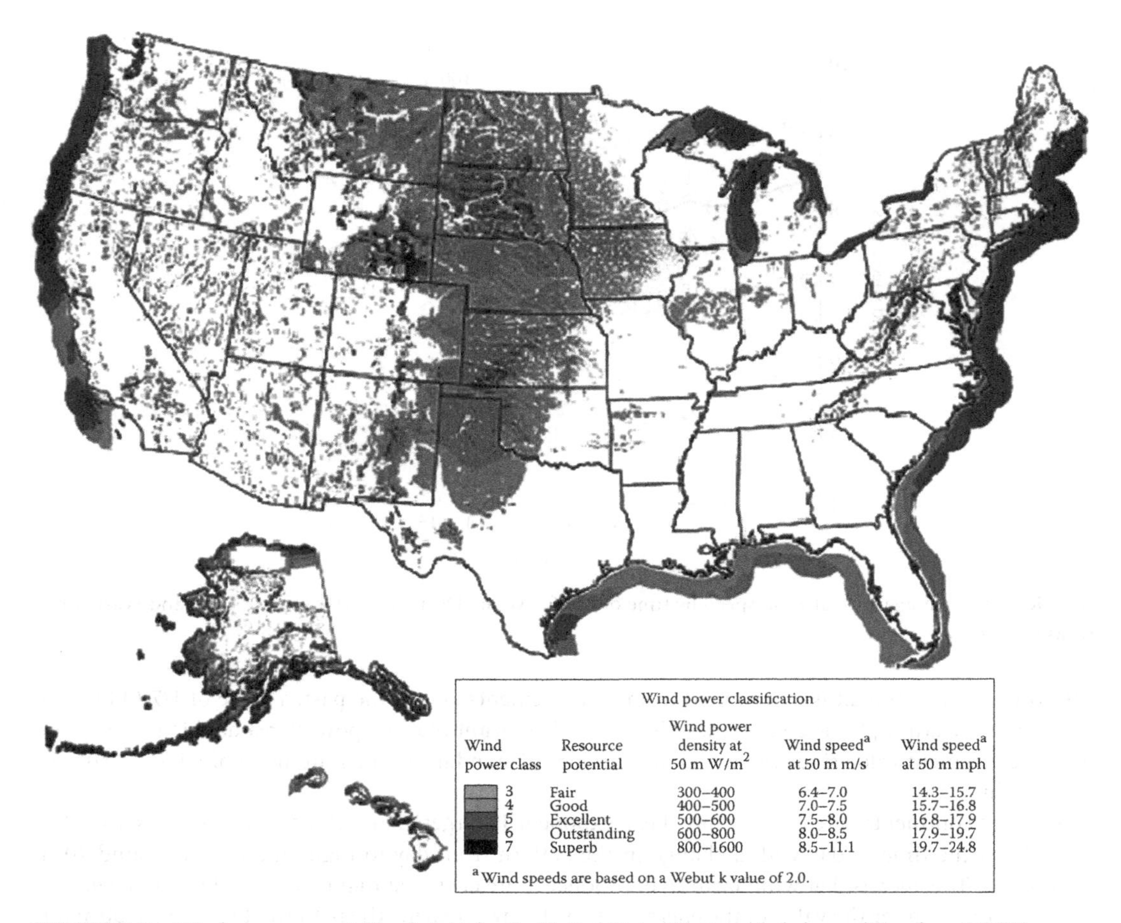

FIGURE 1.2 Wind power map at 50 m height for the United States. Notice wind classes (Map from NREL).

Computer tools for modeling the wind resource have been developed by a number of groups, National Wind Technology Center, National Renewable Energy Laboratory (NREL), United States; RISO in Denmark (WAsP); other government labs, and private industry. WAsP has been employed in over 100 countries and territories around the world. Now revised wind power maps for (50, 80, and even 100 m height) are available, which use terrain, weather balloon data, and computer models [7], and the maps were also verified with available data at 50 m heights. These maps are a good screening tool for wind farm locations, and they show regions of higher-class winds in areas where none was thought to exist. Also because of the larger wind shear than expected, more areas have suitable winds for wind farm development. Remember that 2%–7% accuracy in wind speeds means a 6%–21% error in estimating wind power, so data on site are still needed for locations of wind farms in most areas. Interactive wind speed maps by 3Tier [8] and AWS-Truewind [9] are available online for many locations in the world. Wind Atlases of the World contains links for over 50 countries [10].

Complete coverage of the oceans is now available using reflected microwaves from satellites [11,12]. Ocean wind speed and direction at 10 m are calculated from surface roughness measurements from the daily orbital observations mapped to a 0.25° grid, which are then averaged over 3 days, a week, and a month. Images of the data can be viewed on websites for the world, by region or selected area. Ocean winds are not available within 25 km of the shore, as the radar reflections of the bottom of the ocean skew the data. Ocean winds will indicate onshore winds for islands, coasts, and also some inland regions of higher winds (Figure 1.3). There are now wind farms in the Isthmus of Tehuantepec, Mexico

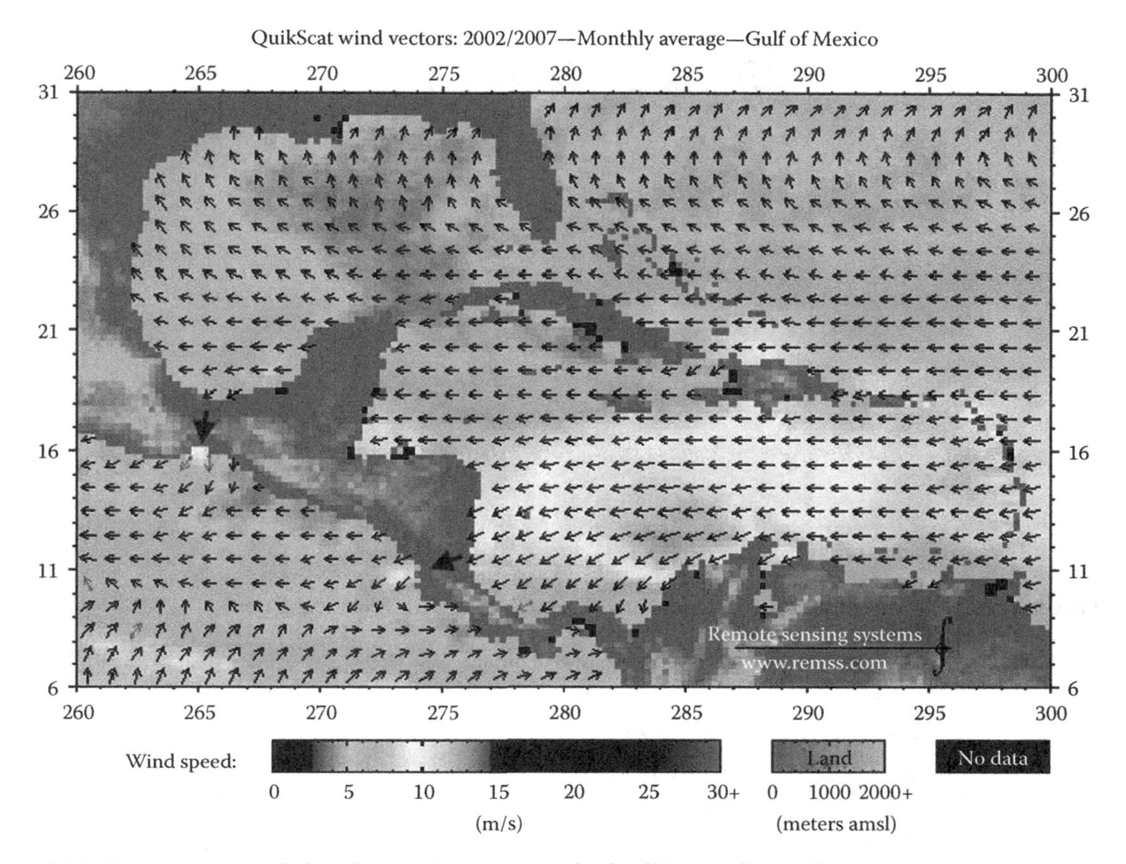

FIGURE 1.3 Ocean winds for July 2002. Two arrows on land indicate excellent onshore wind regions.

and the Arenal region of Costa Rica, where the northeast trade winds (average wind speeds of 10 m/s) are funneled by the land topography.

1.1.3 Wind Turbines

Wind turbines are classified according to the interaction of the blades with the wind, orientation of the rotor axis with respect to the ground and to the tower (upwind, downwind), and innovative or unusual types of machines. The interaction of the blades with the wind is by drag or lift or a combination of the two.

For a drag device, the wind pushes against the blade or sail forcing the rotor to turn on its axis, and drag devices are inherently limited in efficiency since the speed of the device or blades cannot be greater than the wind speed. The maximum theoretical efficiency is 15%. Another major problem is that drag devices have a lot of material in the blades. Although a number of different drag devices (Figure 1.4) have been built, there are essentially no commercial (economically viable) drag devices in production for the generation of electricity.

Most lift devices use airfoils for blades (Figure 1.5), similar to propellers or airplane wings; however, other concepts are Magnus (rotating cylinders) and Savonius wind turbines (Figure 1.6). A Savonius rotor is not strictly a drag device, but it has the same characteristic of large blade area to intercept area. This means more material and problems with the force of the wind on the rotor at high wind speeds, even if the rotor is not turning. An advantage of the Savonius wind turbine is the ease of construction.

FIGURE 1.4 Drag device with cup blades, similar to anemometer. (Courtesy of Charlie Dou.)

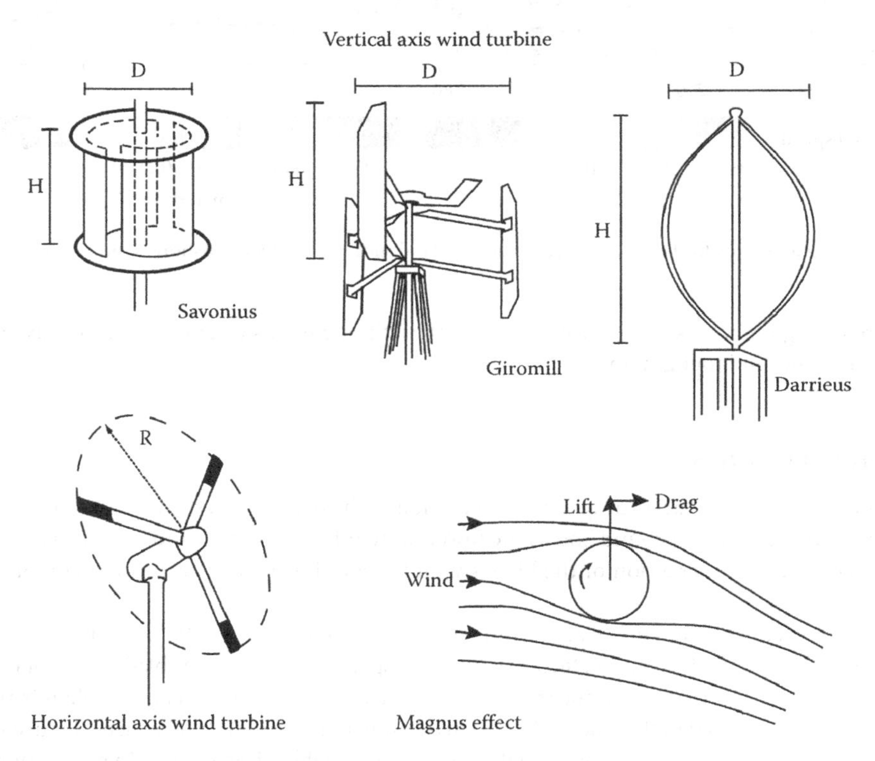

FIGURE 1.5 Diagram of different rotors for horizontal and vertical axis wind turbines.

Using lift, the blades can move faster than the wind and are more efficient in terms of aerodynamics and use of material, a ratio of around 100 to 1 compared to a drag device. The tip speed ratio is the speed of the tip of the blade divided by the wind speed, and lift devices typically have tip speed ratios around seven. There have even been one-bladed wind turbines, which save on material; however, most modern wind turbines have two or three blades.

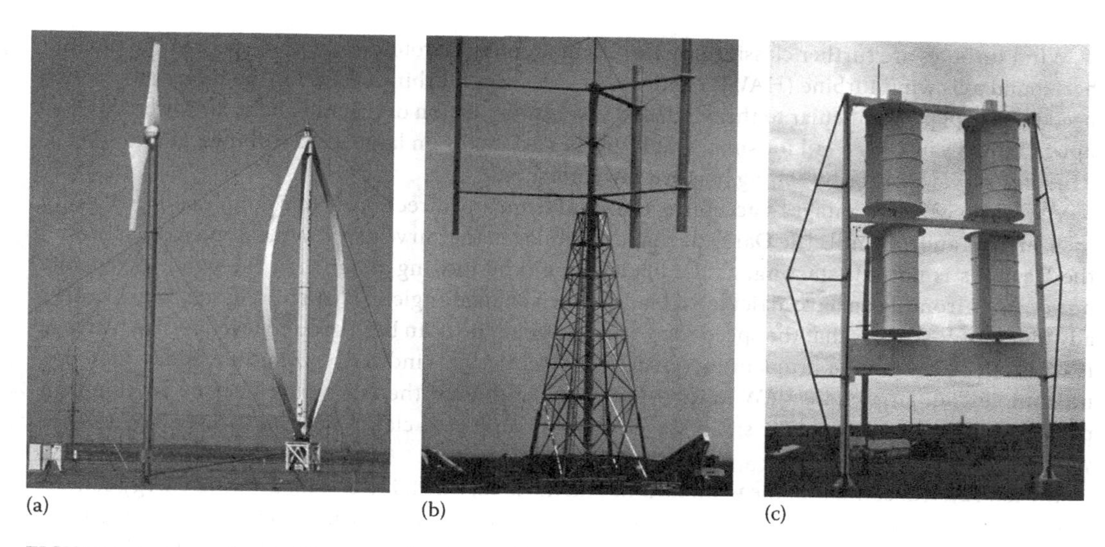

FIGURE 1.6 Examples of different wind turbines. (a) HAWT, diameter = 10 m, 25 kW; Darrieus, diameter = 17 m, 24 m tall rotor, 100 kW. (b) giromill, rotor diameter = 18 m, 12.8 m height, 40 kW. (c) Savonius, 10 kW. (Courtesy of Gary Johnson.)

The power coefficient is the power out or power produced by the wind turbine divided by the power in the wind. From conservation of energy and momentum, the maximum theoretical efficiency of a rotor is 59%. The capacity factor is the average power divided by the rated power. The average power is generally calculated by knowing the energy production divided by the hours in that time period (usually a year or can be calculated for a month or a quarter). For example, if the annual energy production is 4500 MWh for a wind turbine rated at 1.5 MW, then the average power = energy/hours = 4500/8760 = 0.5 MW and the capacity factor would be 0.5 MW/1.5 MW = 0.33 = 33%. So the capacity factor is like an average efficiency. A power curve shows the power produced as a function of wind speed (Figure 1.7). Because there is a large scatter in the measured power versus wind speed, the method of bins (usually 1 m/s bid width suffices) is used.

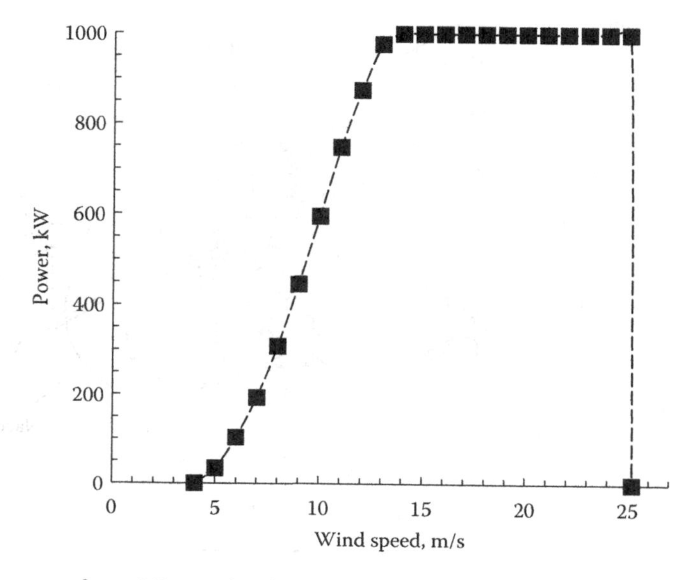

FIGURE 1.7 Power curve for a 1 MW wind turbine.

Wind turbines are further classified by the orientation of the rotor axis with respect to the ground: horizontal axis wind turbine (HAWT) and vertical axis wind turbine (VAWT). The rotors on HAWTs need to be kept perpendicular to the wind, and yaw is this rotation of the unit about the tower axis. For upwind units yaw is by a tail for small wind turbines—a motor on large wind turbines, and for downwind units—yaw may be by coning (passive yaw) or a motor.

VAWT have the advantage of accepting the wind from any direction. Two examples of VAWTs are the Darrieus and giromill. The Darrieus shape is similar to the curve of a moving jump rope; however, the Darrieus is not self-starting, as the blades should be moving faster than the wind to generate power. The giromill can have articulated blades which change angle so it can be self-starting. Another advantage of VAWTs is that the speed increaser and generator can be at ground level. A disadvantage is that taller towers are a problem for VAWTs, especially for wind farm size units. Today there are no commercial, large-scale VAWTs for wind farms, although there are a number of development projects and new companies for small VAWTs. Some companies claim they can scale to MW size for wind farms.

The total system consists of the wind turbine and the load, which is also called a wind energy conversion system (WECS). A typical large wind turbine consists of the rotor (blades and hub), speed increaser (gear box), conversion system, controls and the tower (Figure 1.8). The most common configuration for large wind turbines is three blades, full span pitch control (motors in hub), upwind with yaw motor, speed increaser (gear box), and doubly fed induction generator (allows wider range of rpm for better aerodynamic efficiency). The nacelle is the covering or enclosure of the speed increaser and generator.

The output of the wind turbine, rotational kinetic energy, can be converted to mechanical, electrical, or thermal energy. Generally it is electrical energy. The generators can be synchronous or induction connected directly to the grid, or a variable frequency alternator (permanent magnet alternator) or direct current generator connected indirectly to the grid through an inverter. Most small wind

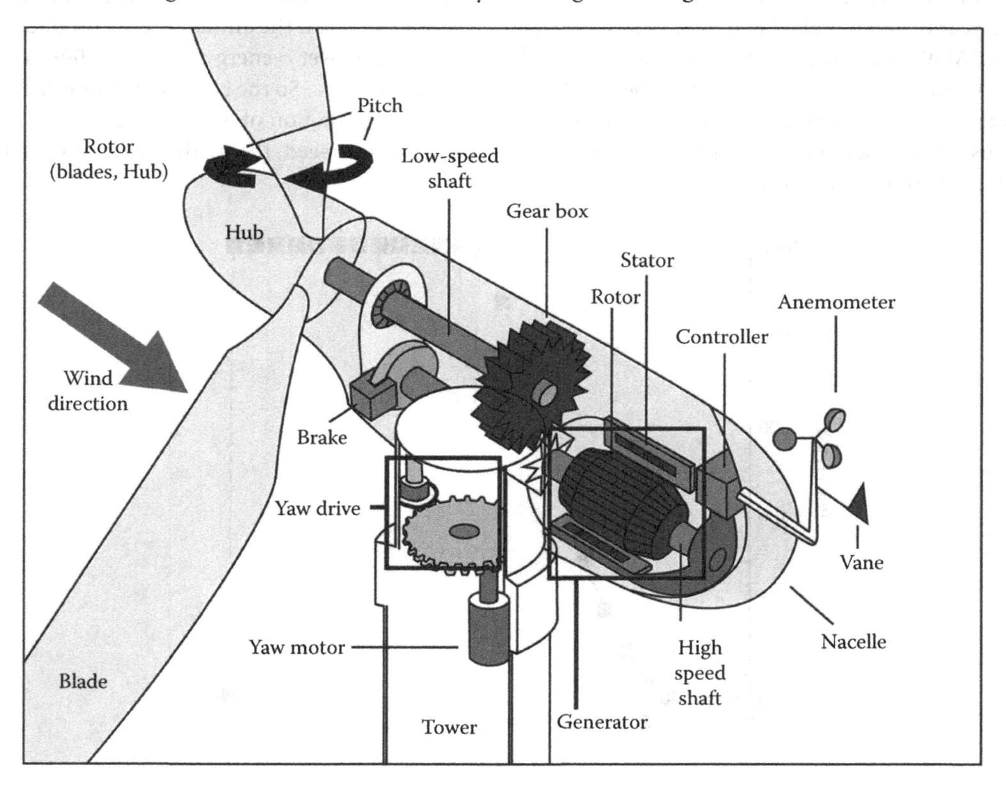

FIGURE 1.8 Diagram of main components of large wind turbine.

turbines are direct drive and no speed increaser and operate at variable rpm. Wind turbines without a gearbox are direct drive units. Enercon has built large wind turbines with huge generators and no speed increaser, which have higher aerodynamic efficiency due to variable rpm operation of the rotor. However, there are some energy losses in the conversion of variable frequency to the constant frequency (50 or 60 Hz) needed for the utility grid.

1.2 Wind Farms

The development of wind farms began in the early 1980s in California with the installation of wind turbines ranging from 20 to 100 kW, as those were the only sizes available in the commercial market. This development of wind farms in California was due to U.S. federal laws and incentives (1980–1985) and due to the avoided costs for energy set by the California Public Utility Commission for electricity generated by those wind farms. As the wind farm market in the world continued, there was a steady progression toward larger size wind turbines due to economies of scale, and today there are commercial multi-megawatt units.

Since then other countries have supported wind energy and by the end of 2010 there were 194.4 GW installed (Figure 1.9) from around 160,000 wind turbines. At 35% capacity factor, the estimated energy production is around $6 * 10^8$ GWh/year. Installation of wind turbines in Europe was led by Denmark in the early days and its manufacturers captured a major share of the world market in the 1980s. Then other European countries installed large numbers of wind turbines and Germany became the world leader. In addition, there was consolidation of manufacturers with both Germany and Spain becoming major players. Then in 2007–2008 the major wind farm installations shifted back to the United States and now in 2010 China is the leader in installed capacity (Figure 1.10) with the United States in second place. Although the United States had a large number of wind turbines installed and the electricity generated by wind was $7.3 * 10^7$ MWh in 2009, wind energy accounted for only 1.3% of the total electricity generated. However, wind power accounted for 35% of new electric power generating capacity in the United States in 2009. Other countries obtain a larger share of their electric demand from wind and Denmark is the leader as 24% of their electricity comes from wind power. In Spain during the early morning on a spring day in 2010, over 50% of the electric demand was provided by wind power.

There are 2939 MW (end of 2010) installed in offshore wind farms in Europe, because of the high cost of land in Europe. Information on European key trends and statistics is available from the European Wind Energy Association and the Global Wind Energy Council. In China, the first 100 MW offshore wind farm was completed in 2010, and in addition, 600 MW offshore and 400 MW intertidal land wind

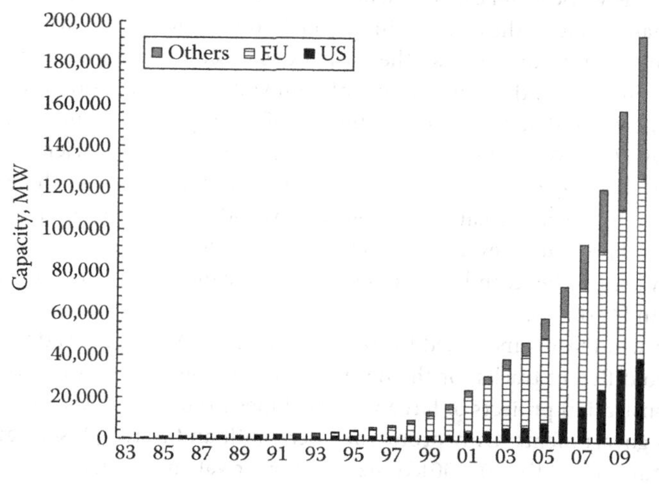

FIGURE 1.9 Wind power installed in the world, 2009, primarily wind farms.

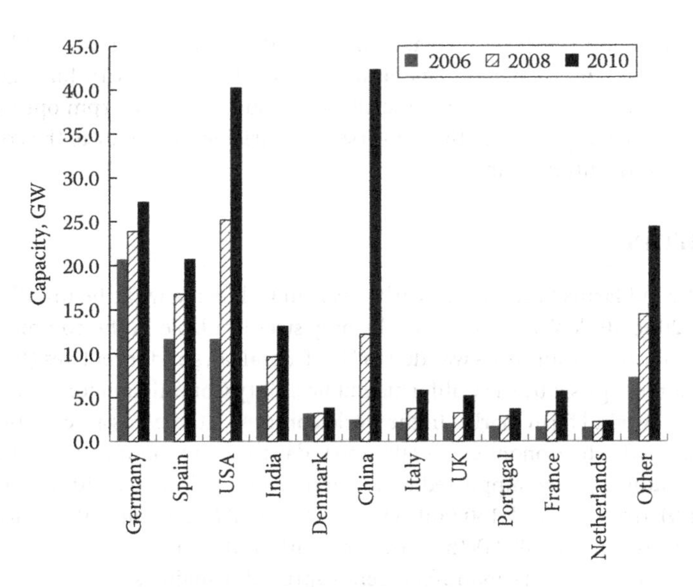

FIGURE 1.10 Cumulative wind power installed by country at the end of the year, primarily wind farms.

power concession projects are under construction. Offshore wind farms are being planned for other parts of the world, for example, in the United States off the East Coast, Texas Gulf Coast, and in the Great Lakes.

The growth of installed capacity in the world has been 20%–30% per year starting in 2005, but with the present world economic recession, some countries did not experience a large growth rate in 2010. Texas surpassed California in installed capacity in 2008 and with over 9700 MW installed by the end of 2010, Texas continues to lead the United States in installed capacity [13]. There have been a number of estimates for the future, one being a world wind installed capacity of 240 GW by 2020, which would be a 100% increase over 2008. However, that projection is now lower than other projections.

Market forecasts for wind power were seen as overly optimistic at the time of the prediction and then were exceeded every time by the actual amount of installations. World wind capacity grew by over 170% over the 5 years from 2005 through 2009. The Global Wind Energy Council forecast in 2009 was for 409 GW by the end of 2014 [14], an increase of 215 GW in the next 4 years.

My estimate is that the world wind capacity will be over 700 GW by 2020, an increase of over 500 GW over 2010. This estimate is due to the changes in national policies promoting wind power, primarily in the United States, China, and Europe. Also the estimate is based on continuing incentives for renewable energy, carbon trading, and the construction of high voltage transmission lines from windy areas to load centers. The new mandate for China is 200 GW of wind power by 2020, and Europe plans an additional 100 GW by 2020. Wind energy could produce 20% of U.S. electricity by 2030 [15], which would require an installed capacity of 300 GW. It is assumed that the rest of the world will install at least 100 GW by the end of 2020. If solar energy, bioenergy, and geothermal energy were included, then renewable energy would provide an even larger percentage of U.S. demand for electricity. The prospects for the wind industry are excellent, and this does not count the increased numbers of distributed, community, and small wind turbines.

Wind turbines for wind farms increased in size from the 100 kW to megawatts due to economies of scale. There were two different tracks for the development of wind turbines for wind farms. The first was R&D plus demonstration projects of large wind turbines for utility power in the 1970s and 1980s, primarily funded by governments. Only prototypes were built and tested [2, see Table 10.10]. The second track was wind turbines in the 50–100 kW size built by private manufacturers [3, Ch. 4] to meet the distributed market in Europe and for wind farms in California. The manufacturers of the second track were successful in developing the modern wind turbine industry, while the units developed primarily by aerospace companies did not make it to the commercial stage.

There were a number of different designs built and sold in the wind farm market in California, including Darrieus wind turbines. In the United States the most common designs were two blades, fixed pitch, rotor downwind, teetered hub, induction generator; and three blades, variable pitch, rotor downwind, induction generator of which U.S. wind power built over 4000 units (late 1980s) for the California wind farm market. In Europe the three blades, fixed pitch, upwind rotor were the predominate design. Now the three blades, rotor upwind, full span variable pitch, and a wider range of rpm are the major type for wind farms. Enercon has a wind turbine with large generator and no gearbox.

Today, wind turbines are available in megawatt sizes with rotor diameters of 60 to over 100 m and installed on towers of 60 to over 100 m. Manufacturers are designing and building wind turbines in the 5–10 MW size, primarily for offshore installations. Out of the top 15 manufacturers in 2007, 10 were from Europe, 2 from China, and 1 each from the United States, Japan, and India. Today other major international companies are buying existing manufacturers of wind turbines or starting manufacturing wind turbines for the wind farm market.

Vestas is the world's leading manufacturer with over 20% of the market and they have installed more than 43,000 wind turbines with a capacity greater than 44,000 MW (2010 data). As an example of a large wind turbine installation, a Vestas, V90, rated at 3 MW, 90 m diameter on 80 m tower is located north of Gruver, Texas (Figure 1.11). Twenty trucks were needed to haul an 800 metric ton crane to the site and another 10 trucks for the turbine and tower. The weight of the components were nacelle = 70 metric tons, rotor = 41 metric tons, and tower = 160 metric tons. The foundation required 460 m^3 of concrete and over 40 metric tons of rebar.

There are economies of scale for installation of wind turbines for wind farms, and in general, most projects need 30–50 MW to reach this level. The spacing for wind turbines is 3–4 rotor diameters within a row and 8–10 rotor diameters from row to row. On ridgelines and mesas there would be one to two rows with a 2-rotor diameter spacing within a row. In general for plains and rolling terrain, the installed capacity could be 5–10 MW per square kilometer and for ridgelines, 8–12 MW per linear kilometer.

FIGURE 1.11 Vestas, V90, 3 MW wind turbine. Notice minivan next to the tower.

FIGURE 1.12 Satellite view of layout of part of the Sweetwater wind farm (south of Sweetwater, Texas). Notice distance between rows is larger than distance between wind turbines within a row. Rows are perpendicular to predominate wind direction.

Satellite images show the layout of wind farms (Figure 1.12); however, the maps may not show the latest installations. In Texas in 2010 there were five wind farms over 500 MW, and the largest was Roscoe at 782 MW.

1.2.1 Small Wind Turbines

There are a number of different configurations and variations in design for small wind turbines (W to 100 kW). Many of the small wind turbines have a tail for both orientation and control in high winds. There are around 700,000 small wind turbines in the world with a capacity around 250–300 MW; however, these are very rough estimates. China has produced around 350,000 small wind turbines, primarily from 50 to 200 W, stand-alone systems with battery storage. Now China is starting to build units in the 1–50 kW range. In the United States and Europe, approximately 25% of the small wind turbines are connected to the grid.

In 2010, there are around 100 manufacturers of small wind turbines with around 40 in Europe and 30 in China, and there is a resurgence of VAWTs designs. The United States is a leading producer of small wind turbines in the 1–50 kW range and the installed capacity of small wind in the United States is around 80 MW. The National Wind Technology Center, NREL has a development program for small wind turbines. Wind energy associations around the world generally have a small wind section and the American Wind Energy Association has Global Market Studies and a U.S. Roadmap [17]. The Roadmap estimates that small wind could provide 3% of U.S. electrical demand by 2020.

1.2.2 Village Power

Village power is another large market for small wind turbines as approximately 1.6 billion people do not have electricity and extension of the grid is too expensive in rural and remote areas with difficult terrain. There are around 2000 village power systems with an installed capacity of 55 MW. Village power systems are mini grids, which can range in size from small micro grids (<100 kWh/day, ~15 kW) to larger communities (tens of MWH/day, hundreds of kW). Today there is an emphasis on systems that

FIGURE 1.13 China village power system (PV/wind/diesel), 54 kW. (Courtesy of Charlie Dou.)

use renewable energy (wind, PV, mini and micro hydro, bioenergy, and hybrid combinations). These systems need to supply reliable, however, limited amount of energy, and in general much of the cost has to be subsidized. The other components of the system are controllers, batteries, inverters, and possibly diesel or gas generators. In windy areas, wind turbines are the least-cost component of the renewable power supply and one or multiple wind turbines may be installed, 10–100 kW range.

China leads the world in installation of renewable village systems of which 100 include wind [2, Ch. 10.5.1]. Their Township Electrification Program in 2002 installed 721 village power systems with a capacity over 15 MW (systems installed; 689 PV, 57 wind/PV and 6 wind). An example is the village power system (54 kW) for Subashi, Zinjiang Province, China (Figure 1.13), which consisted of 20 kW wind, 4 kW PV, 30 kVA diesel, 1000 Ah battery bank, and a 38 kVA inverter. The installed cost was $178,000 for power and mini grid, which is reasonable for a remote site.

1.2.3 Wind Diesel

For remote communities and rural industry, the standard for electric generation is diesel power. Remote electric power is estimated at 12 GW, with 150,000 diesel generator sets ranging in size from 50 to 1000 kW. In many locations these systems are subsidized by regional and national governments.

Diesel generators have low installed costs; however, they are expensive to operate and maintain, especially in remote areas. Even with diesel generators, many small villages only have electricity in the evening. Costs for electricity were in the range from $0.20 to $0.50/kWh; however, as the cost of diesel increases, the cost per kWh increases.

Wind turbines can be installed at existing diesel power plants as a low (fuel saver, diesel does not shut down), medium, or high penetration (wind power supplies more of the load, which results in better economics as diesel engines may be shut down). The wind turbine(s) may be part of a retrofit, an integrated wind-diesel, or wind/PV/diesel hybrid systems for village power. Rough estimates indicate there are over 220 wind-diesel systems in the world, ranging in size from 100 kW to megawatts. Reports on operational experiences from 11 wind-diesel systems are available from the 2004 wind-diesel workshop [18].

At Kotzebue, Alaska, they have six diesel generators (11.2 MW, annual average load = 2.5 MW, peak load = 3.9 MW) and the large reserve capacity is to prevent any loss of load during the winter. The cost of electricity was around $0.50/kWh. Consumption of diesel fuel was around 5.3 million liters per year with an average conversion of 4 kWh/L. There are 17 wind turbines located on a flat windy plain 7 km south of town and 0.8 km from the coast. In 2007 the wind turbines generated 667,500 kWh for a savings of 172,000 L of diesel fuel.

The U.S. Air Force installed four, 225 kW wind turbines connected to two, 1900 kW diesel generators (average load 2.4 MW) on Ascension Island [19] for a low penetration system (14%–24%). Tower height was limited to 30 m due to available crane capacity on the island. In 2003, two additional wind turbines (900 kW each), along with a boiler and advanced controller, were installed and that brought the average wind penetration to 43%–64%. Fuel savings were approximately $1 million/year.

A number of wind turbine manufacturers have wind-diesel, hybrid, and even hydrogen production options. These range from simple, no storage systems to complex, integrated systems with battery storage and dump loads.

1.2.4 Other

There is an overlap of small and large wind turbines installed for the wind-diesel and distributed markets. Distributed systems are the installation of wind turbines on the retail or consumer side of the electric meter for farms, ranches, agribusiness, small industries, and small-scale community wind for schools and other public entities. Examples are as follows:

One 660 kW wind turbine at the American Wind Power Center and Museum, Lubbock, Texas
Ten 1 MW wind turbines at a cotton seed oil plant, Lubbock, Texas
Three school districts in towns near Lubbock, Texas, have install a total of eight 50 kW units
Four 1.5 MW wind turbines supply electricity for the City of Lamar, Colorado

There were approximately 300 MW of community wind projects installed in the United States by 2008. The market in the United States for distributed wind is estimated at 3900 MW by 2020 [17]. In Denmark, individuals or wind turbine cooperatives own around 80% of the 5000 wind turbines and had around 77% of the capacity.

Another market for small wind turbines is power for telecommunication stations, with an estimated 500 having small wind turbines as part of the power supply. This is a growing market due to the increased use of cellular phones, especially in the more remote areas of the world.

Small PV/wind systems for street lighting (Figure 1.14) are now on the market. Even though there may be a transmission line nearby, the cost of the transformer and electricity is more than the cost of electricity from the PV/wind system.

FIGURE 1.14 PV/wind powers streetlight and flashing red lights at stop signs, McCormick Road on I 27 between Canyon and Amarillo, Texas.

Innovative wind turbines have to be evaluated in terms of performance, structural requirements, operation and maintenance, and energy production in relation to constraints and cost of manufacturing. Most innovative wind systems are at the design stage with some even making it to the prototype or demonstration phase. If they become competitive in the market, they would probably be removed from the innovative category. Some examples are tornado type, tethered to reach the high winds of the jet stream, tall tower to use rising hot air, torsion flutter, electrofluid, diffuser augmented, and multiple rotors on same shaft. There have been numerous designs and some prototypes have been built that have different combinations of blades and/or blade shapes. Another innovative example is flying wind turbines [20], now at the design and prototype stage, which would extract power in the winds at 400 m.

There are also companies that are building wind turbines to mount on buildings in the urban environment [2, Ch 9.1]. There is an Internet site for urban wind [21] with downloads available. The wind turbines guideline includes images of flow over buildings and example projects. An unusual design for a building is the incorporation of three wind turbines (225 kW each) on the causeways connecting two skyscrapers in Bahrain [22].

The farm windmill is a long-term application of the conversion of wind energy to mechanical power and it is well designed for pumping small volumes of water at a relatively high lift. It is estimated that there are around 300,000 operating farm windmills in the world and the annual production is around 3,000. The rotor has high solidity, large amount of blade material per rotor swept area, which is similar to drag devices. The tip speed ratio is around 0.8, and the annual average power coefficient is 5%–6% [1].

Different research groups and manufacturers have attempted to improve the performance of the farm windmill and to reduce the cost, especially for developing countries [2, Ch. 10.6]. A wind-electric system is more efficient and can pump enough water for villages or small irrigation. The wind-electric system is a direct connection of the permanent magnet alternator (variable voltage, variable frequency) of the wind turbine to a standard three-phase induction motor driving a centrifugal or submersible pump. Annual power coefficients are around 10%–12%.

1.2.5 Performance

In the final analysis, performance of wind turbines is reduced to energy production and the value or cost of that energy in comparison to other sources of energy. The annual energy production can be estimated by the following methods:

Generator size (rated power)
Rotor area and wind map values
Manufacturer's curve of energy versus annual average wind speed

The generator size method is a rough approximation as wind turbines with the same size rotors (same area) can have different size generators, but it is a fairly good first approximation:

$$\text{AEP} = \text{CF} * \text{GS} * 8760 \text{ kWh/year or MWh/year} \quad (1.3)$$

where
AEP is the annual energy production
CF is the capacity factor
GS is the rated power of wind turbine
8760 is the number of hours in a year

Capacity factors depend on the rated power versus rotor area as wind turbines' models can have different size generators for same size rotor or same size generators for different size rotors for better performance in different wind regimes. For wind farm capacity factors range from 30% to 45% for class 3 wind regimes to class 5 wind regimes.

Availability is the time the wind turbine is available to operate, whether the wind is or is not blowing. Availability of wind turbines is now in the range of 95%–98%. Availability is also an indication of quality or reliability of the wind turbine. So the AEP is reduced due to availability. If the wind turbine is located at higher elevations, then there is also a reduction for change in density (air pressure component) of around 10% for every 1000 m of elevation.

Since the most important factors are the rotor area and the wind regime, the annual energy production can be estimated from

$$\text{AEP} = \text{CF} * \text{Ar} * W_M * 8.76\ \text{kWh/year} \tag{1.4}$$

where

Ar is the area of the rotor, m^2

W_M is the value of power/area for that location from wind map, W/m^2

8.76 h/year, which also converts W to kW

The manufacturer may provide a curve of annual energy production versus annual average wind speed (Figure 1.15), where AEP is calculated from the power curve for that wind turbine and a wind speed histogram calculated from average wind speed using a Rayleigh distribution [2, Ch. 3.11].

The best estimate of annual energy production is the calculated value from measured wind speed data and the power curve of a wind turbine (from measured data). The calculated annual energy production is just the multiplication of the power curve value times the number of hours for each bin (Table 1.2). If the availability is 95% and there is a 10% decrease due to elevation, then the calculate energy production would be around 2600 MWh/year.

The calculated annual energy production is the number from which the economic feasibility of a wind farm project is estimated and the number that is used to justify financing for the project. Wind speed histograms and power curves have to be corrected to the same height and power curves have to be adjusted for air density at that site. In general, wind speed histograms need to be annual averages from 2 to 3 years of data; however, 1 year of data may suffice if it can be compared to a long-term database.

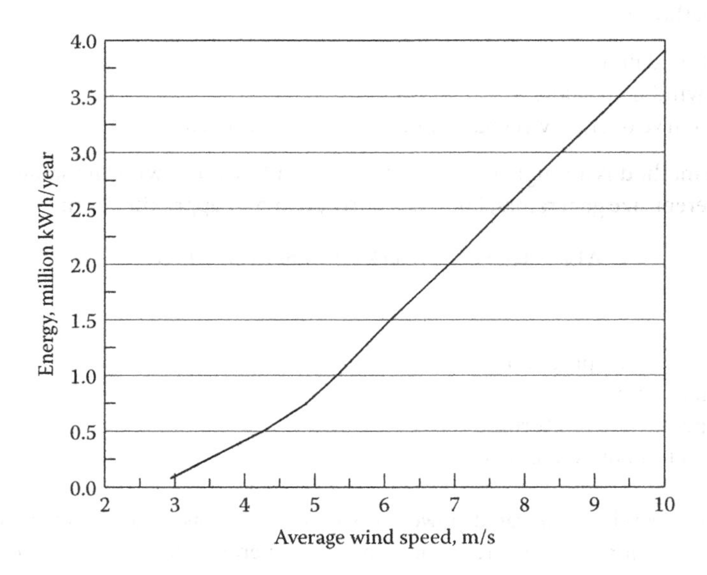

FIGURE 1.15 Manufacturer's curve for estimated annual energy production as function of average wind speed for 1 MW wind turbine.

TABLE 1.2 Calculated Annual Energy Using Power Curve for 1 MW Wind Turbine and Wind Speed Histogram Data for White Deer, Texas (Bin Width = 1 m/s, Data Adjusted to Hub Height of 60 m)

Wind Speed, m/s	Power, kW	Bin, h	Energy, kWh
1	0	119	0
2	0	378	0
3	0	594	0
4	0	760	171
5	34	868	29,538
6	103	914	94,060
7	193	904	174,281
8	308	847	260,760
9	446	756	337,167
10	595	647	384,658
11	748	531	396,855
12	874	419	366,502
13	976	319	311,379
14	1000	234	233,943
15	1000	166	165,690
16	1000	113	113,369
17	1000	75	74,983
18	1000	48	47,964
19	1000	30	29,684
20–24	1000	20	39,540
≥25	0	20	0
	Sum	8760	**3,060,545**

1.3 Institutional Issues

The institutional issues (noninclusive as there are surely others) related to renewable energy contain the following: legislation and regulation; environment; incentives; externalities; world treaties and country responses to greenhouse gas emissions; connection to utility grids (large power generators such as wind farms, large PV arrays, concentrating solar power; large numbers of small systems, distributed and community systems); incentives such as feed-in tariffs, renewable portfolio standards, rebates, and tax credits; and certification standards for equipment and installation of systems. Of course most of these issues are determined by politics and economics.

The interconnection of wind turbines to utility grids, regulations on installation and operation, and environmental concerns are the major issues [2, Ch. 11]. For a few wind turbines on a large utility grid there would be no problems with the amount of power. It would be considered as a negative load, a conservation device that is the same as turning off a load. For large penetration, 20% and greater, other factors such as the variability of the wind and dispatching become important. The utilities are concerned with safety and power quality due to any wind turbines on their grid.

The main environmental issues are visual impact, noise, birds (avian), and bats. The visual impact can be detrimental, especially in locations that are close to scenic areas or parks. It is the familiar story; people are in favor of renewable energy, but not in my backyard. Some people are adamantly opposed to wind farms, most are neutral, and the rest are in favor. In the great plains of the United States, wind turbines are generally seen as favorable due to rural economic impact. For those opposed, generally the visual impact is the most important concern. Noise measurements have shown in general that wind turbines are below the ambient noise; however, the repetitive nose from the blades stands out and one would not want their residence in the middle of a wind farm. The whine from gearboxes on some units is also noticeable. However, with larger wind turbines at higher hub heights and new airfoils, the noise has been reduced.

As with any endeavor, politics enters the situation. To make a change in behavior, especially when the competition is an entrenched industry, you need INCENTIVES, PENALTIES, and EDUCATION. Someone estimated that the amount of each type of energy used is in direct proportion to the amount of subsidies for that type of energy. Subsidies are in the form of taxes, tax breaks, and regulations, all of which generally require legislation. What every entity (industry) wants are incentives for themselves and penalties for their competitors. In addition, they want the government to fund R&D and even commercialization. For the United States, incentives are listed by type, amount, and state [23].

Energy subsidies have generally been in favor of conventional fossil fuels and established energy producers. Subsidies for renewable energy between 1974 and 1997 amounted to $20 billion worldwide. This can be compared with the much larger number for subsidies for conventional energy sources, which was a total of *$300 billion per year* [24], and this number does not even take into account the expenditures for infrastructure, safeguards, and military actions for continued flow of oil and natural gas. The privatization of the electric industry along with the restructuring into generation, transmission, and distribution has opened some doors for renewable energy. The federal production tax credit and renewable portfolio standards have had a major impact on the development of wind farms in the United States.

Net metering is where the electric meter runs in either direction. If the renewable energy system produces more electricity than is needed on site, the utility meter runs backward, and if the load on site is greater, then the meter runs forward. Then the bill is determined at the end of the time period, which is generally 1 month. If the renewable energy system produced more energy over the billing period than was used on site, the utility company pays the avoided cost. Most of the states have net metering that ranges from 10 to 1000 kW, with most in the 10–100 kW range.

Green power is a voluntary consumer decision to purchase electricity supplied from renewable energy sources or to contribute funds for the utility to invest in renewable energy development. Green power is an option in some states' policy and also has been driven by responses of utilities to customer surveys and town meetings. Green power is available to retail or wholesale customers in 22 states [25].

Externalities are defined as social or external costs/benefits, which are attributable to an activity that is not borne by the parties involved in that activity. Externalities are not paid by the producers or consumers and are not included in the market price, although someone at sometime will pay for or be affected by them. Social benefits, generally called subsidies, are paid by someone else and accrue to a group.

1.4 Economics

The most critical factors are (1) initial cost of the installation and (2) the net annual energy production. If the system is connected to the grid, the next important factor is the value of that energy. For wind farms it is the value of the electricity sold to the utility company, for using energy on site, it is generally the value of the electricity displaced, the retail value. In determining economic feasibility, wind energy must compete with the energy available from competing technologies. Natural gas and oil prices have had large fluctuations in the past few years and the future prices for fossil fuels are uncertain, especially when carbon emissions are included. For the United States, if the military costs for insuring the flow of oil from the Middle East were included, that would probably add $0.15–0/30 per liter ($0.50–1.00/gal) to the cost of gasoline. The return from the energy generated should exceed all costs in a reasonable time. For remote locations where there is no electricity, high values for electricity from renewable energy is probably cost competitive with other sources of energy. Of course all values for electricity produced by wind power depend on the resource, so there is a range of values.

Economic analyses [2, Ch. 15] both simple and complicated provide guidelines and simple calculations should be made first. Detailed economic analyses provide information for commercial operations, primarily wind farms. Commonly calculated quantities are (1) simple payback, (2) cost of energy (COE), and (3) cash flow. A simple payback calculation can provide a preliminary judgment of economic feasibility for a wind energy system. The easiest calculation is the cost of the system divided by cost (or value of energy) displace per year.

A life cycle cost (LCC) analysis gives the total cost of the system, including all expenses incurred over the life of the system and salvage value, if any [26,27]. There are two reasons to do an LCC analysis: (1) to compare different power options and (2) to determine the most cost-effective system designs. The competing options to small renewable energy systems are batteries or small diesel generators. For these applications the initial cost of the system, the infrastructure to operate and maintain the system, and the price people pay for the energy are the main concerns. However, even if small renewable systems are the only option, a LCC analysis can be helpful for comparing costs of different designs and/or determining whether a hybrid system would be a cost-effective option. An LCC analysis allows the designer to study the effect of using different components with different reliabilities and lifetimes.

The COE (value of the energy produced by the wind system) gives a levelized value over the life of the system. The lifetime depends on the type of system and is assumed to be 20–25 years for wind turbines. The Electric Power Research Institute (EPRI Tag-Supply method), includes levelized replacement costs (major repairs) and fuel costs [28]; however, for wind energy the fuel costs are zero. The COE is primarily driven by the installed cost and the annual energy production:

$$\text{COE} = \frac{(\text{IC} * \text{FCR} + \text{LRC} + \text{AOM})}{\text{AEP}} \tag{1.5}$$

where

IC is the installed cost
FCR is the fixed charged rate
LRC is the levelized replacement cost
AOM is the annual operation and maintenance cost
AEP is the annual energy production (net)

The COE can be calculated for $/kWh or $/MWh and the last term could be separate as AOM/AEP, again in terms of $/kWh or $/MWh. It may be difficult to obtain good numbers for LCR since repair costs are generally proprietary. One method is to use a 20 year lifetime and estimate LCR as IC/20. That means the major repairs are equal to the initial cost spread over the lifetime.

Example: Wind turbine in a good wind regime. 1 MW, IC = $2,000,000, FCR = 0.07, AEP = 3000 MWh/year, LRC = $100,000/year, AOM = $8/MWh

$$\text{COE} = (\$2{,}000{,}000 * 0.07 + \$100{,}000)/3000 + 8 = 80 + 8 = \$88/\text{MWh}$$

The COE for wind turbines in a wind farm has decreased over the years (Figure 1.16); however, installed costs increased from $1.2 million/MW in 2003 to over $2 million/MW in 2010 (both in $ 2010). Remember these are average values for good wind regimes, while earlier values were for excellent wind regimes.

The COE for wind farms is around $80/MWh (Figure 1.17) without incentives and is competitive with fossil fuels for generating electricity. There are economies of scale for wind and COE for wind farms is less than the COE for small wind systems for residences, agribusiness, and even community and distributed wind (Table 1.3). The annual energy production is estimated by the generator size method in a good wind regime, and capacity factors of 25%–35% were used.

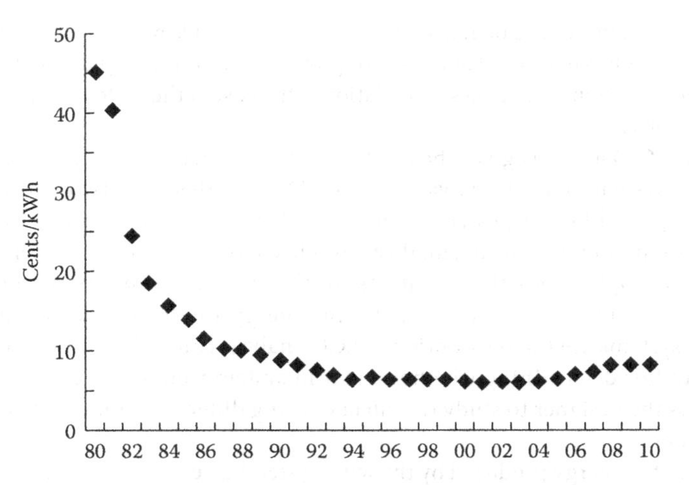

FIGURE 1.16 Cost of energy for wind turbines has declined dramatically since 1980. There is actually a range of values depending on resource, so plotted values are averages for large systems and locations with good to excellent resource.

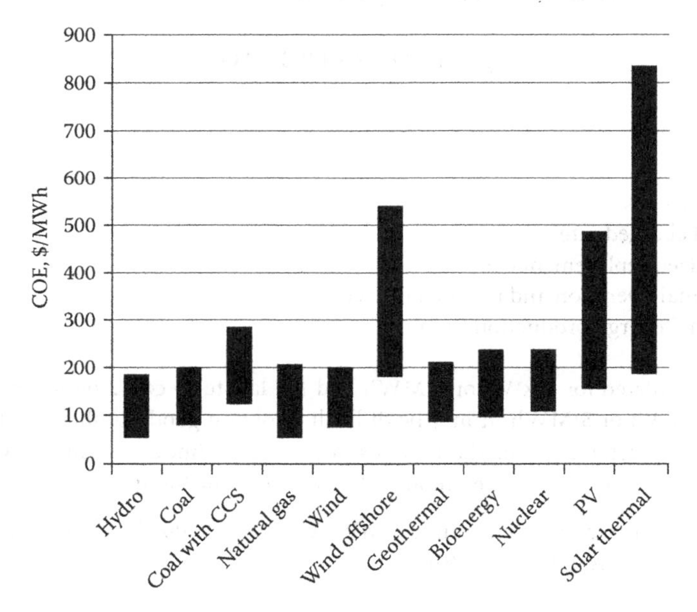

FIGURE 1.17 Estimated levelized cost of new generation resources, 2016 ($ 2009), data from Energy Information Administration. (From Levelized cost of new generation resources in the annual energy outlook 2011, U.S. Energy Information Administration, www.eia.doe.gov/oiaf/aeo/electricity_generation.html.)

The capital cost is the major cost for a project and of that the wind turbine is the major component (Table 1.4). Most renewable energy projects will be similar in that equipment costs are the major item. The installed cost for offshore wind farms is around two times that of wind farms on land.

The value to the landowner can be estimated from the annual energy production and/or MW installed:

1. Royalty on production, 4%–6% with escalation, generally at 10 year periods
2. $4000 to $6000/MW installed per year
3. Combination, (1) or (2), whichever is larger

TABLE 1.3 COE for Wind Turbines in Good Wind Resource Area

Power, kW	IC, $/kW	COE, $/kWh
Remote, Battery	7500–?	0.30–?
0.1–2	5000–6000	0.20–0.30
5–10	5000	0.16–0.25
20–50	4000–5000	0.15–0.18
100	4500	0.13–0.15
1000–	2000	0.07–0.10

Note: Smaller size has higher cost.

TABLE 1.4 Percent Cost for Wind Farm Installation

Component	%
Turbine	74–82
Foundation	1–6
Electric	2–9
Connections to grid	2–9
Land	1–3
Roads, ditching	1–5
Consultants, resource assessment, other	1–3

The return to the landowner is higher than ranching or farming as only 2%–3% of the land is removed from production, primarily for roads (see Figure 1.12).

1.5 Summary

Wind power from wind farms has become a major component for new power installations in many regions of the world. Annual capacity factors range from 0.30 to 0.45 at good to excellent wind locations. One limitation for wind power is that, in general, windy areas are distant from major load centers and wind farms can be installed faster than construction of new transmission lines for the utility grid. As wind power increases beyond 20% penetration into the grid, more strict requirements from power grid operators will be required, such as wind power output forecasting, power quality, fault (low voltage) ride through, etc., similar to conventional power plants.

If storage becomes economical, then renewable energy and especially wind power will supply even more of the world demand for electric energy. At the moment, pumped storage is most economical and should be considered with long-distance extra-high-voltage transmission lines during national electrical power system planning, to accommodate more wind and solar power penetration into large grids.

There is a growing market for small wind turbines for stand-alone and grid connection and midsize wind turbines for the distributed and community market. The market potential for village power is large; however, there are still problems, primarily institutional and economic costs for these communities.

Past estimates of future installation of wind power have been low, so now the planned installations are at least feasible, for example, the proposal, 20% by 2030 for the United States [15]. China will probably become the leader in manufacturing of large wind turbines due to their large increase in installed capacity since 2008 and due to goal of 300 MW by 2020. Ten percent of electricity by wind power for the world by 2020 would require around 1000 GW.

The world faces a tremendous energy problem in terms of supply and in terms of emissions from the use of fossil fuels. The first priority is conservation and energy efficiency and the second is a shift to renewable energy for a sustainable energy future. This shift has started to occur and renewable energy market will grow rapidly over next 30 years.

Links:

Global Wind Energy Council, www.gwec.net (accessed 2/13/2012).
American Wind Energy Association, www.awea.org (accessed on 2/13/2012).
European Wind Energy Association, www.ewea.org (accessed on 2/13/2012).
Alternative Energy Institute, www.windenergy.org (accessed on 2/13/2012).
National Wind Technology Center, NREL, www.nrel.gov/wind (accessed on 2/13/2012).

Many countries have wind energy associations.

References

1. V. Nelson and N. Clark, 2004, *Wind Water Pumping*, CD, Alternative Energy Institute, West Texas A&M University, Canyon, TX.
2. V. Nelson, 2009, *Wind Energy, Renewable Energy and the Environment*, CRC Press, New York.
3. D.G. Sheppard, 2009, Historical development of the windmill. In *Wind Turbine Technology, Fundamental Concepts of Wind Turbine Engineering*, 2nd edn., ed. D.A. Spera, ASME Press, New York.
4. Danish Wind Industry Association, www.windpower.org/en/knowledge.html, great site, check out Guided Tour and Wind with Miller (accessed on 2/13/2012).
5. National Wind Technology Center, NREL, www.nrel.gov/wind/resource_assessment.html (accessed on 2/13/2012).
6. V. Nelson, 2008, Ch. 4 Wind energy. In *Texas Renewable Energy Resource Assessment*, www.seco.cpa.state.tx.us/publications/renewenergy (accessed on 2/13/2012).
7. US Northwest States, wind maps; www.windmaps.org/default.asp (accessed on 2/13/2012).
8. 3Tier, FirstLook Prospecting, www.3tiergroup.com/wind/overview (accessed on 2/13/2012).
9. AWS Truepower Navigator, https://www.windnavigator.com/ (accessed on 2/13/2012).
10. Wind Atlases of the World, www.windatlas.dk/World/About.html (accessed on 2/13/2012).
11. Remote Sensing Systems, www.remss.com (accessed on 2/13/2012).
12. Jet Propulsion Laboratory, Physical Oceanography DAAC, ocean winds, http://podaac.jpl.nasa.gov/index.htm (accessed on 2/13/2012).
13. American Wind Energy Association, Projects, www.awea.org/projects (accessed on 2/13/2012).
14. Global Wind Energy Council, Global Wind 2009 Report, www.gwec.net/index.php?id=167 (accessed on 2/13/2012).
15. 20% Wind Energy by 2030 Report, www1.eere.energy.gov/wind/pdfs/41869.pdf (accessed on 2/13/2012).
16. American Wind Energy Association, www.awea.org/smallwind (accessed on 2/13/2012).
17. T. Forsyth and I. Baring-Gould, 2007, Distributed wind market applications, Technical Report NREL/TP-500-39851, www.nrel.gov/wind/pdfs/forsyth_distributed_market.pdf (accessed on 2/13/2012).
18. Wind Diesel Workshop, Girdwood, AK, 2004 www.eere.energy.gov/windandhydro/pdgfs/workshops/windpowering america/wkshp_2004_wind_diesel.asp (accessed on 2/13/2012).
19. G. Seifert and K. Myers, 2006, Wind-diesel hybrid power, trials and tribulations at Ascension Island, *Proceedings European Energy Conference*, www.ewec 2006 proceedings.info (accessed on 2/13/2012).
20. J. Vlahos March 2011, Blue sky power, *Popular Mechanics*, 51.
21. Wind energy integration in the urban environment, www.urbanwind.org (accessed on 2/13/2012).
22. Bahrain World Trade Center, www.bahrainwtc.com/index.htm (accessed on 2/13/2012).

23. The Database of State Incentives for Renewable Energy (DSIRE), www.dsireusa.org (accessed on 2/13/2012).
24. H. Scheer 1999, Energy subsidies-a basic perspective, *2nd Conference on Financing Renewable Energies, 1999 European Wind Energy Conference*, Bonn, Germany.
25. The Green Power Network, www.eere.energy.gov/greenpower (accessed on 2/13/2012).
26. R.J. Brown and R.R. Yanuck 1980, *Life Cycle Costing, A Practical Guide for Energy Managers*, Fairmont Press, Atlanta, GA.
27. J.M. Sherman, M.S. Gresham, and D.L. Fergason 1982, Wind systems life cycle cost analysis, RFP-3448, UC-60; NREL; Wind energy finance, www.nrel.gov/docs/fy04osti/33638.pdf (accessed on 2/13/2012).
28. J.M. Cohen, T.C. Schweizer, S.M. Hock, and J.B. Cadogan, A methodology for computing wind turbine cost of electricity using utility economic assumptions, *Windpower '89 Proceedings*, NREL/TP-257-3628.
29. Levelized cost of new generation resources in the annual energy outlook 2011, U.S. Energy Information Administration, www.eia.doe.gov/oiaf/aeo/electricity_generation.html (accessed on 2/13/2012).

2

Photovoltaic Fundamentals

Roger A. Messenger
Florida Atlantic University

2.1 Introduction

Becquerel [1] first discovered that sunlight can be converted directly into electricity in 1839, when he observed the photogalvanic effect. After more than a century of theoretical work, the first solar cell was not developed until 1954, by Chapin, Fuller, and Pearson, when sufficiently pure semiconductor material had become available. It had an efficiency of 6%. Only 4 years later, the first solar cells were used on the Vanguard I orbiting satellite.

As is often the case, extraterrestrial use of a technology has led to the terrestrial use of the technology. For extraterrestrial cells, the power:weight ratio was the determining factor for their selection, since weight was the expensive part of the liftoff equation. Reliability, of course, was also of paramount importance. Hence, for extraterrestrial power generation, the cost of the photovoltaic (PV) cells was secondary, provided that they met the primary objectives.

A new objective for PV cells emerged following the organization of the petroleum exporting countries (OPEC) petroleum embargo of the early 1970s, which highlighted the need to seek out alternative

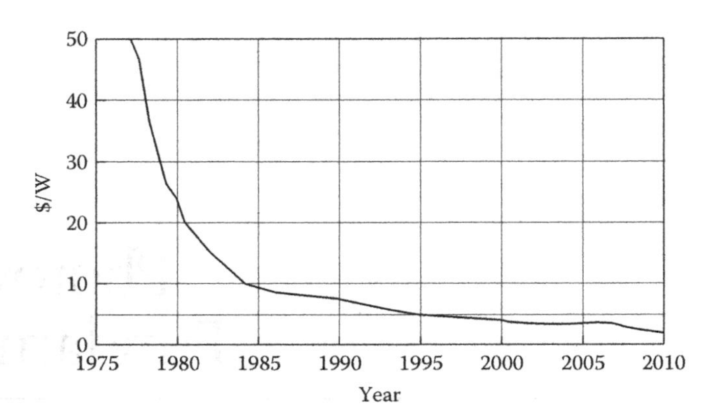

FIGURE 2.1 The decline in cost per watt for photovoltaic modules. (Data from Maycock, P.D., Ed., *Photovoltaic News*, 18(1), 1999; From *International Marketing Data and Statistics*, 22nd edn., Euromonitor PIC, London, U.K., 1998; From Maycock, P.D., *Renewable Energy World*, 5, 147, 2002; From Earth Policy Institute, *Eco-Economy Indicators: Solar Power*—Data www.earth-policy.org/Indicators/Solar/2007_data.htm.)

energy sources. In addition, the observation that fossil fuel sources pollute the atmosphere, water and soil with a number of pollutants such as SO_x, NO_x, particulates, and CO_2 suggested that alternative sources should involve minimal air or water pollution. And, of course, perhaps the most desirable aspect of any new source would be an unlimited supply of fuel, as opposed to the observed dwindling supplies of domestic petroleum.

Solar energy became the obvious choice for alternative energy, either in the form of electricity via the PV effect or via solar thermal heating of water or other substances. But it would be necessary to significantly reduce the cost of capturing the solar resource before it could be used as a replacement for the relatively inexpensive fossil sources.

Since the events of the mid-1970s, the cost per watt of PV cells has declined steadily. Figure 2.1 shows an average decline of approximately 7%/year between 1980 and 2010, which represents a 50% drop every 10 years. In fact, between 2005 and 2010, the price dropped by 50% in a 5 year period. Installations have grown from tens of kilowatts to thousands of megawatts per year, and continue to grow at a rapid rate. Recent rapid growth and accompanying declines in prices can be accounted for by a number of aggressive incentive programs around the world, such as in Germany, Spain, China, and California. For example, Germany now produces more than 20% of its electrical energy by renewable sources. Costa Rica will soon produce nearly 100% of its electrical energy needs from renewable sources, but in this case, hydroelectric and wind will be the major contributors to the renewable portfolio.

2.2 Market Drivers

Price, efficiency, performance, warranties, incentives, interest rates, money availability, space availability, and sometimes aesthetics have been the primary market drivers for current PV technologies. In fact, the decision to install generally involves consideration of all these factors.

Certainly price is important, but obviously if an inexpensive module will cost half as much as another module, but will last only 25% as long, the more expensive module will have a more attractive lifecycle cost. In many instances, where space is limited, the system owner will want to install as much power as possible in the available space in order to generate as much energy as possible. Where space is a consideration, then the metric for PV system selection is based upon kWh/m^2/year/$ for the energy produced by the system. And, in fact, generally the monetary consideration is only secondary to the energy production consideration.

Probably the most important driver among the remaining factors is incentive programs. It has been clearly demonstrated in Germany, Spain, California, New Jersey, Ontario, and a number of other countries, states, and cities that an attractive incentive program will encourage people to install PV systems on their homes and businesses. Incentive programs may take the form of one-time rebates or tax credits or they may take the form of a guaranteed premium price paid over a guaranteed time period for the energy produced by the renewable system. In the final analysis, the owner of a system has a greater incentive as the payback time of the system decreases. Generally, if a system will pay for itself with the value of the energy it produces within a 6 year period, it is a very attractive investment for a homeowner and an acceptable investment for many businesses. The trade-off must be the impact on utility ratepayers, since, generally the funding for the incentive comes either directly from the utility or indirectly via government loans or grants, which, ultimately, are paid for by the taxpayer, who is most likely a ratepayer. Incentive programs must be carefully crafted to ensure acceptance by those who wish to install systems, but not to impose an excessive cost burden on the general population. Just as the energy produced must be sustainable, the incentive program must also be sustainable.

2.3 Optical Absorption

2.3.1 Introduction

When light shines on a material, it is reflected, transmitted, or absorbed. Absorption of light is simply the conversion of the energy contained in the incident photon to some other form of energy, typically heat. Some materials, however, happen to have just the right properties needed to convert the energy in the incident photons to electrical energy.

2.3.2 Semiconductor Materials

Semiconductor materials are characterized as being perfect insulators at absolute zero temperature, with charge carriers being made available for conduction as the temperature of the material is increased. This phenomenon can be explained on the basis of quantum theory, by noting that semiconductor materials have an energy band gap between the valence band and the conduction band. The valence band represents the allowable energies of valence electrons that are bound to host atoms. The conduction band represents the allowable energies of electrons that have received energy from some mechanism and are now no longer bound to specific host atoms.

As temperature of a semiconductor sample is increased, sufficient energy is imparted to a small fraction of the electrons in the valence band for them to move to the conduction band. In effect, these electrons are leaving covalent bonds in the semiconductor host material. When an electron leaves the valence band, an opening is left which may now be occupied by another electron, provided that the other electron moves to the opening. If this happens, of course, the electron that moves in the valence band to the opening, leaves behind an opening in the location from which it moved. If one engages in an elegant quantum mechanical explanation of this phenomenon, it must be concluded that the electron moving in the valence band must have either a negative effective mass and a negative charge or, alternatively, a positive effective mass and a positive charge. The latter has been the popular description, and, as a result, the electron motion in the valence band is called hole motion, where "holes" is the name chosen for the positive charges, since they relate to the moving holes that the electrons have left in the valence band.

What is important to note about these conduction electrons and valence holes is that they have occurred in pairs. Hence, when an electron is moved from the valence band to the conduction band in a semiconductor by whatever means, it constitutes the creation of an electron-hole pair (EHP). Both charge carriers are then free to become a part of the conduction process in the material.

2.3.3 Generation of EHP by Photon Absorption

The energy in a photon is given by the familiar equation,

$$E = h\nu = \frac{hc}{\lambda} \quad (\mathrm{J}), \tag{2.1a}$$

where
 h is Planck's constant ($h = 6.63 \times 10^{-34}$ J s)
 c is the speed of light ($c = 2.998 \times 10^{8}$ m/s)
 ν is the frequency of the photon in Hertz
 λ is the wavelength of the photon in meters

Since energies at the atomic level are typically expressed in electron volts ($1\ \mathrm{eV} = 1.6 \times 10^{-19}$ J) and wavelengths are typically expressed in micrometers, it is possible to express hc in appropriate units, so that if λ is expressed in μm, then E will be expressed in eV. The conversion yields

$$E = \frac{1.24}{\lambda} \quad (\mathrm{eV}). \tag{2.1b}$$

The energy in a photon must exceed the semiconductor bandgap energy, E_g, to be absorbed. Photons with energies at and just above E_g are most readily absorbed because they most closely match bandgap energy and momentum considerations. If a photon has energy greater than the bandgap, it still can produce only a single EHP. The remainder of the photon energy is lost to the cell as heat. It is thus desirable that the semiconductor used for photoabsorption have a bandgap energy such that a maximum percentage of the solar spectrum will be efficiently absorbed.

Now, note that the solar spectrum peaks at $\lambda \approx 0.5$ μm. Equation 2.1b shows that a bandgap energy of approximately 2.5 eV corresponds to the peak in the solar spectrum. In fact, since the peak of the solar spectrum is relatively broad, bandgap energies down to 1.0 eV can still be relatively efficient absorbers, and in certain special cell configurations, even smaller bandgap materials are appropriate.

The nature of the bandgap also affects the efficiency of absorption in a material. A more complete representation of semiconductor bandgaps must show the relationship between bandgap energy as well as bandgap momentum. As electrons make transitions between conduction band and valence band, both energy and momentum transfer normally take place, and both must be properly balanced in accordance with conservation of energy and conservation of momentum laws.

Some semiconducting materials are classified as direct bandgap materials, while others are classified as indirect bandgap materials. Figure 2.2 [6] shows the bandgap diagrams for two materials considering momentum as well as energy. Note that for silicon, the bottom of the conduction band is displaced in the momentum direction from the peak of the valence band. This is an indirect bandgap, while the GaAs diagram shows a direct bandgap, where the bottom of the conduction band is aligned with the top of the valence band.

What these diagrams show is that the allowed energies of a particle in the valence band or the conduction band depend on the particle momentum in these bands. An electron transition from a point in the valence band to a point in the conduction band must involve conservation of momentum as well as energy. For example, in Si, even though the separation of the bottom of the conduction band and the top of the valence band is 1.1 eV, it is difficult for a 1.1 eV photon to excite a valence electron to the conduction band because the transition needs to be accompanied with sufficient momentum to cause

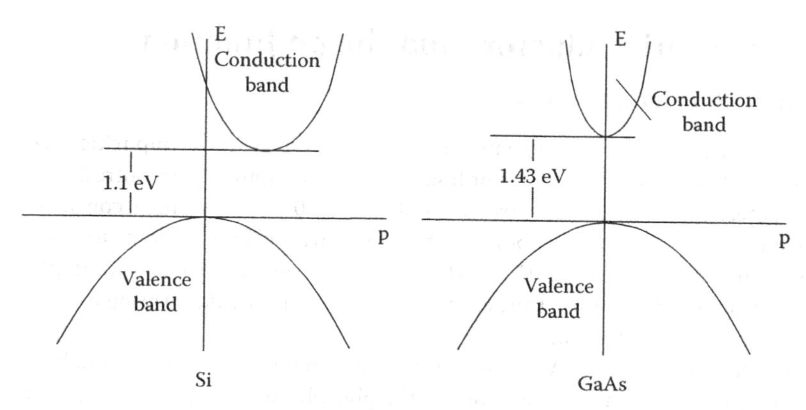

FIGURE 2.2 The energy and momentum diagram for valence and conduction bands in Si and GaAs.

displacement along the momentum axis, and photons carry little momentum. The valence electron must thus simultaneously gain momentum from another source as it absorbs energy from the incident photon. Since such simultaneous events are unlikely, absorption of photons at the Si bandgap energy is several orders of magnitude less likely than absorption of higher energy photons.

Since photons have so little momentum, it turns out that the direct bandgap materials, such as gallium arsenide (GaAs), cadmium telluride (CdTe), copper indium diselenide (CIS), and amorphous silicon (a-Si:H) absorb photons with energy near the material bandgap energy much more readily than do the indirect materials, such as crystalline silicon. As a result, the direct bandgap absorbing material can be several orders of magnitude thinner than indirect bandgap materials and still absorb a significant part of the incident radiation.

The absorption process is similar to many other physical processes, in that the change in intensity with position is proportional to the initial intensity. As an equation, this becomes

$$\frac{dI}{dx} = -\alpha I, \tag{2.2}$$

with the solution

$$I = I_o e^{-\alpha x}, \tag{2.3}$$

where

I is the intensity of the light at a depth x in the material
I_o is the intensity at the surface
α is the absorption constant

The absorption constant depends on the material and on the wavelength. At energies above the bandgap, the absorption constant increases relatively slowly for indirect bandgap semiconductors and increases relatively quickly for direct bandgap materials. Equation 2.3 shows that the thickness of material needed for significant absorption needs to be several times the reciprocal of the absorption constant. This is important information for the designer of a PV cell, since the cell must be sufficiently thick to absorb the incident light.

In any case, when the photon is absorbed, it generates an EHP. The question, then, is what happens to the EHP?

2.4 Extrinsic Semiconductors and the pn Junction

2.4.1 Extrinsic Semiconductors

At T = 0 K, intrinsic semiconductors, i.e., semiconductor materials with no impurities, have all covalent bonds completed with no leftover electrons or holes. If certain impurities are introduced into intrinsic semiconductors, there can be leftover electrons or holes at T = 0 K. For example, consider silicon, which is a group IV element, which covalently bonds with four nearest neighbor atoms to complete the outer electron shells of all the atoms. At T = 0 K, all the covalently bonded electrons are in place, whereas at room temperature, about one in 10^{12} of these covalent bonds will break, forming an EHP, resulting in minimal charge carriers for current flow.

If, however, phosphorous, a group V element, is introduced into the silicon in small quantities, such as one part in 10^6, four of the valence electrons of the phosphorous atoms will covalently bond to the neighboring silicon atoms, while the fifth valence electron will have no electrons with which to covalently bond. This fifth electron remains weakly coupled to the phosphorous atom, readily dislodged by temperature, since it requires only 0.04 eV to excite the electron from the atom to the conduction band [6]. At room temperature, sufficient thermal energy is available to dislodge essentially all of these extra electrons from the phosphorous impurities. These electrons thus enter the conduction band under thermal equilibrium conditions, and the concentration of electrons in the conduction band becomes nearly equal to the concentration of phosphorous atoms, since the impurity concentration is normally on the order of 10^8 times larger than the intrinsic carrier concentration.

Since the phosphorous atoms donate electrons to the material, they are called *donor* atoms and are represented by the concentration, N_D. Note that the phosphorous, or other group V impurities, *do not add holes* to the material. They only add electrons. They are thus designated as n-type impurities.

On the other hand, if group III atoms such as boron are added to the intrinsic silicon, they have only three valence electrons to covalently bond with nearest silicon neighbors. The missing covalent bond appears the same as a hole, which can be released to the material by applying a small amount of thermal energy. Again, at room temperature, nearly all of the available holes from the group III impurity are donated to the conduction process in the host material. Since the concentration of impurities will normally be much larger than the intrinsic carrier concentration, the concentration of holes in the material will be approximately equal to the concentration of p-type impurities.

Historically, group III impurities in silicon have been viewed as electron acceptors, which, in effect, donate holes to the material. Rather than being termed hole donors, however, they have been called *acceptors*. Thus, acceptor impurities donate holes, but no electrons, to the material and the resulting hole density is approximately equal to the density of acceptors, which is represented as N_A.

An interesting property of free electrons and holes is that they like each other. In fact, when they are close to each other, they have a strong tendency to recombine. This observation can be expressed as [6]

$$n_o p_o = n_i^2(T) = KT^3 e^{-E_g/kT}, \tag{2.4}$$

where

n_o represents the thermal equilibrium concentration of free electrons at a point in the semiconductor crystal

p_o represents the thermal equilibrium concentration of holes

n_i represents the concentration of intrinsic charge carriers

T is the temperature in K

K is a constant that depends upon the material

Note that in intrinsic material, $n_o \cong p_o = n_i$ since, by definition, the intrinsic material has no impurities that affect the electrical properties of the material.

In Si, for example, which has approximately 10^{22} covalent bonds/cm^3, approximately 1 in 10^{12} bonds creates an EHP at room temperature. This means that at room temperature for Si, $n_i \cong 10^{10}/\text{cm}^3$.

Equation 2.4 shows that adding a mere one part per million of a donor or acceptor impurity can increase the free charge carrier concentration of the material by a factor of 10^6 for silicon. Equation 2.4 also shows that in thermal equilibrium, if either an n-type or a p-type impurity is added to the host material, that the concentration of the other charge carrier will decrease dramatically, since it is still necessary to satisfy (2.4) under thermal equilibrium conditions.

In extrinsic semiconductors, the charge carrier with the highest concentration is called the *majority carrier* and the charge carrier with the lowest concentration is called the *minority carrier.* Hence, electrons are majority carriers in n-type material, and holes are minority carriers in n-type material. The opposite is true for p-type material.

If both n-type and p-type impurities are added to a material, then whichever impurity has the higher concentration, will become the dominant impurity. However, it is then necessary to acknowledge a net impurity concentration that is given by the difference between the donor and acceptor concentrations. If, for example, $N_D > N_A$, then the net impurity concentration is defined as $N_d = N_D - N_A$. Similarly, if $N_A > N_D$, then $N_a = N_A - N_D$.

2.4.2 pn Junction

2.4.2.1 Junction Formation and Built-In Potential

Although n-type and p-type materials are interesting and useful, the real fun starts when a junction is formed between n-type and p-type materials. The pn junction is treated in gory detail in most semiconductor device theory textbooks. Here, the need is to establish the foundation for the establishment of an electric field across a pn junction and to note the effect of this electric field on photo-generated EHPs.

Figure 2.3 shows a pn junction formed by placing p-type impurities on one side and n-type impurities on the other side. There are many ways to accomplish this structure. The most common is the diffused junction.

When a junction is formed, the first thing to happen is that the conduction electrons on the n-side of the junction notice the scarcity of conduction electrons on the p-side, and the valence holes on the p-side notice the scarcity of valence holes on the n-side. Since both types of charge carrier are undergoing random thermal motion, they begin to diffuse to the opposite side of the junction in search of the wide open spaces. The result is diffusion of electrons and holes across the junction, as indicated in Figure 2.3.

When an electron leaves the n-side for the p-side, however, it leaves behind a positive donor ion on the n-side, right at the junction. Similarly, when a hole leaves the p-side for the n-side, it leaves a negative acceptor ion on the p-side. If large numbers of holes and electrons travel across the junction, large numbers of fixed positive and negative ions are left at the junction boundaries. These fixed ions, as a result of Gauss' law, create an electric field that originates on the positive ions and terminates on the negative ions.

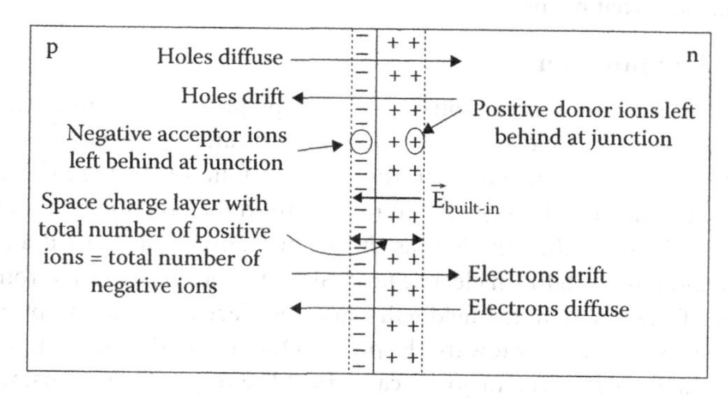

FIGURE 2.3 The pn junction showing electron and hole drift and diffusion.

Hence, the number of positive ions on the n-side of the junction must be equal to the number of negative ions on the p-side of the junction.

The electric field across the junction, of course, gives rise to a drift current in the direction of the electric field. This means that holes will travel in the direction of the electric field and electrons will travel opposite the direction of the field, as shown in Figure 2.3. Notice that for both the electrons and for the holes, the drift current component is opposite the diffusion current component. At this point, one can invoke Kirchhoff's current law to establish that the drift and diffusion components for each charge carrier must be equal and opposite, since there is no net current flow through the junction region. This phenomenon is known as the law of detailed balance.

By setting the sum of the electron diffusion current and the electron drift current equal to zero, it is possible to solve for the potential difference across the junction in terms of the impurity concentrations on either side of the junction. Proceeding with this operation yields

$$V_j = \frac{kT}{q}\ln\frac{n_{no}}{n_{po}}. \tag{2.5a}$$

It is now possible to express the built-in potential in terms of the impurity concentrations on either side of the junction by recognizing that $n_{no} \cong N_D$ and $n_{po} \cong (n_i)^2/N_A$. Substituting these values into (2.5a) yields finally

$$V_j = \frac{kT}{q}\ln\frac{N_A N_D}{n_i^2}. \tag{2.5b}$$

At this point, a word about the region containing the donor ions and acceptor ions is in order. Note first that outside this region, electron and hole concentrations remain at their thermal equilibrium values. Within the region, however, the concentration of electrons must change from the high value on the n-side to the low value on the p-side. Similarly, the hole concentration must change from the high value on the p-side to the low value on the n-side. Considering that the high values are really high, i.e., on the order of $10^{18}/cm^3$, while the low values are really low, i.e., on the order of $10^2/cm^3$, this means that within a short distance of the beginning of the ionized region, the concentration must drop significantly below the equilibrium value. Because the concentrations of charge carriers in the ionized region are so low, this region is often termed the *depletion region*, in recognition of the depletion of mobile charge carriers in the region. Furthermore, because of the charge due to the ions in this region, the depletion region is also often referred to as the *space charge layer*. For the balance of this chapter, this region will simply be referred to as the junction.

The next step in the development of the behavior of the pn junction in the presence of sunlight is to let the sun shine in and see what happens.

2.4.2.2 Illuminated pn Junction

Equation 2.3 governs the absorption of photons at or near a pn junction. Noting that an absorbed photon releases an EHP, it is now possible to explore what happens after the generation of the EHP. Those EHPs generated within the pn junction will be considered first, followed by the EHPs generated outside, but near the junction, and then by EHPs generated further from the junction boundary.

If an EHP is generated within the junction, as shown in Figure 2.4 (points B and C), both charge carriers will be acted upon by the built-in electric field. Since the field is directed from the n-side of the junction to the p-side of the junction, the field will cause the electrons to be swept quickly toward the n-side and the holes to be swept quickly toward the p-side. Once out of the junction region, the optically generated carriers become a part of the majority carriers of the respective regions, with the result that excess concentrations of majority carriers appear at the edges of the junction. These excess majority

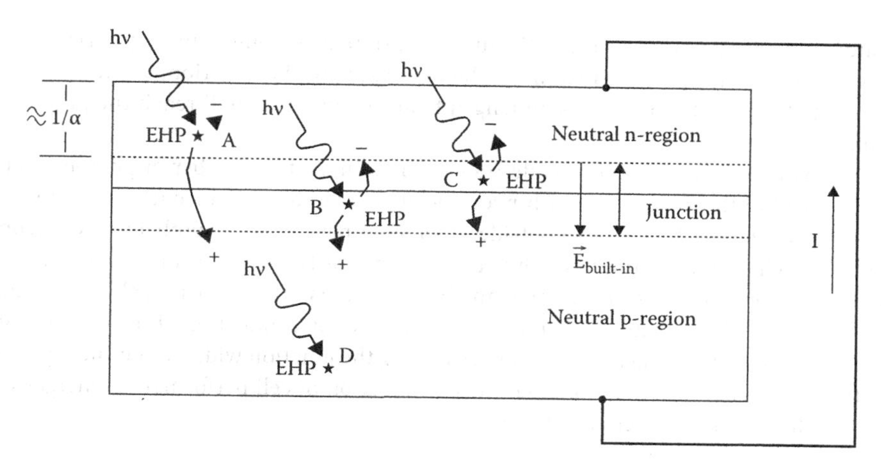

FIGURE 2.4 The illuminated pn junction showing desirable geometry and the creation of electron-hole pairs. *Note:* * At all EHP locations.

carriers then diffuse away from the junction toward the external contacts, since the concentration of majority carriers has been enhanced only near the junction.

The addition of excess majority charge carriers to each side of the junction results in either a voltage between the external terminals of the material or a flow of current in the external circuit or both. If an external wire is connected between the n-side of the material and the p-side of the material, a current, I_l, will flow in the wire from the p-side to the n-side. This current will be proportional to the number of EHPs generated in the junction region, which, in turn, will be proportional to the intensity of the incident light (irradiation).

If an EHP is generated outside the junction region, but close to the junction (with "close" yet to be defined, but shown as point A in Figure 2.4), it is possible that, due to random thermal motion, either the electron, the hole, or both will end up moving into the junction region. Suppose that an EHP is generated in the n-region close to the junction. Then suppose that the hole, which is the minority carrier in the n-region, manages to reach the junction before it recombines. If it can do this, it will be swept across the junction to the p-side and the net effect will be the same as if the EHP had been generated within the junction, since the electron is already on the n-side as a majority carrier. Similarly, if an EHP is generated within the p-region, but close to the junction, and if the minority carrier electron reaches the junction before recombining, it will be swept across to the n-side where it is a majority carrier. So what is meant by close?

Clearly, the minority carriers of the optically generated EHPs outside the junction region must not recombine before they reach the junction. If they do, then, effectively, both carriers are lost from the conduction process, as in point D in Figure 2.4. Since the majority carrier is already on the correct side of the junction, the minority carrier must therefore reach the junction in less than a minority carrier lifetime, τ_n or τ_p.

To convert these times into distances, it is necessary to note that the carriers travel by diffusion once they are created. Since only the thermal velocity has been associated with diffusion, but since the thermal velocity is random in direction, it is necessary to introduce the concept of minority carrier diffusion length, which represents the distance, on the average, which a minority carrier will travel before it recombines. The diffusion length can be shown to be related to the minority carrier lifetime and diffusion constant by the formula [7]

$$L_m = \sqrt{D_m \tau_m}, \tag{2.6}$$

where m has been introduced to represent n for electrons or p for holes. It can also be shown that on the average, if an EHP is generated within a minority carrier diffusion length of the junction, the associated minority carrier will reach the junction. In reality, some minority carriers generated closer than a diffusion length will recombine before reaching the junction, while some minority carriers generated farther than a diffusion length from the junction will reach the junction before recombining.

Hence, to maximize photocurrent, it is desirable to maximize the number of photons that will be absorbed either within the junction or within a minority carrier diffusion length of the junction. The minority carriers of the EHPs generated outside this region have a higher probability of recombining before they have a chance to diffuse to the junction. If a minority carrier from an optically generated EHP recombines before it crosses the junction and becomes a majority carrier, it, along with the opposite carrier with which it recombines, is no longer available for conduction. The engineering design challenge then lies in maximizing α as well as maximizing the junction width and minority carrier diffusion lengths. Additional information about the optimization of cell performance emerges when the performance of the cell under external bias is explored.

2.5 Maximizing Cell Performance

2.5.1 Externally Biased pn Junction

Figure 2.5 shows a pn junction connected to an external battery with the internally generated electric field direction included. If (2.5a) is recalled, taking into account that the externally applied voltage, V, with the exception of any voltage drop in the neutral regions of the material, will appear as opposing the junction voltage, Equation 2.5a becomes

$$V_j - V = \frac{kT}{q}\ln\frac{n_n}{n_p}, \tag{2.7}$$

where now n_n and n_p are the total concentrations of electrons on the n-side of the junction and on the p-side of the junction, no longer in thermal equilibrium. Thermal equilibrium will exist only when the externally applied voltage is zero, meaning that $np = n_i^2$, is only true when $V = 0$. However, under conditions known as low injection levels, it will still be the case that the concentration of electrons on the n-side will remain close to the thermal equilibrium concentration. For this condition, (2.7) becomes

$$V_j - V = \frac{kT}{q}\ln\frac{N_d}{n_p}. \tag{2.8}$$

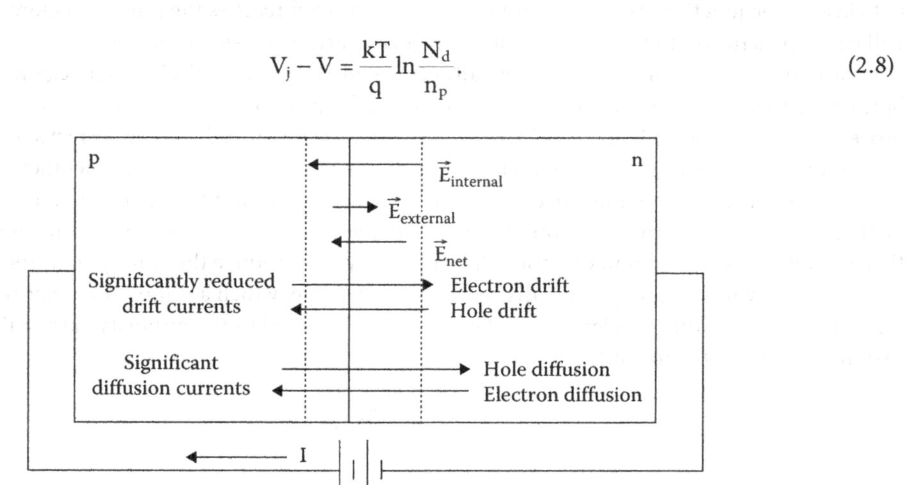

FIGURE 2.5 The pn junction with external bias.

The next step is solving (2.8) for n_p and then going through a detailed analysis of the spatial dependence of n_p between the junction edge and the contact on the p-side of the junction. Then, a similar analysis is done for the spatial dependence of p_n on the p-side of the junction. Finally, it is noted that current flow in the neutral regions of the device is primarily due to diffusion of the higher concentrations of excess charges near the junction toward the contact areas where the minority carrier concentrations are essentially zero. Solving the diffusion equations for n_p and p_n in the neutral regions leads to [8]

$$I = I_n + I_p = qA\left(\frac{D_n n_{po}}{L_n}\,\text{ctnh}\frac{w_p}{L_n} + \frac{D_p p_{no}}{L_p}\,\text{ctnh}\frac{w_n}{L_p}\right)\left(e^{qV/kT} - 1\right), \tag{2.9}$$

where

A is the cross-sectional area of the junction
D_n and D_p are the diffusion coefficients of the minority carriers in the neutral regions
L_n and L_p are the diffusion lengths of the minority carriers in the neutral regions
w_n and w_p are the physical widths of the neutral regions
n_{po} and p_{no} are the equilibrium concentrations of minority carriers in the neutral regions

Note that the current indicated in (2.9) flows in the direction opposite to the optically generated current described earlier. Letting qA (nasty expression) = I_o and incorporating the photocurrent component into (2.9) finally yields the complete equation for the current in the PV cell to be

$$I = I_l - I_o\left(e^{qV/kT} - 1\right), \tag{2.10}$$

which is the familiar diode equation with a photocurrent component. Note that the current of (2.10) is directed out of the positive terminal of the device, so that when the current and voltage are both positive, the device is *delivering* power to the external circuit. The detailed derivation of this expression can be found in Chapter 10 of [8].

The open circuit voltage and the short circuit current of the PV cell can be determined from (2.10). Setting I = 0, solving for V and noting that $I_l >> I_o$ under normal cell illumination conditions gives the cell open circuit voltage

$$V_{OC} = \frac{kT}{q}\ln\frac{I_l + I_o}{I_o} \simeq \frac{kT}{q}\ln\frac{I_l}{I_o}. \tag{2.11}$$

Setting V = 0 and solving for I yields the simple result that $I_{SC} = I_l$. In other words, the cell short circuit current is the photocurrent of the cell.

Even with all the details incorporated in the development of (2.9), what has been left out is the fact that when high current flows in the cell, the ohmic resistance of the neutral regions creates an additional voltage drop beyond the drop across the junction. This series voltage drop in the neutral regions causes the I–V relationship of (2.9) to be modified as shown in Figure 2.6. In this figure, one can observe that for each cell illumination level, only one point on the curve represents maximum power production, P_m, as is shown in the figure. The fill factor of a cell is defined as FF = $P_m/V_{OC}I_{SC}$.

2.5.2 Parameter Optimization

2.5.2.1 Introduction

Equation 2.10 indicates, albeit in a somewhat subtle manner, that to maximize the power output of a PV cell, it is desirable to maximize the open-circuit voltage, short-circuit current, and fill factor of a cell. Noting Figure 2.6, it should be evident that maximizing the open-circuit voltage and the short-circuit

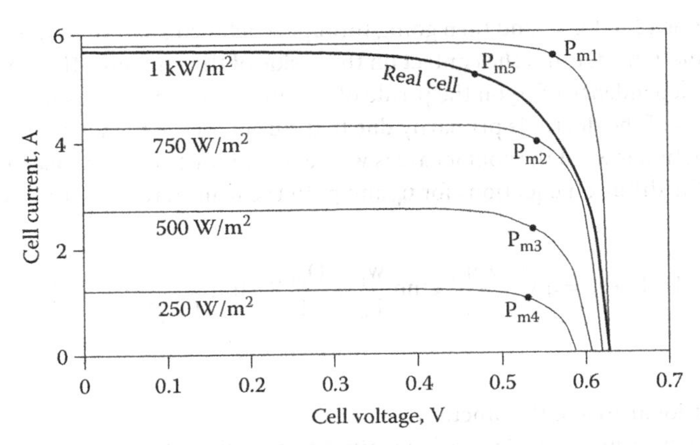

FIGURE 2.6 I–V characteristics of real and ideal PV cells under different illumination levels.

current will maximize the power output for an ideal cell characterized by (2.10). Real cells, of course, have some series resistance, so there will be power dissipated by this resistance, similar to the power loss in a conventional battery due to its internal resistance. In any case, recalling that the open-circuit voltage increases as the ratio of photo current to reverse saturation current increases, a desirable design criteria is to maximize this ratio, provided that it does not proportionally reduce the short-circuit current of the device.

Fortunately, this is not the case, since maximizing the short-circuit current requires maximizing the photocurrent. It is thus instructive to look closely at the parameters that determine both the reverse saturation current and the photocurrent. Techniques for lowering series resistance will then be discussed.

2.5.2.2 Minimizing the Reverse Saturation Current

Beginning with the reverse saturation current as expressed in (2.9), the first observation is that the equilibrium minority carrier concentrations at the edges of the pn junction are related to the intrinsic carrier concentration through (2.4). Hence,

$$p_{no} = \frac{n_i^2}{N_D} \quad \text{and} \quad n_{po} = \frac{n_i^2}{N_A}. \tag{2.12}$$

Thus far, no analytic expression for the constant, K, in (2.4) has been developed. Such an expression can be obtained by considering Fermi levels, densities of states, and other quantities that are discussed in solid-state devices textbooks. Since the goal here is to determine how to minimize the reverse saturation current, and not to go into detail of quantum mechanical proofs, the result is noted here with the recommendation that the interested reader consult a good solid-state devices text for the development of the result. The result is

$$n_i^2 = 4\left(\frac{2\pi kT}{h^2}\right)^3 \left(m_n^* m_p^*\right)^{3/2} e^{-E_g/kT}, \tag{2.13}$$

where

m_n^* and m_p^* are the electron effective mass and hole effective mass in the host material
E_g is the bandgap energy of the host material

These effective masses can be greater than or less than the rest mass of the electron, depending on the degree of curvature of the valence and conduction bands when plotted as energy versus momentum as in Figure 2.2. In fact, the effective mass can also depend on the band in which the carrier resides in a material. For more information on effective mass, the reader is encouraged to consult the references listed at the end of the chapter [6,7].

Now, using (2.6) with (2.12) and (2.13) in (2.9) the following final result for the reverse saturation current is obtained:

$$I_o = \left(4qA\left(\frac{2\pi kT}{h^2}\right)\left(m_n^* m_p^*\right)^{3/2} e^{-E_g/kT}\right) \times \left(\frac{1}{N_A}\sqrt{\frac{D_n}{\tau_n}}\,\mathrm{ctnh}\frac{l_p}{\sqrt{D_n\tau_n}} + \frac{1}{N_D}\sqrt{\frac{D_p}{\tau_p}}\,\mathrm{ctnh}\frac{l_n}{\sqrt{D_p\tau_p}}\right). \tag{2.14}$$

Since the design goal is to minimize I_o while still maximizing the ratio I_ℓ:I_o, the next step is to express the photocurrent in some detail so the values of appropriate parameters can be considered in the design choices.

2.5.2.3 Optimizing Photocurrent

In Section 2.4.2.2, the photocurrent optimization process was discussed qualitatively. In this section the specific parameters that govern the absorption of light and the lifetime of the absorbed charge carriers will be discussed, and a formula for the photocurrent will be presented for comparison with the formula for reverse saturation current. In particular, minimizing reflection of the incident photons, maximizing the minority carrier diffusion lengths, maximizing the junction width, and minimizing surface recombination velocity will be discussed. The PV cell designer will then know exactly what to do to make the perfect cell.

2.5.2.3.1 Minimizing Reflection of Incident Photons

The interface between air and the semiconductor surface constitutes an impedance mismatch, since the electrical conductivities and the dielectric constants of air and a PV cell are different. As a result, part of the incident wave must be reflected in order to meet the boundary conditions imposed by the solution of the wave equation on the electric field, E, and the electric displacement, D.

Those readers who are experts at electromagnetic field theory will recognize that this problem is readily solved by the use of a quarter-wave matching coating on the PV cell. If the coating on the cell has a dielectric constant equal to the geometric mean of the dielectric constants of the cell and of air and if the coating is one-quarter wavelength thick, it will act as an impedance-matching transformer and minimize reflections. Of course, the coating must be transparent to the incident light. This means that it needs to be an insulator with a bandgap that exceeds the energy of the shortest wavelength light to be absorbed by the PV cell. Alternatively, it needs some other property that minimizes the value of the absorption coefficient for the material, such as an indirect bandgap.

It is also important to realize that a quarter wavelength is on the order of 0.1 μm. This is extremely thin, and may pose a problem for spreading a uniform coating of this thickness. And, of course, since it is desirable to absorb a range of wavelengths, the antireflective coating (ARC) will be optimized at only a single wavelength. Despite these problems, coatings have been developed that meet the requirements quite well.

An alternative to ARCs now commonly in use with Si PV cells is to manufacture the cells with textured front and back surfaces, as shown in Figure 2.7. Textured surfaces promote reflection and refraction of the incident photons to extend the photon path lengths within the material, thus enhancing the probability of generating EHPs within a minority carrier diffusion length of the junction. The bottom line is that a textured surface acts to enhance the capture of photons and also acts to prevent the escape

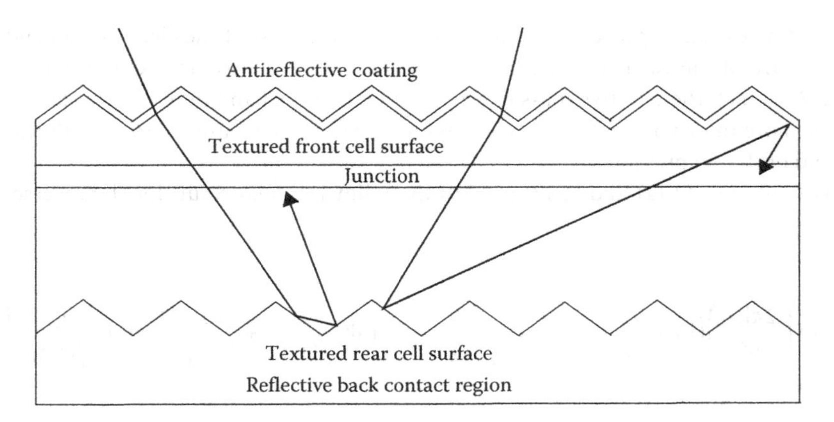

FIGURE 2.7 Maximizing photon capture with textured surfaces.

of captured photons before they can produce EHPs. Furthermore, the textured surface is not wavelength dependent as is the ARC.

2.5.2.3.2 Maximizing Minority Carrier Diffusion Lengths

Since the diffusion lengths are given by (2.6), it is necessary to explore the factors that determine the diffusion constants and minority carrier lifetimes in different materials. It needs to be recognized that changing a diffusion constant may affect the minority carrier lifetime, so the product needs to be maximized.

Diffusion constants depend on scattering of carriers by host atoms as well as by impurity atoms. The scattering process is both material dependent and temperature dependent. In a material at a low temperature with a well-defined crystal structure, scattering of charge carriers is relatively minimal, so they tend to have high mobilities. Figure 2.8a illustrates the experimentally determined dependence of the electron mobility on temperature and on impurity concentration in Si. The Einstein relationship shows that the diffusion constant is proportional to the product of mobility and temperature. This relationship is shown in Figure 2.8b. So, once again, there is a trade-off. While increasing impurity concentrations in the host material *increases* the built-in junction potential, increasing impurity concentrations *decreases* the carrier diffusion constants.

The material of Figure 2.8 is single crystal material. In polycrystalline or amorphous material, the lack of crystal lattice symmetry has a significant effect on the mobility and diffusion constant, causing

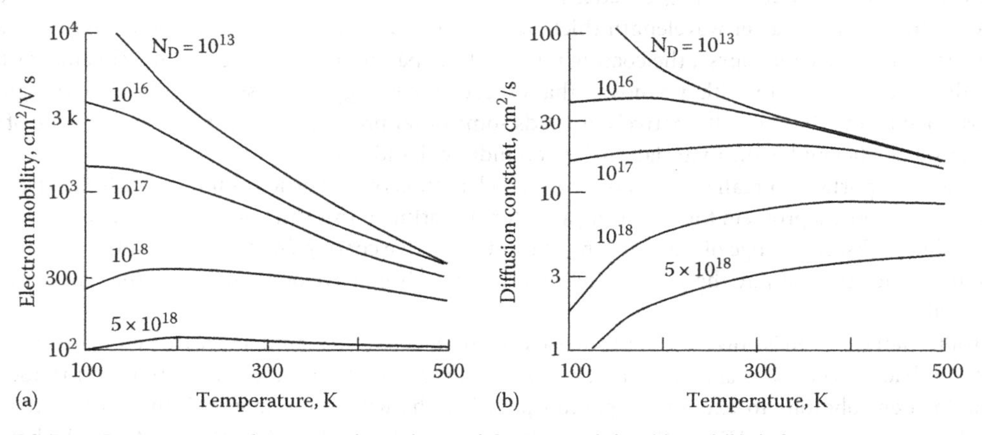

FIGURE 2.8 Temperature and impurity dependence of electron mobility (a), and (b) diffusion constant in silicon. (From Messenger, R.A. and Ventre, J., *Photovoltaic Systems Engineering*, 3rd edn., CRC Press, Boca Raton, FL, 2010.)

significant reduction in these quantities. However, if the absorption constant can be made large enough for these materials, the corresponding decrease in diffusion length may be compensated for by the increased absorption rate.

When an electron and a hole recombine, certain energy and momentum balances must be achieved. Locations in the host material that provide for optimal recombination conditions are known as recombination centers. Hence, the minority carrier lifetime is determined by the density of recombination centers in the host material.

One type of recombination center is a crystal defect, so that as the number of crystal defects increases, the number of recombination centers increases. This means that crystal defects reduce the diffusion constant as well as the minority carrier lifetime in a material.

Impurities also generally make good recombination centers, especially those impurities with energies near the center of the bandgap. These impurities are thus different from the donor and acceptor impurities that are purposefully used in the host material, since donor impurities have energies relatively close to the conduction band and acceptor impurities have energies relatively close to the valence band.

Minority carrier lifetimes also depend on the concentration of charge carriers in the material. An approximation of the dependence of electron minority carrier lifetime on carrier concentration and location of the trapping energy within the energy gap is given by [8]

$$\tau_n = \frac{n'\left[n + p + 2n_i \cosh((E_t - E_i)/kT)\right]}{CN_t(np - n_i^2)}, \tag{2.15a}$$

where

C is the capture cross section of the impurity in cm^3/s
N_t is the density of trapping centers
E_t and E_i are the energies of the trapping center and the intrinsic Fermi level

In most materials, the intrinsic Fermi level is very close to the center of the bandgap. Under most illumination conditions, the hyperbolic term will be negligible compared to the majority carrier concentration and the excess electron concentration, as minority carriers in p-type material will be much larger than the electron thermal equilibrium concentration. Under these conditions, for minority electrons in p-type material, (2.15a) reduces to

$$\tau_n \simeq \frac{1}{CN_t}. \tag{2.15b}$$

Hence, to maximize the minority carrier lifetime, it is necessary to minimize the concentration of trapping centers and to be sure that any existing trapping centers have minimal capture cross sections.

2.5.2.3.3 Maximizing Junction Width

Since it has been determined that it is desirable to absorb photons within the confines of the pn junction, it is desirable to maximize the width of the junction. It is therefore necessary to explore the parameters that govern the junction width.

An expression for the width of a pn junction can be obtained by solving Gauss' law at the junction, since the junction is a region that contains electric charge. Solution of Gauss' law, of course, is dependent upon the ability to express the spatial distribution of the space charge in mathematical, or, at least, in graphical form. Depending on the process used to form the junction, the impurity profile across the junction can be approximated by different expressions. Junctions formed by epitaxial growth or by ion implantation can be controlled to have impurity profiles to meet the discretion of the operator. Junctions grown by diffusion can be reasonably approximated by a linearly graded model. The interested reader is encouraged to consult a reference on semiconductor devices for detailed information on the production of various junction impurity profiles.

The junction with uniform concentrations of impurities is convenient to use to obtain a feeling for how to maximize the width of a junction. Solution of Gauss' law for a junction with uniform concentration of donors on one side and a uniform concentration of acceptors on the other side yields solutions for the width of the space charge layer in the n-type and in the p-type material. The total junction width is then simply the sum of the widths of the two sides of the space charge layer. The results for each side are

$$W_n = \left[\frac{2\varepsilon N_A}{qN_D(N_A + N_D)}\right]^{1/2} (V_j - V)^{1/2} \tag{2.16a}$$

and

$$W_p = \left[\frac{2\varepsilon N_D}{qN_A(N_D + N_A)}\right]^{1/2} (V_j - V)^{1/2}. \tag{2.16b}$$

The overall width of the junction can now be determined by summing Equations 2.16a and 2.16b to get

$$W = \left[\frac{2\varepsilon(N_D + N_A)}{qN_AN_D}\right]^{1/2} (V_j - V)^{1/2}. \tag{2.17}$$

At this point, it should be recognized that the voltage across the junction due to the external voltage across the cell, V, will never exceed the built-in voltage, V_j. The reason is that, as the externally applied voltage becomes more positive, the cell current increases exponentially and causes voltage drops in the neutral regions of the cell, so only a fraction of the externally applied voltage actually appears across the junction. Hence, there is no need to worry about the junction width becoming zero or imaginary. In the case of PV operation, the external cell voltage will hopefully be at the maximum power point, which is generally between 0.5 and 0.6 V for silicon.

Next, observe that, as the external cell voltage increases, the width of the junction decreases. As a result, the absorption of photons decreases. This suggests that it would be desirable to design the cell to have the largest possible built-in potential to minimize the effect of increasing the externally applied voltage. This involves an interesting trade-off, since the built-in junction voltage is logarithmically dependent on the product of the donor and acceptor concentrations (see 2.5b), and the junction width is inversely proportional to the square root of the product of the two quantities. Combining (2.5b) and (2.17) results in

$$W = \left[\frac{2\varepsilon(N_D + N_A)}{qN_DN_A}\right]^{1/2} \left(\frac{kT}{q}\ln\frac{N_AN_D}{n_i^2} - V\right)^{1/2}. \tag{2.18}$$

Note now that maximizing W is achieved by making either $N_A \gg N_D$ or by making $N_D \gg N_A$.

Another way to increase the width of the junction is to include a layer of intrinsic material between the p-side and the n-side as shown in Figure 2.9. In this *pin* junction, there are no impurities to ionize in the intrinsic material, but the ionization still takes place at the edges of the n-type and the p-type material.

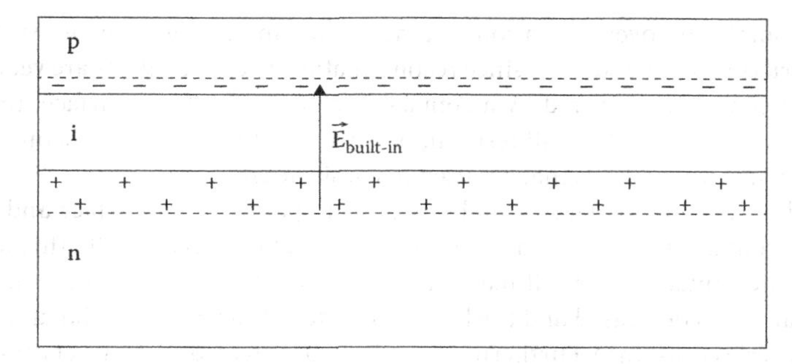

FIGURE 2.9 The pin junction.

As a result, there is still a strong electric field across the junction and there is still a built-in potential across the junction. Since the intrinsic region could conceivably be of any width, it is necessary to determine the limits on the width of the intrinsic region.

The only feature of the intrinsic region that degrades performance is the fact that it has a width. If it has a width, then it takes time for a charge carrier to traverse this width. If it takes time, then there is a chance that the carrier will recombine. Thus, the width of the intrinsic layer simply needs to be kept short enough to minimize recombination. The particles travel through the intrinsic region with a relatively high drift velocity due to the built-in electric field at the junction. Since the thermal velocities of the carriers still exceed the drift velocities by several orders of magnitude, the width of the intrinsic layer needs to be kept on the order of about one diffusion length.

2.5.2.3.4 Minimizing Surface Recombination Velocity

If an EHP is generated near a surface, it becomes more probable that the minority carrier will diffuse to the surface. Since photocurrent depends on minority carriers' diffusing to the junction and ultimately across the junction, surface recombination of minority carriers before they can travel to the junction reduces the available photocurrent. When the surface is within a minority carrier diffusion length of the junction, which is often desirable to ensure that generation of EHPs is maximized near the junction, minority carrier surface recombination can significantly reduce the efficiency of the cell.

Surface recombination depends on the density of excess minority carriers; in this case, as generated by photon absorption, and on the average recombination center density per unit area, N_{sr}, on the surface. The density of recombination centers is very high at contacts and is also high at surfaces in general, since the crystal structure is interrupted at the surface. Imperfections at the surface, whether due to impurities or to crystal defects, all act as recombination centers.

The recombination rate, U, is expressed as number/cm^2/s and is given by [8]

$$U = cN_{sr}m', \tag{2.19}$$

where

m′ is used to represent the excess minority carrier concentration, whether electrons or holes
c is a constant that incorporates the lifetime of a minority carrier at a recombination center

Analysis of the dimensions of the parameters in (2.19) shows that the units of cN_{sr} are cm/s. This product is called the *surface recombination velocity*, S. The total number of excess minority carriers recombining per unit time and subsequent loss of potential photocurrent is thus dependent on the density of recombination centers at the surface and on the area of the surface. Minimizing surface recombination thus may involve reducing the density of recombination centers or reducing the density of minority carriers at the surface.

If the surface is completely covered by a contact, then little can be done to reduce surface recombination if minority carriers reach the surface, since recombination rates at contacts are very high. However, if the surface is not completely covered by a contact, such as at the front surface, then a number of techniques have been discovered that will result in passivation of the surface. Silicon oxide and silicon nitrogen passivation are two methods that are used to passivate silicon surfaces.

Another method of reducing surface recombination is to passivate the surface and then only allow the back contact to contact the cell over a fraction of the total cell area. While this tends to increase series resistance to the contact, if the cell material near the contact is doped more heavily, the ohmic resistance of the material is decreased and the benefit of reduced surface recombination offsets the cost of somewhat higher series resistance. Furthermore, an E-field is created that attracts majority carriers to the contact and repels minority carriers.

2.5.2.3.5 Final Expression for the Photocurrent

An interesting exercise is to calculate the maximum obtainable efficiency of a given PV cell. Equation 2.3 indicates the general expression for photon absorption. Since the absorption coefficient is wavelength dependent, the general formula for overall absorption must take this dependence into account. The challenge in design of the PV cell and selection of appropriate host material is to avoid absorption before the photon is close enough to the junction, but to ensure absorption when the photon is within a minority carrier diffusion length of either side of the junction.

The foregoing discussion can be quantified in terms of cell parameters in the development of an expression for the photocurrent. Considering a monochromatic photon flux incident on the p-side of a p^+n junction (the + indicates strongly doped), the following expression for the hole component of the photocurrent can be obtained. The expression is obtained from the solution of the diffusion equation in the neutral region on the n-side of the junction for the diffusion of the photon-created minority holes to the back contact of the cell [7].

$$\Delta I_{lp} = \frac{qAF_{ph}\alpha L_p}{\alpha^2 L_p^2 - 1}\left[\frac{S\cosh(w_n/L_p) + (D_p/L_p)\sinh(w_n/L_p) + (\alpha D_p - S)e^{-\alpha w_n}}{S\sinh(w_n/L_p) + (D_p/L_p)\cosh(w_n/L_p)} - \alpha L_p\right]. \tag{2.20}$$

It is assumed that the cell has a relatively thin p-side and that the n-side has a width, w_n. In (2.20), F_{ph} represents the number of photons per cm^2 per second per unit wavelength incident on the cell. The effect of the surface recombination velocity on the reduction of photocurrent is more or less clearly demonstrated by (2.20). The mathematical whiz will immediately be able to determine that small values of S maximize the photocurrent, and large values reduce the photocurrent, while one with average math skills may need to plug in some numbers.

Equation 2.20 is thus maximized when α and L_p are maximized and S is minimized. The upper limit of the expression then becomes

$$\Delta I_{lp} = -qAF_{ph}, \tag{2.21}$$

indicating that all photons have been absorbed and all have contributed to the photocurrent of the cell.

Since sunlight is not monochromatic, (2.21) must be integrated over the incident photon spectrum, noting all wavelength-dependent quantities, to obtain the total hole current. An expression must then be developed for the electron component of the current and integrated over the spectrum to yield the total photocurrent as the sum of the hole and electron currents. This mathematical challenge is clearly a member of the nontrivial set of math exercises. Yet, some have persisted at a solution to the problem and have determined the maximum efficiencies that can be expected for cells of various materials. Table 2.1 shows the theoretical optimum efficiencies for several different PV materials.

TABLE 2.1 Theoretical Conversion Efficiency Limits for Several PV Materials at 25°C

Material	E_g	η_{max} (%)
Ge	0.6	13
CIS	1.0	24
Si	1.1	27
InP	1.2	24.5
GaAs	1.4	26.5
CdTe	1.48	27.5
AlSb	1.55	28
a-Si:H	1.65	27
CdS	2.42	18

Source: Rappaport, P., *RCA Review*, 20, 373, 1959; Zweibel, K., *Harnessing Solar Power*, Plenum Press, New York, 1990.

2.5.3 Minimizing Cell Resistance Losses

Any voltage drop in the regions between the junction and the contacts of a PV cell will result in ohmic power losses. In addition, surface effects at the cell edges may result in shunt resistance between the contacts. It is thus desirable to keep any such losses to a minimum by keeping the series resistance of the cell at a minimum and the shunt resistance at a maximum. With the exception of the cell front contacts, the procedure is relatively straightforward.

Most cells are designed with the front layer relatively thin and highly doped, so the conductivity of the layer is relatively high. The back layer, however, is generally more lightly doped in order to increase the junction width and to allow for longer minority carrier diffusion length to increase photon absorption. There must therefore be careful consideration of the thickness of this region in order to maximize the performance of these competing processes.

If the back contact material is allowed to diffuse into the cell, the impurity concentration can be increased at the back side of the cell. This is important for relatively thick cells, commonly fabricated by slicing single crystals into wafers. The contact material must produce either n-type or p-type material if it diffuses into the material, depending on whether the back of the cell is n-type or p-type.

In addition to reducing the ohmic resistance by increasing the impurity concentration, the region near the contact with increased impurity concentration produces an additional electric field that increases the carrier velocity, thus producing a further equivalent reduction in resistance. The electric field is produced in a manner similar to the electric field that is produced at the junction.

For example, if the back material is p-type, holes from the more heavily doped region near the contact diffuse toward the junction, leaving behind negative acceptor ions. Although there is no source of positive ions in the p-region, the holes that diffuse away from the contact create an accumulated positive charge that is distributed through the more weakly doped region. The electric field, of course, causes a hole drift current, which, in thermal equilibrium, balances the hole diffusion current. When the excess holes generated by the photoabsorption process reach the region of the electric field near the contact, however, they are swept more quickly toward the contact. This effect can be viewed as the equivalent of moving the contact closer to the junction, which, in turn, has the ultimate effect of increasing the gradient of excess carriers at the edge of the junction. This increase in gradient increases the diffusion current of holes away from the junction. Since this diffusion current strongly dominates the total current, the total current across the junction is thus increased by the heavily doped layer near the back contact.

At the front contact, another balancing act is needed. Ideally, the front contact should cover the entire front surface. The problem with this, however, is that if the front contact is not transparent to the incident photons, it will reflect them away. In most cases, the front contact is reflecting. Since the front/top layer of the cell is generally very thin, even though it may be heavily doped, the resistance in the transverse direction will be relatively high because of the thin layer. This means that if the contact is placed at the edge of the cell to enable maximum photon absorption, the resistance along the surface to the contact will be relatively large.

The compromise, then, is to create a contact that covers the front surface with many tiny fingers. This network of tiny fingers, which, in turn, are connected to larger and larger fingers, is similar to the configuration of the capillaries that feed veins in a circulatory system. The idea is to maintain more or less constant current density in the contact fingers, so that as more current is collected, the cross-sectional area of the contact must be increased.

Finally, shunt resistance is maximized by ensuring that no leakage occurs at the perimeter of the cell. This can be done by nitrogen passivation or simply by coating the edge of the cell with insulating material to prevent contaminants from providing a current path across the junction at the edges.

2.6 Traditional PV Cells

2.6.1 Introduction

Traditional PV cells are based on the theoretical considerations of Sections 2.3 through 2.5. Cells currently commercially available are based on crystalline, multicrystalline, and amorphous (thin film) silicon (Si-C, Si, a-Si:H); copper indium gallium diselenide (CIGS) thin films; CdTe thin films; and III-V compounds such as gallium arsenide (GaAs). They all have pn junctions and all are subject to the optimization considerations previously discussed. This section will present a brief summary of the structures of each cell along with current (2011) performance information. For the interested reader, reference [8] considers all of the cells in this section in greater detail.

2.6.2 Crystalline Silicon Cells

Crystalline silicon cells can be either monocrystalline or polycrystalline. The monocrystalline cells are generally somewhat more efficient, but are also somewhat more energy intensive to produce. The cells are composed of approximately 200 μm thick slices of p-type single crystal ingots grown from a melt, with their circumferences squared up by slicing the round cross section into an approximately square cross section, similar to the way that lumber is processed from logs. The junction is formed by diffusing n-type impurities to a depth of approximately $1/\alpha$, where α is the absorption constant. The surfaces are then polished, textured and contacts are attached on front and back of the cell. The resulting cell cross section is essentially that of Figure 2.7. An adaptation of the structure of Figure 2.7 involves connecting the top layer of the cell through to the back of the cell so all contacts of the cell will be on the back. This increases cell efficiency by eliminating reflection of incident photons from front contacts. Typical cell conversion efficiencies for cells with front contacts can approach 17% and conversion efficiencies of back contact cells can exceed 20%.

By pouring molten silicon into a quartz crucible with a rectangular or square cross section under carefully controlled temperature conditions, it is possible to form a bar of silicon that consists of crystalline domains, but is not monocrystalline. The advantage of this polycrystalline silicon ingot is that it does not have to be "squared up," thus saving a processing step and some energy as well. The disadvantage is that the crystal boundaries reduce mobilities and provide trapping centers, so the efficiency of the cell is somewhat reduced to the 14%–15% range.

The interesting point, however, is that since monocrystalline silicon cells have rounded corners, when they are assembled into a module to produce more power and more voltage, the module has voids at the corners of the cells, such that no electricity is produced at these locations. The polycrystalline cells, on the other hand, being rectangles, have less wasted module space and the overall module efficiency approximates the efficiency of a monocrystalline module, except for the back contact versions.

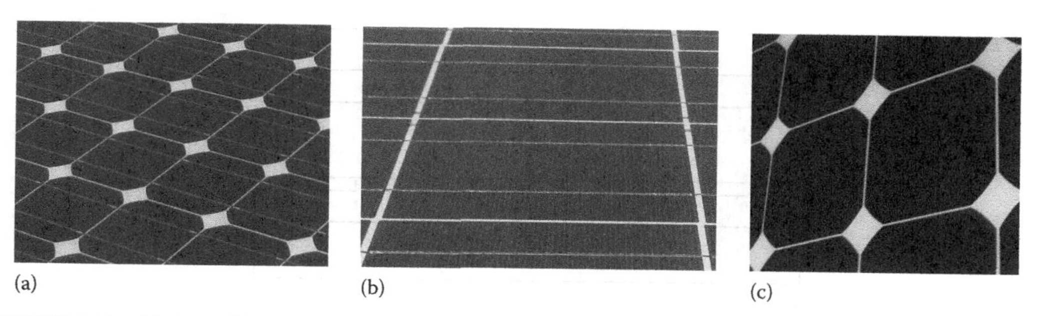

(a) (b) (c)

FIGURE 2.10 (a) Crystalline, (b) multicrystalline, and (c) crystalline back contact Si PV cells.

Figure 2.10 shows photos of front contact monocrystalline cells, front contact polycrystalline cells, and back contact monocrystalline cells.

2.6.3 Amorphous Silicon Cells

Since a-Si:H is not in a crystalline form, it loses the advantages of high mobility and high diffusion constant. The noncrystalline lattice includes a large number of silicon atoms with outer shell electrons that are not covalently bonded to nearest neighbors. These "dangling" electrons create impurity levels in the bandgap and thus affect the lifetimes of the excited charge carriers. Fortunately, it is possible to apply hydrogen to the material such that the hydrogen fills the dangling bonds, thus passivating these sites in the material and improving the electrical properties. The material is thus described as a-Si:H.

Despite the amorphous nature of this material, it has a very favorable direct bandgap, which enables the material to efficiently absorb photons over a short distance. A 2 μm thickness of the material will absorb most of the incident photons, thus the reason why a-Si:H is considered to be a thin-film PV material.

Figure 2.11 shows several different cell structures. Structures b and c are multijunction structures and thus present important challenges to the cell designer. First of all, each layer tends to act as a current source, such that if current sources are connected in series, each layer must generate the same current as every other layer. Secondly, the layers appear as series diodes. This means that although the generating diode is supplying current, this current must flow across a reverse-biased junction between adjacent layers. This implies that significant current will not flow until the voltage across the reverse-biased junction reaches the reverse breakdown potential. Fortunately, with heavy doping on each side of the junction, the reverse breakdown voltage can be reduced to zero and a *tunnel junction* is created that allows current to pass unimpeded.

Commercial a-Si:H PV modules are presently available with flexible structures and conversion efficiencies of 8%–10%. An advantage of the technology and module structure is that the module can be applied directly to an approved surface without the need for any additional mounting components.

2.6.4 Copper Indium Gallium Diselenide Cells

Another popular material for thin-film cells is copper indium gallium diselenide (CIGS). This material is also a direct bandgap that absorbs most photons with energies above the bandgap energy within a thickness of about 2 μm. The material is in commercial production at conversion efficiencies in the neighborhood of 12%.

While it is possible to produce both n-type and p-type CIS, homojunctions in the material are neither stable nor efficient. A good junction can be made, however, by creating a heterojunction with n-type CdS and p-type CIS.

The ideal structure uses near-intrinsic material near the junction to create the widest possible depletion region for collection of generated EHPs. The carrier diffusion length can be as much as 2 μm, which is comparable with the overall film thickness. Figure 2.12 shows a basic ZnO/CdS/CIGS/Mo cell structure, which was in popular use in 2004. Again, CIGS technology is advancing rapidly as a result of the thin-film PV

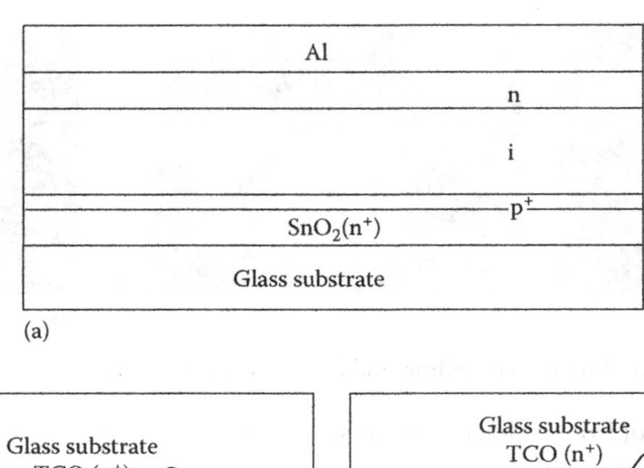

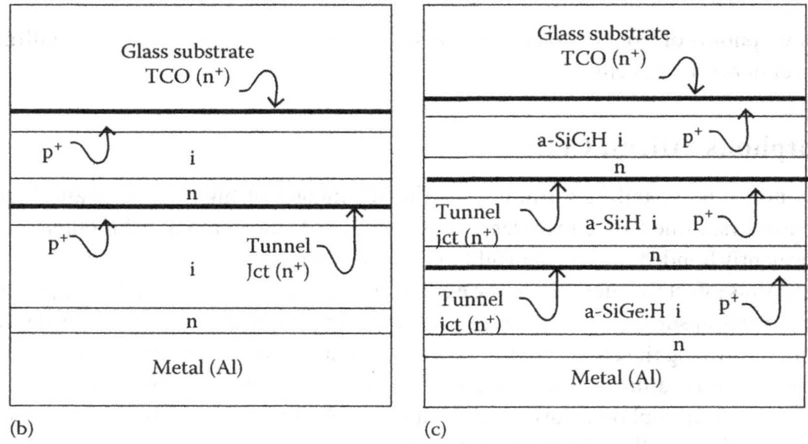

FIGURE 2.11 Three configurations for a-Si:H cells. (a) Basic a-Si:H cell structure, (b) stacked a-Si:H junctions, and (c) SiC-Si-SiGe triple junctions. (From Messenger, R.A. and Ventre, J., *Photovoltaic Systems Engineering*, 3rd edn., CRC Press, Boca Raton, FL, 2010.)

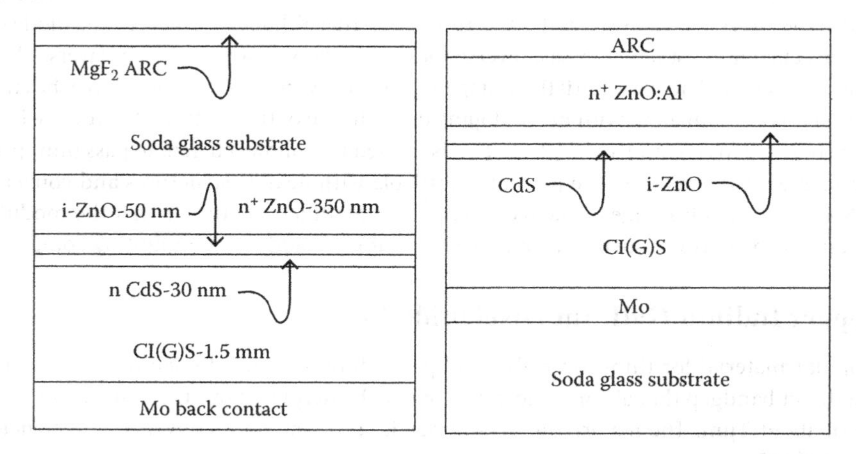

FIGURE 2.12 Typical CIGS thin-film structure. (Adapted from Ullal, H.S. et al., *Proceedings of the 26th IEEE Photovoltaic Specialists Conference*, Anaheim, CA, pp. 301–305, 1997. © IEEE.)

partnership program, so by the time this paragraph is read, the structure of Figure 2.12 may be only suitable for history books and general discussion of the challenges encountered in thin-film cell development.

Nearly a dozen processes have been used to achieve the basic cell structure of Figure 2.12. The processes include radio frequency (rf) sputtering, reactive sputtering, chemical vapor deposition, vacuum evaporation, spray deposition, and electrodeposition. Sometimes these processes are implemented sequentially and sometimes they are implemented concurrently. Recently a novel method of manufacturing CIGS modules using cylindrical tubes with spaces in between to allow photons to pass through the module and be reflected back to the cylindrical tubes from the surface upon which the module is attached has been developed. This structure presents minimal wind loading for the module and consequently for most wind zones it can be simply laid on a flat roof and attached to adjacent modules to form the PV array. This module is becoming popular for use with white membrane flat roofing material.

2.6.5 Cadmium Telluride Cells

In theory, CdTe cells have a maximum efficiency limit close to 25%. The material has a favorable direct bandgap and a large absorption constant, allowing for cells of a few μm thickness. By 2001, efficiencies approaching 17% were being achieved for laboratory cells, and module efficiencies had reached 11% for the best large area (8390 cm^2) module [11]. Efforts were then focused on scaling up the fabrication process to mass produce the modules, with the result of achieving a production cost of less than \$1.00/W in 2008. This cost included an escrow account to be used for recycling the materials at the end of module life.

No fewer than nine companies have shown an interest in commercial applications of CdTe. As of 2001, depending on the fabrication methodology, efficiencies of close to 17% had been achieved for small area cells ($\approx 1\ cm^2$), and 11% on a module with an area of 8390 cm^2 [12].

After it was shown that no degradation was observable after 2 years, production-scale manufacturing began. CdTe modules are now being manufactured and marketed at the rate of more than 1 GW annually for utility scale projects and the magic \$1.00/W production cost barrier has now been broken for these modules, with an announced 4th quarter 2008 production cost of \$0.93/W [8]. Figure 2.13 shows the cross section of a typical CdTe cell. In this figure, antireflective coating (ARC), transparent conducting oxide (TCO) and ethylene vinyl acetate (EVA), which are used to bond the back contact to the glass.

2.6.6 Gallium Arsenide Cells

The 1.43 eV direct bandgap, along with a relatively high absorption constant, makes GaAs an attractive PV material. Historically high production costs, however, have limited the use of GaAs PV cells to extraterrestrial and other special purpose uses, such as in concentrating collectors. Recent advances in concentrating technology, however, enable the use of significantly less active material in a module, such that cost-effective, terrestrial devices may soon be commercially available.

Most modern GaAs cells, however, are prepared by the growth of a GaAs film on a suitable substrate. Figure 2.14 shows one basic GaAs cell structure. The cell begins with the growth of an n-type GaAs layer on a substrate, typically Ge. Then a p-GaAs layer is grown to form the junction and collection region.

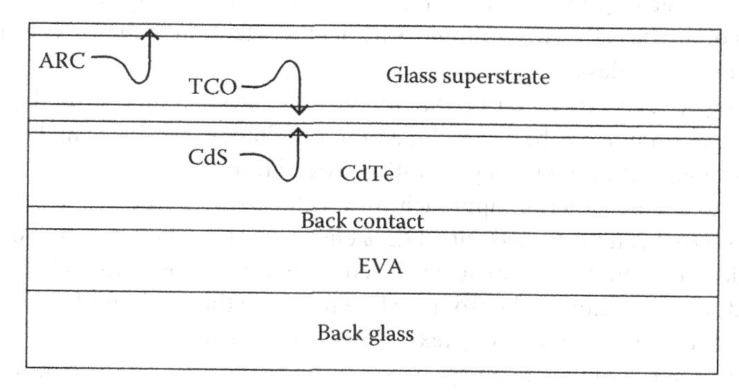

FIGURE 2.13 Basic structure of a CdTe PV cell. (Adapted from Ullal, H.S. et al., *Proceedings of the 26th IEEE Photovoltaic Specialists Conference*, Anaheim, CA, pp. 301, 1997. © IEEE.)

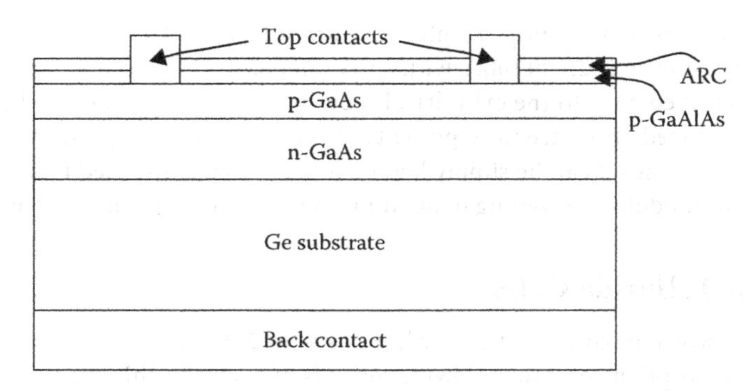

FIGURE 2.14 Structure of a basic GaAs cell with GaAlAs window and passive Ge substrate. (From Messenger, R.A. and Ventre, J., *Photovoltaic Systems Engineering*, 3rd edn., CRC Press, Boca Raton, FL, 2010.)

The top layer of p-type GaAlAs has a bandgap of approximately 1.8 eV. This structure reduces minority carrier surface recombination and transmits photons below the 1.8 eV level to the junction for more efficient absorption.

Cells fabricated with III-V elements are generally extraterrestrial quality. In other words, they are expensive, but they are high-performance units. Efficiencies in excess of 20% are common and efficiencies of cells fabricated on more expensive GaAs substrates have exceeded 34% [11].

An important feature of extraterrestrial quality cells is the need for them to be radiation resistant. Cells are generally tested for their degradation resulting from exposure to healthy doses of 1 MeV or higher energy protons and electrons. Degradation is generally less than 20% for high exposure rates [13].

Extraterrestrial cells are sometimes exposed to temperature extremes, so the cells are also cycled between ~−170°C and +96°C for as many as 1600 cycles. The cells also need to pass a bending test, a contact integrity test, a humidity test, and a high temperature vacuum test, in which the cells are tested at a temperature above 140°C in vacuum for 168 h [13].

Fill factors in excess of 80% have been achieved for GaAs cells. Single cell open-circuit voltages are generally between 0.8 and 0.9 V.

2.7 Emerging Technologies

2.7.1 New Developments in Silicon Technology

While progress continues on conventional Si technology, new ideas are also being pursued for crystalline and amorphous Si cells. The goal of Si technology has been to maintain good transport properties, while improving photon absorption and reducing the material processing cost of the cells. It is likely that several versions of thin Si cells will continue to attract the attention of the PV community, including recent research on thin Si on glass.

Another interesting opportunity for cost reduction in Si cell production is to double up on processing steps. For example, a technique has been developed for simultaneously diffusing boron and phosphorous in a single step, along with growing a passivating oxide layer [14].

As an alternative to the pn junction approach to Si cells, MIS-IL (metal insulator semiconductor inversion layer) cells have been fabricated with 18.5% efficiency [15]. The cell incorporates a point-contacted back electrode to minimize the rear surface recombination, along with Cs beneath the MIS front grid and oxide window passivation of the front surface to define the cell boundaries. Further improvement in cell performance can be obtained by texturing the cell surfaces.

New developments in surface texturing may also simplify the process and result in additional improvement in Si device performance. Discovery of new substrates and methods of growing good quality Si on them is also an interesting possibility for performance improvement and cost reduction for Si cells.

Since new ideas will continue to emerge as interest in Si PV technology continues to grow, the interested reader is encouraged to attend PV conferences and to read the conference publications to stay up-to-date in the field.

2.7.2 CIS-Family-Based Absorbers

Much is yet to be learned about inhomogeneous absorbers and composite absorbers composed of combinations of these various materials. The possibility of multijunction devices is also being explored.

Meanwhile, work is underway to reduce the material usage in the production of CIS modules in order to further reduce production costs. Examples of reduction of material use include halving the width of the Mo contact layer, reduction in the use of H_2S and H_2Se, and a reduction in ZnO, provided that a minimum thickness can be maintained [16].

2.7.3 Other III–V and II–VI Emerging Technologies

It appears that compound tandem cells will receive appreciable emphasis in the III-V family of cells over the next few years. For example, $Ga_{0.84}In_{0.16}As_{0.68}P_{0.32}$, lattice matched to GaAs, has a bandgap of 1.55 eV and may prove to be an ideal material for extraterrestrial use, since it also has good radiation resistance [17]. Cells have been fabricated with $Al_{0.51}In_{0.49}P$ and $Ga_{0.51}In_{0.49}P$ window layers, with the best 1 cm^2 cell having an efficiency of just over 16%, but having a fill factor of 85.4%.

Cell efficiencies can be increased by concentrating sunlight on the cells. Although the homojunction cell efficiency limit under concentration is just under 40%, quantum well (QW) cells have been proposed to increase the concentrated efficiency beyond the 40% level [18]. In QW cells, intermediate energy levels are introduced between the host semiconductor's valence and conduction bands to permit absorption of lower energy photons. These levels must be chosen carefully so that they will not act as recombination centers, however, or the gains of EHPs from lower energy incident photons will be lost to the recombination processes. Laboratory cells have shown higher V_{OC} resulting from a decrease in dark current for these cells.

2.7.4 Other Technologies

2.7.4.1 Thermophotovoltaic Cells

To this point, discussion has been limited to the conversion of visible and near infrared spectrum to EHPs. The reason is simply that the solar spectrum peaks out in the visible range. However, heat sources and incandescent light sources produce radiation in the longer infrared regions, and in some instances, it is convenient to harness radiated heat from these processes by converting it to electricity. This means using semiconductors with smaller bandgaps, such as Ge. More exotic structures, such as InAsSbP, with a bandgap of 0.45–0.48 eV have also been fabricated [19].

2.7.4.2 Intermediate Band Solar Cells

In all cells described to this point, absorption of a photon has resulted in the generation of a single EHP. If an intermediate band material is sandwiched between two ordinary semiconductors, it appears that it may be possible for the material to absorb two photons of relatively low energy to produce a single EHP at the combined energies of the two lower energy photons. The first photon raises an electron from the valence band to the intermediate level, creating a hole in the valence band, and the second photon raises the electron from the intermediate level to the conduction band. The trick is to find such an intermediate band material that will "hold" the electron until another photon of the appropriate energy impinges upon the material. Such a material should have half its states filled with electrons and half empty in order to optimally accommodate this electron transfer process. It appears that III–V compounds may be the best candidates for implementation of this technology. Theoretical maximum efficiency of such a cell is 63.2% [20].

2.7.4.3 Super Tandem Cells

If a large number of cells are stacked with the largest bandgap on top and the bandgap of subsequent cells decreasing, the theoretical maximum efficiency is 86.8% [21]. A 1 cm^2 four-junction cell has been fabricated with an efficiency of 35.4%. The maximum theoretical efficiency of this cell is 41.6% [22]. Perhaps 1 day one of the readers of this paragraph (or one of their great-great grandchildren) will fabricate a cell with the maximum theoretical efficiency.

2.7.4.4 Hot Carrier Cells

The primary loss mechanism in PV cells is the energy lost in the form of heat when an electron is excited to a state above the bottom of the conduction band of a PV cell by a photon with energy greater than the bandgap. The electron will normally drop to the lowest energy available state in the conduction band, with the energy lost in the process being converted to heat. Hence, if this loss mechanism can be overcome, the efficiency of a cell with a single junction should be capable of approaching that of a super tandem cell. One method of preventing the release of this heat energy by the electron is to heat the cell, so the electron will remain at the higher energy state. The process is called *thermoelectronics* and is currently being investigated [21].

2.7.4.5 Optical Up- and Down-Conversion

An alternative to varying the electrical bandgap of a material is to reshape the energies of the incident photon flux. Certain materials have been shown to be capable of absorbing two photons of two different energies and subsequently emitting a photon of the combined energy. Other materials have been shown to be capable of absorbing a single high-energy photon and emitting two lower-energy photons. These phenomena are similar to up-conversion and down-conversion in communications circuits at radio frequencies.

By the use of both types of materials, the spectrum incident on a PV cell can be effectively narrowed to a range that will result in more efficient absorption in the PV cell. An advantage of this process is that the optical up- and down-converters need not be a part of the PV cell. They simply need to be placed between the photon source and the PV cell. In tandem cells, the down-converter would be placed ahead of the top cell and the up-converter would be integrated into the cell structure just ahead of the bottom cell [8].

2.7.4.6 Organic PV Cells

Even more exotic than any of the previously mentioned cells is the organic cell. In the organic cell, electrons and holes are not immediately formed as the photon is absorbed. Instead, the incident photon creates an *exciton*, which is a bound EHP. In order to free the charges, the exciton binding energy must be overcome. This dissociation occurs at the interface between materials of high electron affinity and low ionization potential [22]. Photoluminescence is related to this process. Just to end this section with a little chemistry, the reader will certainly want to know that one material that is a candidate for organic PV happens to be poly{2,5-dimethoxy-1,4-phenylene-1,2-ethenylene-2-methoxy-5-(2-ethyl-hexyloxy)-1,4-phenylene-1,2-ethenylene}, which goes by the nickname M3EH-PPV. Whether M3EH-PPV will dominate the PV market 1 day remains to be seen. So far efficiencies of this very challenging technology have been in the 1% range.

2.8 PV Electronics and Systems

2.8.1 Introduction

Obviously the extent of research and development that has been covered in this chapter may not have been undertaken if a possible market for each technology had not been identified. Prior to 2000, most PV applications were stand-alone applications, such as off-grid cabins. Since that time, however, grid-connected applications have mushroomed and far surpassed stand-alone applications. Grid-connected applications can be either noninteractive or interactive and can be direct grid connect or battery backup. Noninteractive grid-connected systems simply use the grid as a backup source of power when the sun is not shining. Interactive systems are capable of selling energy back to the grid if the host demand is met and excess system output remains available.

Interactive systems require inverters to convert the dc power produced by the PV array into compatible ac power for use at the source as well as for return to the utility. In a utility interactive system, it is necessary for the inverter to shut down if the grid shuts down. However, if a set of batteries are used, then it is possible to energize selected loads when the grid is down as long as the selected loads can be isolated from the grid via switching within the inverter. IEEE Standard 1547 [23] specifies performance parameters for waveform purity and disconnecting from the grid and UL 1741 [24] specifies a testing protocol to ensure compliance with IEEE 1547.

If a system has batteries for backup power, then a battery backup inverter as well as a charge controller will normally be needed.

2.8.2 PV System Electronic Components

2.8.2.1 Inverters

Inverters can have output waveforms ranging from square, for the simplest units, to sine, for the best units. If a unit is to be connected to the grid in a sell mode, it must have no more than 5% total harmonic distortion, with individual harmonic maxima specified by IEEE 1547.

Straight grid-connected, utility interactive inverters are used in the simplest of utility interactive systems. So-called "string inverters" use series/parallel combination of modules that may produce up to 1000 V when open circuited, although installations in the United States must have maximum voltages less than 600 V unless they are on utility-owned property. String inverter power output ranges from the low kW range up to 1 MW, with larger units in the pipeline. String inverters generally have maximum power point tracking (MPPT) circuitry at their inputs, so they can operate the PV array at its maximum power point and thus deliver maximum power to the load. They also incorporate ground fault detection and interruption at their inputs to shut down the PV array if a current-carrying conductor should come in contact with a grounded object. Since they comply with UL1741, they shut down whenever any grid disturbance, such as undervoltage, overvoltage, or frequency error is present on the utility system, even if other inverters are connected to the same system.

Recently, a version of the straight grid-connected inverter, the microinverter, has become very popular. The microinverter is used with either a single module or with a pair of modules. Generally the rated dc module power is between 175 and 240 W. The microinverter is mounted next to the PV module so that ac rather than dc is fed from a rooftop. Furthermore, the ac is connected directly to the utility connection so that if the utility loses power, the microinverter is disabled and no current or voltage is present at its output.

When battery backup is used to create an uninterruptible source, the battery backup inverter must have two separate ac ports. One port is connected to the utility and one is connected to the standby loads. If the grid is operational and the sun is down, the inverter will pass utility power through to the standby loads. If the grid is down, then the inverter provides power to the standby loads via the standby port either directly from the PV array during daytime hours or from the batteries at night, or, for that matter, from a combination of batteries and PV array, depending upon the demand of the standby load. If the grid is down, the grid-connected port is automatically shut down, generally in less than 2 s. Present commercial battery backup inverters are rated at 8 kW or less, with the possibility of connecting series-parallel combinations to deliver up to 80 kW. A 100 kW unit has been developed but is still in the beta testing phase.

2.8.2.2 Charge Controllers

When a battery backup inverter is used in a so-called dc-coupled battery backup configuration, the input of the inverter is connected directly to the system batteries. The PV array is also connected to the batteries, but it is connected through a charge controller. The charge controller is needed for two purposes. The primary purpose is to prevent the batteries from becoming overcharged. The system inverter is set to shut off if the batteries approach the maximum discharge limit. The secondary purpose of the charge controller is to maximize energy transfer from PV array to batteries, assuming that a MPPT charge controller is selected. There is no need for MPPT at the inverter input, since it is always at the battery voltage.

2.9 Conclusions

Regardless of the technology or technologies that may result in low-cost, high-performance PV cells, it must be recognized that the lifecycle cost of a cell depends on the cell's having the longest possible, maintenance-free lifetime. Thus, along with the developments of new technologies for absorbers, development of reliable encapsulants and packaging for the modules will also merit continued research and development activity.

Every year engineers make improvements on products that have been in existence for many years. Automobiles, airplanes, electronic equipment, building materials, and many more common items see improvement every year. Even the yo-yo, a popular children's toy during the 1940s and 1950s, came back with better-performing models. Hence, it should come as no surprise to the engineer to see significant improvements and scientific breakthroughs in the PV industry well into the next millennium. The years ahead promise exciting times for the engineers and scientists working on the development of new photovoltaic cell and system technologies, provided that the massive planning and execution phases can be successfully undertaken. This will certainly be the case as the world of nano-devices is explored.

References

1. Markvart, T., Ed., *Solar Electricity*, John Wiley & Sons, Chichester, U.K., 1994.
2. Maycock, P. D., Ed., *Photovoltaic News*, 18(1), January 1999.
3. *International Marketing Data and Statistics*, 22nd edn., Euromonitor PIC, London, U.K., 1998.
4. Maycock, P. D., The world PV market: Production increases 36%, *Renewable Energy World*, 5, July–August 2002, 147–161.
5. Earth Policy Institute, *Eco-Economy Indicators: Solar Power—Data.* www.earth-policy.org/Indicators/Solar/2007_data.htm (accessed on 2007–2008).
6. Streetman, B. G., *Solid State Electronic Devices*, 4th edn., Prentice Hall, Englewood Cliffs, NJ, 1995.
7. Yang, E. S., *Microelectronic Devices*, McGraw-Hill, New York, 1988.
8. Messenger, R. A. and Ventre, J., *Photovoltaic Systems Engineering*, 3rd edn., CRC Press, Boca Raton, FL, 2010.
9. Rappaport, P., *RCA Review*, 20, September 1959, 373–379.
10. Zweibel, K., *Harnessing Solar Power*, Plenum Press, New York, 1990.
11. Kazmerski, L. L., Photovoltaics R&D in the United States: Positioning for our future, *Proceedings of the 29th IEEE Photovoltaic Specialists Conference*, New Orleans, LA, 2002, pp. 21–27.
12. Ullal, H. S., Zweibel, K., and von Roedern, B., *Proceedings of the 26th IEEE Photovoltaic Specialists Conference*, Anaheim, CA, 1997, pp. 301–305.
13. Brown, M. R. et al., Characterization testing of dual junction GaInP2/GaAs/Ge solar cell assemblies, *Proceedings of the 26th IEEE Photovoltaic Specialists Conference*, Anaheim, CA, 1997, pp. 805–810.
14. Krygowski, T., Rohatgi, A., and Ruby, D., Simultaneous P and B diffusion, in-situ surface passivation, impurity filtering and gettering for high-efficiency silicon solar cells, *Proceedings of the 26th IEEE Photovoltaic Specialists Conference*, Anaheim, CA, 1997, pp. 19–24.
15. Metz, A. et al., *Proceedings of the 26th IEEE Photovoltaic Specialists Conference*, Anaheim, CA, 1997, pp. 31–34.

16. Weiting, R., CIS manufacturing at the megawatt scale, *Proceedings of the 29th IEEE Photovoltaic Specialists Conference*, New Orleans, LA, 2002, pp. 478–483.
17. Jaakkola, R. et al., *Proceedings of the 26th IEEE Photovoltaic Specialists Conference*, Anaheim, CA, 1997, pp. 891–894.
18. Barnham, K. W. J. and Duggan, G., A new approach to high-efficiency multi-band-gap solar cells, *Journal of Applied Physics*, 67, 1990, 3490–3493.
19. Khvostikov, V. P. et al., Zinc-diffused InAsSbP/InAs and Ge TPV cells, *Proceedings of the 29th IEEE Photovoltaic Specialists Conference*, New Orleans, LA, 2002, pp. 943–946.
20. Luque, A. et al., Progress towards the practical implementation of the intermediate band solar cell, *Proceedings of the 29th IEEE Photovoltaic Specialists Conference*, New Orleans, LA, 2002, pp. 1190–1193.
21. Green, M. A., *Proceedings of the 29th IEEE Photovoltaic Specialists Conference*, New Orleans, LA, 2002, pp. 1330–1334.
22. Zahler, J. M. et al., *Proceedings of the 29th IEEE Photovoltaic Specialists Conference*, New Orleans, LA, 2002, pp. 1029–1032.
23. IEEE 1547-2003, IEEE standard for connecting distributed resources to electric power systems, IEEE Standards Coordinating Committee 21, Fuel Cells, Photovoltaics, Distributed Generation and Energy Storage, 2003.
24. UL 1741, *Inverters, Converters and Controllers for Use in Independent Power Systems*, Underwriters Laboratories, Inc., Northbrook, IL, 2005.

3 Advanced Energy Technologies

Saifur Rahman
Virginia Tech

3.1 Storage Systems

Energy storage technologies are of great interest to electric utilities, energy service companies, and automobile manufacturers (for electric vehicle application). The ability to store large amounts of energy would allow electric utilities to have greater flexibility in their operations because with this option the supply and demand do not have to be matched instantaneously. This is especially true when there are intermittent generating sources like wind and solar present in the grid. The availability of the proper battery at the right price will make the electric vehicle a reality, a goal that has eluded the automotive industry thus far. Four types of storage technologies (listed as follows) are discussed in this section, but most emphasis is placed on storage batteries because it is now closest to being commercially viable. The other storage technology widely used by the electric power industry, pumped-storage power plants, is not discussed as this has been in commercial operation for more than 60 years in various countries around the world.

- Flywheel storage
- Compressed air energy storage
- Superconducting magnetic energy storage
- Battery storage

3.1.1 Flywheel Storage

Flywheels store their energy in their rotating mass, which rotates at very high speeds (approaching 100,000 rotations per minute), and are made of composite materials instead of steel because of the composite's ability to withstand the rotating forces exerted on the flywheel. In order to store energy, the flywheel is placed in a sealed container which is then placed in a vacuum to reduce air resistance. Magnets embedded in the flywheel pass near pickup coils. The magnet induces a current in the coil changing the rotational energy into electrical energy. Flywheel storage technologies are finding their way into UPS systems and for data center applications where a bridging power supply is necessary after a power failure and before the back-up diesel engine can be started.

3.1.2 Compressed Air Energy Storage

As the name implies, the compressed air energy storage (CAES) plant uses electricity to compress air, which is stored in underground reservoirs. When electricity is needed, this compressed air is withdrawn, heated with gas or oil, and run through an expansion turbine to drive a generator. The compressed air can be stored in several types of underground structures, including caverns in salt or rock formations, aquifers, and depleted natural gas fields. CAES technology has been in use for over 30 years. A 290 MW CAES plant started operation in Huntorf, Germany in 1978. This plant has demonstrated a 90% availability and 99% starting reliability. A 110 MW CAES plant went into commercial operation in McIntosh, Alabama in the United States in 1991. This plant stores compressed air in a 19-million-cubic-foot cavern mined from a salt dome and has a storage capacity for 26 h operation. For the last 20 years since the Alabama plant went into operation, there has not been any further commercial deployment of this technology. In the last few years there seems to have been a revival of interest in this technology, but no commercial power plant yet.

A typical CAES plant consists of a compressor, an air reservoir, and a combustion turbine generator. During the charging mode, the compressor (usually run by an electric motor) uses electricity to compress air which is stored in the air reservoir. While discharging, the compressed air is released from the air reservoir and fed as input to a gas turbine. This compressed gas is combusted in the turbine to produce electricity. In a CAES system the same synchronous machine can perform the functions of compressing the air and then operate as a generator (single compression-generation train). It may also be designed with two separate machines where one is used as a motor to compress the air and the other is used as a generator. In a combustion turbine, the air that is used to drive the turbine is compressed just prior to combustion and expansion and, as a result, the compressor and the expander must operate at the same time and must have the same air mass flow rate. In the case of a CAES plant, the compressor and the expander can be sized independently to provide the utility-selected "optimal" MW charge and discharge rates which determine the ratio of hours of compression required for each hour of turbine-generator operation. The MW ratings and time ratio are influenced by the utility's load curve, and the price of off-peak power. For example, the CAES plant in Germany requires 4 h of compression per hour of generation. On the other hand, the Alabama plant requires 1.7 h of compression for each hour of generation. At 110 MW net output, the power ratio is 0.818 kW output for each kilowatt input. The heat rate (LHV) is 4122 BTU/kWh with natural gas fuel and 4089 BTU/kWh with fuel oil. Due to the storage option, a partial-load operation of the CAES plant is also very flexible. For example, the heat rate of the expander increases only by 5%, and the airflow decreases nearly linearly when the plant output is turned down to 45% of full load.

3.1.3 Superconducting Magnetic Energy Storage

A third type of advanced energy storage technology is superconducting magnetic energy storage (SMES), which may someday allow electric utilities to store electricity with unparalleled efficiency (90% or more). A simple description of SMES operation follows.

The electricity storage medium is a doughnut-shaped electromagnetic coil of superconducting wire. This coil could be about 1000 m in diameter, installed in a trench, and kept at superconducting temperature by a refrigeration system. Off-peak electricity, converted to direct current (DC), would be fed into this coil and stored for retrieval at any moment. The coil would be kept at a low-temperature superconducting state using liquid helium. The time between charging and discharging could be as little as 20 ms with a round-trip AC–AC efficiency of over 90%.

Developing a commercial-scale SMES plant presents both economic and technical challenges. Due to the high cost of liquid helium, only plants with 1000 MW, 5 h capacity are economically attractive. Even then the plant capital cost can exceed several thousand dollars per kilowatt. As ceramic superconductors, which become superconducting at higher temperatures (maintained by less expensive liquid nitrogen), become more widely available, it may be possible to develop smaller scale SMES plants at a lower price.

3.1.4 Battery Storage

Even though battery storage is the oldest and most familiar energy storage device, significant advances have been made in this technology in recent years to deserve more attention. There has been renewed interest in this technology due to its potential application in nonpolluting electric vehicles. Battery systems are quiet and nonpolluting, and can be installed near load centers and existing suburban substations. These have round-trip AC–AC efficiencies in the range of 85%, and can respond to rapid changes in electrical load (e.g., within 20 ms). Several U.S., European, and Japanese utilities have demonstrated the application of lead–acid batteries for load-following applications. Some of them have been as large as 10 MW with 4 h of storage. The other player in battery development is the automotive industry for electric vehicle applications. The commercial launch of several electric vehicles in 2011 has brought a lot of attention to storage batteries.

3.1.4.1 Battery Types

Chemical batteries are individual cells filled with a conducting medium-electrolyte that, when connected together, form a battery. Multiple batteries connected together form a battery bank. At present, there are two main types of batteries: primary batteries (non-rechargeable) and secondary batteries (rechargeable). Secondary batteries are further divided into two categories based on the operating temperature of the electrolyte. Ambient operating temperature batteries have either aqueous (flooded) or nonaqueous electrolytes. High operating temperature batteries (molten electrodes) have either solid or molten electrolytes. Batteries in EVs are the secondary rechargeable type and are in either of the two sub-categories. A battery for an EV must meet certain performance goals. These goals include quick discharge and recharge capability, long cycle life (the number of discharges before becoming unserviceable), low cost, recyclability, high specific energy (amount of usable energy, measured in watt-hours per pound [lb] or kilogram [kg]), high energy density (amount of energy stored per unit volume), specific power (determines the potential for acceleration), and the ability to work in extreme heat or cold. No battery currently available meets all these criteria.

3.1.4.2 Lead–Acid Batteries

Lead–acid starting batteries (shallow-cycle lead–acid secondary batteries) are the most common battery used in vehicles today. This battery is an ambient temperature, aqueous electrolyte battery. A cousin to this battery is the deep-cycle lead–acid battery, now widely used in golf carts and forklifts. The first electric cars built also used this technology. Although the lead–acid battery is relatively inexpensive, it is very heavy, with a limited usable energy by weight (specific energy). The battery's low specific energy and poor energy density make for a very large and heavy battery pack, which cannot power a vehicle as far as an equivalent gas-powered vehicle. Lead–acid batteries should not be discharged by more than 80% of their rated capacity or depth of discharge (DOD). Exceeding the 80% DOD shortens the life of the battery. Lead–acid batteries are inexpensive, readily available, and are highly recyclable, using the elaborate recycling system already in place. Research continues to try to improve these batteries.

A lead–acid nonaqueous (gelled lead acid) battery uses an electrolyte paste instead of a liquid. These batteries do not have to be mounted in an upright position. There is no electrolyte to spill in an accident. Nonaqueous lead–acid batteries typically do not have as high a life cycle and are more expensive than flooded deep-cycle lead–acid batteries.

3.1.4.3 Nickel Iron and Nickel Cadmium Batteries

Nickel iron (Edison cells) and nickel cadmium (nicad) pocket and sintered plate batteries have been in use for many years. Both of these batteries have a specific energy of around 25 Wh/lb (55 Wh/kg), which is higher than advanced lead–acid batteries. These batteries also have a long cycle life. Both of these batteries are recyclable. Nickel iron batteries are nontoxic, while nicads are toxic. They can also be discharged to 100% DOD without damage. However, these batteries have memory problem, which reduces useful energy recovery if the battery is not fully discharged before recharging. The biggest drawback to these batteries is their cost.

3.1.4.4 Nickel Metal Hydride Batteries

Nickel metal hydride batteries are offered as the best of the next generation of batteries. They have a high specific energy: around 40.8 Wh/lb (90 Wh/kg). According to a U.S. DOE report, the batteries are benign to the environment and are recyclable. They also are reported to have a very long cycle life. Nickel metal hydride batteries have a high self-discharge rate: They lose their charge when stored for long periods of time. They are already commercially available as "AA" and "C" cell batteries, for small consumer appliances and toys. In the past these batteries have been used for EV applications, but most automobile companies are now converging to Lithium ion batteries for electric vehicles.

3.1.4.5 Sodium-Sulfur Batteries

The sodium-sulfur (NaS) battery is a high-temperature battery, with the electrolyte operating at temperatures of 572°F (300°C). The sodium component of this battery explodes on contact with water, which raises certain safety concerns. The materials of the battery must be capable of withstanding the high internal temperatures they create, as well as freezing and thawing cycles. This battery has a very high specific energy: 50 Wh/lb (110 Wh/kg). During the last several years NaS batteries have found applications in power systems where large-scale storage is necessary. There are several examples in the United States and Japan where NaS batteries have been deployed in conjunction with wind farms.

3.1.4.6 Lithium Ion and Lithium Polymer Batteries

These batteries have a very high specific energy: 68 Wh/lb (150 Wh/kg) and store the same charge over their lifetime. These are widely used in laptop computers. Lithium ion (Li-ion) batteries use an anode of carbon, a cathode made of lithium cobalt oxide and an electrolyte gel. Under the stress of rapid charging or heavy use, a Li-ion battery may overheat very quickly with the possibility of fire. As a result, Li-ion batteries have an active protection circuit that prevents the battery from overheating. These batteries allow a vehicle to travel distances and accelerate at a rate comparable to conventional gasoline-powered vehicles. Lithium polymer batteries have the same basic chemistry as Li-ion batteries but use a porous separator which when exposed to the electrolyte, turns into a gel. Because this gel is not flammable, these batteries have a different architecture and are not subject to overheating and fire. Since there are no liquid electrolytes, these batteries can be manufactured in sizes as small as credit cards.

3.1.4.7 Zinc and Aluminum Air Batteries

Zinc air batteries are currently being tested in postal trucks in Germany. These batteries use either aluminum or zinc as a sacrificial anode. As the battery produces electricity, the anode dissolves into the electrolyte. When the anode is completely dissolved, a new anode is placed in the vehicle. The aluminum or zinc and the electrolyte are removed and sent to a recycling facility. These batteries have a specific energy of over 97 Wh/lb (200 Wh/kg). These batteries have been used in German postal vans which carry 80 kWh of energy in their battery, giving them about the same range as 13 gal (49.2 L) of gasoline. In their tests, the vans have achieved a range of 615 miles (990 km) at 25 miles per hour (40 km/h).

3.2 Fuel Cells

In 1839, a British Jurist and an amateur physicist named William Grove first discovered the principle of the fuel cell. Grove utilized four large cells, each containing hydrogen and oxygen, to produce electricity and water which was then used to split water in a different container to produce hydrogen and oxygen. However, it took another 120 years until NASA demonstrated its use to provide electricity and water for some early space flights. Today the fuel cell is the primary source of electricity on the space shuttle. As a result of these successes, industry slowly began to appreciate the commercial value of fuel cells. In addition to stationary power generation applications, there is now a strong push to develop fuel cells for automotive use. Even though fuel cells provide high performance characteristics, reliability, durability, and environmental benefits, a very high investment cost is still the major barrier against large-scale deployments.

3.2.1 Basic Principles

The fuel cell works by processing a hydrogen-rich fuel—usually natural gas or methanol—into hydrogen, which, when combined with oxygen, produces electricity and water. This is the reverse electrolysis process. Rather than burning the fuel, however, the fuel cell converts the fuel to electricity using a highly efficient electrochemical process. A fuel cell has few moving parts, and produces very little waste heat or gas.

A fuel cell power plant is basically made up of three subsystems or sections. In the fuel-processing section, the natural gas or other hydrocarbon fuel is converted to a hydrogen-rich fuel. This is normally accomplished through what is called a steam catalytic reforming process. The fuel is then fed to the power section, where it reacts with oxygen from the air in a large number of individual fuel cells to produce direct current (DC) electricity, and by-product heat in the form of usable steam or hot water. For a power plant, the number of fuel cells can vary from several hundred (for a 40 kW plant) to several thousand (for a multi-megawatt plant). In the final, or third stage, the DC electricity is converted in the power conditioning subsystem to electric utility-grade alternating current (AC).

In the power section of the fuel cell, which contains the electrodes and the electrolyte, two separate electrochemical reactions take place: an oxidation half-reaction occurring at the anode and a reduction half-reaction occurring at the cathode. The anode and the cathode are separated from each other by the electrolyte. In the oxidation half-reaction at the anode, gaseous hydrogen produces hydrogen ions, which travel through the ionically conducting membrane to the cathode. At the same time, electrons travel through an external circuit to the cathode. In the reduction half-reaction at the cathode, oxygen supplied from air combines with the hydrogen ions and electrons to form water and excess heat. Thus, the final products of the overall reaction are electricity, water, and excess heat.

3.2.2 Types of Fuel Cells

The electrolyte defines the key properties, particularly the operating temperature, of the fuel cell. Consequently, fuel cells are classified based on the types of electrolyte used as described next.

1. Polymer electrolyte membrane (PEM)
2. Alkaline fuel cell (AFC)
3. Phosphoric acid fuel cell (PAFC)
4. Molten carbonate fuel cell (MCFC)
5. Solid oxide fuel cell (SOFC)

These fuel cells operate at different temperatures and each is best suited to particular applications. The main features of the five types of fuel cells are summarized in Table 3.1.

3.2.3 Fuel Cell Operation

Basic operational characteristics of the four most common types of fuel cells are discussed in the following sections.

3.2.3.1 Polymer Electrolyte Membrane

The polymer electrolyte membrane (PEM) cell is one in a family of fuel cells that are in various stages of development. It is being considered as an alternative power source for automotive application for electric vehicles. The electrolyte in a PEM cell is a type of polymer and is usually referred to as a membrane, hence the name. PEMs are somewhat unusual electrolytes in that, in the presence of water, which

TABLE 3.1 Comparison of Five Fuel Cell Technologies

Type	Electrolyte	Operating Temperature (°C)	Applications	Advantages
Polymer electrolyte membrane (PEM)	Solid organic polymer poly-perflouro-sulfonic acid	60–100	Electric utility, transportation, portable power	Solid electrolyte reduces corrosion, low temperature, quick start-up
Alkaline (AFC)	Aqueous solution of potassium hydroxide soaked in a matrix	90–100	Military, space	Cathode reaction faster in alkaline electrolyte; therefore high performance
Phosphoric acid (PAFC)	Liquid phosphoric acid soaked in a matrix	175–200	Electric utility, transportation, and heat	Up to 85% efficiency in co-generation of electricity
Molten carbonate (MCFC)	Liquid solution of lithium, sodium, and/or potassium carbonates soaked in a matrix	600–1000	Electric utility	Higher efficiency, fuel flexibility, inexpensive catalysts
Solid oxide (SOFC)	Solid zirconium oxide to which a small amount of yttria is added	600–1000	Electric utility	Higher efficiency, fuel flexibility, inexpensive catalysts Solid electrolyte advantages like PEM

the membrane readily absorbs, the negative ions are rigidly held within their structure. Only the positive (H) ions contained within the membrane are mobile and are free to carry positive charges through the membrane in one direction only, from anode to cathode. At the same time, the organic nature of the PEM structure makes it an electron insulator, forcing it to travel through the outside circuit providing electric power to the load. Each of the two electrodes consists of porous carbon to which very small platinum (Pt) particles are bonded. The electrodes are somewhat porous so that the gases can diffuse through them to reach the catalyst. Moreover, as both platinum and carbon conduct electrons well, they are able to move freely through the electrodes. Chemical reactions that take place inside a PEM fuel cell are presented as follows:

Anode

$$2H_2 \rightarrow 4H^+ + 4e^-$$

Cathode

$$O_2 + 4H^+ + 4e^- \rightarrow 2H_2O$$

$$\text{Net reaction:}\quad 2H_2 + O_2 = 2H_2O$$

Hydrogen gas diffuses through the polymer electrolyte until it encounters a Pt particle in the anode. The Pt catalyzes the dissociation of the hydrogen molecule into two hydrogen atoms (H) bonded to two neighboring Pt atoms. Only then can each H atom release an electron to form a hydrogen ion (H^+) which travels to the cathode through the electrolyte. At the same time, the free electron travels from the anode to the cathode through the outer circuit. At the cathode the oxygen molecule interacts with the hydrogen ion and the electron from the outside circuit to form water. The performance of the PEM fuel cell is limited primarily by the slow rate of the oxygen reduction half-reaction at the cathode, which is 100 times slower than the hydrogen oxidation half-reaction at the anode.

3.2.3.2 Phosphoric Acid Fuel Cell

Phosphoric acid technology has moved from the laboratory research and development to the first stages of commercial application. Turnkey 200kW plants are now available and have been installed at more than 70 sites in the United States, Japan, and Europe. Operating at about 200°C, the phosphoric acid fuel cell (PAFC) plant also produces heat for domestic hot water and space heating, and its electrical efficiency approaches 40%. The principal obstacle against widespread commercial acceptance is cost. Capital costs of about $2500 to $4000/kW must be reduced to $1000 to $1500/kW if the technology is to be accepted in the electric power markets.

The chemical reactions occurring at two electrodes are written as follows:

$$\text{At anode:}\quad 2H_2 \rightarrow 4H^+ + 4e^-$$

$$\text{At cathode:}\quad O_2 + 4H^+ + 4e^- \rightarrow 2H_2O$$

3.2.3.3 Molten Carbonate Fuel Cell

Molten carbonate technology is attractive because it offers several potential advantages over PAFC. Carbon monoxide, which poisons the PAFC, is indirectly used as a fuel in the molten carbonate fuel cell (MCFC). The higher operating temperature of approximately 650°C makes the MCFC a better candidate for combined cycle applications whereby the fuel cell exhaust can be used as input to the intake of a gas turbine or the boiler of a steam turbine. The total thermal efficiency can approach 85%. This technology is at the stage of prototype commercial demonstrations and is estimated to enter the commercial market by 2003 using natural gas, and by 2010 with gas made from coal. Capital costs are expected to be lower than PAFC. MCFCs are now being tested in full-scale demonstration plants. The following equations illustrate the chemical reactions that take place inside the cell:

$$\text{At anode:}\quad 2H_2 + 2CO_3^{2-} \rightarrow 2H_2O + 2CO_2 + 4e^-$$

$$\text{and}\quad 2CO + 2CO_3^{2-} \rightarrow 4CO_2 + 4e^-$$

$$\text{At cathode:}\quad O_2 + 2CO_2 + 4e^- \rightarrow 2O_3^{2-}$$

3.2.3.4 Solid Oxide Fuel Cell

A solid oxide fuel cell (SOFC) is currently being demonstrated at a 100kW plant. Solid oxide technology requires very significant changes in the structure of the cell. As the name implies, the SOFC uses a solid electrolyte, a ceramic material, so the electrolyte does not need to be replenished during the operational life of the cell. This simplifies design, operation, and maintenance, as well as having the potential to reduce costs. This offers the stability and reliability of all solid-state construction and allows higher temperature operation. The ceramic make-up of the cell lends itself to cost-effective fabrication techniques. The tolerance to impure fuel streams make SOFC systems especially attractive for utilizing H_2 and CO from natural gas steam-reforming and coal gasification plants. The chemical reactions inside the cell may be written as follows:

$$\text{At anode:}\quad 2H_2 + 2O^{2-} \rightarrow 2H_2O + 4e^-$$

$$\text{and}\quad 2CO + 2O^{2-} \rightarrow 2CO_2 + 4e^-$$

$$\text{At cathode:}\quad O_2 + 4e^- \rightarrow 2O^{2-}$$

3.3 Summary

Fuel cells can convert a remarkably high proportion of the chemical energy in a fuel to electricity. With the efficiencies approaching 60%, even without co-generation, fuel cell power plants are nearly twice as efficient as conventional power plants. Unlike large steam plants, the efficiency is not a function of the plant size for fuel cell power plants. Small-scale fuel cell plants are just as efficient as the large ones, whether they operate at full load or not. Fuel cells contribute significantly to the cleaner environment; they produce dramatically fewer emissions, and their by-products are primarily hot water and carbon dioxide in small amounts. Because of their modular nature, fuel cells can be placed at or near load centers, resulting in savings of transmission network expansion.

4

Water

Vaughn Nelson
West Texas A&M University

4.1 Introduction

Energy from water is one of the oldest sources of energy, as paddle wheels were used to rotate a millstone to grind grain. A large number of watermills, 200–500 W, for grinding grain are still in use in remote mountains and hilly regions in the developing world. There are an estimated 500,000 in the Himalayas, with around 200,000 in India [1,2]. Of the 25,000–30,000 watermills in Nepal, 2,767 mills were upgraded between 2003 and 2007 [3]. Paddle wheels and buckets powered by moving water were and are still used in some parts of the world for lifting water for irrigation. Water provided mechanical power for the textile and industrial mills of the 1800s as small dams were built, and mill buildings are found along the edges of rivers throughout the United States and Europe. Then, starting in the late 1800s, water stored behind dams was used for the generation of electricity. For example, in Switzerland in the 1920s there were nearly 7000 small-scale hydropower plants.

The energy in water can be potential energy from a height difference, which is what most people think of in terms of hydro; the most common example is the generation of electricity (hydroelectric) from water stored in dams. However, there is also kinetic energy due to water flow in rivers and ocean currents. Finally, there is energy due to tides, which is due to gravitational attraction of the moon and the sun, and energy from waves, which is due to wind. In the final analysis, water energy is just another transformation from solar energy, except for tides.

The energy or work is force * distance, so potential energy due to gravitation is

$$W = F * d = m * g * H \quad \text{(J)} \tag{4.1}$$

The force due to gravity is mass * acceleration, where the acceleration of gravity $g = 9.8\,\text{m/s}^2$ and H = height in meters of the water. For estimations, you may use $g = 10\,\text{m/s}^2$.

For water, generally what is used is the volume, so the mass is obtained from density and volume.

$$\rho = \frac{m}{V} \quad \text{or} \quad m = \rho * V, \quad \text{where } \rho = 1000\,\text{kg/m}^3 \text{ for water}$$

Then, for water Equation 4.1 becomes

$$PE = \rho * g * H * V = 10{,}000 * H * V \tag{4.2}$$

Example 4.1

Find the potential energy for 2000 m^3 of water at a height of 20 m.

$$PE = 10{,}000 * 20 * 2{,}000 = 4 * 10^9\,\text{J} = 4\ \text{GJ}$$

If a mass of water is converted to kinetic energy after falling from a height H, then the velocity can be calculated.

$$KE = PE$$

or

$$0.5\ m * v^2 = m * g * H$$

Then, the velocity of the water is

$$v = (2 * g * H)^{0.5} \quad (\text{m/s}) \tag{4.3}$$

Example 4.2

For data in Example 4.1, find the velocity of that water after falling through 20 m.

$$v = (2 * 10 * 20)^{0.5} = 20\,\text{m/s}$$

Instead of water at some height, there is a flow of water in a river or an ocean current, such as the Gulf Stream. The analysis for energy and power for moving water is similar to wind energy, except there is a large difference in density between water and air. Therefore, for the same amount of power, capture areas for water flow will be a lot smaller.

$$\frac{P}{A} = 0.5 * \rho * v^3 \quad (\text{W/m}^2) \tag{4.4}$$

Example 4.3

Find the power/area for an ocean current that is moving at 1.5 m/s.

$$\frac{P}{A} = 0.5 * (1.5)^3 = 1.7\,\text{kW/m}^2$$

Power is energy/time, and hydraulic power from water or for pumping water from some depth is generally defined in terms of water flow Q and the height. Of course, if you know the time and have either power or energy, then the energy or power can be calculated.

$$P = 10{,}000 * H * \frac{V}{t} = 10 * Q * H \quad (\text{kW}) \tag{4.5}$$

where Q is the flow rate (m^3/s). In terms of pumping smaller volumes of water for residences, livestock, and villages, Q is generally noted as cubic meters per day, so be sure to note what units are used. There will be friction and other losses, so with efficiency ε the power is

$$P = 10{,}000 * \varepsilon * Q * H \tag{4.6}$$

Efficiencies from input to output (generally electric) range from 0.5 to 0.85. Small water turbines have efficiencies up to 80%, so when other losses are included (friction and generator), the overall efficiency is approximately 50%. Maximum efficiency is at the rated design flow and load, which is not always possible as the river flow fluctuates throughout the year or where daily load patterns vary.

The output from the turbine shaft can be used directly as mechanical power, or the turbine can be connected to an electric generator. For many rural industrial applications, shaft power is suitable for grinding grain or oil extraction, sawing wood, small-scale mining equipment, and so on.

4.2 World Resource

Around one-quarter of the solar energy incident on the Earth goes to the evaporation of water; however, as this water vapor condenses, most of the energy goes into the atmosphere as heat. Only 0.06% is rain and snow, and that power and energy of the water flow is the world resource, estimated at around 40,000 TWh/year. The technical potential (Table 4.1) is 15,000 TWh/year, and economic and environmental considerations reduce that potential.

TABLE 4.1 Technical Potential, Hydroelectric Production, and Capacity

	Potential, TWh/year	Production, TWh/year	Capacity, GW
Asia	5,090		
Asia and Oceanic		798	257
Central and South America	2,790	660	136
Europe	2,710	536	166
Eurasia		245	68
Middle East		22	9
Africa	1,890	97	22
North America	1,670	665	164
Oceania	230		
World	14,380	3000	822

Note: Production and capacity data for 2007 or 2008 from U.S. Energy Information Administration.

The classification of hydropower differs by country, authors, and even over time. One classification is large (>30 MW), small (100–30 MW), and micro (≤100 kW). Some examples are as follows: In China, small hydro refers to capacities up to 25 MW, in India up to 15 MW, and in Sweden up to 1.5 MW. Now, in Europe, small hydro means a capacity of up to 10 MW. Today in China, the classifications are large (>30 MW), small (5–30 MW), mini (100 kW to 5 MW), micro (5–100 kW), and pico (<5 kW). Others classify microhydro as 10–100 kW, so be sure to note the range when data are given for capacity and energy for hydropower.

4.3 Hydroelectric

4.3.1 Large (≥30 MW)

In terms of renewable energy, large-scale hydropower (Figure 4.1) is a major contributor to electric generation in the world, over 3000 TWh/year. The world installed capacity for large-scale hydroelectricity has increased around 2% per year, from 462 GW in 1980 to around 850 GW in 2009. However, the hydroelectric percentage of electric power has decreased from 21.5% in 1980 to 16% in 2008 as other sources of electrical energy have increased faster. China is now the leader in installed capacity and generation of electricity (Figure 4.2), with about 14% of their electricity from hydroelectric sources. However, coal in China is the major energy source for the production of electricity, and more coal power has been added than hydroelectric power. In Norway, 98% of the electrical energy is from hydro; Paraguay sells most of its share of electricity from the Itaipu Dam to Argentina. In the United States, the hydroelectric contribution is around 6%. The contribution from small or micro-hydro plants is difficult to estimate but could represent another 5%–10% in terms of world capacity. The capacity factor for hydroelectric power in the world has been fairly consistent at 40%–44% from 1980 to 2008. The capacity factor for hydroelectric power in the United States was 37% in 2008.

Large-scale hydroelectric plants have been constructed all across the world (Table 4.2). The Three Gorges Dam (Figure 4.3) on the Yangtze River is the largest power hydro plant in the world with 18.3 GW and will have a power of 22.5 GW when the rest of the generators are installed in 2011. Previously, the largest project was the Itaipu Dam on the Paraná River between Paraguay and Brazil. The series of dams is 7744 m long and was built from 1975 to 1991. The Aswan High Dam, Egypt (2100 MW), was completed in 1967 and produces more than 10 TWh/year, provides irrigation water for 3.2 million ha, and produces 20,000 ton of fish per year. The entire Temple of Abu Simel had to be moved to higher ground, a major feat in archeology. One of the problems of the Aswan Dam was that farming practices on the banks of the river downstream had to be changed since no yearly floods meant no deposition of fertile silt.

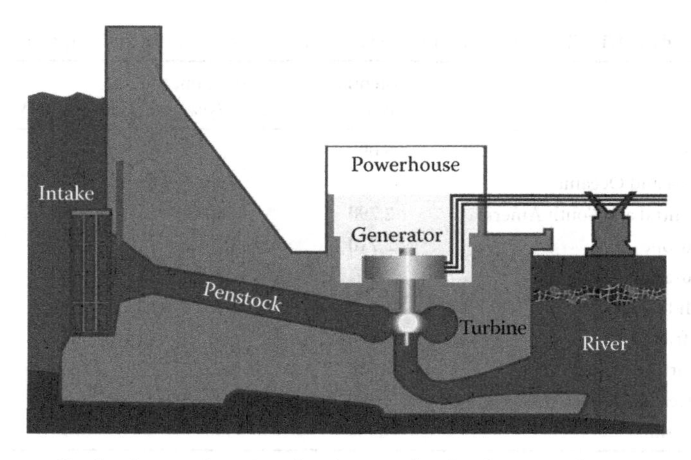

FIGURE 4.1 Diagram of hydroelectric plant. Height of water is level at dam to turbine generator. (From Tennessee Valley Authority, Knoxville, TN.)

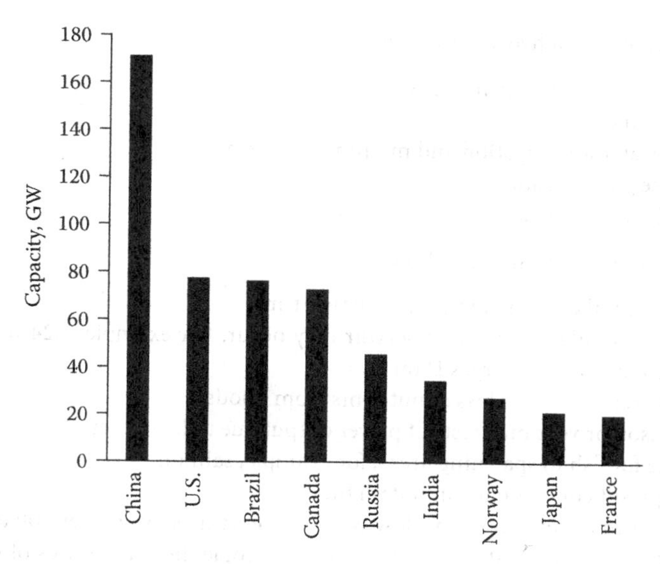

FIGURE 4.2 Installed hydroelectric capacity, 2009, top 10 countries.

TABLE 4.2 Large Hydroelectric Plants in the World, Date Completed, and Capacity

Country	Dam	Year	Capacity, MW
China	Three Gorges	2011	22,500
Brazil–Paraguay	Itaipu	1991	14,700
Venezuela	Guri	1986	10,055
Brazil	Tucurni	1984	8,370
United States	Grand Coulee	1941	6,809
Russia	Sayanao-Shushenskaya	1989	6,500
China	Longtan	2009	6,300
Russia	Krasnoyarsk	1972	6,000
Canada	Robert-Bourassa	1981	5,616
Canada	Churchill Falls	1971	5,429
United States	Hoover	1936	2,079

For photos, see Google images.

FIGURE 4.3 Three Gorges Dam, 22.5 GW, on the Yangtze River, China. (Courtesy of HydroChina, Beijing, China.)

The benefits or advantages of hydropower are as follows:

1. Renewable source, power on demand with reservoirs
2. Long life, 100 years
3. Flood control, water for irrigation and metropolitan areas
4. Low greenhouse gas emission
5. Reservoir for fishing, recreation

Some disadvantages or problems are the following:

1. There is a large initial cost and long construction time.
2. Displacement of population due to reservoir may occur. For example, 1.24 million people were relocated due to the Three Georges Dam.
3. On land downstream, there is loss of nutrients from floods.
4. Drought by season or year may restrict power output due to low water.
5. Lack of passage for fish to spawning areas, for example, salmon.
6. Rivers with high silt content may limit dam life.
7. Dam collapse means many problems downstream. There have been over 200 dam failures in the twentieth century, and it is estimated that 250,000 people died in a series of hydroelectric dam failures in China in 1975.
8. Resource allocation between countries can be a problem [4], especially if a series of dams that use a lot of water for irrigation are built upstream of a country.

In the United States, the first commercial hydroelectric plant (12.5 kW) was built in 1882 on the Fox River in Appleton, Wisconsin. Then, commercial power companies began to install a large number of small hydroelectric plants in mountainous regions near metropolitan areas. The creation of the Federal Power Commission in 1920 increased development of hydroelectric power with regulation and monetary support. The government supported projects for hydroelectric power and for flood control, navigation, and irrigation. The Tennessee Valley Authority was created in 1933 [5], and the Bonneville Power Administration was created in 1937 [6]. Construction of the Hoover Dam (Figure 4.4) started in 1931, and when completed in 1936, it was the largest hydroelectric project in the world at 2 GW [7].

FIGURE 4.4 Hoover Dam, 2 GW, on the Colorado River, United States. (From U.S. Bureau of Reclamation, Washington, DC.)

FIGURE 4.5 Grand Coulee Dam, 6.8 GW, on the Colombia River. (From U.S. Bureau of Reclamation, Washington, DC.)

Hoover Dam was then surpassed in 1941 by the Grand Coulee Dam (Figure 4.5) (6.8 GW) on the Columbia River [8]. The larger power output is due to the higher volume of water available.

A geographic information system (GIS) application, the Virtual Hydropower Prospector, provides maps and information for the United States [9]. The application allows the user to view the plants in the context of hydrography, power and transportation infrastructure, cities and populated areas, and federal land use. Most of the possible sites will be for small hydro.

In the developed countries with significant hydroelectric capacity, many of the best sites for hydroelectric potential already have dams. Many more reservoirs have been built for irrigation, water supply, and flood control than for hydropower as only 3% of the 78,000 dams in the United States have hydropower. Also, the construction of dams has declined in the United States since 1980 (Figure 4.6). So, there is a potential for hydropower by repowering some defunct hydroelectric installations or by new installations of small or medium hydropower at existing dams.

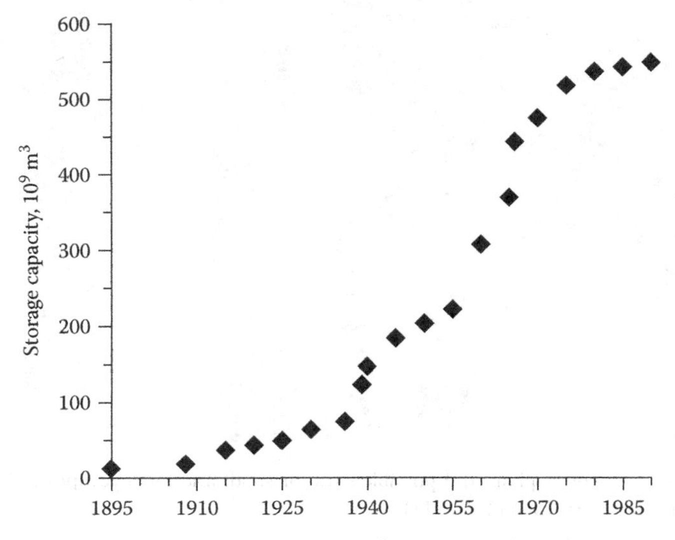

FIGURE 4.6 Reservoir capacity in the United States. (Data from U.S. Geological Survey, Reston, VA.)

4.3.2 Small Hydro (100 kW to 30 MW, 10 MW in Europe)

The definition of small hydro differs by country, so the range in Europe is 100 kW to 10 MW and in other countries is up to 25 or 30 MW. The World Energy Council estimated small hydro (up to 10 MW) was around 25,500 MW in 2006, with the major capacity in Europe, nearly 17,000 MW, and the energy production was estimated at 66 TWh. Now, the World Energy Council estimates the installed capacity of small hydro was around 55 GW in 2010, with China having the largest capacity. The current small hydro electricity generation in Europe (European Union-25, the candidate countries, and Switzerland) is around 47 TWh/year, and the remaining potential is estimated at another 49 TWh/year. This potential consists mainly of low-head sites (below 30 m).

Hydroelectric plants in the United States are predominantly private (69%); however, 75% of the capacity is owned by federal and nonfederal public owners [10], primarily from large power plants. The percentage of low and small hydropower plants in terms of numbers is 86%. This indicates future expansion for hydroelectric power in the United States will be from distributed generation.

A resource assessment of hydropower for 49 states (no resource in Delaware) identified 5667 sites (Figure 4.7) with a potential of around 30,000 MW [11]. The criteria were low power (<1 MW) or small hydro (≥1 MW and ≤30 MW), and the working flow was restricted to half the stream flow rate at the site or sufficient flow to produce 30 MWa (megawatts average), whichever was less. Penstock lengths were limited by the lengths of penstocks of a majority of existing low-power or small hydroelectric plants in the region. The optimum penstock length and location on the stream was determined for the maximum hydraulic head with the minimum length. The number of sites studied was 500,000, with approximately 130,000 sites meeting the feasibility criteria. Then, application of the development model with the limits on working flow and penstock length resulted in a total hydropower potential of 30,000 MWa. The approximately 5,400 sites that could potentially be developed as small hydro plants have a total

FIGURE 4.7 Present hydropower plans and possible sites for small and low hydropower in the United States. (From Idaho National Laboratory, Idaho Falls, ID.)

hydropower potential of 18,000 MWa. Idaho National Lab also developed a probability factor model, Hydropower Evaluation Software, to standardize the environmental assessment.

4.3.3 Microhydro (<100 kW)

Estimation of the number of installations and capacity for microhydro is even more difficult. In general, microhydro does not need dams and a reservoir as water is diverted and then conducted in a penstock to a lower elevation and the water turbine. In most cases, the end production is the generation of electricity.

There are thought to be tens of thousands of microhydro plants in China and significant numbers in Nepal, Sri Lanka, Pakistan, Vietnam, and Peru. The estimate for China was about 500 MW at the end of 2008. China started a program, SDDX [12], in 2003 that installed 146 hydro systems with a capacity of 113.8 MW in remote villages in the Western Provinces and Tibet. Hydropower was the predominant system in terms of capacity compared to wind and photovoltaics (PV), with 721 installations (15.5 MW) for villages and 15,458 installations (1.1 MW) for single households. The average size of the hydropower systems was 780 kW, which is much larger than average for the wind and PV systems (22 kW). Case studies are available for a number of countries [13], and software is available from Microhydro [14].

The advantages of microhydro are the following:

1. Efficient energy source. A small amount of flow (0.5 L/min) with a head of 1 m generates electricity with micro hydro. Electricity can be delivered up to 1.5 km.
2. Reliable. Hydro produces a continuous supply of electrical energy in comparison to other small-scale renewable technologies. Also, backup, whether diesel or batteries (which causes operation and maintenance and cost problems), is not needed.
3. No reservoir required. The water passing through the generator is directed back into the stream with relatively little impact on the surrounding ecology.
4. It is a cost-effective energy solution for remote locations.
5. Power for developing countries. Besides providing power, developing countries can manufacture and implement the technology.

The disadvantages or problems are as follows:

1. Suitable site characteristics are required, including distance from the power source to the load and stream size (flow rate, output, and head).
2. Energy expansion may not be possible.
3. There is low power in the summer months. In many locations, stream size will fluctuate seasonally.
4. Environmental impact is minimal; however, environmental effects must be considered before construction begins.

Impulse turbines are generally more suitable for microhydro applications compared with reaction turbines because of

Greater tolerance of sand and other particles in the water
Better access to working parts
Lack of pressure seals around the shaft
Ease of fabrication and maintenance
Better part-flow efficiency

The major disadvantage of impulse turbines is that they are generally unsuitable for low-head sites. Pelton turbines can be used at heads down to about 10 m; however, they are not used at lower heads because their rotational speeds are too slow, and the runner required is too large. The cross-flow turbine is the best machine for construction by a user.

4.4 Turbines

The two main types of hydro turbines are impulse and reaction. The type selected is based on the head and the flow, or volume of water, at the site (Table 4.3). Other deciding factors include how deep the turbine must be set, efficiency, and cost. Many images are available on the Internet for the different types of turbines.

4.4.1 Impulse Turbines

The impulse turbine uses the velocity of the water to move the runner (rotating part) and discharges to atmospheric pressure. The water stream hits each bucket on the runner, and the water flows out the bottom of the turbine housing. An impulse turbine is generally used for high-head, low-flow applications.

A Pelton turbine (Figure 4.8) has one or more free jets of water impinging on the buckets of a runner. The jet is directed at the centerline of the two buckets. Draft tubes are not required for the impulse turbine since the runner must be located above the maximum tail water to permit operation at atmospheric pressure.

A cross-flow turbine (Figure 4.9) resembles a squirrel cage blower and uses an elongated, rectangular-section nozzle to direct a sheet of water to a limited portion of the runner, about midway on one side. The flow of water crosses through the empty center of the turbine and exits just below the center on the opposite side. A guide vane at the entrance to the turbine directs the flow to a limited portion of the runner. The cross flow was developed to accommodate larger water flows and lower heads than the Pelton turbine.

TABLE 4.3 Classification of Turbine Type

	Head Pressure		
Turbine	High	Medium	Low
Impulse	Pelton	Cross flow	Cross flow
	Turgo	Turgo	
	Multijet pelton	Multijet pelton	
Reaction		Francis	Propeller
			Kaplan

FIGURE 4.8 Pelton runner (cast) showing bucket shape.

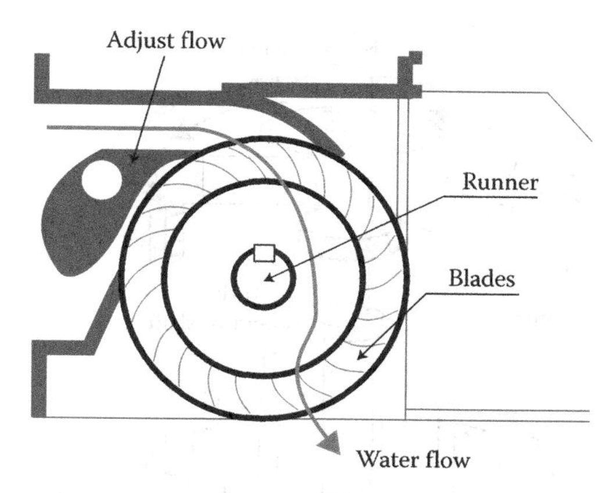

FIGURE 4.9 Diagram of cross-flow turbine.

4.4.2 Reaction Turbines

A reaction turbine develops power from the combined action of pressure and moving water, as the pressure drop across the runner produces power. The runner is in the water stream flowing over the blades rather than striking each individually. Reaction turbines are generally used for sites with lower head and higher flows.

Francis turbines (Figure 4.10) are the most common for hydropower. They are an inward flow turbine that combines radial and axial components. The runner has fixed vanes, usually nine or more. The inlet is spiral shaped with guide vanes to direct the water tangentially to the runner. The guide vanes (or wicket gate) may be adjustable to allow efficient turbine operation for a range of water flow conditions. The other major components are the scroll case, wicket gates, and draft tube (as water speed is reduced, a larger area for the outflow is needed). However, the Francis turbine can be used for heads to 800 m.

A propeller turbine (Figure 4.11) generally has a runner with three to six blades running in a pipe, where the pressure is constant. The pitch of the blades may be fixed or adjustable. The major components besides the runner are a scroll case, wicket gates, and a draft tube.

There are several different types of propeller turbines:

Bulb: The turbine and generator are a sealed unit placed directly in the water stream.

Straflo: The generator is attached directly to the perimeter of the turbine.

Tube: The penstock bends just before or after the runner, allowing a straight-line connection to the generator.

Kaplan: Both the blades and the wicket gates are adjustable, allowing for a wider range of operation.

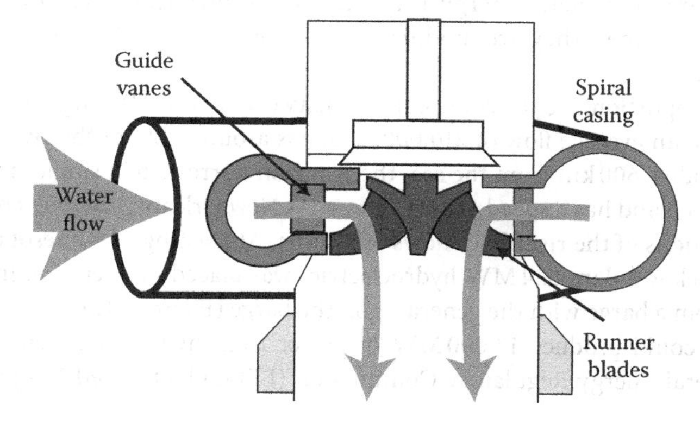

FIGURE 4.10 Diagram of Francis turbine.

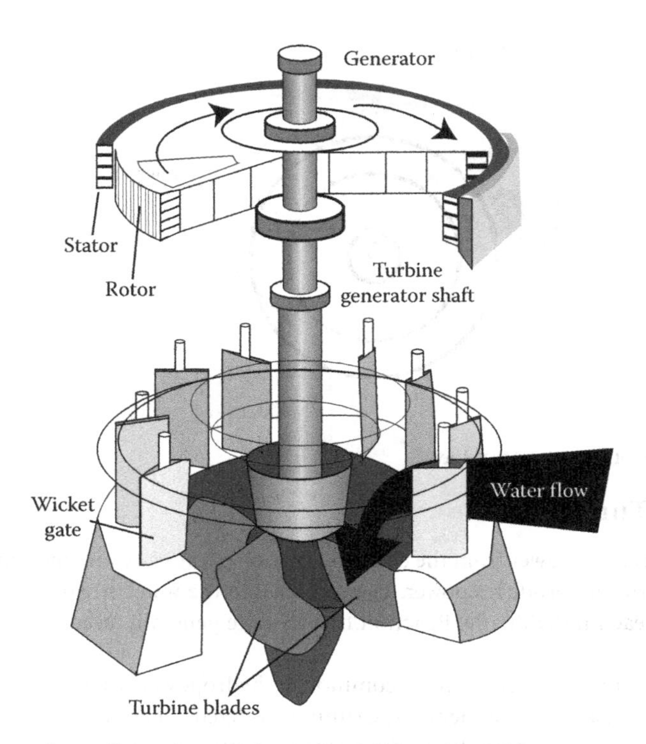

FIGURE 4.11 Diagram of propeller turbine. (From EERE, Washington, DC.)

4.5 Water Flow

Kinetic energy turbines, also called free-flow turbines, generate electricity from the kinetic energy of the flowing water, similar to wind turbines, which generate energy from the flowing air. Systems are also referred to as hydrokinetic, tidal in-stream energy conversion (TISEC), or river in-stream energy conversion (RISEC). The systems may operate in rivers, tides, ocean currents, or even channels or conduits for water. Kinetic systems do not require large civil works, and they can be placed near existing structures such as bridges, tailraces, and channels that increase the natural flow of water. For tidal currents, unidirectional turbines are available; rotation is the same, even though current is from opposite directions. One hydrokinetic system has a hydraulic pump to drive an onshore electric generator. Kinetic energy turbines would have less environmental impact than dams, and like wind turbines, they are modular and can be installed in a short time compared to large civil structures.

The power/area is proportional to the cube of the velocity (Equation 4.4). Large rivers have large flows, and the Amazon, with an average flow of 210,000 m^3/s, has around 20% of the river flow of the world. At the narrows of Óbidos, 600 km from the sea, the Amazon narrows to a single stream that is 1.6 km wide and over 60 m deep and has a speed of 1.8–2.2 m/s. At New Orleans, the speed of the Mississippi is 1.3 m/s, and some sections of the river have flows of 2.2 m/s. At Hastings, Minnesota, a 250-kW hydrokinetic unit located below a dam (4.4 MW hydroelectric) was placed in operation in 2008. The ducted rotor is suspended from a barge with the generator on the barge (Figure 4.12).

The United States could produce 13,000 MW of power from hydrokinetic energy by 2025. As of March 2010, the Federal Energy Regulatory Commission (FERC) had issued 134 preliminary permits

FIGURE 4.12 In-river system, 250 kW, on Mississippi River, Hastings, Minnesota. (Courtesy of Hydro Green, Houston, TX.)

for hydrokinetic projects (Figure 4.13) with a total capacity of 9864 MW. Notice that many of the permits are on the Mississippi River.

The FERC requires consideration of any environmental effects of the proposed construction, installation, operation, and removal of the project. The description should include

1. Any physical disturbance (vessel collision or other project-related risks for fish, marine mammals, seabirds, and other wildlife as applicable)
2. Species-specific habitat creation or displacement
3. Increased vessel traffic
4. Exclusion or disturbance of recreational, commercial, industrial, or other uses of the waterway and changes in navigational safety
5. Any above- or below-water noise disturbance, including estimated decibel levels during project construction, installation, operation, and removal
6. Any electromagnetic field disturbance
7. Any changes in river or tidal flow, wave regime, or coastal or other geomorphic processes
8. Any accidental contamination from device failures, vessel collisions, and storm damage
9. Chemical toxicity of any component of, or biofouling coating on, the project devices or transmission line
10. Any socioeconomic effects on the commercial fishing industry from potential loss of harvest or effect on access routes to fishing grounds

An important factor for water flow is that, at good locations, power will not vary like that of wind turbines, especially for in-river locations, so capacity factors can be much higher. One manufacturer stated that capacity factors should be at least 30% for tides and 50% for in-river systems. As always, the final result for comparison is the cost per kilowatt hour, which should be life-cycle costs.

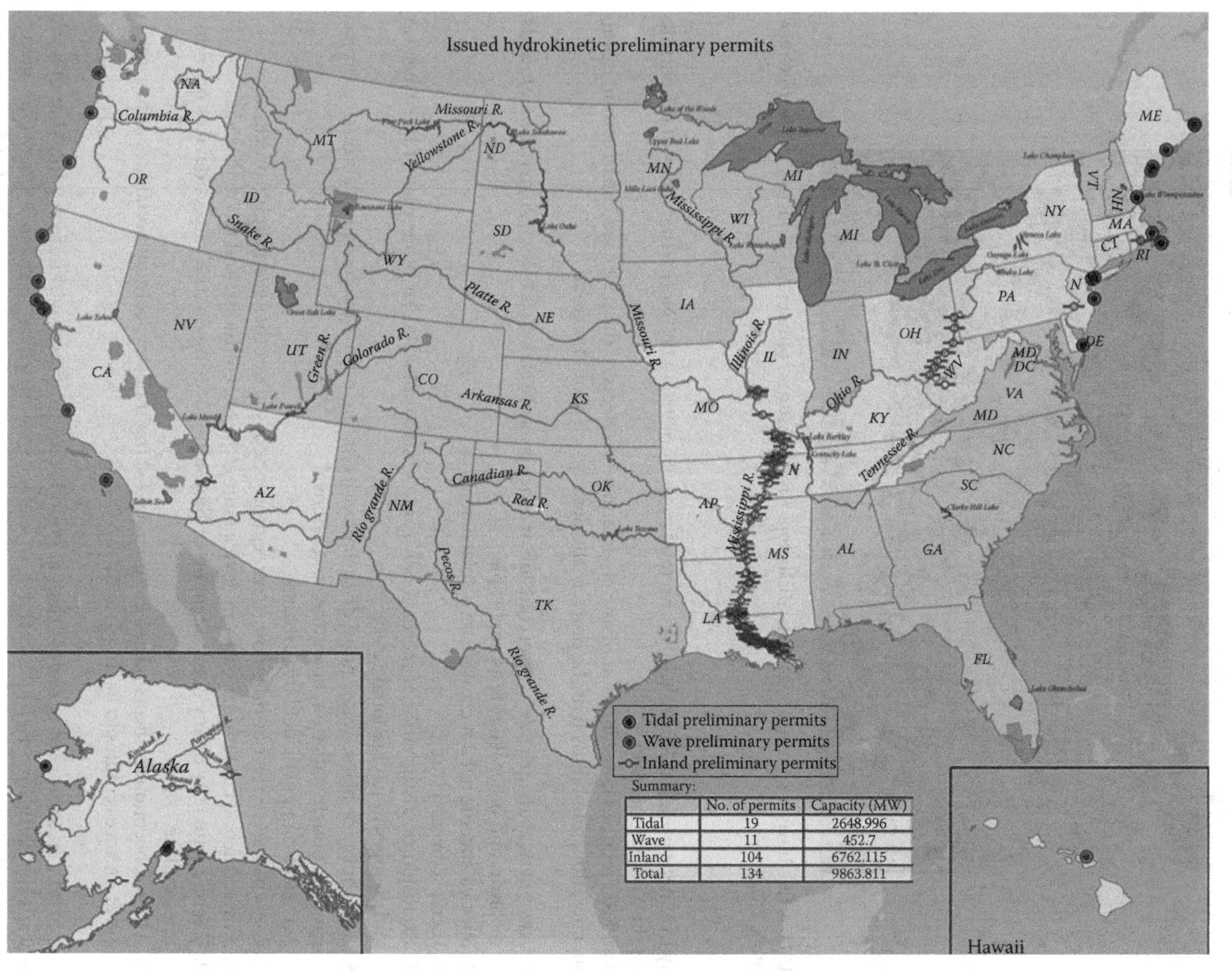

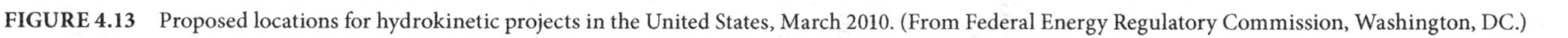
FIGURE 4.13 Proposed locations for hydrokinetic projects in the United States, March 2010. (From Federal Energy Regulatory Commission, Washington, DC.)

4.6 Tides

Tides are due to the gravitational attraction of the moon and the sun at the surface of the Earth. The effect of the moon on the Earth in terms of tides is larger than the effect of the sun, even through the gravitational force of the sun is larger. To find how the gravitational force of the moon distorts any volume of the material body of the Earth, the gradient of the gravitational force of the moon on that volume must be found (a gradient is how force changes with distance; in calculus, it is the differentiation with respect to length). The tidal effects (Figure 4.14) are superimposed on the near-spherical Earth, and there will be two tides per day due to the spin of Earth. When the tidal effects of the sun and moon are aligned, the tides are higher, spring tides. When the continents are added, the ocean bulges reflect from shorelines, which causes currents, resonant motions, and standing waves, so there are some places in the oceans where the tidal variations are nearly zero. In other locations, the coastal topography can intensify water heights with respect to the land. The largest tidal ranges in the world are the Bay of Fundy (11.7 m), Ungava Bay (9.75 m), Bristol Channel (9.6 m), and the Turnagain Arm of Cook Inlet, Alaska (9.2 m). The potential world tidal current energy is on about 2200 TWh/year.

Small mills were used on tidal sections of rivers in the Middle Ages for grinding grain. Today, there are only a few tidal systems installed in the world: the French installed a tidal system on the Rance Estuary (constructed from 1961 to 1967) with a power of 240 MW; an 18-MW rim generator at Annapolis Royal, Nova Scotia, Canada (1984); a 400-kW unit in the Bay of Kislaya, Russia (1968); and a 500-kW unit at Jangxia Creek, East China Sea.

The simplest system for generation of electricity is an ebb system, which involves a dam, known as a barrage, across an estuary. Barrages make use of the potential energy in the difference in height between high and low tides. Sluice gates on the barrage allow the tidal basin to fill on high tides (flood tide) and to generate power on the outgoing tide (ebb tide). Flood generating systems generate power from both tides but are less favored than ebb generating systems. Barrages across the full width of a tidal estuary have high civil infrastructure costs, there is a worldwide shortage of viable sites, and there are more environmental issues.

Tidal lagoons are similar to barrages but can be constructed as self-contained structures, not fully across an estuary, and generally have much lower cost and environmental impact. Furthermore, they can be configured to generate continuously, which is not the case with barrages. Different tidal systems, installed and proposed plants, and prototype and demonstration projects are given in Ref. [15].

The potential for tidal in-stream systems was estimated at 692 MW for five states in the United States [16]. A kinetic energy demonstration project (Figure 4.15) is installed in the East River, New York City, and consists of two 35-kW turbines, 5-m diameter, with passive yaw. In 9000 h of operation, the system generated 70 MWh. Another prototype, SeaGen, is installed in Strangford Narrows, Northern Ireland, with rated power of 1.2 MW at a current velocity of 2.4 m/s and with twin 16-m diameter rotors (Figure 4.16). The rotor blades can be pitched through 180° to generate power on both ebb and food tides. The twin power units are mounted on winglike extensions on a tubular steel monopole, and the system can be raised above sea level for maintenance. Kinetic energy systems are being considered because of

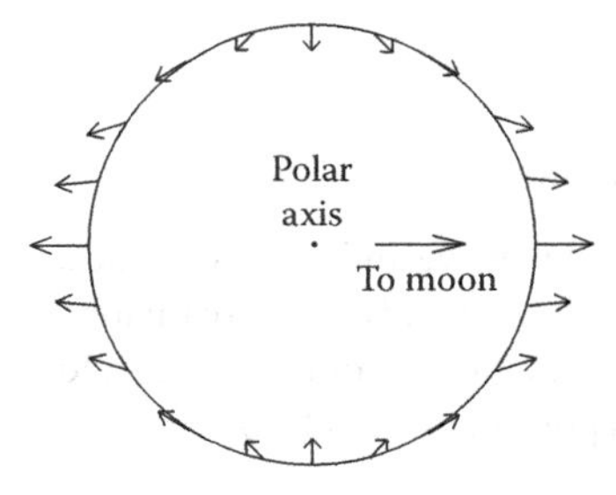

FIGURE 4.14 Tidal forces on the Earth due to the moon.

FIGURE 4.15 Tidal turbine, 35 kW, being installed in East River, New York City. (Courtesy of Verdant Power, New York.)

FIGURE 4.16 Tidal system, 1.2 MW, in Strangford Narrows, Northern Ireland. (Courtesy of Sea Generation, Bristol, U.K.)

the lower cost, lower ecological impact, increased availability of sites compared to barrages, and shorter time for installation.

Advantages for tidal systems are as follows:

1. Renewable
2. Predictable

Disadvantages or problems are as follows:

1. A barrage across an estuary is expensive and affects a wide area.
2. The environment is changed upstream and downstream for some distance. Many birds rely on the tide uncovering the mudflats so that they can feed. Fish ladders are needed.
3. There is intermittent power as power is available for around 10 h each day when the tide is moving in or out.
4. There are few suitable sites for tidal barrages.

4.7 Ocean

As with other renewable resources, the ocean energy is large [17]. The global technical resource exploitable with today's technology is estimated to be 20,000 TWh/year for ocean currents, 45,000 TWh/year for wave energy, 33,000 TWh/year for ocean thermal energy conversion (OTEC), and 20,000 TWh/year for salinity gradient energy. Of course, economics and other factors will greatly reduce the potential production, and future actual energy production will be even smaller.

Besides the environmental considerations mentioned, there are a number of technical challenges for ocean energy to be utilized at a commercial scale:

Avoidance of cavitations (bubble formation)
Prevention of marine growth buildup
Reliability (since maintenance costs may be high)
Corrosion resistance

4.7.1 Currents

There are large currents in the ocean (Figure 4.17), and detailed information on surface currents by ocean is available [18, only Atlantic and Polar at this time]. For example, the Gulf Stream transports a significant amount of warm water toward the North Atlantic and the coast of Europe. The core of the Gulf Stream current is about 90 km wide and has peak velocities greater than 2 m/s. The relatively constant extractable energy density near the surface of the Gulf Stream, the Florida Straits Current, is about 1 kW/m^2. Although the volume and velocity are adequate for in-stream hydro-kinetic systems, an ocean current would need to be close to the shore.

The total world power in ocean currents has been estimated to be about 5000 GW, with power densities of up to 15 kW/m^2 [19]. The European Union, Japan, and China are interested in and pursuing the application of ocean current energy systems.

4.7.2 Waves

Waves are created by the progressive transfer of energy from the wind as it blows over the surface of the water. Once created, waves can travel large distances without much reduction in energy. The energy in

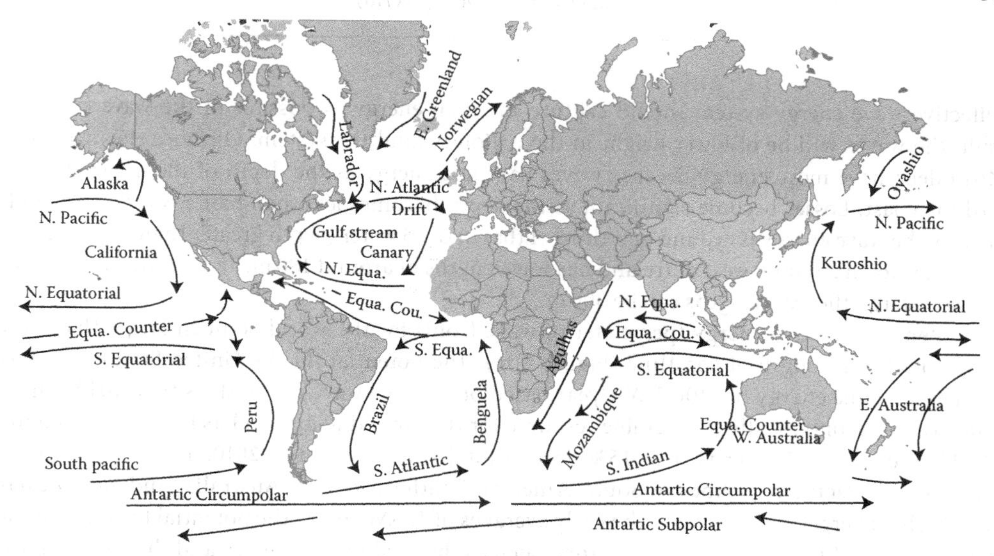

FIGURE 4.17 Major ocean currents in the world. (With permission from Michael Pidwirney.)

a wave is proportional to the height squared. In data for wave heights, be sure to note that height is for crest to trough, and amplitude is midpoint to crest.

$$E = 0.5 * \rho * g * \frac{H^2}{16} \quad \text{(J)} \tag{4.7}$$

where H is wave height. This is for a single wave, but in the ocean, there is superposition of waves, and the energy transported is by group velocity. The speed of the wave, wave length, and frequency (or period, which is 1/frequency) are related by

$$\text{Speed} = \text{Wavelength}\,(\lambda) * \text{Frequency}\,(f)$$

In deep water where the water depth is larger than half the wavelength, the power per length (meter) of the wave front is given by

$$\frac{P}{L} = \rho * g^2 * H^2 * \frac{T}{(64 * \pi)} \sim 0.5 * A^2 * T \quad \text{(kW/m)} \tag{4.8}$$

where T is the period of the wave (time it takes for successive crests to pass one point). In major storms, the largest waves offshore are about 15 m high and have a period of about 15 s, so the power is large, around 1.7 MW/m.

Example 4.4

Calculate power/length for waves off New Zealand if the average wave height is 7 m with a wave period of 8 s. The power/length is

$$\frac{P}{L} = 0.5 * 7^2 * 8 = 196 \quad \text{(kW/m)}$$

An effective wave energy system should capture as much energy as possible of the wave energy. As a result, the waves will be of lower height in the region behind the system. Offshore sites with water 25–40 m deep have more energy because waves have less energy as the depth of the ocean decreases toward the coast. Losses become significant as the depth becomes less than half a wavelength, and at 20 m deep, the wave energy is around one-third of that in deep water (depth greater than one-half wavelength). The North Atlantic west of Ireland has wavelengths of around 180 m, and off the West Coast of the United States, the wavelengths can be 300 m.

The potential for wave energy (per meter of wave front) for the world is much larger than ocean currents due to the length of coastline (Figure 4.18). The potential for the United States is 240 GW, with an extractable energy of 2100 TWh/year based on average wave power density of 10 kW/m. The technically and economically recoverable resource for the United Kingdom has been estimated to be 50–90 TWh of electricity per year or 15%–25% of total U.K. demand in 2010. The western coast of Europe and the Pacific coastlines of South America, Southern Africa, Australia, and New Zealand are also highly energetic. Any area with yearly averages of 15 kW/m has the potential to generate wave energy. Note that this threshold excludes areas such as the Mediterranean Sea and the Great Lakes of North America.

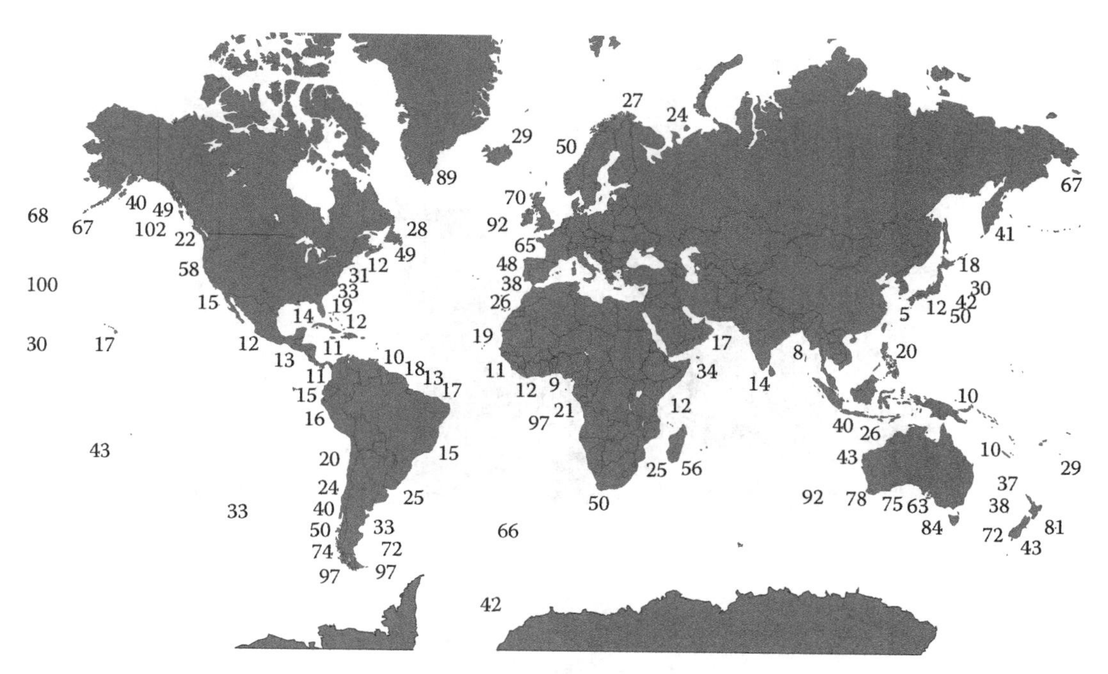

FIGURE 4.18 Average wave energy (kW/m) for coastlines around the world; values are for deep-water sites. (Courtesy of Pelamis Wave Energy.)

The resource or wave climate can be obtained from recorded data, and satellites now provide current worldwide data and are used for prediction of wave heights. For wave energy systems, it is also important to determine the statistical occurrence of the extreme waves that can be expected at the site over the lifetime of the system since the system should be designed to survive peak waves.

Once the general area of the wave farm site has been determined, more analysis is needed to pick the best site within that area, for example, by examining the mean wave direction, variability, and the possibility of local focusing of waves. Another essential task includes the calculation of calm periods that allow sufficient time for maintenance and other operations. However, as noted, large waves have lots of power and could damage or destroy the system, so design and construction must take these large waves into account.

The mechanisms for capture of wave energy are point absorber, reservoir, attenuator, oscillating water column, and other mechanisms. There are a number of prototypes and demonstration projects but few commercial projects. A point absorber has a small dimension in relation to the wavelength (Figure 4.19).

The reservoir system is where the waves are forced to higher heights by channels or ramps, and the water is captured in a reservoir (Figure 4.20). Locations for land installations for reservoir and oscillating water column systems will be much more limited than offshore systems; however, land installations are easier to construct and maintain. The Wave Dragon is a floating offshore platform (Figures 4.21 and 4.22).

The Pelamis Wave Energy Converter [20], an attenuator, is a semisubmerged, articulated cylindrical attenuator linked by hinged joints (Figure 4.23). The wave-induced motion of these joints drives hydraulic rams, which pump high-pressure fluid through hydraulic motors via smoothing accumulators. The hydraulic motors drive an electrical generator, and the power from all the joints is fed down a single cable to a junction on the seabed. Several devices (Figure 4.24) can then be linked to shore through a single seabed cable. Current production machines have four power conversion modules: 750-kW rated power, 180 m long, 4-m diameter. The power table and the wave climate are combined to give the electrical power response over time and, from that, its average level and its variability. Depending on the wave resource, the capacity factor is 25%–40%.

(a)

(b)

FIGURE 4.19 PowerBuoy prototype, 40 kW, 14.6 m long, 3.5 m diameter; floats 4.25 m above surface of water. (Courtesy of Ocean Power Technology, Pennington, NJ.)

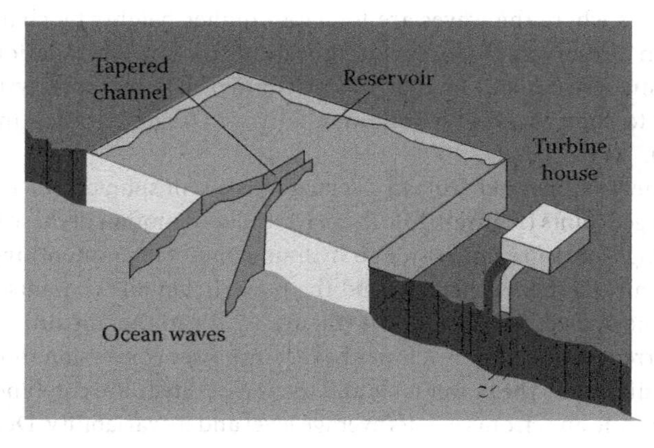

FIGURE 4.20 Diagram of a reservoir system on land.

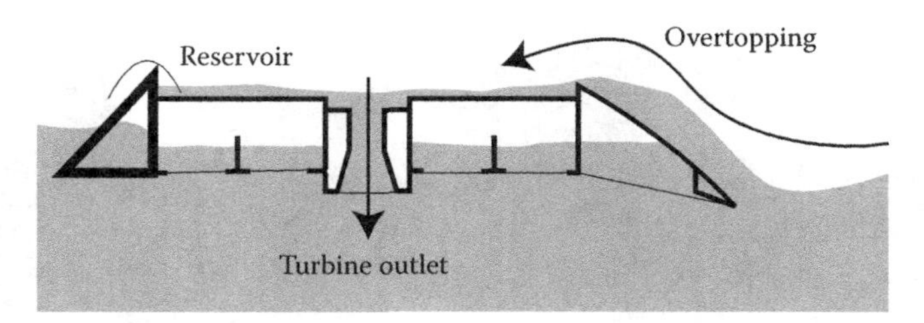

FIGURE 4.21 Diagram of floating reservoir system. (Diagram from Wave Dragon, Ottawa, Ontario, Canada, http://www.wavedragon.net.)

FIGURE 4.22 Prototype floating reservoir system, Nissum Bredning, Denmark. (Courtesy of Wave Dragon, Ottawa, Ontario, Canada, http://www.wavedragon.net.)

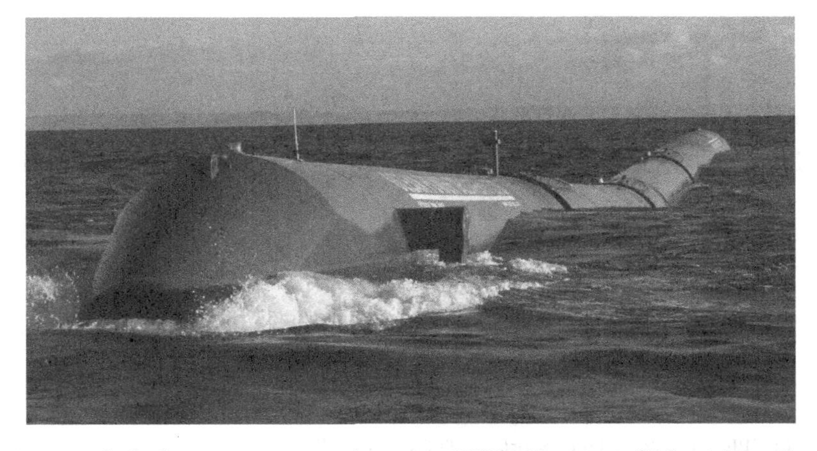

FIGURE 4.23 Sea trial of Pelamis Wave Energy Converter, 750 kW. (Courtesy of Pelamis Wave Energy.)

In an oscillating water column, as a wave enters the column, the air pressure within the column is increased, and as the wave retreats, air pressure is reduced (Figure 4.25). The Wells turbine turns in the same direction irrespective of the airflow direction. The land-installed marine power energy transmitter (LIMPET) unit on Isle of Islay, Scotland [21], has an inclined oscillating water column, with an inlet width of 21 m (Figure 4.26) with the mean water depth at the entrance at 6 m. The system (rated power is 500 kW) has three water columns contained within concrete tubes, 6 m by 6 m, inclined at 40° to the

FIGURE 4.24 Installation of three units at Aguçadoura, Portugal, 2.25 MW total power. (Courtesy of Pelamis Wave Energy.)

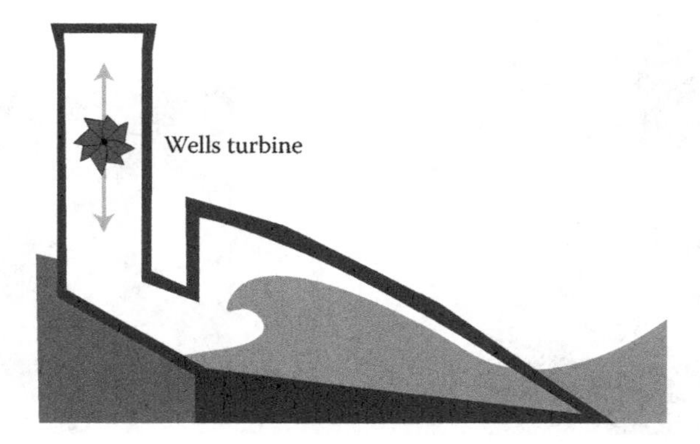

FIGURE 4.25 Diagram of oscillating water column system.

FIGURE 4.26 LIMPET on Islay Island, Scotland, 500 kW; installed 2000. (Courtesy of Voith Hydro Wavgen, Inverness-Shire, U.K.)

horizontal, giving a total water surface area of 169 m^2. The upper parts of the tubes are connected to a single tube, which contains a Wells generator.

The design of the air chamber is important to maximize the conversion of wave energy to pneumatic power, and the turbines need to be matched to the air chamber. The performance has been optimized for annual average wave intensities of between 15 and 25 kW/m.

FIGURE 4.27 Oyster hydroelectric wave energy converter, 315 kW; unit installed at Billa Croo, Orkney, Scotland. (Courtesy of Acquamarine Power, Edinburgh, U.K.)

In another system, waves drive a hinged flap connected (Figure 4.27) to the seabed at around 10 m depth, which then drives hydraulic pistons to deliver high-pressure water via a pipeline to an onshore electrical turbine.

4.7.3 Ocean Thermal Energy Conversion

OTEC for producing electricity is the same as solar ponds, for which the thermal difference between surface water and deep water drives a Rankine cycle. There is one major difference: The deep ocean water is rich in nutrients, which can be used for mariculture. In both systems, there is the production of freshwater.

An OTEC system needs a temperature difference of 20°C from cold water within 1000 m of the surface, which occurs across vast areas of the world (Figure 4.28). The systems can be on or near the shore. The three general types of OTEC processes are closed cycle, open cycle, and hybrid cycle.

In the closed-cycle system, heat transferred from the warm surface seawater causes a working fluid to turn to vapor, and the expanding vapor drives a turbine attached to an electric generator. Cold seawater passing through a condenser containing the vaporized working fluid turns the vapor back into a liquid, which is then recycled through the system.

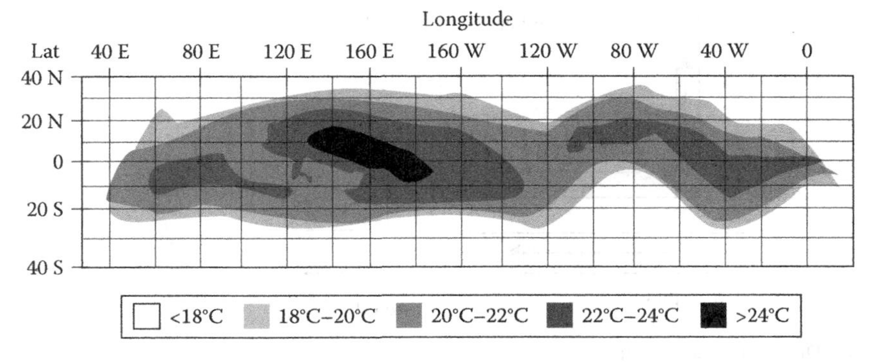

FIGURE 4.28 Ocean thermal differences, surface to depth of 1000 m. (Courtesy of *NREL*, Golden, CO.)

An open-cycle system uses the warm surface water itself as the working fluid. The water vaporizes in a near vacuum at surface water temperatures. The expanding vapor drives a low-pressure turbine attached to an electrical generator. The vapor, which is almost pure freshwater, is condensed into a liquid by exposure to cold temperatures from deep ocean water. If the condenser keeps the vapor from direct contact with seawater, the water can be used for drinking water, irrigation, or aquaculture. A direct contact condenser produces more electricity, but the vapor is mixed with cold seawater, and the mixture is discharged to the ocean. Hybrid systems use parts of both open- and closed-cycle systems to optimize production of electricity and freshwater.

The first prototype OTEC project (22 kW) was installed at Matanzas Bay, Cuba, in 1930 [22]. Then, in the latter part of the twentieth century, experimental systems were installed in Hawaii and Japan. An experimental, open-cycle, onshore system was operated intermittently between 1992 and 1998 at the Keahole Point Facility, National Energy Laboratory, Hawaii. Surface water is 26°C, and the deep-water temperature is 6°C (depth of 823 m); the system produced a maximum power of 250 kW. However, the power requirements for pumping the surface (36.3 m^3/min) and deep (24.6 m^3/min) seawater were around 200 kW. A small fraction (10%) of the steam produced was diverted to a surface condenser for the production of freshwater, about 22 L/min. In 1981, Japan demonstrated a shore-based, 100-kWe closed-cycle plant in the Republic of Nauru in the Pacific Ocean. The cold-water pipe was laid on the seabed at a depth of 580 m. The plant produced 31.5 kWe of net power during continuous operating tests.

4.7.4 Salinity Gradient

Salinity gradient energy is derived from the difference in the salt concentration between seawater and river water. Two practical methods for this are reverse electrodialysis and pressure-retarded osmosis; both rely on osmosis with ion-specific membranes. A small prototype (4 kW) started operation in 2009 in Tofte, Norway. The pressure generated is equal to a water column of 120 m, which is used to drive a turbine to generate electricity.

4.8 Other

Another application for water flow is ram pumps, where the pressure from water over a drop of a few meters is used to lift a small percentage of that water through a much greater height for water for people or irrigation. Ram pumps were developed over 200 year ago and can be made locally [23–25]. The operation of a ram pump (Figure 4.29) is as follows:

1. Water from a stream flows down the drive pipe and out of the waste valve.
2. As the flow of water accelerates, the waste valve is forced shut, causing a pressure surge (or water hammer) as the moving water is brought to a halt.

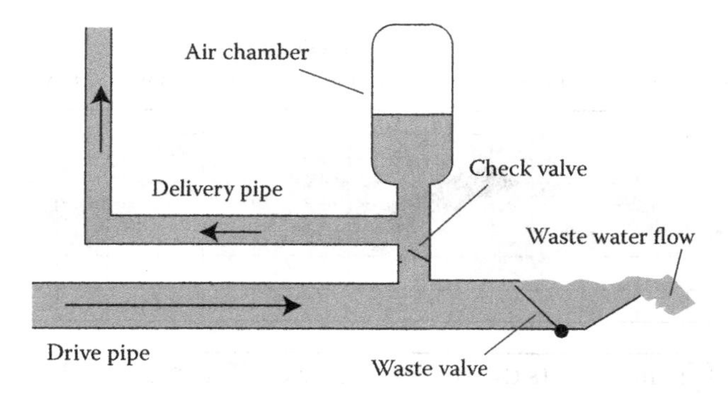

FIGURE 4.29 Diagram of ram pump.

3. The pressure surge causes the check valve to open, allowing high-pressure water to enter the air chamber and delivery pipe. The pressurized air in the air chamber helps to smooth out the pressure surges to give a continuous flow through the delivery pipe.
4. As the pressure surge subsides, the pressurized air in the air chamber causes the check valve to close. The sudden closure of the check valve reduces the pressure in the drive pipe so that the waste valve opens, and the pump is returned to start the cycle again. Most ram pumps operate at 30–100 cycles a minute.

The Alternative Indigenous Development Foundation in the Philippines developed durable ram pumps, and the maintenance is done locally on the moving parts that need regular replacement. The five different size ram pumps can deliver between 1,500 and 72,000 L/day up to a height of 200 m. The 98 ram pumps installed by 2007 were delivering over 900 m^3/day of water, serving over 15,000 people and irrigating large areas of land.

References

1. Water wheels of India. http://www.goodnewsindia.com/Pages/content/discovery/waterwheels.html (accessed on 2/13/2012).
2. Nepal Ghatta Project. Water mills in Nepal. http://www.nathaneagle.com/ghatta (accessed on 2/13/2012).
3. Centre for Rural Technology, Nepal. http://crtnepal.org (accessed on 2/13/2012), see gallery photos of watermills.
4. M. T. Klare. 2001. *Resource Wars, the New Landscape of Global Conflict*. Metropolitan Books, New York.
5. From the new deal to a new century. http://www.tva.com/abouttva/history.htm (accessed on 2/13/2012).
6. Booneville Power Administration, History. http://www.bpa.gov/corporate/About_BPA/history.cfm (accessed on 2/13/2012).
7. Hoover Dam. http://www.usbr.gov/lc/hooverdam/History/storymain.html (accessed on 2/13/2012), www.arizona-leisure.com/hoover-dam.html (accessed on 2/13/2012).
8. Grand Coulee Dam. http://www.usbr.gov/pn/grandcoulee/ (accessed on 2/13/2012).
9. Virtual Hydropower Prospector. http://hydropower.inl.gov/prospector (accessed on 2/13/2012).
10. D. G. Hall and K. S. Reeves. 2006. *A Study of United States Hydroelectric Plant Ownership*. Idaho National Laboratory, Idaho Falls, ID. http://hydropower.inel.gov/hydrofacts/pdfs/a_study_of_united_states_hydroelectric_plant_ownership.pdf (accessed on 2/13/2012).
11. Energy Efficiency and Renewable Energy Network. 2006. Feasibility assessment of the water energy resources of the United States for new low power and small hydro classes of hydroelectric plants. http://hydropower.inel.gov/resourceassessment/index.shtml (accessed on 2/13/2012).
12. S. Jingli, C. Dou, and R. Dongming. 2008. *Renewable Energy Based Chinese un-Electrified Region Electrification*. Chemical Industry Press, Beijing, China, Chapter 6.
13. S. Khennas and A. Barnett. 2000. Best practices for sustainable development of micro hydro projects in developing countries. http://www.microhydropower.net/download/bestpractsynthe.pdf
14. Microhydro. http://www.microhydropower.net (accessed on 2/13/2012).
15. University of Stratclyde, http://www.esru.strath.ac.uk/EandE/Web_sites/01-02/RE_info/Tidal%20Power.htm (accessed on 2/13/2012).
16. R. Bedard, M. Previsic, B. Polagye, G. Hagerman, and A. Casavant. 2006. North American tidal in-stream energy conversion technology feasibility. EPRI TP-008-NA. http://oceanenergy.epri.com/streamenergy.html (accessed on 2/13/2012); (http://oceanenergy.epri.com/attachments/streamenergy/reports/008_Summary_Tidal_Report_06-10-06.pdf (accessed on 2/13/2012).
17. World Energy Council. Survey of Energy Resources. 2010. See Hydropower chapter. http://www.worldenergy.org/publications/3040.asp (accessed on 2/13/2012).

18. Cooperative Institute for Marine and Atmospheric Sciences, University of Miami. Miami, FL. http://oceancurrents.rsmas.miami.edu (accessed on 2/13/2012).
19. Technology white paper on ocean current energy potential on U.S. continental shelf. http://ocsenergy.anl.gov/documents/docs/OCS_EIS_WhitePaper_Current.pdf (accessed on 2/13/2012).
20. Pelamis Wave Power. http://www.pelamiswave.com/index.php (accessed on 2/13/2012). Gallery and videos are available.
21. Islay LIMPET project monitoring final report. ETSU V/06/00180/00/Rep, 2002. http://www.wavegen.co.uk/pdf/art.1707.pdf (accessed on 2/13/2012).
22. National Renewable Energy Laboratory. Ocean thermal energy conversion. http://www.nrel.gov/otec/ (accessed on 2/13/2012).
23. Clemson University. Home-made hydraulic ram pump. http://www.clemson.edu/irrig/Equip/ram.htm (accessed on 2/13/2012).
24. Case Study, Ashden Awards. http://www.ashdenawards.org/winners/aidfoundation (accessed on 2/13/2012).
25. Local manufacture and installation of hydraulic ram pumps for village water supply. http://www.ashdenawards.org/files/reports/AIDFI_2007_technical_report.pdf (accessed on 2/13/2012).

Recommended Resources

Links

Acaua Marine Power, Ocean Power Technologies. http://www.oceanpowertechnologies.com (accessed on 2/13/2012).

Bonneville Power Administration. 2008. Renewable energy technology roadmap (wind, ocean wave, in-stream tidal and solar).

EPRI tidal in-stream energy conversion (TISEC) project. http://oceanenergy.epri.com/streamenergy.html (accessed on 2/13/2012).

European Ocean Energy Association. http://www.eu-oea.com (accessed on 2/13/2012).

Hydropower Research Foundation. http://www.hydrofoundation.org/index.html (accessed on 2/13/2012).

International Energy Agency, Ocean Energy Systems. http://www.iea-oceans.org (accessed on 2/13/2012).

International Hydropower Association. http://www.hydropower.org (accessed on 2/13/2012).

International Network on Small Hydro Power. http://www.inshp.org/main.asp (accessed on 2/13/2012).

International Small-Hydro Atlas. http://www.small-hydro.com (accessed on 2/13/2012).

Microhydro Power. http://practicalaction.org/energy/micro_hydro_expertise (accessed on 2/13/2012).

Microhydro Power. Links to case studies. http://www.microhydropower.net/index.php (accessed on 2/13/2012).

Micro hydro Solomon Islands. http://www.pelena.com.au/pelton_turbine.htm (accessed on 2/13/2012).

National Hydropower Association. http://www.hydro.org (accessed on 2/13/2012).

Oceanweather, current significant wave height and direction by regions of the world. http://www.oceanweather.com/data (accessed on 2/13/2012).

United States, Water Power Technology Program. http://www1.eere.energy.gov/water/about.html (accessed on 2/13/2012).

Wavebob. http://www.wavebob.com (accessed on 2/13/2012).

Wave Dragon. http://www.wavedragon.net (accessed on 2/13/2012).

II

Electric Power Generation: Conventional Methods

Rama Ramakumar

Rama Ramakumar has been a professor of electrical and computer engineering at Oklahoma State University, Stillwater, Oklahoma, since 1976 and the PSO/Albrecht Naeter Professor since 1991. He has been the director of the OSU Engineering Energy Laboratory since 1987. Dr. Ramakumar holds memberships in several professional and honorary societies. He was granted the title of regents professor at the June 2008 meeting of the OSU Board of Regents. He has been primarily involved in renewable energy and energy systems research. His research and teaching interests are in energy conversion, renewable energy, power electronics, and system reliability. He has authored a textbook entitled *Engineering Reliability: Fundamentals and Applications* published by Prentice Hall in 1993. He was named a fellow of IEEE in 1994 for his "contributions to renewable energy systems and leadership in power engineering education."

5

Hydroelectric Power Generation

Steven R. Brockschink (retired)
Stantec Consulting

James H. Gurney (retired)
BC Hydro

Douglas B. Seely
Stantec Consulting

Hydroelectric power generation involves the storage of a hydraulic fluid, water, conversion of the hydraulic (potential) energy of the fluid into mechanical (kinetic) energy in a hydraulic turbine, and conversion of the mechanical energy to electrical energy in an electric generator.

The first hydroelectric power plants came into service in the 1880s and now comprise approximately 20% (875 GW) of the world's installed generation capacity (World Energy Council, 2010). Hydroelectricity is an important source of renewable energy and provides significant flexibility in base loading, peaking, and energy storage applications. While initial capital costs are high, the inherent simplicity of hydroelectric plants, coupled with their low operating and maintenance costs, long service life, and high reliability, makes them a very cost-effective and flexible source of electricity generation. Especially valuable is their operating characteristic of fast response for start-up, loading, unloading, and following of system load variations. Other useful features include their ability to start without the availability of power system voltage (black start capability), ability to transfer rapidly from generation mode to synchronous-condenser mode, and pumped storage application.

Hydroelectric units have been installed in capacities ranging from a few kilowatts to nearly 1 GW. Multiunit plant sizes range from a few kilowatts to a maximum of 22.5 GW.

5.1 Planning of Hydroelectric Facilities

5.1.1 Siting

Hydroelectric plants are located in geographic areas where they will make economic use of hydraulic energy sources. Hydraulic energy is available wherever there is a flow of liquid and accumulated head. Head represents potential energy and is the vertical distance through which the fluid falls in the energy conversion process. The majority of sites utilize the head developed by freshwater; however, other liquids such as saltwater and treated sewage have been utilized. The siting of a prospective hydroelectric plant requires careful evaluation of technical, economic, environmental, and social factors. A significant portion of the project cost may be required for mitigation of environmental effects on fish and wildlife and relocation of infrastructure and population from flooded areas.

5.1.2 Hydroelectric Plant Schemes

There are three main types of hydroelectric plant arrangements, classified according to the method of controlling the hydraulic flow at the site:

1. Run-of-the-river plants, having small amounts of water storage and thus little control of the flow through the plant
2. Storage plants, having the ability to store water and thus control the flow through the plant on a daily or seasonal basis
3. Pumped storage plants, in which the direction of rotation of the turbines is reversed during off-peak hours, pumping water from a lower reservoir to an upper reservoir, thus "storing energy" for later production of electricity during peak hours

5.1.3 Selection of Plant Capacity, Energy, and Other Design Features

The generating capacity of a hydroelectric plant is a function of the head and flow rate of water discharged through the hydraulic turbines, as shown in the following equation:

$$P = 9.8\eta QH \tag{5.1}$$

where
- P is the power (kW)
- η is the plant efficiency
- Q is the discharge flow rate (m^3/s)
- H is the head (m)

Flow rate and head are influenced by reservoir inflow, storage characteristics, plant and equipment design features, and flow restrictions imposed by irrigation, minimum downstream releases, or flood control requirements. Historical daily, seasonal, maximum (flood), and minimum (drought) flow conditions are carefully studied in the planning stages of a new development. Plant capacity, energy, and physical features such as the dam and spillway structures are optimized through complex economic studies that consider the hydrological data, planned reservoir operation, performance characteristics of plant equipment, construction costs, the value of capacity and energy, and financial discount rates. The costs of substation, transmission, telecommunications, and off-site control facilities are also important considerations in the economic analysis. If the plant has storage capability, then societal benefits from flood control may be included in the economic analysis.

Another important planning consideration is the selection of the number and size of generating units installed to achieve the desired plant capacity and energy, taking into account installed unit costs, unit availability, and efficiencies at various unit power outputs (American Society of Mechanical Engineers–Hydropower Technical Committee, 1996).

5.2 Hydroelectric Plant Features

Figures 5.1 and 5.2 illustrate the main components of a hydroelectric generating unit. The generating unit may have its shaft oriented in a vertical, horizontal, or inclined direction depending on the physical conditions of the site and the type of turbine applied. Figure 5.1 shows a typical vertical shaft Francis turbine unit and Figure 5.2 shows a horizontal shaft propeller turbine unit. The following sections will

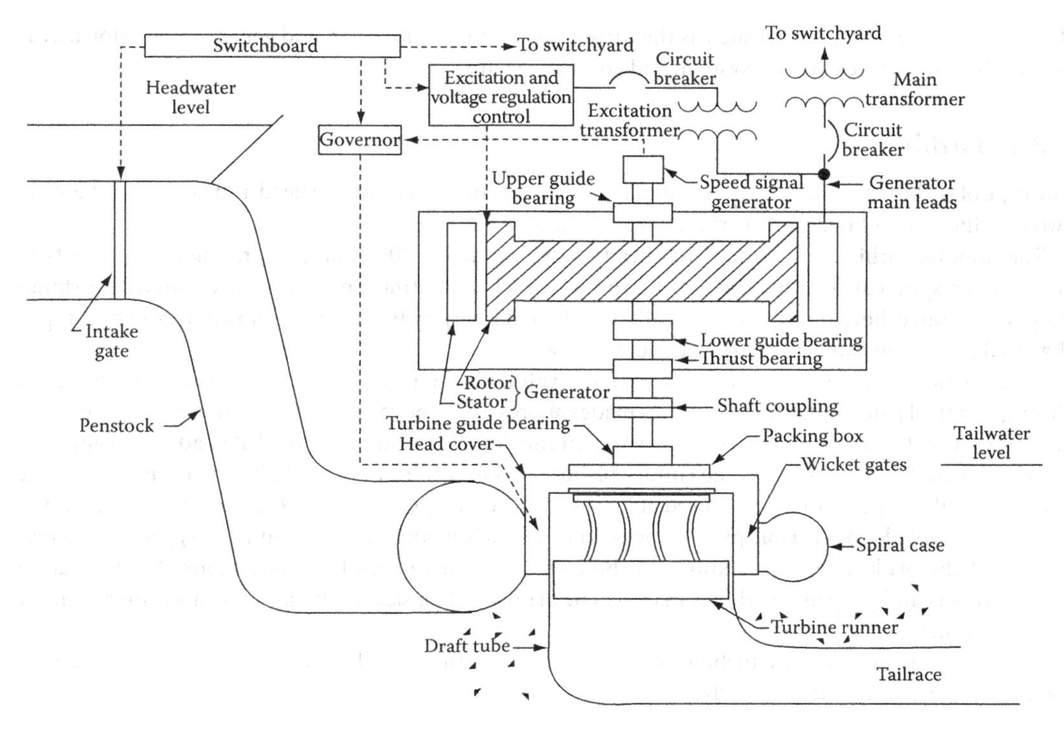

FIGURE 5.1 Vertical Francis unit arrangement. (From IEEE Standard 1020, IEEE Guide for Control of Small Hydroelectric Power Plants. Copyright IEEE. All rights reserved.)

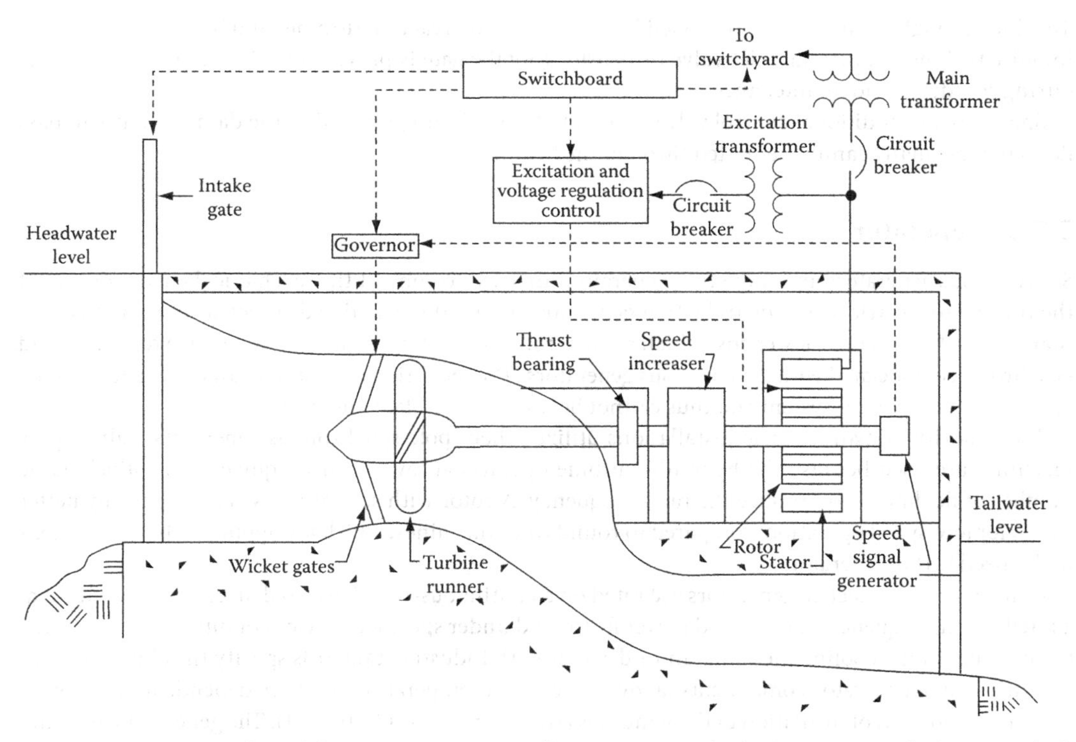

FIGURE 5.2 Horizontal axial-flow unit arrangement. (From IEEE Standard 1020, IEEE Guide for Control of Small Hydroelectric Power Plants. Copyright IEEE. All rights reserved.)

describe the main components such as the turbine, generator, switchgear, and generator transformer, as well as the governor, excitation system, and control systems.

5.2.1 Turbine

The type of turbine selected for a particular application is influenced by the head and flow rate. There are two classifications of hydraulic turbines: impulse and reaction.

The impulse turbine is used for high heads—approximately 300 m or greater. High-velocity jets of water strike spoon-shaped buckets on the runner which is at atmospheric pressure. Impulse turbines may be mounted horizontally or vertically and include perpendicular jets (known as a Pelton type), diagonal jets (known as a Turgo type), or cross-flow types.

In a reaction turbine, the water passes from a spiral casing through stationary radial guide vanes, through control gates and onto the runner blades at pressures above atmospheric. There are two categories of reaction turbine—Francis and propeller. In the Francis turbine, installed at heads up to approximately 360 m, the water impacts the runner blades tangentially and exits axially. The propeller turbine uses a propeller-type runner and is used at low heads—below approximately 45 m. The propeller runner may use fixed blades or variable pitch blades—known as a Kaplan or double regulated type—that allows control of the blade angle to maximize turbine efficiency at various hydraulic heads and generation levels. Francis and propeller turbines may also be arranged in a slant, tubular, bulb, and rim generator configurations.

Water discharged from the turbine is directed into a draft tube where it exits to a tailrace channel, lower reservoir, or directly to the river.

5.2.2 Flow Control Equipment

The flow through the turbine is controlled by wicket gates on reaction turbines and by needle nozzles on impulse turbines. A turbine inlet valve or penstock intake gate is provided for isolation of the turbine during shutdown and maintenance.

Spillways and additional control valves and outlet tunnels are provided in the dam structure to pass flows that normally cannot be routed through the turbines.

5.2.3 Generator

Synchronous generators and induction generators are used to convert the mechanical energy output of the turbine to electrical energy. Induction generators are used in small hydroelectric applications (less than 5 MVA) due to their lower cost which results from elimination of the exciter, voltage regulator, and synchronizer associated with synchronous generators. The induction generator draws its excitation current from the electrical system and thus cannot be used in an isolated power system.

The majority of hydroelectric installations utilize salient pole synchronous generators. Salient pole machines are used because the hydraulic turbine operates at low speeds, requiring a relatively large number of field poles to produce the rated frequency. A rotor with salient poles is mechanically better suited for low-speed operation, compared to round rotor machines, which are applied in horizontal axis high-speed turbo-generators.

Generally, hydroelectric generators are rated on a continuous-duty basis to deliver net kVA output at a rated speed, frequency, voltage, and power factor and under specified service conditions including the temperature of the cooling medium (air or direct water). Industry standards specify the allowable temperature rise of generator components (above the coolant temperature) that are dependent on the voltage rating and class of insulation of the windings (IEEE, C50.12; IEC, 60034-1). The generator capability curve (Figure 5.3) describes the maximum real and reactive power output limits at rated voltage within which the generator rating will not be exceeded with respect to stator and rotor heating and other limits.

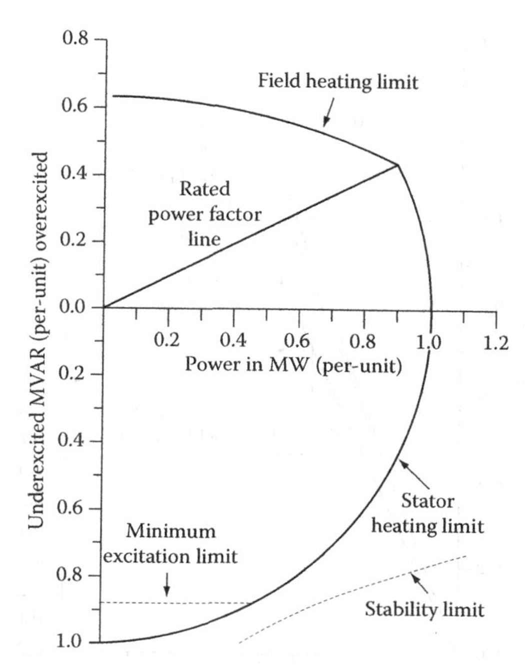

FIGURE 5.3 Typical hydro-generator capability curve (0.9 power factor, rated voltage). (From IEEE Standard 492, IEEE Guide for Operation and Maintenance of Hydro-Generators. Copyright IEEE. All rights reserved.)

Standards also provide guidance on short-circuit capabilities and continuous and short-time current unbalance requirements (IEEE, C50.12; IEEE, 492).

Synchronous generators require direct current (DC) field excitation to the rotor, provided by the excitation system described in Section 5.2.7. The generator saturation curve (Figure 5.4) describes the relationship of terminal voltage, stator current, and field current.

While the generator may be vertical or horizontal, the majority of new installations are vertical. The basic components of a vertical generator are the stator (frame, magnetic core, and windings), rotor (shaft, thrust block, spider, rim, and field poles with windings), thrust bearing, one or two guide bearings, upper and lower brackets for the support of bearings and other components, and sole plates which are bolted to the foundation. Other components may include a direct connected exciter, speed signal generator, rotor brakes, rotor jacks, and ventilation systems with surface air coolers (IEEE, 1095).

The stator core is composed of stacked steel laminations attached to the stator frame. The stator winding may consist of single-turn or multiturn coils or half-turn bars, connected in series to form a three phase circuit. Double layer windings, consisting of two coils per slot, are most common. One or more circuits are connected in parallel to form a complete phase winding. The stator winding is normally connected in wye configuration, with the neutral grounded through one of a number of alternative methods that depend on the amount of phase-to-ground fault current that is permitted to flow (IEEE, C62.92.2, C37.101). Generator output voltages range from approximately 480 VAC to 22 kVAC line-to-line, depending on the MVA rating of the unit. Temperature detectors are installed between coils in a number of stator slots.

The rotor is normally comprised of a spider frame attached to the shaft, a rim constructed of solid steel or laminated rings, and field poles attached to the rim. The rotor construction will vary significantly depending on the shaft and bearing system, unit speed, ventilation type, rotor dimensions, and characteristics of the driving hydraulic turbine. Damper windings or amortisseurs in the form of copper or brass rods are embedded in the pole faces for damping rotor speed oscillations.

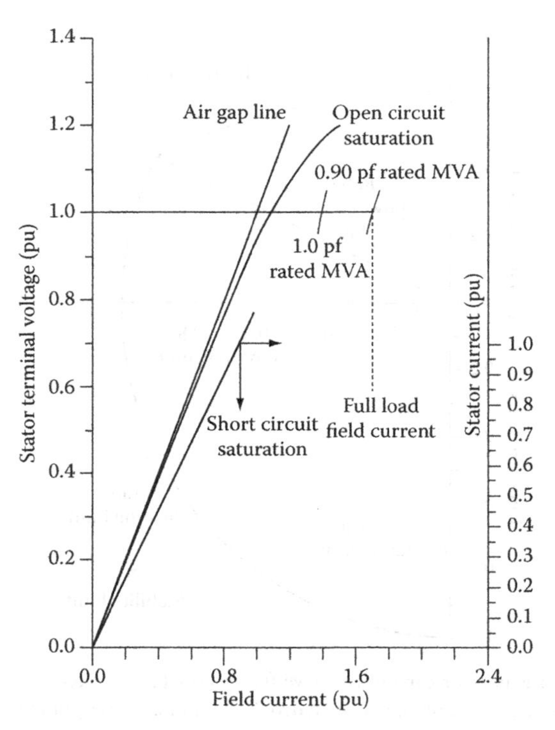

FIGURE 5.4 Typical hydro-generator saturation curves. (From IEEE Standard 492, IEEE Guide for Operation and Maintenance of Hydro-Generators. Copyright IEEE. All rights reserved.)

The thrust bearing supports the mass of both the generator and turbine plus the hydraulic thrust imposed on the turbine runner and is located either above the rotor (suspended unit) or below the rotor (umbrella unit). Thrust bearings are constructed of oil-lubricated, segmented, babbitt-lined shoes. One or two oil-lubricated generator guide bearings are used to restrain the radial movement of the shaft.

Fire protection systems are normally installed to detect combustion products in the generator enclosure, initiate rapid de-energization of the generator, and release extinguishing material. Carbon dioxide and water are commonly used as the fire quenching medium.

Excessive unit vibrations may result from mechanical or magnetic unbalance. Vibration monitoring devices such as proximity probes to detect shaft run out are provided to initiate alarms and unit shutdown.

The choice of generator inertia is an important consideration in the design of a hydroelectric plant. The speed rise of the turbine-generator unit under load rejection conditions, caused by the instantaneous disconnection of electrical load, is inversely proportional to the combined inertia of the generator and turbine. Turbine inertia is normally about 5% of the generator inertia. During the design of the plant, unit inertia, effective wicket gate or nozzle closing and opening times, and penstock dimensions are optimized to control the pressure fluctuations in the penstock and speed variations of the turbine-generator during load rejection and load acceptance. Speed variations may be reduced by increasing the generator inertia at added cost. Inertia can be added by increasing the mass of the generator, adjusting the rotor diameter, or by adding a flywheel. The unit inertia also has a significant effect on the transient stability of the electrical system, as this factor influences the rate at which energy can be moved in or out of the generator to control the rotor angle acceleration during system fault conditions. (See Kundur [1994], and Part II of Power System Stability and Control of this handbook.)

5.2.4 Generator Terminal Equipment

The generator output is connected to terminal equipment via cable, busbar, or isolated phase bus. The terminal equipment comprises current transformers (CTs), voltage transformers (VTs), and surge suppression devices. The CTs and VTs are used for unit protection, metering and synchronizing, and for governor and excitation system functions. The surge protection devices, consisting of surge arresters and capacitors, protect the generator and low-voltage windings of the step-up transformer from lightning and switching-induced surges.

5.2.5 Generator Switchgear

The generator circuit breaker and associated isolating disconnect switches are used to connect and disconnect the generator to and from the power system. The generator circuit breaker may be located on either the low-voltage or high-voltage side of the generator step-up transformer. In some cases, the generator is connected to the system by means of circuit breakers located in the switchyard of the generating plant. The generator circuit breaker may be of the oil filled, air magnetic, air blast, or compressed gas insulated type, depending on the specific application. The circuit breaker is closed as part of the generator synchronizing sequence and is opened (tripped) either by operator control, as part of the automatic unit stopping sequence, or by operation of protective relay devices in the event of unit fault conditions.

5.2.6 Generator Step-Up Transformer

The generator transformer steps up the generator terminal voltage to the voltage of the power system or plant switchyard. Generator transformers are generally specified and operated in accordance with international standards for power transformers, with the additional consideration that the transformer will be operated close to its maximum rating for the majority of its operating life. Various types of cooling systems are specified depending on the transformer rating and physical constraints of the specific application. In some applications, dual low-voltage windings are provided to connect two generating units to a single bank of step-up transformers. Also, transformer tertiary windings are sometimes provided to serve the AC station service requirements of the power plant.

5.2.7 Excitation System

The excitation system fulfills two main functions:

1. It produces DC voltage (and power) to force current to flow in the field windings of the generator. There is a direct relationship between the generator terminal voltage and the quantity of current flowing in the field windings as described in Figure 5.4.
2. It provides a means for regulating the terminal voltage of the generator to match a desired setpoint and to provide damping for power system oscillations.

Prior to the 1960s, generators were generally provided with rotating exciters that fed the generator field through a slip ring arrangement, a rotating pilot exciter feeding the main exciter field, and a regulator controlling the pilot exciter output. Since the 1960s, the most common arrangement is thyristor bridge rectifiers fed from a transformer connected to the generator terminals, referred to as a "potential source controlled rectifier high initial response exciter" or "bus-fed static exciter" (IEEE, 421.1, 421.2, 421.4, 421.5). Another system used for smaller high-speed units is a brushless exciter with a rotating AC generator and rotating rectifiers.

Modern static exciters have the advantage of providing extremely fast response times and high field ceiling voltages for forcing rapid changes in the generator terminal voltage during system faults. This is necessary to overcome the inherent large time constant in the response between terminal voltage and field voltage (referred to as T'_{do}, typically in the range of 5–10 s). Rapid terminal voltage forcing is necessary to maintain transient stability of the power system during and immediately after system faults. Power system stabilizers are also applied to static exciters to cause the generator terminal voltage to vary in phase with the speed deviations of the machine, for damping power system dynamic oscillations. (See Kundur (1994), and Part II of Power System Stability and Control of this handbook.)

Various auxiliary devices are applied to the static exciter to allow remote setting of the generator voltage and to limit the field current within rotor thermal and under excited limits. Field flashing equipment is provided to build up generator terminal voltage during starting to the point at which the thyristor can begin gating. Power for field flashing is provided either from the station battery or alternating current (AC) station service.

5.2.8 Governor System

The governor system is the key element of the unit speed and power control system (IEEE, 125, 1207; IEC, 61362; ASME, PTC 29). It consists of control and actuating equipment for regulating the flow of water through the turbine, for starting and stopping the unit, and for regulating the speed and power output of the turbine generator. The governor system includes setpoint and sensing equipment for speed, power and actuator position, compensation circuits, and hydraulic power actuators which convert governor control signals to mechanical movement of the wicket gates (Francis and Kaplan turbines), runner blades (Kaplan turbine), and nozzle jets (Pelton turbine). The hydraulic power actuator system includes high-pressure oil pumps, pressure tanks, oil sump, actuating valves, and servomotors.

Older governors are of the mechanical-hydraulic type, consisting of ballhead speed sensing, mechanical dashpot and compensation, gate limit, and speed droop adjustments. Modern governors are of the electro-hydraulic type where the majority of the sensing, compensation, and control functions are performed by electronic or microprocessor circuits. Compensation circuits utilize proportional plus integral (PI) or proportional plus integral plus derivative (PID) controllers to compensate for the phase lags in the penstock–turbine–generator–governor control loop. PID settings are normally adjusted to ensure that the hydroelectric unit remains stable when serving an isolated electrical load. These settings ensure that the unit contributes to the damping of system frequency disturbances when connected to an integrated power system. Various techniques are available for modeling and tuning the governor (IEEE, 1207).

A number of auxiliary devices are provided for remote setting of power, speed, and actuator limits and for electrical protection, control, alarming, and indication. Various solenoids are installed in the hydraulic actuators for controlling the manual and automatic start-up and shutdown of the turbine-generator unit.

5.2.9 Control Systems

Detailed information on the control of hydroelectric power plants is available in industry standards (IEEE, 1010, 1020, 1249). A general hierarchy of control is illustrated in Table 5.1. Manual controls, normally installed adjacent to the device being controlled, are used during testing and maintenance, and as a backup to the automatic control systems. Figure 5.5 illustrates the relationship of control locations and typical functions available at each location. Details of the control functions available at each location are described in IEEE 1249. Automatic sequences implemented for starting, synchronizing, and shutdown of hydroelectric units are detailed in IEEE 1010.

Modern hydroelectric plants and plants undergoing rehabilitation and life extension are incorporating higher levels of computer automation (IEEE, 1249, 1147). The relative simplicity of hydroelectric plant control allows most plants to be operated in an unattended mode from off-site control centers.

TABLE 5.1 Summary of Control Hierarchy for Hydroelectric Plants

Control Category	Subcategory	Remarks
Location	Local	Control is local at the controlled equipment or within sight of the equipment
	Centralized	Control is remote from the controlled equipment, but within the plant
	Off-site	Control location is remote from the project
Mode	Manual	Each operation needs a separate and discrete initiation; could be applicable to any of the three locations
	Automatic	Several operations are precipitated by a single initiation; could be applicable to any of the three locations
Operation (supervision)	Attended	Operator is available at all times to initiate control action
	Unattended	Operation staff is not normally available at the project site

Source: IEEE Standard 1249, IEEE Guide for Computer-Based Control for Hydroelectric Power Plant Automation. With permission.

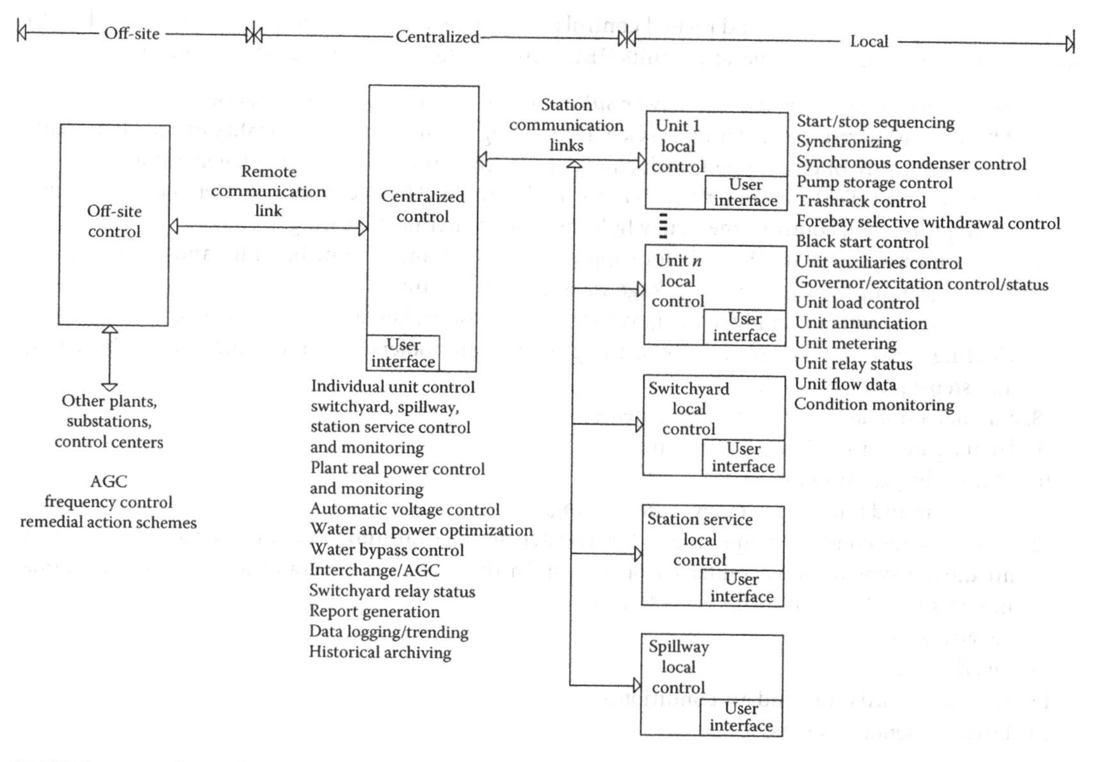

FIGURE 5.5 Relationship of local, centralized, and off-site control. (From IEEE Standard 1249, IEEE Guide for Computer-Based Control for Hydroelectric Power Plant Automation.)

The current trend is to apply automated condition monitoring systems for hydroelectric plant equipment. Condition monitoring systems, coupled with expert system computer programs, allow plant owners and operators to more fully utilize the capacity of plant equipment and water resources, make better maintenance and replacement decisions, and maximize the value of installed assets.

5.2.10 Protection Systems

The turbine-generator unit and related equipment are protected against mechanical, electrical, hydraulic, and thermal damage that may occur as a result of abnormal conditions within the plant or on the power system to which the plant is connected. Abnormal conditions are detected automatically by means of protective relays and other devices and measures are taken to isolate the faulty equipment as quickly as possible while maintaining the maximum amount of equipment in service. Typical protective devices include electrical fault detecting relays, temperature, pressure, level, speed, and fire sensors, and vibration monitors associated with the turbine, generator, and related auxiliaries. The protective devices operate in various isolation and unit shutdown sequences, depending on the severity of the fault.

The type and extent of protection will vary depending on the size of the unit, manufacturer's recommendations, owner's practices, and industry standards.

Specific guidance on application of protection systems for hydroelectric plants is provided in IEEE 1010, 1020, C37.102, C37.91.

5.2.11 Plant Auxiliary Equipment

A number of auxiliary systems and related controls are provided throughout the hydroelectric plant to support the operation of the generating units (IEEE, 1010, 1020). These include the following:

1. Switchyard systems (see *Electric Power Substations Engineering* of this handbook).
2. Alternating current (AC) station service. Depending on the size and criticality of the plant, multiple sources are often supplied, with emergency backup provided by a diesel generator.
3. Direct current (DC) station service. It is normally provided by one or more battery banks, for supply of protection, control, emergency lighting, and exciter field flashing.
4. Lubrication systems, particularly for supply to generator and turbine bearings and bushings.
5. Drainage pumps, for removing leakage water from the plant.
6. Air compressors, for supply to the governors, generator brakes, and other systems.
7. Cooling water systems, for supply to the generator air coolers, generator and turbine bearings, and step-up transformer.
8. Fire detection and extinguishing systems.
9. Intake gate or isolation valve systems.
10. Draft tube gate systems.
11. Reservoir and tailrace water level monitoring.
12. Synchronous condenser equipment, for dewatering the draft tube to allow the runner to spin in air during synchronous condenser operation. In this case, the generator acts as a synchronous motor, supplying or absorbing reactive power.
13. Service water systems.
14. Overhead crane.
15. Heating, ventilation, and air conditioning.
16. Environmental systems.

5.3 Special Considerations Affecting Pumped Storage Plants

A pumped storage unit is one in which the turbine and generator are operated in the reverse direction to pump water from the lower reservoir to the upper reservoir. The generator becomes a motor, drawing its energy from the power system, and supplies mechanical power to the turbine which acts as a pump. The motor is started with the wicket gates closed and the draft tube water depressed with compressed air. The motor is accelerated in the pump direction and when at full speed and connected to the power system, the depression air is expelled, the pump is primed, and the wicket gates are opened to commence pumping action.

5.3.1 Pump Motor Starting

Various methods are utilized to accelerate the generator/motor in the pump direction during starting (IEEE, 1010). These include the following:

1. Full voltage, across the line starting—Used primarily on smaller units, the unit breaker is closed and the unit is started as an induction generator. Excitation is applied near rated speed and the machine reverts to synchronous motor operation.
2. Reduced voltage, across the line starting—A circuit breaker connects the unit to a starting bus tapped from the unit step-up transformer at one-third to one-half rated voltage. Excitation is applied near rated speed and the unit is connected to the system by means of the generator circuit breaker. Alternative methods include the use of a series reactor during starting and energization of partial circuits on multiple circuit machines.
3. Pony motor starting—A variable speed wound-rotor motor attached to the AC station service and coupled to the motor/generator shaft is used to accelerate the machine to synchronous speed.
4. Synchronous starting—A smaller generator, isolated from the power system, is used to start the motor by connecting the two in parallel on a starting bus, applying excitation to both units, and opening the wicket gates on the smaller generator. When the units reach synchronous speed, the motor unit is disconnected from the starting bus and connected to the power system.
5. Semisynchronous (reduced frequency, reduced voltage) starting—An isolated generator is accelerated to about 80% rated speed and paralleled with the motor unit by means of a starting bus. Excitation is applied to the generating unit and the motor unit starts as an induction motor. When the speed of the two units is approximately equal, excitation is applied to the motor unit, bringing it into synchronism with the generating unit. The generating unit is then used to accelerate both units to rated speed and the motor unit is connected to the power system.
6. Static starting—A static converter/inverter connected to the AC station service is used to provide variable frequency power to accelerate the motor unit. Excitation is applied to the motor unit at the beginning of the start sequence and the unit is connected to the power system when it reaches synchronous speed. The static starting system can be used for dynamic braking of the motor unit after disconnection from the power system, thus extending the life of the unit's mechanical brakes.

5.3.2 Phase Reversing of the Generator/Motor

It is necessary to reverse the direction of rotation of the generator/motor by interchanging any two of the three phases. This is achieved with multipole motor operated switches or with circuit breakers.

5.3.3 Draft Tube Water Depression

Water depression systems using compressed air are provided to lower the level of the draft tube water below the runner to minimize the power required to accelerate the motor unit during the transition to pumping mode. Water depression systems are also used during motoring operation of a conventional hydroelectric unit while in synchronous-condenser mode. Synchronous-condenser operation is used to provide voltage support for the power system and to provide spinning reserve for rapid loading response when required by the power system.

5.4 Construction and Commissioning of Hydroelectric Plants

The construction and commissioning of a new hydroelectric plant, rehabilitation of an existing plant, or replacement of existing equipment require rigorous attention to the evaluation, design, installation, inspection, testing and commissioning of equipment and systems (IEEE, 1095, 1147, 1248).

References

American Society of Mechanical Engineers–Hydropower Technical Committee, *The Guide to Hydropower Mechanical Design*, HCI Publications, Kansas City, MO, 1996.

ASME PTC 29, Speed Governing Systems for Hydraulic Turbine Generator Units.

IEC Standard 60034-1, Rotating Electrical Machines—Part 1: Rating and Performance.

IEC Standard 61362, Guide to Specification of Hydraulic Turbine Control Systems.

IEEE Standard C37.91, IEEE Guide for Protecting Power Transformers.

IEEE Standard C37.101, IEEE Guide for Generator Ground Protection.

IEEE Standard C37.102, IEEE Guide for AC Generator Protection.

IEEE Standard C50.12, IEEE Standard for Salient-Pole 50 Hz and 60 Hz Synchronous Generators and Generator/Motors for Hydraulic Turbine Applications Rated 5 MVA and Above.

IEEE Standard C62.92.2, IEEE Guide for the Application of Neutral Grounding in Electrical Utility Systems, Part II—Grounding of Synchronous Generator Systems.

IEEE Standard 125, IEEE Recommended Practice for Preparation of Equipment Specifications for Speed-Governing of Hydraulic Turbines Intended to Drive Electric Generators.

IEEE Standard 421.1, IEEE Standard Definitions for Excitation Systems for Synchronous Machines.

IEEE Standard 421.2, IEEE Guide for Identification, Testing and Evaluation of the Dynamic Performance of Excitation Control Systems.

IEEE Standard 421.4, IEEE Guide for the Preparation of Excitation System Specifications.

IEEE Standard 421.5, IEEE Recommended Practice for Excitation Systems for Power Stability Studies.

IEEE Standard 492, IEEE Guide for Operation and Maintenance of Hydro-Generators.

IEEE Standard 1010, IEEE Guide for Control of Hydroelectric Power Plants.

IEEE Standard 1020, IEEE Guide for Control of Small Hydroelectric Power Plants.

IEEE Standard 1095, IEEE Guide for Installation of Vertical Generators and Generator/Motors for Hydroelectric Applications.

IEEE Standard 1147, IEEE Guide for the Rehabilitation of Hydroelectric Power Plants.

IEEE Standard 1207, IEEE Guide for the Application of Turbine Governing Systems for Hydroelectric Generating Units.

IEEE Standard 1248, IEEE Guide for the Commissioning of Electrical Systems in Hydroelectric Power Plants.

IEEE Standard 1249, IEEE Guide for Computer-Based Control for Hydroelectric Power Plant Automation.

Kundur, P., Power System Stability and Control, McGraw-Hill, New York, 1994.

World Energy Council, Survey of Energy Resources, 2010.

6 Synchronous Machinery

Paul I. Nippes
Magnetic Products and Services, Inc.

6.1 General

Synchronous motors convert electrical power to mechanical power; synchronous generators convert mechanical power to electrical power; and synchronous condensers supply only reactive power to stabilize system voltages.

Synchronous motors, generators, and condensers perform similarly, except for a heavy cage winding on the rotor of motors and condensers for self-starting.

A rotor has physical magnetic poles, arranged to have alternating north and south poles around the rotor diameter which are excited by electric current, or uses permanent magnets, having the same number of poles as the stator electromagnetic poles.

The rotor RPM = 120 × Electrical System Frequency/Poles.

The stator winding, fed from external AC multi-phase electrical power, creates rotating electromagnetic poles.

At speed, rotor poles turn in synchronism with the stator rotating electromagnetic poles, torque being transmitted magnetically across the "air gap" power angle, lagging in generators and leading in motors.

Synchronous machine sizes range from fractional watts, as in servomotors, to 1500 MW, as in large generators.

Voltages vary, up to 25,000 V AC stator and 1,500 V DC rotor.

Installed horizontal or vertical at speed ranges up to 130,000 RPM, normally from 40 RPM (waterwheel generators) to 3,600 RPM (turbine generators).

Frequency at 60 or 50 Hz mostly, 400 Hz military; however, synthesized variable frequency electrical supplies are increasingly common and provide variable motor speeds to improve process efficiency.

Typical synchronous machinery construction and performance are described; variations may exist on special smaller units.

This document is intentionally general in nature. Should the reader want specific application information, refer to standards: NEMA MG-1; IEEE 115, C50-10 and C50-13; IEC 600034: 1–11, 14–16, 18, 20, 44, 72, and 136, plus other applicable specifications.

6.2 Construction (see Figure 6.1)

6.2.1 Stator

6.2.1.1 Frame

The exterior frame, made of steel, either cast or a weldment, supports the laminated stator core and has feet, or flanges, for mounting to the foundation. Frame vibration from core magnetic forcing or rotor unbalance is minimized by resilient mounting the core and/or by designing to avoid frame resonance with forcing frequencies. If bracket type bearings are employed, the frame must support the bearings, oil seals, and gas seals when cooled with hydrogen or gas other than air. The frame also provides protection from the elements and channels cooling air, or gas, into and out of the core, stator windings, and rotor. When the unit is cooled by gas contained within the frame, heat from losses is removed by coolers having water circulating through finned pipes of a heat exchanger mounted within the frame. Where cooling water is unavailable and outside air cannot circulate through the frame because of its dirty or toxic condition, large air-to-air heat exchangers are employed, the outside air being forced through the cooler by an externally shaft-mounted blower.

6.2.1.2 Stator Core Assembly

The stator core assembly of a synchronous machine is almost identical to that of an induction motor. A major component of the stator core assembly is the core itself, providing a high permeability path

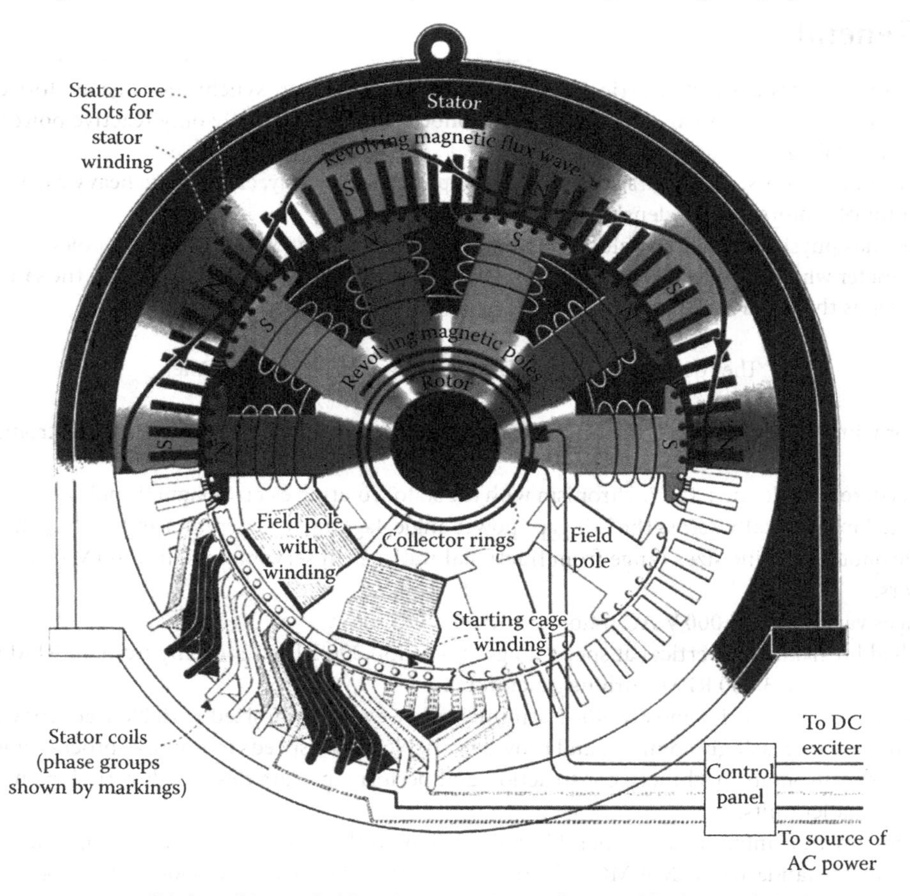

FIGURE 6.1 Magnetic "skeleton" (upper half) and structural parts (lower half) of a 10 pole (720 RPM at 60 cycles) synchronous motor. (From The Electric Machinery Company, Inc., *The ABC's of Synchronous Motors*, 7(1), 5, 1944. With permission.)

for magnetism. The stator core is comprised of thin silicon steel laminations and insulated by a surface coating minimizing eddy current and hysteresis losses generated by alternating magnetism. The laminations are stacked as full rings or segments, in accurate alignment, either in a fixture or in the stator frame, having ventilation spacers inserted periodically along the core length. The completed core is compressed and clamped axially to about 10 kg/cm^2 using end fingers and heavy clamping plates. Core end heating from stray magnetism is minimized, especially on larger machines, by using non-magnetic materials at the core end or by installing a flux shield of either tapered laminations or copper shielding.

A second major component is the stator winding made up of insulated coils placed in axial slots of the stator core inside diameter. The coil make-up, pitch, and connections are designed to produce rotating stator electromagnetic poles in synchronism with the rotor magnetic poles. The stator coils are retained into the slots by slot wedges driven into grooves in the top of the stator slots. Coil end windings are bound together and to core-end support brackets. If the synchronous machine is a generator, the rotating rotor pole magnetism generates voltage in the stator winding which delivers power to an electric load. If the synchronous machine is a motor, its electrically powered stator winding generates rotating electromagnetic poles and the attraction of the rotor magnets, operating in synchronism, produces torque and delivery of mechanical power to the drive shaft.

6.2.2 Rotor

6.2.2.1 The Rotor Assembly

The rotor of a synchronous machine is a highly engineered unitized assembly capable of rotating satisfactorily at synchronous speed continuously according to standards or as necessary for the application. The central element is the shaft, having journals to support the rotor assembly in bearings. Located at the rotor assembly axial mid-section is the rotor core embodying magnetic poles. When the rotor is round it is called "non-salient pole," or turbine generator type construction and when the rotor has protruding pole assemblies, it is called "salient pole" construction.

The non-salient pole construction, used mainly on turbine generators (and also as wind tunnel fan drive motors), has two or four magnetic poles created by direct current in coils located in slots at the rotor outside diameter. Coils are restrained in the slots by slot wedges and at the ends by retaining rings on large high-speed rotors, and fiberglass tape on other units where stresses permit. This construction is not suited for use on a motor requiring self-starting as the rotor surface, wedges, and retaining rings overheat and melt from high currents of self-starting.

A single piece forging is sometimes used on salient pole machines, usually with four or six poles. Salient poles can also be integral with the rotor lamination and can be mounted directly to the shaft or fastened to an intermediate rotor spider. Each distinct pole has an exciting coil around it carrying excitation current or else it employs permanent magnets. In a generator, a moderate cage winding in the face of the rotor poles, usually with pole-to-pole connections, is employed to dampen shaft torsional oscillation and to suppress harmonic variation in the magnetic waveform. In a motor, heavy bars and end connections are required in the pole face to minimize and withstand the high heat of starting duty.

Direct current excites the rotor windings of salient, and non-salient pole motors and generators, except when permanent magnets are employed. The excitation current is supplied to the rotor from either an external DC supply through collector rings or a shaft-mounted brushless exciter. Positive and negative polarity bus bars or cables pass along and through the shaft as required to supply excitation current to the windings of the field poles.

When supplied through collector rings, the DC current could come from a shaft-driven DC or AC exciter rectified output, from an AC–DC motor-generator set, or from plant power. DC current supplied by a shaft-mounted AC generator is rectified by a shaft-mounted rectifier assembly.

As a generator, excitation current level is controlled by the voltage regulator. As a motor, excitation current is either set at a fixed value, or is controlled to regulate power factor, motor current, or system stability.

In addition, the rotor also has shaft-mounted fans or blowers for cooling and heat removal from the unit plus provision for making balance weight additions or corrections.

6.2.2.2 Bearings and Couplings

Bearings on synchronous machinery are anti-friction, grease, or oil-lubricated on smaller machines, journal type oil-lubricated on large machines, and tilt-pad type on more sophisticated machines, especially where rotor dynamics are critical. Successful performance of magnetic bearings, proving to be successful on turbo-machinery, may also come to be used on synchronous machinery as well.

As with bearings on all large electrical machinery, precautions are taken with synchronous machines to prevent bearing damage from stray electrical shaft currents. An elementary measure is the application of insulation on the outboard bearing, if a single-shaft end unit, and on both bearing and coupling at the same shaft end for double-shaft end drive units. Damage can occur to bearings even with properly applied insulation, when solid-state controllers of variable frequency drives, or excitation, cause currents at high frequencies to pass through the bearing insulation as if it were a capacitor. Shaft grounding and shaft voltage and grounding current monitoring can be employed to predict and prevent bearing and other problems.

6.3 Performance

6.3.1 Synchronous Machines, in General

This section covers performance common to synchronous motors, generators, and condensers.

Saturation curves (Figure 6.2) are either calculated or obtained from test and are the basic indicators of machine design suitability. From these the full load field, or excitation, amperes for either motors or generators are determined as shown, on the rated voltage line, as "*Rated Load*." For synchronous

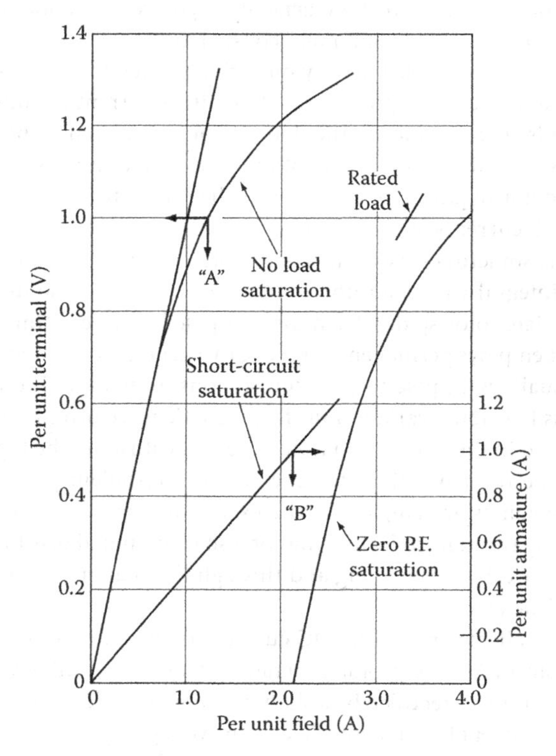

FIGURE 6.2 Saturation curves.

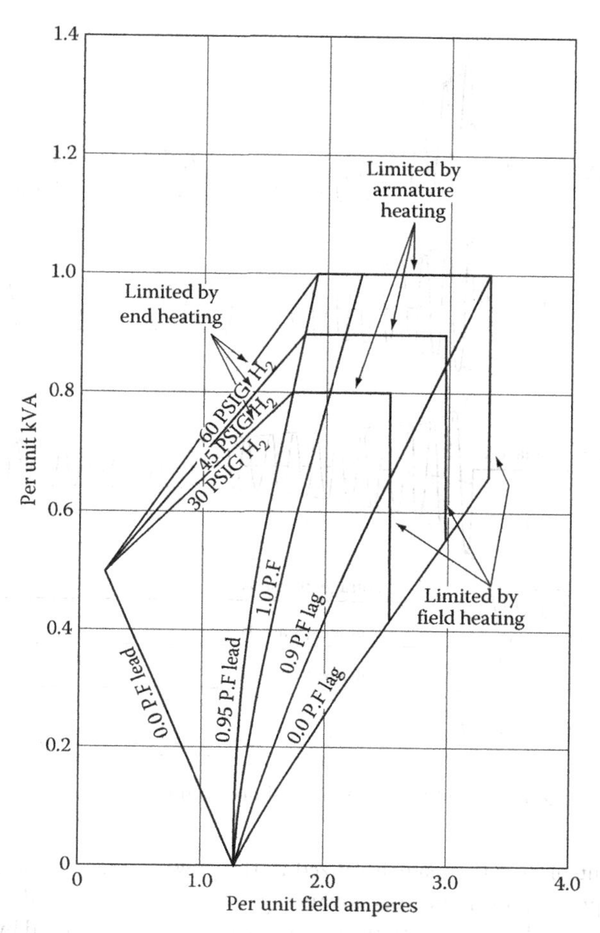

FIGURE 6.3 Vee curves.

condensers, the field current is at the crossing of the zero P.F. saturation line at 1.0 V. As an approximate magnetic figure of merit, the no-load saturation curve should not exceed its extrapolated straight line by more than 25%, unless of a special design. From these criteria, and the knowledge of the stator current and cooling system effectiveness, the manufacturer can project the motor component heating, and thus insulation life, and the efficiency of the machine at different loads.

Vee curves (Figure 6.3) show overall loading performance of a synchronous machine for different loads and power factors, but more importantly show how heating and stability limit loads. For increased hydrogen pressures in a generator frame, the load capability increases markedly.

The characteristics of all synchronous machines when their stator terminals are short-circuited are similar (see Figure 6.4). There is an initial subtransient period of current increase of 8–10 times rated, with one phase offsetting an equal amount. These decay in a matter of milliseconds to a transient value of three to five times rated, decaying in tenths of a second to a relatively steady value. Coincident with this, the field current increases suddenly by three to five times, decaying in tenths of a second. The stator voltage on the shorted phases drops to zero and remains so until the short circuit is cleared.

6.3.2 Synchronous Generator Capability

The synchronous generator normally has easy starting duty as it is brought up to speed by a prime mover. Then the rotor excitation winding is powered with DC current, adjusted to rated voltage, and transferred

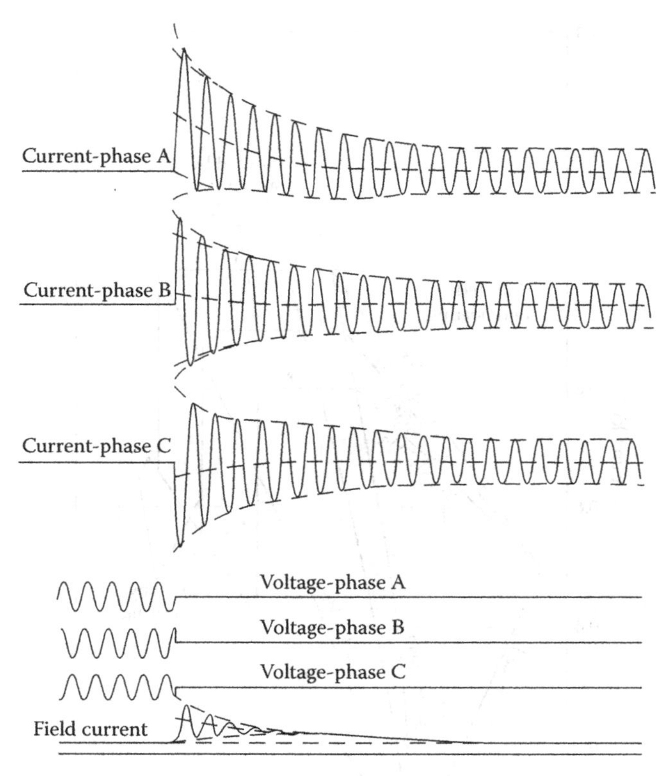

FIGURE 6.4 Typical oscillogram of a sudden three-phase short circuit.

to voltage regulator control. It is then synchronized to the power system, closing the interconnecting circuit breaker as the prime mover speed is advancing, at a snail's pace, leading the electric system. Once on line, its speed is synchronized with the power system and KW is raised by increasing the prime mover KW input. The voltage regulator adjusts excitation current to hold voltage. Increasing the voltage regulator set point increases KVAR input to the system, reducing the power factor toward lagging and vice versa. Steady operating limits are provided by its Reactive Capability Curve (see Figure 6.5). This curve shows the possible kVA reactive loading, lagging, or leading, for given KW loading. Limitations consist of field heating, armature heating, stator core end heating, and operating stability over different regions of the reactive capability curve.

6.3.3 Synchronous Motor and Condenser Starting

The duty on self-starting synchronous motors and condensers is severe, as there are large induction currents in the starting cage winding once the stator winding is energized (see Figure 6.6). These persist as the motor comes up to speed, similar to but not identical to starting an induction motor. Similarities exist to the extent that extremely high torque impacts the rotor initially and decays rapidly to an average value, increasing with time. Different from the induction motor is the presence of a large oscillating torque. The oscillating torque decreases in frequency as the rotor speed increases. This oscillating frequency is caused by the saliency effect of the protruding poles on the rotor. Meanwhile, the stator current remains constant until 80% speed is reached. The oscillating torque at decaying frequency may excite train torsional natural frequencies during acceleration, a serious train design consideration. An anomaly occurs at half speed as a dip in torque and current due to the coincidence of line frequency torque with oscillating torque frequency. Once the rotor is close to rated speed, excitation is applied

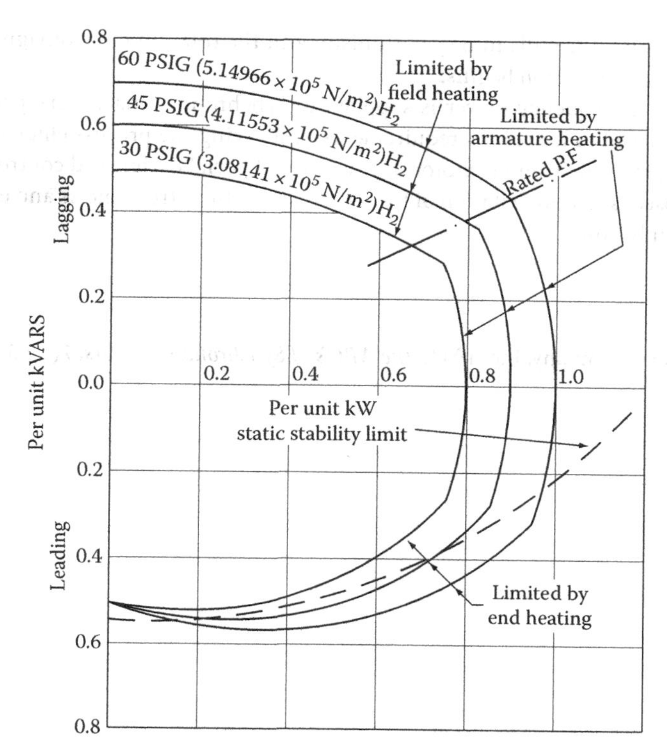

FIGURE 6.5 Typical reactive capability curve.

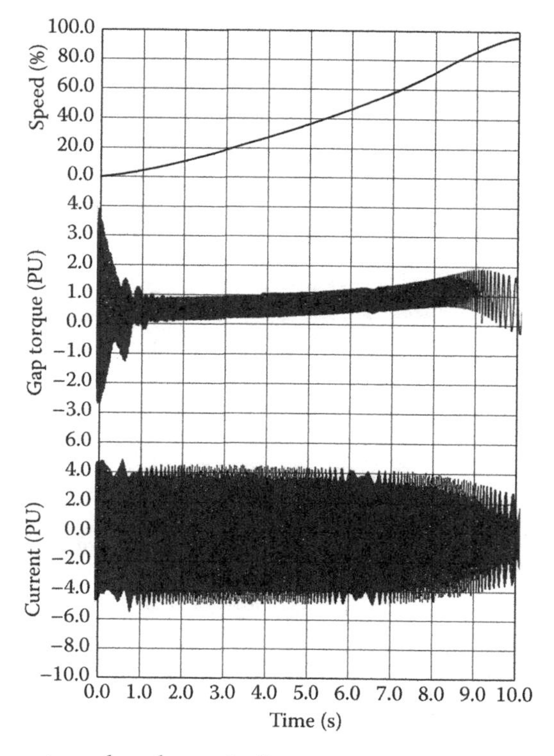

FIGURE 6.6 Synchronous motor and condenser starting.

to the field coils and the rotor pulls into synchronism with the rotating electromagnetic poles. At this point, stable steady-state operation begins.

Increasingly, variable frequency power is supplied to synchronous machinery primarily to deliver the optimum motor speed to meet load requirements, improving the process efficiency. It can also be used for soft-starting the synchronous motor or condenser. Special design and control are employed to avert problems imposed, such as excitation of train torsional natural frequencies and extra heating from harmonics of the supply power.

Reference

The Electric Machinery Company, Inc. 1944. *The ABC's of Synchronous Motors*, 7(1): 5.

7

Thermal Generating Plants

Kenneth H. Sebra
Baltimore Gas and Electric Company

Thermal generating plants are designed and constructed to convert energy from fuel (coal, oil, gas, or radiation) into electric power. The actual conversion is accomplished by a turbine-driven generator. Thermal generating plants differ from industrial plants in that the nature of the product never changes. The plant will always produce electric energy. The things that may change are the fuel used (coal, oil, or gas) and environmental requirements. Many plants that were originally designed for coal were later converted to oil, converted back to coal, and then converted to gas. Environmental requirements have changed, which has required the construction of air and water emissions control systems.

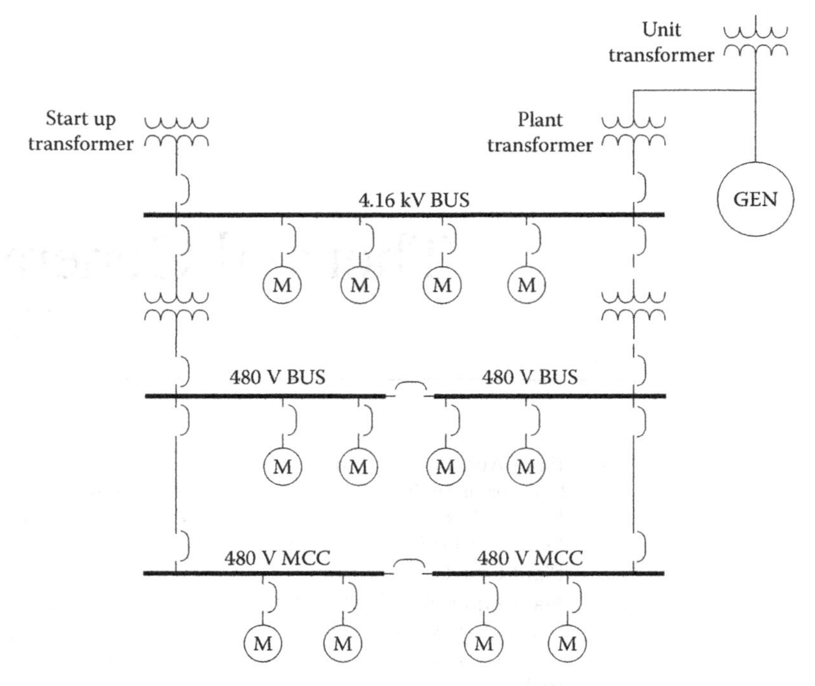

FIGURE 7.1 Typical plant layout.

Plant electrical systems should be designed to allow for further growth. Sizing of transformers and buses is at best a matter of guesswork. The plant electrical system should be sized at 5%–10% the size of the generating unit depending on the plant configuration and number of units at the plant site. The layout of a typical system is seen in Figure 7.1.

7.1 Plant Auxiliary System

7.1.1 Selection of Auxiliary System Voltages

The most common plant auxiliary system voltages are 13,800, 6,900, 4,160, 2,400, and 480 V. The highest voltage is determined by the largest motor. If motors of 4,000 hp or larger are required, one should consider using 13,800 V. If the largest motor required is less than 4000 hp, then 4160 V should be satisfactory.

7.1.2 Auxiliary System Loads

Auxiliary load consists of motors and transformers. Transformers supply lower level buses which supply smaller motors and transformers which supply lower voltage buses. Generation plants built before 1950 may have an auxiliary generator that is connected to the main generator shaft. The auxiliary generator will supply plant loads when the plant is up and running.

7.1.3 Auxiliary System Power Sources

The power sources for a generating plant consist of one or more off-site sources and one or more on-site sources. The on-site sources are the generator and, in some cases, a black start diesel generator or a gas turbine generator which may be used as a peaker.

7.1.4 Auxiliary System Voltage Regulation Requirements

Most plants will not require voltage regulation. A load flow study will indicate if voltage regulation is required. Transformers with tap changers, static var compensators, or induction regulators may be used to keep plant bus voltages within acceptable limits. Switched capacitor banks and overexcited synchronous motors may also be used to regulate bus voltage.

7.2 Plant One-Line Diagram

The one-line diagram is the most important document you will use. Start with a conceptual one-line and add detail as it becomes available. The one-line diagram will help you think about your design and make it easier to discuss with others. Do not be afraid to get something on paper very early and modify as you get more information about the design. Consider how the plant will be operated. Will there be a start-up source and a running source? Are there on-site power sources?

7.3 Plant Equipment Voltage Ratings

Establish at least one bus for each voltage rating in the plant. Two or more buses may be required depending on how the plant will be operated.

7.4 Grounded vs. Ungrounded Systems

A method of grounding must be determined for each voltage level in the plant.

7.4.1 Ungrounded

Most systems will be grounded in some manner with the exception for special cases of 120 V control systems which may be operated ungrounded for reliability reasons. An ungrounded system may be allowed to continue to operate with a single ground on the system. Ungrounded systems are undesirable because ground faults are difficult to locate. Also, ground faults can result in system overvoltage, which can damage equipment that is connected to the ungrounded system.

7.4.2 Grounded

Most systems 480 V and lower will be solidly grounded.

7.4.3 Low-Resistance Grounding

Low-resistance grounding systems are used at 2400 V and above. This system provides enough ground fault current to allow relay coordination and limits ground fault current to a value low enough to prevent equipment damage.

7.4.4 High-Resistance Grounding

High-resistance grounding systems limit ground fault current to a very low value but make relay coordination for ground faults difficult.

7.5 Miscellaneous Circuits

7.5.1 Essential Services

Essential services such as critical control required for plant shutdown, fire protection, and emergency lighting should be supplied by a battery-backed inverter. This is equipment that must continue to operate after a loss of off-site power. Some of these loads may be supplied by an on-site diesel generator or gas turbine if a delay after loss of off-site power is acceptable.

7.5.2 Lighting Supply

Lighting circuits should be designed with consideration to emergency lighting to the control room and other vital areas of the plant. Consideration should be given to egress lighting and lighting requirements for plant maintenance.

7.6 DC Systems

The plant will require at least one DC system for control and operation of essential systems when off-site power is lost. The required operating time for the emergency equipment that will be operated from the DC systems must be established in order to size the batteries. The installation of a diesel generator may reduce the size of the battery.

7.6.1 125 V DC

A 125 V DC system is supplied for circuit breaker and protective relaying. The system voltage may collapse to close to zero during fault conditions and would not be capable of supplying relay control and breaker trip current when it is needed to operate.

7.6.2 250 V DC

A 250 V DC system may be required to supply turbine generator emergency motors such as turning gear motors and emergency lube oil motors.

7.7 Power Plant Switchgear

7.7.1 High-Voltage Circuit Breakers

High-voltage circuit breakers of 34.5 kV and above may be used in the switchyard associated with the generating plant, but are rarely used in a generating plant.

7.7.2 Medium-Voltage Switchgear

Medium-voltage breakers are 2.4–13.8 kV. Breakers in this range are used for large motors in the plant. The most prevalent is 4.16 kV.

7.7.2.1 Medium-Voltage Air Circuit Breakers

Air circuit breakers were the most common type of breaker until about 1995. Due to large size and high maintenance requirements of air circuit breakers, they have been replaced by vacuum breakers.

7.7.2.2 Medium-Voltage Vacuum Circuit Breakers

Vacuum circuit breakers are the most common type of circuit breaker used in new installations. Vacuum circuit breakers are being used to replace air circuit breakers. Vacuum breakers are smaller and can provide additional space if the plant needs to be expanded to meet new requirements. Before using vacuum circuit breakers, a transient analysis study should be performed to determine if there is a need for surge protection. If required, surge protection can be supplied by the installation of capacitors and/or surge suppressors can be used to eliminate voltage surge problems.

7.7.2.3 Medium-Voltage SF6 Circuit Breakers

SF6 circuit breakers have the same advantages as vacuum circuit breakers but there is some environmental concern with the SF6 gas.

7.7.3 Low-Voltage Switchgear

Low voltage is 600 V and below. The most common voltage used is 480 V.

7.7.3.1 Low-Voltage Air Circuit Breakers

Air circuit breakers are used in load centers that may include a power transformer. Air circuit breakers are used for motors greater than 200 hp and less than about 600 hp. Low-voltage circuit breakers are self-contained in that fault protection is an integral part of the breaker. Low-voltage devices, which do not contain fault protection devices, are referred to as low-voltage switches. Low-voltage breakers may be obtained with various combinations of trip elements. Long time, short time, and ground trip elements may be obtained in various combinations.

Low-voltage breakers manufactured before 1970 will contain oil dashpot time delay trip elements. Breakers manufactured after the mid-1970s until about 1990 will contain solid-state analog trip elements. Breakers manufactured after 1990 will contain digital trip elements. The digital elements provide much more flexibility.

A circuit that may be large enough for a load center circuit breaker but is operated several times a day should not be put on a load center circuit breaker. The circuit breaker would be put through its useful life in a very short time. A motor starter would be a better choice.

7.7.4 Motor Control Centers

Motor control centers are self-contained and may include molded case breakers or combination starters. Molded case breakers are available as either magnetic or thermal-magnetic. The magnetic trip breakers are instantaneous trip only and the thermal-magnetic trip breakers are time delay with instantaneous trip. Magnetic breakers can be used with a contactor to make a combination starter. Time delay trip is provided by overload relays mounted on the contactor. Solid-state equipment is available to use in motor control centers and allows much greater flexibility.

7.7.5 Circuit Interruption

The purpose of a circuit breaker is to provide a method of interrupting the circuit either to turn the load on and off or to interrupt fault current. The first requirement is based on the full load current of the load. The second requirement is based on the maximum fault current as determined by the fault current study. There is no relationship between the load current and the fault current. If modifications are made to the electric power system, the fault interrupting current requirement may increase. Failure to recognize this could result in the catastrophic failure of a breaker.

7.8 Auxiliary Transformers

7.8.1 Selection of Percent Impedance

The transformer impedance is always compromised. High transformer impedance will limit fault current and reduce the required interrupting capability of switchgear and, therefore, reduce the cost. Low impedance will reduce the voltage drop through the transformer and therefore improve voltage regulation. A careful analysis using a load flow study will help in arriving at the best compromise.

7.8.2 Rating of Voltage Taps

Transformers should be supplied with taps to allow adjustment in bus voltage. Optimum tap settings can be determined using a load flow study.

7.9 Motors

7.9.1 Selection of Motors

Many motors are required in a thermal generating plant and range in size from fractional horsepower to several thousand horsepower. These motors may be supplied with the equipment they drive or they may be specified by the electrical engineer and purchased separately. The small motors are usually supplied by the equipment supplier and the large motors specified by the electrical engineer. How this will be handled must be resolved very early in the project. The horsepower cut-off point for each voltage level must be decided. The maximum plant voltage level must be established. A voltage of 13.8 kV may be required if very large horsepower motors are to be used. This must be established very early in the plant design so that a preliminary one-line diagram may be developed.

7.9.2 Types of Motors

7.9.2.1 Squirrel Cage Induction Motors

The squirrel cage induction motor is the most common type of large motor used in a thermal generating plant. Squirrel cage induction motors are very rugged and require very little maintenance.

7.9.2.2 Wound Rotor Induction Motors

The wound rotor induction motor has a rotor winding which is brought out of the motor through slip rings and brushes. While more flexible than a squire cage induction motor, the slip rings and brushes are an additional maintenance item. Wound rotor motors are only used in special applications in a power plant.

7.9.2.3 Synchronous Motors

Synchronous motors may be required in some applications. Large slow-speed, 1800 rpm or less may require a synchronous motor. A synchronous motor may used to supply VARs and improve voltage regulation. If the synchronous motor is going to be used as a VAR source, the field supply must be sized large enough to over-excite the field.

7.9.2.4 Direct Current Motors

Direct current motors are used primarily on emergency systems such as turbine lube oil and turbine turning gear. Direct current motors may also be used on some control valves.

7.9.2.5 Single-Phase Motors

Single-phase motors are fractional horsepower motors and are usually supplied with the equipment.

7.9.2.6 Motor Starting Limitations

The starting current for induction motors is about six times full load current. This must be taken into account when sizing transformers and should be part of the load flow analysis. If the terminal voltage is allowed to drop too low, below 80%, the motor will stall. Methods of reduced voltage starting are available, but should be avoided if possible. The most reliable designs are the simplest.

7.10 Main Generator

The turbine generator will be supplied as a unit. The size and characteristics are usually determined by the system planners as a result of system load requirements and system stability requirements.

7.10.1 Associated Equipment

7.10.1.1 Exciters and Excitation Equipment

The excitation system will normally be supplied with the generator.

7.10.2 Electronic Exciters

Modern excitation systems are solid state and, in recent years, most have digital control systems.

7.10.3 Generator Neutral Grounding

The generator neutral is never connected directly to ground. The method used to limit the phase to ground fault current to a value equal to or less than the three-phase fault current is determined by the way the generator is connected to the power system. If the generator is connected directly to the power system, a resistor or inductor connected between the neutral of the generator and ground will be used to limit the ground fault current. If the generator is connected to the power system through a transformer in a unit configuration, the neutral of the generator may be connected to ground through a distribution transformer with a resistor connected across the secondary of the transformer. The phase-to-ground fault current can be limited to 5–10 A using this method.

7.10.4 Isolated Phase Bus

The generator is usually connected to the step-up transformer through an isolated phase bus. This separated phase greatly limits the possibility of a phase-to-phase fault at the terminals of the generator.

7.11 Cable

Large amounts of cable are required in a thermal generating plant. Power, control, and instrumentation cable should be selected carefully with consideration given to long life. Great care should be given in the installation of all cable. Cable replacement can be very expensive.

7.12 Electrical Analysis

All electrical studies should be well-documented for use in plant modifications. These studies will be of great value in evaluating plant problems.

7.12.1 Load Flow

A load flow study should be performed as early in the design as possible even if the exact equipment is not known. The load flow study will help in getting an idea of transformer size and potential voltage drop problems.

A final load flow study should be performed to document the final design and will be very helpful if modifications are required in the future.

7.12.2 Short-Circuit Analysis

Short-circuit studies must be performed to determine the requirements for circuit breaker interrupting capability. Relay coordination should be studied as well.

7.12.3 Surge Protection

Surge protection may be required to limit transient overvoltage caused by lightning and circuit switching. A surge study should be performed to determine the needs of each plant configuration. Surge arrestors and/or capacitors may be required to limit transient voltages to acceptable levels.

7.12.4 Phasing

A phasing diagram should be made to determine correct transformer connections. An error here could prevent busses from being paralleled.

7.12.5 Relay Coordination Studies

Relay coordination studies should be performed to ensure proper coordination of the relay protection system. The protective relay system may include overcurrent relays, bus differential relays, transformer differential relays, voltage relays, and various special function relays.

7.13 Maintenance and Testing

A good plant design will take into account maintenance and testing requirements. Equipment must be accessible for maintenance and provisions should be made for test connections.

7.14 Start-Up

A start-up plan should be developed to ensure equipment will perform as expected. This plan should include insulation testing. Motor starting current magnitude and duration should be recorded and relay coordination studies verified. Voltage level and load current readings should be taken to verify load flow studies. This information will also be very helpful when evaluating plant operating conditions and problems.

References

General

Beeman, D., Ed., *Industrial Power Systems Handbook*, McGraw-Hill, New York, 1955.

Central Station Engineers of the Westinghouse Corporation, *Electrical Transmission and Distribution Reference Book*, Westinghouse, Butler, PA.

IEEE Standard 666-1991, IEEE Design Guide for Electric Power Service Systems for Generating Stations.

Grounding

IEEE 665-1955, IEEE Guide for Generating Station Grounding.

IEEE 1050-1996, IEEE Guide for Instrumentation and Control Grounding in Generating Stations.

DC Systems

IEEE 485-1997, IEEE Recommended Practice for Sizing Lead-Acid Batteries for Station Applications.
IEEE 946-1992, IEEE Recommended Practice for the Design of DC Auxiliary Power Systems for Generating Stations.

Switchgear

IEEE Standards Collection: Power and Energy-Switchgear, 1998 Edition.

Auxiliary Transformers

IEEE Distribution, Power and Regulating Transformers Standards Collection, 1998 Edition.

Motors

IEEE Electric Machinery Standards Collection, 1997 Edition.

Cable

IEEE 835-1994, IEEE Standard Power Cable Ampacity Tables.

Electrical Analysis

Clarke, E., *Circuit Analysis of AC Power Systems*, General Electric Company, New York, 1961.
Stevenson, W.D., *Elements of Power Systems Analysis*, McGraw-Hill, New York, 1962.
Wager, C.F. and Evans, R.D., *Symmetrical Components*, McGraw-Hill, New York, 1933.

8

Distributed Utilities

John R. Kennedy
Georgia Power Company

Rama Ramakumar
Oklahoma State University

Distributed utilities (DU), as the name indicates, employ distributed generation (DG) sources and distributed storage (DS) of energy throughout the service area. DG and DS devices can be operating individually or together and in stand-alone mode or in conjunction with an existing electric utility grid.

Increasing interest in the environmental aspects of electric power generation and utilization, the need to maintain a very high level of reliability and power quality to serve the evolving digital society, and the desire to harness renewable energy resources have combined to sharpen the focus on DG and DS in recent years.

Continuing technological advances in materials, system control and operation, and forecasting techniques have enabled DG with stochastic inputs such as insolation and wind to join the ranks of predictable generation technologies involving fuel cells, microturbines, gas turbines, and battery storage. In most cases, outputs of these devices are small in comparison with traditional utility-scale quantities. However, strategic deployment of DG and DS can significantly improve the overall operation of the electric power system from the economic, reliability, and environmental points of view.

Ratings of DG can range from a few kilowatts to several megawatts as in the case of large wind farms, utility-scale photovoltaic (PV), and solar-thermal systems. Typically they enter at the secondary voltage level, 120 V single-phase to 480 V three-phase, and at medium voltage levels, 2.4–25 kV, depending on the technology and the geographic parameters involved. Some of the recent developments envisage the entry of DC microgrids as components of DU to effectively harness renewable energy resources at local levels.

In this section, an overview of the different issues associated with DU (DG and DS), including available technologies, interfacing, brief discussion on economic and possible regulatory treatment, applications, and some practical examples are included. Emerging technologies discussed will include fuel

TABLE 8.1 Distributed Generation Technology Chart

Technology	Size	Fuel Sources	AC Interface Type	Applications
Fuel cells	0.5 kW—larger units with stacking	Natural gas, hydrogen, petroleum products, land-fill gas	Inverter type	Continuous
Microturbines	10–100 kW larger sizes	Natural gas, petroleum products	Inverter type	Continuous standby
Batteries	0.1 kW–2 MW+	Storage	Inverter type	PQ, peaking
Flywheel	>0.1–0.5 kW	Storage	Inverter type	PQ, peaking
PV	A few kW to MWs	Insolation	Inverter type	Peaking
Gas turbine	10 kW–5 MW+	Natural gas, petroleum products	Conventional	Continuous peaking
WECS	Tens to hundreds of MWs	Wind	Direct/inverter	Intermittent/energy source
Solar thermal	Tens to hundreds of MWs	Insolation	Conventional	Continuous with gas backup, peaking

cells, microturbines, small gas turbines, PV systems, wind farms, and solar-thermal systems. Energy storage technologies are briefly discussed. Interfacing issues include general protection, overcurrent protection, islanding issues, communication and control, voltage regulation, frequency control, fault detection, safety issues, and synchronization. In the section on applications, some economic issues, auxiliary services, peak shaving, reliability, and power quality issues are briefly covered.

8.1 Available Technologies

Many of the "new" technologies have been around for several years, but the relative cost per kilowatt of these technologies compared to conventional power plants has made their use limited until now. Utility rules and interconnect requirements have also limited the use of small generators and storage devices to mostly emergency, standby, and power quality applications. The scenario of deregulation coupled with environmental consciousness has changed all that. Utilities are no longer assured that they can recover the costs of large base generation plants, and stranded investment of transmission and distribution facilities is constantly undergoing shifting considerations. This, coupled with improvements in the cost and reliability of DU technologies, has opened an ever-increasing market for small and "green" power plants. In the near future (already in some cases), these new technologies should be competitive with conventional plants and provide high reliability with less investment risk. Some of the technologies are listed in the following. Energy storage technologies are enjoying rapid improvements in the small capacity range, specifically for use in portable electronic devices. At the same time, large-scale storage using flow batteries and similar technologies are also rapidly advancing. Power electronic technologies have experienced dramatic improvements to enable effective interfacing with ac grid, starting with dc outputs or rectified variable frequency outputs. Table 8.1 lists different technologies, their size ranges, input sources, interface types, and most likely applications.

8.2 Fuel Cells

Fuel cell technology has been around since its invention by William Grove in 1839. From the 1960s to the present, fuel cells have been the power source used for space flight missions. Unlike other generation technologies, fuel cells act like continuously fueled batteries, producing direct current (DC) by using an electrochemical process. The basic design of all fuel cells consists of an anode, electrolyte, and cathode. Hydrogen or a hydrogen-rich fuel gas is passed over the anode, and oxygen or air is passed over the cathode.

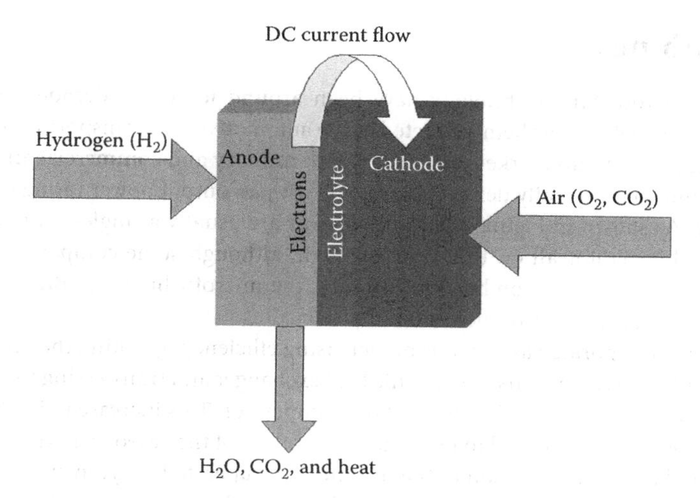

FIGURE 8.1 Basic fuel cell operation.

A chemical combination then takes place producing a constant supply of electrons (DC current) with by-products of water, carbon dioxide, and heat. The DC power can be used directly or it can be fed to a power conditioner and converted to AC power (see Figure 8.1).

Most of the present technologies have a fuel reformer or processor that can take most hydrocarbon-based fuels, separate out the hydrogen, and produce high-quality power with negligible emissions. This would include gasoline, natural gas, coal, methanol, light oil, or even landfill gas. In addition, fuel cells can be more efficient than conventional generators since they operate isothermally and, as such, are not subject to Carnot limitations. Theoretically they can obtain efficiencies as high as 85% when the excess heat produced in the reaction is used in a combined cycle mode. These features, along with relative size and weight, have also made the fuel cell attractive to the automotive industry as an alternative to battery power for electric vehicles. The major differences in fuel cell technology concern the electrolyte composition. The major types are the proton exchange membrane fuel cell (PEFC) also called the PEM, the phosphoric acid fuel cell (PAFC), the molten carbonate fuel cell (MCFC), and the solid oxide fuel cell (SOFC) (Table 8.2).

Fuel cell power plants can come in sizes ranging from a few watts to several megawatts with stacking. The main disadvantage to the fuel cell is the initial high cost of installation. With the interest in efficient and environmentally friendly generation, coupled with the automotive interest in an EV alternative power source, improvements in the technology and lower costs are expected. As with all new technologies, volume of sales should also lower the unit price.

TABLE 8.2 Comparison of Fuel Cell Types

	PAFC	MCFC	SOFC	PEMFC
Electrolyte	Phosphoric acid	Molten carbonate salt	Ceramic	Polymer
Operating temperature	375°F (190°C)	1200°F (650°C)	1830°F (1000°C)	175°F (80°C)
Fuels	Hydrogen (H_2)	H_2/CO	H_2/CO_2/CH_4	H_2
	Reformate	Reformate	Reformate	Reformate
Reforming	External	External	External	External
Oxidant	O_2/air	CO_2/O_2/air	O_2/air	O_2/air
Efficiency (HHV)	40%–50%	50%–60%	45%–55%	40%–50%

Source: Li, X. *Principles of Fuel Cells,* Section 1.4.5, Taylor and Francis, Boca Raton, Florida, 2006.

8.3 Microturbines

Experiments with microturbine technology have been around for many decades, with the earliest attempts of wide-scale applications being targeted at the automotive and transportation markets. These experiments later expanded into markets associated with military and commercial aircraft and mobile systems. Microturbines are typically defined as systems with an output power rating of between 10 kW up to a few 100 kW. As shown in Figure 8.2, these systems are usually a single-shaft design with compressor, turbine, and generator all on the common shaft, although some companies are engineering dual-shaft systems. Like the large combustion turbines, the microturbines are Brayton cycle systems, and will usually have a recuperator in the system.

The recuperator is incorporated as a means of increasing efficiency by taking the hot turbine exhaust through a heavy (and relatively expensive) metallic heat exchanger and transferring the heat to the input air, which is also passed through parallel ducts of the recuperator. This increase in inlet air temperature helps reduce the amount of fuel needed to raise the temperature of the gaseous mixture during combustion to levels required for total expansion in the turbine. A recuperated Brayton cycle microturbine can operate at efficiencies of approximately 30%, while these aeroderivative systems operating without a recuperator would have efficiencies in the 15–18% range.

Another requirement of microturbine systems is that the shaft must spin at very high speeds, in excess of 50,000 RPM and in some cases doubling that rate, due to the low inertia of the shaft and connected components. This high speed is used to keep the weight of the system low and increase the power density over other generating technologies. Although many of the microturbines are touted as having only a single moving part, there are numerous ancillary devices required that do incorporate moving parts such as cooling fans, fuel compressors, and pumps.

Since the turbine requires extremely high speeds for optimal performance, the generator cannot operate as a synchronous generator. Typical microturbines have a permanent magnet motor/generator incorporated onto the shaft of the system. The high rotational speed gives an AC output in excess of 1000 Hz, depending on the number of poles and actual rotational speed of the microturbine. This high-frequency AC source is rectified, forming a common DC bus voltage that is then converted to a 60 Hz AC output by an onboard inverter.

The onboard electronics are also used to start the microturbine, either in a stand-alone mode or in grid parallel applications. Typically, the utility voltage will be rectified and the electronics are used to convert this DC voltage into a variable frequency AC source. This variable frequency drive will power the permanent magnet motor/generator (which is operating as a motor), and will ramp the turbine speed up to a preset RPM, a point where stabile combustion and control can be maintained. Once this preset

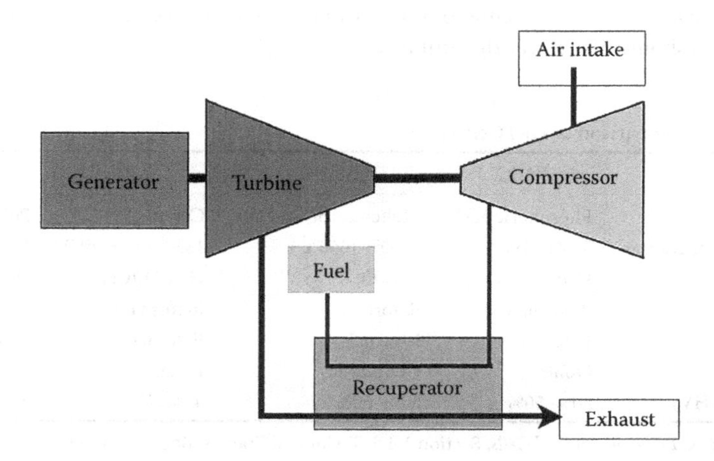

FIGURE 8.2 Turbine block diagram configuration with recuperator.

speed is obtained and stabile combustion is taking place, the drive shuts down and the turbine speed increases until the operating point is maintained and the system operates as a generator. The time from a "shaft stop" to full load condition is anywhere from 30 s to 3 min, depending on manufacturer recommendations and experiences.

Although microturbine products are in the early stages of commercialization, there are cost targets that have been announced from all of the major manufacturers of these products. The early market entry price of these systems is in excess of $600 per kilowatt, more than comparably sized units of alternative generation technologies, but all of the major suppliers have indicated that costs will fall as the number of units being put into the field increases.

The microturbine family has a very good environmental rating, due to natural gas being a primary choice for fuel and the inherent operating characteristics, which puts these units at an advantage over diesel generation systems.

8.4 Combustion Turbines

There are two basic types of combustion turbines (CTs) other than the microturbines: the heavy frame industrial turbines and the aeroderivative turbines. The heavy frame systems are derived from similar models that were steam turbine designs. As can be identified from the name, they are of very heavy construction. The aeroderivative systems have a design history from the air flight industry, and are of a much lighter and higher speed design. These types of turbines, although similar in operation, do have some significant design differences in areas other than physical size. These include areas such as turbine design, combustion areas, rotational speed, and air flows.

Although these units were not originally designed as a "distributed generation" technology, but more so for central station and large co-generation applications, the technology is beginning to economically produce units with ratings in the hundreds of kilowatts and single-digit megawatts. These turbines operate as Brayton cycle systems and are capable of operating with various fuel sources. Most applications of the turbines as DG will operate on either natural gas or fuel oil. The operating characteristics between the two systems can best be described in tabular form as shown in Table 8.3.

The CT unit consists of three major mechanical components: a compressor, a combustor, and a turbine. The compressor takes the input air and compresses it, which will increase the temperature and decrease the volume per the Brayton cycle. The fuel is then added and the combustion takes place in the combustor, which increases both the temperature and volume of the gaseous mixture, but leaves the pressure as a constant. This gas is then expanded through the turbine where the power is extracted through the decrease in pressure and temperature and the increase in volume.

If efficiency is the driving concern, and the capital required for the increased efficiency is available, the Brayton cycle systems can have either co-generation systems, heat recovery steam generators, or simple recuperators added to the CT unit. Other equipment modifications and improvements can be incorporated into these types of combustion turbines such as multistage turbines with fuel re-injection, inter-cooler between multistage compressors, and steam/water injection.

Typical heat rates for simple cycle combustion turbines vary across manufacturers, but are in a range from 11,000 to 20,000 BTU/kWh. However, these numbers decrease as recuperation and co-generation are added. CTs typically have a starting reliability in the 99% range and operating reliability approaching 98%.

TABLE 8.3 Basic Combustion Turbine Operating Characteristics

	Heavy Frame	Aeroderivative
Size (same general rating)	Large	Compact
Shaft speed	Synchronous	Higher speed (coupled through a gear box)
Air flow	High (lower compression)	Lower (high compression)
Start-up time	15 min	2–3 min

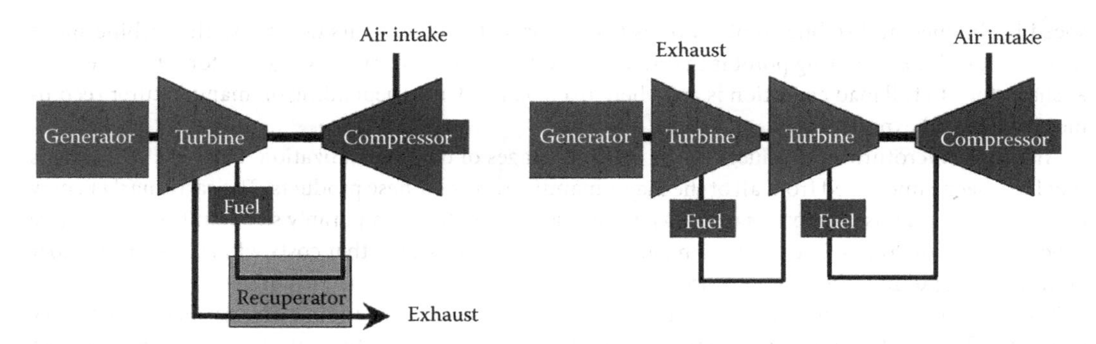

FIGURE 8.3 Basic combustion turbine designs.

The operating environment has a major effect on the performance of combustion turbines. The elevation at which the CT is operating has a degradation factor of around 3.5% per 1000 ft of increased elevation and the ambient temperature has a similar degradation per 10° increase.

Figure 8.3 shows a block diagram of a simple cycle combustion turbine with a recuperator (left) and a combustion turbine with multistage turbine and fuel re-injection (right).

8.5 Photovoltaics

PV devices, commonly known as solar cells, convert incident solar radiation (insolation) directly into dc electrical energy. PV technology is highly flexible and versatile in terms of size (milliwatts to megawatts) and siting (from watches and calculators to utility-scale central station systems). Starting with its discovery in Bell Labs in 1954, PV has progressed from an expensive but critical option for space programs to utility-scale terrestrial use.

Insolation is an abundant resource. It arrives on the earth's surface at a rate of 120 million GW (about 20,000 kW for each and every 1 of the 6 billion humans inhabiting the earth). In comparison, the total rate of energy consumed by humans is estimated to be 15,000 GW, about 8,000 times smaller than insolation.

Among the many attributes of PV are silent, simple, and environmentally benign operation, no moving parts (other than trackers, if used), modular with no serious efficiency and cost penalties, and high power output to weight ratio. However, cost continues to be a barrier for large-scale generation of electricity. Cost of PV modules is in the range of \$3–\$4 per peak watt, cost of energy output is still about four to six times the generation cost realized with modern large wind-electric conversion systems. Absence of the resource at night times and variability introduced by moving clouds necessitate some form of energy storage and reconversion system or backup generation (including a conventional power grid). Many niche applications have been developed that are cost-effective at present. PV's siting flexibility will allow installation in locations ranging from rooftops and village-level systems all the way to large MW-scale central station configurations.

With an average overall efficiency of 10% and a peak insolation of 1 kW/m^2, an aggregated cell area of 1 ha (10,000 m^2 or 2.471 acres) will generate an electrical output of 1 MW. However, the need to have tilted arrays of small modules spaced to avoid inter-array shadowing will result in a land area requirement of approximately 3–4 ha per MW with fixed-tilt flat-plate arrays. If tracking is employed, the land area required will be even larger. Concentrating plants will require about 2–2.5 times the land area to avoid shadowing as compared to fixed-tilt flat-plate arrays. The need to improve overall efficiency is obvious to reduce the land area required.

There are no serious technical problems for large-scale generation of electricity using PV. The first 1 MW plant was built in 9 months and started operation near Hesperia, California in 1982, and generated about 3 million kWh per year. By 2010, large-scale generation of electricity using PV had spread to more than 100 countries. At 97 MW, the Sarnia PV plant in Canada is the largest plant at present. Other large PV plants are Montalto di Castro plant in Italy rated 84.2 MW, Finsterwalde Solar Park in

Germany rated at 80.7 MW, Rovigo plant in Italy rated at 70 MW, Olmedilla PV Park in Spain rated at 60 MW, Strasskirchen Solar Park in Germany rated at 54 MW, and Lieberose PV Park, also in Germany rated at 53 MW. Many more are in the planning and construction stages all around the world.

Some of the other developments include transparent building-integrated PV (BIPV) systems, development of ac modules employing small power electronic systems with each module and improvements in the large-scale manufacturing of PV cells, panels, and arrays. Most recent advances are aimed at capturing a larger fraction of the incident solar spectrum and increasing the efficiencies using titanium-based nanotechnologies. All along, technologies based on silicon in its various forms (single crystal, poly crystal, amorphous, etc.) have exhibited dramatic efficiency improvements ranging all the way to just above 24%, a new world record.

8.6 Solar-Thermal-Electric Systems

Solar-thermal-electric systems harness insolation in the form of thermal energy and utilize the solar heat as the input for a conventional thermal cycle with a suitable working fluid, which could be simply water. The temperatures that can be obtained on the hot side depend on the type of collector employed. Coupled with a supplementary heater (typically burning natural gas) and/or a thermal energy storage and retrieval system, schedulable power output can be obtained on demand on a 24/7 basis.

There are two basic options available for collecting solar thermal energy:

1. Distributed collector with central or DG
2. Central collector with central generation system

Solar thermal collectors are classified based on the collection temperature as low, medium, or high. For conversion to electrical energy, only medium and high temperature collectors are viable and practical. Medium temperatures at around 400°C are obtained using parabolic trough systems. Parabolic dishes can concentrate insolation with ratios in the range of 600–2000 and temperatures in excess of 1500°C can be obtained at the focal point. Central receiver systems operate at concentration ratios in the range of 300–1500 and collect solar energy in thermal form at temperatures of 500°C–1500°C.

Distributed collectors can be a flat-plate type (not suitable for electric power generation) or a parabolic trough type resulting in a line-focus system or a point-focusing parabolic dish. With parabolic dishes, one generator can be placed in the focus of each dish and aggregation of energy is accomplished on the electrical side. With line-focus systems, an array of collectors (requiring approximately 2 ha per MWe) is employed to collect and transport the medium-grade thermal energy to a central location for use in a thermodynamic cycle.

In a central collector (receiver) plant, heliostats concentrate insolation onto a central receiver (in the 300–1500 ratio) where the energy is transferred to a working fluid (typically water) or to molten salt for storage and retrieval. The thermal energy thus collected is used to generate steam, which drives a turbine/generator system to produce electricity. Central receiver plants must be built in the scale of tens or hundreds of MW to be economically feasible.

Thermal energy storage and retrieval is far easier than storage in electrical form. It can be supplemented by fossil fuel (typically natural gas) burning subsystems to obtain schedulable and economically viable outputs. This is one of the primary reasons for the recent upsurge in construction and operation of solar-thermal-electric systems worldwide.

Solar thermal power industry is growing rapidly around the world with 1.2 GW of capacity under construction as of April 2009 and another 13.9 GW announced through 2014. Spain was the epicenter of solar thermal power development in 2010 with 22 projects aggregating to 1037 MW. In the United States, 5600 MW of solar thermal power projects have been announced. Solar Millennium, LLC and Chevron Energy Solutions, joint developers of a project, propose to construct, own, and operate the Blythe Solar Power Project in Southern California. The project is a concentrated solar thermal electric generating facility with four adjacent, independent, and identical solar plants of 250 MW nominal capacity each for a total capacity of 1000 MW. eSolar and Penglai Electric, a privately owned Chinese electrical power equipment manufacturer, have reached a master licensing agreement to build at least 2 GW of solar thermal power plants in China over the next 10 years. In India, the Jawaharlal Nehru National Solar Mission project is aimed at supplementing up to 20 GW of solar generated energy by 2022 across India. The prime phase of the project is to establish off-grid solar PV as well as solar thermal by 2013.

8.7 Wind Electric Conversion Systems

Wind is moving air, resulting from the uneven heating of the earth's atmosphere by the incident solar energy. Thus, wind energy is an indirect form of solar energy, which is fully renewable. Wind energy is abundant (about 1670 trillion kWh/year over the land area of the earth). Including the offshore resources, a comprehensive study undertaken in 2005 estimated the potential of wind power on land and near-shore to be 72 TW, equivalent to 54,000 MTOE (million tons of oil equivalent) per year, which is over five times the world's current energy use in all forms. This abundant resource can be easily converted to rotary mechanical energy for coupling to an electrical generator to generate electricity. Since the collection area is perpendicular to ground surface, they pose minimal burden on land area and can coexist with many farming and other activities. It is estimated that over the land area, large WECS require about 6 ha per MW. Smaller unit sizes will require larger land areas. The blades occupy only a small fraction of the collector area, thus promising cost-effective conversion. Concerns about avian mortality have been largely mitigated by employing large diameter (60–120 m) turbines operating at slow rotational speeds in the range of 15–20 RPM.

Wind turbines have progressed from the early multibladed farm windmills that dotted the Midwest for pumping water for livestock in the United States to technologically complex large units employing sophisticated electromechanical energy converters operating in conjunction with advanced power electronic systems.

After several unsuccessful attempts to operate large wind turbines at constant speeds coupled to synchronous generators, detailed computer models revealed that such operation incurred undue stresses on the blades, tower, etc. and that variable speed operation alleviated this problem and extended the life of the system. Consequently, all the modern large WECS operate in the variable speed mode and employ suitable techniques on the electrical side to obtain constant frequency output. At present, the most widely used approach involves a double output induction generator (DFIG) that feeds the grid directly from the stator and through a power electronic frequency converter from the rotor. With a 20% variation in the range of wind turbine speeds, the rating of the power electronic system need be only 20% of the overall system rating. One of the promising new technologies under development is to eliminate the gear box altogether and employ direct drive large diameter (around 6 m) permanent magnet generators with lightweight rotating magnet systems and fixed armatures. The variable frequency output from the stationary armature is processed through a full power converter and fed to the grid.

Selecting a suitable location to site wind turbines is very critical to the success of the system. In order to aid this process, wind regimes are classified into several classes as listed in Table 8.4. Even small differences in mean wind speeds can lead to significant changes in the energy generated on an annual basis and capacity factors, which typically range from 20% to 30% or above.

In the year 2000, wind electrical conversion was the fastest growing and least cost electric generation option in the world. Almost 4000 MW of installed capacity was added during 2000, bringing the total to slightly above 17,500 MW globally. At the start of 2009, aggregated worldwide nameplate capacity of wind power plants was approximately 122 GW and energy production was around 260 TWh, which was about 1.5% of the worldwide electricity usage. By May 2009, around 80 countries were using wind power on a commercial basis. Countries such as Denmark and Spain have considerable share of wind in energy generation sector with shares of 19% and 13%, respectively. China had originally set a generation target of 30,000 MW by 2020 from renewable energy sources and reached 22,500 MW by the end of 2009 with a potential to surpass 30,000 MW by the end of 2010.

TABLE 8.4 Wind Power Density Classes at 10 and 50 m Heights

Class #	Power Density W/m² @ 10 m	Wind Speed m/s @ 10 m	Power Density W/m² @ 50 m	Wind Speed m/s @ 50 m
1	0–100	0–4.4	0–200	0–5.6
2	100–150	4.4–5.1	200–300	5.6–6.4
3	150–200	5.1–5.6	300–400	6.4–7.0
4	200–250	5.6–6.0	400–500	7.0–7.5
5	250–300	6.0–6.4	500–600	7.5–8.0
6	300–400	6.4–7.0	600–800	8.0–8.8
7	Over 400	Over 7.0	Over 800	Over 8.8

In recent years, United States has added more wind energy to its grid than any other country by installing 35,159 MW of wind power. According to the U.S. Department of Energy, wind power is capable of becoming a major contributor of electricity to America in three decades. A scenario of 20% of electricity from wind by 2030 is proclaimed as a national goal. The annual energy generated by these installations will depend on their capacity factors, which are strongly dependent on the prevailing wind regimes.

8.8 Storage Technologies

Storage technologies include batteries, flywheels, ultracapacitors, and to some extent photovoltaics. Most of these technologies are best suited for power quality and reliability enhancement applications, due to their relative energy storage capabilities and power density characteristics, although some large battery installations could be used for peak shaving. All of the storage technologies have a power electronic converter interface and can be used in conjunction with other DU technologies to provide "seamless" transitions when power quality is a requirement.

8.9 Interface Issues

A whole chapter could be written just about interface issues, but this discussion will touch on the highlights. Most of the issues revolve around safety and quality of service. We will discuss some general guidelines and the general utility requirements and include examples of different considerations. In addition to the interface issues, the DU installation must also provide self-protection to prevent short circuit or other damage to the unit. Self-protection will not be discussed here. The most important issues are listed in Table 8.5.

In addition to the interface issues identified in Table 8.5, there are also operating limits that must be considered. These are listed in Table 8.6.

Utility requirements vary but generally depend on the application of a distributed source. If the unit is being used strictly for emergency operation, open transition peak shaving, or any other stand-alone type operation, the interface requirements are usually fairly simple, since the units will not be operating in parallel with the utility system. When parallel operation is anticipated or required, the interface requirements become more complex. Protection, safety, power quality, and system coordination become issues that must be addressed. In the case of parallel operation, there are generally three major factors that determine the degree of protection required. These would include the size and type of the generation, the location on the system, and how the installation will operate (one-way vs. two-way). Generator sizes are generally classified as follows:

Large: Greater than 3 MVA or possibility of "islanding" a portion of the system
Small: Between large and extremely small
Extremely small: Generation less than 100 kVA

TABLE 8.5 Interface Issues

Issue	Definition	Concern
Automatic reclosing	Utility circuit breakers can test the line after a fault	If a generator is still connected to the system, it may not be in synchronization, thus damaging the generator or causing another trip
Faults	Short circuit condition on the utility system	Generator may contribute additional current to the fault, causing a miss operation of relay equipment
Islanding	A condition where a portion of the system continues to operate isolated from the utility system	Power quality, safety, and protection may be compromised in addition to possible synchronization problems
Protection	Relays, instrument transformers, circuit breakers	Devices must be utility grade rather than industrial grade for better accuracy
		Devices must also be maintained on a regular schedule by trained technicians
Communication	Devices necessary for utility control during emergency conditions	Without control of the devices, islanding and other undesirable operation of devices

TABLE 8.6 Operating Limits

1. Voltage—The operating range for voltage must maintain a level of ±15% of nominal for service voltage (ANSI C84.1), and have a means of automatic separation if the level gets out of the acceptable range within a specified time
2. Flicker—Flicker must be within the limits as specified by the connecting utility. Methods of controlling flicker are discussed in IEEE Std. 519-1992, 10.5
3. Frequency—Frequency must be maintained within ±0.5 Hz of 60 Hz and have an automatic means of disconnecting if this is not maintained. If the system is small and isolated, there might be a larger frequency window. Larger units may require an adjustable frequency range to allow for clock synchronization
4. Power factor—The power factor should be within 0.85 lagging or leading for normal operation. Some systems that are designed for compensation may operate outside these limits
5. Harmonics—Both voltage and current harmonics must comply with the values for generators as specified in IEEE Std. 519-1992 for both total and individual harmonics

Location on the system and individual system characteristics determine impedance of a distribution line, which in turn determines the available fault current and other load characteristics that influence "islanding" and make circuit protection an issue. This will be discussed in more detail later.

The type of operation is the other main issue and is one of the main determinants in the amount of protection required. One-way power flow where power will not flow back into the utility has a fairly simple interface, but is dependent on the other two factors, while two-way interfaces can be quite complex. An example is shown in Figure 8.4. Smaller generators and "line-commutated" units would have less stringent requirements. Commutation methods will be discussed later. Reciprocating engines such as diesel and turbines with mass, and "self-commutating" units, which could include microturbines and fuel cells, would require more stringent control packages due to their islanding and reverse power capabilities.

Most of the new developing technologies are inverter based and there are efforts now in IEEE to revise the old Standard P929 Recommended Practice for Utility Interface of Photovoltaic (PV) Systems to include other inverter-based devices. The standards committee is looking at the issues with inverter-based devices in an effort to develop a standard interface design that will simplify and reduce the cost, while not sacrificing the safety and operational concerns. Inverter interfaces generally fall into two classes: line-commutated inverters and self-commutated inverters.

8.9.1 Line-Commutated Inverters

These inverters require a switching signal from the line voltage in order to operate. Therefore, they will cease operation if the line signal, i.e., utility voltage, is abnormal or interrupted. These are not as popular

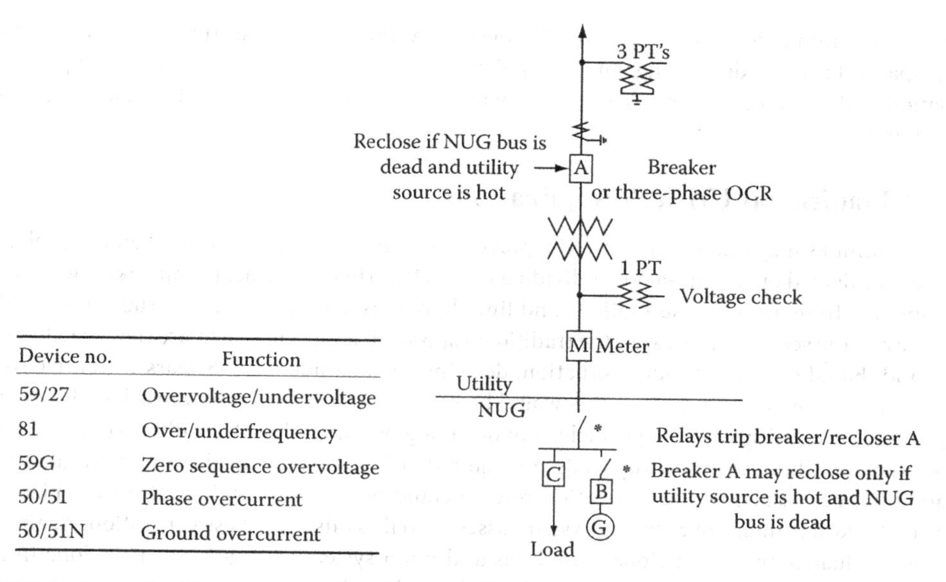

Device no.	Function
59/27	Overvoltage/undervoltage
81	Over/underfrequency
59G	Zero sequence overvoltage
50/51	Phase overcurrent
50/51N	Ground overcurrent

FIGURE 8.4 Example of large generator interface requirements for distribution. (From *Georgia Power Bulletin*, 18–8, generator interface requirements.)

today for single-phase devices due to the filtering elements required to meet the harmonic distortion requirements, but are appearing in some of the three-phase devices where phase cancellation minimizes the use of the additional components.

8.9.2 Self-Commutated Inverters

These inverters, as implied by the name, are self-commutating. All stand-alone units are self-commutated, but not all self-commutated inverters are stand-alone. They can be designed as either voltage or current sources and most that are now being designed to be connected to the utility system are designed to be current sources. These units still use the utility voltage signal as a comparison and produce current at that voltage and frequency. A great deal of effort has gone into the development of nonislanding inverters that are of this type.

8.10 Applications

Applications vary and will become more diverse as utilities unbundle. Listed in the following are some examples of the most likely.

8.10.1 Ancillary Services

Ancillary services support the basic electrical services and are essential for the reliability and operation of the electric power system. The electrical services that are supported include generating capacity, energy supply, and the power delivery system. FERC requires six ancillary services, including system control, regulation (frequency), contingency reserves (both spinning and supplemental), voltage control, and energy imbalance. In addition, load following, backup supply, network stability, system "blackstart," loss replacement, and dynamic scheduling are necessary for the operation of the system. Utilities have been performing these functions for decades, but as vertically integrated regulated monopoly organizations. As these begin to disappear, and a new structure with multiple competing parties emerges, DU might be able to supply several of these.

The DU providing these services could be owned by the former traditional utility, customers, or third-party brokers, depending on the application. The main obstacles to this approach are aggregation and communication when dealing with many small resources rather than large central station sources.

8.10.2 "Traditional Utility" Applications

Traditional utilities may find the use of DU a practical way to solve loading and reliability problems if each case is evaluated on a stand-alone individual basis. Deferring investment is one likely way that DU can be applied. In many areas, substations and lines have seasonal peaks that are substantially higher than the rest of the year. In these cases, the traditional approach has been to increase the capacity to meet the demand. Based on the individual situation, delaying the upgrade for 2–5 years with a DU system could be a more economical solution. This would be especially true if different areas had different seasonal peaks and the DU system was portable, thus deferring two upgrades. DU could also be used instead of conventional facilities when backup feeds are required or to improve reliability or power quality.

In addition, peak shaving and generation reserve could be provided with strategically placed DU systems that take advantage of reducing system losses as well as offsetting base generation. Again, these have to be evaluated on an individual case basis and not a system average basis as is done in many economic studies. The type of technology used will depend on the particular requirements. In general, storage devices such as flywheels and batteries are better for power quality applications due to their fast response time, in many cases half a cycle. Generation devices are better suited for applications that require more than 30 min of supply, such as backup systems, alternate feeds, peak shaving, and demand deferrals. Generation sources can also be used instead of conventional facilities in certain cases.

8.10.3 Customer Applications

Individual customers with special requirements may find DU technologies that meet their needs. Customers who require "enhanced" power quality and reliability of service already utilize UPS systems with battery backup to condition the power to sensitive equipment, and many hospitals, waste treatment plants, and other emergency services providers have emergency backup systems supplied by standby generator systems. As barriers go down and technologies improve, customer-sited DU facilities could provide many of the ancillary services as well as sell excess power into the grid. Fuel cell and even diesel generators could be especially attractive for customers with requirements of heat and steam. Many of the fuel cell technologies are now looking at the residential market with small units that would be connected to the grid but supply the additional requirements for customers with special power quality needs.

8.10.4 Third-Party Service Providers

Third-party service providers could provide all the services listed earlier for the utilities and customers, in addition to selling power across the grid. In many cases, an end user does not have the expertise to operate and maintain generation systems and would prefer to purchase the services.

8.11 Conclusions

Significant portions of the evolving "smart grid" will contain both DG and DS to enable the integration of renewable energy technologies and improve the overall economics, reliability, and power quality. Strong desire to decrease the carbon footprint will accelerate this process. Advances in material technologies (in particular the use of nanotechnologies to improve PV devices), power electronic devices and systems, use of sensors and two-way communication, and the merging of internet and power grid to replace "copper and steel" by "silicon and glass" all point to an expanding role for DU.

References

Ackermann, T., *Wind Power in Power Systems*, John Wiley and Sons Ltd., Royale Institute of Technology, Stockholm, Sweden, 2005.

ANSI/IEEE Std. 1001–1998, *IEEE Guide for Interfacing Dispersed Storage and Generation Facilities with Electric Utility Systems*, IEEE Standards Coordinating Committee 23, February 9, 1989.

Archer, C.L. and Jacobson, M.Z., Evaluation of global wind power, 2005, Retrieved April 21, 2006.

Chiradeja, P. and Ramakumar, R., An approach to quantify the technical benefits of distributed generation, *IEEE Transactions on Energy Conversion*, 19(4): 764–773, 2004.

Davis, M.W., *Microturbines—An Economic and Reliability Evaluation for Commercial, Residential, and Remote Load Applications*, IEEE Transactions PE-480-PWRS-0-10-1998.

Delmerico, R.W., Miller, N.W., and Owen, E.L., *Power System Integration Strategies for Distributed Generation*, Power Systems Energy Consulting GE International, Inc., *Distributed Electricity Generation Conference*, Denver, CO, January 25, 1999.

Department of Defense Website, www.dodfuelcell.com/fcdescriptions.html

Energy Information Administration, Official Energy Statistics from the US Government, http://www.eia.doe.gov (accessed on February, 2012).

Goldstein, H.L., *Small Turbines in Distributed Utility Application Natural Gas Pressure Supply Requirements*, NREL/SP-461-21073, Golden, CO, May 1996.

Hirschenhofer, J.H., DOE forum on fuel cell technologies, *IEEE Winter Power Meeting*, New York, Parsons Corporation Presentation, February 4, 1999.

http://windenergyelectricity.com/wind-energy-electricity-production.htm

http://www.energy.ca.gov

http://www.renewableenergyworld.com/rea/news/article/2010/01/esolar-to-build-2-gw-of-solar-thermal-in-china

http://www.renewableenergyworld.com/rea/news/article/2010/06/sunpower-sets-solar-cell-efficiency-record-at-24-2

http://www.thesolarguide.com

Hurley, B., Where does the wind come from and how much is there, *Claverton Energy Conference*, Bath, U.K., October 24, 2008.

IEEE Standard P929, Draft 10, *Recommended Practice for Utility Interface of Photovoltaic (PV) Systems*, February 1999.

Jewell, W.T. and Ramakumar, R., The history of utility- interactive photovoltaic generation, *IEEE Transactions on Energy Conversion*, 3(3): 583–588, 1988.

Kirby, B., Distributed generation: A natural for ancillary services, *Distributed Electric Generation Conference*, Denver, CO, January 25, 1999.

Lenardic, D., Large-scale photovoltaic power plants ranking 1–50 *PVresources.com*, 2010.

Manning, P., With green power comes great responsibility, *Sydney Morning Herald*, October 10, 2009, Retrieved October 12, 2009.

Markman, J., It's solar power's time to shine, MSN Money, Retrieved 2008-06-0.

Oplinger, J.L., *Methodology to Assess the Market Potential of Distributed Generation*, Power Systems Energy Consulting GE International, Inc., *Distributed Electric Generation Conference*, Denver, CO, January 25, 1999.

Price, T.J., James Blyth—Britain's first modern wind power engineer, *Wind Engineering*, 29(3): 191–200, May 3, 2005. doi:10.1260/030952405774354921.

Quirky old-style contraptions make water from wind on the mesas of West Texas, Mysanantonio.com, September 23, 2007, Retrieved August 29, 2010.

Ramakumar, R., Renewable energy sources and developing countries, *IEEE Transactions on Power Apparatus and Systems*, PAS-102(2): 502–510, 1983.

Ramakumar, R., Electricity from renewable energy: A timely option, *National Seminar on Renewable Energy Sources-Electrical Power Generation*, Pune, Maharashtra, India, August 31–September 01, 2001.

Ramakumar, R. and Bigger, J.E., Photovoltaic systems, *Proceedings of the IEEE—Special Issue on Advanced Power Generation Technologies*, 81(3): 365–377, 1993.

Ramakumar, R. et al., Prospects for tapping solar energy on a large scale, *Solar Energy*, 16(2): 107–115, 1974.

Ramakumar, R. et al., Economic aspects of advanced generation systems, *Proceedings of the IEEE- Special Issue on Advanced Power Generation Technologies*, 81(3): 318–332, 1993.

REN21, Renewables 2010 Global Status Report, Paris, France, p. 19, 2010.

Smil, V., Energy at the crossroads, *Global Science Forum Conference on Scientific Challenges for Energy Research*, Paris, France, May 17–18, 2006.

Solar thermal energy: An industry report, http://www.solar-thermal.com/solar-thermal.pdf, 2008.

Southern Company Parallel Operation Requirements, Protection and Control Committee, August 4, 1998.

Technology Overviews, DOE Forum on Fuel Cell Technology, *IEEE Winter Power Meeting*, New York, February 4, 1999.

World Wind Energy Report 2009, www.wwindea.org

III

Transmission System

George G. Karady

George G. Karady received his MSEE and DE in electrical engineering from Technical University of Budapest, Budapest, Hungary. He is a power system chair professor at Arizona State University, Arizona, where he teaches electrical power and conducts research in power electronics, high-voltage techniques, and electric power systems. Previously, he served as chief consulting electrical engineer, manager of electrical systems, and chief engineer of computer technology at EBASCO Services. He was also an electrical task supervisor for the Tokomak Fusion Test Reactor project at Princeton. Prior to this, he worked for the Hydro Quebec Institute of Research as a program manager. Dr. Karady started his career at Budapest University of Technology and Economics, where he has progressed from postdoctoral student to deputy department head. He is a registered also a professional engineer in New York, New Jersey, and Quebec and has authored more than 200 technical papers.

Dr. Karady is active in IEEE. He was the chairman of Chapter/Membership's Award Committee, Education Committee's Award Subcommittee, and Working Group (WG) on Non-ceramic Insulators. He also served in the U.S. National Committee of CIGRE as vice president and secretary treasurer. He was a member of the Canadian Electrical Engineering Association and the Electrical Engineering Association of Hungary.

9

Concept of Energy Transmission and Distribution

George G. Karady
Arizona State University

The purpose of the electric transmission system is the interconnection of the electric-energy-producing power plants or generating stations with the loads. A three-phase AC system is used for most transmission lines. The operating frequency is 60 Hz in the United States and 50 Hz in Europe, Australia, and part of Asia. The three-phase system has three phase conductors. The system voltage is defined as the rms voltage between the conductors, also called line-to-line voltage. The voltage between the phase conductor and ground, called line-to-ground voltage, is equal to the line-to-line voltage divided by the square root of three. Figure 9.1 shows a typical system.

The figure shows the Phoenix area 230 kV system, which interconnects the local power plants and the substations supplying different areas of the city. The circles are the substations and the squares are the generating stations. The system contains loops that assure that each load substation is supplied by at least two lines. This assures that the outage of a single line does not cause loss of power to any customer. For example, the Aqua Fria generating station (marked: power plant) has three outgoing lines. Three high-voltage cables supply the Country Club Substation (marked: substation with cables). The Pinnacle Peak Substation (marked: substation with transmission lines) is a terminal for six transmission lines. This example shows that the substations are the node points of the electric system. The system is interconnected with the neighboring systems. As an example, one line goes to Glen Canyon and the other to Cholla from the Pinnacle Peak substation.

In the middle of the system, which is in a congested urban area, high-voltage cables are used. In open areas, overhead transmission lines are used. The cost per mile of overhead transmission lines is 6%–10% less than underground cables. The major components of the electric system, the transmission lines, and cables are described briefly later [1].

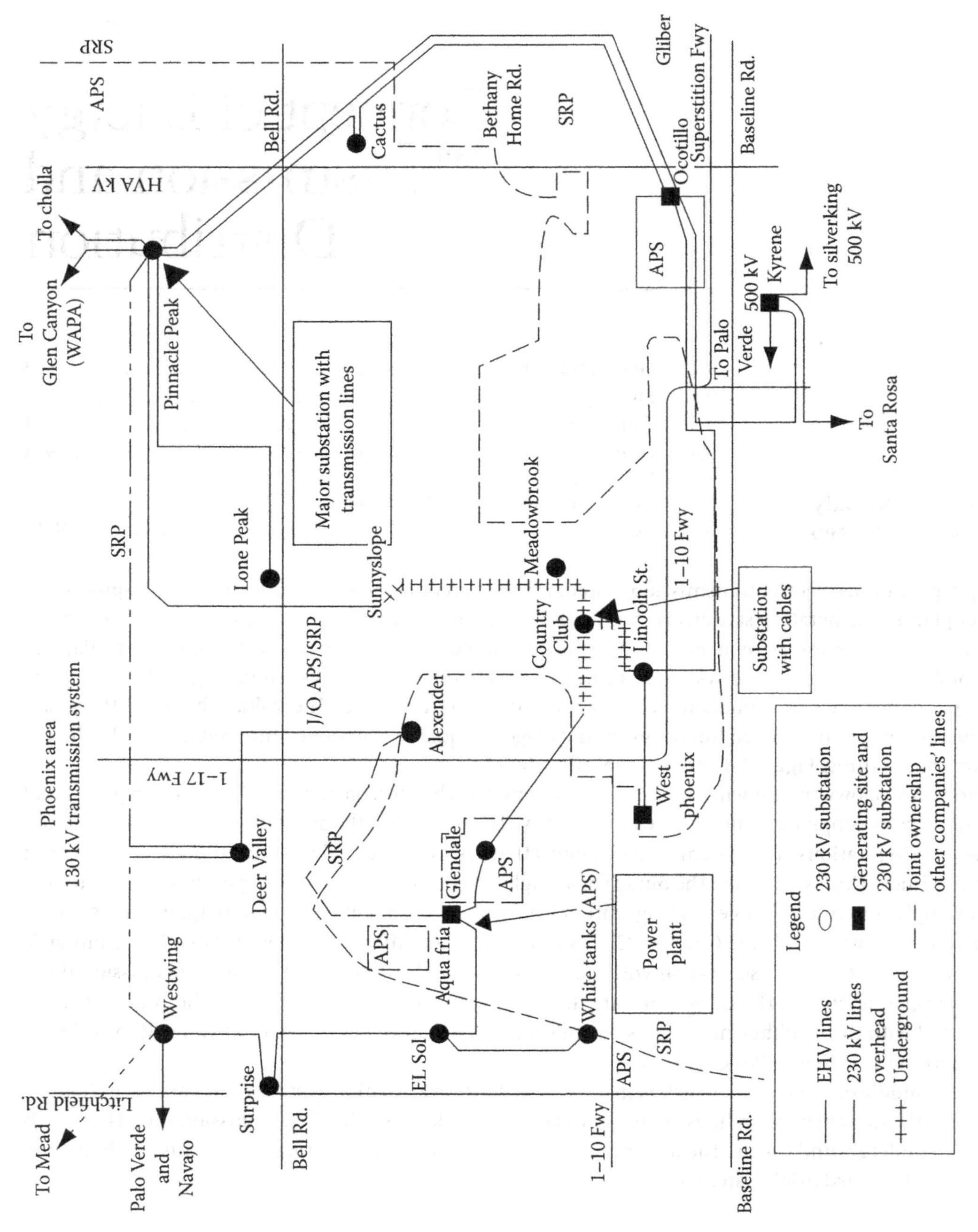

FIGURE 9.1 One line diagram of a high-voltage electric transmission system.

9.1 Generation Stations

The generating station converts the stored energy of gas, oil, coal, nuclear fuel, or water position to electric energy. The most frequently used power plants are as follows:

Thermal power plant. The fuel is pulverized coal or natural gas. Older plants may use oil. The fuel is mixed with air and burned in a boiler that generates steam. The high-pressure and high-temperature steam drives the turbine, which turns the generator that converts the mechanical energy to electric energy.

Nuclear power plant. Enriched uranium produces atomic fission that heats water and produces steam. The steam drives the turbine and generator.

Hydro power plants. A dam increases the water level on a river, which produces fast water flow to drive a hydro-turbine. The hydro-turbine drives a generator that produces electric energy.

Gas turbine. Natural gas is mixed with air and burned. This generates a high-speed gas flow that drives the turbine, which turns the generator.

Combined cycle power plant. This plant contains a gas turbine that generates electricity. The exhaust from the gas turbine is high-temperature gas. The gas supplies a heat exchanger to preheat the combustion air to the boiler of a thermal power plant. This process increases the efficiency of the combined cycle power plant. The steam drives a second turbine, which drives the second generator. This two-stage operation increases the efficiency of the plant.

9.2 Switchgear

The safe operation of the system requires switches to open lines automatically in case of a fault, or manually when the operation requires it. Figure 9.2 shows the simplified connection diagram of a generating station.

The generator is connected directly to the low-voltage winding of the main transformer. The transformer high-voltage winding is connected to the bus through a circuit breaker (CB), disconnect switch, and current transformer (CT). The generating station auxiliary power is supplied through an auxiliary transformer through a CB, disconnect switch, and CT. Generator CBs, connected between the generator and transformer, are frequently used in Europe. These breakers have to interrupt the very large short-circuit current of the generators, which results in high cost.

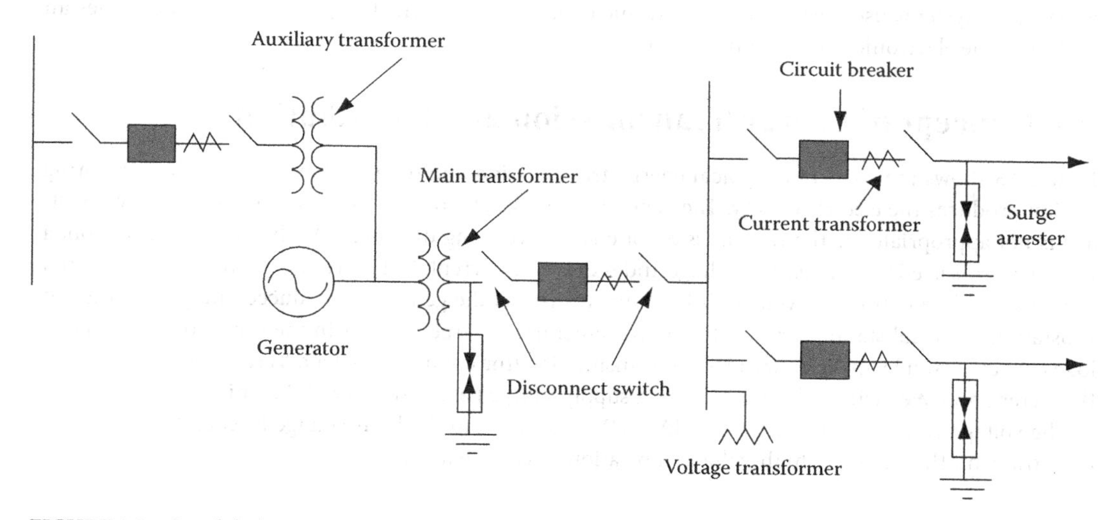

FIGURE 9.2 Simplified connection diagram of a generating station.

The high-voltage bus supplies two outgoing lines. The station is protected from lightning and switching surges by a surge arrester.

Circuit breaker (CB) is a large switch that interrupts the load and fault current. Fault detection systems automatically open the CB, but it can be operated manually.

Disconnect switch provides visible circuit separation and permits CB maintenance. It can be operated only when the CB is open, in no-load condition.

Potential transformers (PTs) and CTs reduce the voltage to 120 V and the current to 5 A and insulate the low-voltage circuit from the high voltage. These quantities are used for metering and protective relays. The relays operate the appropriate CB in case of a fault.

Surge arresters are used for protection against lightning and switching overvoltages. They are voltage dependent, nonlinear resistors.

9.3 Control Devices

In an electric system, the voltage and current can be controlled. The voltage control uses parallel connected devices, while the flow or current control requires devices connected in series with the lines.

Tap-changing transformers are frequently used to control the voltage. In this system, the turns-ratio of the transformer is regulated, which controls the voltage on the secondary side. The ordinary tap changer uses a mechanical switch. A thyristor-controlled tap changer has recently been introduced.

A shunt capacitor connected in parallel with the system through a switch is the most frequently used voltage control method. The capacitor reduces lagging-power-factor reactive power and improves the power factor. This increases voltage and reduces current and losses. Mechanical and thyristor switches are used to insert or remove the capacitor banks.

The frequently used static var compensator (SVC) consists of a switched capacitor bank and a thyristor-controlled inductance. This permits continuous regulation of reactive power.

The current of a line can be controlled by a capacitor connected in series with the line. The capacitor reduces the inductance between the sending and receiving points of the line. The lower inductance increases the line current if a parallel path is available.

In recent years, electronically controlled series compensators have been installed in a few transmission lines. This compensator is connected in series with the line, and consists of several thyristor-controlled capacitors in series or parallel, and may include thyristor-controlled inductors. Medium- and low-voltage systems use several other electronic control devices. The last part in this section gives an outline of the electronic control of the system.

9.4 Concept of Energy Transmission and Distribution

Figure 9.3 shows the concept of typical energy transmission and distribution systems. The generating station produces the electric energy. The generator voltage is around 15–25 kV. This relatively low voltage is not appropriate for the transmission of energy over long distances. At the generating station, a transformer is used to increase the voltage and reduce the current. In Figure 9.3, the voltage is increased to 500 kV and an extra-high-voltage (EHV) line transmits the generator-produced energy to a distant substation. Such substations are located on the outskirts of large cities or in the center of several large loads. As an example, in Arizona, a 500 kV transmission line connects the Palo Verde Nuclear Station to the Kyrene and Westwing substations, which supply a large part of the city of Phoenix.

The voltage is reduced at the 500/220 kV EHV substation to the high-voltage level and high-voltage lines transmit the energy to high-voltage substations located within cities.

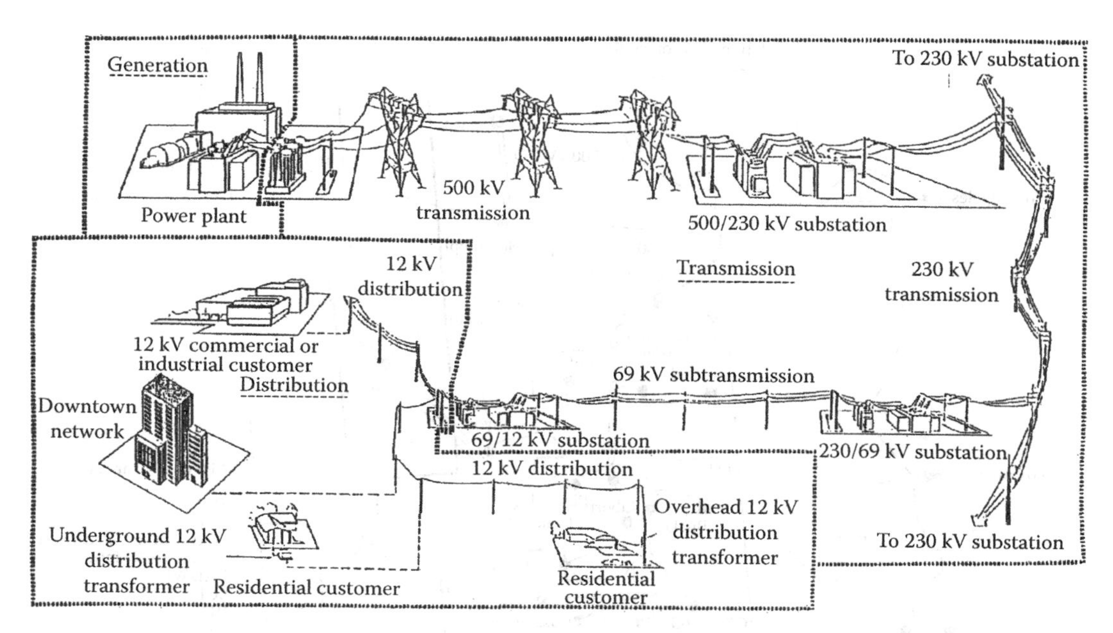

FIGURE 9.3 Concept of electric energy transmission.

At the high-voltage substation, the voltage is reduced to 69 kV. Subtransmission lines connect the high-voltage substation to many local distribution stations located within cities. Subtransmission lines are frequently located along major streets [2,3].

The voltage is reduced to 12 kV at the distribution substation. Several distribution lines emanate from each distribution substation as overhead or underground lines. Distribution lines distribute the energy along streets and alleys. Each line supplies several step-down transformers distributed along the line. The distribution transformer reduces the voltage to 230/115 V, which supplies houses, shopping centers, and other local loads. The large industrial plants and factories are supplied directly by a subtransmission line or a dedicated distribution line as shown in Figure 9.3.

The overhead transmission lines are used in open areas such as interconnections between cities or along wide roads within the city. In congested areas within cities, underground cables are used for electric energy transmission. The underground transmission system is environmentally preferable but has a significantly higher cost. In Figure 9.3 the 12 kV line is connected to a 12 kV cable, which supplies commercial or industrial customers [4]. The figure also shows 12 kV cable networks supplying downtown areas in a large city. Most newly developed residential areas are supplied by 12 kV cables through pad-mounted step-down transformers as shown in Figure 9.3.

9.4.1 High-Voltage Transmission Lines

High-voltage and extra-high-voltage (EHV) transmission lines interconnect power plants and loads, and form an electric network. Figure 9.4 shows a typical high-voltage and EHV system.

This system contains 500, 345, 230, and 115 kV lines. The figure also shows that the Arizona (AZ) system is interconnected with transmission systems in California, Utah, and New Mexico. These interconnections provide instantaneous help in case of lost generation in the AZ system. This also permits the export or import of energy, depending on the needs of the areas.

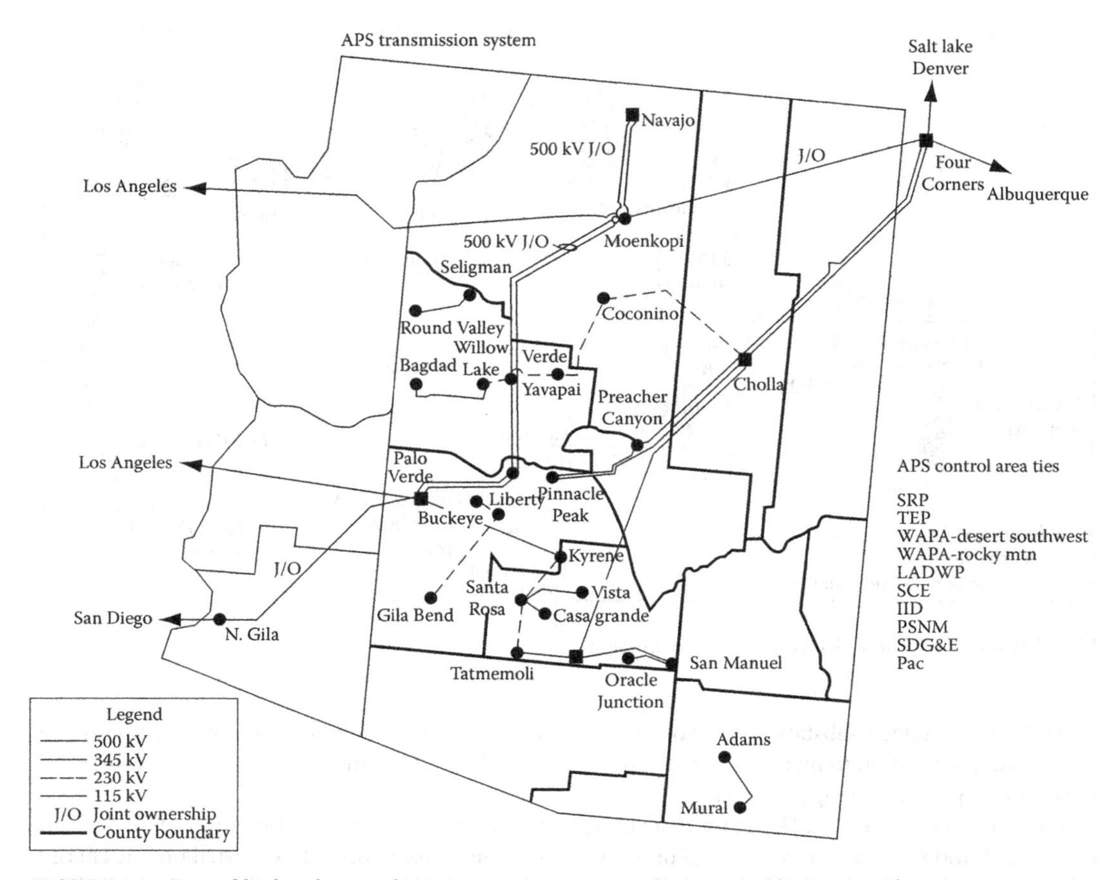

FIGURE 9.4 Typical high-voltage and EHV transmission system (Arizona Public Service, Phoenix area system).

Presently, synchronous ties (AC lines) interconnect all networks in the eastern United States and Canada. Synchronous ties also (AC lines) interconnect all networks in the western United States and Canada. Several nonsynchronous ties (DC lines) connect the East and the West. These interconnections increase the reliability of the electric supply systems.

In the United States, the nominal voltage of the high-voltage lines is between 100 and 230 kV. The voltage of the EHV lines is above 230 kV and below 800 kV. The voltage of an ultrahigh-voltage line is above 800 kV. The maximum length of high-voltage lines is around 200 miles. EHV transmission lines generally supply energy up to 400–500 miles without intermediate switching and var support. Transmission lines are terminated at the bus of a substation.

The physical arrangement of most EHV lines is similar. Figure 9.5 shows the major components of an EHV, which are as follows:

1. Tower: The figure shows a lattice, steel tower.
2. Insulator: V strings hold four bundled conductors in each phase.
3. Conductor: Each conductor is a stranded, steel-reinforced aluminum cable.
4. Foundation and grounding: Steel-reinforced concrete foundation and grounding electrodes placed in the ground.
5. Shield conductors: Two grounded shield conductors protect the phase conductors from lightning.

At lower voltages, the appearance of lines can be improved by using more aesthetically pleasing steel tubular towers. Steel tubular towers are made out of a tapered steel tube equipped with banded arms. The arms hold the insulators and the conductors. Figure 9.6 shows typical 230 kV steel tubular and

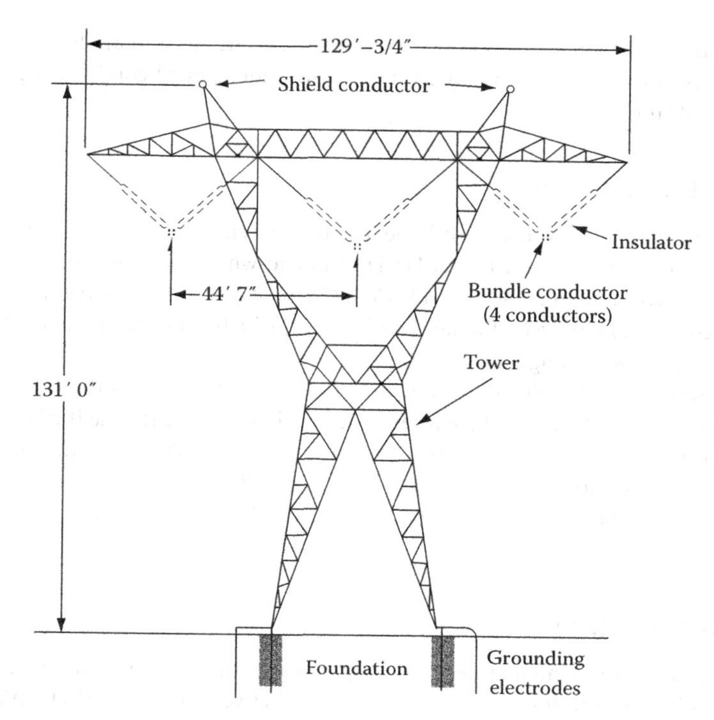

FIGURE 9.5 Typical high-voltage transmission line. (From Fink, D.G. and Beaty, H.W., *Standard Handbook for Electrical Engineering*, 11th edn., McGraw-Hill, New York, Sec. 18, 1978.)

FIGURE 9.6 Typical 230 kV constructions.

lattice double-circuit towers. Both lines carry two three-phase circuits and are built with two conductor bundles to reduce corona and radio and TV noise. Grounded shield conductors protect the phase conductors from lightning [1].

9.4.2 High-Voltage DC Lines

High-voltage DC lines are used to transmit large amounts of energy over long distances or through waterways. One of the best known is the Pacific HVDC Intertie, which interconnects southern California with Oregon. Another DC system is the ±400 kV Coal Creek-Dickenson lines. Another famous HVDC system is the interconnection between England and France, which uses underwater cables. In Canada, Vancouver Island is supplied through a DC cable.

In an HVDC system, the AC voltage is rectified and a DC line transmits the energy. At the end of the line, an inverter converts the DC voltage to AC. A typical example is the Pacific HVDC Intertie that operates with ±500 kV voltage and interconnects Southern California with the hydro stations in Oregon.

Figure 9.7 shows a guyed tower arrangement used on the Pacific HVDC Intertie. Four guy wires balance the lattice tower. The tower carries a pair of two-conductor bundles supported by suspension insulators.

9.4.3 Subtransmission Lines

Typical subtransmission lines interconnect the high-voltage substations with distribution stations within a city. The voltage of the subtransmission system is between 46, 69, and 115 kV. The maximum length of sub-transmission lines is in the range of 50–60 miles. Most subtransmission lines are located along streets and alleys. Figure 9.8 shows a typical subtransmission system.

This system operates in a looped mode to enhance continuity of service. This arrangement assures that the failure of a line will not interrupt the customer's power.

Figure 9.9 shows a typical double-circuit subtransmission line, with a wooden pole and post-type insulators. Steel tube or concrete towers are also used. The line has a single conductor in each phase. Postinsulators hold the conductors without metal cross arms. One grounded shield conductor on the top of the tower shields the phase conductors from lightning. The shield conductor is grounded at each tower. Plate or vertical tube electrodes (ground rod) are used for grounding.

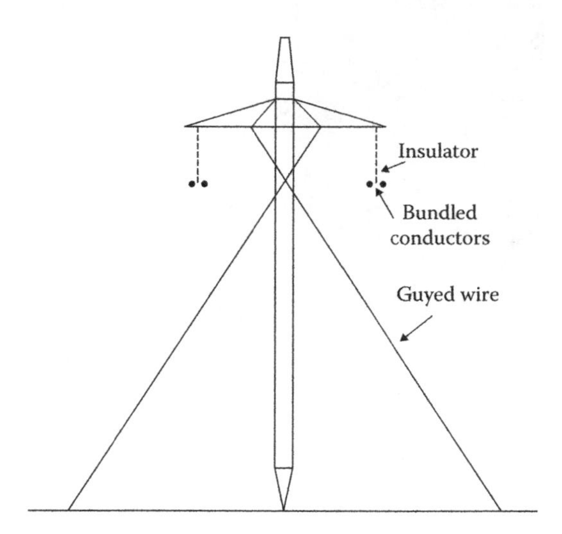

FIGURE 9.7 HVDC tower arrangement. (From Fink, D.G. and Beaty, H.W., *Standard Handbook for Electrical Engineering*, 11th edn., McGraw-Hill, New York, Sec. 18, 1978.)

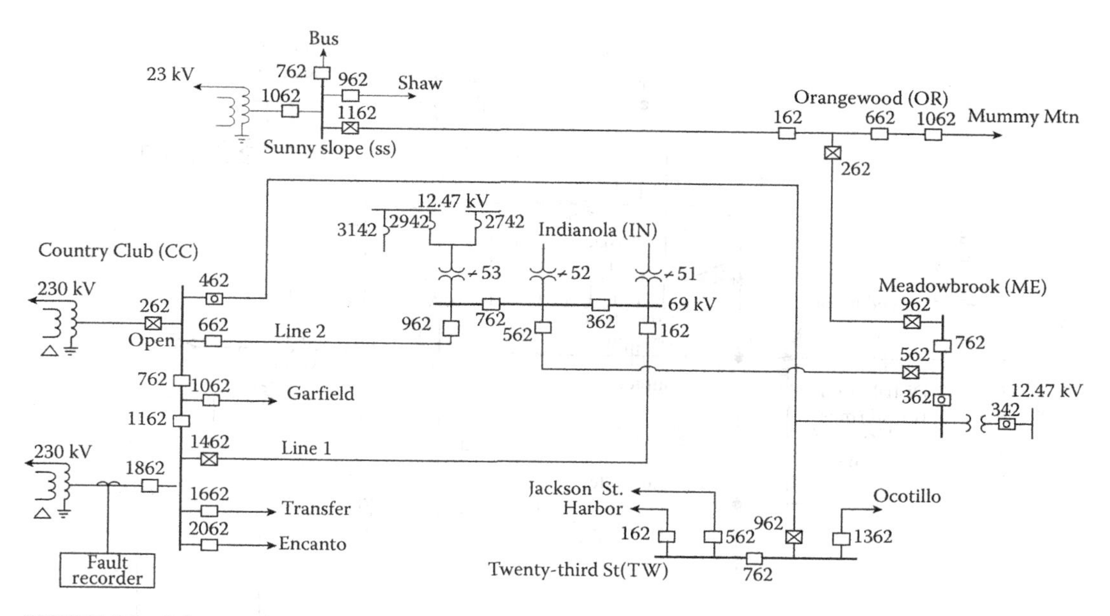

FIGURE 9.8 Subtransmission system.

FIGURE 9.9 Typical subtransmission line.

9.4.4 Distribution Lines

The distribution system is a radial system. Figure 9.10 shows the concept of a typical urban distribution system. In this system, a main three-phase feeder goes through the main street. Single-phase subfeeders supply the crossroads. Secondary mains are supplied through transformers. The consumer's service drops supply the individual loads. The voltage of the distribution system is between 4.6 and 25 kV. Distribution feeders can supply loads up to 20–30 miles.

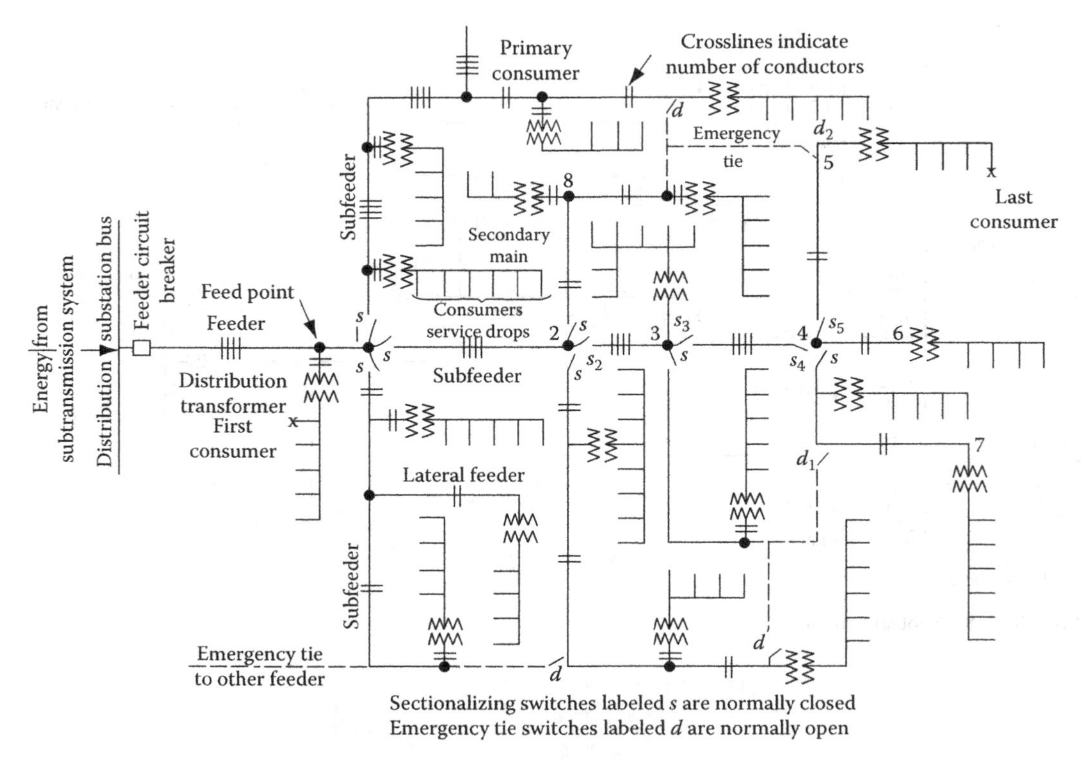

FIGURE 9.10 Concept of radial distribution system.

Many distribution lines in the United States have been built with a wood pole and cross arm. The wood is treated with an injection of creosote or other wood preservative that protects the wood from rotting and termites. Most poles are buried in a hole without foundation. Lines built recently may use a simple concrete block foundation. Small porcelain or nonceramic, pin-type insulators support the conductors. The insulator pin is grounded to eliminate leakage current, which can cause burning of the wood tower. A simple vertical copper rod is used for grounding. Shield conductors are seldom used. Figure 9.11 shows typical distribution line arrangements.

Because of the lack of space in urban areas, distribution lines are often installed on the subtransmission line towers. This is referred to as underbuild. A typical arrangement is shown in Figure 9.12.

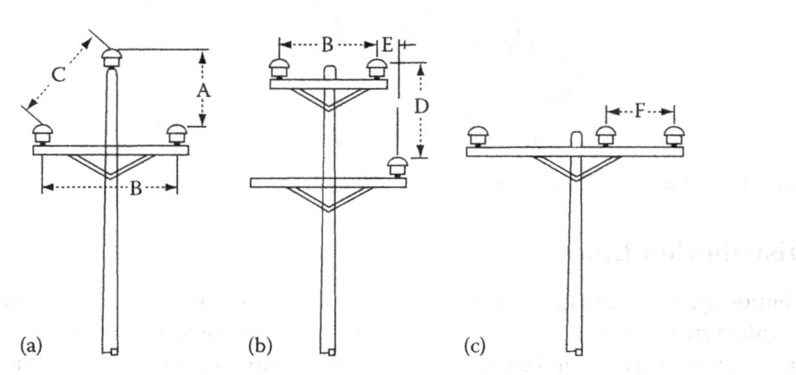

FIGURE 9.11 Distribution line arrangements: (a) pole top, (b) two arm, and (c) single arm.

FIGURE 9.12 Distribution line installed under the subtransmission line.

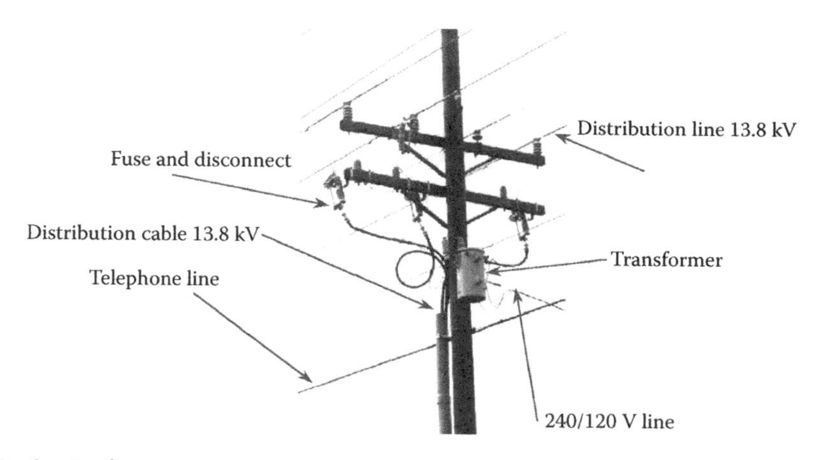

FIGURE 9.13 Service drop.

The figure shows that small porcelain insulators support the conductors. The insulators are installed on metal brackets that are bolted onto the wood tower. This arrangement reduces the right-of-way requirement and saves space.

Transformers mounted on distribution poles frequently supply individual houses or groups of houses. Figure 9.13 shows a typical transformer pole, consisting of a transformer that supplies a 240/120 V service drop, and a 13.8 kV distribution cable. The latter supplies a nearby shopping center, located on the other side of the road. The 13.8 kV cable is protected by a cut-off switch that contains a fuse mounted on a pivoted insulator. The lineman can disconnect the cable by pulling the cut-off open with a long insulated rod (hot stick).

References

1. Electric Power Research Institute, *Transmission Line Reference Book, 345 kV and Above*, Electric Power Research Institute, Palo Alto, CA, 1987.
2. Fink, D.G. and Beaty, H.W., *Standard Handbook for Electrical Engineering*, 11th edn., McGraw-Hill, New York, Sec. 18, 1978.
3. Gonen, T., *Electric Power Distribution System Engineering*, Wiley, New York, 1986.
4. Gonen, T., *Electric Power Transmission System Engineering*, Wiley, New York, 1986.

10 Transmission Line Structures

Joe C. Pohlman
Consultant

An overhead transmission line (OHTL) is a very complex, continuous, electrical/mechanical system. Its function is to transport power safely from the circuit breaker on one end to the circuit breaker on the other. It is physically composed of many individual components made up of different materials having a wide variety of mechanical properties, such as

- Flexible vs. rigid
- Ductile vs. brittle
- Variant dispersions of strength
- Wear and deterioration occurring at different rates when applied in different applications within one micro-environment or in the same application within different micro-environments

This discussion will address the nature of the structures which are required to provide the clearances between the current-carrying conductors, as well as their safe support above the earth. During this discussion, reference will be made to the following definitions:

Capability: Capacity (×) availability

Reliability level: Ability of a line (or component) to perform its expected capability

Security level: Ability of a line to restrict progressive damage after the failure of the first component

Safety level: Ability of a line to perform its function safely

10.1 Traditional Line Design Practice

Present line design practice views the support structure as an isolated element supporting half span of conductors and overhead ground wires (OHGWs) on either side of the structure. Based on the voltage level of the line, the conductors and OHGWs are configured to provide, at least, the minimum clearances mandated by the National Electrical Safety Code (NESC) (IEEE, 1990), as well as other applicable codes. This configuration is designed to control the separation of

- Energized parts from other energized parts
- Energized parts from the support structure of other objects located along the r-o-w
- Energized parts above ground

The NESC divides the United States into three large global loading zones: heavy, medium, and light and specifies radial ice thickness/wind pressure/temperature relationships to define the minimum load levels that must be used within each loading zone. In addition, the Code introduces the concept of an Overload Capacity Factor (OCF) to cover uncertainties stemming from the

- Likelihood of occurrence of the specified load
- Dispersion of structure strength
- Grade of construction
- Deterioration of strength during service life
- Structure function (suspension, dead-end, angle)
- Other line support components (guys, foundations, etc.)

Present line design practice normally consists of the following steps:

1. The owning utility prepares an agenda of loading events consisting of
   a. Mandatory regulations from the NESC and other codes
   b. Climatic events believed to be representative of the line's specific location
   c. Contingency loading events of interest; i.e., broken conductor
   d. Special requirements and expectations
   Each of these loading events is multiplied by its own OCF to cover uncertainties associated with it to produce an agenda of final ultimate design loads (see Figure 10.1).
2. A ruling span is identified based on the sag/tension requirements for the preselected conductor.
3. A structure type is selected based on past experience or on recommendations of potential structure suppliers.
4. Ultimate design loads resulting from the ruling span are applied statically as components in the longitudinal, transverse, and vertical directions, and the structure deterministically designed.
5. Using the loads and structure configuration, ground line reactions are calculated and used to accomplish the foundation design.
6. The ruling span line configuration is adjusted to fit the actual r-o-w profile.
7. Structure/foundation designs are modified to account for variation in actual span lengths, changes in elevation, and running angles.
8. Since most utilities expect the tangent structure to be the weakest link in the line system, hardware, insulators, and other accessory components are selected to be stronger than the structure.

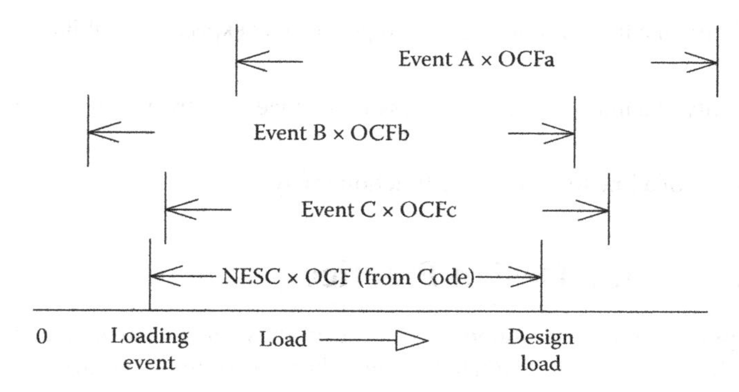

FIGURE 10.1 Development of a loading agenda.

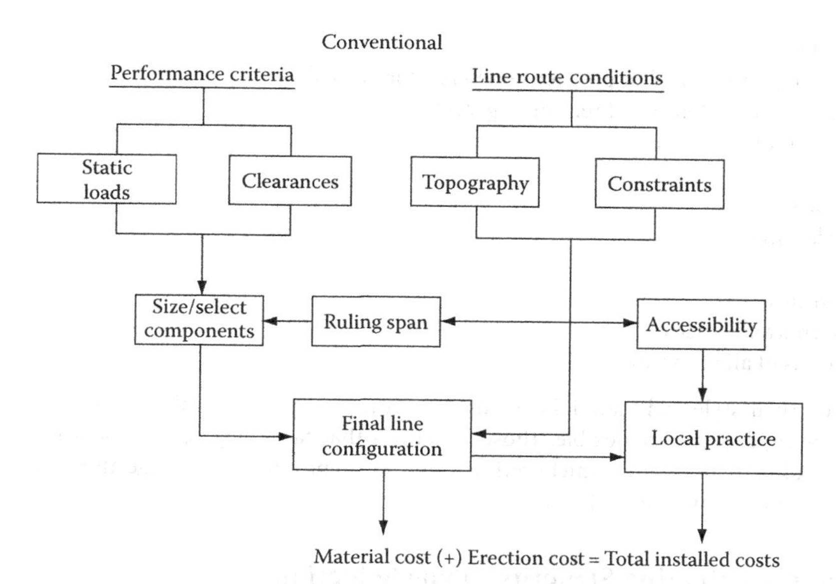

FIGURE 10.2 Search for cost effectiveness.

In as much as structure types are available in a wide variety of concepts, materials, and costs, several iterations would normally be attempted in search of the most cost effective line design based on total installed costs (see Figure 10.2).

While deterministic design using static loads applied in quadrature is a convenient mathematical approach, it is obviously not representative of the real-world exposure of the structural support system. OHTLs are tens of yards wide and miles long and usually extend over many widely variant microtopographical and microclimatic zones, each capable of delivering unique events consisting of magnitude of load at a probability-of-occurrence. That component along the r-o-w that has the highest probability of occurrence of failure from a loading event becomes the weak link in the structure design and establishes the reliability level for the total line section. Since different components are made from different materials that have different response characteristics and that wear, age, and deteriorate at different rates, it is to be expected that the weak link:

- Will likely be different in different line designs
- Will likely be different in different site locations within the same line
- Can change from one component to another over time

10.1.1 Structure Types in Use

Structures come in a wide variety of styles:

- Lattice towers
- Cantilevered or guyed poles and masts
- Framed structures
- Combinations of the above

They are available in a wide variety of materials:

- Metal
  Galvanized steel and aluminum rods, bars and rolled shapes
  Fabricated plate
  Tubes

- Concrete
  Spun with pretensioned or post-tensioned reinforcing cable
  Statically cast nontensioned reinforcing steel
  Single or multiple piece
- Wood
  As grown
  Glued laminar
- Plastics
- Composites
- Crossarms and braces
- Variations of all of the above

Depending on their style and material contents, structures vary considerably in how they respond to load. Some are rigid. Some are flexible. Those structures that can safely deflect under load and absorb energy while doing so, provide an ameliorating influence on progressive damage after the failure of the first element (Pohlman and Lummis, 1969).

10.1.2 Factors Affecting Structure Type Selection

There are usually many factors that impact on the selection of the structure type for use in an OHTL. Some of the more significant are briefly identified below.

Erection technique: It is obvious that different structure types require different erection techniques. As an example, steel lattice towers consist of hundreds of individual members that must be bolted together, assembled, and erected onto the four previously installed foundations. A tapered steel pole, on the other hand, is likely to be produced in a single piece and erected directly on its previously installed foundation in one hoist. The lattice tower requires a large amount of labor to accomplish the considerable number of bolted joints, whereas the pole requires the installation of a few nuts applied to the foundation anchor bolts plus a few to install the crossarms. The steel pole requires a large-capacity crane with a high reach which would probably not be needed for the tower. Therefore, labor needs to be balanced against the need for large, special equipment and the site's accessibility for such equipment.

Public concerns: Probably the most difficult factors to deal with arise as a result of the concerns of the general public living, working, or coming in proximity to the line. It is common practice to hold public hearings as part of the approval process for a new line. Such public hearings offer a platform for neighbors to express individual concerns that generally must be satisfactorily addressed before the required permit will be issued. A few comments demonstrate this problem.

The general public usually perceives transmission structures as "eyesores" and distractions in the local landscape. To combat this, an industry study was made in the late 1960s (Dreyfuss, 1968) sponsored by the Edison Electric Institute and accomplished by Henry Dreyfuss, the internationally recognized industrial designer. While the guidelines did not overcome all the objections, they did provide a means of satisfying certain very highly controversial installations (Pohlman and Harris, 1971).

Parents of small children and safety engineers often raise the issue of lattice masts, towers, and guys, constituting an "attractive challenge" to determined climbers, particularly youngsters.

Inspection, assessment, and maintenance: Depending on the owning utility, it is likely their in-house practices will influence the selection of the structure type for use in a specific line location. Inspections and assessment are usually made by human inspectors who use diagnostic technologies to augment their personal senses of sight and touch. The nature and location of the symptoms of critical interest are such that they can be most effectively examined from specific perspectives. Inspectors must work from the most advantageous location when making inspections. Methods can include observations from ground or fly-by patrol, climbing, bucket trucks, or helicopters. Likewise, there are certain maintenance activities that are known or believed to be required for particular structure types. The equipment necessary to maintain the structure should be taken into consideration during the structure type selection process to assure there will be no unexpected conflict between maintenance needs and r-o-w restrictions.

Future upgrading or uprating: Because of the difficulty of procuring r-o-w's and obtaining the necessary permits to build new lines, many utilities improve their future options by selecting structure types for current line projects that will permit future upgrading and/or uprating initiatives.

10.2 Current Deterministic Design Practice

Figure 10.3 shows a loading agenda for a double-circuit, 345-kV line built in the upper Midwest region of the United States on steel lattice towers. Over and above the requirements of the NESC, the utility had specified these loading events:

- A heavy wind condition (Pohlman and Harris, 1971)
- A wind on bare tower (Carton and Peyrot, 1992)
- Two maximum vertical loads on the OHGW and conductor supports (CIGRE, 1995; Ostendorp, 1998)
- Two broken wire contingencies (Dreyfuss, 1968; Pohlman and Lummis, 1969)

TANGENT AND LIGHT ANGLE SUSPENSION TOWER – 345 DOUBLE CIRCUIT

OHGW:	Two 7/16″ diameter galvanized steel strand
Conductors:	Six twin conductor bundles of 1431 KCM 45/7 ACSR
Weight span:	1650 ft
Wind span:	1100 ft
Line angle:	0° to 2°

Load Case	Load Event	Radial Ice (″)	Wind Pressure Wire (psf)	Wind Pressure Structure (psf)	Load Direction	OCF
1	NESC Heavy	1/2	4	5.1	T	2.54
					L	1.65
					V	1.27
2	One broken OHGW combined with wind and ice	1/2	8	13.0	T	1.0
					L	1.0
					V	1.0
3	One broken conductor bundle combined with wind and ice	1/2	8	13.0	T	1.0
					L	1.0
					V	1.0
4	Heavy wind	0	16	42.0	T	1.0
					L	1.0
					V	1.0
5	Wind on bare tower (no conductors or OHGW)	0	0	46.2	T	1.0
					L	1.0
					V	1.0
6	Vertical load at any OHGW support of 3,780 lbs. (not simultaneously)	0	0	0	V	1.0
7	Vertical load at any conductor support of 17,790 lbs. (not simultaneously)	0	0	0	V	1.0

FIGURE 10.3 Example of loading agenda.

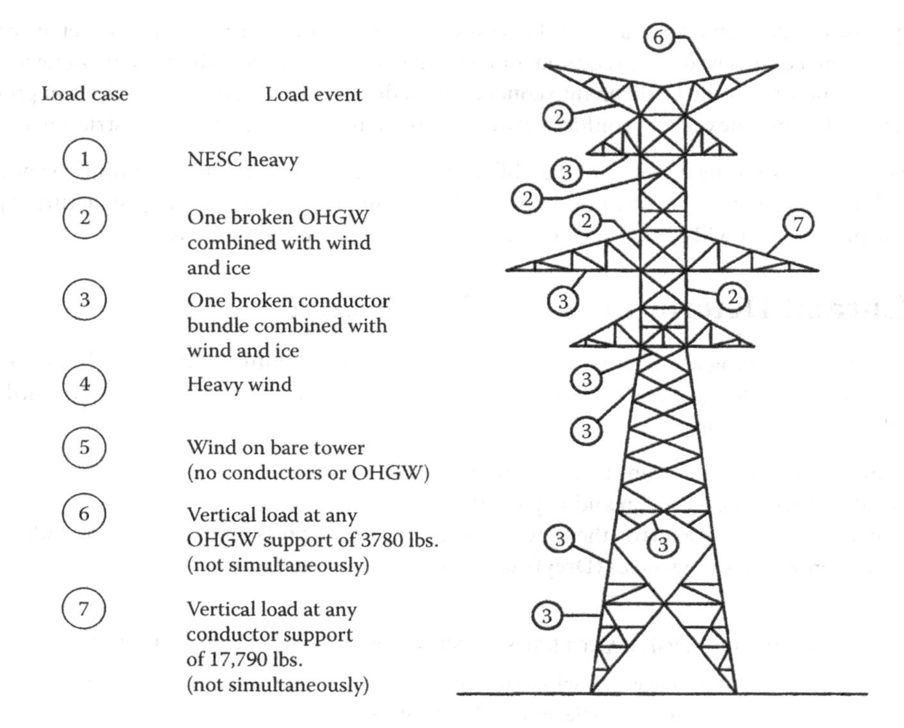

FIGURE 10.4 Results of deterministic design.

It was expected that this combination of loading events would result in a structural support design with the capability of sustaining 50-year recurrence loads likely to occur in the general area where the line was built. Figure 10.4 shows that different members of the structure, as designed, were under the control of different loading cases from this loading agenda. While interesting, this does not

- Provide a way to identify weak links in the support structure
- Provide a means for predicting performance of the line system
- Provide a framework for decision-making

10.2.1 Reliability Level

The shortcomings of deterministic design can be demonstrated by using 3D modeling/simulation technology which is in current use (Carton and Peyrot, 1992) in forensic investigation of line failures. The approach is outlined in Figure 10.5. After the structure (as designed) is properly modeled, loading events of increasing magnitude are analytically applied from different directions until the actual critical capacity for each key member of interest is reached. The probability of occurrence for those specific loading events can then be predicted for the specific location of that structure within that line section by professionals skilled in the art of micrometeorology.

Figure 10.6 shows a few of the key members in the example for Figure 10.4:

- The legs had a probability of failure in that location of once in 115 years.
- Tension chords in the conductor arm and OHGW arm had probabilities of failure of 110 and 35 years, respectively.
- A certain wind condition at an angle was found to be critical for the foundation design with a probability of occurrence at that location of once in 25 years.

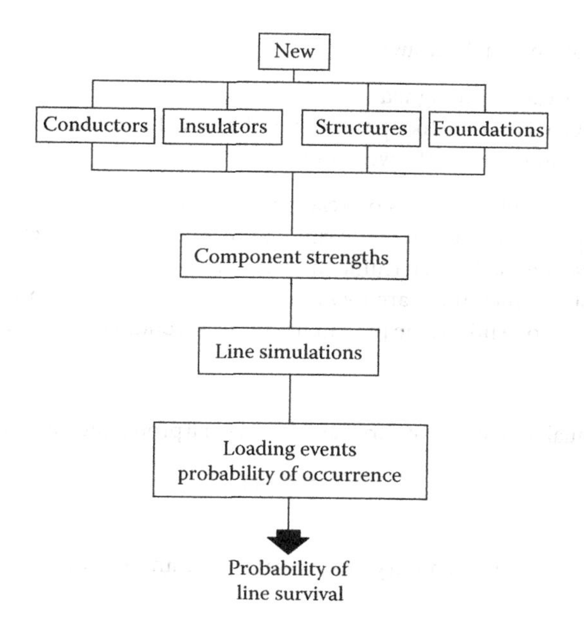

FIGURE 10.5 Line simulation study.

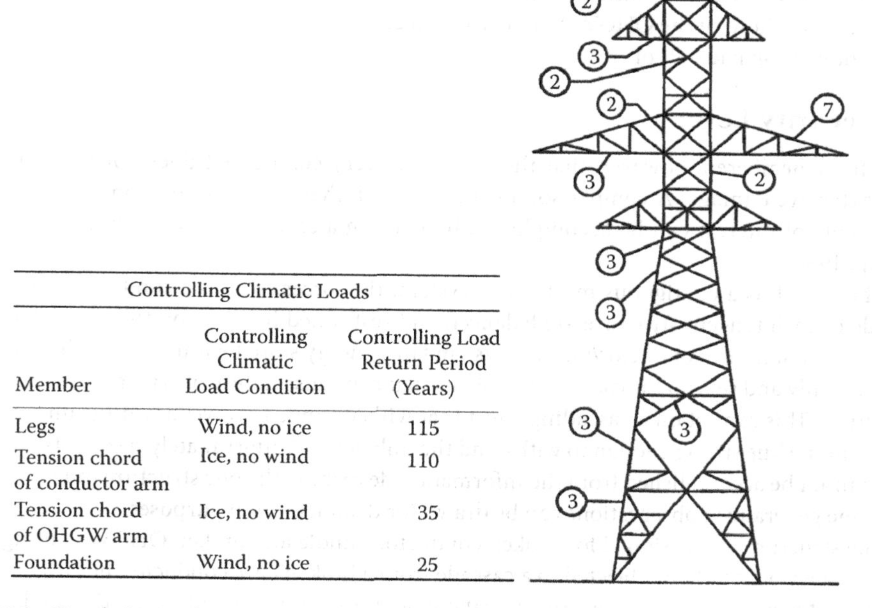

Controlling Climatic Loads		
Member	Controlling Climatic Load Condition	Controlling Load Return Period (Years)
Legs	Wind, no ice	115
Tension chord of conductor arm	Ice, no wind	110
Tension chord of OHGW arm	Ice, no wind	35
Foundation	Wind, no ice	25

FIGURE 10.6 Simulation study output.

Some interesting observations can be drawn:

- The legs were conservatively designed.
- The loss of an OHGW is a more likely event than the loss of a conductor.
- The foundation was found to be the weak link.

In addition to the interesting observations on relative reliability levels of different components within the structural support system, the output of the simulation study also provides the basis for a decision-making process which can be used to determine the cost effectiveness of management initiatives. Under the simple laws of statistics, when there are two independent outcomes to an event, the probability of the first outcome is equal to one minus the probability of the second. When these outcomes are survival and failure:

$$\text{Annual probability of survival} = 1 - \text{Annual probability of failure}$$

$$\text{Ps} = 1 - \text{Pf} \tag{10.1}$$

If it is desired to know what the probability of survival is over an extended length of time, i.e., n years of service life:

$$[\text{Ps1} \times \text{Ps2} \times \text{Ps3} \times \cdots \text{Psn}] = (\text{ps})\text{n} \tag{10.2}$$

Applying this principle to the components in the deterministic structure design and considering a 50-year service life as expected by the designers:

- The legs had a Ps of 65%
- The tension chord in the conductor arm had a Ps of 63%
- The tension chord of the OHGW arm had a Ps of 23%
- The foundation had a Ps of 13%

10.2.2 Security Level

It should be remembered, however, that the failure of every component does not necessarily progress into extensive damage. A comparison of the total risk that would result from the initial failure of components of interest can be accomplished by making a security-level check of the line design (Osterdorp, 1998).

Since the OHTL is a contiguous mechanical system, the forces from the conductors and OHGWs on one side of each tangent structure are balanced and restrained by those on the other side. When a critical component in the conductor/OHGW system fails, energy stored within the conductor system is released suddenly and sets up unbalanced transients that can cause failure of critical components at the next structure. This can set off a cascading effect that will continue to travel downline until it encounters a point in the line strong enough to withstand the unbalance. Unfortunately, a security check of the total line cannot be accomplished from the information describing the one structure in Figure 10.4; but perhaps some generalized observations can be drawn for demonstration purposes.

Since the structure was designed for broken conductor bundle and broken OHGW contingencies, it appears the line would not be subjected to a cascade from a broken bare conductor, but what if the conductor was coated with ice at the time? Since ice increases the energy trapped within the conductor prior to release, it might be of interest to determine how much ice would be "enough." Three-dimensional modeling would be employed to simulate ice coating of increasing thicknesses until the critical amount is defined. A proper micrometeorological study could then identify the probability of occurrence of a storm system capable of delivering that amount of ice at that specific location.

In the example, a wind condition with no ice was identified that would be capable of causing foundation failure once every 25 years. A security-level check would predict the amount of resulting losses and damages that would be expected from this initiating event compared to the broken-conductor-under-ice-load contingencies.

10.3 Improved Design Approaches

The above discussion indicates that technologies are available today for assessing the true capability of an OHTL that was created using the conventional practice of specifying ultimate static loads and designing a structure that would properly support them. Because there are many different structure types made from different materials, this was not always straightforward. Accordingly, many technical societies prepared guidelines on how to design the specific structure needed. These are listed in the accompanying references. The interested reader should realize that these documents are subject to periodic review and revision and should, therefore, seek the most current version.

While the technical fraternity recognizes that the mentioned technologies are useful for analyzing existing lines and determining management initiatives, something more direct for designing new lines is needed. There are many efforts under way. The most promising of these is *Improved Design Criteria of OHTLs Based on Reliability Concepts* (Ostendorp, 1998), currently under development by CIGRE Study Committee 22: Recommendations for Overhead Lines. Appendix 10.A outlines the methodology involved in words and in a diagram. The technique is based on the premise that loads and strengths are stochastic variables and the combined reliability is computable if the statistical functions of loads and strength are known. The referenced report has been circulated internationally for trial use and comment. It is expected that the returned comments will be carefully considered, integrated into the report, and the final version submitted to the International Electrotechnical Commission (IEC) for consideration as an International Standard.

10.A Appendix A: General Design Criteria—Methodology

The recommended methodology for designing transmission line components is summarized in Figure 10.7 and can be described as follows:

a. Gather preliminary line design data and available climatic data.*
b1. Select the reliability level in terms of return period of design loads. (**Note:** Some national regulations and/or codes of practice sometimes impose design requirements, directly or indirectly, that may restrict the choice offered to designers.)
b2. Select the security requirements (failure containment).
b3. List safety requirements imposed by mandatory regulations and construction and maintenance loads.
c. Calculate climatic variables corresponding to selected return period of design loads.
d1. Calculate climatic limit loadings on components.
d2. Calculate loads corresponding to security requirements.
d3. Calculate loads related to safety requirements during construction and maintenance.
e. Determine the suitable strength coordination between line components.
f. Select appropriate load and strength factors applicable to load and strength equations.
g. Calculate the characteristic strengths required for components.
h. Design line components for the above strength requirements.

* In some countries, design wind speed, such as the 50-year return period, is given in National Standards.

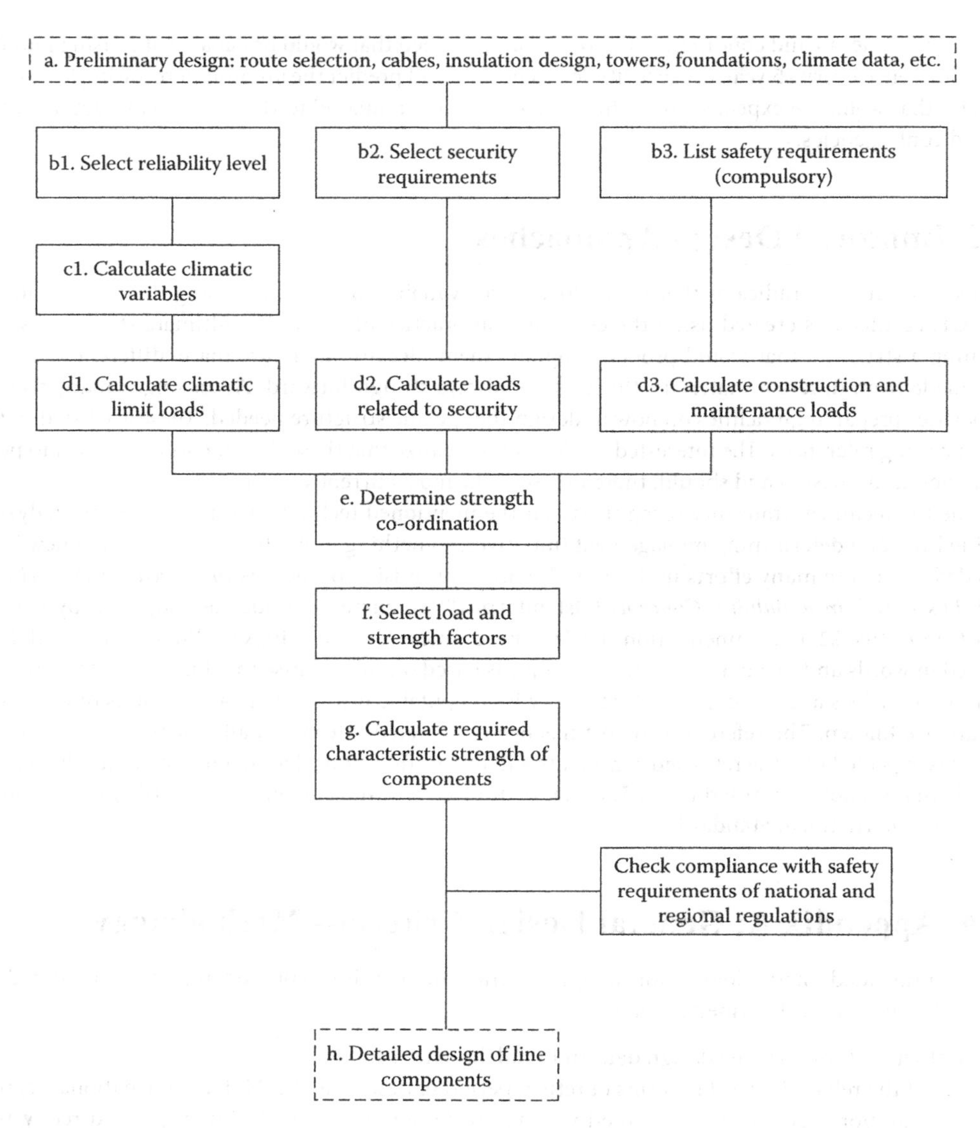

FIGURE 10.7 Methodology.

This document deals with items (b) through (g). Items (a) and (h) are not part of the scope of this document. They are identified by a dotted frame in Figure 10.7.

Source: Improved design criteria of overhead transmission lines based on reliability concepts, *CIGRE SC22 Report*, October, 1995.

References

Carton, T. and Peyrot, A., Computer aided structural and geometric design of power lines, *IEEE Transactions on Power Line System*, 7(1), 438–443, 1992.

Dreyfuss, H., *Electric Transmission Structures*, Edison Electric Institute, Publication No. 67-61, 1968.

Guide for the design and use of concrete poles, ASCE 596-6, 1987.

Guide for the design of prestressed concrete poles, ASCE/PCI Joint Commission on Concrete Poles, February, 1992. Draft.

Guide for the design of transmission towers, *ASCE Manual on Engineering Practice*, 52, 1988.

Guide for the design steel transmission poles, *ASCE Manual on Engineering Practice*, 72, 1990.

IEEE, *National Electrical Safety Code*, ANSI C-2, New York, 1990.

IEEE standard 751, *IEEE Trial-Use Design Guide for Wood Transmission Structures*, Piscataway, NJ, February 1991.

Improved design criteria of overhead transmission lines based on reliability concepts, CIGRE SC-22 Report, October 1995.

Ostendorp, M., Longitudinal loading and cascading failure assessment for transmission line upgrades, *ESMO Conference* '98, Orlando, FL, April 26–30, 1998.

Pohlman, J. and Harris, W., Tapered steel H-frames gain acceptance through scenic valley, *Electric Light and Power Magazine*, 48(vii), 55–58, 1971.

Pohlman, J. and Lummis, J., Flexible structures offer broken wire integrity at low cost, *Electric Light and Power*, 46(V, 144–148.4), 1969.

11

Insulators and Accessories

George G. Karady
Arizona State University

Richard G. Farmer
Arizona State University

Electric insulation is a vital part of an electrical power system. Although the cost of insulation is only a small fraction of the apparatus or line cost, line performance is highly dependent on insulation integrity. Insulation failure may cause permanent equipment damage and long-term outages. As an example, a short circuit in a 500kV system may result in a loss of power to a large area for several hours. The potential financial losses emphasize the importance of a reliable design of the insulation.

The insulation of an electric system is divided into two broad categories:

1. Internal insulation
2. External insulation

Apparatus or equipment has mostly internal insulation. The insulation is enclosed in a grounded housing, which protects it from the environment. External insulation is exposed to the environment. A typical example of internal insulation is the insulation for a large transformer where insulation between turns and between coils consists of solid (paper) and liquid (oil) insulation protected by a steel tank. An overvoltage can produce internal insulation breakdown and a permanent fault.

External insulation is exposed to the environment. Typical external insulation is the porcelain insulators, supporting transmission line conductors. An overvoltage caused by flashover produces only a temporary fault. The insulation is self-restoring.

This section discusses external insulation used for transmission lines and substations.

11.1 Electrical Stresses on External Insulation

The external insulation (transmission line or substation) is exposed to electrical, mechanical, and environmental stresses. The applied voltage of an operating power system produces electrical stresses. The weather and the surroundings (industry, rural dust, oceans, etc.) produce additional environmental stresses.

The conductor weight, wind, and ice can generate mechanical stresses. The insulators must withstand these stresses for long periods of time. It is anticipated that a line or substation will operate for more than 20–30 years without changing the insulators. However, regular maintenance is needed to minimize the number of faults per year. The typical number of insulation failure–caused faults is 0.5–10 per year, per 100 mile of line.

11.1.1 Transmission Lines and Substations

Transmission line and substation insulation integrity is one of the most dominant factors in power system reliability. We will describe typical transmission lines and substations to demonstrate the basic concept of external insulation application.

Figure 11.1 shows a high-voltage transmission line. The major components of the line are

1. Conductors
2. Insulators
3. Support structure tower

The insulators are attached to the tower and support the conductors. In a suspension tower, the insulators are in a vertical position or in a V-arrangement. In a dead-end tower, the insulators are in a horizontal position. The typical transmission line is divided into sections and two dead-end towers terminate each section. Between 6 and 15 suspension towers are installed between the two dead-end towers. This sectionalizing prevents the propagation of a catastrophic mechanical fault beyond each section. As an example, a tornado-caused collapse of one or two towers could create a domino effect, resulting in the collapse of many miles of towers, if there are no dead ends.

Figure 11.2 shows a lower voltage line with post-type insulators. The rigid, slanted insulator supports the conductor. A high-voltage substation may use both suspension and post-type insulators. References [1,2] give a comprehensive description of transmission lines and discuss design problems.

FIGURE 11.1 A 500 kV suspension tower with V string insulators.

FIGURE 11.2 69 kV transmission line with post insulators.

11.1.2 Electrical Stresses

The electrical stresses on insulation are created by

1. Continuous power frequency voltages
2. Temporary overvoltages
3. Switching overvoltages
4. Lightning overvoltages

11.1.2.1 Continuous Power Frequency Voltages

The insulation has to withstand normal operating voltages. The operating voltage fluctuates from changing load. The normal range of fluctuation is around ±10%. The line-to-ground voltage causes the voltage stress on the insulators. As an example, the insulation requirement of a 220 kV line is at least

$$1.1 \times \frac{220\text{ kV}}{\sqrt{3}} \cong 140\text{ kV} \tag{11.1}$$

This voltage is used for the selection of the number of insulators when the line is designed. The insulation can be laboratory tested by measuring the dry flashover voltage of the insulators. Because the line insulators are self-restoring, flashover tests do not cause any damage. The flashover voltage must be larger than the operating voltage to avoid outages. For a porcelain insulator, the required dry flashover voltage is about 2.5–3 times the rated voltage. A significant number of the apparatus standards recommend dry withstand testing of every kind of insulation to be two (2) times the rated voltage plus 1 kV for 1 min of time. This severe test eliminates most of the deficient units.

TABLE 11.1 Expected Amplitude of Temporary Overvoltages

Type of Overvoltage	Expected Amplitude	Duration
Fault overvoltages		
Effectively grounded	1.3 per unit	1 s
Resonant grounded	1.73 per unit or greater	10 s
Load rejection		
System substation	1.2 per unit	1–5 s
Generator station	1.5 per unit	3 s
Resonance	3 per unit	2–5 min
Transformer energization	1.5–2.0 per unit	1–20 s

11.1.2.2 Temporary Overvoltages

Ground faults, switching, load rejection, line energization, or resonance generates relatively long duration power frequency or close to power frequency overvoltages. The duration is from 5 s to several minutes. The expected peak amplitudes and duration are listed in Table 11.1.

The base is the crest value of the rated voltage. The dry withstand test, with two times the maximum operating voltage plus 1 kV for 1 min, is well-suited to test the performance of insulation under temporary overvoltages.

11.1.2.3 Switching Overvoltages

The opening and closing of circuit breakers causes switching overvoltages. The most frequent causes of switching overvoltages are fault or ground fault clearing, line energization, load interruption, interruption of inductive current, and switching of capacitors.

Switching produces unidirectional or oscillatory impulses with durations of 5,000–20,000 μs. The amplitude of the overvoltage varies between 1.8 and 2.5 per unit. Some modern circuit breakers use pre-insertion resistance, which reduces the overvoltage amplitude to 1.5–1.8 per unit. The base is the crest value of the rated voltage.

Switching overvoltages are calculated from computer simulations that can provide the distribution and standard deviation of the switching overvoltages. Figure 11.3 shows typical switching impulse voltages. Switching surge performance of the insulators is determined by flashover tests. The test is

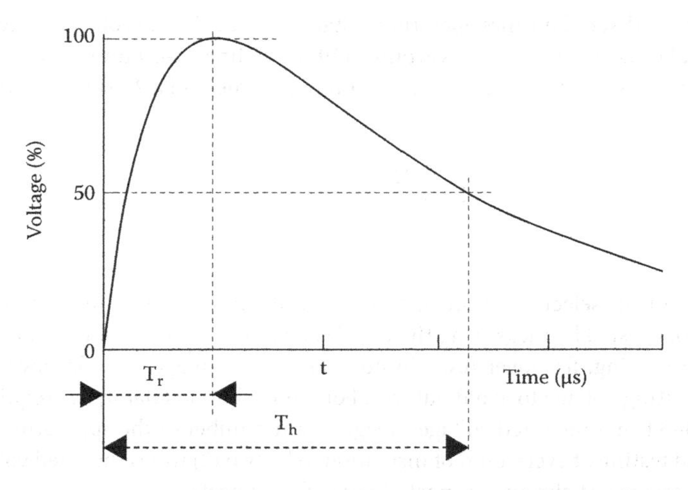

FIGURE 11.3 Switching overvoltages. T_r = 20–5000 μs, T_h < 20,000 μs, where T_r is the time-to-crest value and T_h is the time-to-half value.

performed by applying a standard impulse with a time-to-crest value of 250 μs and time-to-half value of 5000 μs. The test is repeated 20 times at different voltage levels and the number of flashovers is counted at each voltage level. These represent the statistical distribution of the switching surge impulse flashover probability. The correlation of the flashover probability with the calculated switching impulse voltage distribution gives the probability, or risk, of failure. The measure of the risk of failure is the number of flashovers expected by switching surges per year.

11.1.2.4 Lightning Overvoltages

Lightning overvoltages are caused by lightning strikes

1. To the phase conductors
2. To the shield conductor (the large current-caused voltage drop in the grounding resistance may cause flashover to the conductors [back flash])
3. To the ground close to the line (the large ground current induces voltages in the phase conductors)

Lighting strikes cause a fast-rising, short-duration, unidirectional voltage pulse. The time-to-crest value is between 0.1 and 20 μs. The time-to-half value is 20–200 μs.

The peak amplitude of the overvoltage generated by a direct strike to the conductor is very high and is practically limited by the subsequent flashover of the insulation. Shielding failures and induced voltages cause somewhat less overvoltage. Shielding failure–caused overvoltage is around 500–2000 kV. The lightning-induced voltage is generally less than 400 kV. The actual stress on the insulators is equal to the impulse voltage.

The insulator basic insulation level (BIL) is determined by using standard lightning impulses with a time-to-crest value of 1.2 μs and time-to-half value of 50 μs. This is a measure of the insulation strength for lightning. Figure 11.4 shows a typical lightning pulse.

When an insulator is tested, peak voltage of the pulse is increased until the first flashover occurs. Starting from this voltage, the test is repeated 20 times at different voltage levels and the number of flashovers is counted at each voltage level. This provides the statistical distribution of the lightning impulse flashover probability of the tested insulator.

11.1.3 Environmental Stresses

Most environmental stress is caused by weather and by the surrounding environment, such as industry, sea, or dust in rural areas. The environmental stresses affect both mechanical and electrical (M&E) performance of the line.

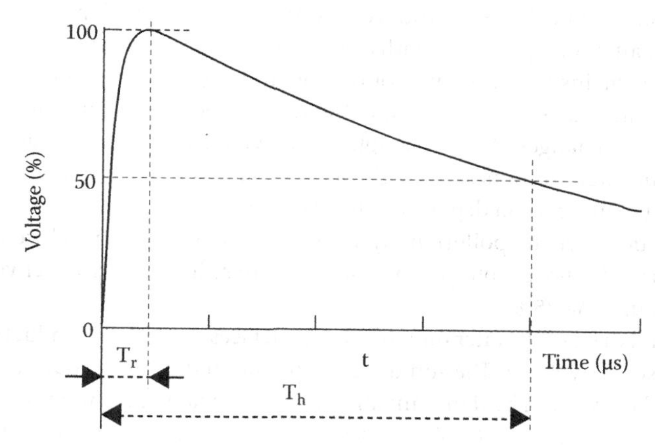

FIGURE 11.4 Lightning overvoltages. T_r = 0.1–20 μs, T_h = 20–200 μs, where T_r is the time-to-crest value and T_h is the time-to-half value.

11.1.3.1 Temperature

The temperature in an outdoor station or line may fluctuate between −50°C and +50°C, depending upon the climate. The temperature change has no effect on the electrical performance of outdoor insulation. It is believed that high temperatures may accelerate aging. Temperature fluctuation causes an increase of mechanical stresses; however, it is negligible when well-designed insulators are used.

11.1.3.2 UV Radiation

UV radiation accelerates the aging of nonceramic composite insulators, but has no effect on porcelain and glass insulators. Manufacturers use fillers and modified chemical structures of the insulating material to minimize the UV sensitivity.

11.1.3.3 Rain

Rain wets porcelain insulator surfaces and produces a thin conducting layer most of the time. This reduces the flashover voltage of the insulators. As an example, a 230 kV line may use an insulator string with 12 standard ball-and-socket-type insulators. Dry flashover voltage of this string is 665 kV and the wet flashover voltage is 502 kV. The percentage reduction is about 25%.

Nonceramic polymer insulators have a water-repellent hydrophobic surface that reduces the effects of rain. As an example, with a 230 kV composite insulator, dry flashover voltage is 735 kV and wet flashover voltage is 630 kV. The percentage reduction is about 15%. The insulator's wet flashover voltage must be higher than the maximum temporary overvoltage.

11.1.3.4 Icing

In industrialized areas, conducting water may form ice due to water-dissolved industrial pollution. An example is the ice formed from acid rain water. Ice deposits form bridges across the gaps in an insulator string that result in a solid surface. When the sun melts the ice, a conducting water layer will bridge the insulator and cause flashover at low voltages. Melting ice–caused flashover has been reported in the Quebec and Montreal areas.

11.1.3.5 Pollution

Wind drives contaminant particles into insulators. Insulators produce turbulence in airflow, which results in the deposition of particles on their surfaces. The continuous depositing of the particles increases the thickness of these deposits. However, the natural cleaning effect of wind, which blows loose particles away, limits the growth of deposits. Occasionally, rain washes part of the pollution away. The continuous depositing and cleaning produces a seasonal variation of the pollution on the insulator surfaces. However, after a long time (months, years), the deposits are stabilized and a thin layer of solid deposit will cover the insulator. Because of the cleaning effects of rain, deposits are lighter at the top of the insulators and heavier at the bottom. The development of a continuous pollution layer is compounded by chemical changes. As an example, in the vicinity of a cement factory, the interaction between the cement and water produces a tough, very sticky layer. Around highways, the wear of car tires produces a slick, tar-like carbon deposit on the insulator's surface.

Moisture, fog, and dew wet the pollution layer, dissolve the salt, and produce a conducting layer, which in turn reduces the flashover voltage. The pollution can reduce the flashover voltage of a standard insulator string by about 20%–25%.

Near the ocean, wind drives salt water onto insulator surfaces, forming a conducting salt-water layer, which reduces the flashover voltage. The sun dries the pollution during the day and forms a white salt layer. This layer is washed off even by light rain and produces a wide fluctuation in pollution levels.

The equivalent salt deposit density (ESDD) describes the level of contamination in an area. ESDD is measured by periodically washing down the pollution from selected insulators using distilled water. The resistivity of the water is measured and the amount of salt that produces the same resistivity is calculated.

TABLE 11.2 Site Severity (IEEE Definitions)

Description	ESDD (mg/cm^2)
Very light	0–0.03
Light	0.03–0.06
Moderate	0.06–0.1
Heavy	<0.1

TABLE 11.3 Typical Sources of Pollution

Pollution Type	Source of Pollutant	Deposit Characteristics	Area
Rural areas	Soil dust	High resistivity layer, effective rain washing	Large areas
Desert	Sand	Low resistivity	Large areas
Coastal area	Sea salt	Very low resistivity, easily washed by rain	10–20 km from the sea
Industrial	Steel mill, coke plants, chemical plants, generating stations, quarries	High conductivity, extremely difficult to remove, insoluble	Localized to the plant area
Mixed	Industry, highway, desert	Very adhesive, medium resistivity	Localized to the plant area

The obtained mg value of salt is divided by the surface area of the insulator. This number is the ESDD. The pollution severity of a site is described by the average ESDD value, which is determined by several measurements.

Table 11.2 shows the criteria for defining site severity.

The contamination level is light or very light in most parts of the United States and Canada. Only the seashores and heavily industrialized regions experience heavy pollution. Typically, the pollution level is very high in Florida and on the southern coast of California. Heavy industrial pollution occurs in the industrialized areas and near large highways. Table 11.3 gives a summary of the different sources of pollution.

The flashover voltage of polluted insulators has been measured in laboratories. The correlation between the laboratory results and field experience is weak. The test results provide guidance, but insulators are selected using practical experience.

11.1.3.6 Altitude

The insulator's flashover voltage is reduced as altitude increases. Above 1500 ft, an increase in the number of insulators should be considered. A practical rule is a 3% increase of clearance or insulator strings' length per 1000 ft as the elevation increases.

11.1.4 Mechanical Stresses

Suspension insulators need to carry the weight of the conductors and the weight of occasional ice and wind loading.

In northern areas and in higher elevations, insulators and lines are frequently covered by ice in the winter. The ice produces significant mechanical loads on the conductor and on the insulators. The transmission line insulators need to support the conductor's weight and the weight of the ice in the adjacent spans. This may increase the mechanical load by 20%–50%.

The wind produces a horizontal force on the line conductors. This horizontal force increases the mechanical load on the line. The wind-force-produced load has to be added vectorially to the weight-produced forces. The design load will be the larger of the combined wind and weight, or ice and wind load.

The dead-end insulators must withstand the longitudinal load, which is higher than the simple weight of the conductor in the half span.

A sudden drop in the ice load from the conductor produces large-amplitude mechanical oscillations, which cause periodic oscillatory insulator loading (stress changes from tension to compression and back).

The insulator's 1 min tension strength is measured and used for insulator selection. In addition, each cap-and-pin or ball-and-socket insulator is loaded mechanically for 1 min and simultaneously energized. This M&E value indicates the quality of insulators. The maximum load should be around 50% of the M&E load.

The Bonneville Power Administration uses the following practical relation to determine the required M&E rating of the insulators:

1. M&E > 5 * Bare conductor weight/span
2. M&E > Bare conductor weight + Weight of 3.81 cm (1.5 in.) of ice on the conductor (3 lb/ft^2)
3. M&E > 2 * Bare conductor weight + Weight of 0.63 cm (1/4 in.) of ice on the conductor and loading from a wind of 1.8 kg/ft^2 (4 lb/ft^2)

The required M&E value is calculated from all equations above and the largest value is used.

11.2 Ceramic (Porcelain and Glass) Insulators

11.2.1 Materials

Porcelain is the most frequently used material for insulators. Insulators are made of wet, processed porcelain. The fundamental materials used are a mixture of feldspar (35%), china clay (28%), flint (25%), ball clay (10%), and talc (2%). The ingredients are mixed with water. The resulting mixture has the consistency of putty or paste and is pressed into a mold to form a shell of the desired shape. The alternative method is formation by extrusion bars that are machined into the desired shape. The shells are dried and dipped into a glaze material. After glazing, the shells are fired in a kiln at about 1200°C. The glaze improves the mechanical strength and provides a smooth, shiny surface. After a cooling-down period, metal fittings are attached to the porcelain with Portland cement. Reference [3] presents the history of porcelain insulators and discusses the manufacturing procedure.

Toughened glass is also frequently used for insulators [4]. The melted glass is poured into a mold to form the shell. Dipping into hot and cold baths cools the shells. This thermal treatment shrinks the surface of the glass and produces pressure on the body, which increases the mechanical strength of the glass. Sudden mechanical stresses, such as a blow by a hammer or bullets, will break the glass into small pieces. The metal end fitting is attached by alumina cement.

11.2.2 Insulator Strings

Most high-voltage lines use ball-and-socket-type porcelain or toughened glass insulators. These are also referred to as "cap and pin." The cross section of a ball-and-socket-type insulator is shown in Figure 11.5.

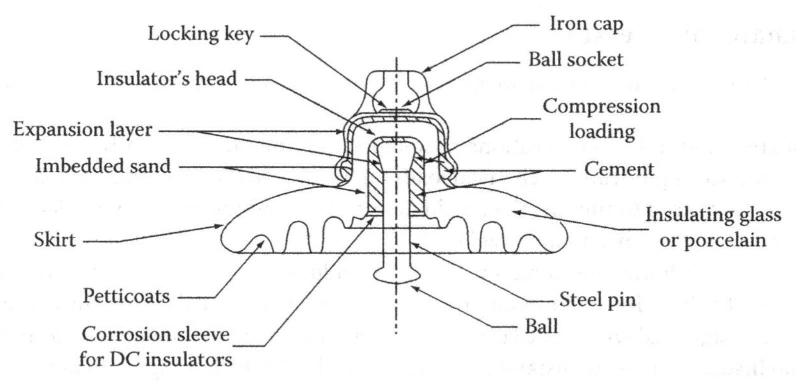

FIGURE 11.5 Cross section of a standard ball-and-socket insulator.

TABLE 11.4 Technical Data of a Standard Insulator

Diameter	25.4 cm	(10 in.)
Spacing	14.6 cm	(5-3/4 in.)
Leakage distance	305 cm	(12 ft)
Typical operating voltage	10 kV	
Mechanical strength	75 kN	(15 klb)

Table 11.4 shows the basic technical data of these insulators.

The porcelain skirt provides insulation between the iron cap and steel pin. The upper part of the porcelain is smooth to promote rain washing and cleaning of the surface. The lower part is corrugated, which prevents wetting and provides a longer protected leakage path. Portland cement attaches the cup and pin. Before the application of the cement, the porcelain is sandblasted to generate a rough surface. A thin expansion layer (e.g., bitumen) covers the metal surfaces. The loading compresses the cement and provides high mechanical strength.

The metal parts of the standard ball-and-socket insulator are designed to fail before the porcelain fails as the mechanical load increases. This acts as a mechanical fuse protecting the tower structure.

The ball-and-socket insulators are attached to each other by inserting the ball in the socket and securing the connection with a locking key. Several insulators are connected together to form an insulator string. Figure 11.6 shows a ball-and-socket insulator string and the clevis-type string, which is used less frequently for transmission lines.

Fog-type, long leakage distance insulators are used in polluted areas, close to the ocean, or in industrial environments. Figure 11.7 shows representative fog-type insulators, the mechanical strength of

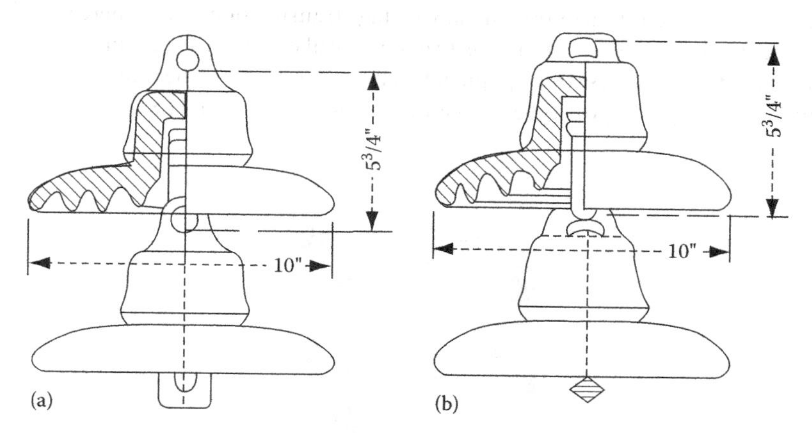

FIGURE 11.6 Insulator string: (a) clevis type and (b) ball-and-socket type.

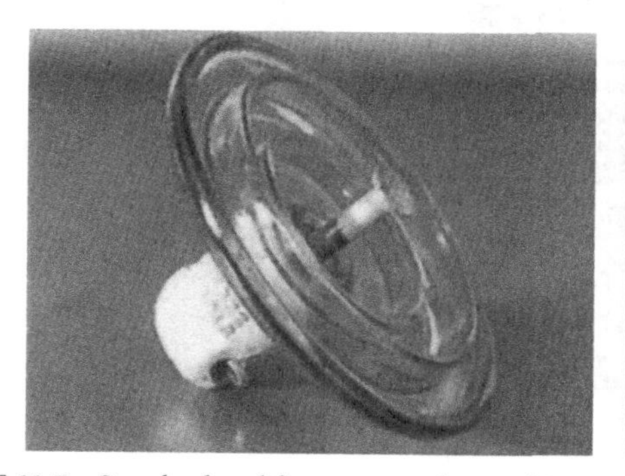

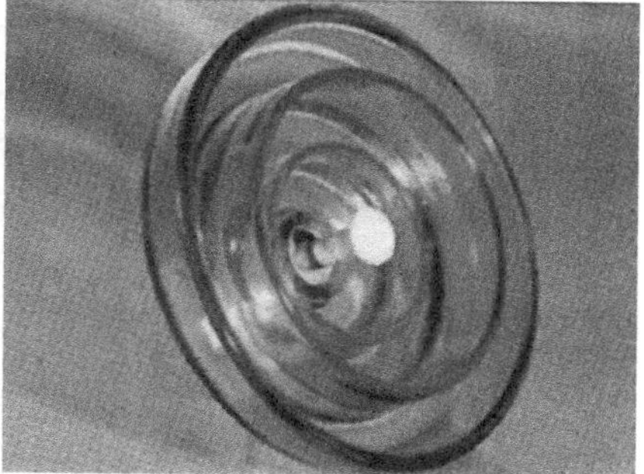

FIGURE 11.7 Standard and fog-type insulators. (Courtesy of Sediver, Inc., Nanterre, France.)

TABLE 11.5 Typical Number of Standard (5-1/4 ft × 10 in.) Insulators at Different Voltage Levels

Line Voltage (kV)	Number of Standard Insulators
69	4–6
115	7–9
138	8–10
230	12
287	15
345	18
500	24
765	30–35

which is higher than standard insulator strength. As an example, a 6 1/2 × 12 1/2 fog-type insulator is rated to 180 kN (40 klb) and has a leakage distance of 50.1 cm (20 in.).

Insulator strings are used for high-voltage transmission lines and substations. They are arranged vertically on support towers and horizontally on dead-end towers. Table 11.5 shows the typical number of insulators used by utilities in the United States and Canada in lightly polluted areas.

11.2.3 Post-Type Insulators

Post-type insulators are used for medium- and low-voltage transmission lines, where insulators replace the cross-arm (Figure 11.3). However, the majority of post insulators are used in substations where insulators support conductors, bus bars, and equipment. A typical example is the interruption chamber of a live tank circuit breaker. Typical post-type insulators are shown in Figure 11.8.

FIGURE 11.8 Post insulators.

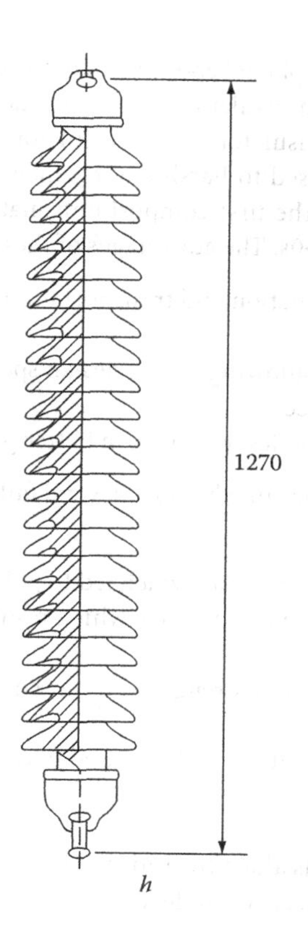

FIGURE 11.9 Long rod insulator.

Older post insulators are built somewhat similar to cap-and-pin insulators, but with hardware that permits stacking of the insulators to form a high-voltage unit. These units can be found in older stations. Modern post insulators consist of a porcelain column, with weather skirts or corrugation on the outside surface to increase leakage distance. For indoor use, the outer surface is corrugated. For outdoor use, a deeper weather shed is used. The end-fitting seals the inner part of the tube to prevent water penetration. Figure 11.8 shows a representative unit used at a substation. Equipment manufacturers use the large post-type insulators to house capacitors, fiber-optic cables and electronics, current transformers, and operating mechanisms. In some cases, the insulator itself rotates and operates disconnect switches.

Post insulators are designed to carry large compression loads, smaller bending loads, and small tension stresses.

11.2.4 Long Rod Insulators

The long rod insulator is a porcelain rod with an outside weather shed and metal end fittings. The long rod is designed for tension load and is applied on transmission lines in Europe. Figure 11.9 shows a typical long rod insulator. These insulators are not used in the United States because vandals may shoot the insulators, which will break and cause outages. The main advantage of the long rod design is the elimination of metal parts between the units, which reduces the insulator's length.

11.3 Nonceramic (Composite) Insulators

Nonceramic insulators use polymers instead of porcelain. High-voltage composite insulators are built with mechanical load-bearing fiberglass rods, which are covered by polymer weather sheds to assure high electrical strength.

The first insulators were built with bisphenol epoxy resin in the mid-1940s and are still used in indoor applications. Cycloaliphatic epoxy resin insulators were introduced in 1957. Rods with weather sheds were molded and cured to form solid insulators. These insulators were tested and used in England for several years. Most of them were exposed to harsh environmental stresses and failed. However, they have been successfully used indoors. The first composite insulators, with fiberglass rods and rubber weather sheds, appeared in the mid-1960s. The advantages of these insulators are as follows [5–7]:

- Lightweight, which lowers construction and transportation costs
- More vandalism resistant
- Higher strength-to-weight ratio, allowing longer design spans
- Better contamination performance
- Improved transmission line aesthetics, resulting in better public acceptance of a new line

However, early experiences were discouraging because several failures were observed during operation. Typical failures experienced were

- Tracking and erosion of the shed material, which led to pollution and caused flashover
- Chalking and crazing of the insulator's surface, which resulted in increased contaminant collection, arcing, and flashover
- Reduction of contamination flashover strength and subsequent increased contamination-induced flashover
- Deterioration of mechanical strength, which resulted in confusion in the selection of mechanical line loading
- Loosening of end fittings
- Bonding failures and breakdowns along the rod–shed interface
- Water penetration followed by electrical failure

As a consequence of reported failures, an extensive research effort led to second- and third-generation nonceramic transmission line insulators. These improved units have tracking-free sheds, better corona resistance, and slip-free end fittings. A better understanding of failure mechanisms and of mechanical strength–time dependency has resulted in newly designed insulators that are expected to last 20–30 years [8,9]. Increased production quality control and automated manufacturing technology has further improved the quality of these third-generation nonceramic transmission line insulators.

11.3.1 Composite Suspension Insulators

A cross section of a third-generation composite insulator is shown in Figure 11.10. The major components of a composite insulator are

- End fittings
- Corona ring(s)
- Fiberglass-reinforced plastic rod
- Interface between shed and sleeve
- Weather shed

11.3.1.1 End Fittings

End fittings connect the insulator to a tower or conductor. It is a heavy metal tube with an oval eye, socket, ball, tongue, and a clevis ending. The tube is attached to a fiberglass rod. The duty of the end fitting is to provide a reliable, nonslip attachment without localized stress in the fiberglass rod. Different manufacturers use different technologies. Some methods are as follows:

1. The ductile galvanized iron-end fitting is wedged and glued with epoxy to the rod.
2. The galvanized forged steel-end fitting is swaged and compressed to the rod.

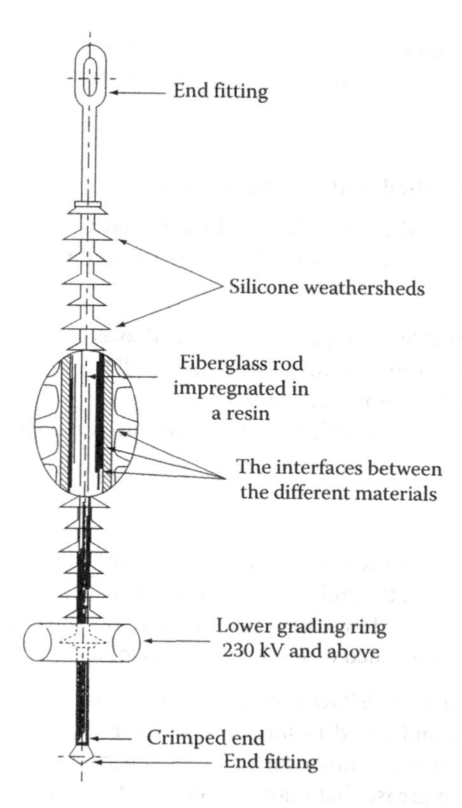

FIGURE 11.10 Cross section of a typical composite insulator. (From *Toughened Glass Insulators*, Sediver, Inc., Nanterre, France, 1993. With permission.)

3. The malleable cast iron, galvanized forged steel, or aluminous bronze-end fitting is attached to the rod by controlled swaging. The material is selected according to the corrosion resistance requirement. The end-fitting coupling zone serves as a mechanical fuse and determines the strength of the insulator.
4. High-grade forged steel or ductile iron is crimped to the rod with circumferential compression.

The interface between the end fitting and the shed material must be sealed to avoid water penetration. Another technique, used mostly in distribution insulators, involves the weather shed overlapping the end fitting.

11.3.1.2 Corona Ring(s)

Electrical field distribution along a nonceramic insulator is nonlinear and produces very high electric fields near the end of the insulator. High fields generate corona and surface discharges, which are the source of insulator aging. Above 230 kV, each manufacturer recommends aluminum corona rings be installed at the line end of the insulator. Corona rings are used at both ends at higher voltages (>500 kV).

11.3.1.3 Fiberglass-Reinforced Plastic Rod

The fiberglass is bound with epoxy or polyester resin. Epoxy produces better-quality rods but polyester is less expensive. The rods are manufactured in a continuous process or in a batch mode, producing the required length. The even distribution of the glass fibers assures equal loading, and the uniform impregnation assures good bonding between the fibers and the resin. To improve quality, some manufacturers use E-glass to avoid brittle fractures. Brittle fracture can cause sudden shattering of the rod.

11.3.1.4 Interfaces between Shed and Fiberglass Rod

Interfaces between the fiberglass rod and weather shed should have no voids. This requires an appropriate interface material that assures bonding of the fiberglass rod and weather shed. The most frequently used techniques are as follows:

1. The fiberglass rod is primed by an appropriate material to assure the bonding of the sheds.
2. Silicon rubber or ethylene propylene diene monomer (EPDM) sheets are extruded onto the fiberglass rod, forming a tube-like protective covering.
3. The gap between the rod and the weather shed is filled with silicon grease, which eliminates voids.

11.3.1.5 Weather Shed

All high-voltage insulators use rubber weather sheds installed on fiberglass rods. The interface between the weather shed, fiberglass rod, and the end fittings is carefully sealed to prevent water penetration. The most serious insulator failure is caused by water penetration to the interface.

The most frequently used weather shed technologies are as follows:

1. Ethylene propylene copolymer (EPM) and silicon rubber alloys, where hydrated-alumina fillers are injected into a mold and cured to form the weather sheds. The sheds are threaded to the fiberglass rod under vacuum. The inner surface of the weather shed is equipped with O-ring-type grooves filled with silicon grease that seals the rod–shed interface. The gap between the end fittings and the sheds is sealed by axial pressure. The continuous slow leaking of the silicon at the weather shed junctions prevents water penetration.
2. High-temperature vulcanized (HTV) silicon rubber sleeves are extruded on the fiberglass surface to form an interface. The silicon rubber weather sheds are injection-molded under pressure and placed onto the sleeved rod at a predetermined distance. The complete subassembly is vulcanized at high temperatures in an oven. This technology permits the variation of the distance between the sheds.
3. The sheds are directly injection molded under high pressure and high temperature onto the primed rod assembly. This assures simultaneous bonding to both the rod and the end fittings. Both EPDM and silicon rubber are used. This one-piece molding assures reliable sealing against moisture penetration.
4. One piece of silicon or EPDM rubber shed is molded directly to the fiberglass rod. The rubber contains fillers and additive agents to prevent tracking and erosion.

11.3.2 Composite Post Insulators

The construction and manufacturing method of post insulators is similar to that of suspension insulators. The major difference is in the end fittings and the use of a larger diameter fiberglass rod. The latter is necessary because bending is the major load on these insulators. The insulators are flexible, which permits bending in case of sudden overload. A typical post-type insulator used for 69 kV lines is shown in Figure 11.11.

Post-type insulators are frequently used on transmission lines. Development of station-type post insulators has just begun. The major problem is the fabrication of high strength, large diameter fiberglass tubes and sealing of the weather shed.

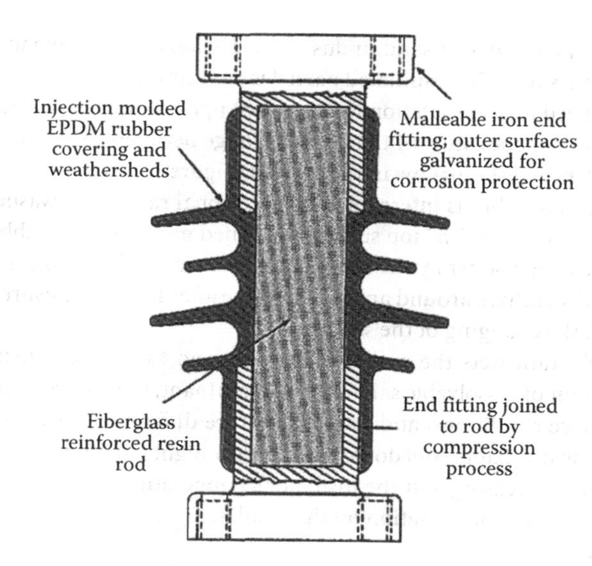

FIGURE 11.11 Post-type composite insulator. (From *Toughened Glass Insulators*, Sediver, Inc., Nanterre, France, 1993. With permission.)

11.4 Insulator Failure Mechanism

11.4.1 Porcelain Insulators

Cap-and-pin porcelain insulators are occasionally destroyed by direct lightning strikes, which generate a very steep wave front. Steep-front waves break down the porcelain in the cap, cracking the porcelain. The penetration of moisture results in leakage currents and short circuits of the unit.

Mechanical failures also crack the insulator and produce short circuits. The most common cause is water absorption by the Portland cement used to attach the cap to the porcelain. Water absorption expands the cement, which in turn cracks the porcelain. This reduces the mechanical strength, which may cause separation and line dropping.

Short circuits of the units in an insulator string reduce the electrical strength of the string, which may cause flashover in polluted conditions.

Glass insulators use alumina cement, which reduces water penetration and the head-cracking problem. A great impact, such as a bullet, can shatter the shell, but will not reduce the mechanical strength of the unit.

The major problem with the porcelain insulators is pollution, which may reduce the flashover voltage under the rated voltages. Fortunately, most areas of the United States are lightly polluted. However, some areas with heavy pollution experience flashover regularly.

11.4.2 Insulator Pollution

Insulation pollution is a major cause of flashovers and of long-term service interruptions. Lightning-caused flashovers produce short circuits. The short-circuit current is interrupted by the circuit breaker and the line is reclosed successfully. The line cannot be successfully reclosed after pollution-caused flashover because the contamination reduces the insulation's strength for a long time. Actually, the insulator must dry before the line can be reclosed.

11.4.2.1 Ceramic Insulators

Pollution-caused flashover is an involved process that begins with the pollution source. Some sources of pollution are salt spray from an ocean, salt deposits in the winter, dust and rubber particles during the summer from highways and desert sand, industrial emissions, engine exhaust, fertilizer deposits, and generating station emissions. Contaminated particles are carried in the wind and deposited on the insulator's surface. The speed of accumulation is dependent upon wind speed, line orientation, particle size, material, and insulator shape. Most of the deposits lodge between the insulator's ribs and behind the cap because of turbulence in the airflow in these areas (Figure 11.12).

The deposition is continuous, but is interrupted by occasional rain. Rain washes the pollution away and high winds clean the insulators. The top surface is cleaned more than the ribbed bottom. The horizontal and V strings are cleaned better by the rain than the I strings. The deposit on the insulator forms a well-dispersed layer and stabilizes around an average value after longer exposure times. However, this average value varies with the changing of the seasons.

Fog, dew, mist, or light rain wets the pollution deposits and forms a conductive layer. Wetting is dependent upon the amount of dissolvable salt in the contaminant, the nature of the insoluble material, duration of wetting, surface conditions, and the temperature difference between the insulator and its surroundings. At night, the insulators cool down with the low night temperatures. In the early morning, the air temperature begins increasing, but the insulator's temperature remains constant. The temperature difference accelerates water condensation on the insulator's surface. Wetting of the contamination layer starts leakage currents.

Leakage current density depends upon the shape of the insulator's surface. Generally, the highest current density is around the pin. The current heats the conductive layer and evaporates the water at the areas with high current density. This leads to the development of dry bands around the pin. The dry bands modify the voltage distribution along the surface. Because of the high resistance of the dry bands, it is across them that most of the voltages will appear. The high voltage produces local arcing. Short arcs (Figure 11.13) will bridge the dry bands.

Leakage current flow will be determined by the voltage drop of the arcs and by the resistance of the wet layer in series with the dry bands. The arc length may increase or decrease, depending on the layer resistance. Because of the large layer resistance, the arc first extinguishes, but further wetting reduces the resistance, which leads to increases in arc length. In adverse conditions, the level of contamination is high and the layer resistance becomes low because of intensive wetting. After several arcing periods, the length of the dry band will increase and the arc will extend across the insulator. This contamination causes flashover.

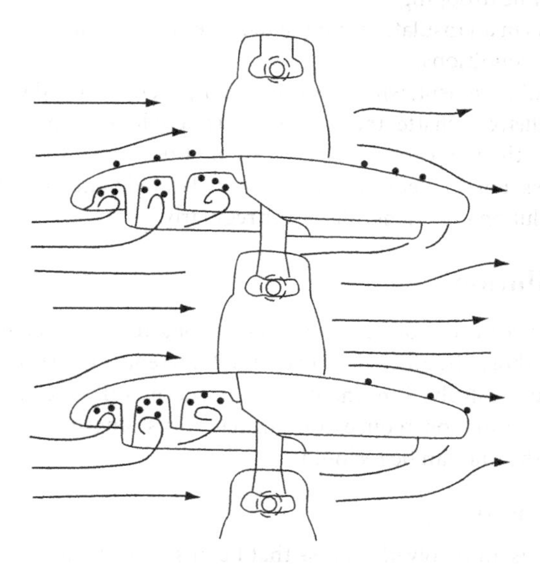

FIGURE 11.12 Deposit accumulation. (From *Application Guide for Composite Suspension Insulators*, Sediver, Inc., York, SC, 1993. With permission.)

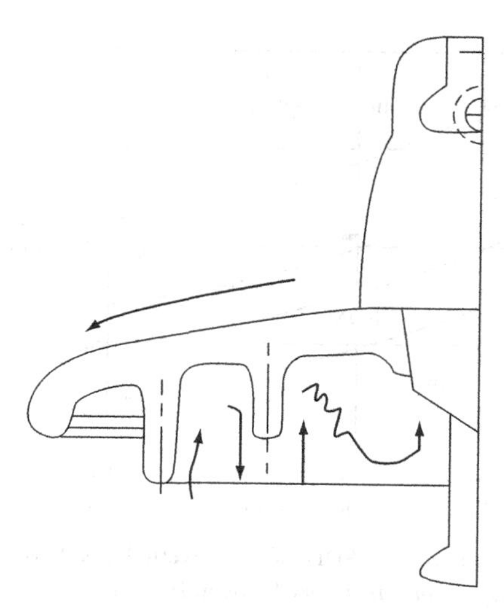

FIGURE 11.13 Dry-band arcing. (From *Application Guide for Composite Suspension Insulators*, Sediver, Inc., York, SC, 1993. With permission.)

In favorable conditions when the level of contamination is low, layer resistance is high and arcing continues until the sun or wind dries the layer and stops the arcing. Continuous arcing is harmless for ceramic insulators, but it ages nonceramic and composite insulators.

The mechanism described above shows that heavy contamination and wetting may cause insulator flashover and service interruptions. Contamination in dry conditions is harmless. Light contamination and wetting causes surface arcing and aging of nonceramic insulators.

11.4.2.2 Nonceramic Insulators

Nonceramic insulators have a dirt- and water-repellent (hydrophobic) surface that reduces pollution accumulation and wetting. The different surface properties slightly modify the flashover mechanism.

Contamination buildup is similar to that in porcelain insulators. However, nonceramic insulators tend to collect less pollution than ceramic insulators. The difference is that in a composite insulator, the diffusion of low-molecular-weight silicone oil covers the pollution layer after a few hours. Therefore, the pollution layer will be a mixture of the deposit (dust, salt) and silicone oil. A thin layer of silicone oil, which provides a hydrophobic surface, will also cover this surface.

Wetting produces droplets on the insulator's hydrophobic surface. Water slowly migrates to the pollution and partially dissolves the salt in the contamination. This process generates high resistivity in the wet region. The connection of these regions starts leakage current. The leakage current dries the surface and increases surface resistance. The increase of surface resistance is particularly strong on the shaft of the insulator where the current density is higher.

Electrical fields between the wet regions increase. These high electrical fields produce spot discharges on the insulator's surface. The strongest discharge can be observed at the shaft of the insulator. This discharge reduces hydrophobicity, which results in an increase of wet regions and an intensification of the discharge. At this stage, dry bands are formed at the shed region. In adverse conditions, this phenomenon leads to flashover. However, most cases of continuous arcing develop as the wet and dry regions move on the surface.

The presented flashover mechanism indicates that surface wetting is less intensive in nonceramic insulators. Partial wetting results in higher surface resistivity, which in turn leads to significantly higher flashover voltage. However, continuous arcing generates local hot spots, which cause aging of the insulators.

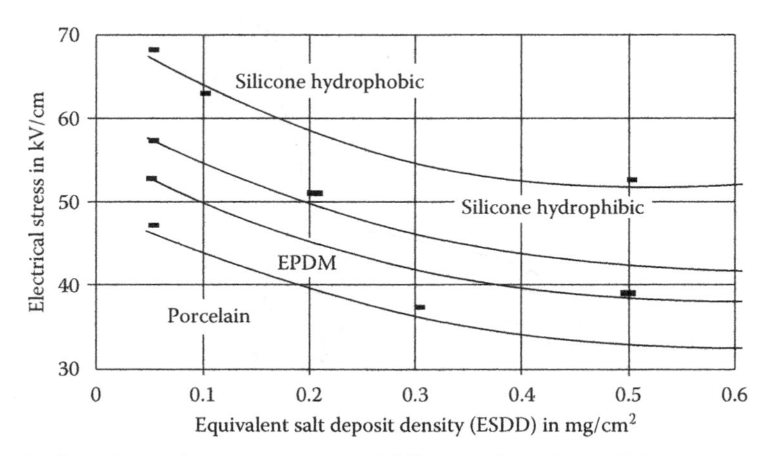

FIGURE 11.14 Surface electrical stress vs. ESDD of fully wetted insulators (laboratory test results). (From *Application Guide for Composite Suspension Insulators*, Sediver, Inc., York, SC, 1993. With permission.)

TABLE 11.6 Number of Standard Insulators for Contaminated Areas

	Level of Contamination			
System Voltage KV	Very Light	Light	Moderate	Heavy
138	6/6	8/7	9/7	11/8
230	11/10	14/12	16/13	19/15
345	16/15	21/17	24/19	29/22
500	25/22	32/27	37/29	44/33
765	36/32	47/39	53/42	64/48

Note: First number is for I-string; second number is for V-string.

11.4.3 Effects of Pollution

The flashover mechanism indicates that pollution reduces flashover voltage. The severity of flashover voltage reduction is shown in Figure 11.14. This figure shows the surface electrical stress (field), which causes flashover as a function of contamination, assuming that the insulators are wet. This means that the salt in the deposit is completely dissolved. The ESDD describes the level of contamination.

These results show that the electrical stress, which causes flashover, decreases by increasing the level of pollution on all of the insulators. This figure also shows that nonceramic insulator performance is better than ceramic insulator performance. The comparison between EPDM and silicone shows that flashover performance is better for the latter.

Table 11.6 shows the number of standard insulators required in contaminated areas. This table can be used to select the number of insulators, if the level of contamination is known.

Pollution and wetting cause surface discharge arcing, which is harmless on ceramic insulators, but produces aging on composite insulators. Aging is a major problem and will be discussed in the next section.

11.4.4 Composite Insulators

The Electric Power Research Institute (EPRI) conducted a survey analyzing the cause of composite insulator failures and operating conditions. The survey was based on the statistical evaluation of failures reported by utilities.

Results show that a majority of insulators (48%) are subjected to very light pollution and only 7% operate in heavily polluted environments. Figure 11.15 shows the typical cause of composite insulator failures. The majority of failures are caused by deterioration and aging. Most electrical failures are

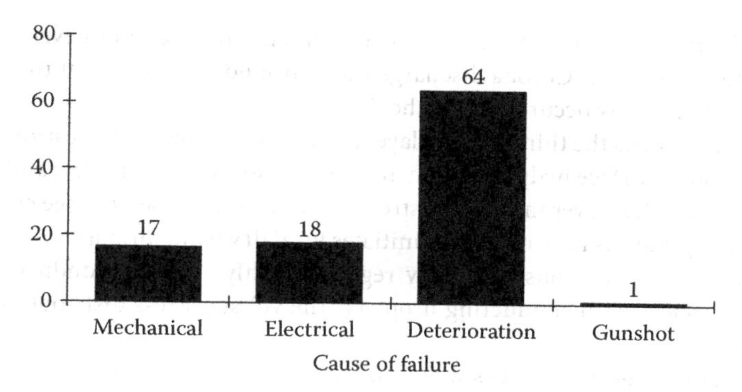

FIGURE 11.15 Cause of composite insulator failure. (From Schneider, H. et al., *IEEE Trans. Power Del.* 4(4), 2214, 1989.)

caused by water penetration at the interface, which produces slow tracking in the fiberglass rod surface. This tracking produces a conduction path along the fiberglass surface and leads to internal breakdown of the insulator. Water penetration starts with corona or erosion-produced cuts, holes on the weather shed, or mechanical load-caused separation of the end-fitting and weather shed interface.

Most of the mechanical failures are caused by breakage of the fiberglass rods in the end fitting. This occurs because of local stresses caused by inappropriate crimping. Another cause of mechanical failures is brittle fracture. Brittle fracture is initiated by the penetration of water containing slight acid from pollution. The acid may be produced by electrical discharge, initiate chemical reactions which attracts bonds in the glass-fiber. This cutting of the bonds causes smooth fracture of the glass-fiber rod. The brittle fractures start at high mechanical stress points, many times in the end fitting.

11.4.5 Aging of Composite Insulators

Most technical work concentrates on the aging of nonceramic insulators and the development of test methods that simulate the aging process. Transmission lines operate in a polluted atmosphere. Inevitably, insulators will become polluted after several months in operation. Fog and dew cause wetting and produce uneven voltage distribution, which results in surface discharge. Observations of transmission lines at night by a light magnifier show that surface discharge occurs in nearly every line in wet conditions. UV radiation and surface discharge cause some level of deterioration after long-term operation. These are the major causes of aging in composite insulators which also lead to the uncertainty of an insulator's life span. If the deterioration process is slow, the insulator can perform well for a long period of time. This is true of most locations in the United States and Canada. However, in areas closer to the ocean or areas polluted by industry, deterioration may be accelerated and insulator failure may occur after a few years of exposure [10,11]. Surveys indicate that some insulators operate well for 18–20 years and others fail after a few months. An analysis of laboratory data and literature surveys permits the formulation of the following aging hypothesis:

1. Wind drives dust and other pollutants into the composite insulator's water-repellent surface. The combined effects of mechanical forces and UV radiation produce slight erosion of the surface, increasing surface roughness and permitting the slow buildup of contamination.
2. Diffusion drives polymers out of the bulk skirt material and embeds the contamination. A thin layer of polymer will cover the contamination, assuring that the surface maintains hydrophobicity.
3. High humidity, fog, dew, or light rain produces droplets on the hydrophobic insulator surface. Droplets may roll down from steeper areas. In other areas, contaminants diffuse through the thin polymer layer and droplets become conductive.
4. Contamination between the droplets is wetted slowly by the migration of water into the dry contaminant. This generates a high resistance layer and changes the leakage current from capacitive to resistive.

5. The uneven distribution and wetting of the contaminant produces an uneven voltage stress distribution along the surface. Corona discharge starts around the droplets at the high stress areas. Additional discharge may occur between the droplets.
6. The discharge consumes the thin polymer layer around the droplets and destroys hydrophobicity.
7. The deterioration of surface hydrophobicity results in dispersion of droplets and the formation of a continuous conductive layer in the high stress areas. This increases leakage current.
8. Leakage current produces heating, which initiates local dry band formation.
9. At this stage, the surface consists of dry regions, highly resistant conducting surfaces, and hydrophobic surfaces with conducting droplets. The voltage stress distribution will be uneven on this surface.
10. Uneven voltage distribution produces arcing and discharges between the different dry bands. These cause further surface deterioration, loss of hydrophobicity, and the extension of the dry areas.
11. Discharge and local arcing produces surface erosion, which ages the insulator's surface.
12. A change in the weather, such as the sun rising, reduces the wetting. As the insulator dries, the discharge diminishes.
13. The insulator will regain hydrophobicity if the discharge-free dry period is long enough. Typically, silicon rubber insulators require 6–8 h; EPDM insulators require 12–15 h to regain hydrophobicity.
14. Repetition of the described procedure produces erosion on the surface. Surface roughness increases and contamination accumulation accelerates aging.
15. Erosion is due to discharge-initiated chemical reactions and a rise in local temperature. Surface temperature measurements, by temperature indicating point, show local hot-spot temperatures between 260°C and 400°C during heavy discharge.

The presented hypothesis is supported by the observation that the insulator life spans in dry areas are longer than in areas with a wetter climate. Increasing contamination levels reduce an insulator's life span. The hypothesis is also supported by observed beneficial effects of corona rings on insulator life.

DeTourreil and Lambeth [9] reported that aging reduces the insulator's contamination flashover voltage. Different types of insulators were exposed to light natural contamination for 36–42 months at two different sites. The flashover voltage of these insulators was measured using the "quick flashover salt fog" technique, before and after the natural aging. The quick flashover salt fog procedure subjects the insulators to salt fog (80 kg/m^3 salinity). The insulators are energized and flashed over 5–10 times. Flashover was obtained by increasing the voltage in 3% steps every 5 min from 90% of the estimated flashover value until flashover. The insulators were washed, without scrubbing, before the salt fog test. The results show that flashover voltage on the new insulators was around 210 kV and the aged insulators flashed over around 184–188 kV. The few years of exposure to light contamination caused a 10%–15% reduction of salt fog flashover voltage.

Natural aging and a follow-up laboratory investigation indicated significant differences between the performance of insulators made by different manufacturers. Natural aging caused severe damage on some insulators and no damage at all on others.

11.5 Methods for Improving Insulator Performance

Contamination caused flashovers produce frequent outages in severely contaminated areas. Lines closer to the ocean are in more danger of becoming contaminated. Several countermeasures have been proposed to improve insulator performance. The most frequently used methods are as follows:

1. *Increasing leakage distance by increasing the number of units or by using fog-type insulators.* The disadvantages of the larger number of insulators are that both the polluted and the impulse flashover voltages increase. The latter jeopardizes the effectiveness of insulation coordination because of the increased strike distance, which increases the overvoltages at substations.

2. *Application insulators are covered with a semiconducting glaze.* A constant leakage current flows through the semiconducting glaze. This current heats the insulator's surface and reduces the moisture of the pollution. In addition, the resistive glaze provides an alternative path when dry bands are formed. The glaze shunts the dry bands and reduces or eliminates surface arcing. The resistive glaze is exceptionally effective near the ocean.
3. *Periodic washing of the insulators with high-pressure water.* The transmission lines are washed by a large truck carrying water and pumping equipment. Trained personnel wash the insulators by aiming the water spray toward the strings. Substations are equipped with permanent washing systems. High-pressure nozzles are attached to the towers and water is supplied from a central pumping station. Safe washing requires spraying large amounts of water at the insulators in a short period of time. Fast washing prevents the formation of dry bands and pollution-caused flashover. However, major drawbacks of this method include high installation and operational costs.
4. *Periodic cleaning of the insulators by high-pressure-driven abrasive material, such as ground corn cobs or walnut shells.* This method provides effective cleaning, but cleaning of the residual from the ground is expensive and environmentally undesirable.
5. *Replacement of porcelain insulators with nonceramic insulators.* Nonceramic insulators have better pollution performance, which eliminates short-term pollution problems at most sites. However, insulator aging may affect the long-term performance.
6. *Covering the insulators with a thin layer of room-temperature vulcanized (RTV) silicon rubber coating.* This coating has a hydrophobic and dirt-repellent surface, with pollution performance similar to nonceramic insulators. Aging causes erosion damage to the thin layer after 5–10 years of operation. When damage occurs, it requires surface cleaning and a reapplication of the coating. Cleaning by hand is very labor intensive. The most advanced method is cleaning with high-pressure-driven abrasive materials like ground corn cobs or walnut shells. The coating is sprayed on the surface using standard painting techniques.
7. *Covering the insulators with a thin layer of petroleum or silicon grease.* Grease provides a hydrophobic surface and absorbs the pollution particles. After 1 or 2 years of operation, the grease saturates the particles and it must be replaced. This requires cleaning of the insulator and application of the grease, both by hand. Because of the high cost and short life span of the grease, it is not used anymore.

11.6 Accessories

Most high-voltage transmission lines use aluminum cable steel–reinforced (ACSR) conductors or all aluminum conductors (AAC). These conductors are described in more details in Chapter 22. The conductor must be attached to the insulators at each tower. The attachment must prevent slipping, but must be flexible to minimize the mechanical stress on insulators and permit free movement of the conductors. Figure 11.16 shows a suspension unit. The figure shows that this unit permits small conductor movement in all directions.

Extra-high-voltage lines use bundle conductors. Each phase contains two, three, or four conductors connected in parallel. The use of bundle conductors reduces the line-generated TV and radio interference, conductor impedance, and increases the maximum permitted phase current. As an example, the two bundle conductors require a suspension holder shown in Figure 11.17.

Similar holders are available for three and four bundle conductors.

At the dead-end towers, the conductors are terminated at the insulators at both sides of the tower and a flexible conductor connects the two insulator ends together assuring the current flow, as shown in Figure 11.18.

The hardware used for the line termination is shown in Figure 11.19. This is a compression-type termination used for ACSR conductors up to 500 kV.

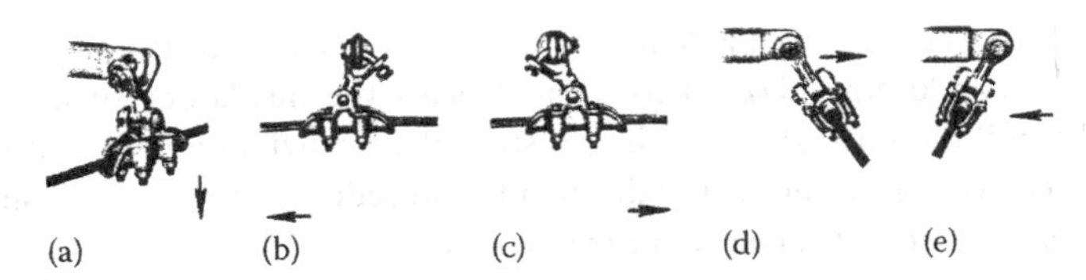

FIGURE 11.16 Suspension-type conductor holder. (a) through (e) shows the flexibility of the holder, permitting movement of the conductor in all direction.

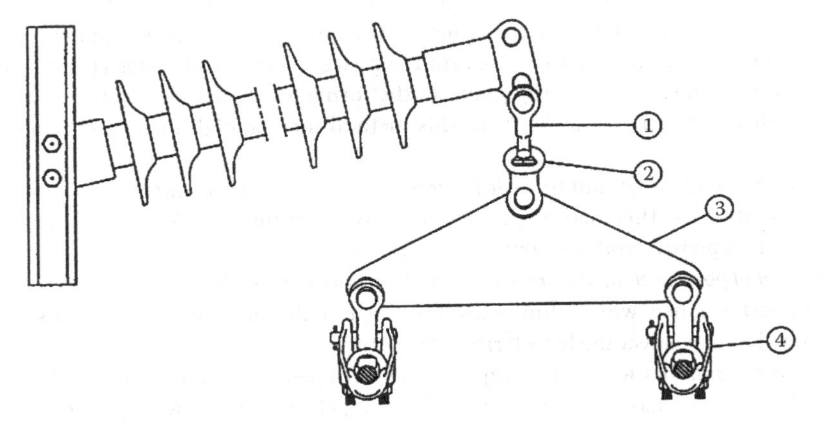

FIGURE 11.17 Suspension-type conductor holder for two bundle conductors.

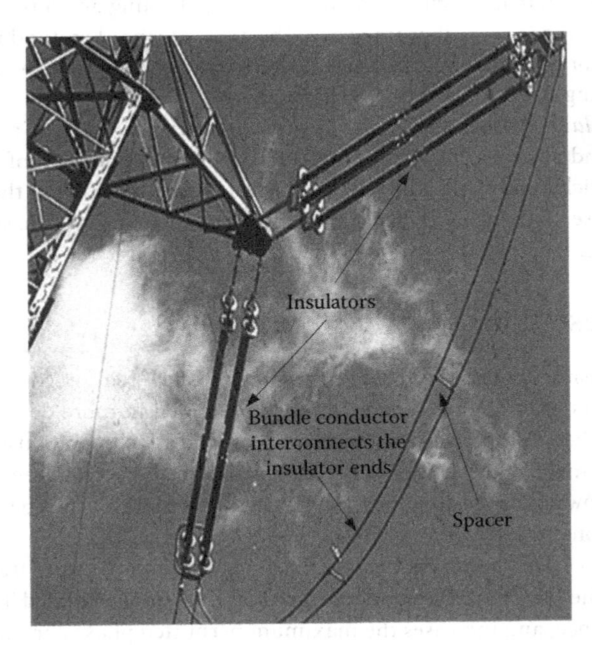

FIGURE 11.18 Line termination on dead-end tower with two bundle conductors.

The conductor bundle requires spacers preventing the tangling of the conductors. Figure 11.18 shows spacers on the interconnection at a dead-end tower. Figure 11.20a shows dimensions of a spacer for two bundle conductors and the photograph in Figure 11.20b shows a spacer used for three bundle conductors.

The wind generates Aeolian vibration on the transmission line conductors, which produces small amplitude (typically only a few centimeters) vertical movement of the conductor. Vortices on the

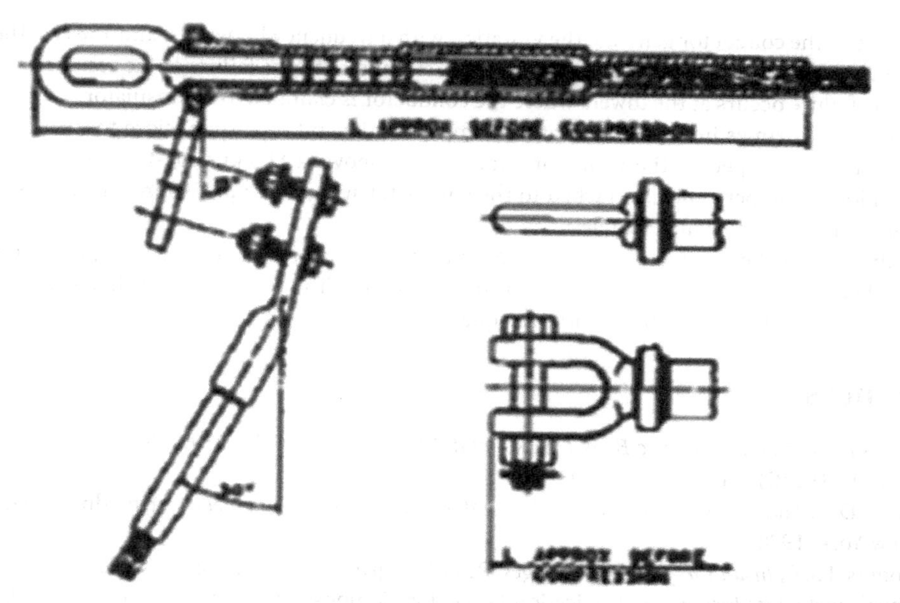

FIGURE 11.19 Compression dead end for ASCR conductors. (From AFL Telecommunication website: Conductor accessories: http://www.acasolutions.com/resource_center/brochures. With permission.)

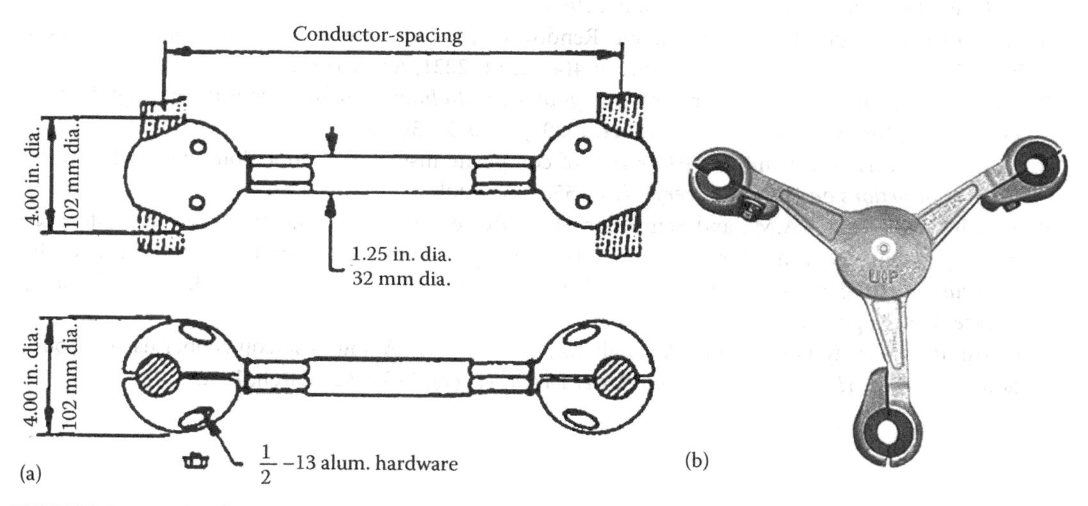

FIGURE 11.20 Conductor spacers. (a) Spacer for two conductors. (b) Spacer for three conductors. (From AFL Telecommunication website: Conductor accessories: http://www.acasolutions.com/resource_center/brochures. With permission.)

FIGURE 11.21 Stockbridge conductor vibration damper. (From AFL Telecommunication website: Conductor accessories: http://www.acasolutions.com/resource_center/brochures. With permission.)

leeward side of the conductor generate the vibration with a frequency between 5 and 150 Hz. The vibration produces periodic bending of the conductor, which causes fatigue failure of the conductor strands. Most of the failure occurs at the towers where the conductor is clamped to the insulator.

The power companies install two vibration dampers at the end each span, close to the point where the conductor is clamped to the insulator. Figure 11.21 shows a Stockbridge-type damper. A short (30–80 cm long) damper cable is attached to the conductor with a clamp. Two metal weights are connected at each end of the damper cable.

The vibration of the conductor will initiate swinging motion of the damper weights. The weights periodically hit the cable, which greatly damps the oscillation. The weights, the stiffness, and length of the damper cable are tuned to the vibration frequency.

References

1. *Transmission Line Reference Book (345 kV and Above)*, 2nd edn., EL 2500, Electric Power Research Institute (EPRI), Palo Alto, CA, 1987.
2. Fink, D.G. and Beaty, H.W., *Standard Handbook for Electrical Engineers*, 11th edn., McGraw-Hill, New York, 1978.
3. Looms, J.S.T., *Insulators for High Voltages*, Peter Peregrinus Ltd., London, U.K., 1988.
4. *Toughened Glass Insulators*, Application Guide for Composite Suspension Insulators, Sediver Inc., Nanterre, France, 1993.
5. Hall, J.F., History and bibliography of polymeric insulators for outdoor application, *IEEE Transactions on Power Delivery*, 8(1), 376–385, January, 1993.
6. Schneider, H., Hall, J.F., Karady, G., and Rendowden, J., Nonceramic insulators for transmission lines, *IEEE Transactions on Power Delivery*, 4(4), 2214–2221, April, 1989.
7. Karady, G.G., Outdoor insulation, *Proceedings of the Sixth International Symposium on High Voltage Engineering*, New Orleans, LA, September, 1989, pp. 30-01–30-08.
8. DeTourreil, C.H. and Lambeth, P.J., Aging of composite insulators: Simulation by electrical tests, *IEEE Transactions on Power Delivery*, 5(3), 1558–1567, July 1990.
9. Karady, G.G., Rizk, F.A.M., and Schneider, H.H., Review of CIGRE and IEEE research into pollution performance of nonceramic insulators: Field aging effect and laboratory test techniques, in *International Conference on Large Electric High Tension Systems (CIGRE)*, Group 33, (33–103), Paris, France, 1–8, August, 1994.
10. Gorur, R.S., Karady, G.G., Jagote, A., Shah, M., and Yates, A., Aging in silicon rubber used for outdoor insulation, *IEEE Transactions on Power Delivery*, 7(2), 525–532, March, 1992.

12

Transmission Line Construction and Maintenance

Jim Green
Salt River Project

Daryl Chipman
Salt River Project

Yancy Gill
Salt River Project

12.1 Introduction

Electric transmission lines are constructed to provide a path for electricity to flow from a generation source to a specific service area or to intertie with the transmission grid. Transmission of electricity is categorized by voltage level and can range from 69 through 765 kV and higher. The majority of new transmission line construction is being done at higher voltages commonly referred to as extra high voltage (EHV) or voltages greater than 230 kV. The higher voltages reduce power losses inherent in long-distance transmission of electricity, thus allowing the system to operate more efficiently. EHV line construction requires more specific construction procedures and constructor expertise than construction of lower voltage lines.

Transmission lines are engineered and designed to meet National Electric Safety Code Standards and the criteria of regulating authorities such as North American Electric Reliability Council (NERC) overseen by Federal Energy Regulatory Commission (FERC). Design and construction must also comply with regulations, rules, and laws set forth by federal, state, and local authorities. The engineering and design must also consider constructability of the line design and future maintenance requirements. Additional information regarding transmission line construction can be found in IEEE Standards 524–2003 and 1441–2004.

12.2 Transmission Line Siting

Historically, transmission line easements or rights-of-way (ROW) followed as direct a route as possible from the generation source to the service territory or grid interconnection. Today, consideration also has to be given to the proximity of populated areas, archeological, geological, and environmental concerns. EHV lines ROW require wider easements not only to facilitate construction and future maintenance but to insure that NESC rules are not violated. These requirements must be met or mitigated before a permit or license to construct and operate will generally be issued by the regulatory authorities.

12.3 Sequence of Line Construction

Prior to construction, decisions as to structure types, foundation requirements, conductor size and type, insulation, and line hardware have all been determined. The ROW has been surveyed and the transmission centerline, structure locations, and edge of ROW have been marked. Next, the ROW is made ready for construction activities by constructing access roads, clearing obstacles, and removing vegetation that may hinder construction efforts. Once this is accomplished, line construction can begin. A typical line construction sequence is as follows:

- Foundations installed as applicable
- Structures and hardware delivered to the designated construction sites
- Structure assembly and erection
- Hardware and insulators assembled and installed
- Conductor travelers installed
- Install pulling lines
- Pull conductor
- Sag and clip conductors
- Install vibration dampers or spacer dampers as applicable

Factors to consider during the construction process include the following:

- Foundations are constructed according to engineered design but should be checked for correctness regarding location, type, orientation, and the bisector verified prior to structure erection. Figure 12.1 illustrates a stub angle braced and set in a foundation pier.
- Structure assembly should be as complete as possible prior to the structure erection, which reduces the amount of aerial work required prior to conductor stringing. This may include all

FIGURE 12.1 Stub angle braced and set in foundation pier.

FIGURE 12.2 Tower assembled from layback to bridge on ground and lifted onto the tower body and legs for final assembly.

hardware, insulator assemblies, and stringing blocks. Attention must be paid to total weight of the assembled structure so that the lifting capacity of the equipment used in the erection process is not exceeded. An example of this is illustrated in Figure 12.2 where the tower from the layback to the bridge was assembled on the ground, lifted, and set to the tower body and legs.

- Consideration needs to be given to how damaged and incorrectly fabricated steel members and gussets are going to be addressed. Incorrect bolt hole placement or patterns can be corrected in the field using a "fill and drill" process, as long as the new hole spacing and edge distance conforms to the engineering design or applicable structural steel standard(s) for bolted connections. Bent lattice needs to be assessed on an individual basis to determine if a field repair is possible. Figure 12.3 shows a gusset with incorrectly placed holes.
- Aluminum conductors with steel reinforcement (ACSR conductors) have been used extensively in the past, but conductors with higher operating temperatures and current carrying capabilities are becoming more prevalent with today's EHV transmission line construction. The design of these types of conductor employs aluminum stranding that is softer than conventional ACSR and requires considerably more care in handling during installation. Careful planning and a thorough understanding of stringing procedures are needed to prevent damage to the conductor during stringing operations.
- Conductor reels are supplied by the conductor manufacturer and can be either the nonreturnable (NR) wood or the returnable metal (RM, RMT) type. Reel size is dependent on type and size of conductor specified by the engineering design or conductor manufacturer. The most commonly used reels are the RMT 96.60 size often referred to as 8 ft reels. Wooden reels are not recommended for use in transmission stringing.
- A worksite grounding plan is necessary to provide a safe electrical environment for all on-site personnel. This is accomplished by evaluating all electrical hazards and soil resistivity within the worksites with respect to the placement of equipment and personnel. Typically, the greatest electrical hazard faced by transmission line construction workers is induced current when working in congested transmission line corridors or when working near energized distribution or transmission lines.

FIGURE 12.3 Gusset with incorrect hole placement.

12.4 Conductor Pulling Plan

A conductor pulling plan must be developed for the entire line prior to the start of conductor stringing. The conductor pulling plan is dependent on a variety of factors:

- Conductor size and lengths on reels
- Span lengths
- Structure heights
- Ground clearance during stringing operations
- Obstructions, such as railroad crossings, road crossings, other line crossings
- Line angles (individual and cumulative)
- Allowable stringing tensions
- Terrain and elevation changes
- Dead-end structure locations
- Equipment setup locations
- Midline splice location access
- Capacities of pulling and tensioning equipment

With all of these factors taken into consideration, it is desirable to make the pull section as long as allowable but generally not more than the length of the wire on two conductor reels. Also a pulling plan should incorporate dead-end structures whenever possible as the termination of a pull section.

12.5 Conductor Stringing Methods

12.5.1 Slack or Layout Method

With this stringing method, the conductor is dragged along the ground by means of a pulling vehicle, or the reel is carried along the line on a vehicle from which the conductor is laid on the ground. Usually, a braking device is provided to control conductor payout. When the conductor is dragged past a supporting structure, pulling is stopped and the conductor is placed in travelers attached to the structure before proceeding to the next structure. This method is only applicable to the construction of new lines and for cases in which the conductor surface condition is not critical. This method is not usually economical in urban locations where hazards exist from traffic or where there is danger of contact with energized circuits.

12.5.2 Tension Stringing

Tension stringing is preferred for all transmission conductor installations. Using this method, the conductor is kept under tension during the stringing process, which keeps the conductor off the ground, minimizing the possibility of conductor surface damage and facilitates overcoming obstacles such as road crossings and also for maintaining clearance from other overhead lines. In a typical tension stringing operation, a pilot line is pulled through travelers installed on the structures for the length of the conductor pull section. The pilot line is then used to pull in the heavier pulling line. The pulling line is then attached to the conductor with a swivel and a woven grip, commonly referred to as a pulling sock, and then pulled in from the wire setup to the puller.

Pulling speed is an important factor in achieving a smooth stringing operation. Speeds of 3–5 miles/h usually provide for smooth passage of the running board over the stringing travelers. Slower speeds tend to result in unnecessary traveler and insulator hardware swing as the pulling grip and running board pass over the traveler. Faster speeds reduce the time to react in the event of an equipment malfunction.

Stringing tensions should be kept low as possible during the stringing process to minimize conductor creep. Conductor creep is a function of time, temperature, and conductor tension which results in permanent elongation of the conductor. Conductors that have been subjected to excessively high stringing tensions or that have been allowed to remain in travelers for an extended period of time will experience an abnormal amount of creep. If this occurs, the sag tables should be corrected to compensate for the additional creep elongation. If not, the initial sag tensions will be higher than designed and could result in conductor damage due to vibration.

Major equipment and tools required for tension stringing include the following:

Reel stands: A device designed to support one or more conductor or ground wire reels and can be skid, trailer, or truck mounted. Reel stands can accommodate conductor reels of varying sizes and should be equipped with reel brakes to prevent the reels from turning when pulling is stopped. They are used for either slack or tension stringing.

Tensioner: A device designed to hold tension against a pulling line or conductor during stringing operations. The tensioner consists of one or more pairs of urethane or neoprene-lined single or multiple groove bullwheels in which each pair is arranged in tandem. Tension is accomplished by friction generated against the conductor that is reeved around the grooves of a pair of the bullwheels. Some tensioners are equipped with their own engines, which retard the bullwheels mechanically, hydraulically, or through a combination of both.

Puller: A device designed to pull a conductor during stringing operations. The puller can be either the drum or bullwheel type. It can be truck or trailer mounted and is normally equipped with its own engine, which drives the drum mechanically, hydraulically, or through a combination of both. The pulling line can be either synthetic fiber or wire rope.

Pilot line winder: Pilot line winders usually have multiple drums to provide pilot lines for several phase or ground wire positions. They have operating characteristics similar to drum-type pullers. Pilot line winders are used to pull in the larger pulling lines that will in turn be connected to the conductors to complete the pulling sequence. They can also be used to pull in overhead ground wires if the capacity rating is sufficient.

Pulling vehicle: A suitably configured vehicle used to install pilot lines or pulling lines in accessible ROW.

Helicopter: Helicopters are sometimes used to install pilot lines, especially in rough terrain or where vehicle traffic is restricted on the ROW. Helicopters have proven to be efficient and cost effective when compared with traditional methods of pilot line installation. When a helicopter is used to install pilot lines, it is necessary to have the travelers equipped with outrigger arms that guide the pilot line into the throat area of the traveler. Spring-loaded gates keep the line from being pulled out of the traveler throat as the helicopter continues the installation to the adjacent travelers. Bundle conductor travelers may have additional guides to deposit the lines into the pulling line sheave of the traveler. Bundle travelers are directional as the guides or gates must open toward the puller and away from the wire setup location.

Helicopters are also used in tower erection to lift structural elements and assemblies as well as transporting men and materials in difficult terrain.

Travelers or conductor blocks: Travelers must be sized correctly for the size and type of conductor being installed. It is recommended that the sheave be at least 20 times the conductor diameter as measured from the bottom of the conductor groove. The radius of the conductor groove should be 1.10 times the radius of the conductor. The flare of the groove should be between 12° and 20° from the vertical to facilitate the passage of swivels, pulling grips and to contain the conductor within the groove, particularly at line angles. Travelers are available with single sheave or with multiple sheave combinations to accommodate different conductor bundle configurations. Sheaves should be lined with neoprene or urethane material to prevent damage to the conductor. Travelers must run freely or they will adversely affect stringing and sagging operations. Also traveler efficiency must be considered when determining pulling/sagging section lengths. Finally, always ensure that the manufacturer's safe working load for the traveler does not exceed the traveler stringing loads.

Grounded travelers: The same as a standard traveler but have additional unlined rollers that make electrical contact between the conductor and the assembly which is connected to ground. They are usually placed at several locations along the pull section with a set relatively close to each end of the pull and at both sides of any energized line crossings.

Running grounds: These consist of spring tensioned unlined rollers that ensure constant contact with the conductor. The unit is connected directly to a suitable ground. Running grounds should be placed between the reel stands and the Tensioner, between the Tensioner and the first structure out and also on the pulling cable/wire rope between the puller and the first structure out on the pull section. Running grounds are also known as rolling grounds.

Pulling lines: Pulling lines can be either wire or synthetic rope as long as the rope is of a sufficient rated strength with appropriate safety factor to withstand the applied stringing tensions. Pulling lines should also be non-rotating, i.e., the rope will not imply twist or torque to the conductor.

Swivel: A device used to connect pulling lines to the conductor or from conductor to conductor. It is constructed so that each end will spin or rotate independently, thus reducing the transfer of rotational torque from one line to the other.

Woven grip: Also referred to as Kellems grip or most commonly as a sock. It is similar to a Chinese finger grip and constricts to grip the conductor when tension is applied. One end is open to allow the conductor to be inserted and the other end is fitted with an eye that facilitates attaching a swivel.

Running board: A device used for pulling multiple conductors with only one pulling line. A running board is also referred to as an alligator or gator.

12.6 Equipment Setup

Once pull sections are identified, then the pulling, tensioning, and reel stand placement can be done. The physical location of each piece of equipment integral to the wire and puller setups is of the utmost importance. Distance from the first structure out to the puller or tensioner must take into consideration the loading capacities designed into the structure and traveler supports. The general rule is that the distance from the structure be a minimum of 3 ft horizontal for every 1 ft vertical to the traveler attachment point. This is also applicable to the conductor snubbing location which will be between the tensioner or puller and the structure. The snubbing locations are sites where the conductors are temporarily fixed/anchored to allow for sagging, conductor splicing, and serve as the pull site for the next sag section.

The reel stands should be located a sufficient distance from the rear of the tensioner to allow for enough fleet angle of the conductor leaving the reel and entering the bullwheels of the tensioner so that no damage or scuffing of the conductor can occur. Also all equipment at each location must be grounded and bonded together to ensure that no difference in ground potential exists.

Examples of conductor stringing and stringing equipment are presented in Figures 12.4 through 12.11.

FIGURE 12.4 Flying in pilot line, note distribution line crossing and crossing structure.

FIGURE 12.5 Five-drum pilot line puller (two statics and three conductors).

FIGURE 12.6 Drum-type line puller.

FIGURE 12.7 Tensioner and reel stands set up for conductor pull.

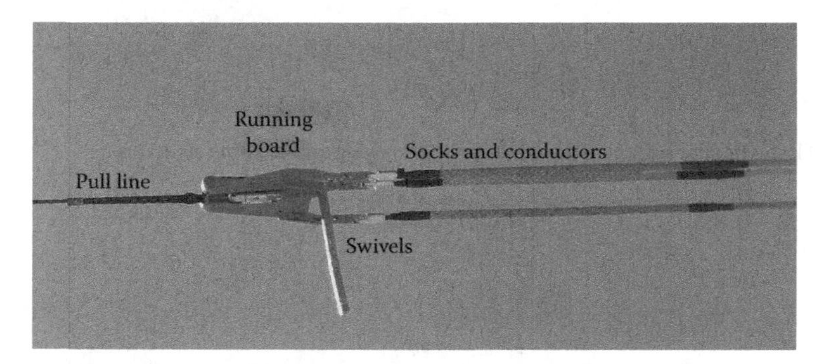

FIGURE 12.8 Running board with three conductors.

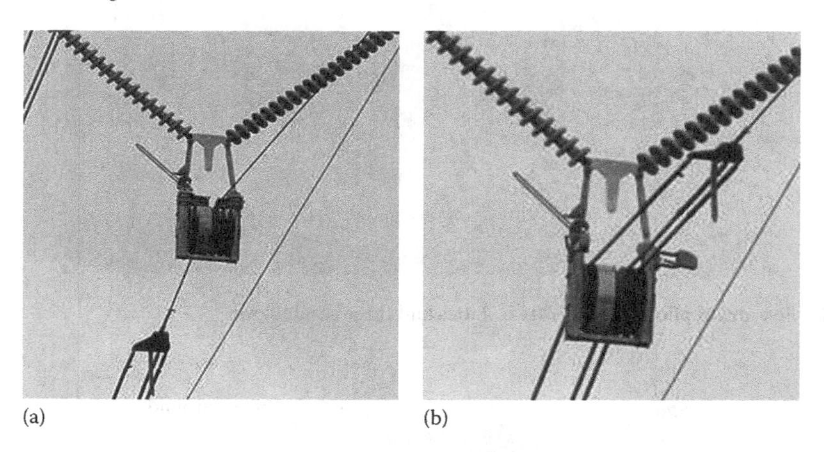

FIGURE 12.9 Running board with conductors approaching, (a) passing, and (b) the traveler.

12.7 Sagging

Overhead conductors are flexible and uniform in weight and when suspended between two supports form the shape of a caternary. Sagging transmission conductors is the process of obtaining the proper caternary shape based upon the sag-tension design requirements. The two most common methods used to sag transmission conductor are the transit method and stop watch or time of return method. The transit method is typically preferred as it provides for greater accuracy and control of the sagging process.

The sagging process begins by identifying the sag spans where the measurements are made to obtain the desired caternary shape. Sag spans should be located near each end of the sag section with preference

FIGURE 12.10 Pressing outer sleeve of splice.

FIGURE 12.11 Snub location being prepared for wire letup.

given to longer more level spans. Long sag sections will require additional sag spans in the middle of the sag section. Sagging conductor in hilly terrain presents the additional difficulty of balancing the horizontal conductor tensions. The imbalance of the horizontal tension is due to gravity pulling the conductor through the travelers to the downhill end of the sag section. This increases the sag and lowers the tensions in the downhill spans. To restore the horizontal tension or pull the conductor uphill, a clipping offset is used. The clipping offset is a calculated distance, measured along the conductor, from the plum mark to a point on the conductor at which the center of the suspension clamp is to be placed.

Conductor is usually progressively sagged from the tensioner end to the puller end of the sag section. Prior to sagging, bundle travelers at structures with line angles of more than three degrees should be exchanged for single sheaves for each conductor and suspended as closely as possible to their respective clipped in position to allow for more accurate sagging. Also, when sagging bundled conductors, all subconductors in the bundle should be sagged at the same time in order to maintain their mechanical characteristics relative to each other. Conductor sagging should be initiated as soon as possible after all conductors are pulled in. Should the sagging operation be delayed, the conductor tension should be reduced as much as possible to avoid causing excessive conductor creep prior to sagging. Once sagging is complete, all conductors should be marked at the suspension point on each structure prior to any conductors being clipped in the section. Clipping-in is the process of installing the conductor in the permanent suspension clamps that replace the stringing travelers on the support structure. Finally, if dampers, spacers, or spacer dampers are required, they should be installed on the conductor immediately after clipping to prevent vibration damage due to the wind. Installation locations for this type of line hardware are important for optimal conductor damping and/or spacing and are usually provided by the manufacturer.

12.8 Overhead Transmission Line Maintenance

12.8.1 Introduction

Overhead transmission line maintenance is the management of all transmission line assets (structures, hardware, conductor, and insulators) with the goal to optimize the risk, reliability, and operation over their respective life cycles. The typical overhead transmission line maintenance program is comprised of

- Transmission line inspections
- Transmission line fault investigations and corrective action(s)
- Transmission line work
- Vegetation management
- Data/Information management and analysis
- Emergency restoration of transmission structures

The organization and performance of these tasks is driven by a utility's work practices, system age and build, voltage, environment, and regulatory requirements. Regardless of how a utility chooses to organize and perform these basic maintenance tasks, it is imperative that they are all integrated so that information derived from or technological advancements to any individual task can drive process changes to the others. Not integrating these maintenance tasks, i.e., treating them as separate entities, will result in less than optimal maintenance practices.

12.8.2 Overhead Transmission Line Inspections

One of the primary sources from which transmission line maintenance work is derived is by inspection of the transmission line assets. Inspection methods work since the majority of overhead transmission line failure modes are time or cyclically dependent and typically exhibit visible or measureable distress prior to failure. Examples of the type of failure modes encountered include; wear, fatigue, corrosion, loosing of mechanical fasteners, electrical breakdown/tracking, in addition to fungal and insect attack on wood assets. Random damage events, vandalism, for example, typically do not result in immediate failures and are often identified within the inspection cycle. Severe structural loading events, such as extreme weather or a vehicle/airplane striking the transmission line or structure, often result in immediate failure and/or operational loss. These types of events cannot be prevented by transmission line inspections and are better addressed by an emergency restoration plan which is discussed later in this chapter.

Transmission line inspection plans are mainly developed from experience by compiling information such as past failures, material/environment degradation mechanisms, voltage, risk to transmission system and grid, public risk, and regulatory requirements. Once the inspection plan has been established, it is important to follow the plan while continually assessing its effectiveness. Any changes to the inspection practices and plan should be well documented as to why the process is being changed. Failure to follow your established inspection plan and adequately document your findings could result in regulatory fines and increase the risk of litigation.

FIGURE 12.12 Examples of basic types of visual inspections: (a) ground, (b) climb and shake, and (c) air.

Visual inspections are the primary means of identifying overhead transmission line problems from which maintenance work is derived. They are normally conducted at predetermined time intervals and can be via ground, climb and shake, or air (Figure 12.12).

Visual inspections can be general, looking at all the transmission line assets, or they can be specific to a particular asset or component due to a previous identified failure mode. Table 12.1 lists the types of findings that usually result from visual inspections. It is not uncommon to supplement or combine visual inspections with other inspection technologies such as radio frequency (RF), infrared and corona cameras, corrosion potential measurements, Light detection and ranging (LiDAR) mapping, and acoustic/ultrasonic methods

TABLE 12.1 Typical Transmission Line Findings from Visual Inspections

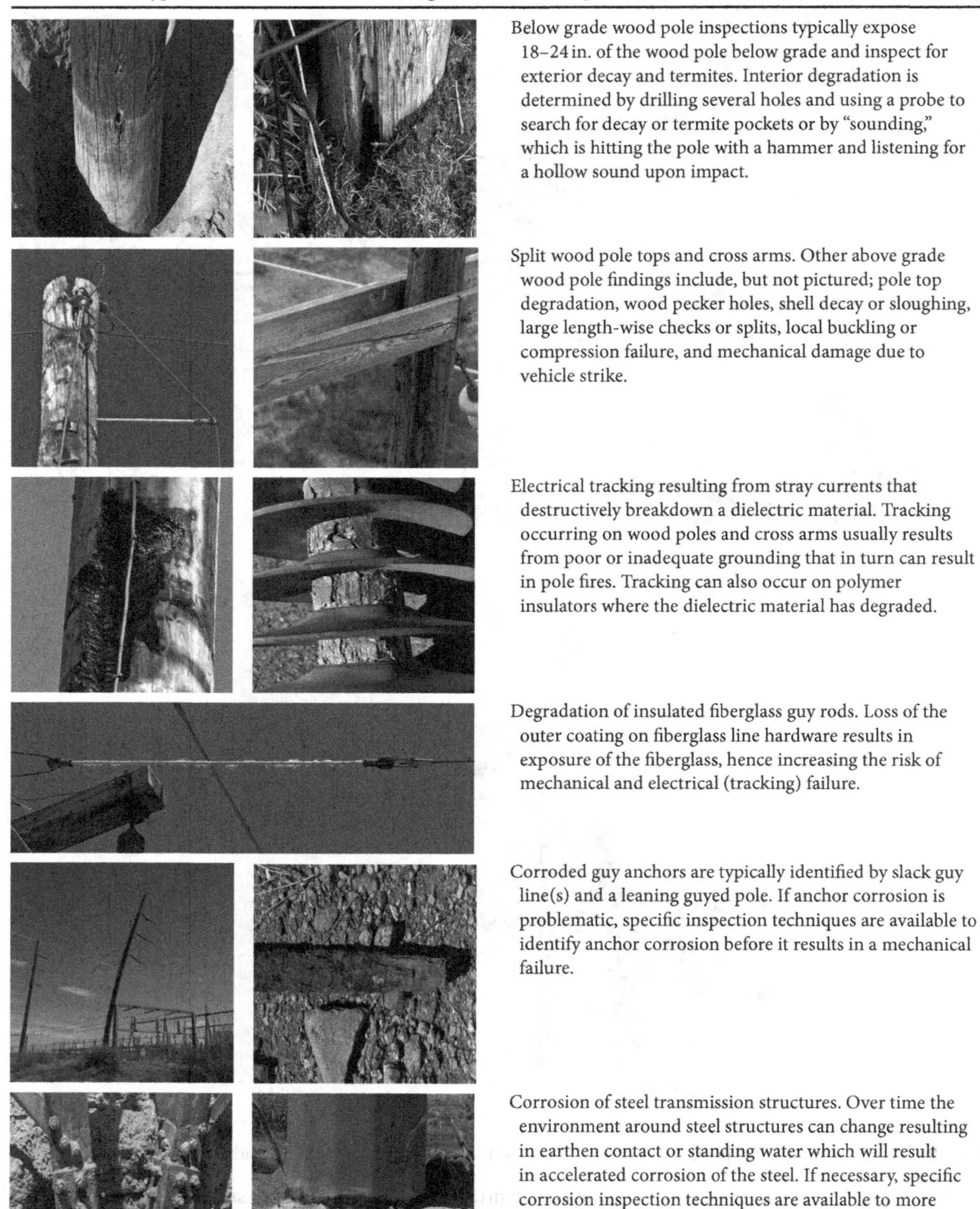

Below grade wood pole inspections typically expose 18–24 in. of the wood pole below grade and inspect for exterior decay and termites. Interior degradation is determined by drilling several holes and using a probe to search for decay or termite pockets or by “sounding,” which is hitting the pole with a hammer and listening for a hollow sound upon impact.

Split wood pole tops and cross arms. Other above grade wood pole findings include, but not pictured; pole top degradation, wood pecker holes, shell decay or sloughing, large length-wise checks or splits, local buckling or compression failure, and mechanical damage due to vehicle strike.

Electrical tracking resulting from stray currents that destructively breakdown a dielectric material. Tracking occurring on wood poles and cross arms usually results from poor or inadequate grounding that in turn can result in pole fires. Tracking can also occur on polymer insulators where the dielectric material has degraded.

Degradation of insulated fiberglass guy rods. Loss of the outer coating on fiberglass line hardware results in exposure of the fiberglass, hence increasing the risk of mechanical and electrical (tracking) failure.

Corroded guy anchors are typically identified by slack guy line(s) and a leaning guyed pole. If anchor corrosion is problematic, specific inspection techniques are available to identify anchor corrosion before it results in a mechanical failure.

Corrosion of steel transmission structures. Over time the environment around steel structures can change resulting in earthen contact or standing water which will result in accelerated corrosion of the steel. If necessary, specific corrosion inspection techniques are available to more thoroughly evaluate steel corrosion.

TABLE 12.1 (continued) Typical Transmission Line Findings from Visual Inspections

Transmission line hardware failures are often the result of loose or missing bolts and cotter keys, corrosion, fatigue, and improper or poor installation.

Flashed insulators typically result from a lightning strike or foreign debris, such as a string of balloons, shorting the insulator to ground. Flashed insulators usually remain functional but should be replaced at the first available opportunity.

Broken porcelain insulators result from the propagation of microcracks, cement expansion, lightning puncture, excessive corona discharge around the insulator cap, and vandalism (gunshot). Industrial and coastal areas are also prone to cap and pin corrosion. Similar problems are also encountered for toughened glass insulators; however, the material failure mechanisms differ.

Insulator contamination can compromise the dry arc distance of an insulator or insulator string resulting in a flash over. Common contamination sources include birds, industrial sources, and salt fog in coastal environments. If natural processes cannot clean the insulator(s) sufficiently, protective screening, bird deterrents, or insulator cleaning could become necessary.

Mechanical failure of polymer insulators is often due to handling damage, binding hardware, vandalism, manufacturing issues, and exposure of the fiberglass rod to the environment as a result of excessive corona cutting of the seal or sheath, sheath damage due to lightning, and electrical tracking under the sheath.

Conductor damage can be the result of vandalism (gunshot), line hardware wearing against the conductor, incidental contact such as from a crane boom, or during transmission line construction.

(continued)

TABLE 12.1 (continued) Typical Transmission Line Findings from Visual Inspections

	Design issues such as conductor uplift, left photo, and missing or improperly installed hardware such as the absence of a standoff for the optical ground wire, right photo.
	Foundation issues such as erosion, burying of foundation caps, or foundation cap cracking.
	Encroachments due to construction, stock piling of materials, and vegetation within the transmission line ROW. Also blocking or limiting access to the ROW is considered encroachment and should be noted in the inspection report.
	Bent tower lattice can be due to buckling from high wind loads, foundation settling/movement, impact from vehicles or earth, and linemen climbing/stepping on smaller members. Failed steel members are usually the result of fatigue at bolted connections, corrosion, and major loading events such as vehicle impact and extreme weather.

for a more comprehensive and in-depth evaluation of the transmission line structures, conductor, insulators, and hardware. Examples of the uses for the aforementioned inspection technologies include the following:

- LiDAR mapping for line rating and vegetation management.
- Ultrasonic techniques and corrosion potential measurements for steel poles, tower stub angles, and anchor rods for corrosion management.
- Corona camera evaluation of polymer insulators to identify excessive corona discharge that can damage the seal at the end fittings and/or the silicon sheath resulting in exposure of the fiberglass rod. Exposing the fiberglass rod to the environment significantly increases the risk of a mechanical or electrical failure.
- Infrared temperature measurements to evaluate poor electrical connections in connectors, switches, and splices.

An excellent resource covering all aspects of transmission line inspection methodologies and technologies is EPRI's "Overhead Transmission Inspection and Assessment Guidelines" also known as the EPRI Yellow Book.

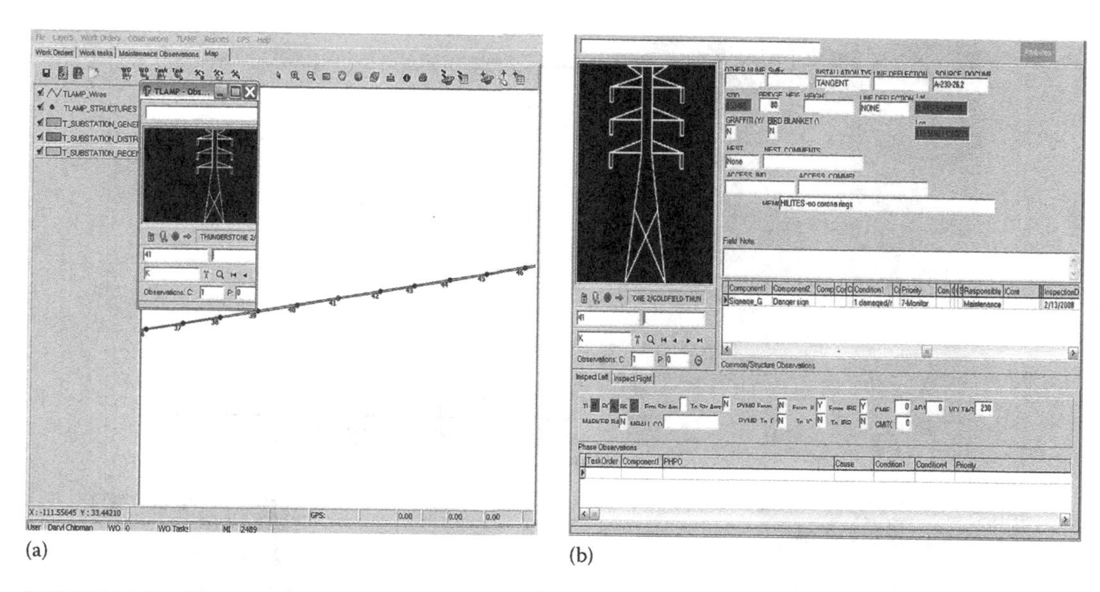

FIGURE 12.13 Example of inspection software entry screen(s). (a) The software is GPS enabled and highlights the closest structure, and (b) allowing for menu-driven inspection entries/findings.

12.8.3 Transmission Line Inspection Software

Most of today's transmission line inspection software is built on a geographic information system (GIS) platform in addition to being global positioning system (GPS) enabled. The reporting format of transmission line problems or issues within the software is usually standardized or menu driven, Figure 12.13. Reporting standardization gives consistency to the inspection data, thus making the data more searchable and easier to analyze. Comment sections, while valuable, should only be used to document problems not found within the standardized report or to enhance the report with specific information related to the problem or transmission asset.

12.8.4 Transmission Line Fault Investigations and Corrective Action(s)

One of the primary objectives of any transmission line maintenance program is to prevent line faults. Therefore, it is important that when a fault does occur, it is investigated to determine if it can be effectively addressed by preventative maintenance measures. One method for investigating transmission line faults and assessing the potential for preventing future faults is the root cause analysis (RCA) process. RCA is usually systematic and standardized with regards to conducting and documenting the investigative process. Upon completion of the investigation, work is initiated on identifying the root cause(s) and potential corrective actions. Finally, the potential corrective actions are evaluated for their cost and potential to prevent future fault occurrences. Figure 12.14 illustrates the RCA process for a lightning investigation.

Another method by which faults can be evaluated is by control charting. Control charting of transmission line faults can be useful in assessing transmission line maintenance and inspection programs by determining the statistical characteristics of faulting associated with a particular transmission line, voltage, or fault type. An example of this is illustrated in Figure 12.15 for a 115 kV transmission system located in the Southwestern United States in which all faults that occurred from 1997 to 2009 are charted. By knowing the stable faulting characteristics, one is better able to assess the maintenance impact on the transmission system and thus distinguish between changes driven by special variation versus those that are inherent.

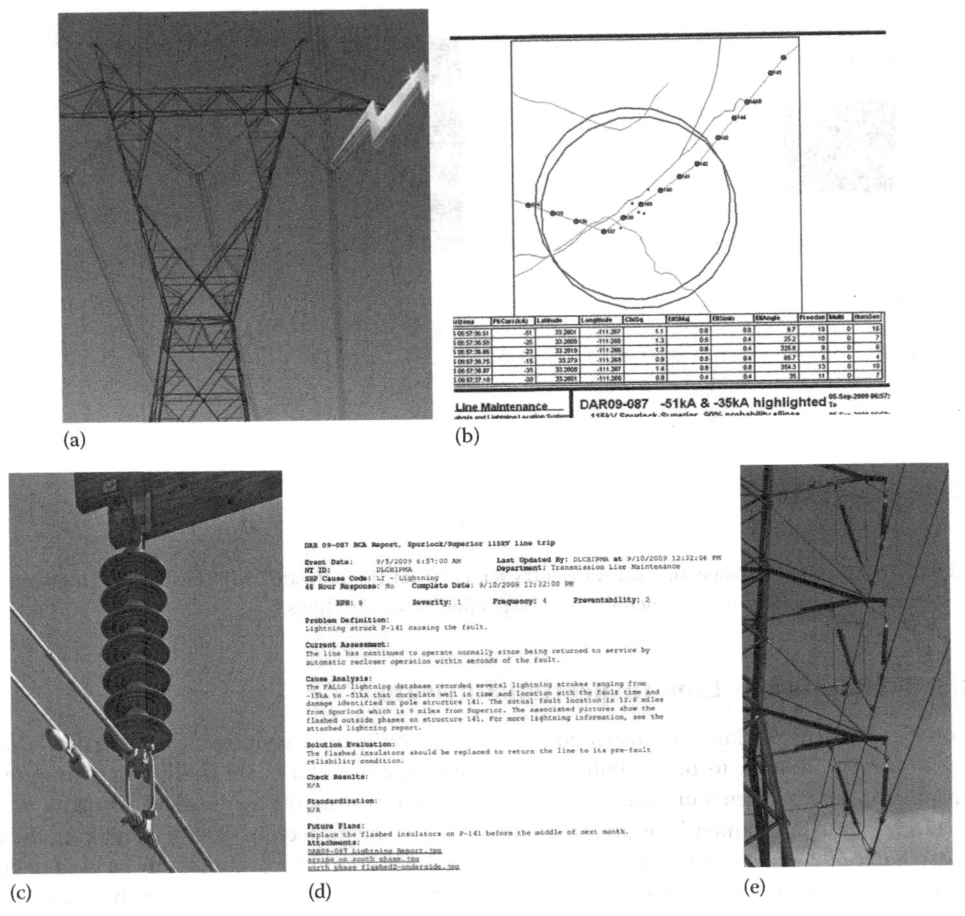

DAR 09-087 RCA Report, Spurlock/Superior 115kV line trip

Event Date: 9/5/2009 6:57:00 AM **Last Updated By:** DLCHIPMA at 9/10/2009 12:32:06 PM
NT ID: DLCHIPMA **Department:** Transmission Line Maintenance
SRP Cause Code: LI - Lightning
48 Hour Response: No **Complete Date:** 9/10/2009 12:32:00 PM

RPN: 8 **Severity:** 1 **Frequency:** 4 **Preventability:** 2

Problem Definition:
Lightning struck P-141 causing the fault.

Current Assessment:
The line has continued to operate normally since being returned to service by automatic recloser operation within seconds of the fault.

Cause Analysis:
The FALLS lightning database recorded several lightning strokes ranging from -15kA to -51kA that correlate well in time and location with the fault time and damage identified on pole structure 141. The actual fault location is 12.6 miles from Spurlock which is 9 miles from Superior. The associated pictures show the flashed outside phases on structure 141. For more lightning information, see the attached lightning report.

Solution Evaluation:
The flashed insulators should be replaced to return the line to its pre-fault reliability condition.

Check Results:
N/A

Standardization:
N/A

Future Plans:
Replace the flashed insulators on P-141 before the middle of next month.
Attachments:
DAR09-087 Lightning Report.jpg
arcing on south phase.jpg
north phase flashed2-underside.jpg

FIGURE 12.14 Example of the RCA process for a lightning investigation. (a) Lightning strikes a transmission structure resulting in a line fault, (b) fault location analysis estimates the probable location from relay data and this is compared to the lightning strike data from the National Lightning Detection Network (NLDN), (c) a field patrol verifies the fault location, assesses damage, and investigates the root cause, (d) the root cause report is created and stored for future trending purposes, and (e) as appropriate, line improvements are made, such as the installation of lightning arrestors.

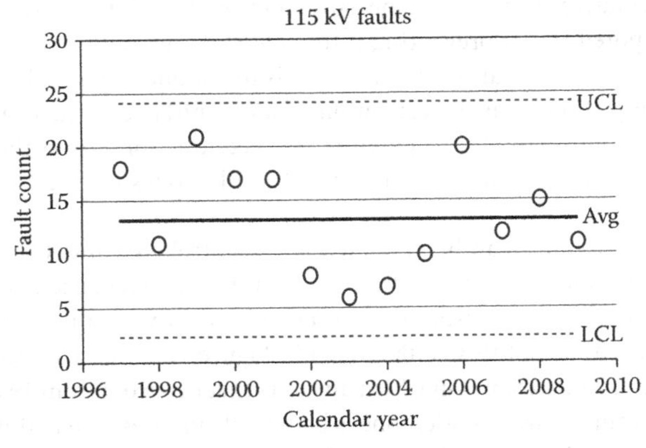

FIGURE 12.15 "C"-type control chart for annual faults on 115 kV transmission system located in the Southwest United States.

12.9 Transmission Line Work

Scheduling the necessary maintenance repairs starts by assigning all maintenance findings with a work priority based on a timeline for repairs to be completed. Most utilities have multiple work priority levels ranging from repair immediately to monitor. Work priority durations are developed based upon a utility's work practices to address the particular transmission line problem and the problem's impact to public safety, as well as the transmission system and grid. It is important to be realistic in developing and assigning work priorities as a poorly designed and/or applied work priority methodology will result in work inefficiencies and could create unintended regulatory issues. Table 12.2 presents typical transmission line maintenance repair work.

12.9.1 Live Line Work

The term "live line work" refers to working with the conductors in the energized state by either the bare-hand technique or insulated hot stick. This method of transmission line maintenance is preventative and preferred for transmission lines where de-energizing is cost prohibitive, adversely affects reliability, or results in extensive customer outages (radial feed). Live line maintenance is highly specialized work and utilities that perform live line work have the following:

- Clear and consistent guidelines for the performance of such work.
- Equipment and vehicles specifically designed for the energized environment.
- Regular live line training to maintain qualified/certified personnel.
- A transmission system that allows for the minimum approach distance (MAD) to be maintained while working under energized conditions. A simple definition of MAD is the distance upon which the air gap provides sufficient insulation from electrical sparkover to ground due to a potential overvoltage. Figure 12.16 is a simple illustration of the work envelope and MAD for live line work. The distance, D, in Figure 12.16 must be equal to or greater than the MAD which is a function of phase voltage, potential overvoltage, and altitude.

Additional information regarding live line working methods, tools and terminology can be found in IEEE Standards 516–2009 and 935–1989 as well as EPRI's *Live Work Reference Book* also known as the *EPRI Tan Book*.

12.9.2 Worksite Grounding

The purpose behind worksite grounding is to protect transmission line personnel while working on electrically isolated (de-energized) lines in the event the lines become accidently energized or due to induction from adjacent parallel lines. Two fundamental rules are universal to worksite grounding:

1. The installation of protective grounds is considered to be energized work. The circuit is always treated as energized until properly grounded.
2. The line worker should never place himself or herself in series with the grounding system.

Voltage differences commonly referred to as step, touch, and transfer-touch, occur when there is a difference in potential between two points as a result of accidental energization or induction, Figure 12.17. This difference in voltage can result in potentially life threatening current flow for line personnel within the voltage gradient zone. Worksite grounding is designed to reduce the voltage difference within the work zone, thereby minimizing the risk to line personnel working on electrically isolated lines. Information on the temporary protective grounds used for grounding can be found in ASTM F855–2009.

TABLE 12.2 Typical Transmission Line Repair Work

Steel pole repair due to impact. Repairs involve welding a similar thickness patch over the dented/damaged area, if damaged beyond repair the pole is replaced.

Replacement of wood poles, H-structures, and associated members such as cross arms and bracing are usually due to wood degradation or damage/failure as the result of vehicle impact or severe weather.

Wood pole ground line reinforcement usually involves driving a steel C-channel down the side of the pole and attaching it with steel bands. The preferred placement of the steel C-channel is with the open end of the "C" in line with the conductor.

Replacing broken or flashed porcelain or toughened glass insulators. The photos to the left are examples of live line or energized insulator replacement. Polymer insulators are also replaced for similar reasons, however, the appearance and failure mechanisms differ from those of porcelain and toughened glass insulators.

Conductors that have broken strands are typically repaired with patch rod. Patch rod consists of a set of helical wound rods that are formed around the damaged area to provide mechanical strength and electrical continuity.

Conductor splicing is required for damage that is beyond the mechanical capacity of patch rod or for situations where additional conductor needs to be added to existing conductor. The photos to the left show the circumcision of the aluminum to the steel and the completed compression splice.

TABLE 12.2 (continued) Typical Transmission Line Repair Work

Replacement of dampers, spacers, spacer dampers, and other line hardware. The photos to the left are examples of live line spacer-damper replacement from a conductor buggy and boom truck.

Insulator cleaning is sometimes necessary in areas of high contamination. The dry cleaning process uses compressed air to force suitable abrasive media such as corn cobs out of a nozzle, mechanically removing the contamination. The dry process can be done while energized. The wet cleaning process uses pressurized water and must be done under de-energized conditions.

Foundation caps and aboveground retaining structures are used to protect the structure at the foundation–structure interface near and above the grade line. Foundation cap cracking and deterioration are normally the result of poor concrete, environmental degradation, or impact damage.

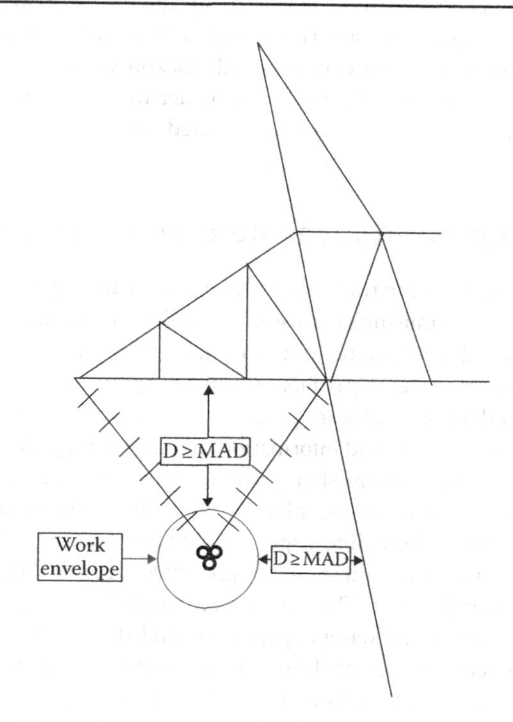

FIGURE 12.16 Illustration of the work envelope, which is the ergonomic work area for line personnel to operate and the distance, D, to grounded conductive elements. The distance, D, must be greater than or equal to the MAD as determined from the phase voltage, potential over voltage, and altitude.

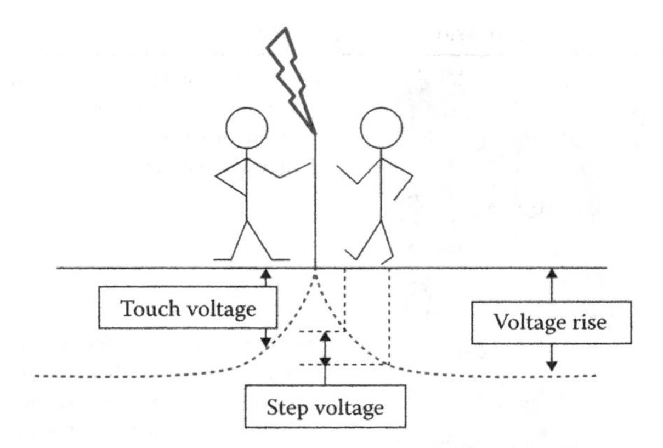

FIGURE 12.17 Illustration of an electrical fault or induced voltage and the resulting voltage rise curve. Touch and step voltages occur when there is a voltage difference between two points. Transfer-touch is similar to touch voltage except the voltage is passed through a conductive element prior to the touch.

In addition to grounding, two other approaches are available for protecting against voltage differences. They are insulation and isolation. Insulation is achieved through the use of insulating (non-conductive) mats, footwear, and gloves. Isolation involves limiting physical access to the work zone usually by barricading or fencing.

12.9.3 Vegetation Management

Transmission line vegetation management is regulated by the North American Electric Reliability Corporation (NERC) via Standard FAC-003-Transmission Vegetation Management Program. This standard was developed in response to massive regional blackouts that were the direct result of vegetation related faults. NERC requires the transmission owner to prepare and keep current all objectives, practices, procedures, and work specifications associated with preventing outages from vegetation located on and adjacent to the transmission line ROW.

12.10 Data/Information Management and Analysis

Effective and efficient management of the transmission line assets can only occur by integrating the data and information from all aspects of transmission inspection and work management programs with that of the GIS or asset management database(s). In today's environment this means that all the software platforms must be capable of interfacing to the necessary databases and perhaps to each other. Software that is not well integrated often results in less than optimal work practices and missed data opportunities that make any follow up analysis difficult. However, data and information that is well integrated provides the opportunity for better understanding of the entire transmission system and analysis of this combined data often results in improved transmission line maintenance operations and asset life cycle performance. Guidance on the collection and management of transmission line inspection data can be found in IEEE Standard 1808–2011.

An example of analyzing data from several sources is presented for 500 kV transmission lines with failing tri-bundle spring-type spacer dampers. The failure mechanism for this type of spacer damper is two body mechanical wear between the steel damping spring(s) and the aluminum body which eventually results in separation of the spacer damper. Data from transmission line visual inspections and installation dates from the GIS database was used to construct the Weibull failure probabilities. Using the Weibull probabilities, a spacer-damper failure forecast was developed which resulted in implementation of a replacement plan suited to the exposure risk and failure characteristics of the spacer dampers. The failure mechanism and analysis process for the tri-bundle spring type spacer damper is given in Figure 12.18.

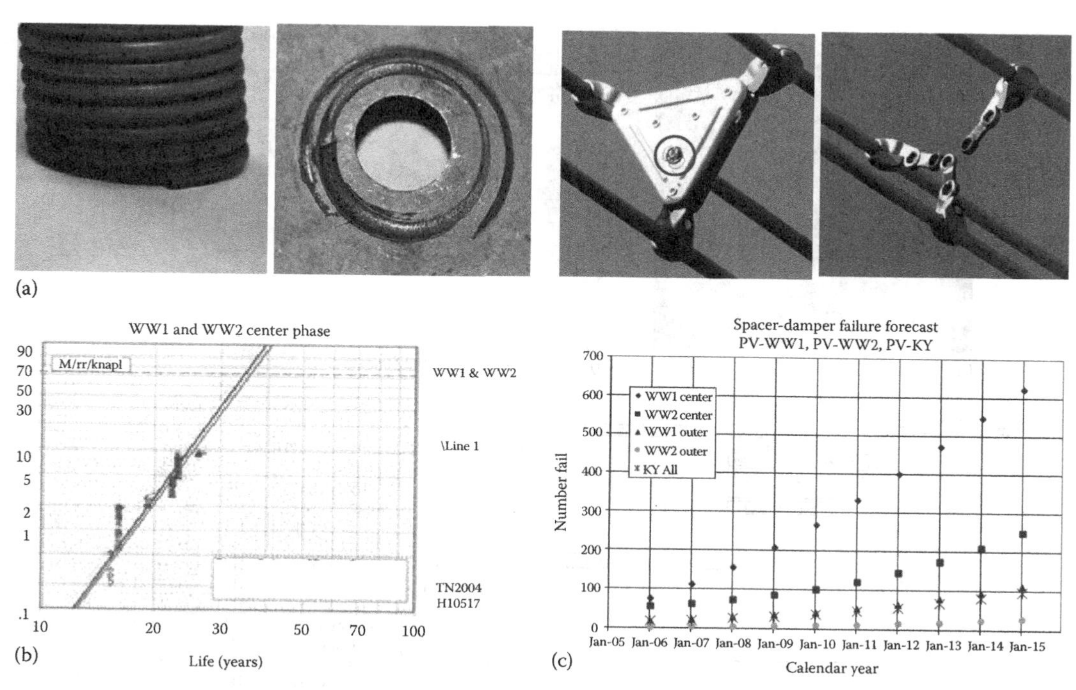

FIGURE 12.18 Spring type spacer-damper failures and analysis. (a) Failure is due to two body wear between the steel damping spring and the aluminum housing, (b) Weibull failure probabilities were determined from inspection and GIS information, and (c) used to develop a failure forecast.

Another example of analysis utilizing data and information from several different sources is presented for the asset management of 69 kV tangent wood poles. This analysis combines information from the GIS database such as pole material, treatment, dimensions, framing, the presence of under built distribution, span lengths, and age with that from the ground line inspection and local wind gust data to provide a structural risk assessment of all tangent wood poles in the circuit. This type of analytical approach to evaluating 69 kV wood poles is much more effective in regards to establishing replacement budgets and managing 69 kV system risk rather than solely basing decisions on the wood pole inspection data alone. Figure 12.19 illustrates the analysis process.

12.11 Emergency Restoration of Transmission Structures

Planning for high impact, low risk of occurrence events that result in loss of transmission structure(s) requires evaluating the economic and operational impact of the downed transmission line to the local transmission system and regional grid. If conditions are such that the line can remain out of service until construction of new structure(s) are complete, then the restoration plan needs to consider maintaining a reasonable number of structures in inventory necessary for line restoration. However, if the economic and operational impacts are such that a prolonged outage is intolerable, the restoration plan's main focus should be on the quick and efficient restoration of the line. Planning for the rapid restoration of transmission line structures at voltages 345 kV and higher usually requires the use of modular emergency restoration structures. These structures consist of column sections fabricated from lightweight, high-strength aluminum alloy, which are easy to transport, and once on-site can be arranged to construct a variety of guyed structures. An example of this is presented in Figure 12.20 for a 500 kV tower which was collapsed due to extremely high localized winds (microburst). This restoration took approximately 4 days from mobilization to line operation.

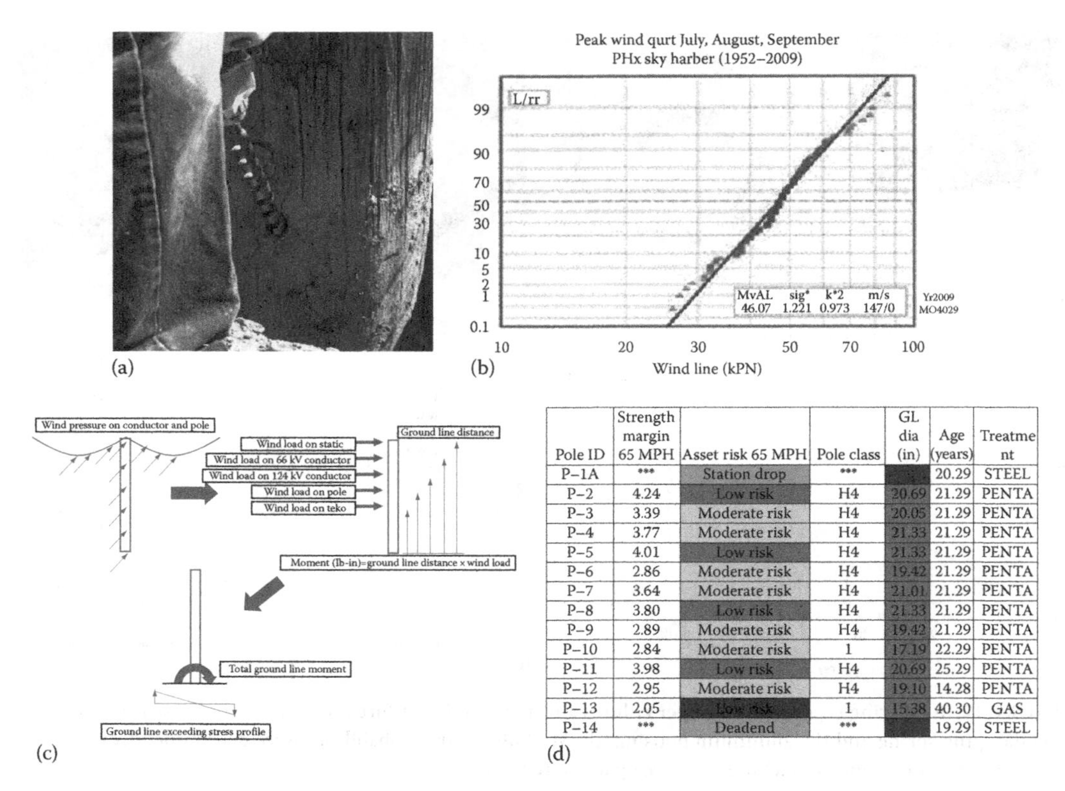

Pole ID	Strength margin 65 MPH	Asset risk 65 MPH	Pole class	GL dia (in)	Age (years)	Treatment
P-1A	***	Station drop	***		20.29	STEEL
P-2	4.24	Low risk	H4	20.69	21.29	PENTA
P-3	3.39	Moderate risk	H4	20.05	21.29	PENTA
P-4	3.77	Moderate risk	H4	21.33	21.29	PENTA
P-5	4.01	Low risk	H4	21.33	21.29	PENTA
P-6	2.86	Moderate risk	H4	19.42	21.29	PENTA
P-7	3.64	Moderate risk	H4	21.01	21.29	PENTA
P-8	3.80	Low risk	H4	21.33	21.29	PENTA
P-9	2.89	Moderate risk	H4	19.42	21.29	PENTA
P-10	2.84	Moderate risk	1	17.19	22.29	PENTA
P-11	3.98	Low risk	H4	20.69	25.29	PENTA
P-12	2.95	Moderate risk	H4	19.10	14.28	PENTA
P-13	2.05	Low risk	1	15.38	40.30	GAS
P-14	***	Deadend	***		19.29	STEEL

FIGURE 12.19 The structural risk assessment process for individual 69 kV tangent wood poles is accomplished by combining and analyzing data from multiple sources. (a) Wood pole inspection data, (b) is combined with loading data, (c) and (d) analyzed to assess wood pole and circuit risk.

FIGURE 12.20 Collapsed 500 kV tower from a microburst and subsequent installation of emergency restoration towers.

References

Transmission line construction references

IEEE Standard 524-2003, *IEEE Guide to the Installation of Overhead Transmission Line Conductors.*

IEEE Standard 1441-2004, *IEEE Guide for the Inspection of Overhead Transmission Line Construction.*

Transmission line maintenance references

ASTM F 855-97, Standard Specifications for Temporary Grounding Systems to be Used on De-Energized Electrical Power Lines and Equipment.

EPRI Live Working Reference Book. EPRI, Palo Alto, CA, 2009. 1018974.

*EPRI Overhead Transmission Inspection and Assessment Guidelines—*2009. EPRI, Palo Alto, CA, 2009. 1017693.

IEEE Standard 516-2009, *IEEE Guide for Maintenance Methods on Energized Power Lines.*

IEEE Standard 935-1989, *IEEE Guide on Terminology for Tools and Equipment to Be Used in Live Line Working.*

IEEE Standard 1808-2011, *IEEE Guide for Collecting and Managing Transmission Line Inspection and Maintenance Data.*

North American Electric Reliability Corporation (NERC) Reliability Standards for the Bulk Electric System of North America, Standard FAC-003-Transmission Vegetation Management Program.

13

Insulated Power Cables Used in Underground Applications

Michael L. Dyer
Salt River Project

Aesthetics is primarily the major reason for installing power cables underground, providing open views of the landscape free of poles and wires. One could also argue that underground lines are more reliable than overhead lines as they are not susceptible to weather and tree caused outages, common to overhead power lines. This is particularly true of temporary outages caused by wind, which represents approximately 80% of all outages occurring on overhead systems. However, underground lines are susceptible to being damaged by excavations (reason behind "call before digging" locating programs implemented by many states in the United States). The time required to repair a damaged underground line may be considerably longer than an overhead line. Underground lines are typically 10 times more expensive to install than overhead lines. The ampacity, current carrying capacity, of an underground line is less than an equivalent sized overhead line. Underground lines require a higher degree of planning than overhead, because it is costly to add or change facilities in an existing system. Underground cables do not have an infinite life, because the dielectric insulation is subjected to aging; therefore, systems should be designed with future replacement or repair as a consideration.

13.1 Underground System Designs

There are two types of underground systems (Figure 13.1).

1. Radial—Where the transformers are served from a single source.
2. Looped—Where the transformers are capable of being served from one of two sources. During normal operation an open is located at one of the transformers, usually the midpoint.

A radial system has the lowest initial cost, because a looped system requires the additional facilities to the second source. Outage restoration on a radial system requires either a cable repair or replacement, whereas on a looped system, switching to the alternate source is all that is required.

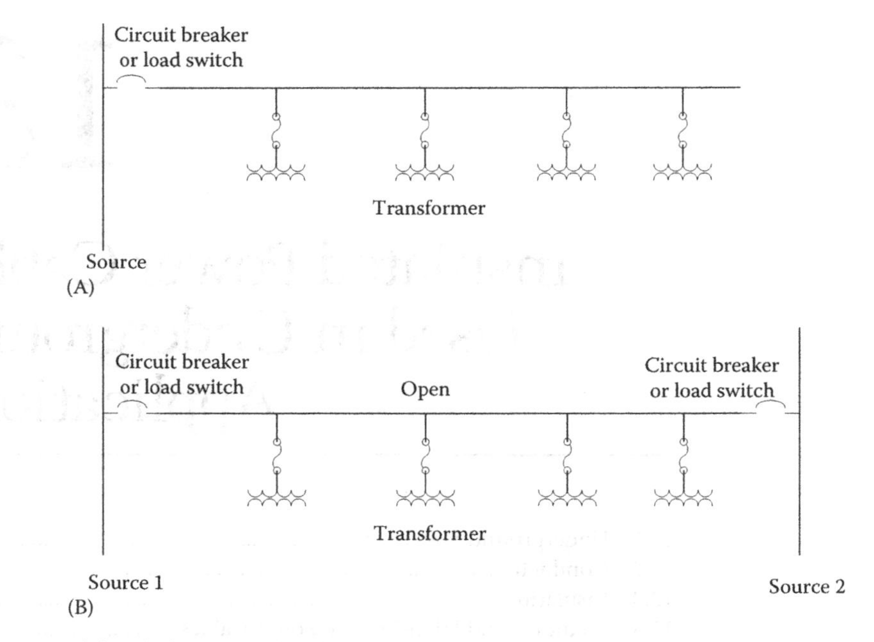

FIGURE 13.1 (A) Radial system and (B) looped system.

Underground cable can be directly buried in earth, which is the lowest initial cost, allows splicing at the point of failure as a repair option and allows for maximum ampacity. Cables may also be installed in conduit, which is an additional cost, requires replacement of a complete section as the repair option, reduces the ampacity, because the conduit wall and surrounding air are additional thermal resistances, but provides protection to the cable.

Underground power cables have three classifications.

1. Low voltage—Limited to 2 kV. Primarily used as service cables
2. Medium voltage—2–46 kV. Primarily used to supply distribution transformers
3. High voltage—Above 46 kV. Primarily used to supply substation transformers

American Standards Testing Material (ASTM), Insulated Cable Engineering Association (ICEA), National Electrical Manufacturing Association (NEMA), and Association of Edison Illuminating Companies (AEIC) have published standards for the various types of power cables.

13.2 Conductor

Common among all classes in function is the central conductor, the purpose of which is to conduct power (current and voltage) to serve a load. The metals of choice are either copper or aluminum. This central conductor may be composed of a single element (solid) or composed of multiple elements (stranded), on the basis of a geometric progression of 6, 12, 18, etc., of individual strands for each layer. Each layer is helically applied in the opposite direction of the underlying layer.

There are three common types of stranding available.

1. Concentric round
2. Compressed round (97% of the diameter of concentric)
3. Compact round (90%–91% of the diameter of concentric)

Note: Some types of connectors may be suitable for stranded types 1 and 2 but not type 3 for the same size.

To improve manufacturing, 19 wire combination unilay stranding (helically applied in one direction one operation) has become popular in low-voltage applications, where some of the outer strands are of a smaller diameter, but the same outside diameter as compressed round is retained. Another stranding method which retains the same overall diameter is single input wire (SIW) compressed, which can be used to produce a wide range of conductors using a smaller range of the individual strands.

Conductors used at transmission voltages may have exotic stranding to reduce the voltage stress.

Cables requiring greater flexibility such as portable power cable utilize very fine strands with a rope type stranding.

Typical sizes for power conductors are #6 American wire gage (AWG) through 1000 kcmil. One cmil is defined as the area of a circle having a diameter of 1 mile (0.0001 in.). Solid conductors are usually limited to a maximum of #1/0 because of flexibility.

The metal type and size determines the ampacity and losses (I^2R). Copper having a higher intrinsic conductivity will have a greater ampacity and lower resistance than an equivalent size aluminum conductor. Aluminum 1350 alloy medium hardness is typical for power cable use.

13.3 Insulation

In order to install power cables underground, the conductor must be insulated. For low-voltage applications, a layer of insulation is extruded onto the conductor. Many types of insulation compounds have been used from natural or synthetic rubber, polyvinyl chloride (PVC), high molecular weight polyethylene (HMWPE), and cross-linked polyethylene (XLPE) to name a few. Although each insulation type has various characteristics, operating temperature and durability are probably the most important. XLPE is probably the most widely used insulation for low-voltage cables. XLPE is a thermoset plastic with its hydrocarbon molecular chains cross-linked. Cross-linking is a curing process, which occurs under heat and pressure, or as used for low-voltage cables, moisture and allows an operating temperature of 90°C.

Multiple layer cable insulation composed of a softer compound under a harder compound, a single layer harder insulation, or a self-healing insulation are used to address protection of the conductor, typically for direct buried low-voltage power cables.

13.4 Medium- and High-Voltage Power Cables

Medium- and high-voltage power cables, in addition to being insulated, are shielded to contain and evenly distribute the electric field within the insulation.

The components and function of a medium- and high-voltage cable are as follows (Figure 13.2A and B):

1. The center conductor—Metallic path to carry power.
2. The conductor shield—A semiconducting layer placed over the conductor to provide a smooth conducting cylinder around the conductor. Typical of today's cables, this layer is a semiconducting plastic, polymer with a carbon filler, extruded directly over the conductor. This layer represents a very smooth surface, which, because of direct contact with the conductor, is elevated to the applied voltage on the conductor.
3. The insulation—A high dielectric material to isolate the conductor. The two basic types used today are XLPE or ethylene propylene rubber (EPR). Because of an aging effect known as treeing (Figure 13.3), on the basis of its visual appearance, caused by moisture in the presence of an electric field, a modified version of XLPE designated tree retardant (TRXLPE) has replaced the use of XLPE for medium-voltage applications. High-voltage transmission cables still utilize XLPE, but they usually have a moisture barrier. TRXLPE is a very low loss dielectric that is reasonably flexible and has an operating temperature limit of 90°C or 105°C depending on type. TRXLPE because it is cross-linked, does not melt at high operating temperatures but softens. EPR is a

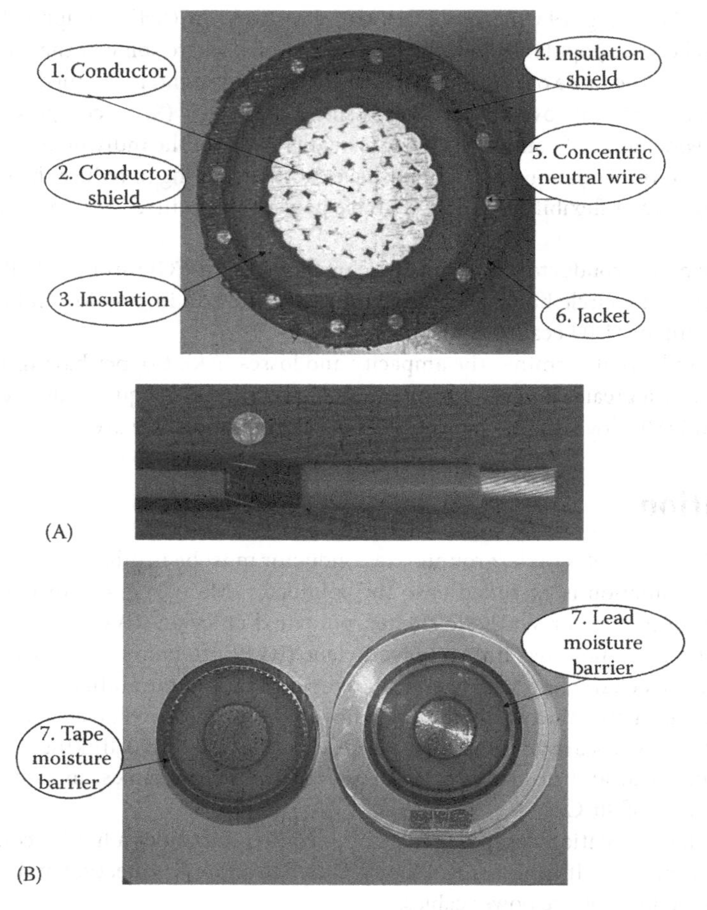

FIGURE 13.2 (A) Medium-voltage cable components and (B) high-voltage cable components.

FIGURE 13.3 Tree in XLPE.

rubber-based insulation having higher losses than TRXLPE and is very flexible and has an operating temperature limit of 105°C. EPR does not melt or soften as much as TRXLPE at high operating temperatures, because of its high filler content.

4. The insulation shield—A semiconducting layer to provide a smooth cylinder around the outside surface of the insulation. Typical shield compound is a polymer with a carbon filler that is extruded directly over the insulation. This layer, for medium-voltage applications, is not fully bonded to the insulation (strippable) to allow relatively easy removal for the installation of cable accessories. Transmission cables have this layer bonded to the insulation, which requires shaving tools to remove.
5. The metallic shield—A metallic layer, which may be composed of wires, tapes, or corrugated tube. This shield is connected to the ground, which keeps the insulation shield at ground potential and provides a return path for fault current. Medium-voltage cables can utilize the metallic shield as the neutral return conductor if sized accordingly. Typical metallic shield sizing criteria:
   a. Equal in ampacity to the central conductor for one phase applications.
   b. One-third the ampacity for three-phase applications.
   c. Fault duty for three-phase feeders and transmission applications.
6. Overall jacket—A plastic layer applied over the metallic shield for physical protection. This polymer layer may be extruded as a loose tube or directly over the metallic shield (encapsulated). Although both provide physical protection, the encapsulated jacket removes the space present in a loose tube design, which may allow longitudinal water migration. The typical compound used for jackets is linear low density polyethylene (LLDPE), because of its ruggedness and relatively low water vapor transmission rate. Jackets can be specified insulating (most common) or semiconducting (when jointly buried and randomly laid with communication cables).
7. Moisture barrier—A sealed metallic barrier applied either over or under the overall jacket. Typically used for transmission cables, this barrier may be a sealed tape, corrugated tube, or lead sheath.

Cable components 1–4 comprise the cable core, which in cross-section, is a capacitor with the conductor shield and insulation shield making up the plates on each side of a dielectric. These plates evenly distribute the electric field radially in all directions within the insulation; however, until the metallic shield is added and effectively grounded, the insulation shield is subject to capacitive charging and presents a shock hazard. To be considered effectively grounded, the National Electrical Safety Code (NESC) requires a minimum of four ground connections per mile of line or eight grounds per mile when jointly buried with communication cables for insulating jackets. Semiconducting jackets are considered grounded when in contact with earth.

Because medium- and high-voltage cables are shielded, special methods are required to connect them to devices or other cables. Since the insulation shield is conductive and effectively grounded, it must be carefully removed a specific distance from the conductor end, on the basis of the operating voltage. Once the insulation shield has been removed, the electric field will no longer be contained within the insulation and the highest electrical stress will be concentrated at the end of the insulation shield (Figure 13.4). Premolded, cold or heat shrink, or special tapes are applied at this location to control this stress, allowing the cable to be connected to various devices (Figure 13.5).

13.5 Shield Bonding Practice

Generally, the metallic shields on distribution circuits are grounded at every device. Transmission circuits, however, may use one of the following configurations.

Multiple ground connections (multigrounded) (Figure 13.6A): The metallic shield will carry an induced current because they surround the alternating current in the central conductor. This circulating current results in an I^2R heating loss, which adversely affects the ampacity of the cable.

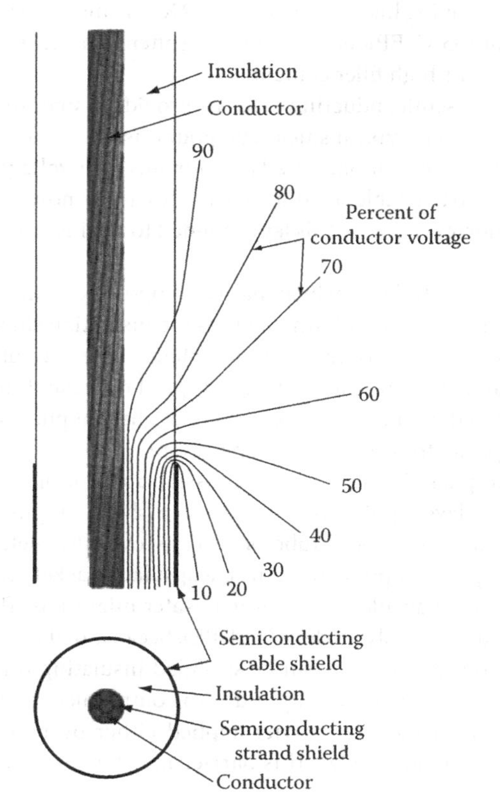

FIGURE 13.4 Voltage distribution in the insulation with the cable shield removed.

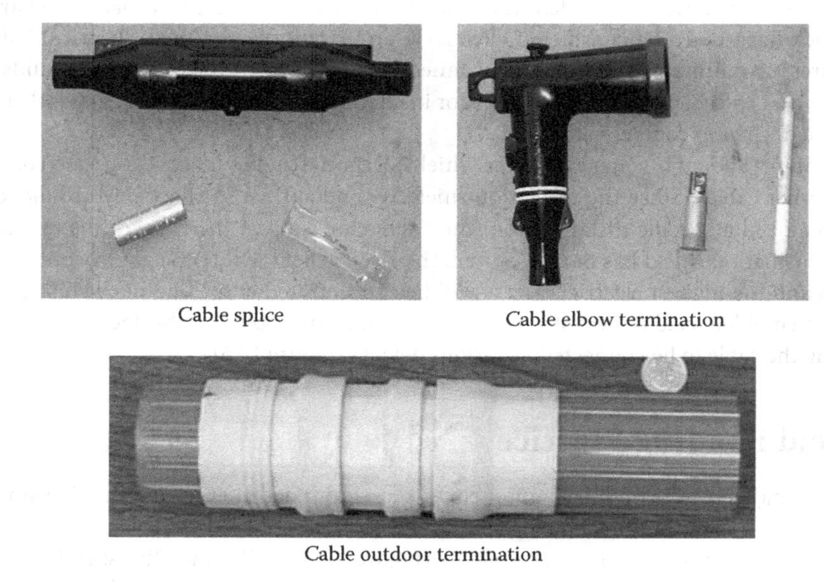

FIGURE 13.5 Cable accessories.

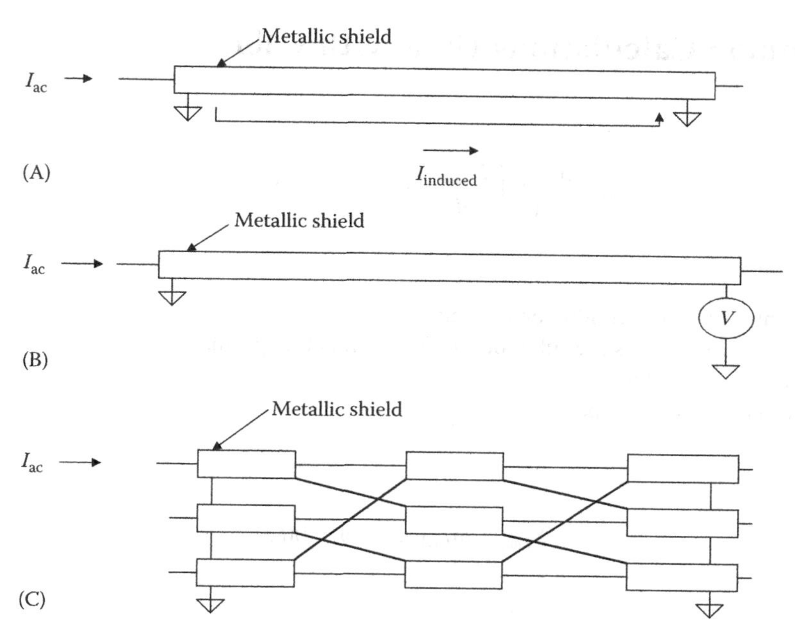

FIGURE 13.6 (A) Multigrounded shield, (B) single point grounded shield, and (C) cross-bonding shields.

Single point grounded (Figure 13.6B): The metallic shield is grounded at a single point and no current can flow in the metallic shield because there is no closed circuit. This configuration allows the maximum ampacity rating for the cable; however, a voltage will be present on the open end, which may be a hazard. This voltage is dependent on the cable spacing, current, and cable length.

Cross-bonding (Figure 13.6C): The three-phase circuit is divided into three equal segments. The metallic shield between each segment is connected to an adjacent phase using insulated conductor. Splices at these segments must interrupt the insulation shield to be effective.

13.6 Installation Practice

When cables are directly buried in earth, the trench bottom may require bedding sand or select backfill free from rocks that could damage the cable over time. When the cable is installed in conduit, the pulling tension must be limited so as not to damage the conductor, insulation, or shields. Typical value when using a wire basket grip is 3000 lbs. When the cable is pulled around a bend, the pulling tension results in a side-wall bearing force against the inside surface of the elbow. This force must be limited to avoid crushing the cable components. Cables also have a minimum bending radius limit that prevents distortion of the cable components.

13.7 System Protection Devices

Two types of protecting devices are used on cable systems.

1. Overcurrent—Fuses or circuit breakers. These devices isolate the cable from its source, preventing the flow of damaging levels of current during an overload, or remove a faulted cable from the system allowing restoration of the unfaulted parts.
2. Overvoltage—Surge arrester. This device prevents damaging overvoltages caused by lightning or switching surges from entering the cable by clamping the voltage to a level tolerated by the cable insulation.

13.8 Common Calculations Used with Cable

Inductance

$$L_{\text{cable}} = \frac{\mu_o}{2\pi}\left(\ln\left(\frac{2s_{\text{cable}}}{d}\right) + \frac{1}{4}\right) \quad \mu_o = 4\pi 10^{-7}\,\frac{\text{H}}{\text{m}},$$

where

s_{cable} is the center-to-center conductor spacing
for three single cables s_{cable} is the cube root of each conductor spacing
d is the conductor diameter
μ_o is the permeability of free space

Inductive reactance

$$X_{\text{cable}} = \omega L_{\text{cable}} L \quad \omega = 2\pi f,$$

where

f is the frequency
L_{cable} is the inductance
L is the length

Capacitance

$$C_{\text{cable}} = \frac{2\pi\varepsilon_o\varepsilon}{\ln(D/d)} \quad \varepsilon_o = \frac{10^{-9}}{36\pi}\,\frac{\text{F}}{\text{m}},$$

where

ε is the relative dielectric constant of the insulation (2.4—XLPE, 2.9—EPR)
ε_o is the free space permittivity
D is the diameter of insulation under insulation shield when present
d is the diameter of the conductor in inches over the conductor shield when present

Charging current

$$I_{\text{cap}} = V_n(\omega C_{\text{cable}} L),$$

where

C_{cable} is the capacitance
V_n is the voltage line to neutral
L is the length

Ampacity

$$I_{\text{amp}} = \sqrt{\frac{T_c - T_a}{R_{\text{ac}} R_{\text{th}}}},$$

where
T_c is the conductor temperature
T_a is the ambient temperature
R_{ac} is the ac resistance at the operating temperature
R_{th} is the thermal resistance of surrounding environment

Voltage drop

$$\text{Voltage drop} = I_{cable}\left(R_{cable}\cos(\phi) + X_{cable}\sin(\phi)\right),$$

where
I_{cable} is the current in conductor
R_{cable} is the total ac resistance of the cable
X_{cable} is the total ac reactance of the cable
ϕ is the phase angle between supply voltage and current

For single-phase calculations the values of the main and the return conductors must be used.

Pulling tension single cable in straight conduit

$$T = \mu WL,$$

where
μ is the coefficient of dynamic friction (0.2–0.7 dependent on cable exterior and type of conduit)
W is the cable weight per unit length
L is the length

Pulling tension single cable through conduit bend

$$T_{out} = T_{in}e^{\mu\phi}(\text{lbs}),$$

where
T_{in} is the tension entering the bend
μ is the coefficient of dynamic friction (0.2–0.7 dependent on cable exterior and type of conduit)
ϕ is the bend angle in radians

The pulling tensions of each segment of the conduit path are added together to determine the total pulling tension.

When multiple single cables are installed in a conduit, a multiplier must be applied to the cable weight, accounting for configuration as follows:

For three cables with a triangular configuration the weight multiplier is

$$W_{\text{multiplier triangular}} = \frac{2}{\sqrt{1-(d/(D-d))^2}}.$$

For three cables with a cradled configuration

$$W_{\text{multiplier triangular}} = 1 + \frac{4}{3}\left(\frac{d}{D-d}\right)^2,$$

where

d is the single cable outside diameter

D is the conduit inside diameter

References

ANSI/IEEE 575-1988. IEEE Guide for the Application of Sheath-Bonding Methods for Single-Conductor Cables and the Calculation of Induced Voltages and Currents in Cable Sheaths.

Arnold, T. P. (ed.). 1997. *Southwire Company Power Cable Manual*, 2nd edn. Southwire Company, Carrollton, GA.

Association of Edison Illuminating Companies, AEIC CS8-2005. Specification for Extruded Dielectric, Shielded Power Cables Rated 5 through 46 kV.

ICEA Standard S-81-570-2005. 600 Volt Rated Cables of Ruggedized Design for Direct Burial Installations as Single Conductors or Assemblies of Single Conductors.

ICEA Standard S-94-694-2004. Concentric Neutral Cables Rated 5 through 46 kV.

ICEA Standard S-97-682-2000. Utility Shielded Power Cables Rated 5 through 46 kV.

ICEA Standard S-105-692-2004. 600 Volt Single layer Thermoset Insulated Utility Underground Distribution Cables.

ICEA Standard S-108-720-2004. Extruded Insulation Power Cables Rated above 46 through 345 kV.

IEEE 48-1996. IEEE Standard Test Procedures and Requirements for Alternating Current Cable Terminations 2.5 kV through 765 kV.

IEEE 386-2005. IEEE Standard for Separable Insulated Connector Systems for Power Distribution Systems above 600 V.

IEEE 404-1993. IEEE Standard for Cable Joints for use with Extruded Dielectric Cable Rated 5000–138,000 V and Cable Joints for use with Laminated Dielectric Cable Rated 2500–500,000 V.

IEEE 1215-2001. IEEE Guide for the Application of Separable Insulated Connectors.

Insulated Cable Engineering Association, ICEA, Standard P-53-426. Ampacities, 15–69 kV 1/c Power Cable Including Effect of Shield Losses (Solid Dielectric).

Walker, M. 1982. *Aluminum Electrical Conductor Handbook*. The Aluminum Association, New York.

14

Transmission Line Parameters

Manuel
Reta-Hernández
Universidad Autónoma de Zacatecas

14.1 Transmission Line Parameters

The power transmission line is one of the main components of an electric power system. Its major function is to transport electric energy, with minimal losses, from the power sources to the load centers, usually separated by long distances. The design of a transmission line depends on four electrical parameters:

1. Series resistance
2. Series inductance
3. Shunt capacitance
4. Shunt conductance

The series resistance relies basically on the physical composition of the conductor at a given temperature. The series inductance and shunt capacitance are produced by the presence of magnetic and electric fields around the conductors, and depend on their geometrical arrangement. The shunt conductance is due to leakage currents flowing across insulators and air. As leakage current is considerably small compared to nominal current, it is usually neglected, and therefore, the shunt conductance is normally not considered for the transmission line modeling.

14.1.1 Series Resistance

AC resistance of a conductor in a transmission line is based on the calculation of its DC resistance. If DC current is flowing along a round cylindrical conductor, the current is uniformly distributed over its cross-section area and its DC resistance, at a given temperature, is evaluated by Equation 14.1:

$$R_{DC} = \frac{\rho l}{A} \quad (\Omega) \tag{14.1}$$

where

ρ is the conductor resistivity at a given temperature (Ω m)
l is the conductor length (m)
A is the conductor cross-section area (m^2)

If AC current is flowing, rather than DC current, the conductor effective resistance is higher due to frequency or skin effect.

14.1.1.1 Frequency Effect

Frequency of AC voltages produces a second effect on the conductor resistance due to the non-uniform distribution of the current. This phenomenon is known as *skin effect*. As frequency increases, the current tends to go toward the surface of the conductor and the current density decreases at the center. Skin effect reduces the effective cross-section area used by the current, and thus, the effective resistance increases. Also, although in very small amount, a further resistance increase occurs when other current-carrying conductors are present in the immediate vicinity. A skin correction factor k, obtained by differential equations and Bessel functions, is considered to reevaluate the AC series resistance as shown in Equation 14.2. For a 60 Hz frequency, it is estimated $k = 1.02$:

$$R_{AC} = R_{DC}k \tag{14.2}$$

Other variations in the series AC resistance are caused by

- Temperature
- Spiraling of stranded conductors
- Bundle conductors' arrangement

14.1.1.2 Temperature Effect

The resistivity of any conductive material varies linearly over an operating temperature, and therefore, the resistance of any conductor suffers the same variations. As temperature rises, the conductor resistance increases linearly, over normal operating temperatures, according to Equation 14.3:

$$R_2 = R_1\left(\frac{T+t_2}{T+t_1}\right) \quad (\Omega) \tag{14.3}$$

where

R_2 is the resistance at second temperature t_2
R_1 is the resistance at initial temperature t_1
T is the temperature coefficient for a specific material (°C)

Resistivity (ρ) and temperature coefficient (T) constants depend upon the particular conductor material. Table 14.1 lists resistivity and temperature coefficients of some typical conductor materials.

14.1.1.3 Spiraling and Bundle Conductor Effect

There are two types of transmission line conductors: overhead and underground. Overhead conductors, made of naked metal and suspended on insulators, are preferred over underground conductors because

TABLE 14.1 Resistivity and Temperature Coefficient of Some Conductors

Material	Resistivity at 20°C (Ω m)	Temperature Coefficient (°C)
Silver	1.59×10^{-8}	243.0
Annealed copper	1.72×10^{-8}	234.5
Hard-drawn copper	1.77×10^{-8}	241.5
Aluminum	2.83×10^{-8}	228.1

of the low cost and easy maintenance. Also, overhead transmission lines use aluminum conductors, because of the lower cost and lighter weight compared to copper conductors, although more cross-section area is needed to conduct the same amount of current.

Among different types of commercially available aluminum conductors, aluminum-conductor-steel-reinforced (ACSR) conductor is one of the most used conductors in overhead transmission lines. It consists of alternate layers of stranded conductors, spiraled in opposite directions to hold the strands together, surrounding a core of steel strands.

The purpose of introducing a steel core inside the stranded aluminum conductors is to obtain a high strength-to-weight ratio. A stranded conductor offers more flexibility and is easier to manufacture than a solid large conductor. However, the total resistance is increased because the outside strands are larger than the inside strands on account of the spiraling [1,2]. The resistance of each wound conductor at any layer, per-unit length, is based on its total length as shown in Equation 14.4:

$$R_{cond} = \frac{\rho}{A}\sqrt{1+\left(\pi\frac{1}{p}\right)^2} \quad (\Omega/\text{m}) \tag{14.4}$$

where

R_{cond} is the resistance of wound conductor (Ω)

$\sqrt{1+(\pi(1/p))^2}$ is the length of wound conductor (m)

$p = (l_{turn}/2r_{layer})$ is the relative pitch of wound conductor, l_{turn} is the length of one turn of the spiral (m), and $2r_{layer}$ is the diameter of the layer (m)

The parallel combination of n conductors, with same diameter per layer, gives the resistance per layer as follows:

$$R_{layer} = \frac{1}{\sum_{i=1}^{n}(1/R_i)} \quad (\Omega/\text{m}) \tag{14.5}$$

Similarly, the total resistance of the stranded conductor is evaluated by the parallel combination of resistances per layer.

In high-voltage overhead transmission lines, there may be more than one conductor per phase (bundle configuration) to increase the current capability and to reduce corona effect discharge. Corona effect occurs when the surface potential gradient of a conductor exceeds the dielectric strength of the surrounding air (30 kV/cm during fair weather), producing ionization in the area close to the conductor, with consequent corona losses, audible noise, and radio interference. As corona effect is a function of conductor diameter, line configuration, and conductor surface condition, meteorological conditions play a key role in its evaluation. Corona losses under rain or snow, for instance, are much higher than in dry weather.

Corona, however, can be reduced by increasing the total conductor surface. Although corona losses rely on meteorological conditions, their evaluation takes into account the conductance between conductors and between conductors and ground. By increasing the number of conductors per phase, the total cross-section area increases, the current capacity increases, and the total series AC resistance decreases proportionally to the number of conductors per bundle. Conductor bundles may be applied to any voltage but are always used at 345 kV and above to limit corona [2,3]. Spacers made of steel or aluminum bars are used to maintain the distance between bundle conductors along the line. Figure 14.1 shows some typical arrangement of stranded bundle configurations.

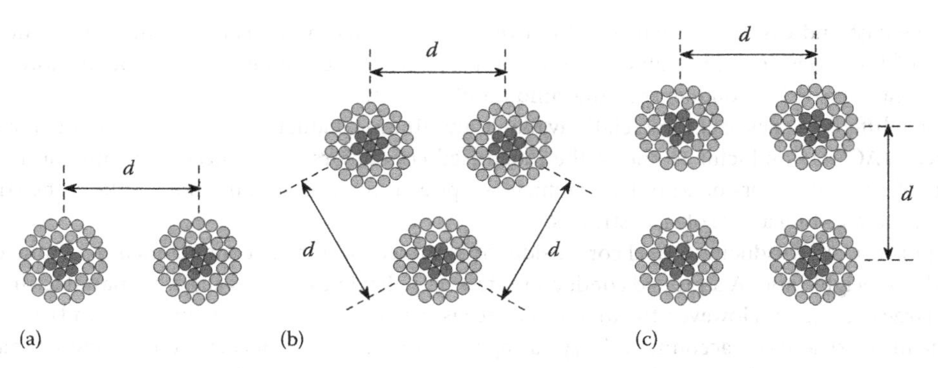

FIGURE 14.1 Stranded conductors arranged in bundles per phase of (a) two, (b) three, and (c) four.

14.1.1.4 Current-Carrying Capacity (Ampacity)

In overhead transmission lines, the current-carrying capacity is determined mostly by the conductor resistance and the heat dissipated from its surface. The heat generated in a conductor (Joules's effect) is dissipated from its surface area by convection and radiation given by Equation 14.6:

$$I^2R = S(w_c + w_r) \quad \text{(W)} \tag{14.6}$$

where
- R is the conductor resistance (Ω)
- I is the conductor current-carrying (A)
- S is the conductor surface area (sq. in.)
- w_c is the convection heat loss (W/sq. in.)
- w_r is the radiation heat loss (W/sq. in.)

Heat dissipation by convection is defined in Equation 14.7 as

$$w_c = \frac{0.0128\sqrt{pv}}{T_{air}^{0.123}\sqrt{d_{cond}}}\Delta t \quad \text{(W)} \tag{14.7}$$

where
- p is the atmospheric pressure (atm)
- v is the wind velocity (ft/s)
- d_{cond} is the conductor diameter (in.)
- T_{air} is the air temperature (°C)
- $\Delta t = T_c - T_{air}$ is the temperature rise of the conductor (°C)

Heat dissipation by radiation is obtained from Stefan–Boltzman law and is defined as

$$w_r = 36.8E\left[\left(\frac{T_c}{1000}\right)^4 - \left(\frac{T_{air}}{1000}\right)^4\right] \quad \text{(W/sq. in.)} \tag{14.8}$$

where
- w_r is the radiation heat loss (W/sq. in.)
- E is the emissivity constant (1 for the absolute black body and 0.5 for oxidized copper)
- T_c is the conductor temperature (°C)
- T_{air} is the ambient temperature (°C)

Substituting Equations 14.7 and 14.8 in 14.6 we can obtain the conductor ampacity at given temperatures, as shown in Equations 14.9 and 14.10:

$$I = \sqrt{\frac{S(w_c + w_r)}{R}} \quad \text{(A)} \tag{14.9}$$

$$I = \sqrt{\frac{S}{R}\left(\frac{\Delta t\left(0.0128\sqrt{pv}\right)}{T_{air}^{0.123}\sqrt{d_{cond}}} + 36.8\,E\left(\frac{T_c^4 - T_{air}^4}{1000^4}\right)\right)} \quad \text{(A)} \tag{14.10}$$

Some approximated current-carrying capacity for overhead ACSR and AAC conductors are presented in Section 14.1.5 [1–3].

14.1.2 Series Inductance and Series Inductive Reactance

A current-carrying conductor produces concentric magnetic flux lines around the conductor. If the current varies with the time, the magnetic flux changes and a voltage is induced. Therefore, an inductance is present, defined as the ratio of the magnetic flux linkage and the current. The magnetic flux produced by the current in transmission line conductors produces a total inductance whose magnitude depends on the line configuration [4–7]. To determine the inductance of the line, it is necessary to calculate, as in any magnetic circuit with permeability μ, the following factors:

1. Magnetic field intensity H
2. Magnetic field density B
3. Flux linkage λ

14.1.2.1 Inductance of a Solid, Round, Infinitely Long Conductor

Consider an infinitely long, solid cylindrical conductor with radius r, carrying-current I as shown in Figure 14.2. If the conductor is made of a nonmagnetic material, and the current is assumed uniformly distributed (no skin effect), then the generated internal and external magnetic field lines are concentric circles around the conductor with direction defined by the right-hand rule.

14.1.2.2 Internal Inductance due to Internal Magnetic Flux

To obtain the internal inductance, a magnetic field with radius x inside the conductor of length l is chosen, as shown in Figure 14.3.

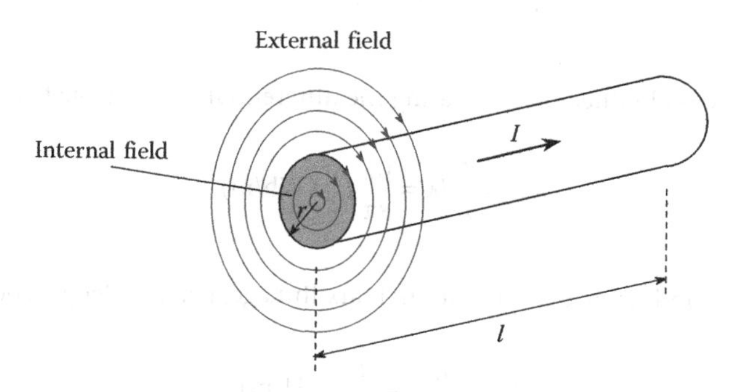

FIGURE 14.2 External and internal concentric magnetic flux lines around the conductor.

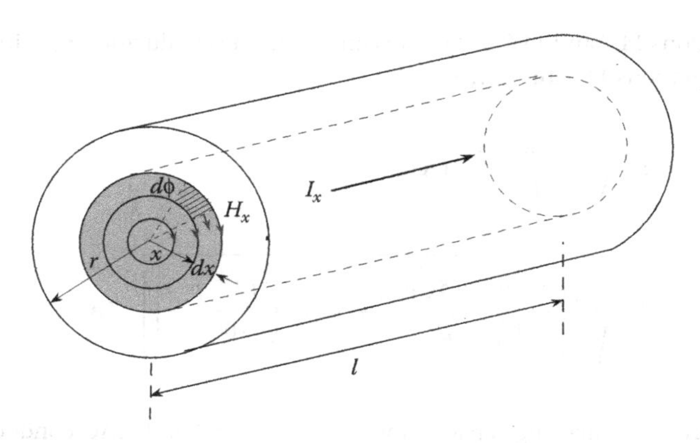

FIGURE 14.3 Internal magnetic flux.

The fraction of the current I_x enclosed in the area of the circle chosen is determined by

$$I_x = I\frac{\pi x^2}{\pi r^2} \quad \text{(A)} \tag{14.11}$$

Ampere's law determines the magnetic field intensity H_x, constant at any point along the circle contour as

$$H_x = \frac{I_x}{2\pi x} = \frac{I}{2\pi r^2}\,x \quad \text{(A/m)} \tag{14.12}$$

The magnetic flux density B_x is obtained by

$$B_x = \mu H_x = \frac{\mu_0}{2\pi}\left(\frac{Ix}{r^2}\right) \quad \text{(T)} \tag{14.13}$$

where $\mu = \mu_0 = 4\pi \times 10^{-7}$ H/m is the magnetic permeability for a nonmagnetic material.

The differential flux $d\phi$ enclosed in a ring of thickness dx for a 1 m length of conductor, and the differential flux linkage $d\lambda$ in the respective area are defined in Equations 14.14 and 14.15:

$$d\phi = B_x\,dx = \frac{\mu_0}{2\pi}\left(\frac{Ix}{r^2}\right)dx \quad \text{(Wb/m)} \tag{14.14}$$

$$d\lambda = \frac{\pi x^2}{\pi r^2}\,d\phi = \frac{\mu_0}{2\pi}\left(\frac{Ix^3}{r^4}\right)dx \quad \text{(Wb/m)} \tag{14.15}$$

The internal flux linkage is obtained by integrating the differential flux linkage from $x = 0$ to $x = r$:

$$\lambda_{int} = \int_0^r d\lambda = \frac{\mu_0}{8\pi}I \quad \text{(Wb/m)} \tag{14.16}$$

Therefore, the conductor inductance due to internal flux linkage, per-unit length, becomes

$$L_{int} = \frac{\lambda_{int}}{I} = \frac{\mu_0}{8\pi} \quad \text{(H/m)} \tag{14.17}$$

14.1.2.3 External Inductance

The external inductance is evaluated assuming that the total current I is concentrated at the conductor surface (maximum skin effect). At any point, on an external magnetic field circle of radius y (Figure 14.4), the magnetic field intensity H_y and the magnetic field density B_y, per-unit length, are defined by Equations 14.18 and 14.19:

$$H_y = \frac{I}{2\pi y} \quad \text{(A/m)} \tag{14.18}$$

$$B_y = \mu\, H_y = \frac{\mu_0}{2\pi}\frac{I}{y} \quad \text{(T)} \tag{14.19}$$

The differential flux $d\phi$ enclosed in a ring of thickness dy, from point D_1 to point D_2, for a 1 m length of conductor is

$$d\phi = B_y\, dy = \frac{\mu_0}{2\pi}\frac{I}{y}dy \quad \text{(Wb/m)} \tag{14.20}$$

As the total current I flows in the surface conductor, then the differential flux linkage $d\lambda$ has the same magnitude as the differential flux $d\phi$:

$$d\lambda = d\phi = \frac{\mu_0}{2\pi}\frac{I}{y}dy \quad \text{(Wb/m)} \tag{14.21}$$

The total external flux linkage enclosed by the ring is obtained by integrating from D_1 to D_2 becomes

$$\lambda_{1-2} = \int_{D_1}^{D_2} d\lambda = \frac{\mu_0}{2\pi} I \int_{D_1}^{D_2} \frac{dy}{y} = \frac{\mu_0}{2\pi} I \ln\left(\frac{D_1}{D_2}\right) \quad \text{(Wb/m)} \tag{14.22}$$

In general, the total external flux linkage from the surface of the conductor to any point D, per-unit length, is

$$\lambda_{ext} = \int_{r}^{D} d\lambda = \frac{\mu_0}{2\pi} I \ln\left(\frac{D}{r}\right) \quad \text{(Wb/m)} \tag{14.23}$$

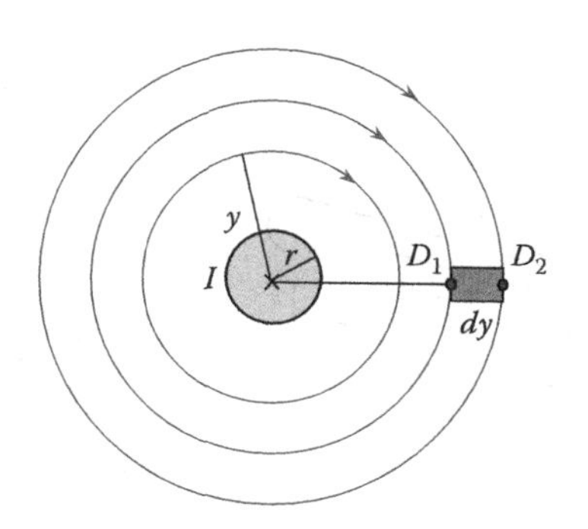

FIGURE 14.4 External magnetic field.

The summation of the internal and external flux linkage at any point D permits evaluation of the total inductance of the conductor L_{tot}, per-unit length, as follows:

$$\lambda_{int} + \lambda_{ext} = \frac{\mu_0}{2\pi} I \left[\frac{1}{4} + \ln\left(\frac{D}{r}\right) \right] = \frac{\mu_0}{2\pi} I \left[\ln\left(e^{1/4}\right) + \ln\left(\frac{D}{r}\right) \right] = \frac{\mu_0}{2\pi} I \ln\left(\frac{D}{e^{-1/4} r}\right) \quad \text{(Wb/m)} \tag{14.24}$$

Therefore, the expression of total inductance is given by Equation 14.25:

$$L_{tot} = \frac{\lambda_{int} + \lambda_{ext}}{I} = \frac{\mu_0}{2\pi} \ln\left(\frac{D}{GMR}\right) \quad \text{(H/m)} \tag{14.25}$$

where GMR (geometric mean radius) $= e^{-1/4} r = 0.7788\, r$.

GMR can be considered as the radius of a fictitious conductor assumed to have no internal flux but with the same inductance as the actual conductor with radius r.

14.1.2.4 Inductance of a Two-Wire, Single-Phase Line

Now, consider a two-wire single-phase line with solid cylindrical conductors A and B with the same radius r, same length l, and separated by a distance D, where $D > r$, and conducting the same current I, as shown in Figure 14.5. The current flows from the source to the load in conductor A and returns in conductor B ($I_A = -I_B$).

The magnetic flux generated by one conductor links the second conductor. The total flux linking conductor A, for instance, has two components: (a) the flux generated by conductor A and (b) the flux generated by conductor B which links conductor A.

As shown in Figure 14.6, the total flux linkage from conductors A and B at point P are

$$\lambda_{AP} = \lambda_{AAP} + \lambda_{ABP} \tag{14.26}$$

$$\lambda_{BP} = \lambda_{BBP} + \lambda_{BAP} \tag{14.27}$$

where

λ_{AAP} is the flux linkage from magnetic field of conductor A on conductor A at point P
λ_{ABP} is the flux linkage from magnetic field of conductor B on conductor A at point P
λ_{BBP} is the flux linkage from magnetic field of conductor B on conductor B at point P
λ_{BAP} is the flux linkage from magnetic field of conductor A on conductor B at point P

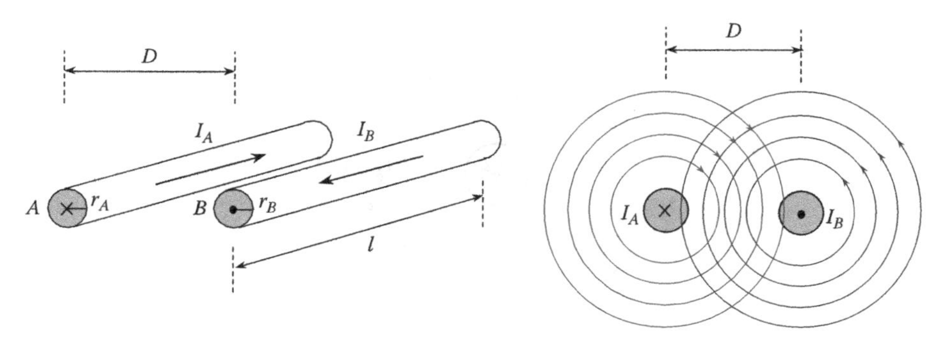

FIGURE 14.5 External magnetic flux around conductors in a two-wire, single-phase line.

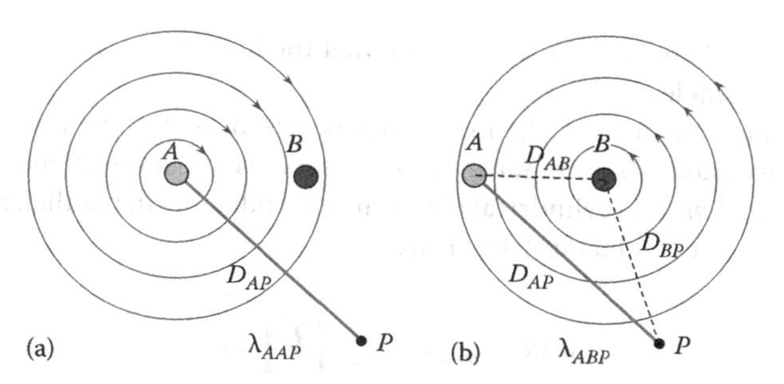

FIGURE 14.6 Flux linkage of (a) conductor *A* at point *P* and (b) conductor *B* on conductor *A* at point *P*. Single-phase system.

The expressions of the aforementioned flux linkages, per-unit length, are

$$\lambda_{AAP} = \frac{\mu_0}{2\pi} I \ln\left(\frac{D_{AP}}{GMR_A}\right) \quad \text{(Wb/m)} \tag{14.28}$$

$$\lambda_{ABP} = \int_{D}^{D_{BP}} B_{BP}\, dP = -\frac{\mu_0}{2\pi} I \ln\left(\frac{D_{BP}}{D}\right) \quad \text{(Wb/m)} \tag{14.29}$$

$$\lambda_{BAP} = \int_{D}^{D_{AP}} B_{AP}\, dP = -\frac{\mu_0}{2\pi} I \ln\left(\frac{D_{AP}}{D}\right) \quad \text{(Wb/m)} \tag{14.30}$$

$$\lambda_{BBP} = \frac{\mu_0}{2\pi} I \ln\left(\frac{D_{BP}}{GMR_B}\right) \quad \text{(Wb/m)} \tag{14.31}$$

The total flux linkage of the system at point *P* is the algebraic summation of λ_{AP} and λ_{BP}:

$$\lambda_P = \lambda_{AP} + \lambda_{BP} = (\lambda_{AAP} + \lambda_{ABP}) + (\lambda_{BAP} + \lambda_{BBP}) \tag{14.32}$$

$$\lambda_P = \frac{\mu_0}{2\pi} I \ln\left[\left(\frac{D_{AP}}{GMR_A}\right)\left(\frac{D}{D_{AP}}\right)\left(\frac{D_{BP}}{GMR_B}\right)\left(\frac{D}{D_{BP}}\right)\right] = \frac{\mu_0}{2\pi} I \ln\left(\frac{D^2}{GMR_A GMR_B}\right) \quad \text{(Wb/m)} \tag{14.33}$$

If the conductors have the same radius, $r_A = r_B = r$, and the point *P* is shifted to infinity, then the total flux linkage of the system becomes

$$\lambda = \frac{\mu_0}{\pi} I \ln\left(\frac{D}{GMR}\right) \quad \text{(Wb/m)} \tag{14.34}$$

and the total inductance per-unit length is

$$L_{1\text{-phase system}} = \frac{\lambda}{I} = \frac{\mu_0}{\pi} \ln\left(\frac{D}{GMR}\right) \quad \text{(H/m)} \tag{14.35}$$

Comparing Equations 14.25 and 14.35, it can be seen that the inductance of the single-phase system is twice the inductance of a single conductor.

For a line with stranded conductors, the inductance is determined using a new GMR value named $GMR_{stranded}$, evaluated according to the number of conductors. If conductors A and B in the single-phase system are formed by n and m solid cylindrical identical subconductors in parallel, respectively, then the expressions or $GMR_{stranded}$ for conductor A and B are

$$GMR_{A\_stranded} = \sqrt[n^2]{\prod_{i=1}^{n}\prod_{j=1}^{n} D_{ij}} \tag{14.36}$$

$$GMR_{B\_stranded} = \sqrt[m^2]{\prod_{i=1}^{m}\prod_{j=1}^{m} D_{ij}} \tag{14.37}$$

Generally, the $GMR_{stranded}$ for a particular cable can be found in conductors tables given by the manufacturer.

If the line conductor is composed of bundle conductors, the inductance is reevaluated taking into account the number of bundle conductors and the separation among them. The GMR_{bundle} is introduced to determine the final inductance value. Assuming the same separation among bundle conductors, the equation for GMR_{bundle}, up to three conductors per bundle, is defined as

$$GMR_{n\,bundle\,conductors} = \sqrt[n]{d^{n-1} GMR_{stranded}} \tag{14.38}$$

where

n is the number of conductors per bundle
$GMR_{stranded}$ is the GMR of the stranded conductor
d is the distance between bundle conductors

In four conductors per bundle with the same separation between consecutive conductors, the GMR_{bundle}, is evaluated, according to their geometry, as

$$\begin{aligned} GMR_{4\,bundle\,conductors} &= \sqrt[16]{\left(GMR_{stranded}\, d\, d\, d\sqrt{2}\right)^4} \\ &= \sqrt[16]{4 GMR_{stranded}^4\, d^{12}} \\ &= 4^{1/16} GMR_{stranded}^{1/4}\, d^{3/4} \\ &= 1.09\sqrt[4]{GMR_{stranded}\, d^3} \end{aligned} \tag{14.39}$$

14.1.2.5 Inductance of Three-Phase Transmission Line in Asymmetrical Arrangement

Derivations of inductance in a single-phase system can be extended to obtain the inductance per phase in a three-phase system. Consider a three-phase, three-conductor system with solid cylindrical conductors with identical radius r_A, r_B, and r_C, placed horizontally with separation D_{AB}, D_{BC}, and D_{CA} (where $D > r$) among them. Corresponding currents I_A, I_B, and I_C flow along each conductor as shown in Figure 14.7.

The total magnetic flux enclosing conductor A at a point P away from the conductors, is the sum of the flux produced by conductors A, B, and C as indicated in Equation 14.40:

$$\phi_{AP} = \phi_{AAP} + \phi_{ABP} + \phi_{ACP} \tag{14.40}$$

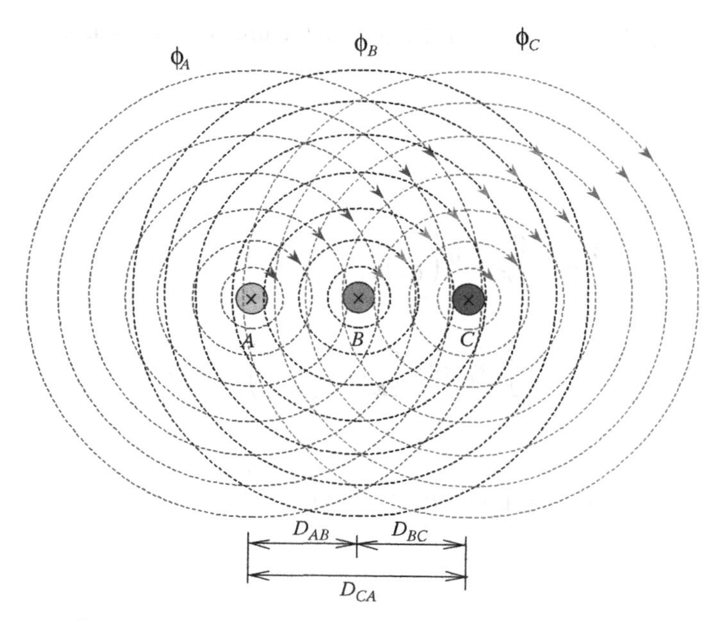

FIGURE 14.7 Magnetic flux produced by each conductor in a three-phase system.

where

ϕ_{AAP} is the flux produced by current I_A on conductor A at point P
ϕ_{ABP} is the flux produced by current I_B on conductor A at point P
ϕ_{ACP} is the flux produced by current I_C on conductor A at point P

Considering 1 m length for each conductor, the expressions for the fluxes are

$$\phi_{AAP} = \frac{\mu_0}{2\pi} I_A \ln\left(\frac{D_{AP}}{GMR_A}\right) \quad \text{(Wb/m)} \tag{14.41}$$

$$\phi_{ABP} = \frac{\mu_0}{2\pi} I_B \ln\left(\frac{D_{BP}}{D_{AB}}\right) \quad \text{(Wb/m)} \tag{14.42}$$

$$\phi_{ACP} = \frac{\mu_0}{2\pi} I_C \ln\left(\frac{D_{CP}}{D_{AC}}\right) \quad \text{(Wb/m)} \tag{14.43}$$

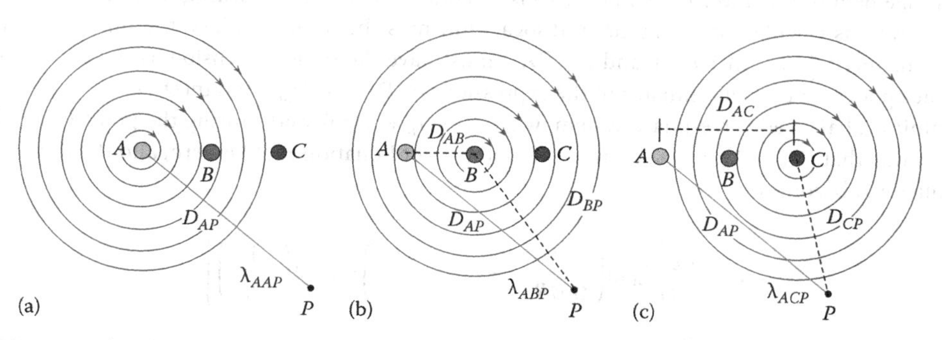

FIGURE 14.8 Flux linkage of (a) conductor A at point P, (b) conductor B on conductor A at point P, and (c) conductor C on conductor A at point P. Three-phase system.

The corresponding flux linkage of conductor A at point P (Figure 14.8), is evaluated as

$$\lambda_{AP} = \lambda_{AAP} + \lambda_{ABP} + \lambda_{ACP} \tag{14.44}$$

having

$$\lambda_{AAP} = \frac{\mu_0}{2\pi} I_A \ln\left(\frac{D_{AP}}{GMR_A}\right) \quad \text{(Wb/m)} \tag{14.45}$$

$$\lambda_{ABP} = \int_{D_{AB}}^{D_{BP}} B_{BP}\, dP = \frac{\mu_0}{2\pi} I_B \ln\left(\frac{D_{BP}}{D_{AB}}\right) \quad \text{(Wb/m)} \tag{14.46}$$

$$\lambda_{ACP} = \int_{D_{AC}}^{D_{CP}} B_{CP}\, dP = \frac{\mu_0}{2\pi} I_C \ln\left(\frac{D_{CP}}{D_{AC}}\right) \quad \text{(Wb/m)} \tag{14.47}$$

where

λ_{AP} is the total flux linkage of conductor A at point P
λ_{AAP} is the flux linkage from magnetic field of conductor A on conductor A at point P
λ_{ABP} is the flux linkage from magnetic field of conductor B on conductor A at point P
λ_{ACP} is the flux linkage from magnetic field of conductor C on conductor A at point P

Substituting Equations 14.45 through 14.47 in Equation 14.44 and rearranging, according to natural logarithms law, we have

$$\lambda_{AP} = \frac{\mu_0}{2\pi}\left[I_A \ln\left(\frac{D_{AP}}{GMR_A}\right) + I_B \ln\left(\frac{D_{BP}}{D_{AB}}\right) + I_C \ln\left(\frac{D_{CP}}{D_{AC}}\right)\right] \quad \text{(Wb/m)} \tag{14.48}$$

$$\lambda_{AP} = \frac{\mu_0}{2\pi}\left[I_A \ln\left(\frac{1}{GMR_A}\right) + I_B \ln\left(\frac{1}{D_{AB}}\right) + I_C \ln\left(\frac{1}{D_{AC}}\right)\right] + \frac{\mu_0}{2\pi}[I_A \ln(D_{AP}) + I_B \ln(D_{BP}) + I_C \ln(D_{CP})] \quad \text{(Wb/m)} \tag{14.49}$$

The arrangement of Equation 14.48 into 14.49 is algebraically correct according to natural logarithms law. However, as calculation of any natural logarithm must be dimensionless, the numerator in the expressions $\ln(1/GMR_A)$, $\ln(1/D_{AB})$, and $\ln(1/D_{AC})$ must have the same dimension as the denominator. The same applies for the denominator in the expressions $\ln(D_{AP})$, $\ln(D_{BP})$, and $\ln(D_{CP})$.

Assuming a balanced three-phase system, where $I_A + I_B + I_C = 0$, and shifting the point P to infinity in such a way that $D_{AP} = D_{BP} = D_{CP}$, then the second part of Equation 14.49 is zero, and the flux linkage of conductor A becomes

$$\lambda_A = \frac{\mu_0}{2\pi}\left[I_A \ln\left(\frac{1}{GMR_A}\right) + I_B \ln\left(\frac{1}{D_{AB}}\right) + I_C \ln\left(\frac{1}{D_{AC}}\right)\right] = \frac{\mu_0}{2\pi}[I_A L_{AA} + I_B L_{AB} + I_C L_{AC}] \quad \text{(Wb/m)} \tag{14.50}$$

Similarly, the flux linkage expressions for conductors B and C are

$$\lambda_B = \frac{\mu_0}{2\pi}\left[I_A \ln\left(\frac{1}{D_{BA}}\right) + I_B \ln\left(\frac{1}{GMR_B}\right) + I_C \ln\left(\frac{1}{D_{BC}}\right)\right]$$

$$= \frac{\mu_0}{2\pi}[I_A L_{BA} + I_B L_{BB} + I_C L_{BC}] \quad \text{(Wb/m)} \tag{14.51}$$

$$\lambda_C = \frac{\mu_0}{2\pi}\left[I_A \ln\left(\frac{1}{D_{CA}}\right) + I_B \ln\left(\frac{1}{D_{CB}}\right) + I_C \ln\left(\frac{1}{GMR_C}\right)\right]$$

$$= \frac{\mu_0}{2\pi}[I_A L_{CA} + I_B L_{CB} + I_C L_{CC}] \quad \text{(Wb/m)} \tag{14.52}$$

where

$\lambda_A, \lambda_B, \lambda_C$ are the total flux linkages of conductors A, B, and C

L_{AA}, L_{BB}, L_{CC} are the self-inductances of conductors A, B, and C

$L_{AB}, L_{BC}, L_{CA}, L_{BA}, L_{CB}, L_{AC}$ are the mutual inductance among conductors A, B, and C

The flux linkage of each phase conductor depends on the three currents, and therefore, the inductance per phase is not only one, as in the single-phase system. Instead, there are self and mutual conductor inductances, as shown in Equations 14.50 through 14.52.

14.1.2.6 Inductance of Balanced Three-Phase Transmission Line in Symmetrical Arrangement

If the three-phase transmission line has an asymmetrical arrangement, as in the horizontal arrangement, there are nine different inductances (mutual and self-inductances). However, a single inductance per phase can be obtained if the three conductors are arranged with the same separation among them (symmetrical arrangement), where $D = D_{AB} = D_{BC} = D_{CA}$. For a balanced three-phase system ($I_A + I_B + I_C = 0$, or $I_A = -I_B - I_C$), the flux linkage of each conductor, per-unit length, will be the same. Substituting these values in Equation 14.50, the expression of λ_A is simplified as follows:

$$\lambda_A = \frac{\mu_0}{2\pi}\left[(-I_B - I_C)\ln\left(\frac{1}{GMR_A}\right) + I_B \ln\left(\frac{1}{D}\right) + I_C \ln\left(\frac{1}{D}\right)\right]$$

$$= \frac{\mu_0}{2\pi}\left[-I_B \ln\left(\frac{D}{GMR_A}\right) - I_C \ln\left(\frac{D}{GMR_A}\right)\right]$$

$$= \frac{\mu_0}{2\pi}\left[I_A \ln\left(\frac{D}{GMR_A}\right)\right] \quad \text{(Wb/m)} \tag{14.53}$$

If GMR value is the same for all conductors (either single or bundle GMR), the total flux linkage expression is the same for all phases. Therefore, the equivalent inductance per phase is

$$L_{phase} = \frac{\mu_0}{2\pi}\ln\left(\frac{D}{GMR_{phase}}\right) \quad \text{(H/m)} \tag{14.54}$$

14.1.2.7 Inductance of Transposed Three-Phase Transmission Lines

In actual transmission lines, the phase conductors cannot maintain symmetrical arrangement along the whole length because of construction considerations, even when bundle conductors spacers are used. With asymmetrical spacing, the inductance will be different for each phase, with a corresponding unbalanced voltage drop on each conductor. Therefore, the single-phase equivalent circuit to represent the power system cannot be used.

However, it is possible to assume symmetrical arrangement in the transmission line by transposing the phase conductors. In a transposed system, each phase conductor occupies the location of the other two phases for one-third of the total line length as shown in Figure 14.9.

The flux linkage of conductor A in each section along the total length is obtained considering the distances related to conductors of phase B and C:

$$\lambda_{A\,section\,1} = \frac{\mu_0}{2\pi}\left[I_A \ln\left(\frac{1}{GMR_A}\right) + I_B \ln\left(\frac{1}{D_{21}}\right) + I_C \ln\left(\frac{1}{D_{31}}\right)\right] \tag{14.55}$$

$$\lambda_{A\,section\,2} = \frac{\mu_0}{2\pi}\left[I_A \ln\left(\frac{1}{GMR_A}\right) + I_B \ln\left(\frac{1}{D_{32}}\right) + I_C \ln\left(\frac{1}{D_{12}}\right)\right] \tag{14.56}$$

$$\lambda_{A\,section\,3} = \frac{\mu_0}{2\pi}\left[I_A \ln\left(\frac{1}{GMR_A}\right) + I_B \ln\left(\frac{1}{D_{13}}\right) + I_C \ln\left(\frac{1}{D_{23}}\right)\right] \tag{14.57}$$

where

$\lambda_{A\,section\,1}$ is the flux linkage of conductor A in section 1
$\lambda_{A\,section\,2}$ is the flux linkage of conductor A in section 2
$\lambda_{A\,section\,3}$ is the flux linkage of conductor A in position 3
D_{21} is the distance between conductor B and A when conductor A is in section 1
D_{31} is the distance between conductor C and A when conductor A is in section 1

Then, the average value of flux linkage in phase A is expressed in Equation 14.58:

$$\begin{aligned}\lambda_A &= \frac{\lambda_{A\,section\,1}(l/3) + \lambda_{A\,section\,2}(l/3) + \lambda_{A\,section\,3}(l/3)}{l}\\ &= \frac{\lambda_{A\,section\,1} + \lambda_{A\,section\,2} + \lambda_{A\,section\,3}}{3}\\ &= \frac{\mu_0}{6\pi}\left[3I_A \ln\left(\frac{1}{GMR_A}\right) + I_B \ln\left(\frac{1}{D_{12}D_{23}D_{31}}\right) + I_C \ln\left(\frac{1}{D_{13}D_{21}D_{32}}\right)\right]\end{aligned} \tag{14.58}$$

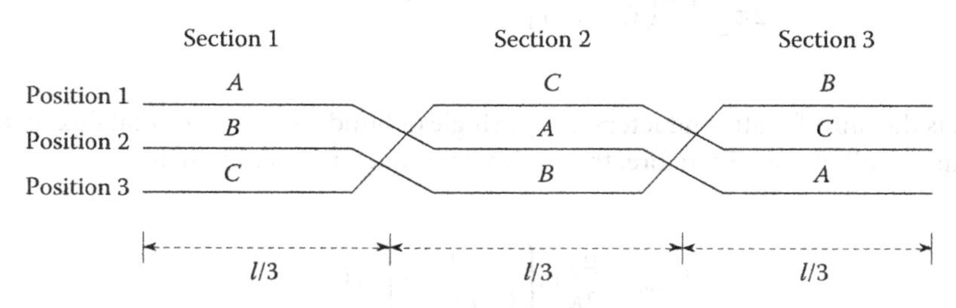

FIGURE 14.9 Arrangement of conductors in a transposed three-phase transmission line.

As $D_{13} = D_{31}$, $D_{23} = D_{32}$, and $D_{12} = D_{21}$, and $-I_A = I_B + I_C$ in a balanced three-phase system, then λ_A is simplified in Equation 14.59:

$$\begin{aligned}\lambda_A &= \frac{\mu_0}{6\pi}\left[3I_A \ln\left(\frac{1}{GMR_A}\right) - I_A \ln\left(\frac{1}{D_{12}D_{23}D_{31}}\right)\right]\\ &= \frac{\mu_0}{6\pi}\left[3I_A \ln\left(\frac{1}{GMR_A}\right) + I_A \ln(D_{12}D_{23}D_{31})\right]\\ &= \frac{\mu_0}{2\pi} I_A \ln\left(\frac{(D_{12}D_{23}D_{31})^{1/3}}{GMR_A}\right)\end{aligned} \tag{14.59}$$

Flux linkages λ_B and λ_B are obtained in similar procedure. The expression $(D_{12}D_{23}D_{31})^{1/3}$ is defined as the average distance *GMD* (geometrical mean distance) and substitutes the distance *D* of the symmetrical arrangement case. As conductor of phase A is assumed the same as conductor of phase B and C, then $GMR_A = GMR_B = GMR_C$.

Therefore, calculation of phase inductance derived for symmetrical arrangement is valid for transposed transmission lines. The expression of inductance per phase per unit-length becomes

$$L_{phase} = \frac{\mu_0}{2\pi} \ln\left(\frac{GMD}{GMR_{phase}}\right) \quad \text{(H/m)} \tag{14.60}$$

Once the inductance per phase is obtained, the inductive reactance per unit-length is expressed as follows:

$$X_{L_{phase}} = 2\pi f\, L_{phase} = \mu_0 f \ln\left(\frac{GMD}{GMR_{phase}}\right) \quad (\Omega/\text{m}) \tag{14.61}$$

For bundle conductors, the GMR_{bundle} value is determined, as in the single-phase transmission line case, by the number of conductors, and by the number of conductors per bundle and the separation among them. The expression for the total inductive reactance per phase yields

$$X_{L_{phase}} = \mu_0 f \ln\left(\frac{GMD}{GMR_{bundle}}\right) \quad (\Omega/\text{m}) \tag{14.62}$$

where

$GMR_{bundle} = (d^{n-1}\, GMR_{phase})^{1/n}$ up to three conductors per bundle

$GMR_{bundle} = 1.09(d^3\, GMR_{phase})^{1/4}$ for four conductors per bundle. GMR_{phase} is the geometric mean radius of phase conductor, either solid or stranded.

14.1.3 Shunt Capacitance and Capacitive Reactance

Capacitance exists among transmission line conductors due to their potential difference. To evaluate the capacitance between conductors in a surrounding medium with permittivity ε, it is necessary to determine the voltage between the conductors, and the electric field strength of the surrounding [4–7].

14.1.3.1 Capacitance of a Single Solid Conductor

Consider a solid, cylindrical, long conductor with radius r, in a free space with permittivity ε_0, and with a charge of $q+$ coulombs per meter, uniformly distributed on the surface. There is constant electric field strength on the surface of cylinder (Figure 14.10). The resistivity of the conductor is assumed to be zero (perfect conductor), which results in zero internal electric field due to the charge on the conductor.

The charge $q+$ produces an electric field radial to the conductor with equipotential surfaces concentric to the conductor. According to Gauss's law, the total electric flux leaving a closed surface is equal to the total charge inside the volume enclosed by the surface. Therefore, at an outside point P separated x meters from the center of the conductor, the electric field flux density, and the electric field intensity are

$$Density_P = \frac{q}{A} = \frac{q}{2\pi x} \quad \text{(C)} \tag{14.63}$$

$$E_P = \frac{Density_P}{\varepsilon} = \frac{q}{2\pi\varepsilon_0 x} \quad \text{(V/m)} \tag{14.64}$$

where

$Density_P$ is the electric flux density at point P (C)
E_P is the electric field intensity at point P (V/m)
A is the surface of a concentric cylinder with 1 m length and radius x (m^2)
$\varepsilon = \varepsilon_0 = 10^{-9}/36\pi$ is the permittivity of free space assumed for the conductor (F/m)

The potential difference or voltage difference between two outside points P_1 and P_2 with corresponding distances x_1 and x_2 from the conductor center is defined by integrating the electric field intensity from x_1 to x_2:

$$V_{P1-P2} = \int_{x_1}^{x_2} E_P \frac{dx}{x} = \int_{x_1}^{x_2} \frac{q}{2\pi\varepsilon_0}\frac{dx}{x} = \frac{q}{2\pi\varepsilon_0}\ln\left[\frac{x_2}{x_1}\right] \quad \text{(V)} \tag{14.65}$$

Then, the capacitance between points P_1 and P_2 is evaluated as

$$C_{P1-P2} = \frac{q}{V_{P1-P2}} = \frac{2\pi\varepsilon_0}{\ln[x_2/x_1]} \quad \text{(F/m)} \tag{14.66}$$

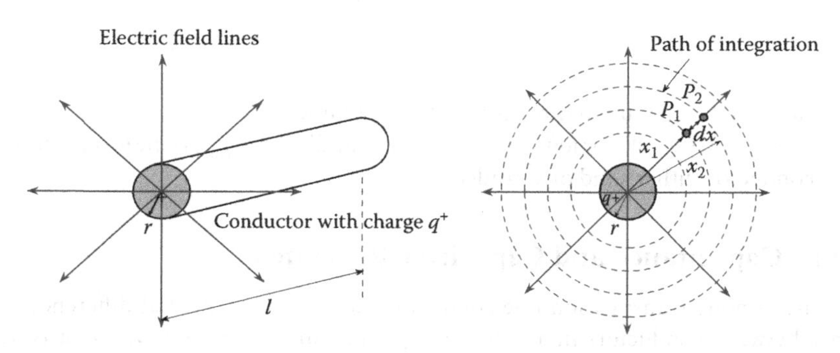

FIGURE 14.10 Electric field produced from a single conductor.

If point P_1 is located at the conductor surface ($x_1 = r$), and point P_2 is located at ground surface below the conductor ($x_2 = h$), then the voltage of the conductor, and the capacitance between the conductor and ground are

$$V_{cond} = \frac{q}{2\pi\varepsilon_0}\ln\left[\frac{h}{r}\right] \quad \text{(V)} \tag{14.67}$$

$$C_{cond\text{-}ground} = \frac{q}{V_{cond}} = \frac{2\pi\varepsilon_0}{\ln[h/r]} \quad \text{(F/m)} \tag{14.68}$$

14.1.3.2 Capacitance of a Single-Phase Line with Two Wires

Consider a two-wire single-phase line with conductors *A* and *B* with the same radius *r*, separated by a distance $D > r_A$ and r_B. The conductors are energized by a voltage source such that conductor *A* has a charge q^+ and conductor *B* a charge q^- as shown in Figure 14.11.

The charge of each conductor generates independent electric fields. Charge q^+ of conductor *A* generates a voltage V_{AB-A} between both conductors. Similarly, charge q^- of conductor *B* generates a voltage V_{AB-B} between conductors.

V_{AB-A} is calculated by integrating the electric field intensity, due to the charge of conductor *A* on conductor *B* from r_A to *D*:

$$V_{AB-A} = \int_{r_A}^{D} E_A dx = \frac{+q}{2\pi\varepsilon_0}\ln\left[\frac{D}{r_A}\right] \tag{14.69}$$

V_{AB-B} is calculated by integrating the electric field intensity due to the charge of conductor *B* from *D* to r_B:

$$V_{AB-B} = \int_{D}^{r_B} E_B\, dx = \frac{-q}{2\pi\varepsilon_0}\ln\left[\frac{r_B}{D}\right] = \frac{q}{2\pi\varepsilon_0}\ln\left[\frac{D}{r_B}\right] \tag{14.70}$$

The total voltage is the sum of the generated voltages V_{AB-A} and V_{AB-B}:

$$V_{AB} = V_{AB-A} + V_{AB-B} = \frac{q}{2\pi\varepsilon_0}\ln\left[\frac{D}{r_A}\right] + \frac{q}{2\pi\varepsilon_0}\ln\left[\frac{D}{r_B}\right] = \frac{q}{2\pi\varepsilon_0}\ln\left[\frac{D^2}{r_A r_B}\right] \tag{14.71}$$

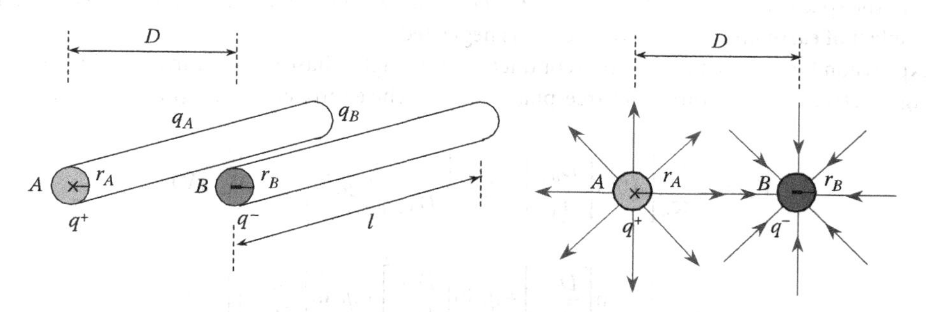

FIGURE 14.11 Electric field produced from a two-wire, single-phase system.

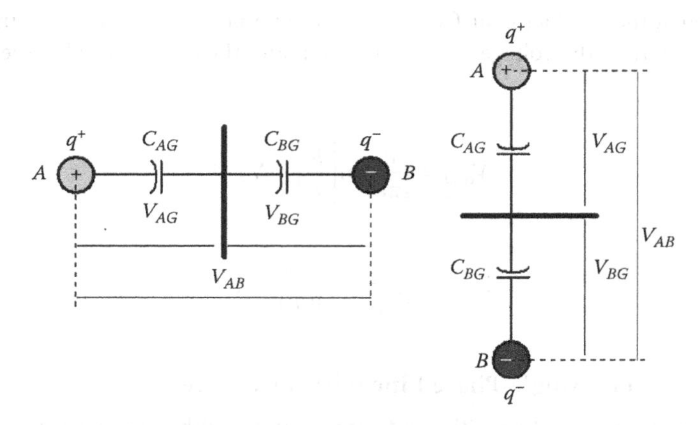

FIGURE 14.12 Capacitance between line-to-ground in a two-wire, single-phase line.

If the conductors have the same radius, $r_A = r_B = r$, then the voltage between conductors V_{AB}, and the capacitance between conductors C_{AB}, for a 1 m line length are

$$V_{AB} = \frac{q}{\pi\varepsilon_0}\ln\left[\frac{D}{r}\right] \quad \text{(V)} \tag{14.72}$$

$$C_{AB} = \frac{\pi\varepsilon_0}{\ln[D/r]} \quad \text{(F/m)} \tag{14.73}$$

The voltage between each conductor and ground (Figure 14.12) is one-half of the voltage between the two conductors. Therefore, the capacitance from either line to ground is twice the capacitance between lines:

$$V_{AG} = V_{BG} = \frac{V_{AB}}{2} \quad \text{(V)} \tag{14.74}$$

$$C_{AG} = \frac{q}{V_{AG}} = \frac{2\pi\varepsilon_0}{\ln[D/r]} \quad \text{(F/m)} \tag{14.75}$$

14.1.3.3 Capacitance of Three-Phase Transmission Line in Asymmetrical Arrangement

Consider a three-phase line with the same voltage magnitude between phases, and assuming a balanced three-phase system with *abc* (positive) sequence such that $q_A + q_B + q_C = 0$. The conductors have radii r_A, r_B, and r_C, and the space between conductors are D_{AB}, D_{BC}, and D_{AC} (where D_{AB}, D_{BC}, and $D_{AC} > r_A$, r_B, and r_C). Also, the effect of earth and neutral conductors is neglected.

The expression for voltage between two conductors in a single-phase system can be extended to obtain the voltages between conductors in a three-phase system. The expressions for V_{AB} and V_{AC} are

$$V_{AB} = \frac{1}{2\pi\varepsilon_0}\left[q_A \ln\left[\frac{D_{BA}}{r_A}\right] + q_B \ln\left[\frac{r_B}{D_{AB}}\right] + q_C \ln\left[\frac{D_{BC}}{D_{AC}}\right]\right] \quad \text{(V)} \tag{14.76}$$

$$V_{AC} = \frac{1}{2\pi\varepsilon_0}\left[q_A \ln\left[\frac{D_{CA}}{r_A}\right] + q_B \ln\left[\frac{D_{CB}}{D_{AB}}\right] + q_C \ln\left[\frac{r_C}{D_{AC}}\right]\right] \quad \text{(V)} \tag{14.77}$$

With all conductors with the same radii $r_A = r_B = r_C = r$, and assuming a balanced system line-to-line voltages with sequence *abc*, where $V_{AB} = \sqrt{3}\,V_{AN}\angle 30^\circ$ and $V_{AC} = -V_{CA} = \sqrt{3}\,V_{AN}\angle -30^\circ$, the line-to-neutral voltage V_{AN} can be expressed in terms of V_{AB} and V_{AC} as

$$V_{AN} = \frac{V_{AB} + V_{AC}}{3} \tag{14.78}$$

$$\begin{aligned} V_{AN} &= \frac{1}{6\pi\varepsilon_0}\left[q_A \ln\left[\frac{D_{BA}}{r_A}\right] + q_B \ln\left[\frac{r_B}{D_{AB}}\right] + q_C \ln\left[\frac{D_{BC}}{D_{AC}}\right] + q_A \ln\left[\frac{D_{CA}}{r_A}\right] + q_B \ln\left[\frac{D_{CB}}{D_{AB}}\right] + q_C \ln\left[\frac{r_C}{D_{AC}}\right]\right] \\ &= \frac{1}{6\pi\varepsilon_0}\left[q_A \ln\left[\frac{D_{BA}D_{CA}}{r^2}\right] + q_B \ln\left[\frac{rD_{BC}}{D_{AB}^2}\right] + q_C \ln\left[\frac{rD_{BC}}{D_{AC}^2}\right]\right] \end{aligned} \tag{14.79}$$

Similar procedure can be followed to obtain V_{BN} and V_{CN}. However, it can be observed from Equation 14.79 that, for asymmetrical arrangement, as in the inductance evaluation case, there is no a single-phase capacitance value.

14.1.3.4 Capacitance of Three-Phase Transmission Line in Symmetrical Arrangement

If the three-phase system has triangular arrangement with equidistant conductors (symmetrical arrangement) such that $D_{AB} = D_{BC} = D_{AC} = D$, with the same radii in all conductors such that $r_A = r_B = r_C = r$ (where $D > r$), then the expressions for V_{AB} and V_{AC} from Equations 14.76 and 14.77, are

$$\begin{aligned} V_{AB} &= \frac{1}{2\pi\varepsilon_0}\left[q_A \ln\left[\frac{D}{r}\right] + q_B \ln\left[\frac{r}{D}\right] + q_C \ln\left[\frac{D}{D}\right]\right] \\ &= \frac{1}{2\pi\varepsilon_0}\left[q_A \ln\left[\frac{D}{r}\right] + q_B \ln\left[\frac{r}{D}\right]\right] \quad \text{(V)} \end{aligned} \tag{14.80}$$

$$\begin{aligned} V_{AC} &= \frac{1}{2\pi\varepsilon_0}\left[q_A \ln\left[\frac{D}{r}\right] + q_B \ln\left[\frac{D}{D}\right] + q_C \ln\left[\frac{r}{D}\right]\right] \\ &= \frac{1}{2\pi\varepsilon_0}\left[q_A \ln\left[\frac{D}{r}\right] + q_C \ln\left[\frac{r}{D}\right]\right] \quad \text{(V)} \end{aligned} \tag{14.81}$$

Assuming again balanced line-to-line voltages with sequence *abc*, expressed in terms of the line-to-neutral voltage, where $V_{AN} = (V_{AB} + V_{AC})/3$, and substituting V_{AB} and V_{AC} from Equations 14.80 and 14.81, we have

$$\begin{aligned} V_{AN} &= \frac{1}{6\pi\varepsilon_0}\left[\left[q_A \ln\left[\frac{D}{r}\right] + q_B \ln\left[\frac{r}{D}\right]\right] + \left[q_A \ln\left[\frac{D}{r}\right] + q_C \ln\left[\frac{r}{D}\right]\right]\right] \\ &= \frac{1}{6\pi\varepsilon_0}\left[2q_A \ln\left[\frac{D}{r}\right] + (q_B + q_C)\ln\left[\frac{r}{D}\right]\right] \quad \text{(V)} \end{aligned} \tag{14.82}$$

Under balanced conditions $q_A + q_B + q_C = 0$, or $-q_A = (q_B + q_C)$ then, the final expression for the line-to-neutral voltage is

$$V_{AN} = \frac{1}{2\pi\varepsilon_0} q_A \ln\left[\frac{D}{r}\right] \quad \text{(V)} \tag{14.83}$$

The positive sequence capacitance per unit-length between phase A and neutral can now be obtained. The same result is obtained for capacitance between phases B and C to neutral:

$$C_{AN} = \frac{q_A}{V_{AN}} = \frac{2\pi\varepsilon_0}{\ln[D/r]} \quad \text{(F/m)} \tag{14.84}$$

14.1.3.5 Capacitance of Stranded Bundle Conductors

The calculation of the capacitance in Equation 14.84 is based on

1. Solid conductors with zero resistivity (zero internal electric field)
2. Charge uniformly distributed
3. Equilateral spacing of phase conductors

In actual transmission lines, the resistivity of the conductors produces a small internal electric field, and therefore, the electric field at the conductor surface is smaller than the estimated. However, the difference is negligible for practical purposes.

Because of the presence of other charged conductors, the charge distribution is nonuniform, and therefore the estimated capacitance is different. However, this effect is negligible for most practical calculation. In a line with stranded conductors, the capacitance is evaluated assuming a solid conductor with the same radius as the outside radius of the stranded conductor. This produces a negligible difference.

Most transmission lines do not have equilateral spacing of phase conductors. This causes differences between the line-to-neutral capacitances of the three phases. However, transposing the phase conductors balances the system resulting in equal line-to-neutral capacitance for each phase. Consider a transposed three-phase line with conductors having the same radius r, and with space between conductors D_{AB}, D_{BC}, and D_{AC}, where D_{AB}, D_{BC}, and $D_{AC} > r$.

Assuming *abc* positive sequence, the expressions for V_{AB} on the first, second, and third sections of the transposed line (refer to Figure 14.9) are

$$V_{AB\,\text{section 1}} = \frac{1}{2\pi\varepsilon_0}\left[q_A \ln\left[\frac{D_{AB}}{r}\right] + q_B \ln\left[\frac{r}{D_{AB}}\right] + q_C \ln\left[\frac{D_{BC}}{D_{AC}}\right]\right] \quad \text{(V)} \tag{14.85}$$

$$V_{AB\,\text{section 2}} = \frac{1}{2\pi\varepsilon_0}\left[q_A \ln\left[\frac{D_{BC}}{r}\right] + q_B \ln\left[\frac{r}{D_{BC}}\right] + q_C \ln\left[\frac{D_{AC}}{D_{AB}}\right]\right] \quad \text{(V)} \tag{14.86}$$

$$V_{AB\,\text{section 3}} = \frac{1}{2\pi\varepsilon_0}\left[q_A \ln\left[\frac{D_{AC}}{r}\right] + q_B \ln\left[\frac{r}{D_{AC}}\right] + q_C \ln\left[\frac{D_{AB}}{D_{BC}}\right]\right] \quad \text{(V)} \tag{14.87}$$

Similarly, the expressions for V_{AC} on the first, second, and third sections of the transposed line are

$$V_{AC\,\text{section 1}} = \frac{1}{2\pi\varepsilon_0}\left[q_A \ln\left[\frac{D_{AC}}{r}\right] + q_B \ln\left[\frac{D_{BC}}{D_{AB}}\right] + q_C \ln\left[\frac{r}{D_{AC}}\right]\right] \quad \text{(V)} \tag{14.88}$$

$$V_{AC\,section\,2} = \frac{1}{2\pi\varepsilon_0}\left[q_A \ln\left[\frac{D_{AB}}{r}\right] + q_B \ln\left[\frac{D_{AC}}{D_{BC}}\right] + q_C \ln\left[\frac{r}{D_{AB}}\right]\right] \text{ (V)} \tag{14.89}$$

$$V_{AC\,section\,3} = \frac{1}{2\pi\varepsilon_0}\left[q_A \ln\left[\frac{D_{BC}}{r}\right] + q_B \ln\left[\frac{D_{AB}}{D_{AC}}\right] + q_C \ln\left[\frac{r}{D_{BC}}\right]\right] \text{ (V)} \tag{14.90}$$

Taking the average value of the three sections, we have the final expressions of V_{AB} and V_{AC} of the transposed line in Equations 14.91 and 14.92:

$$\begin{aligned} V_{AB\ transp} &= \frac{V_{AB\ section\ 1} + V_{AB\ section\ 2} + V_{AB\ section\ 3}}{3} \\ &= \frac{1}{6\pi\varepsilon_0}\left[q_A \ln\left[\frac{D_{AB}D_{BC}D_{AC}}{r^3}\right] + q_B \ln\left[\frac{r^3}{D_{AB}D_{BC}D_{AC}}\right] + q_C \ln\left[\frac{D_{BC}D_{AC}D_{AB}}{D_{AC}D_{AB}D_{BC}}\right]\right] \text{ (V)} \end{aligned} \tag{14.91}$$

$$\begin{aligned} V_{AC\ transp} &= \frac{V_{AC\ section\ 1} + V_{AC\ section\ 2} + V_{AC\ section\ 3}}{3} \\ &= \frac{1}{6\pi\varepsilon_0}\left[q_A \ln\left[\frac{D_{AC}D_{AB}D_{BC}}{r^3}\right] + q_B \ln\left[\frac{D_{BC}D_{AC}D_{AB}}{D_{AB}D_{BC}D_{AC}}\right] + q_C \ln\left[\frac{r^3}{D_{AC}D_{AB}D_{BC}}\right]\right] \text{ (V)} \end{aligned} \tag{14.92}$$

For a balanced system where $-q_A = (q_B + q_C)$, the phase-to-neutral voltage V_{AN} (phase voltage) becomes

$$\begin{aligned} V_{AN\ transp} &= \frac{V_{AB\ transp} + V_{AC\ transp}}{3} \\ &= \frac{1}{18\pi\varepsilon_0}\left[2q_A \ln\left[\frac{D_{AB}D_{BC}D_{AC}}{r^3}\right] + (q_B + q_C)\ln\left[\frac{r^3}{D_{AC}D_{AB}D_{BC}}\right]\right] \\ &= \frac{1}{18\pi\varepsilon_0}\left[2q_A \ln\left[\frac{D_{AB}D_{BC}D_{AC}}{r^3}\right] - q_A \ln\left[\frac{r^3}{D_{AC}D_{AB}D_{BC}}\right]\right] \\ &= \frac{1}{6\pi\varepsilon_0} q_A \ln\left[\frac{D_{AB}D_{BC}D_{AC}}{r^3}\right] \\ &= \frac{1}{2\pi\varepsilon_0} q_A \ln\left[\frac{GMD}{r}\right] \text{ (V)} \end{aligned} \tag{14.93}$$

where $GMD = \sqrt[3]{D_{AB}D_{BC}D_{CA}}$ is the geometrical mean distance for a three-phase line.

For bundle conductors, an equivalent radius r_e replaces the radius r of a single conductor and is determined by the number of conductors per bundle and the spacing of conductors. The expression of r_e is similar to GMR_{bundle} used in the calculation of the inductance per phase, except that the actual outside radius of the conductor is used instead of the GMR_{phase}. Therefore, the expression for V_{AN} is

$$V_{AN\ transp} = \frac{1}{2\pi\varepsilon_0} q_A \ln\left[\frac{GMD}{r_e}\right] \text{ (V)} \tag{14.94}$$

where

$r_e = \sqrt[n]{d^{n-1}r}$ is the equivalent radius for up to three conductors per bundle (m)

$r_e = 1.09\sqrt[4]{d^3 r}$ is the equivalent radius for four conductors per bundle (m), d is the distance between bundle conductors (m), and n is the number of conductor per bundle

Finally, the capacitance and capacitive reactance, per-unit length, from phase to neutral can be evaluated as

$$C_{AN\ transp} = \frac{q_A}{V_{AN\ transp}} = \frac{2\pi\varepsilon_0}{\ln[GMD/r_e]} \quad \text{(F/m)} \tag{14.95}$$

$$X_{AN\ transp} = \frac{1}{2\pi f C_{AN\ transp}} = \frac{1}{4\pi f \varepsilon_0}\ln\left[\frac{GMD}{r_e}\right] \quad (\Omega/\text{m}) \tag{14.96}$$

14.1.3.6 Capacitance due to Earth's Surface

Considering a single overhead conductor with a return path through the earth, separated a distance H from earth's surface, the charge of the earth would be equal in magnitude to that on the conductor but of opposite sign. If the earth is assumed as a perfectly conductive horizontal plane with infinite length, then the electric field lines will go from the conductor to the earth, perpendicular to the earth's surface (Figure 14.13).

To calculate the capacitance, the negative charge of the earth can be replaced by an equivalent charge of an image conductor with the same radius as the overhead conductor, lying just below the overhead conductor (Figure 14.14).

The same principle can be extended to calculate the capacitance per phase of a three-phase system. Figure 14.15 shows an equilateral arrangement of identical single conductors for phase A, B, and C carrying the charges q_A, q_B, and q_C and their respective image conductors A', B', and C'.

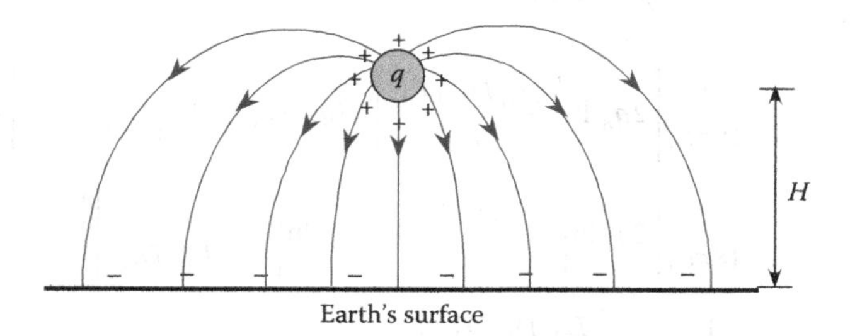

FIGURE 14.13 Distribution of electric field lines from an overhead conductor to earth's surface.

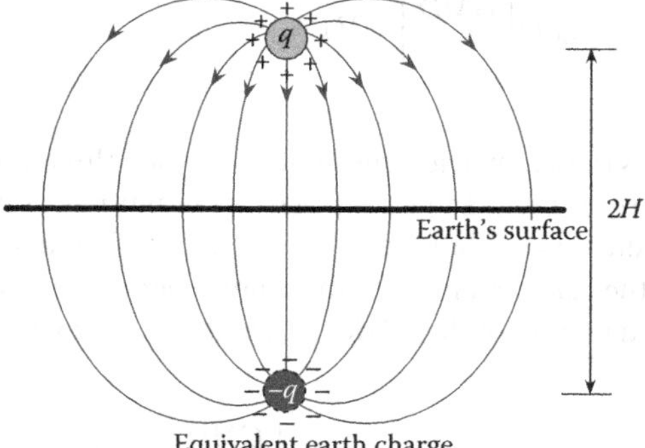

FIGURE 14.14 Equivalent image conductor representing the charge of the earth.

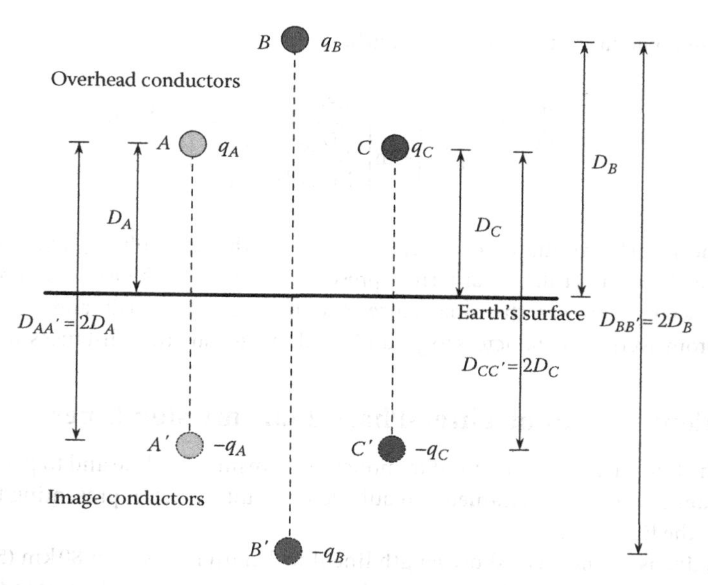

FIGURE 14.15 Arrangement of image conductors in a three-phase transmission line.

D_A, D_B, and D_C are perpendicular distances from phases A, B, and C to earth's surface. $D_{AA'}$, $D_{BB'}$, and $D_{CC'}$ are perpendicular distances from phases A, B, and C to image conductors A', B', and C'. Voltage V_{AB} can be obtained as

$$V_{AB} = \frac{1}{2\pi\varepsilon_0}\begin{bmatrix} q_A \ln\,[D_{AB}/r_A] + q_B \ln\,[r_B/D_{AB}] + q_C \ln\,[D_{BC}/D_{AC}] - \\ -q_A \ln\,[D_{AB'}/D_{AA'}] - q_B \ln\,[D_{BB'}/D_{AB'}] - q_C \ln\,[D_{BC'}/D_{AC'}] \end{bmatrix} \text{ (V)} \tag{14.97}$$

With identical overhead conductors ($r = r_A = r_B = r_C$) in equilateral arrangement, $D = D_{AB} = D_{BC} = D_{CA}$, the expression for V_{AB} becomes

$$V_{AB} = \frac{1}{2\pi\varepsilon_0}\left[q_A\left(\ln\left[\frac{D}{r}\right] - \ln\left[\frac{D_{AB'}}{D_{AA'}}\right]\right) + q_B\left(\ln\left[\frac{r}{D}\right] - \ln\left[\frac{D_{BB'}}{D_{AB'}}\right]\right) - q_C \ln\left[\frac{D_{BC'}}{D_{AC'}}\right]\right] \text{ (V)} \tag{14.98}$$

Similarly, the expressions for V_{BC} and V_{AC} are

$$V_{BC} = \frac{1}{2\pi\varepsilon_0}\left[-q_A \ln\left[\frac{D_{CA'}}{D_{BA'}}\right] + q_B\left(\ln\left[\frac{D}{r}\right] - \ln\left[\frac{D_{CB'}}{D_{BB'}}\right]\right) + q_C\left(\ln\left[\frac{r}{D}\right] - \ln\left[\frac{D_{CC'}}{D_{BC'}}\right]\right)\right] \text{ (V)} \tag{14.99}$$

$$V_{AC} = \frac{1}{2\pi\varepsilon_0}\left[q_A\left(\ln\left[\frac{D}{r}\right] - \ln\left[\frac{D_{CA'}}{D_{AA'}}\right]\right) - q_B \ln\left[\frac{D_{CB'}}{D_{AB'}}\right] + q_C\left(\ln\left[\frac{r}{D}\right] - \ln\left[\frac{D_{CC'}}{D_{AC'}}\right]\right)\right] \text{ (V)} \tag{14.100}$$

The phase voltage V_{AN} becomes, through algebraic reduction,

$$\begin{aligned} V_{AN} &= \frac{V_{AB} + V_{AC}}{3} \\ &= \frac{1}{2\pi\varepsilon_0} q_A \left(\ln\left[\frac{D}{r}\right] - \ln\left[\frac{\sqrt[3]{D_{AB'}D_{BC'}D_{CA'}}}{\sqrt[3]{D_{AA'}D_{BB'}D_{CC'}}}\right]\right) \text{ (V)} \end{aligned} \tag{14.101}$$

Therefore, the phase capacitance C_{AN}, per-unit length, is

$$C_{AN} = \frac{q_A}{V_{AN}} = \frac{2\pi\varepsilon_0}{\ln\left[\frac{D}{r}\right] - \ln\left[\frac{\sqrt[3]{D_{AB'}D_{BC'}D_{CA'}}}{\sqrt[3]{D_{AA'}D_{BB'}D_{CC'}}}\right]} \quad \text{(F/m)} \tag{14.102}$$

Equations 14.84 and 14.102 have similar expressions, except for the term $\ln[(D_{AB'}, D_{BC'}, D_{CA'})^{1/3}/(D_{AA'}, D_{BB'}, D_{CC'})^{1/3}]$ included in Equation 14.102. That term represents the effect of the earth on phase capacitance, increasing its total value. However, the capacitance increment is really small, and is usually neglected, because distances from overhead conductors to ground are always greater than distances among conductors.

14.1.4 Equivalent Circuit of Three-Phase Transmission Lines

Once evaluated, the line parameters are used to model the transmission line and to perform design calculations. The arrangement of the parameters (equivalent circuit model) representing the line depends upon the length of the line [4–9].

A transmission line is defined as a short-length line if its length is less than 80 km (50 miles). In this case, the shut capacitance effect is negligible and only the series resistance and series inductive reactance are considered. Assuming balanced conditions, the line can be represented by the equivalent circuit of a single phase with resistance R, and inductive reactance X_L in series, as shown in Figure 14.16.

Sending voltage and current V_S and I_S can be expressed in terms of receiving voltage and current V_R and I_R at the load side as

$$V_S = V_R + ZI_R \tag{14.103}$$

$$I_S = I_R \tag{14.104}$$

where $Z = R + jX_L$ is the total series impedance of the transmission line (Ω).

If the transmission line has a length between 80 km (50 miles) and 240 km (150 miles), the line is considered a medium-length line and its single-phase equivalent circuit can be represented by a nominal π circuit configuration. The shunt capacitance of the line is divided into two equal parts, each placed at the sending and receiving ends of the line. Figure 14.17 shows the equivalent circuit for a medium-length line.

Sending voltage and current V_S and I_S can be expressed in terms of receiving voltage and current V_R and I_R at the load side as

$$V_S = V_R + V_Z = V_R + Z[I_R + V_R(Y_R)] = V_R\left[1 + Z\left(\frac{Y}{2}\right)\right] + I_R Z \tag{14.105}$$

$$I_S = V_{Y_S} + I_Z = V_S(Y_S) + [I_R + V_R(Y_R)] = V_R\left[Y\left(1 + Z\left(\frac{Y}{4}\right)\right)\right] + I_R\left[1 + Z\left(\frac{Y}{2}\right)\right] \tag{14.106}$$

In matrix form, we have

$$\begin{bmatrix} V_S \\ I_S \end{bmatrix} = \begin{bmatrix} 1 + Z(Y/2) & Z \\ Y[1 + Z(Y/4)] & 1 + Z(Y/2) \end{bmatrix} \begin{bmatrix} V_R \\ I_R \end{bmatrix} = \begin{bmatrix} A & B \\ C & D \end{bmatrix} \begin{bmatrix} V_R \\ I_R \end{bmatrix} \tag{14.107}$$

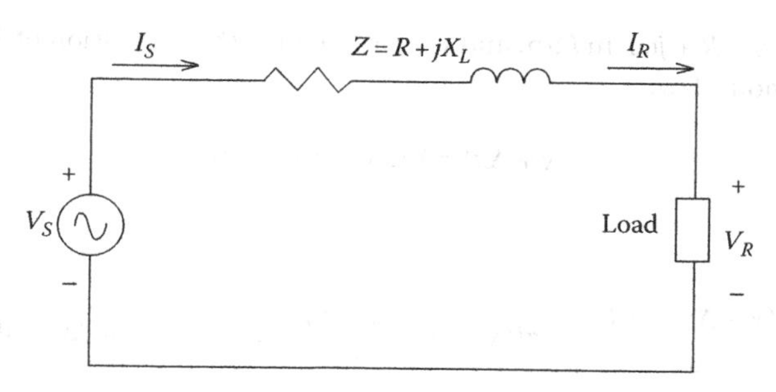

FIGURE 14.16 Equivalent circuit of a short-length transmission line.

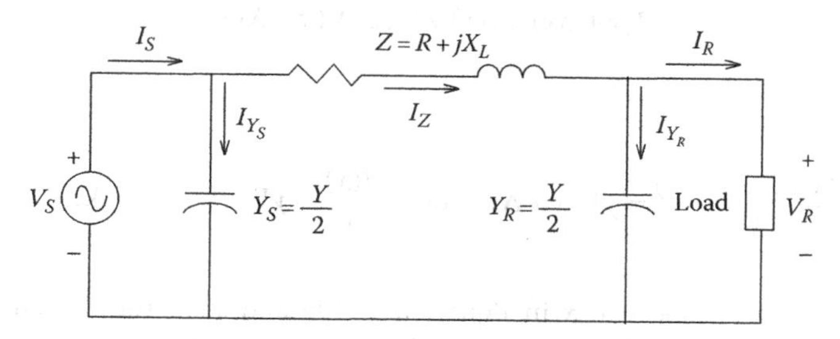

FIGURE 14.17 Equivalent circuit of a medium-length transmission line.

where

$A = 1 + Z\,(Y/2)$ pu

$B = Z\,\Omega$

$C = Y\,[1 + Z(Y/4)]$ S

$D = 1 + Z(Y/2)$ pu

A, B, C, and D are defined as approximated lumped parameters models, and are used in both short-length and medium-length transmission lines. However, if the line is larger than 240 km, the model must consider parameters uniformly distributed along the line. The appropriate series impedance and shunt capacitance are found by solving the corresponding differential equations, where voltages and currents are described as a function of distance and time. Figure 14.18 shows the distributed parameters along the line, neglecting the shunt conductance.

In Figure 14.18, the total length l of the transmission line is taken from the receiving end ($x = 0$) to the sending end ($x = l$). It is shown a small length Δx in which there are a series impedance $z\Delta x$ and a shunt

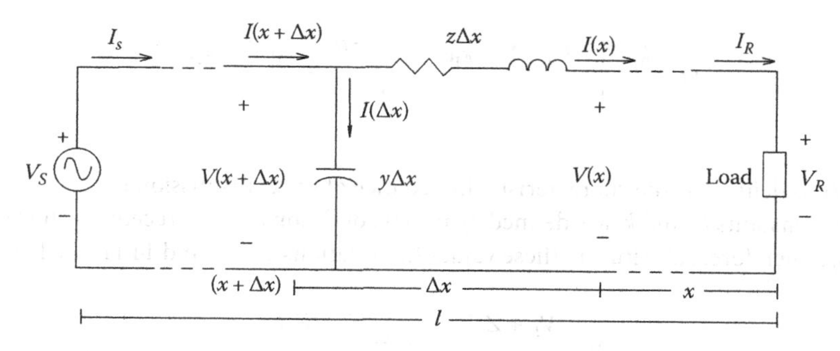

FIGURE 14.18 Distributed parameters of a long-length transmission line model.

admittance $y\Delta x$, where $z = R + j\omega L$ in Ω/m and $y = j\omega C$ in S/m. The expression of $V(x + \Delta x)$ is obtained applying voltage Kirchhoff's law:

$$V(x+\Delta x) = V(x) + (z\Delta x)I(x) \tag{14.108}$$

Rearranging terms,

$$\frac{V(x+\Delta x) - V(x)}{\Delta x} = zI(x) \quad \text{or} \quad \frac{dV(x)}{dx} = zI(x) \quad \text{if } \lim \Delta x \to 0 \tag{14.109}$$

Similarly, the expression for current $I(x + \Delta x)$ is obtained applying current Kirchhoff's law:

$$I(x+\Delta x) = I(x) + (y\Delta x)V(x+\Delta x) \tag{14.110}$$

Rearranging terms,

$$\frac{I(x+\Delta x) - I(x)}{\Delta x} = yV(x+\Delta x) \quad \text{or} \quad \frac{dI(x)}{dx} = yV(x) \quad \text{if } \lim \Delta x \to 0 \tag{14.111}$$

Taking the derivative with respect to x in Equation 14.109 and substituting Equation 14.111 gives Equation 14.112. Similarly, taking the derivative with respect to x in Equation 14.110 and substituting Equation 14.109 gives Equation 14.113:

$$\frac{d^2V(x)}{dx^2} = zyV(x) = \gamma^2 V(x) \quad \text{or} \quad \frac{d^2V(x)}{dx^2} - \gamma^2 V(x) = 0 \tag{14.112}$$

$$\frac{d^2I(x)}{dx^2} = zyI(x) = \gamma^2 I(x) \quad \text{or} \quad \frac{d^2I(x)}{dx^2} - \gamma^2 I(x) = 0 \tag{14.113}$$

where $\gamma = \sqrt{zy}$ is defined as the propagation constant.

Equations 14.112 and 14.113 are second-order linear homogenous differential equations. Solving Equation 14.112 for $V(x)$ we have

$$V(x) = k_1 e^{\gamma x} + k_2 e^{-\gamma x} \tag{14.114}$$

The solution for $I(x)$ is obtained by taking the derivative of Equation 14.114 with respect to x and substituting the value in Equation 14.109:

$$I(x) = \frac{k_1 e^{\gamma x} - k_2 e^{-\gamma x}}{(z/y)} = \frac{k_1 e^{\gamma x} - k_2 e^{-\gamma x}}{\sqrt{z/y}} = \frac{k_1 e^{\gamma x} - k_2 e^{-\gamma x}}{Z_c} \tag{14.115}$$

where $Z_c = \sqrt{z/y}$ is defined as the characteristic impedance of the transmission line.

The values of constants k_1 and k_2 are defined by initial conditions. At the receiving end ($x = 0$), $V(0) = V_R$ and $I(0) = I_R$. Therefore, substituting these values in Equations 14.114 and 14.115 we have

$$k_1 = \frac{V_R + Z_c I_R}{2}, \quad k_2 = \frac{V_R - Z_c I_R}{2} \tag{14.116}$$

The reduced expressions of $V(x)$ and $I(x)$ along any point of the transmission lines become

$$\begin{aligned} V(x) &= \left(\frac{V_R + Z_c I_R}{2}\right)e^{\gamma x} + \left(\frac{V_R - Z_c I_R}{2}\right)e^{-\gamma x} \\ &= \left(\frac{e^{\gamma x} + e^{-\gamma x}}{2}\right)V_R + Z_c\left(\frac{e^{\gamma x} - e^{-\gamma x}}{2}\right)I_R \\ &= \cosh(\gamma x)V_R + Z_c \sinh(\gamma x)I_R \end{aligned} \tag{14.117}$$

$$\begin{aligned} I(x) &= \left(\frac{V_R + Z_c I_R}{2Z_c}\right)e^{\gamma x} + \left(\frac{V_R - Z_c I_R}{2Z_c}\right)e^{-\gamma x} \\ &= \frac{1}{Z_c}\left(\frac{e^{\gamma x} - e^{-\gamma x}}{2}\right)V_R + \left(\frac{e^{\gamma x} + e^{-\gamma x}}{2}\right)I_R \\ &= \frac{1}{Z_c}\sinh(\gamma x)V_R + \cosh(\gamma x)I_R \end{aligned} \tag{14.118}$$

Describing Equations 14.117 and 14.118 in matrix form,

$$\begin{bmatrix} V(x) \\ I(x) \end{bmatrix} = \begin{bmatrix} A(x) & B(x) \\ C(x) & D(x) \end{bmatrix}\begin{bmatrix} V_R \\ I_R \end{bmatrix} \tag{14.119}$$

where

$A(x) = \cosh(\gamma x)$ pu
$B(x) = Z_c \sinh(\gamma x)$ Ω
$C(x) = (1/Z_c) \sinh(\gamma x)$ S
$D(x) = \cosh(\gamma x)$ pu

$A(x)$, $B(x)$, $C(x)$, and $D(x)$ are known as exact *ABCD* parameters for a transmission line in steady state.

$V(x)$ and $I(x)$ can give the exact value of voltage and current at any distance x from the receiving end, no matter what is the total length of the transmission line. However, in most of the cases, the only points of interest are at the receiving and sending ends. If V_R and I_R are known, then the matrix form of equations for V_S and I_S, where $x = l$, are

$$\begin{bmatrix} V_S \\ I_S \end{bmatrix} = \begin{bmatrix} A & B \\ C & D \end{bmatrix}\begin{bmatrix} V_R \\ I_R \end{bmatrix} \tag{14.120}$$

where

$A = \cosh(\gamma l)$ pu
$B = Z_c \sinh(\gamma l)$ Ω
$C = (1/Z_c) \sinh(\gamma l)$ S
$D = \cosh(\gamma l)$ pu

In this case, a long-length transmission line can be represented with an equivalent π circuit model, similar to the nominal π circuit representing the medium-length transmission line, as shown in Figure 14.19, where Z' is the modified value of total series impedance (Ω), and Y' is the modified value of total shunt admittance (S).

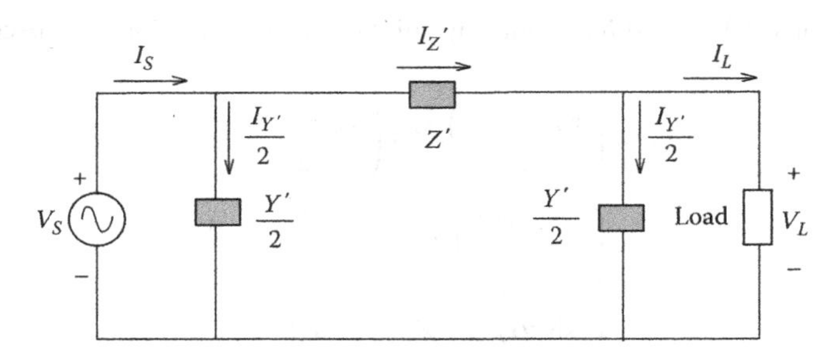

FIGURE 14.19 Equivalent π circuit model of a long-length transmission line.

From the equivalent π circuit model, the corresponding matrix array for V_S and I_S, following the same procedure as in the medium-length transmission line, are

$$\begin{bmatrix} V_S \\ I_S \end{bmatrix} = \begin{bmatrix} 1+Z'\left(\frac{Y'}{2}\right) & Z' \\ Y'\left[1+Z'\left(\frac{Y'}{4}\right)\right] & 1+Z'\left(\frac{Y'}{2}\right) \end{bmatrix} \begin{bmatrix} V_R \\ I_R \end{bmatrix} = \begin{bmatrix} A & B \\ C & D \end{bmatrix} \begin{bmatrix} V_R \\ I_R \end{bmatrix} \tag{14.121}$$

As parameters A, B, C, and D from Equations 14.120 and 14.121 are the same, we have

$$A = \cosh(\gamma l) = 1 + Z'\left(\frac{Y'}{2}\right) \quad \text{(pu)} \tag{14.122}$$

$$B = Z_c \sinh(\gamma l) = Z' \quad (\Omega) \tag{14.123}$$

$$C = \frac{1}{Z_c}\sinh(\gamma l) = Y'\left[1 + Z'\left(\frac{Y'}{4}\right)\right] \quad \text{(S)} \tag{14.124}$$

$$D = \cosh(\gamma l) = 1 + Z'\left(\frac{Y'}{2}\right) \quad \text{(pu)} \tag{14.125}$$

The equivalent expression of $Y'/2$ is obtained from Equations 14.122 through 14.124 as

$$\frac{Y'}{2} = \frac{\tanh(\gamma l)}{Z_c} = \frac{Y}{2}\frac{\tanh(\gamma l)}{\gamma l/2} \tag{14.126}$$

Example 14.1

A 60 Hz, 350 km, three-phase, transposed transmission line, is designed with three bobolink conductors per phase, in horizontal arrangement. Distance between consecutive phases is $D = 9.5$ m and distance among bundle conductors is $d = 0.45$ m. The transmission line is connected to a 250 MVA load at 0.8 lagging power factor and at a line-to-line voltage of 400 kV. Determine (a) the series impedance and shunt admittance, (b) the $ABCD$ parameters of the π equivalent circuit, and (c) the voltage and current at the sending end.

Data

Conductor
Bobolink (Table 14.2): *Diam* = 36.25 mm, $R_{AC,75°C}$ = 0.0503 Ω/km, GMR_c = 14.39 mm, l = 350 km

Arrangement
Horizontal, Cond. per phase = 3, D = 9.5 m, d = 0.45 m

Load
$S_{3\text{-}phase}$ = 250 MVA, V_{LL} = 400 kV, PF = 0.8 lagging, f = 60 Hz

Solution

a. *Series impedance and shunt admittance*

$$GMD = \sqrt[3]{D \times D \times 2D} = \sqrt[3]{9.5 \times 9.5 \times 9.5} = 11.969\,\text{m}$$

$$GMR_{bundle} = \sqrt[3]{d^2 GMR_c} = \sqrt[3]{0.45^2 \times 0.01439} = 0.143\,\text{m}$$

$$r_{e\ bundle} = \sqrt[3]{d^2 r_c} = \sqrt[3]{0.45^2 \times 0.018} = 0.154\,\text{m}$$

$$R_{phase} = \frac{R_c}{\text{Cond per phase}} = \frac{0.0503\,\Omega/\text{km}}{3} = 1.677 \times 10^{-5}\,\Omega/\text{m}$$

$$X_{L\,phase} = j\omega L_{phase} = j2\pi f\left(\frac{\mu_0}{2\pi}\ln\left(\frac{GMD}{GMR_{bundle}}\right)\right) = j377 \times \frac{4\pi \times 10^{-7}}{2\pi} \times \ln\left(\frac{11.969}{0.143}\right) = j3.33 \times 10^{-4}\,\Omega/\text{m}$$

$$Y_{C\,phase} = j\omega C_{phase} = j2\pi f\left(\frac{2\pi\varepsilon_0}{\ln\left(\dfrac{GMD}{r_{e\ bundle}}\right)}\right) = j377 \times \frac{8.85 \times 10^{-7}}{\ln(11.69/0.154)} = j4.817 \times 10^{-9}\,\text{S/m}$$

$$z_{phase} = R_{phase} + jX_{L\,phase} = 1.677 \times 10^{-5} + j3.33 \times 10^{-4}\,\Omega/\text{m}$$

$$y_{phase} = G_{phase} + jY_{C\,phase} = 0 + j4.817 \times 10^{-9}\,\text{S/m}$$

b. *ABCD parameters of the π equivalent circuit*

$$Z_C = \sqrt{\frac{z}{y}} = \sqrt{\frac{1.677 \times 10^{-5} + j3.33 \times 10^{-4}}{0 + j4.817 \times 10^{-9}}} = 263.364 + j6.61\,\Omega$$

$$\gamma = \sqrt{zy} = \sqrt{(1.677 \times 10^{-5} + j3.33 \times 10^{-4})(0 + j4.817 \times 10^{-9})} = 3.184 \times 10^{-8} + j1.269 \times 10^{-6}\,1/\text{m}$$

$$A = \cosh(\gamma l) = \cosh((3.184 \times 10^{-8} + j1.269 \times 10^{-6})(3{,}50{,}000)) = 0.903 + j4.787 \times 10^{-3}\ \text{pu}$$

$$B = Z_C \sinh(\gamma l) = (263.364 + j6.61) \times \sinh((3.184 \times 10^{-8} + j1.269 \times 10^{-6})(350{,}000))$$
$$= 5.49 + j113.074\ \Omega$$

$$C = \frac{1}{Z_C}\sinh(\gamma l) = \frac{1}{263.364 + j6.61}\sinh((3.184 \times 10^{-8} + j1.269 \times 10^{-6})(350{,}000))$$
$$= -2.726 \times 10^{-6} + j1.631 \times 10^{-3}\ \text{S}$$

$$D = \cosh(\gamma l) = \cosh((3.184 \times 10^{-8} + j1.269 \times 10^{-6})(3{,}50{,}000)) = 0.903 + j4.787 \times 10^{-3}\ \text{pu}$$

c. *Voltage and current at the sending end*

$$V_{R\,LL} = 400\angle 30^\circ\ \text{kV}$$

$$V_{R\,phase} = \frac{V_{R\,LL}}{\sqrt{3}}\angle -30^\circ = 230.94\angle 0^\circ\ \text{kV}$$

$$S_{Load\ 3\text{-}phase} = 250\angle -\text{acos(PF)} = 250\angle \text{acos}(0.8) = 250\angle 36.87^\circ\ \text{MVA}$$

$$S_{Load\ phase} = \frac{S_{Load\ 3\text{-}phase}}{3} = \frac{250\angle 25.842^\circ}{3} = 83.333\angle 36.87^\circ\ \text{MVA}$$

$$I_{R\ phase} = I_{Load} = \overline{\frac{S_{Load\ phase}}{V_{R\ phase}}} = \overline{\frac{83.333\angle 36.87^\circ}{230.94\angle 0^\circ}} = \frac{83.333\angle -36.87^\circ}{230.94\angle 0^\circ} = 360.844\angle -36.87^\circ\ \text{A}$$

$$\begin{bmatrix} V_S \\ I_S \end{bmatrix} = \begin{bmatrix} A & B \\ C & D \end{bmatrix}\begin{bmatrix} V_R \\ I_R \end{bmatrix} = \begin{bmatrix} 0.903 + j4.787 \times 10^{-3} & 5.49 + j113.074 \\ -2.726 \times 10^{-6} + j1.631 \times 10^{-3} & 0.903 + j4.787 \times 10^{-3} \end{bmatrix}\begin{bmatrix} 230.94\angle 0^\circ \\ 360.844\angle -36.87^\circ \end{bmatrix}$$

$$= \begin{bmatrix} 236.874\angle 7.9^\circ \\ 318.592\angle 34.959^\circ \end{bmatrix}$$

$$V_{S\ phase} = 236.874\angle 7.9^\circ\ \text{kV}$$

$$I_{S\ phase} = 318.592\angle 34.959^\circ\ \text{A}$$

$$V_{S\,LL} = \sqrt{3}V_{S\ phase}\angle 30^\circ = \sqrt{3} \times 236.874\ \angle 7.9^\circ = 410.277\angle 30^\circ\ \text{kV}$$

14.1.5 Characteristics of Overhead Conductors

Table 14.2 presents typical values of resistance, inductive reactance, and capacitance reactance, per-unit length, of ACSR conductors. The size of the conductors (cross-section area) is specified in square millimeters and *kcmil*, where a *cmil* is the cross-section area of a circular conductor with a diameter of 1/1000 in. The tables include also the approximate current-carrying capacity of the conductors assuming 60 Hz, wind speed of 1.4 miles/h, and conductor and air temperatures of 75°C and 25°C, respectively. Table 14.3 presents the corresponding characteristics of AAC conductors.

TABLE 14.2 Characteristics of Aluminum-Cable-Steel Reinforced Conductors

Code	Cross-Section Area			Stranding	Diameter		Layers	Approx. Current-Carrying Capacity	Resistance (mΩ/km)				GMR	60-Hz Reactances (Dm = 1 m)	
	Total	Aluminum							DC	AC (60 Hz)					
	(mm²)	(kcmil)	(mm²)	Al/Steel	Conductor (mm)	Core (mm)		(A)	25°C	25°C	50°C	75°C	(mm)	X_L (Ω/km)	X_C (MΩ/km)
—	1521	2776	1407	84/19	50.80	13.87	4		21.0	24.5	26.2	28.1	20.33	0.294	0.175
Joree	1344	2515	1274	76/19	47.75	10.80	4		22.7	26.0	28.0	30.0	18.93	0.299	0.178
Thrasher	1235	2312	1171	76/19	45.77	10.34	4		24.7	27.7	30.0	32.2	18.14	0.302	0.180
Kiwi	1146	2167	1098	72/7	44.07	8.81	4		26.4	29.4	31.9	34.2	17.37	0.306	0.182
Bluebird	1181	2156	1092	84/19	44.75	12.19	4		26.5	29.0	31.4	33.8	17.92	0.303	0.181
Chukar	976	1781	902	84/19	40.69	11.10	4		32.1	34.1	37.2	40.1	16.28	0.311	0.186
Falcon	908	1590	806	54/19	39.24	13.08	3	1380	35.9	37.4	40.8	44.3	15.91	0.312	0.187
Lapwing	862	1590	806	45/7	38.20	9.95	3	1370	36.7	38.7	42.1	45.6	15.15	0.316	0.189
Parrot	862	1510	765	54/19	38.23	12.75	3	1340	37.8	39.2	42.8	46.5	15.48	0.314	0.189
Nuthatch	818	1510	765	45/7	37.21	9.30	3	1340	38.7	40.5	44.2	47.9	14.78	0.318	0.190
Plover	817	1431	725	54/19	37.21	12.42	3	1300	39.9	41.2	45.1	48.9	15.06	0.316	0.190
Bobolink	775	1431	725	45/7	36.25	9.07	3	1300	35.1	42.6	46.4	50.3	14.39	0.320	0.191
Martin	772	1351	685	54/19	36.17	12.07	3	1250	42.3	43.5	47.5	51.6	14.63	0.319	0.191
Dipper	732	1351	685	45/7	35.20	8.81	3	1250	43.2	44.9	49.0	53.1	13.99	0.322	0.193
Pheasant	726	1272	645	54/19	35.10	11.71	3	1200	44.9	46.1	50.4	54.8	14.20	0.321	0.193
Bittern	689	1272	644	45/7	34.16	8.53	3	1200	45.9	47.5	51.9	56.3	13.56	0.324	0.194
Grackle	681	1192	604	54/19	34.00	11.33	3	1160	47.9	49.0	53.6	58.3	13.75	0.323	0.194
Bunting	646	1193	604	45/7	33.07	8.28	3	1160	48.9	50.4	55.1	59.9	13.14	0.327	0.196
Finch	636	1114	564	54/19	32.84	10.95	3	1110	51.3	52.3	57.3	62.3	13.29	0.326	0.196
Bluejay	603	1113	564	45/7	31.95	8.00	3	1110	52.4	53.8	58.9	64.0	12.68	0.329	0.197
Curlew	591	1033	523	54/7	31.62	10.54	3	1060	56.5	57.4	63.0	68.4	12.80	0.329	0.198
Ortolan	560	1033	525	45/7	30.78	7.70	3	1060	56.5	57.8	63.3	68.7	12.22	0.332	0.199
Merganser	596	954	483	30/7	31.70	13.60	2	1010	61.3	61.8	67.9	73.9	13.11	0.327	0.198
Cardinal	546	954	483	54/7	30.38	10.13	3	1010	61.2	62.0	68.0	74.0	12.31	0.332	0.200

(continued)

TABLE 14.2 (continued) Characteristics of Aluminum-Cable-Steel Reinforced Conductors

	Cross-Section Area				Diameter			Approx. Current-Carrying Capacity	Resistance (mΩ/km)					60-Hz Reactances ($Dm = 1$ m)	
		Aluminum								AC (60 Hz)					
Code	Total (mm²)	(kcmil)	(mm²)	Stranding Al/Steel	Conductor (mm)	Core (mm)	Layers	(A)	DC 25°C	25°C	50°C	75°C	GMR (mm)	X_L (Ω/km)	X_C (MΩ/km)
Rail	517	954	456	45/7	29.59	7.39	3	1010	61.2	62.4	68.3	74.3	11.73	0.335	0.201
Baldpate	562	900	456	30/7	30.78	13.21	2	960	65.0	65.5	71.8	78.2	12.71	0.329	0.199
Canary	515	900	456	54/7	29.51	9.83	3	970	64.8	65.5	72.0	78.3	11.95	0.334	0.201
Ruddy	478	900	443	45/7	28.73	7.19	3	970	64.8	66.0	72.3	78.6	11.40	0.337	0.202
Crane	501	875	443	54/7	29.11	9.70	3	950	66.7	67.5	74.0	80.5	11.80	0.335	0.202
Willet	474	874	403	45/7	28.32	7.09	3	950	66.7	67.9	74.3	80.9	11.25	0.338	0.203
Skimmer	479	795	403	30/7	29.00	12.40	2	940	73.5	74.0	81.2	88.4	11.95	0.334	0.202
Mallard	495	795	403	30/19	28.96	12.42	2	910	73.5	74.0	81.2	88.4	11.95	0.334	0.202
Drake	469	795	403	26/7	28.14	10.36	2	900	73.3	74.0	81.2	88.4	11.43	0.337	0.203
Condor	455	795	403	54/7	27.74	9.25	3	900	73.4	74.1	81.4	88.6	11.22	0.339	0.204
Cuckoo	455	795	403	24/7	27.74	9.25	2	900	73.4	74.1	81.4	88.5	11.16	0.339	0.204
Tern	431	795	403	45/7	27.00	6.76	3	900	73.4	74.4	81.6	88.8	10.73	0.342	0.205
Coot	414	795	403	36/1	26.42	3.78	3	910	73.0	74.4	81.5	88.6	10.27	0.345	0.206
Buteo	447	715	362	30/7	27.46	11.76	2	840	81.8	82.2	90.2	98.3	11.34	0.338	0.204
Redwing	445	715	362	30/19	27.46	11.76	2	840	81.8	82.2	90.2	98.3	11.34	0.338	0.204
Starling	422	716	363	26/7	26.7	9.82	2	840	81.5	82.1	90.1	98.1	10.82	0.341	0.206
Crow	409	715	362	54/7	26.31	8.76	3	840	81.5	82.2	90.2	98.2	10.67	0.342	0.206
Stilt	410	716	363	24/7	26.31	8.76	2	840	81.5	82.2	90.2	98.1	10.58	0.343	0.206
Grebe	388	716	363	45/7	25.63	6.4	3	840	81.5	82.5	90.4	98.4	10.18	0.346	0.208
Gannet	393	666	338	26/7	25.76	9.5	2	800	87.6	88.1	96.6	105.3	10.45	0.344	0.208
Gull	382	667	338	54/7	25.4	8.46	3	800	87.5	88.1	96.8	105.3	10.27	0.345	0.208
Flamingo	382	667	338	24/7	25.4	8.46	2	800	87.4	88.1	96.7	105.3	10.21	0.346	0.208
Scoter	397	636	322	30/7	25.88	11.1	2	800	91.9	92.3	101.4	110.4	10.70	0.342	0.207
Egret	396	636	322	30/19	25.88	11.1	2	780	91.9	92.3	101.4	110.4	10.70	0.342	0.207
Grosbeak	375	636	322	26/7	25.15	9.27	2	780	91.7	92.2	101.2	110.3	10.21	0.346	0.209
Goose	364	636	322	54/7	24.82	8.28	3	770	91.8	92.4	101.4	110.4	10.06	0.347	0.208
Rook	363	636	322	24/7	24.82	8.28	2	770	91.7	92.3	101.3	110.3	10.06	0.347	0.209

Kingbird	340	636	322	18/1	23.88	4.78	2	780	91.2	92.2	101.1	110.0	9.27	0.353	0.211
Swirl	331	636	322	36/1	23.62	3.38	3	780	91.3	92.4	101.3	110.3	9.20	0.353	0.212
Wood Duck	378	605	307	30/7	25.25	10.82	2	760	96.7	97.0	106.5	116.1	10.42	0.344	0.208
Teal	376	605	307	30/19	25.25	10.82	2	770	96.7	97.0	106.5	116.1	10.42	0.344	0.208
Squab	356	605	356	26/7	25.54	9.04	2	760	96.5	97.0	106.5	116.0	9.97	0.347	0.208
Peacock	346	605	307	24/7	24.21	8.08	2	760	96.4	97.0	106.4	115.9	9.72	0.349	0.210
Duck	347	606	307	54/7	24.21	8.08	3	750	96.3	97.0	106.3	115.8	9.81	0.349	0.210
Eagle	348	557	282	30/7	24.21	10.39	2	730	105.1	105.4	115.8	126.1	10.00	0.347	0.210
Dove	328	556	282	26/7	23.55	8.66	2	730	104.9	105.3	115.6	125.9	9.54	0.351	0.212
Parakeet	319	557	282	24/7	23.22	7.75	2	730	104.8	105.3	115.6	125.9	9.33	0.352	0.212
Osprey	298	556	282	18/1	22.33	4.47	2	740	104.4	105.2	115.4	125.7	8.66	0.358	0.214
Hen	298	477	242	30/7	22.43	9.6	2	670	122.6	122.9	134.9	147.0	9.27	0.353	0.214
Hawk	281	477	242	26/7	21.79	8.03	2	670	122.4	122.7	134.8	146.9	8.84	0.357	0.215
Flicker	273	477	273	24/7	21.49	7.16	2	670	122.2	122.7	134.7	146.8	8.63	0.358	0.216
Pelican	255	477	242	18/1	20.68	4.14	2	680	121.7	122.4	134.4	146.4	8.02	0.364	0.218
Lark	248	397	201	30/7	20.47	8.76	2	600	147.2	147.4	161.9	176.4	8.44	0.360	0.218
Ibis	234	397	201	26/7	19.89	7.32	2	590	146.9	147.2	161.7	176.1	8.08	0.363	0.220
Brant	228	398	201	24/7	19.61	6.53	2	590	146.7	147.1	161.6	176.1	7.89	0.365	0.221
Chickadee	213	397	201	18/1	18.87	3.78	2	590	146.1	146.7	161.0	175.4	7.32	0.371	0.222
Oriole	210	336	170	30/7	18.82	8.08	2	530	173.8	174.0	191.2	208.3	7.77	0.366	0.222
Linnet	198	336	170	26/7	18.29	6.73	2	530	173.6	173.8	190.9	208.1	7.41	0.370	0.224
Widgeon	193	336	170	24/7	18.03	6.02	2	530	173.4	173.7	190.8	207.9	7.25	0.371	0.225
Merlin	180	336	170	18/1	16.46	3.48	2	530	173.0	173.1	190.1	207.1	6.74	0.377	0.220
Piper	187	300	152	30/7	17.78	7.62	2	500	195.0	195.1	214.4	233.6	7.35	0.370	0.225
Ostrich	177	300	152	26/7	17.27	6.38	2	490	194.5	194.8	214.0	233.1	7.01	0.374	0.227
Gadwall	172	300	152	24/7	17.04	5.69	2	490	194.5	194.8	213.9	233.1	6.86	0.376	0.227
Phoebe	160	300	152	18/1	16.41	3.28	2	490	193.5	194.0	213.1	232.1	6.37	0.381	0.229
Junco	167	267	135	30/7	16.76	7.19	2	570	219.2	219.4	241.1	262.6	6.92	0.375	0.228
Partridge	157	267	135	26/7	16.31	5.99	2	460	218.6	218.9	240.5	262.0	6.61	0.378	0.229
Waxwing	143	267	135	18/1	15.47	3.1	2	460	217.8	218.1	239.7	261.1	6.00	0.386	0.232

Sources: Electric Power Research Institute, *Transmission Line Reference Book 345 kV and Above*, EPRI, Palo Alto, CA, 1987; Glover, J.D. and Sarma, M.S., *Power System Analysis and Design*, 3rd edn., Brooks/Cole, Pacific Grove, CA, 2002. With permission.

Current capacity evaluated at 75°C conductor temperature, 25°C air temperature, wind speed of 1.4 miles/h, and frequency of 60 Hz.

TABLE 14.3 Characteristics of All-Aluminum Conductors (AAC)

Code	Cross-Section Area		Stranding	Diameter	Layers	Approx. Current-Carrying Capacity (A)	Resistance (mΩ/km)				GMR	60-Hz Reactances (Dm = 1 m)	
							DC	AC (60 Hz)					
	(mm²)	kcmil or AWG		(mm)			25°C	25°C	50°C	75°C	(mm)	X_L (Ω/km)	X_C (MΩ/km)
Coreopsis	806.2	1591	61	36.93	4	1380	36.5	39.5	42.9	46.3	14.26	0.320	0.190
Gladiolus	765.8	1511	61	35.99	4	1340	38.4	41.3	44.9	48.5	13.90	0.322	0.192
Carnation	725.4	1432	61	35.03	4	1300	40.5	43.3	47.1	50.9	13.53	0.322	0.193
Columbine	865.3	1352	61	34.04	4	1250	42.9	45.6	49.6	53.6	13.14	0.327	0.196
Narcissus	644.5	1272	61	33.02	4	1200	45.5	48.1	52.5	56.7	12.74	0.329	0.194
Hawthorn	604.1	1192	61	31.95	4	1160	48.7	51.0	55.6	60.3	12.34	0.331	0.197
Marigold	564.2	1113	61	30.89	4	1110	52.1	54.3	59.3	64.3	11.92	0.334	0.199
Larkspur	524	1034	61	29.77	4	1060	56.1	58.2	63.6	69.0	11.49	0.337	0.201
Bluebell	524.1	1034	37	29.71	3	1060	56.1	58.2	63.5	68.9	11.40	0.337	0.201
Goldenrod	483.7	955	61	28.6	4	1010	60.8	62.7	68.6	74.4	11.03	0.340	0.203
Magnolia	483.6	954	37	28.55	3	1010	60.8	62.7	68.6	74.5	10.97	0.340	0.203
Crocus	443.6	875	61	27.38	4	950	66.3	68.1	74.5	80.9	10.58	0.343	0.205
Anemone	443.5	875	37	27.36	3	950	66.3	68.1	74.5	80.9	10.49	0.344	0.205
Lilac	403.1	796	61	26.11	4	900	73.0	74.6	81.7	88.6	10.09	0.347	0.207
Arbutus	402.9	795	37	26.06	3	900	73.0	74.6	81.7	88.6	10.00	0.347	0.207
Nasturtium	362.5	715	61	24.76	4	840	81.2	82.6	90.5	98.4	9.57	0.351	0.209
Violet	362.8	716	37	24.74	3	840	81.1	82.5	90.4	98.3	9.48	0.351	0.209
Orchid	322.2	636	37	23.32	3	780	91.3	92.6	101.5	110.4	8.96	0.356	0.212
Mistletoe	281.8	556	37	21.79	3	730	104 4	105.5	115 8	126.0	8.38	0.361	0.215
Dahlia	281.8	556	19	21.72	2	730	104.4	105.5	115 8	125.9	8.23	0.362	0.216
Syringa	241.5	477	37	20.193	3	670	121.8	122.7	134.7	146.7	7.74	0.367	0.219
Cosmos	241.9	477	19	20.142	2	670	121.6	122.6	134.5	146.5	7.62	0.368	0.219
Canna	201.6	398	19	18.36	2	600	145.9	146.7	161.1	175.5	6.95	0.375	0.224
Tulip	170.6	337	19	16.92	2	530	172.5	173.2	190.1	207.1	6.40	0.381	0.228
Laurel	135.2	267	19	15.06	2	460	217.6	218.1	239.6	261.0	5.70	0.390	0.233
Daisy	135.3	267	7	14.88	1	460	217.5	218	239.4	260.8	5 39	0.394	0.233
Oxlip	107.3	212 or (4/0)	7	13.26	1	340	274.3	274.7	301.7	328.8	4.82	0.402	0.239
Phlox	85	168 or (3/0)	7	11.79	1	300	346.4	346.4	380.6	414.7	4.27	0.411	0.245

Aster	67.5	133 or (2/0)	7	10.52	1	270	436.1	439.5	479.4	522.5	3.81	0.4 0	0.25
Poppy	53.5	106 or (1/0)	7	9.35	1	230	550	550.2	604.5	658.8	3.38	0.429	0.256
Pansy	42.4	#1 AWG	7	8.33	1	200	694.2	694.2	763.2	831.6	3.02	0.438	0.261
Iris	33.6	#2 AWG	7	7.42	1	180	874.5	874.5	960.8	1047.9	2.68	0.446	0.267
Rose	21.1	#3 AWG	7	5.89	1	160	1391.5	1391.5	1528.9	1666.3	2.13	0.464	0.278
Peachbell	13.3	#4 AWG	7	4.67	1	140	2214.4	2214.4	2443.2	2652	1.71	0.481	0.289
Even sizes													
Bluebonnet	1773.3	3500	7	54.81	6		16.9	22.2	23.6	25.0	21.24	0.290	0.172
Trillium	1520.2	3000	127	50.75	6		19.7	24.6	26.2	27.9	19.69	0.296	0.175
Lupine	1266.0	2499	91	46.30	5		23.5	27.8	29.8	31.9	17.92	0.303	0.180
Cowslip	1012.7	1999	91	41.40	5		29.0	32.7	35.3	38.0	16.03	0.312	0.185
Jessamine	887.0	1750	61	38.74	4		33.2	36.5	39.5	42.5	14.94	0.317	0.188
Hawkweed	506.7	1000	37	29.24	3	1030	58.0	60.0	65.5	71.2	11.22	0.339	0.201
Camelia	506.4	999	61	29.26	4	1030	58.1	60.1	65.5	71.2	11.31	0.338	0.201
Snapdragon	456.3	900	61	27.74	4	970	64.4	66.3	72.5	78.7	10.73	0.342	0.204
Cockscomb	456.3	900	37	27.74	3	970	64.4	66.3	72.5	78.7	10.64	0.343	0.204
Cattail	380.1	750	61	25.35	4	870	77.4	78.9	86.4	93.9	9.78	0.349	0.208
Petunia	380.2	750	37	23.85	3	870	77.4	78.9	86.4	93.9	9.72	0.349	0.208
Flag	354.5	700	61	24.49	4	810	83.0	84.4	92.5	100.6	9.45	0.352	0.210
Verbena	354.5	700	37	24.43	3	810	83.0	84.4	92.5	100.6	9.39	0.352	0.210
Meadowsweet	303.8	600	37	2.63	3	740	96.8	98.0	107.5	117.0	8.69	0.358	0.214
Hyacinth	253.1	500	37	20.65	3	690	116.2	117.2	128.5	140.0	7.92	0.365	0.218
Zinnia	253.3	500	19	20.60	2	690	116.2	117.2	128.5	139.9	7.80	0.366	0.218
Golden tuft	228.0	450	19	19.53	2	640	129.0	129.9	142.6	155.3	7.41	0.370	0.221
Daffodil	177.3	350	19	17.25	2	580	165.9	166.6	183.0	199.3	6.52	0.379	0.227
Peony	152.1	300	19	15.98	2	490	193.4	194.0	213.1	232.1	6.04	0.385	0.230
Valerian	126.7	250	19	14.55	2	420	232.3	232.8	255.6	278.6	5.52	0.392	0.235
Sneezewort	126.7	250	7	14.40	1	420	232.2	232.7	255.6	278.4	5.21	0.396	0.235

Sources: Electric Power Research Institute, *Transmission Line Reference Book 345 kV and Above*, EPRI, Palo Alto, CA, 1987; Glover, J.D. and Sarma, M.S., *Power System Analysis and Design*, 3rd edn., Brooks/Cole, Pacific Grove, CA, 2002. With permission.

Current capacity evaluated at 75°C conductor temperature, 25°C air temperature, wind speed of 1.4 miles/h, and frequency of 60 Hz.

References

1. Electric Power Research Institute, *Transmission Line Reference Book 345 kV and Above*, 2nd edn., EPRI, Palo Alto, CA, 1987.
2. Barnes, C. C., *Power Cables. Their Design and Installation*, 2nd edn., Chapman and Hall Ltd., London, U.K., 1966.
3. Glover, J. D. and Sarma, M. S., *Power System Analysis and Design*, 3rd edn., Brooks/Cole, Pacific Grove, CA, 2002.
4. Stevenson W. D. Jr., *Elements of Power System Analysis*, 4th. edn., McGraw-Hill, New York, 1982.
5. Saadat, H., *Power System Analysis*, McGrawHill, New York, 1999.
6. Gross, C. A., *Power System Analysis*, John Wiley and Sons, Inc., New York, 1979.
7. Yamayee, Z. A. and Bala J. L. Jr., *Electromechanical Energy Devices and Power Systems*, John Wiley and Sons, Inc., New York, 1994.
8. Gungor, B. R., *Power Systems*, Harcourt Brace Jovanovich, Orlando, FL, 1988.
9. Zaborszky, J. and Rittenhouse, J. W., *Electric Power Transmission. The Power System in the Steady State*, The Ronald Press Company, New York, 1954.

15

Sag and Tension of Conductor

Dale A. Douglass
Power Delivery Consultants, Inc.

F. Ridley Thrash
Southwire Company

The energized conductors of transmission and distribution lines must be placed to totally eliminate the possibility of injury to people. Overhead conductors, however, elongate with time, temperature, and tension, thereby changing their original positions after installation. Despite the effects of weather and loading on a line, the conductors must remain at safe distances from buildings, objects, and people or vehicles passing beneath the line at all times. To ensure this safety, the shape of the terrain along the right-of-way, the height and lateral position of the conductor support points, and the position of the conductor between support points under all wind, ice, and temperature conditions must be known.

Bare overhead transmission or distribution conductors are typically quite flexible and uniform in weight along their length. Because of these characteristics, they take the form of a catenary (Ehrenberg, 1935; Winkelmann, 1959) between support points. The shape of the catenary changes with conductor temperature, ice and wind loading, and time. To ensure adequate vertical and horizontal clearance under all weather and electrical loadings, and to ensure that the breaking strength of the conductor is not exceeded, the behavior of the conductor catenary under all conditions must be known before the line is designed. The future behavior of the conductor is determined through calculations commonly referred to as sag-tension calculations.

Sag-tension calculations predict the behavior of conductors based on recommended tension limits under varying loading conditions. These tension limits specify certain percentages of the conductor's rated breaking strength that are not to be exceeded upon installation or during the life of the line. These conditions, along with the elastic and permanent elongation properties of the conductor, provide the basis for determinating the amount of resulting sag during installation and long-term operation of the line.

Accurately determined initial sag limits are essential in the line design process. Final sags and tensions depend on initial installed sags and tensions and on proper handling during installation. The final sag shape of conductors is used to select support point heights and span lengths so that the minimum clearances will be maintained over the life of the line. If the conductor is damaged or the initial sags are incorrect, the line clearances may be violated or the conductor may break during heavy ice or wind loadings.

15.1 Catenary Cables

A bare-stranded overhead conductor is normally held clear of objects, people, and other conductors by periodic attachment to insulators. The elevation differences between the supporting structures affect the shape of the conductor catenary. The catenary's shape has a distinct effect on the sag and tension of the conductor, and therefore, must be determined using well-defined mathematical equations.

15.1.1 Level Spans

The shape of a catenary is a function of the conductor weight per unit length, w, the horizontal component of tension, H, span length, S, and the maximum sag of the conductor, D. Conductor sag and span length are illustrated in Figure 15.1 for a level span.

The exact catenary equation uses hyperbolic functions. Relative to the low point of the catenary curve shown in Figure 15.1, the height of the conductor, $y(x)$, above this low point is given by the following equation:

$$y(x) = \frac{H}{w}\cosh\left(\left(\frac{w}{H}x\right) - 1\right) = \frac{w(x^2)}{2H} \tag{15.1}$$

Note that x is positive in either direction from the low point of the catenary. The expression to the right is an approximate parabolic equation based upon a MacLaurin expansion of the hyperbolic cosine.

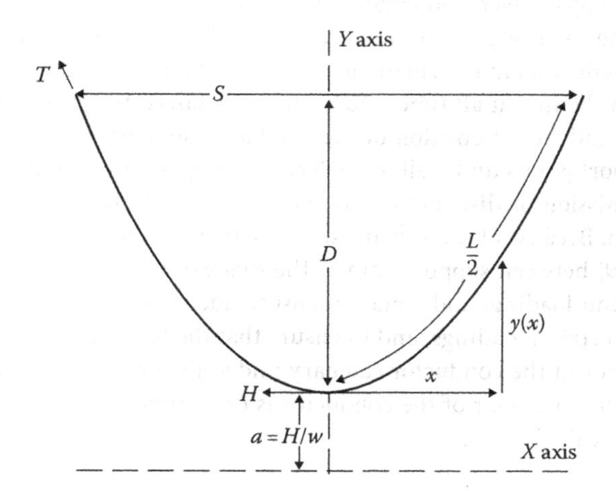

FIGURE 15.1 Catenary curve for level spans.

For a level span, the low point is in the center, and the sag, D, is found by substituting $x = S/2$ in the preceding equations. The exact and approximate parabolic equations for sag become the following:

$$D = \frac{H}{w}\left(\cosh\left(\frac{wS}{2H}\right) - 1\right) = \frac{w(S^2)}{8H} \tag{15.2}$$

The ratio, H/w, which appears in all of the preceding equations, is commonly referred to as the catenary constant. An increase in the catenary constant, having the units of length, causes the catenary curve to become shallower and the sag to decrease. Although it varies with conductor temperature, ice, and wind loading, and time, the catenary constant typically has a value in the range of several thousand feet for most transmission-line catenaries.

The approximate or parabolic expression is sufficiently accurate as long as the sag is <5% of the span length. As an example, consider a 1000 ft span of Drake conductor ($w = 1.096$ lb/ft) installed at a tension of 4500 lb. The catenary constant equals 4106 ft. The calculated sag is 30.48 and 30.44 ft using the hyperbolic and approximate equations, respectively. Both estimates indicate a sag-to-span ratio of 3.4% and a sag difference of only 0.5 in.

The horizontal component of tension, H, is equal to the conductor tension at the point in the catenary where the conductor slope is horizontal. For a level span, this is the midpoint of the span length. At the ends of the level span, the conductor tension, T, is equal to the horizontal component plus the conductor weight per unit length, w, multiplied by the sag, D, as shown in the following:

$$T = H + wD \tag{15.3}$$

Given the conditions in the preceding example calculation for a 1000 ft level span of Drake ACSR, the tension at the attachment points exceeds the horizontal component of tension by 33 lb. It is common to perform sag-tension calculations using the horizontal tension component, but the average of the horizontal and support point tension is usually listed in the output.

15.1.2 Conductor Length

Application of calculus to the catenary equation allows the calculation of the conductor length, $L(x)$, measured along the conductor from the low point of the catenary in either direction.

The resulting equation becomes

$$L(x) = \frac{H}{w}\sinh\left(\frac{wx}{H}\right) = x\left(1 + \frac{x^2(w^2)}{6H^2}\right) \tag{15.4}$$

For a level span, the conductor length corresponding to $x = S/2$ is half of the total conductor length and the total length, L, is

$$L = \left(\frac{2H}{w}\right)\sinh\left(\frac{Sw}{2H}\right) = S\left(1 + \frac{S^2(w^2)}{24H^2}\right) \tag{15.5}$$

The parabolic equation for conductor length can also be expressed as a function of sag, D, by substitution of the sag parabolic equation, giving

$$L = S + \frac{8D^2}{3S} \tag{15.6}$$

15.1.3 Conductor Slack

The difference between the conductor length, L, and the span length, S, is called slack. The parabolic equations for slack may be found by combining the preceding parabolic equations for conductor length, L, and sag, D:

$$L - S = S^3\left(\frac{w^2}{24H^2}\right) = D^2\left(\frac{8}{3S}\right) \tag{15.7}$$

While slack has units of length, it is often expressed as the percentage of slack relative to the span length. Note that slack is related to the cube of span length for a given H/w ratio and to the square of sag for a given span. For a series of spans having the same H/w ratio, the total slack is largely determined by the longest spans. It is for this reason that the ruling span (RS) is nearly equal to the longest span rather than the average span in a series of suspension spans.

Equation 15.7 can be inverted to obtain a more interesting relationship showing the dependence of sag, D, upon slack, $L - S$:

$$D = \sqrt{\frac{3S(L-S)}{8}} \tag{15.8}$$

As can be seen from the preceding equation, small changes in slack typically yield large changes in conductor sag.

15.1.4 Inclined Spans

Inclined spans may be analyzed using essentially the same equations that were used for level spans. The catenary equation for the conductor height above the low point in the span is the same. However, the span is considered to consist of two separate sections, one to the right of the low point and the other to the left as shown in Figure 15.2 (Winkelmann, 1959). The shape of the catenary relative to the low point is unaffected by the difference in suspension point elevation (span inclination).

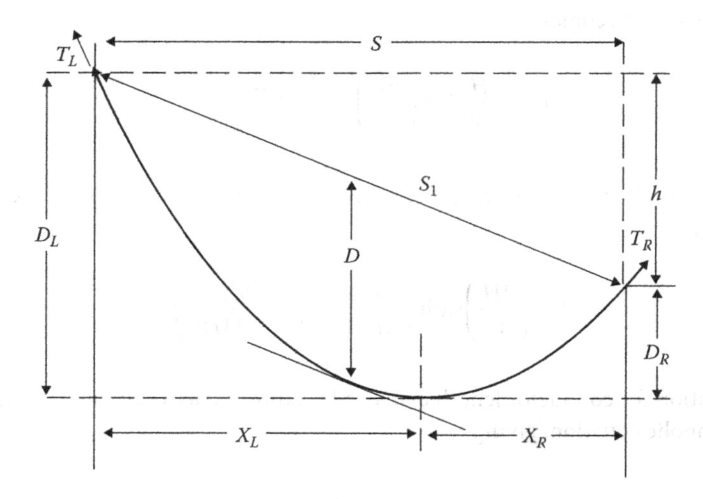

FIGURE 15.2 Inclined catenary span.

In each direction from the low point, the conductor elevation, $y(x)$, relative to the low point is given by

$$y(x) = \frac{H}{w}\cosh\left(\left(\frac{w}{H}x\right) - 1\right) = \frac{w(x^2)}{2H} \tag{15.9}$$

Note that x is considered positive in either direction from the low point.

The horizontal distance, x_L, from the left support point to the low point in the catenary is

$$x_L = \frac{S}{2}\left(1 + \frac{h}{4D}\right) \tag{15.10}$$

The horizontal distance, x_R, from the right support point to the low point of the catenary is

$$x_R = \frac{S}{2}\left(1 - \frac{h}{4D}\right) \tag{15.11}$$

where

- S is the horizontal distance between support points
- h is the vertical distance between support points
- D is the sag measured vertically from a line through the points of conductor support to a line tangent to the conductor

The midpoint sag, D, is approximately equal to the sag in a horizontal span equal in length to the inclined span, S_l.

Knowing the horizontal distance from the low point to the support point in each direction, the preceding equations for $y(x)$, L, D, and T can be applied to each side of the inclined span.

The total conductor length, L, in the inclined span is equal to the sum of the lengths in the x_R and x_L sub-span sections:

$$L = S + \left(x_R^3 + x_L^3\right)\left(\frac{w^2}{6H^2}\right) \tag{15.12}$$

In each sub-span, the sag is relative to the corresponding support point elevation:

$$D_R = \frac{wx_R^2}{2H} \quad D_L = \frac{wx_L^2}{2H} \tag{15.13}$$

or in terms of sag, D, and the vertical distance between support points:

$$D_R = D\left(1 - \frac{h}{4D}\right)^2 \quad D_L = D\left(1 + \frac{h}{4D}\right)^2 \tag{15.14}$$

and the maximum tension is

$$T_R = H + wD_R \quad T_L = H + wD_L \tag{15.15}$$

or in terms of upper and lower support points:

$$T_u = T_l + wh \tag{15.16}$$

where

D_R is the sag in right sub-span section
D_L is the sag in left sub-span section
T_R is the tension in right sub-span section
T_L is the tension in left sub-span section
T_u is the tension in conductor at upper support
T_l is the tension in conductor at lower support

The horizontal conductor tension is equal at both supports. The vertical component of conductor tension is greater at the upper support and the resultant tension, T_u, is also greater.

15.1.5 Ice and Wind Conductor Loads

When a conductor is covered with ice and/or is exposed to wind, the effective conductor weight per unit length increases. During occasions of heavy ice and/or wind load, the conductor catenary tension increases dramatically along with the loads on angle and dead-end structures. Both the conductor and its supports can fail unless these high-tension conditions are considered in the line design.

The National Electric Safety Code (NESC) suggests certain combinations of ice and wind corresponding to heavy, medium, and light loading regions of the United States. Figure 15.3 is a map of the United States indicating those areas (NESC, 2007). The combinations of ice and wind corresponding to loading region are listed in Table 15.1.

The NESC also suggests that increased conductor loads due to high wind loads without ice be considered. Figure 15.4 shows the suggested wind pressure as a function of geographical area for the United States (ASCE Std. 7–88).

Certain utilities in very heavy ice areas use glaze ice thicknesses of as much as 2 in. to calculate iced conductor weight. Similarly, utilities in regions where hurricane winds occur may use wind loads as high as 34 lb/ft^2.

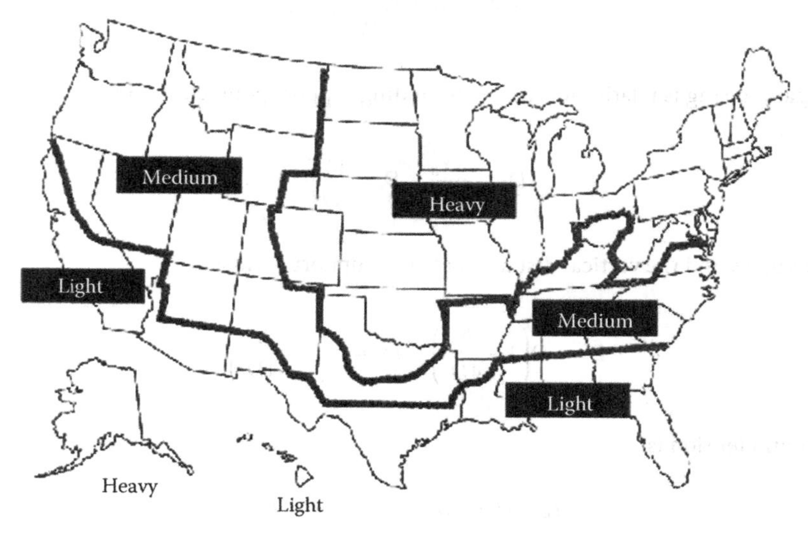

FIGURE 15.3 Ice and wind load areas of the United States.

TABLE 15.1 Definitions of Ice and Wind Load for NESC Loading Areas

	Loading Districts			
	Heavy	Medium	Light	Extreme Wind Loading
Radial thickness of ice				
(in.)	0.50	0.25	0	0
(mm)	12.5	6.5	0	0
Horizontal wind pressure				
(lb/ft²)	4	4	9	See Figure 15.4
(Pa)	190	190	430	
Temperature				
(°F)	0	+15	+30	+60
(°C)	−20	−10	−1	+15
Constant to be added to the resultant for all conductors				
(lb/ft)	0.30	0.20	0.05	0.0
(N/m)	4.40	2.50	0.70	0.0

As the NESC indicates, the degree of ice and wind loads varies with the region. Some areas may have heavy icing, whereas some areas may have extremely high winds. The loads must be accounted for in the line design process so they do not have a detrimental effect on the line. Some of the effects of both the individual and combined components of ice and wind loads are discussed in the following.

15.1.5.1 Ice Loading

The formation of ice on overhead conductors may take several physical forms (glaze ice, rime ice, or wet snow). The impact of lower density ice formation is usually considered in the design of line sections at high altitudes.

The formation of ice on overhead conductors has the following influence on line design:

- Ice loads determine the maximum vertical conductor loads that structures and foundations must withstand.
- In combination with simultaneous wind loads, ice loads also determine the maximum transverse loads on structures.
- In regions of heavy ice loads, the maximum sags and the permanent increase in sag with time (difference between initial and final sags) may be due to ice loadings.

Ice loads for use in designing lines are normally derived on the basis of past experience, code requirements, state regulations, and analysis of historical weather data. Mean recurrence intervals for heavy ice loadings are a function of local conditions along various routings. The impact of varying assumptions concerning ice loading can be investigated with line design software.

The calculation of ice loads on conductors is normally done with an assumed glaze ice density of 57 lb/ft³. The weight of ice per unit length is calculated with the following equation:

$$w_{ice} = 1.244t(D_c + t) \tag{15.17}$$

where

t is the thickness of ice, in.

D_c is the conductor outside diameter, in.

w_{ice} is the resultant weight of ice, lb/ft

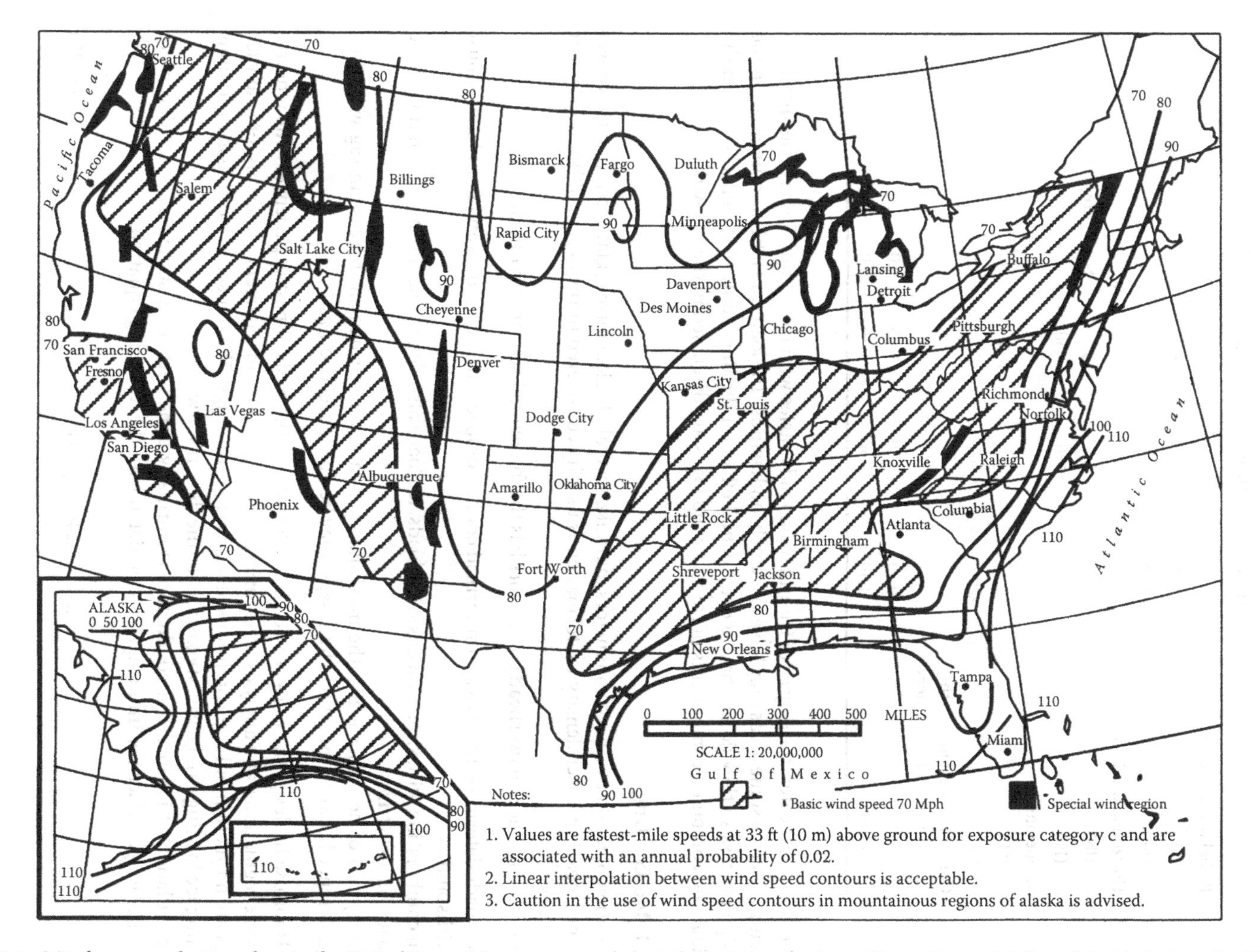

FIGURE 15.4 Wind pressure design values in the United States. Maximum recorded wind speed in miles/hour. (From Overend, P.R. and Smith, S., *Impulse Time Method of Sag Measurement*, American Society of Civil Engineers, Reston, VA, 1986. With permission.)

TABLE 15.2 Ratio of Iced to Bare Conductor Weight

ACSR Conductor	D_c, in.	W_{bare}, lb/ft	W_{ice}, lb/ft	$\frac{W_{bare}+W_{ice}}{W_{bare}}$
#1/0 AWG-6/1 "Raven"	0.398	0.1451	0.559	4.8
477 kcmil-26/7 "Hawk"	0.858	0.6553	0.845	2.3
1590 kcmil-54/19 "Falcon"	1.545	2.042	1.272	1.6

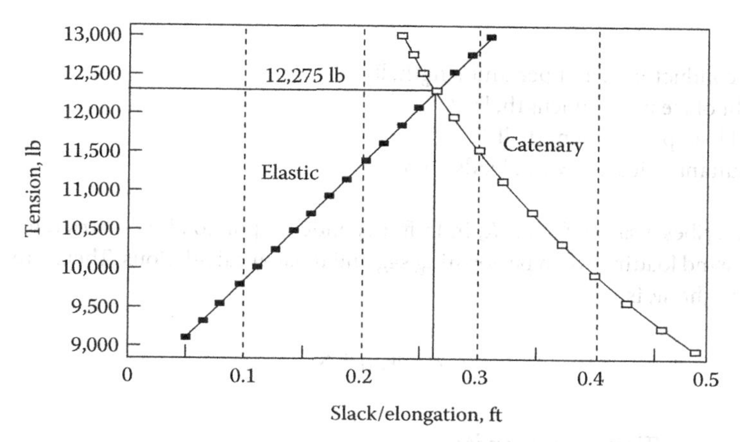

FIGURE 15.5 Sag-tension solution for 600 ft span of Drake at 0°F and 0.5 in. ice.

The ratio of iced weight to bare weight depends strongly upon conductor diameter. As shown in Table 15.2 for three different conductors covered with 0.5 in. radial glaze ice, this ratio ranges from 4.8 for #1/0 AWG to 1.6 for 1590 kcmil conductors. As a result, small diameter conductors may need to have a higher elastic modulus and higher tensile strength than large conductors in heavy ice and wind loading areas to limit sag (Figure 15.5).

15.1.5.2 Wind Loading

Wind loadings on overhead conductors influence line design in a number of ways:

- The maximum span between structures may be determined by the need for horizontal clearance to edge of right-of-way during moderate winds.
- The maximum transverse loads for tangent and small angle suspension structures are often determined by infrequent high wind-speed loadings.
- Permanent increases in conductor sag may be determined by wind loading in areas of light ice load.

Wind pressure load on conductors, P_w, is commonly specified in lb/ft^2. The relationship between P_w and wind velocity is given by the following equation:

$$P_w = 0.0025(V_w)^2 \tag{15.18}$$

where V_w is the wind speed in miles per hour.

The wind load per unit length of conductor is equal to the wind pressure load, P_w, multiplied by the conductor diameter (including radial ice of thickness t, if any), is given by the following equation:

$$W_w = P_w \frac{(D_c + 2t)}{12} \tag{15.19}$$

15.1.5.3 Combined Ice and Wind Loading

If the conductor weight is to include both ice and wind loading, the resultant magnitude of the loads must be determined vectorially. The weight of a conductor under both ice and wind loading is given by the following equation:

$$w_{w+i} = \sqrt{(w_b + w_i)^2 + (w_w)^2} \tag{15.20}$$

where

w_b is the bare conductor weight per unit length, lb/ft
w_i is the weight of ice per unit length, lb/ft
w_w is the wind load per unit length, lb/ft
w_{w+i} is the resultant of ice and wind loads, lb/ft

The NESC prescribes a safety factor, K, in lb/ft, dependent upon loading district, to be added to the resultant ice and wind loading when performing sag and tension calculations. Therefore, the total resultant conductor weight, w, is

$$w = w_{w+i} + K \tag{15.21}$$

15.1.6 Conductor Tension Limits

The NESC recommends limits on the tension of bare overhead conductors as a percentage of the conductor's rated breaking strength. The tension limits are 60% under maximum ice and wind load, 33.3% initial unloaded (when installed) at 60°F, and 25% final unloaded (after maximum loading has occurred) at 60°F. It is common, however, for lower unloaded tension limits to be used. Except in areas experiencing severe ice loading, it is not unusual to find tension limits of 60% maximum, 25% unloaded initial, and 15% unloaded final. This set of specifications could easily result in an actual maximum tension on the order of only 35%–40%, an initial tension of 20% and a final unloaded tension level of 15%. In this case, the 15% tension limit is said to govern.

Transmission-line conductors are normally not covered with ice, and winds on the conductor are usually much lower than those used in maximum load calculations. Under such everyday conditions, tension limits are specified to limit aeolian vibration to safe levels. Even with everyday lower tension levels of 15%–20%, it is assumed that vibration control devices will be used in those sections of the line that are subject to severe vibration. Aeolian vibration levels, and thus appropriate unloaded tension limits, vary with the type of conductor, the terrain, span length, and the use of dampers. Special conductors, such as ACSS, SDC, and VR, exhibit high self-damping properties and may be installed to the full code limits, if desired.

15.2 Approximate Sag-Tension Calculations

Sag-tension calculations, using exacting equations, are usually performed with the aid of a computer; however, with certain simplifications, these calculations can be made with a handheld calculator. The latter approach allows greater insight into the calculation of sags and tensions than is possible with complex computer programs. Equations suitable for such calculations, as presented in the preceding section, can be applied to the following example:

It is desired to calculate the sag and slack for a 600 ft level span of 795 kcmil-26/7 ACSR "Drake" conductor. The bare conductor weight per unit length, w_b, is 1.094 lb/ft. The conductor is installed with a horizontal tension component, H, of 6,300 lb, equal to 20% of its rated breaking strength of 31,500 lb.

By the use of Equation 15.2, the sag for this level span is

$$D = \frac{1.094(600^2)}{(8)6300} = 7.81\,\text{ft}\ (2.38\,\text{m})$$

The length of the conductor between the support points is determined using Equation 15.6:

$$L = 600 + \frac{8(7.81)^2}{3(600)} = 600.27\,\text{ft}\ (182.96\,\text{m})$$

Note that the conductor length depends solely on span and sag. It is not directly dependent on conductor tension, weight, or temperature. The conductor slack is the conductor length minus the span length; in this example, it is 0.27 ft (0.0826 m).

15.2.1 Sag Change with Thermal Elongation

ACSR and AAC conductors elongate with increasing conductor temperature (Table 15.3). The rate of linear thermal expansion for the composite ACSR conductor is less than that of the AAC conductor because the steel strands in the ACSR elongate at approximately half the rate of aluminum. The effective linear thermal expansion coefficient of a non-homogenous conductor, such as Drake ACSR, may be found from the following equations (Fink and Beatty, 1993):

$$E_{AS} = E_{AL}\left(\frac{A_{AL}}{A_{TOTAL}}\right) + E_{ST}\left(\frac{A_{ST}}{A_{TOTAL}}\right) \tag{15.22}$$

$$\alpha_{AS} = \alpha_{AL}\left(\frac{E_{AL}}{E_{AS}}\right)\left(\frac{A_{AL}}{A_{TOTAL}}\right) + \alpha_{ST}\left(\frac{E_{ST}}{E_{AS}}\right)\left(\frac{A_{ST}}{A_{TOTAL}}\right) \tag{15.23}$$

where
E_{AL} is the Elastic modulus of aluminum, psi
E_{ST} is the Elastic modulus of steel, psi
E_{AS} is the Elastic modulus of aluminum-steel composite, psi
A_{AL} is the Area of aluminum strands, square units
A_{ST} is the Area of steel strands, square units
A_{TOTAL} is the Total cross-sectional area, square units
α_{AL} is the Aluminum coefficient of linear thermal expansion, per °F
α_{ST} is the Steel coefficient of thermal elongation, per °F
α_{AS} is the Composite aluminum-steel coefficient of thermal elongation, per °F

TABLE 15.3 Iterative Solution for Increased Conductor Temperature

Iteration #	Length, L_n, ft	Sag, D_n, ft	Tension, H_n, lb	New Trial Tension, lb
ZTL	600.550	11.1	4435	—
1	600.836	13.7	3593	$\frac{4435+3593}{2} = 4014$
2	600.809	13.5	3647	$\frac{3647+4014}{2} = 3831$
3	600.797	13.4	3674	$\frac{3674+3831}{2} = 3753$
4	600.792	13.3	3702	$\frac{3702+3753}{2} = 3727$

The elastic modulus for solid aluminum wire is 10 million psi and for steel wire is 30 million psi. The elastic modulus for stranded wire is reduced. The modulus for stranded aluminum is assumed to be 8.6 million psi for all strandings. The modulus for the steel core of ACSR conductors varies with stranding as follows:

- 27.5×10^6 for single-strand core
- 27.0×10^6 for 7-strand core
- 26.5×10^6 for 19-strand core

Using elastic moduli of 8.6 and 27.0 million psi for aluminum and steel, respectively, the elastic modulus for Drake ACSR is

$$E_{AS} = (8.6 \times 10^6)\left(\frac{0.6247}{0.7264}\right) + (27.0 \times 10^6)\left(\frac{0.1017}{0.7264}\right) = 11.2 \times 10^6 \text{ psi}$$

and the coefficient of linear thermal expansion is

$$\alpha_{AS} = 12.8 \times 10^{-6}\left(\frac{8.6\times10^6}{11.2\times10^6}\right)\left(\frac{0.6247}{0.7264}\right) + 6.4 \times 10^{-6}\left(\frac{27.0\times10^6}{11.2\times10^6}\right)\left(\frac{0.1017}{0.7264}\right)$$
$$= 10.6 \times 10^{-6}/°\text{F}$$

If the conductor temperature changes from a reference temperature, T_{REF}, to another temperature, T, the conductor length, L, changes in proportion to the product of the conductor's effective thermal elongation coefficient, α_{AS}, and the change in temperature, $T - T_{REF}$, as shown in the following:

$$L_T = L_{T_{REF}}(1 + \alpha_{AS}(T - T_{REF})) \tag{15.24}$$

For example, if the temperature of the Drake conductor in the preceding example increases from 60°F (15°C) to 167°F (75°C), then the length at 60°F increases by 0.68 ft (0.21 m) from 600.27 ft (182.96 m) to 600.95 ft (183.17 m):

$$L_{(167°\text{F})} = 600.27(1 + (10.6\times10^{-6})(167 - 60)) = 600.95 \text{ ft}$$

Ignoring for the moment any change in length due to change in tension, the sag at 167°F (75°C) may be calculated for the conductor length of 600.95 ft (183.17 m) using Equation 15.8:

$$D = \sqrt{\frac{3(600)(0.95)}{8}} = 14.62 \text{ ft}$$

Using a rearrangement of Equation 15.2, this increased sag is found to correspond to a decreased tension of

$$H = \frac{w(S^2)}{8D} = \frac{1.094(600^2)}{8(14.62)} = 3367 \text{ lb}$$

If the conductor were inextensible, that is, if it had an infinite modulus of elasticity, then these values of sag and tension for a conductor temperature of 167°F would be correct. For any real conductor, however, the elastic modulus of the conductor is finite and changes in tension do change the conductor length. The use of the preceding calculation, therefore, will overstate the increase in sag (Figure 15.6).

The preceding approximate tension calculations could have been more accurate with the use of actual stress–strain curves and graphic sag-tension solutions, as described in detail in *Graphic Method for Sag*

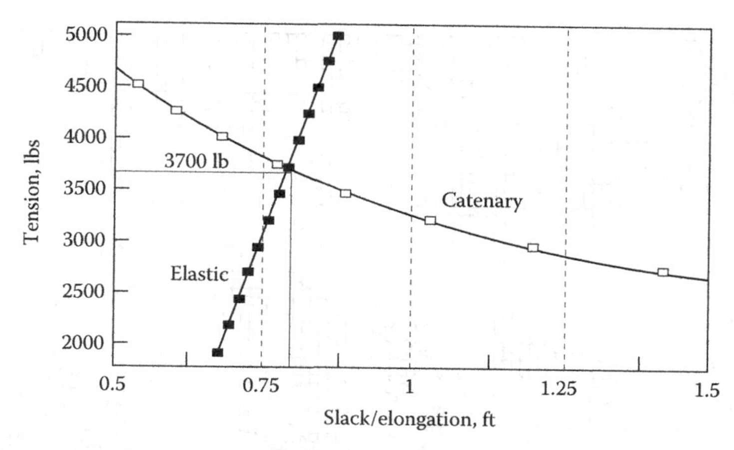

FIGURE 15.6 Sag-tension solution for 600 ft span of Drake at 167°F.

Tension Calculations for ACSR and Other Conductors (Aluminum Company of America, 1961). This method, although accurate, is very slow and has been replaced completely by computational methods.

15.3 Numerical Sag-Tension Calculations

Sag-tension calculations are normally done numerically and allow the user to enter many different loading and conductor temperature conditions. Both initial and final conditions are calculated and multiple tension constraints can be specified. The complex stress–strain behavior of ACSR-type conductors can be modeled numerically, including both temperature, and elastic and plastic effects.

15.3.1 Stress–Strain Curves

Stress–strain curves for bare overhead conductor include a minimum of an initial curve and a final curve over a range of elongations from 0% to 0.45%. For conductors consisting of two materials, an initial and final curve for each is included. Creep curves for various lengths of time are typically included as well.

Overhead conductors are not purely elastic. They stretch with tension, but when the tension is reduced to zero, they do not return to their initial length. That is, conductors are plastic; the change in conductor length cannot be expressed with a simple linear equation, as for the preceding hand calculations. The permanent length increase that occurs in overhead conductors yields the difference in initial and final sag-tension data found in most computer programs.

Figure 15.7 shows a typical stress–strain curve for a 26/7 ACSR conductor (Aluminum Association, 1974); the curve is valid for conductor sizes ranging from 266.8 to 795 kcmil. A 795 kcmil-26/7 ACSR "Drake" conductor has a breaking strength of 31,500 lb (14,000 kg) and an area of 0.7264 in.^{2} (46.9 mm^2) so that it fails at an average stress of 43,000 psi (30 kg/mm^2). The stress–strain curve illustrates that when the percent of elongation at a stress is equal to 50% of the conductor's breaking strength (21,500 psi), the elongation is less than 0.3% or 1.8 ft (0.55 m) in a 600 ft (180 m) span.

Note that the component curves for the steel core and the aluminum stranded outer layers are separated. This separation allows for changes in the relative curve locations as the temperature of the conductor changes.

For the preceding example, with the Drake conductor at a tension of 6300 lb (2860 kg), the length of the conductor in the 600 ft (180 m) span was found to be 0.27 ft longer than the span. This tension corresponds to a stress of 8600 psi (6.05 kg/mm^2). From the stress–strain curve in Figure 15.7, this corresponds to an initial elongation of 0.105% (0.63 ft). As in the preceding hand calculation, if the conductor is reduced to zero tension, its unstressed length would be less than the span length.

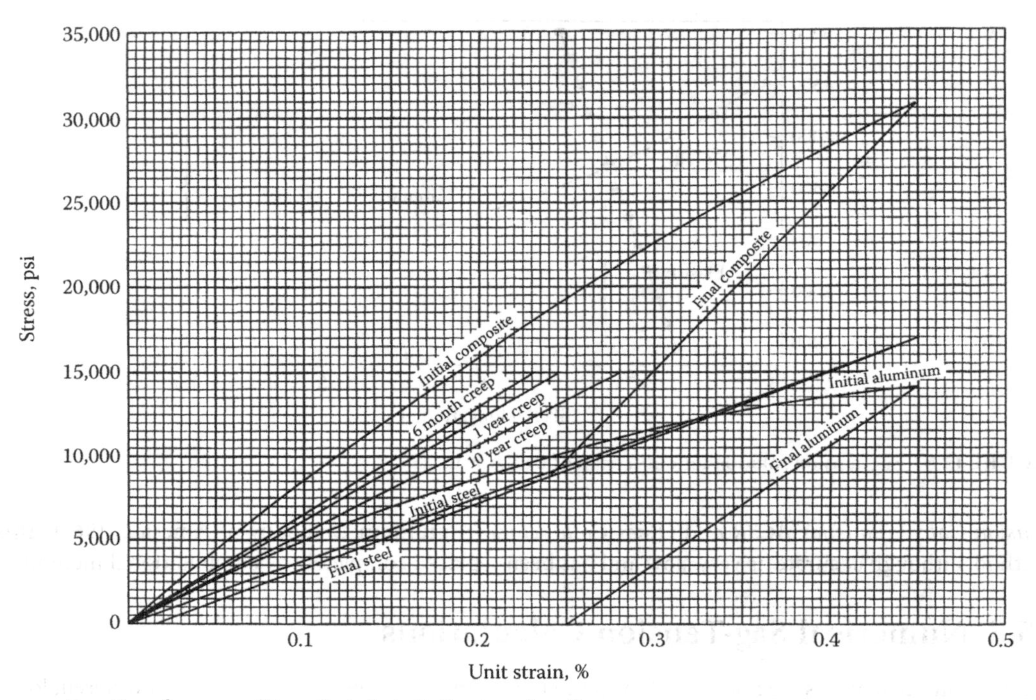

Equations for curves (X = unit strain in %; Y = stress in psi):

Initial composite: $X = 4.07 \times 10^{-3} + (1.28 \times 10^{-5})Y - (1.18 \times 10^{-10})Y^2 + (5.64 \times 10^{-15})Y^3$
$Y = -512 + (8.617 \times 10^4)X - (1.18 \times 10^4)X^2 - (5.76 \times 10^{-4})X^3$
Initial steel: $Y = (37.15 \times 10^3)X$
Initial aluminum: $Y = -512 = (4.902 \times 10^4)X - (1.18 \times 10^4)X^2 - (5.76 \times 10^4)X^3$
Final composite: $Y = (107.55X - 17.65) \times 10^3$
Final Steel: $Y = (38.60X - 0.65) \times 10^3$
Final aluminum: $Y = (68.95X - 17.00) \times 10^3$
6 month creep: $Y = (68.75 \times 10^3)X$
1 year creep: $Y = (60.60 \times 10^3)X$
10 year creep: $Y = (53.45 \times 10^3)X$

Test temperature 70°F–75°F

FIGURE 15.7 Stress–strain curves for 26/7 ACSR.

Figure 15.8 is a stress–strain curve (Aluminum Association, 1974) for an all-aluminum 37-strand conductor ranging in size from 250 to 1033.5 kcmil. Because the conductor is made entirely of aluminum, there is only one initial and final curve.

15.3.1.1 Permanent Elongation

Once a conductor has been installed at an initial tension, it can elongate further. Such elongation results from two phenomena: permanent elongation due to high tension levels resulting from ice and wind loads, and creep elongation under everyday tension levels. These types of conductor elongation are discussed in the following sections.

15.3.1.2 Permanent Elongation due to Heavy Loading

Both Figures 15.7 and 15.8 indicate that when the conductor is initially installed, it elongates following the initial curve that is not a straight line. If the conductor tension increases to a relatively high level under ice and wind loading, the conductor will elongate. When the wind and ice loads abate, the conductor elongation will reduce along a curve parallel to the final curve, but the conductor will never return to its original length.

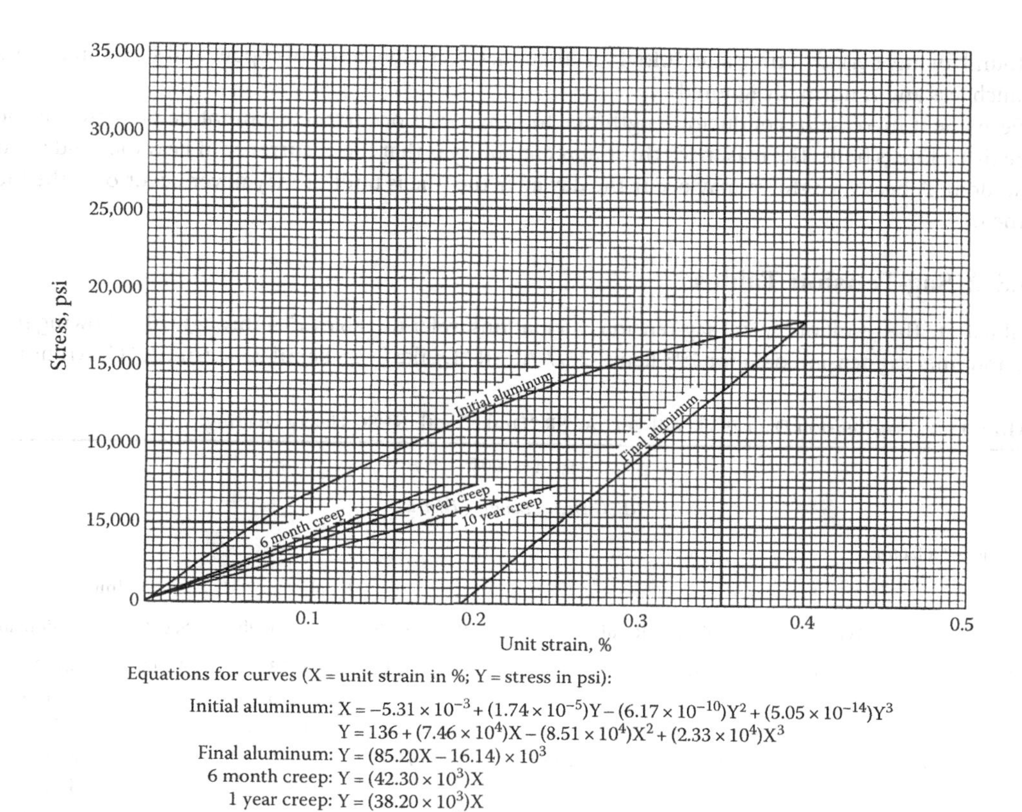

Equations for curves (X = unit strain in %; Y = stress in psi):

Initial aluminum: $X = -5.31 \times 10^{-3} + (1.74 \times 10^{-5})Y - (6.17 \times 10^{-10})Y^2 + (5.05 \times 10^{-14})Y^3$

$Y = 136 + (7.46 \times 10^4)X - (8.51 \times 10^4)X^2 + (2.33 \times 10^4)X^3$

Final aluminum: $Y = (85.20X - 16.14) \times 10^3$

6 month creep: $Y = (42.30 \times 10^3)X$

1 year creep: $Y = (38.20 \times 10^3)X$

10 year creep: $Y = (30.60 \times 10^3)X$

Test temperature 70°F–75°F

FIGURE 15.8 Stress–strain curves for 37-strand AAC.

For example, refer to Figure 15.8 and assume that a newly strung 795 kcmil-37 strand AAC "Arbutus" conductor has an everyday tension of 2780 lb. The conductor area is 0.6245 in.^{2}, so the everyday stress is 4450 psi and the elongation is 0.062%. Following an extremely heavy ice and wind load event, assume that the conductor stress reaches 18,000 psi. When the conductor tension decreases back to everyday levels, the conductor elongation will be permanently increased by more than 0.2%. Also the sag under everyday conditions will be correspondingly higher, and the tension will be less. In most numerical sag-tension methods, final sag-tensions are calculated for such permanent elongation due to heavy loading conditions.

15.3.1.3 Permanent Elongation at Everyday Tensions (Creep Elongation)

Conductors permanently elongate under tension even if the tension level never exceeds everyday levels. This permanent elongation caused by everyday tension levels is called creep (Aluminum Company of America, 1961). Creep can be determined by long-term laboratory creep tests, the results of which are used to generate creep curves. On stress–strain graphs, creep curves are usually shown for 6 month, 1 year, and 10 year periods. Figure 15.8 shows these typical creep curves for a 37 strand 250.0 through 1033.5 kcmil AAC. In Figure 15.8, assume that the conductor tension remains constant at the initial stress of 4450 psi. At the intersection of this stress level and the initial elongation curve, 6 month, 1 year, and 10 year creep curves, the conductor elongation from the initial elongation of 0.062% increases to 0.11%, 0.12%, and 0.15%, respectively. Because of creep elongation, the resulting final sags are greater and the conductor tension is less than the initial values.

Creep elongation in aluminum conductors is quite predictable as a function of time and obeys a simple exponential relationship. Thus, the permanent elongation due to creep at everyday tension can be found for any period of time after initial installation. Creep elongation of copper and steel conductors is much less and is normally ignored.

Permanent increase in conductor length due to heavy load occurrences cannot be predicted at the time that a line is built. The reason for this unpredictability is that the occurrence of heavy ice and wind is random. A heavy ice storm may occur the day after the line is built or may never occur over the life of the line.

15.3.2 Sag-Tension Tables

To illustrate the result of typical sag-tension calculations, refer to Tables 15.4 through 15.9 showing initial and final sag-tension data for 795 kcmil-26/7 ACSR "Drake," 795 kcmil-37 strand AAC "Arbutus,"

TABLE 15.4 Sag-Tension Data 795 kcmil-26/7 ACSR "Drake" with NESC Heavy Loading

Span = 600 ft

NESC Heavy Loading District

Creep is *not* a factor

					Final		Initial	
Temp, °F	Ice, in.	Wind, lb/ft²	*K*, lb/ft	Resultant Weight, lb/ft	Sag, ft	Tension, lb	Sag, ft	Tension, lb
0	0.50	4.00	0.30	2.509	11.14	10,153	11.14	10,153
						5,415 Al		5,415 Al
						4,738 St		4,738 St
32	0.50	0.00	0.00	2.094	44.54	8,185	11.09	8,512
						3,819 Al		4,343 Al
						4,366 St		4,169 St
−20	0.00	0.00	0.00	1.094	6.68	7,372	6.27	7,855
						3,871 Al		4,465 Al
						3,501 St		3,390 St
0	0.00	0.00	0.00	1.094	7.56	6,517	6.89	7,147
						3,111 Al		3,942 Al
						3,406 St		3,205 St
30	0.00	0.00	0.00	1.094	8.98	5,490	7.95	6,197
						2,133 Al		3,201 Al
						3,357 St		2,996 St
60	0.00	0.00	0.00	1.094	10.44	4,725[a]	9.12	5,402
						1,321 Al		2,526 Al
						3,404 St		2,875 St
90	0.00	0.00	0.00	1.094	11.87	4,157	10.36	4,759
						634 Al		1,922 Al
						3,522 St		2,837 St
120	0.00	0.00	0.00	1.094	13.24	3,727	11.61	4,248
						35 Al		1,379 Al
						3,692 St		2,869 St
167	0.00	0.00	0.00	1.094	14.29	3,456	13.53	3,649
						0 Al		626 Al
						3,456 St		3,022 St
212	0.00	0.00	0.00	1.094	15.24	3,241	15.24	3,241
						0 Al		0 Al
						3,241 St		3,239 St

[a] Design condition.

TABLE 15.5 Tension Differences in Adjacent Dead-End Spans

Conductor: Drake
795 kcmil-26/7 ACSR
Area = 0.7264 in.^{2}
Creep *is* a factor

Span = 700 ft

NESC Heavy Loading District

Temp, °F	Ice, in.	Wind, lb/ft^2	*K*, lb/ft	Resultant Weight, lb/ft	Final Sag, ft	Final Tension, lb	Initial Sag, ft	Initial Tension, lb
0	0.50	4.00	0.30	2.509	13.61	11,318	13.55	11,361
32	0.50	0.00	0.00	2.094	13.93	9,224	13.33	9,643
−20	0.00	0.00	0.00	1.094	8.22	8,161	7.60	8,824
0	0.00	0.00	0.00	1.094	9.19	7,301	8.26	8,115
30	0.00	0.00	0.00	1.094	10.75	6,242	9.39	7,142
60	0.00	0.00	0.00	1.094	12.36	5,429	10.65	6,300[a]
90	0.00	0.00	0.00	1.094	13.96	4,809	11.99	5,596
120	0.00	0.00	0.00	1.094	15.52	4,330	13.37	5,020
167	0.00	0.00	0.00	1.094	16.97	3,960	15.53	4,326
212	0.00	0.00	0.00	1.094	18.04	3,728	17.52	3,837

Conductor: Drake
795 kcmil-26/7 ACSR
Area = 0.7264 in.^{2}
Creep is *not* a factor

Span = 1000 ft

NESC Heavy Loading District

Temp, °F	Ice, in.	Wind, lb/ft^2	*K*, lb/ft	Resultant Weight, lb/ft	Final Sag, ft	Final Tension, lb	Initial Sag, ft	Initial Tension, lb
0	0.50	4.00	0.30	2.509	25.98	12,116	25.98	12,116
32	0.50	0.00	0.00	2.094	26.30	9,990	25.53	10,290
−20	0.00	0.00	0.00	1.094	18.72	7,318	17.25	7,940
0	0.00	0.00	0.00	1.094	20.09	6,821	18.34	7,469
30	0.00	0.00	0.00	1.094	22.13	6,197	20.04	6,840
60	0.00	0.00	0.00	1.094	24.11	5,689	21.76	6,300[a]
90	0.00	0.00	0.00	1.094	26.04	5,271	23.49	5,839
120	0.00	0.00	0.00	1.094	27.89	4,923	25.20	5,444
167	0.00	0.00	0.00	1.094	30.14	4,559	27.82	4,935
212	0.00	0.00	0.00	1.094	31.47	4,369	30.24	4,544

[a] Design condition.

and 795 kcmil Type 16 "Drake/SDC" conductors in NESC light and heavy loading areas for spans of 1000 and 300 ft. Typical tension constraints of 15% final unloaded at 60°F, 25% initial unloaded at 60°F, and 60% initial at maximum loading are used.

With most sag-tension calculation methods, final sags are calculated for both heavy ice/wind load and for creep elongation. The final sag-tension values reported to the user are those with the greatest increase in sag.

15.3.2.1 Initial vs. Final Sags and Tensions

Rather than calculate the line sag as a function of time, most sag-tension calculations are determined based on initial and final loading conditions. Initial sags and tensions are simply the sags and tensions

TABLE 15.6 Sag and Tension Data for 795 kcmil-26/7 ACSR "Drake" 600 ft Ruling Span

Conductor: Drake
795 kcmil-26/7 ACSR — Span = 600 ft
Area = 0.7264 in.^{2}
Creep is *not* a factor — *NESC Heavy Loading District*

					Final		Initial	
Temp, °F	Ice, in.	Wind, lb/ft^2	*K*, lb/ft	Resultant Weight, lb/ft	Sag, ft	Tension, lb	Sag, ft	Tension, lb
0	0.50	4.00	0.30	2.509	11.14	10,153	11.14	10,153
32	0.50	0.00	0.00	2.094	11.54	8,185	11.09	8,512
−20	0.00	0.00	0.00	1.094	6.68	7,372	6.27	7,855
0	0.00	0.00	0.00	1.094	7.56	6,517	6.89	7,147
30	0.00	0.00	0.00	1.094	8.98	5,490	7.95	6,197
60	0.00	0.00	0.00	1.094	10.44	4,725[a]	9.12	5,402
90	0.00	0.00	0.00	1.094	11.87	4,157	10.36	4,759
120	0.00	0.00	0.00	1.094	13.24	3,727	11.61	4,248
167	0.00	0.00	0.00	1.094	14.29	3,456	13.53	3,649
212	0.00	0.00	0.00	1.094	15.24	3,241	15.24	3,241

[a] Design condition.

at the time the line is built. Final sags and tensions are calculated if (1) the specified ice and wind loading has occurred, and (2) the conductor has experienced 10 years of creep elongation at a conductor temperature of 60°F at the user-specified initial tension.

15.3.2.2 Special Aspects of ACSR Sag-Tension Calculations

Sag-tension calculations with ACSR conductors are more complex than such calculations with AAC, AAAC, or ACAR conductors. The complexity results from the different behavior of steel and aluminum strands in response to tension and temperature. Steel wires do not exhibit creep elongation or plastic elongation in response to high tensions. Aluminum wires do creep and respond plastically to high stress levels. Also, they elongate twice as much as steel wires do in response to changes in temperature.

Table 15.10 presents various initial and final sag-tension values for a 600 ft span of a Drake ACSR conductor under heavy loading conditions. Note that the tension in the aluminum and steel components is shown separately. In particular, some other useful observations are as follows:

1. At 60°F, without ice or wind, the tension level in the aluminum strands decreases with time as the strands permanently elongate due to creep or heavy loading.
2. Both initially and finally, the tension level in the aluminum strands decreases with increasing temperature reaching zero tension at 212°F and 167°F for initial and final conditions, respectively.
3. At the highest temperature (212°F), where all the tension is in the steel core, the initial and final sag-tensions are nearly the same, illustrating that the steel core does not permanently elongate in response to time or high tension.

15.4 Ruling Span Concept

Transmission lines are normally designed in line sections with each end of the line section terminated by a strain structure that allows no longitudinal (along the line) movement of the conductor (Winkelman, 1959). Structures within each line section are typically suspension structures that support the conductor vertically, but allow free movement of the conductor attachment point either longitudinally or transversely.

TABLE 15.7 Stringing Sag Table for 795 kcmil-26/7 ACSR "Drake" 600 ft Ruling Span

600 ft Ruling Span

Controlling Design Condition:

15% RBS at 60°F, No Ice or Wind, Final

NESC Heavy Load District

Horizontal	6493	6193	5910	5645	5397	5166	4952	4753	4569
Tension, lb	20	30	40	50	60	70	80	90	100
Temp, °F Spans	Sag, ft-in.	Sag, ft-in.	Sag, ft-in.	Sag, ft-in.	Sag, ft-in.	Sag, ft-in.	Sag, ft-in.	Sag, ft-in.	Sag, ft-in.
400	3–4	3–6	3–8	3–11	4–1	4–3	4–5	4–7	4–9
410	3–6	3–9	3–11	4–1	4–3	4–5	4–8	4–10	5–0
420	3–9	3–11	4–1	4–3	4–6	4–8	4–10	5–1	5–3
430	3–11	4–1	4–3	4–6	4–8	4–11	5–1	5–4	5–6
440	4–1	4–3	4–6	4–8	4–11	5–2	5–4	5–7	5–10
450	4–3	4–6	4–8	4–11	5–2	5–4	5–7	5–10	6–1
460	4–5	4–8	4–11	5–2	5–4	5–7	5–10	6–1	6–4
470	4–8	4–11	5–1	5–4	5–7	5–10	6–1	6–4	6–7
480	4–10	5–1	5–4	5–7	5–10	6–1	6–4	6–8	6–11
490	5–1	5–4	5–7	5–10	6–1	6–4	6–8	6–11	7–2
500	5–3	5–6	5–9	6–1	6–4	6–7	6–11	7–2	7–6
510	5–6	5–9	6–0	6–4	6–7	6–11	7–2	7–6	7–9
520	5–8	6–0	6–3	6–7	6–10	7–2	7–6	7–9	8–1
530	5–11	6–2	6–6	6–10	7–1	7–5	7–9	8–1	8–5
540	6–2	6–5	6–9	7–1	7–5	7–9	8–1	8–5	8–9
550	6–4	6–8	7–0	7–4	7–8	8–0	8–4	8–8	9–1
560	6–7	6–11	7–3	7–7	7–11	8–4	8–8	9–0	9–5
570	6–10	7–2	7–6	7–10	8–3	8–7	9–0	9–4	9–9
580	7–1	7–5	7–9	8–2	8–6	8–11	9–4	9–8	10–1
590	7–4	7–8	8–1	8–5	8–10	9–3	9–7	10–0	10–5
600	7–7	7–11	8–4	8–9	9–1	9–6	9–11	10–4	10–9
610	7–1	8–3	8–7	9–0	9–5	9–10	10–3	10–9	11–2
620	8–1	8–6	8–11	9–4	9–9	10–2	10–7	11–1	11–6
630	8–	8–9	9–2	9–7	10–1	10–6	11–0	11–5	11–11
640	8–8	9–1	9–6	9–11	10–5	10–10	11–4	11–9	12–3
650	8–11	9–4	9–9	10–3	10–9	11–2	11–8	12–2	12–8
660	9–2	9–7	10–1	10–7	11–1	11–6	12–0	12–6	13–1
670	9–5	9–11	10–5	10–11	11–5	11–11	12–5	12–11	13–5
680	9–9	10–3	10–8	11–2	11–9	12–3	12–9	13–4	13–10
690	10–0	10–6	11–0	11–6	12–1	12–7	13–2	13–8	14–3
700	10–4	10–10	11–4	11–11	12–5	13–0	13–6	14–1	14–8

TABLE 15.8 Time-Sag Table for Stopwatch Method

Return of Wave											
Sag, in.	Third Time, s	Fifth Time, s	Sag, in.	Third Time, s	Fifth Time, s	Sag, in.	Third Time, s	Fifth Time, s	Sag, in.	Third Time, s	Fifth Time, s
5	1.9	3.2	55	6.4	10.7	105	8.8	14.7	155	10.7	17.9
6	2.1	3.5	56	6.5	10.8	106	8.9	14.8	156	10.8	18.0
7	2.3	3.8	57	6.5	10.9	107	8.9	14.9	157	10.8	18.0
8	2.4	4.1	58	6.6	11.1	109	9.0	15.0	158	10.9	18.1
9	2.6	4.3	59	6.6	11.1	109	9.0	15.0	159	10.9	18.1
10	2.7	4.6	60	6.7	11.1	110	9.1	15.1	160	10.9	18.2
11	2.9	4.8	61	6.7	11.2	111	9.1	15.2	161	11.0	18.2
12	3.0	5.0	62	6.8	11.3	112	9.1	15.2	162	11.0	18.2
13	3.1	5.2	63	6.9	11.4	113	9.2	15.3	163	11.0	18.4
14	3.2	5.4	64	6.9	11.5	114	9.2	15.4	164	11.1	18.4
15	3.3	5.6	65	7.0	11.6	115	9.3	15.4	165	11.1	18.5
16	3.5	5.8	66	7.0	11.7	116	9.3	15.5	166	11.1	18.5
17	3.6	5.9	67	7.1	11.8	117	9.3	15.6	167	11.2	18.6
18	3.7	6.1	68	7.1	11.9	118	9.4	15.6	168	11.2	18.7
19	3.8	6.3	69	7.2	12.0	119	9.4	15.7	169	11.2	18.7
20	3.9	6.4	70	7.2	12.0	120	9.5	15.8	170	11.3	18.8
21	4.0	6.6	71	7.3	12.1	121	9.5	15.8	171	11.3	18.8
22	4.0	6.7	72	7.3	12.2	122	9.5	15.9	172	11.3	18.9
23	4.1	6.9	73	7.4	12.3	123	9.6	16.0	173	11.4	18.9
24	4.2	7.0	74	7.4	12.4	124	9.6	16.0	174	11.4	19.0
25	4.3	7.2	75	7.5	12.5	125	9.7	16.1	175	11.4	19.0
26	4.4	7.3	76	7.5	12.5	126	9.7	16.2	176	11.4	19.1
27	4.5	7.5	77	7.6	12.6	127	9.7	16.2	177	11.5	19.1
28	4.6	7.6	78	7.6	12.7	128	9.8	16.3	178	11.5	19.2
29	4.6	7.7	79	7.7	12.8	129	9.8	16.3	179	11.5	19.3
30	4.7	7.9	80	7.7	12.9	130	9.8	16.4	180	11.6	19.3
31	4.8	8.0	81	7.8	13.0	131	9.9	16.5	181	11.6	19.4
32	4.9	8.1	82	7.8	13.0	132	9.9	16.5	182	11.6	19.4
33	5.0	8.3	83	7.9	13.1	133	10.0	16.6	183	11.7	19.5
34	5.0	8.4	84	7.9	13.2	134	10.0	16.7	184	11.7	19.5
35	5.1	8.5	85	8.0	13.3	135	10.0	16.7	185	11.7	19.6
36	5.2	8.6	86	8.0	13.3	136	10.1	16.8	186	11.8	19.6
37	5.3	8.8	87	8.1	13.4	137	10.1	16.8	187	11.8	19.7
38	5.3	8.9	88	8.1	13.5	138	10.1	16.9	188	11.8	19.7
39	5.4	9.0	89	8.1	13.6	139	10.2	17.0	189	11.9	19.8
40	5.5	9.1	90	8.2	13.7	140	10.2	17.0	190	11.9	19.8
41	5.5	9.2	91	8.2	13.7	141	10.3	17.1	191	11.9	19.9
42	5.6	9.3	92	8.3	13.8	142	10.3	17.1	192	12.0	19.9
43	5.7	9.4	93	8.3	13.9	143	10.3	17.2	193	12.0	20.0
44	5.7	9.5	94	8.4	14.0	144	10.4	17.3	194	12.0	20.0
45	5.8	9.7	95	8.4	14.0	145	10.4	17.3	195	12.1	20.1
46	5.9	9.8	96	8.5	14.1	146	10.4	17.4	196	12.1	20.1
47	5.9	9.9	97	8.5	14.2	147	10.5	17.4	197	12.1	20.2
48	6.0	10.0	98	8.5	14.2	148	10.5	17.5	198	12.1	20.0
49	6.0	10.1	99	8.6	14.3	149	10.5	17.6	199	12.2	20.3
50	6.1	10.2	100	8.6	14.4	150	10.6	17.6	200	12.2	20.3
51	6.2	10.3	101	8.7	14.5	151	10.6	17.7	201	12.2	20.4
52	6.2	10.4	102	8.7	14.5	152	10.6	17.7	202	12.3	20.5

TABLE 15.8 (continued) Time-Sag Table for Stopwatch Method

Return of Wave											
Sag, in.	Third Time, s	Fifth Time, s	Sag, in.	Third Time, s	Fifth Time, s	Sag, in.	Third Time, s	Fifth Time, s	Sag, in.	Third Time, s	Fifth Time, s
53	6.3	10.5	103	8.8	14.6	153	10.7	17.8	203	12.3	20.5
54	6.3	10.6	104	8.8	14.7	154	10.7	17.9	204	12.3	20.6

Note: To calculate the time of return of other waves, multiply the time in seconds for one wave return by the number of wave returns or, more simply, select the combination of values from the table that represents the number of wave returns desired. For example, the time of return of the 8th wave is the sum of the 3rd and 5th, while for the 10th wave it is twice the time of the 5th.

The approximate formula giving the relationship between sag and time is given as follows:

$$D = 12.075\left(\frac{T}{N}\right)^2 \text{ (in.)}$$

where

D is the sag, in.
T is the time, s
N is the number of return waves counted

TABLE 15.9 Sag-Tension Data 795 kcmil-26/7 ACSR "Drake" with NESC Light Loading 300 and 1000 ft Spans

Conductor: Drake
795 kcmil-26/7 ACSR
Area = 0.7264 in.^{2}
Creep *is* a factor

Span = 300 ft

NESC Light Loading District

Temp, °F	Ice, in.	Wind, lb/ft^2	*K*, lb/ft	Weight, lb/ft	Final Sag, ft	Final Tension, lb	Initial Sag, ft	Initial Tension, lb
30	0.00	9.00	0.05	1.424	2.37	6769	2.09	7664
30	0.00	0.00	0.00	1.094	1.93	6364	1.66	7404
60	0.00	0.00	0.00	1.094	2.61	4725[a]	2.04	6033
90	0.00	0.00	0.00	1.094	3.46	3556	2.57	4792
120	0.00	0.00	0.00	1.094	1.00	3077	3.25	3785
167	0.00	0.00	0.00	1.094	4.60	2678	4.49	2746
212	0.00	0.00	0.00	1.094	5.20	2371	5.20	2371

Conductor: Drake
795 kcmil-26/7 ACSR
Area = 0.7264 in.^{2}
Creep *is* a factor

Span = 1000 ft

NESC Light Loading District

Temp, °F	Ice, in.	Wind, lb/ft^2	*K*, lb/ft	Weight, lb/ft	Final Sag, ft	Final Tension, lb	Initial Sag, ft	Initial Tension, lb
30	0.00	9.00	0.05	1.424	28.42	6290	27.25	6558
30	0.00	0.00	0.00	1.094	27.26	5036	25.70	5339
60	0.00	0.00	0.00	1.094	29.07	4725[a]	27.36	5018
90	0.00	0.00	0.00	1.094	30.82	4460	28.98	4740
120	0.00	0.00	0.00	1.094	32.50	4232	30.56	4498
167	0.00	0.00	0.00	1.094	34.49	3990	32.56	4175
212	0.00	0.00	0.00	1.094	35.75	3851	35.14	3917

Note: Calculations based on (1) NESC light loading district and (2) tension limits: (a) initial loaded—60% RBS @ 30°F; (b) initial unloaded—25% RBS @ 60°F; and (c) final unloaded—15% RBS @ 60°F.

[a] Design condition.

TABLE 15.10 Sag-Tension Data 795 kcmil-26/7 ACSR "Drake" NESC Heavy Loading 300 and 1000 ft Spans

Conductor: Drake
795 kcmil-26/7 ACSR/SD
Area = 0.7264 in.^{2}
Creep *is* a factor

Span = 300 ft

NESC Heavy Loading District

					Final		Initial	
Temp, °F	Ice, in.	Wind, lb/ft^2	K, lb/ft	Weight, lb/ft	Sag, ft	Tension, lb	Sag, ft	Tension, lb
0	0.50	4.00	0.30	2.509	2.91	9,695	2.88	9,802
32	0.50	0.00	0.00	2.094	3.13	7,528	2.88	8,188
−20	0.00	0.00	0.00	1.094	1.26	9,733	1.26	9,756
0	0.00	0.00	0.00	1.094	1.48	8,327	1.40	8,818
30	0.00	0.00	0.00	1.094	1.93	6,364	1.66	7,404
60	0.00	0.00	0.00	1.094	2.61	4,725[a]	2.04	6,033
90	0.00	0.00	0.00	1.094	3.46	3,556	2.57	4,792
120	0.00	0.00	0.00	1.094	4.00	3,077	3.25	3,785
167	0.00	0.00	0.00	1.094	4.60	2,678	4.49	2,746
212	0.00	0.00	0.00	1.094	5.20	2,371	5.20	2,371

Conductor: Drake
795 kcmil-26/7 ACSR
Area = 0.7264 in.^{2}
Creep is *not* a factor

Span = 1000 ft

NESC Heavy Loading District

					Final		Initial	
Temp, °F	Ice, in.	Wind, lb/ft^2	K, lb/ft	Weight, lb/ft	Sag, ft	Tension, lb	Sag, ft	Tension, lb
0	0.50	4.00	0.30	2.509	30.07	10,479	30.07	10,479
32	0.50	0.00	0.00	2.094	30.56	8,607	29.94	8,785
−20	0.00	0.00	0.00	1.094	24.09	5,694	22.77	6,023
0	0.00	0.00	0.00	1.094	25.38	5,406	23.90	5,738
30	0.00	0.00	0.00	1.094	27.26	5,036	25.59	5,362
60	0.00	0.00	0.00	1.094	29.07	4,725[a]	27.25	5,038
90	0.00	0.00	0.00	1.094	30.82	4,460	28.87	4,758
120	0.00	0.00	0.00	1.094	32.50	4,232	30.45	4,513
167	0.00	0.00	0.00	1.094	34.36	4,005	32.85	4,187
212	0.00	0.00	0.00	1.094	35.62	3,865	35.05	3,928

Note: Calculations based on (1) NESC heavy loading district and (2) tension limits: (a) initial loaded—60% RBS @ 0°F; (b) initial unloaded—25% RBS @ 60°F; and (c) final unloaded—15% RBS @ 60°F.

[a] Design condition.

TABLE 15.11 Sag-Tension Data 795 kcmil-Type 16 ACSR/SD with NESC Light Loading 300 and 1000 ft Spans

Conductor: Drake
795 kcmil-Type 16 ACSR/SD
Area = 0.7261 in.^{2}
Creep *is* a factor

Span = 300 ft

NESC Light Loading District

					Final		Initial	
Temp, °F	Ice, in.	Wind, lb/ft^2	*K*, lb/ft	Weight, lb/ft	Sag, ft	Tension, lb	Sag, ft	Tension, lb
30	0.00	9.00	0.05	1.409	1.59	9,980	1.31	12,373
30	0.00	0.00	0.00	1.093	1.26	9,776	1.03	11,976
60	0.00	0.00	0.00	1.093	1.60	7,688	1.16	10,589[a]
90	0.00	0.00	0.00	1.093	2.12	5,806	1.34	9,159
120	0.00	0.00	0.00	1.093	2.69	4,572	1.59	7,713
167	0.00	0.00	0.00	1.093	3.11	3,957	2.22	5,545
212	0.00	0.00	0.00	1.093	3.58	3,435	3.17	3,877

Conductor: Drake
795 kcmil-Type 16 ACSR/SD
Area = 0.7261 in.^{2}
Creep *is* a factor

Span = 1000 ft

NESC Light Loading District

					Final		Initial	
Temp, °F	Ice, in.	Wind, lb/ft^2	*K*, lb/ft	Weight, lb/ft	Sag, ft	Tension, lb	Sag, ft	Tension, lb
30	0.00	9.00	0.05	1.409	17.21	10,250	15.10	11,676
30	0.00	0.00	0.00	1.093	15.22	8,988	12.69	10,779
60	0.00	0.00	0.00	1.093	17.21	7,950[a]	13.98	9,780
90	0.00	0.00	0.00	1.093	19.26	7,108	15.44	8,861
120	0.00	0.00	0.00	1.093	21.31	6,428	17.03	8,037
167	0.00	0.00	0.00	1.093	24.27	5,647	19.69	6,954
212	0.00	0.00	0.00	1.093	25.62	5,352	22.32	6,136

Note: Calculations based on (1) NESC light loading district and (2) tension limits: (a) initial loaded—60% RBS @ 30°F; (b) initial unloaded—25% RBS @ 60°F; and (c) final unloaded—15% RBS @ 60°F.

[a] Design condition.

15.4.1 Tension Differences for Adjacent Dead-End Spans

Tables 15.11 and 15.12 contains initial and final sag-tension data for a 700 ft and a 1000 ft dead-end span when a Drake ACSR conductor is initially installed to the same 6300 lb tension limits at 60°F. Note that the difference between the initial and final limits at 60°F is approximately 460 lb. Even the initial tension (equal at 60°F) differs by almost 900 lb at −20°F and 600 lb at 167°F.

15.4.2 Tension Equalization by Suspension Insulators

At a typical suspension structure, the conductor is supported vertically by a suspension insulator assembly, but allowed to move freely in the direction of the conductor axis. This conductor movement is possible due to insulator swing along the conductor axis. Changes in conductor tension between spans, caused by changes in temperature, load, and time, are normally equalized by insulator swing, eliminating horizontal tension differences across suspension structures.

TABLE 15.12 Sag-Tension Data 795 kcmil-Type 16 ACSR/SD with NESC Heavy Loading 300 and 1000 ft Spans

Conductor: Drake
795 kcmil-Type 16 ACSR/SD — Span = 300 ft
Area = 0.7261 in.^{2}
Creep *is* a factor — *NESC Heavy Loading District*

					Final		Initial	
Temp, °F	Ice, in.	Wind, lb/ft^2	*K*, lb/ft	Weight, lb/ft	Sag, ft	Tension, lb	Sag, ft	Tension, lb
0	0.50	4.00	0.30	2.486	2.19	12,774	2.03	13,757
32	0.50	0.00	0.00	2.074	2.25	10,377	1.90	12,256
−20	0.00	0.00	0.00	1.093	.91	13,477	.87	14,156
0	0.00	0.00	0.00	1.093	1.03	11,962	.92	13,305
30	0.00	0.00	0.00	1.093	1.26	9,776	1.03	11,976
60	0.00	0.00	0.00	1.093	1.60	7,688	1.16	10,589[a]
90	0.00	0.00	0.00	1.093	2.12	5,806	1.34	9,159
120	0.00	0.00	0.00	1.093	2.69	4,572	1.59	7,713
167	0.00	0.00	0.00	1.093	3.11	3,957	2.22	5,545
212	0.00	0.00	0.00	1.093	3.58	3,435	3.17	3,877

Conductor: Drake
795 kcmil-Type 16 ACSR/SD — Span = 1000 ft
Area = 0.7261 in.^{2}
Creep *is* a factor — *NESC Heavy Loading District*

					Final		Initial	
Temp, °F	Ice, in.	Wind, lb/ft^2	*K*, lb/ft	Weight, lb/ft	Sag, ft	Tension, lb	Sag, ft	Tension, lb
0	0.50	4.00	0.30	2.486	20.65	15,089	20.36	15,299
32	0.50	0.00	0.00	2.074	20.61	12,607	19.32	13,445
−20	0.00	0.00	0.00	1.093	12.20	11,205	10.89	12,552
0	0.00	0.00	0.00	1.093	13.35	10,244	11.56	11,832
30	0.00	0.00	0.00	1.093	15.22	8,988	12.69	10,779
60	0.00	0.00	0.00	1.093	17.21	7,950[a]	13.98	9,780
90	0.00	0.00	0.00	1.093	19.26	7,108	15.44	8,861
120	0.00	0.00	0.00	1.093	21.31	6,428	17.03	8,037
167	0.00	0.00	0.00	1.093	24.27	5,647	19.69	6,954
212	0.00	0.00	0.00	1.093	25.62	5,352	22.32	6,136

Note: Calculations based on (1) NESC heavy loading district and (2) tension limits: (a) initial loaded—60% RBS @ 0°F; (b) initial unloaded—25% RBS @ 60°F; (c) final unloaded—15% RBS @ 60°F.

[a] Design condition.

15.4.3 Ruling Span Calculation

Sag-tension can be found for a series of suspension spans in a line section by the use of the RS concept (Ehrenberg, 1935; Winkelman, 1959). The RS for the line section is defined by the following equation:

$$RS = \sqrt{\frac{S_1^3 + S_2^3 + \cdots + S_n^3}{S_1 + S_2 + \cdots + S_n}} \tag{15.25}$$

where

RS is the ruling span for the line section containing n suspension spans
S_1 is the Span length of the first suspension span
S_2 is the Span length of the second suspension span
S_n is the Span length of the nth suspension span

Alternatively, a generally satisfactory method for estimating the RS is to take the sum of the average suspension span length plus two-thirds of the difference between the maximum span and the average span. However, some judgment must be exercised in using this method because a large difference between the average and maximum span may cause a substantial error in the RS value.

As discussed, suspension spans are supported by suspension insulators that are free to move in the direction of the conductor axis. This freedom of movement allows the tension in each suspension span to be assumed to be the same and equal to that calculated for the RS. This assumption is valid for the suspension spans and RS under the same conditions of temperature and load, for both initial and final sags. For level spans, sag in each suspension span is given by the parabolic sag equation:

$$D_i = \frac{w\left(S_i^2\right)}{8H_{RS}} \tag{15.26}$$

where

D_i is the sag in the ith span
S_i is the span length of the ith span
H_{RS} is the tension from RS sag-tension calculations

The sag in level suspension spans may also be calculated using the ratio:

$$D_i = \frac{S_i}{S_{RS}} D_{RS} \tag{15.27}$$

where D_{RS} is the sag in RS.

Suspension spans vary in length, though typically not over a large range. Conductor temperature during sagging varies over a range considerably smaller than that used for line design purposes.

If the sag in any suspension span exceeds approximately 5% of the span length, a correction factor should be added to the sags obtained from the previous equation or the sag should be calculated using catenary Equation 15.29. This correction factor may be calculated as follows:

$$Correction = D^2 \frac{w}{6H} \tag{15.28}$$

where

D is the sag obtained from parabolic equation
w is the weight of conductor, lb/ft
H is the horizontal tension, lb

TABLE 15.13 Sag-Tension Data 795 kcmil-37 Strand AAC "Arbutus" with NESC Light Loading 300 and 1000 ft Spans

Conductor: Arbutus
795 kcmil-37 Strands AAC
Area = 0.6245 in.^{2}
Creep *is* a factor

Span = 300 ft

NESC Light Loading District

					Final		Initial	
Temp, °F	Ice, in.	Wind, lb/ft^2	*K*, lb/ft	Weight, lb/ft	Sag, ft	Tension, lb	Sag, ft	Tension, lb
30	0.00	9.00	0.05	1.122	3.56	3546	2.82	4479
30	0.00	0.00	0.00	0.746	2.91	2889	2.06	4075
60	0.00	0.00	0.00	0.746	4.03	2085[a]	2.80	2999
90	0.00	0.00	0.00	0.746	5.13	1638	3.79	2215
120	0.00	0.00	0.00	0.746	6.13	1372	4.86	1732
167	0.00	0.00	0.00	0.746	7.51	1122	6.38	1319
212	0.00	0.00	0.00	0.746	8.65	975	7.65	1101

Conductor: Arbutus
795 kcmil-37 Strands AAC
Area = 0.6245 in.^{2}
Creep *is* a factor

Span = 1000 ft

NESC Light Loading District

					Final		Initial	
Temp, °F	Ice, in.	Wind, lb/ft^2	*K*, lb/ft	Weight, lb/ft	Sag, ft	Tension, lb	Sag, ft	Tension, lb
30	0.00	9.00	0.05	1.122	44.50	3185	42.85	3305
30	0.00	0.00	0.00	0.746	43.66	2158	41.71	2258
60	0.00	0.00	0.00	0.746	45.24	2085[a]	43.32	2175
90	0.00	0.00	0.00	0.746	46.76	2018	44.89	2101
120	0.00	0.00	0.00	0.746	48.24	1958	46.42	2033
167	0.00	0.00	0.00	0.746	50.49	1873	48.72	1939
212	0.00	0.00	0.00	0.746	52.55	1801	50.84	1860

Note: Calculations based on (1) NESC light loading district and (2) tension limits: (a) initial loaded—60% RBS @ 30°F; (b) initial unloaded—25% RBS @ 60°F; and (c) final unloaded—15% RBS @ 60°F.

[a] Design condition.

The catenary equation for calculating the sag in a suspension or stringing span is

$$Sag = \frac{H}{w}\left(\cosh\frac{Sw}{2H} - 1\right) \tag{15.29}$$

where

S is the span length, ft
H is the horizontal tension, lb
w is the resultant weight, lb/ft

15.4.4 Stringing Sag Tables

Conductors are typically installed in line section lengths consisting of multiple spans. The conductor is pulled from the conductor reel at a point near one strain structure progressing through travelers attached to each suspension structure to a point near the next strain structure. After stringing, the conductor tension is increased until the sag in one or more suspension spans reaches the appropriate stringing sags based on the RS for the line section. The calculation of stringing sags is based on the preceding sag equation.

Table 15.13 shows a typical stringing sag table for a 600 ft RS of Drake ACSR with suspension spans ranging from 400 to 700 ft and conductor temperatures of 20°F–100°F. All values in this stringing table are calculated from RS initial tensions, shown in Table 15.12 using the parabolic sag equation.

15.5 Line Design Sag-Tension Parameters

In laying out a transmission line, the first step is to survey the route and draw up a plan-profile of the selected right-of-way. The plan-profile drawings serve an important function in linking together the various stages involved in the design and construction of the line. These drawings, prepared based on the route survey, show the location and elevation of all natural and man-made obstacles to be traversed by, or adjacent to, the proposed line. These plan-profiles are drawn to scale and provide the basis for tower spotting and line design work.

Once the plan-profile is completed, one or more estimated RSs for the line may be selected. Based on these estimated RSs and the maximum design tensions, sag-tension data may be calculated providing initial and final sag values. From these data, sag templates may be constructed to the same scale as the plan-profile for each RS, and used to graphically spot structures.

15.5.1 Catenary Constants

The sag in a RS is equal to the weight per unit length, w, times the span length, S, squared, divided by eight times the horizontal component of the conductor tension, H. The ratio of conductor horizontal tension, H, to weight per unit length, w, is the catenary constant, H/w. For a RS sag-tension calculation using eight loading conditions, a total of 16 catenary constant values could be defined, one for initial and final tension under each loading condition.

Catenary constants can be defined for each loading condition of interest and are used in any attempt to locate structures. Some typical uses of catenary constants for locating structures are to avoid overloading, assure ground clearance is sufficient at all points along the right-of-way, and minimize blowout or uplift under cold weather conditions. To do this, catenary constants are typically found for (1) the maximum line temperature, (2) heavy ice and wind loading, (3) wind blowout, and (4) minimum conductor temperature. Under any of these loading conditions, the catenary constant allows sag calculation at any point within the span.

15.5.2 Wind Span

The maximum wind span of any structure is equal to the distance measured from center to center of the two adjacent spans supported by a structure. The wind span is used to determine the maximum horizontal force a structure must be designed to withstand under high wind conditions. Wind span is not dependent on conductor sag or tension, only on horizontal span length.

15.5.3 Weight Span

The weight span of a structure is a measure of the maximum vertical force a structure must be designed to withstand. The weight span is equal to the horizontal distance between the low points and the vertex of two adjacent spans. The maximum weight span for a structure is dependent on the loading condition being a minimum for heavy ice and wind load. When the elevations of adjacent structures are the same, the wind and weight spans are equal.

15.5.4 Uplift at Suspension Structures

Uplift occurs when the weight span of a structure is negative. On steeply inclined spans, the low point of sag may fall beyond the lower support. This indicates that the conductor in the uphill span is exerting a

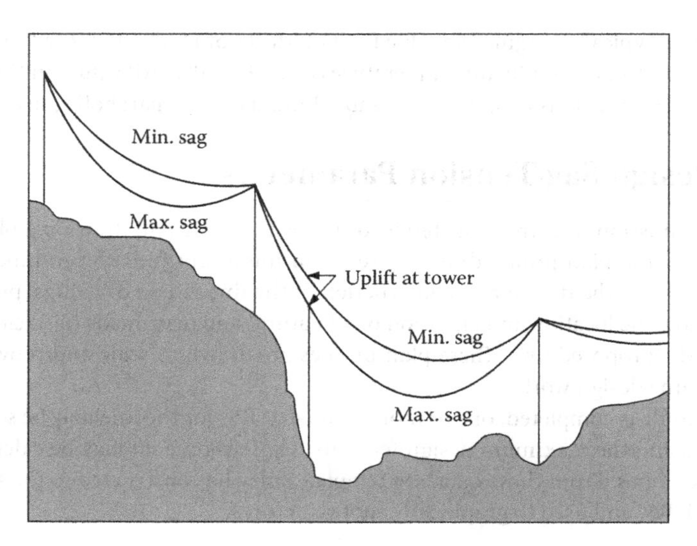

FIGURE 15.9 Conductor uplift.

negative or upward force on the lower tower. The amount of this upward force is equal to the weight of the conductor from the lower tower to the low point in the sag. If the upward pull of the uphill span is greater than the downward load of the next adjacent span, actual uplift will be caused and the conductor will swing free of the tower. This usually occurs under minimum temperature conditions and must be dealt with by adding weights to the insulator suspension string or using a strain structure (Figure 15.9).

15.5.5 Tower Spotting

Given sufficiently detailed plan-profile drawings, structure heights, wind/weight spans, catenary constants, and minimum ground clearances, structure locations can be chosen such that ground clearance is maintained and structure loads are acceptable. This process can be done by hand using a sag template, plan-profile drawing, and structure heights, or numerically by one of several commercial programs.

15.6 Conductor Installation

Installation of a bare overhead conductor can present complex problems. Careful planning and a thorough understanding of stringing procedures are needed to prevent damage to the conductor during the stringing operations. The selection of stringing sheaves, tensioning method, and measurement techniques are critical factors in obtaining the desired conductors sagging results. Conductor stringing and sagging equipment and techniques are discussed in detail in the *IEEE Guide to the Installation of Overhead Transmission Line Conductors*, IEEE Std. 524–1992. Some basic factors concerning installation are covered in this section. Because the terminology used for equipment and installation procedures for overhead conductors varies throughout the utility industry, a limited glossary of terms and equipment definitions excerpted from IEEE Std. 524–1992 is provided in the chapter appendix. A complete glossary is presented in the *IEEE Guide to the Installation of Overhead Transmission Line Conductors.*

15.6.1 Conductor Stringing Methods

There are two basic methods of stringing conductors, categorized as either slack or tension stringing. There are as many variations of these methods as there are organizations installing conductors. The selected method, however, depends primarily on the terrain and conductor surface damage requirements.

15.6.1.1 Slack or Layout Stringing Method

Slack stringing of conductor is normally limited to lower voltage lines and smaller conductors. The conductor reel(s) is placed on reel stands or "jack stands" at the beginning of the stringing location. The conductor is unreeled from the shipping reel and dragged along the ground by means of a vehicle or pulling device. When the conductor is dragged past a supporting structure, pulling is stopped and the conductor placed in stringing sheaves attached to the structure. The conductor is then reattached to the pulling equipment and the pull continued to the next structure.

This stringing method is typically used during construction of new lines in areas where the right-of-way is readily accessible to vehicles used to pull the conductor. However, slack stringing may be used for repair or maintenance of transmission lines where rugged terrain limits the use of pulling and tensioning equipment. It is seldom used in urban areas or where there is any danger of contact with high-voltage conductors.

15.6.1.2 Tension Stringing

A tension stringing method is normally employed when installing transmission conductors. Using this method, the conductor is unreeled under tension and is not allowed to contact the ground. In a typical tension stringing operation, travelers are attached to each structure. A pilot line is pulled through the travelers and is used, in turn, to pull in heavier pulling line. This pulling line is then used to pull the conductor from the reels and through the travelers. Tension is controlled on the conductor by the tension puller at the pulling end and the bullwheel tension retarder at the conductor payout end of the installation. Tension stringing is preferred for all transmission installations. This installation method keeps the conductor off the ground, minimizing the possibility of surface damage and limiting problems at roadway crossings. It also limits damage to the right-of-way by minimizing heavy vehicular traffic.

15.6.2 Tension Stringing Equipment and Setup

Stringing equipment typically includes bullwheel or drum pullers for back-tensioning the conductor during stringing and sagging; travelers (stringing blocks) attached to every phase conductor and shield wire attachment point on every structure; a bullwheel or crawler tractor for pulling the conductor through travelers; and various other special items of equipment. Figure 15.10 illustrates a typical stringing and sagging setup for a stringing section and the range of stringing equipment required. Provision for conductor splicing during stringing must be made at tension site or midspan sites to avoid pulling splices through the travelers.

During the stringing operation, it is necessary to use proper tools to grip the strands of the conductor evenly to avoid damaging the outer layer of wires. Two basic types or categories of grips are normally used in transmission construction. The first is a type of grip referred to as a pocketbook, suitcase, bolted, etc., that hinges to completely surround the conductor and incorporates a bail for attaching to the pulling line. The second type is similar to a Chinese finger grip and is often referred to as a basket or "Kellem" grip. Such a grip, shown in Figure 15.11, is often used because of its flexibility and small size, making it easily pulled through sheaves during the stringing operation. Whatever type of gripping device is used, a swivel should be installed between the pulling grip and pulling line or running board to allow free rotation of both the conductor and the pulling line.

A traveler consists of a sheave or pulley wheel enclosed in a frame to allow it to be suspended from structures or insulator strings. The frame must have some type of latching mechanism to allow insertion and removal of the conductor during the stringing operation. Travelers are designed for a maximum safe working load. Always ensure that this safe working load will not be exceeded during the stringing operation. Sheaves are often lined with neoprene or urethane materials to prevent scratching of conductors in high-voltage applications; however, unlined sheaves are also available for special applications.

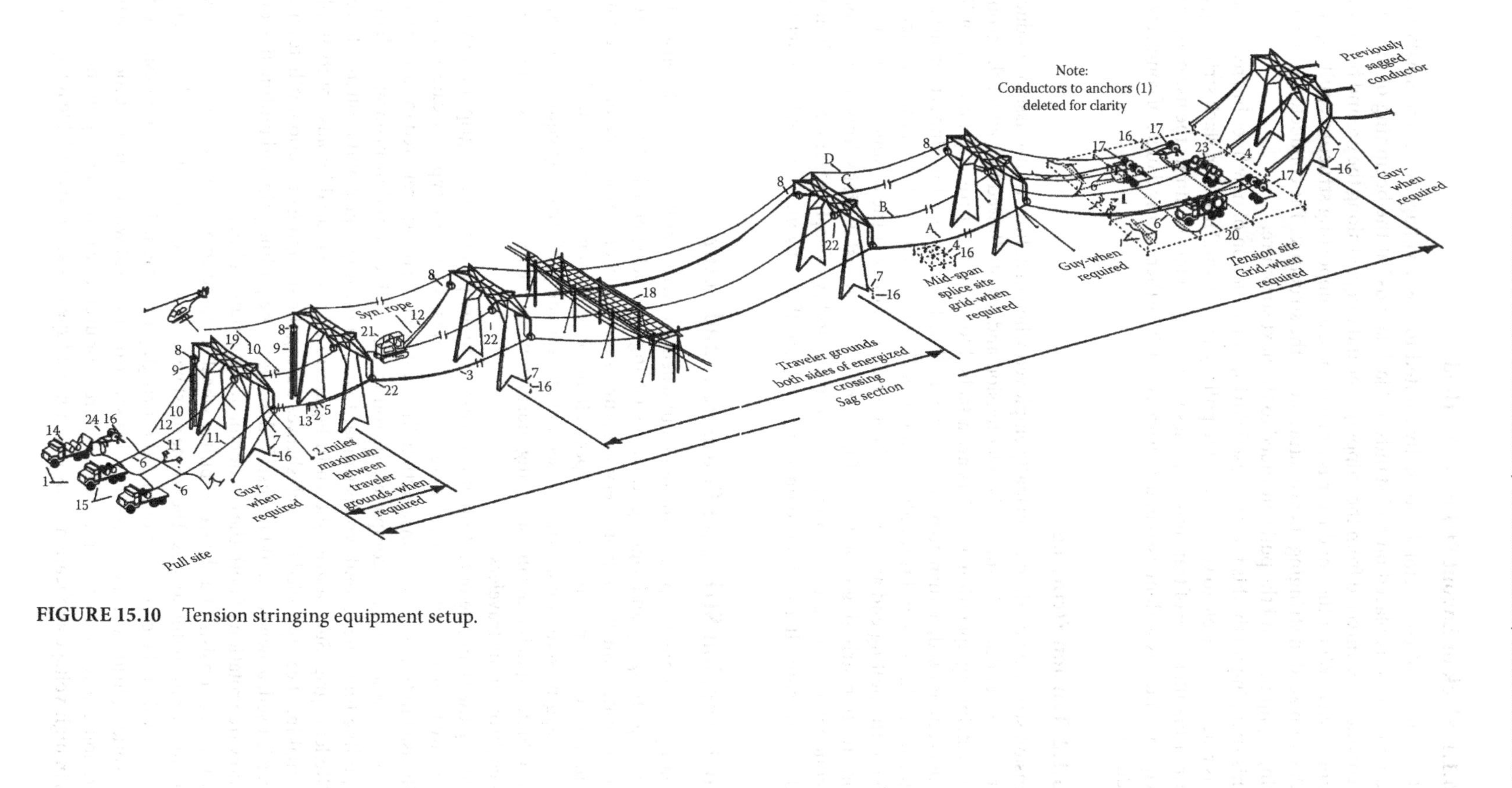

FIGURE 15.10 Tension stringing equipment setup.

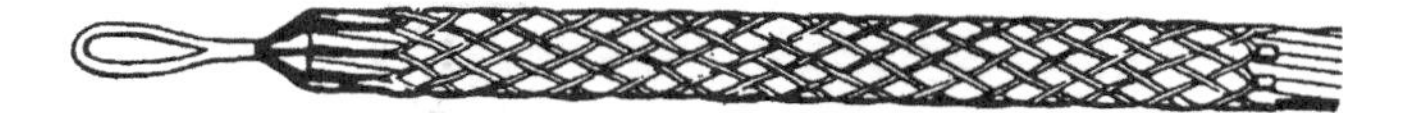

FIGURE 15.11 Basket grip pulling device.

Travelers used in tension stringing must be free rolling and capable of withstanding high running or static loads without damage. Proper maintenance is essential. Very high longitudinal tension loads can develop on transmission structures if a traveler should "freeze" during tension stringing, possibly causing conductor and/or structure damage. Significant levels of rotation resistance will also yield tension differences between spans, resulting in incorrect sag.

Proper selection of travelers is important to assure that travelers operate correctly during tension stringing and sagging. The sheave diameter and the groove radius must be matched to the conductor.

Figure 15.12 illustrates the minimum sheave diameter for typical stringing and sagging operations. Larger diameter sheaves may be required where particularly severe installation conditions exist.

15.6.3 Sagging Procedure

It is important that the conductors be properly sagged at the correct stringing tension for the design RS. A series of several spans, a line section, is usually sagged in one operation. To obtain the correct sags and to insure the suspension insulators hang vertically, the horizontal tension in all spans must be equal. Figures 15.13 through 15.18 depict typical parabolic methods and computations required for sagging conductors. Factors that must be considered when sagging conductors are creep elongation during stringing and prestressing of the conductor.

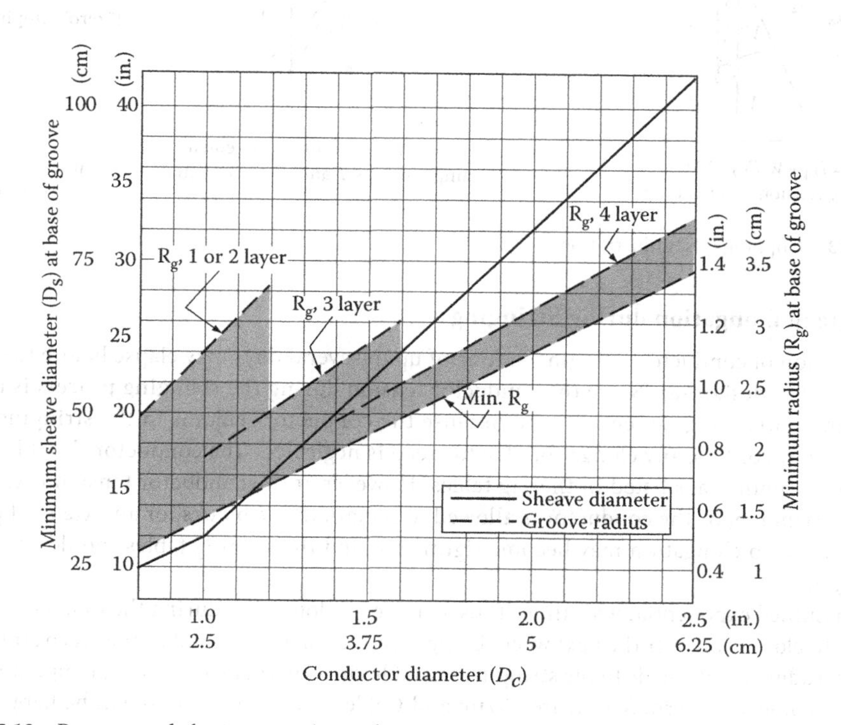

FIGURE 15.12 Recommended minimum sheave dimensions.

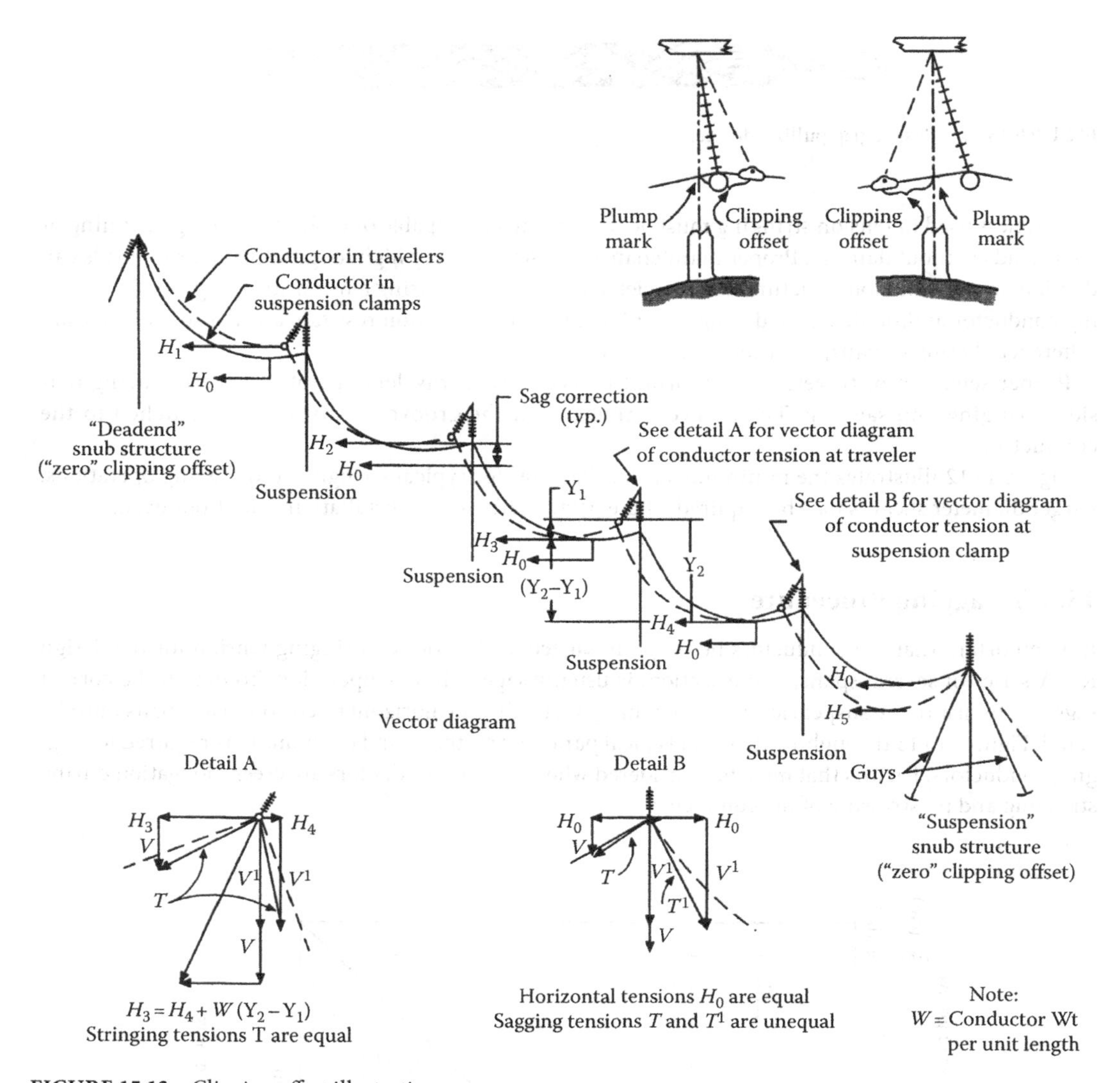

FIGURE 15.13 Clipping offset illustration.

15.6.3.1 Creep Elongation during Stringing

Upon completion of conductor stringing, a time of up to several days may elapse before the conductor is tensioned to design sag. Since the conductor tension during the stringing process is normally well below the initial sagging tension, and because the conductor remains in the stringing sheaves for only a few days or less, any elongation due to creep is neglected. The conductor should be sagged to the initial stringing sags listed in the sag tables. However, if the conductor tension is excessively high during stringing, or the conductor is allowed to remain in the blocks for an extended period of time, then the creep elongation may become significant and the sagging tables should be corrected prior to sagging.

Creep is assumed exponential with time. Thus, conductor elongation during the first day under tension is equal to elongation over the next week. Using creep estimation formulas, the creep strain can be estimated and adjustments made to the stringing sag tables in terms of an equivalent temperature. Also, should this become a concern, Southwire's Wire and Cable Technology Group will be happy to work with you to solve the problem.

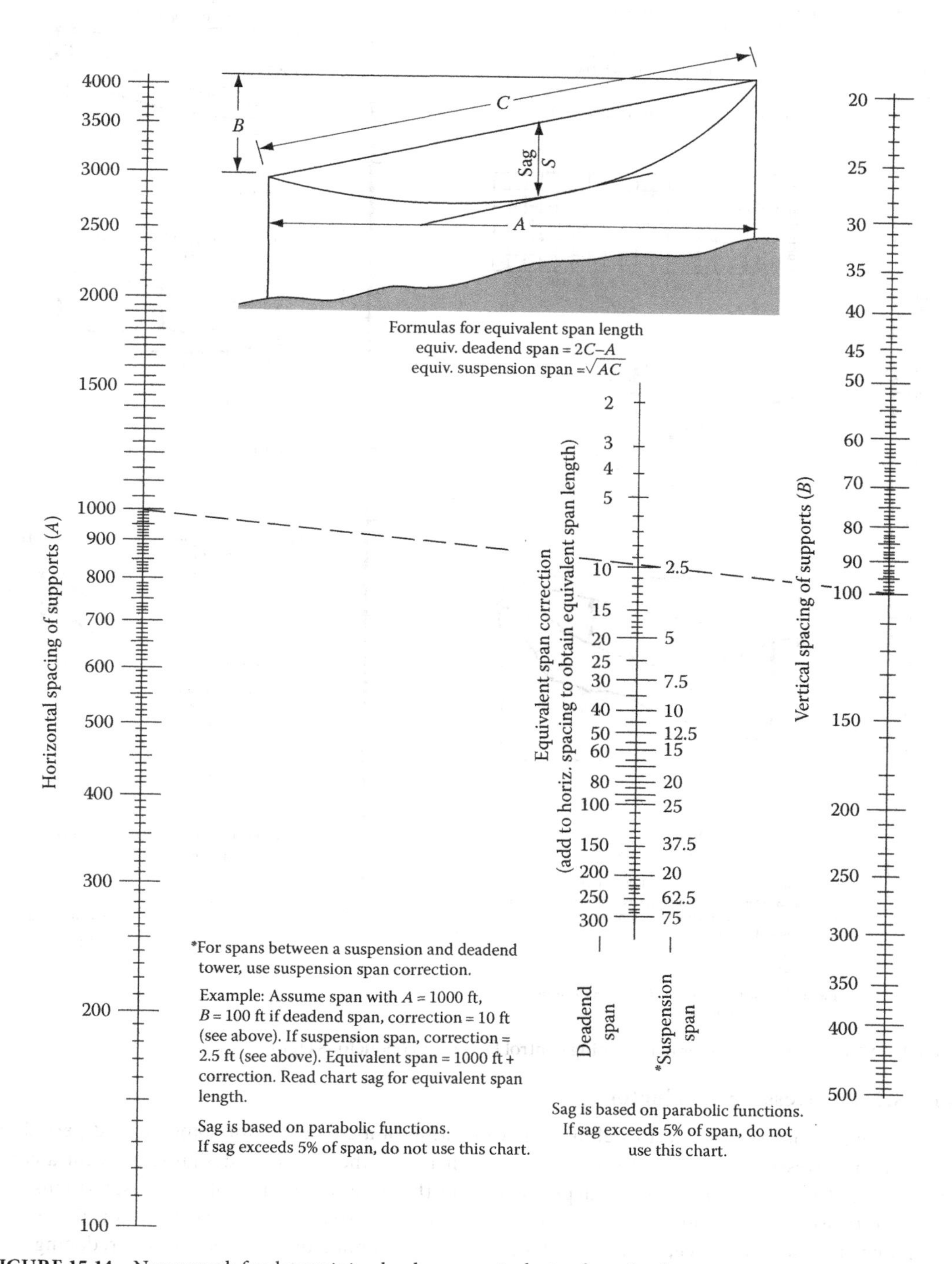

FIGURE 15.14 Nomograph for determining level span equivalents of non-level spans.

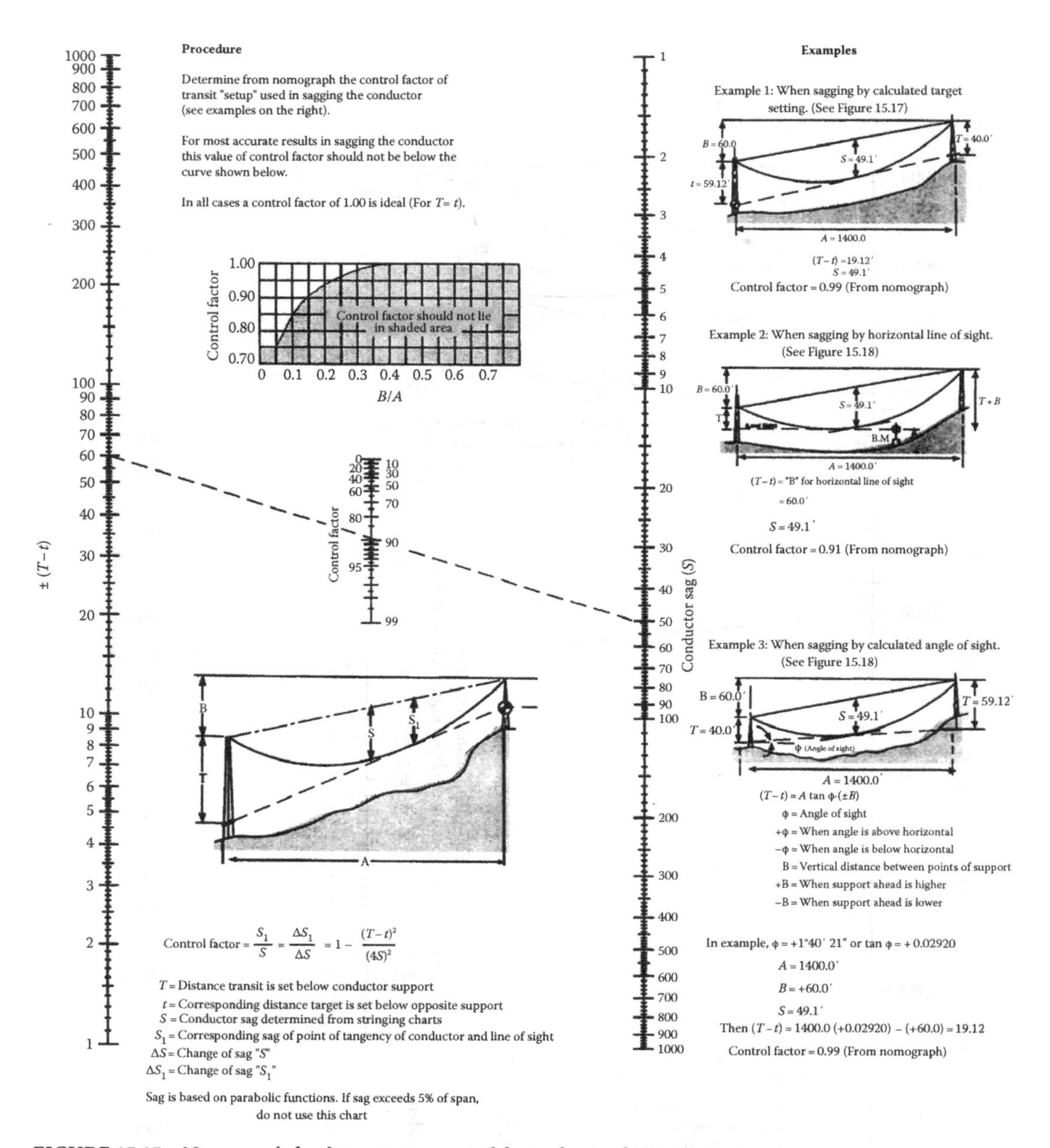

FIGURE 15.15 Nomograph for determining control factor for conductor sagging.

15.6.3.2 Prestressing Conductor

Prestressing is sometimes used to stabilize the elongation of a conductor for some defined period of time. The prestressing tension is normally much higher than the unloaded design tension for a conductor. The degree of stabilization is dependent upon the time maintained at the prestress tension. After prestressing, the tension on the conductor is reduced to stringing or design tension limits. At this reduced tension, the creep or plastic elongation of the conductor has been slowed, reducing the permanent elongation due to strain and creep for a defined period of time. By tensioning a conductor to levels approaching 50% of its breaking strength for times on the order of a day, creep elongation will be temporarily halted (Cahill, 1973). This simplifies concerns about creep during subsequent installation but presents both equipment and safety problems.

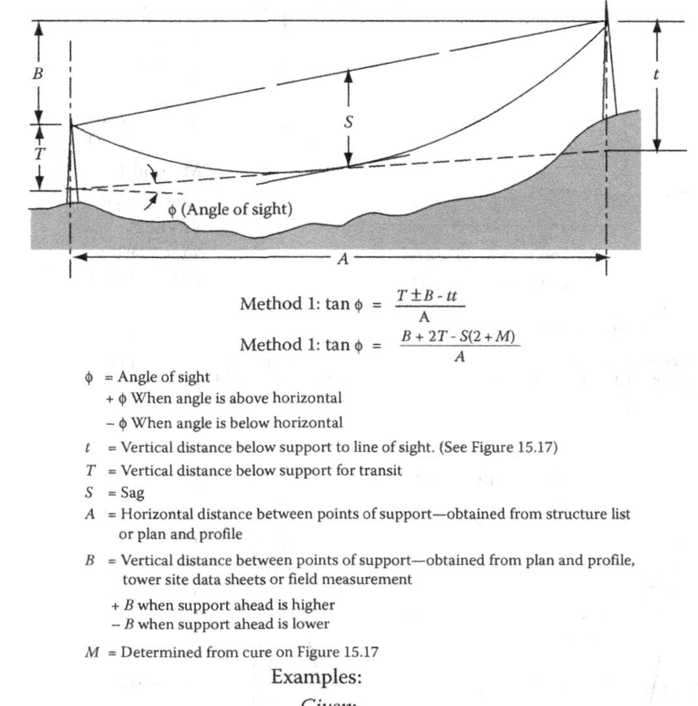

Method 1: $\tan\phi = \dfrac{T \pm B - tt}{A}$

Method 1: $\tan\phi = \dfrac{B + 2T - S(2 + M)}{A}$

ϕ = Angle of sight
 + ϕ When angle is above horizontal
 − ϕ When angle is below horizontal

t = Vertical distance below support to line of sight. (See Figure 15.17)

T = Vertical distance below support for transit

S = Sag

A = Horizontal distance between points of support—obtained from structure list or plan and profile

B = Vertical distance between points of support—obtained from plan and profile, tower site data sheets or field measurement
 + B when support ahead is higher
 − B when support ahead is lower

M = Determined from cure on Figure 15.17

Examples:

Given:

$A = 1400.0'$	$S = 49.1'$	@ 60°F
$B = 160.0'$	$S = 51.2'$	@ 90°F
$T = 40.0'$	$T = 59.12'$	@ 60°F
	$T = 63.76'$	@ 90°F

Method 1

$$\text{Tan}\,\phi = \frac{T \pm B - t}{A}$$

$$\text{Tan}\,\phi_{60°F} = \frac{40.0 - 60.0 - 59.12}{1400.0} = 0.02920$$

$\phi_{60°F} = +1°\,40'21''$

$$\text{Tan}\,\phi_{90°F} = \frac{40.0 - 60.0 - 63.76}{1400.0} = 0.02589$$

$\phi_{90°F} = +1°\,28'\,59''$

Change in angle ϕ for
$5°F = (1°\,40'\,21'' - 1°\,28'\,55'')\left(\frac{5}{30}\right) = 0°\,1'\,54''$

Method 2

$$\text{Tan}\,\phi = \frac{B + 2T - S(2 + M)}{A}$$

$$\text{Tan}\,\phi_{60°F} = \frac{60.0 + (40.0)(2) - (49.1)(2+0.019)}{1400.0} = 0.02919$$

$\phi_{60°F} = +1°\,40'\,19''$

$$\text{Tan}\,\phi_{90°F} = \frac{60.0 + (40.0)(2) - (51.2)(2+0.027)}{1400.0} = 0.02587$$

$\phi_{90°F} = +1°\,28'\,55''$

Change in angle ϕ for
$5°F = (1°\,40'\,19'' - 1°\,28'\,55'')\left(\frac{5}{30}\right) = 0°\,1'\,54''$

FIGURE 15.16 Conductor sagging by calculated angle of sight.

15.6.3.3 Sagging by Stopwatch Method

A mechanical pulse imparted to a tensioned conductor moves at a speed proportional to the square root of tension divided by weight per unit length. By initiating a pulse on a tensioned conductor and measuring the time required for the pulse to move to the nearest termination, the tension, and thus the sag of the conductor, can be determined. This stopwatch method (Overend and Smith, 1986) has come into wide use even for long spans and large conductors.

The conductor is struck a sharp blow near one support and the stopwatch is started simultaneously. A mechanical wave moves from the point where the conductor was struck to the next support point at which it will be partially reflected. If the initiating blow is sharp, the wave will travel up and down the span many times before dying out. Time-sag tables such as the one shown in Table 15.14 are available from many sources. Specially designed sagging stopwatches are also available.

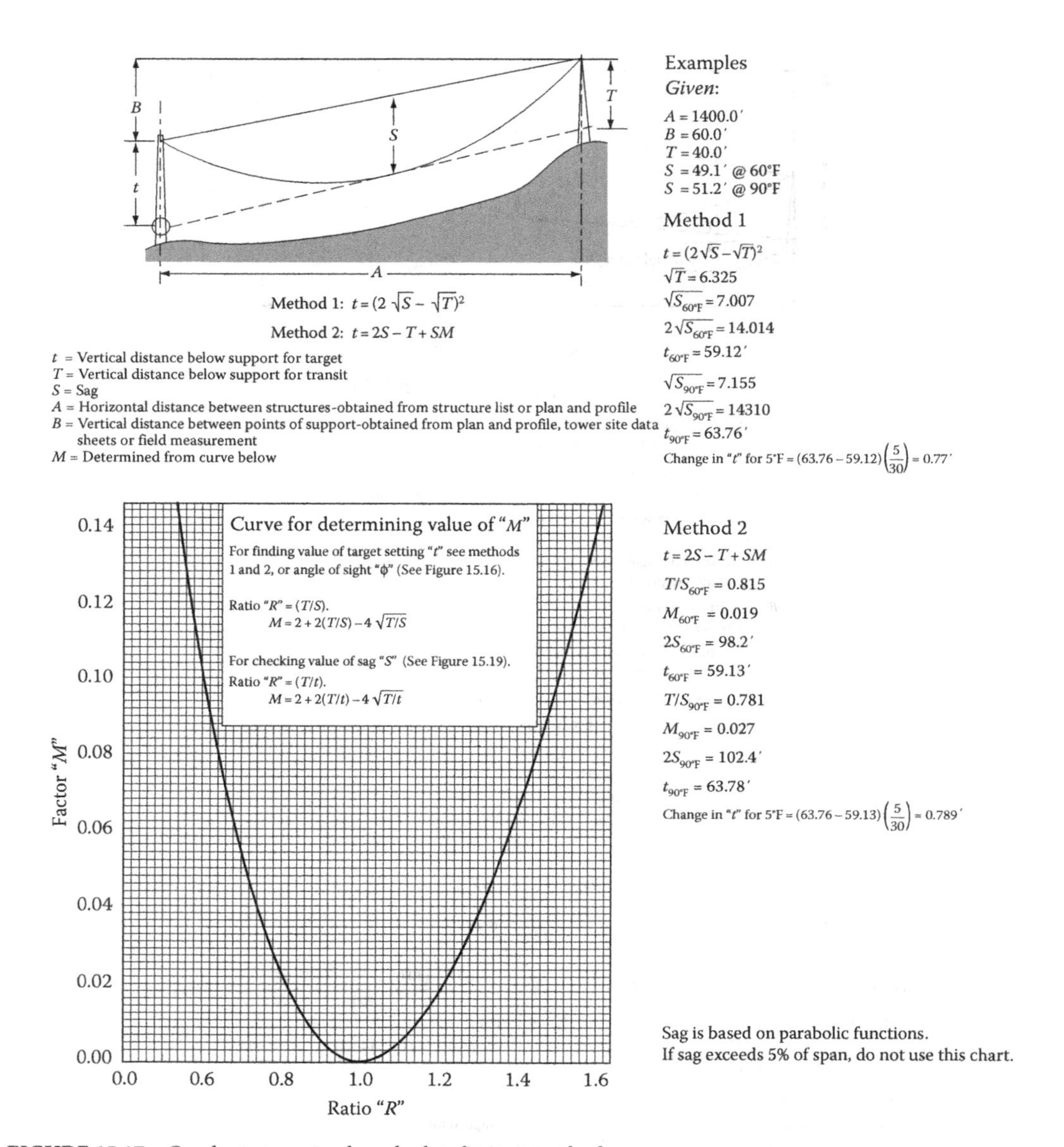

FIGURE 15.17 Conductor sagging by calculated target method.

The reflected wave can be detected by lightly touching the conductor but the procedure is more likely to be accurate if the wave is both initiated and detected with a light rope over the conductor. Normally, the time for the return of the third or fifth wave is monitored.

Traditionally, a transit sagging method has been considered to be more accurate for sagging than the stopwatch method. However, many transmission-line constructors use the stopwatch method exclusively, even with large conductors.

15.6.3.4 Sagging by Transit Methods

IEEE Guide Std. 524–1993 lists three methods of sagging conductor with a transit: "Calculated Angle of Sight," "Calculated Target Method," and "Horizontal Line of Sight." The method best suited to a particular line sagging situation may vary with terrain and line design.

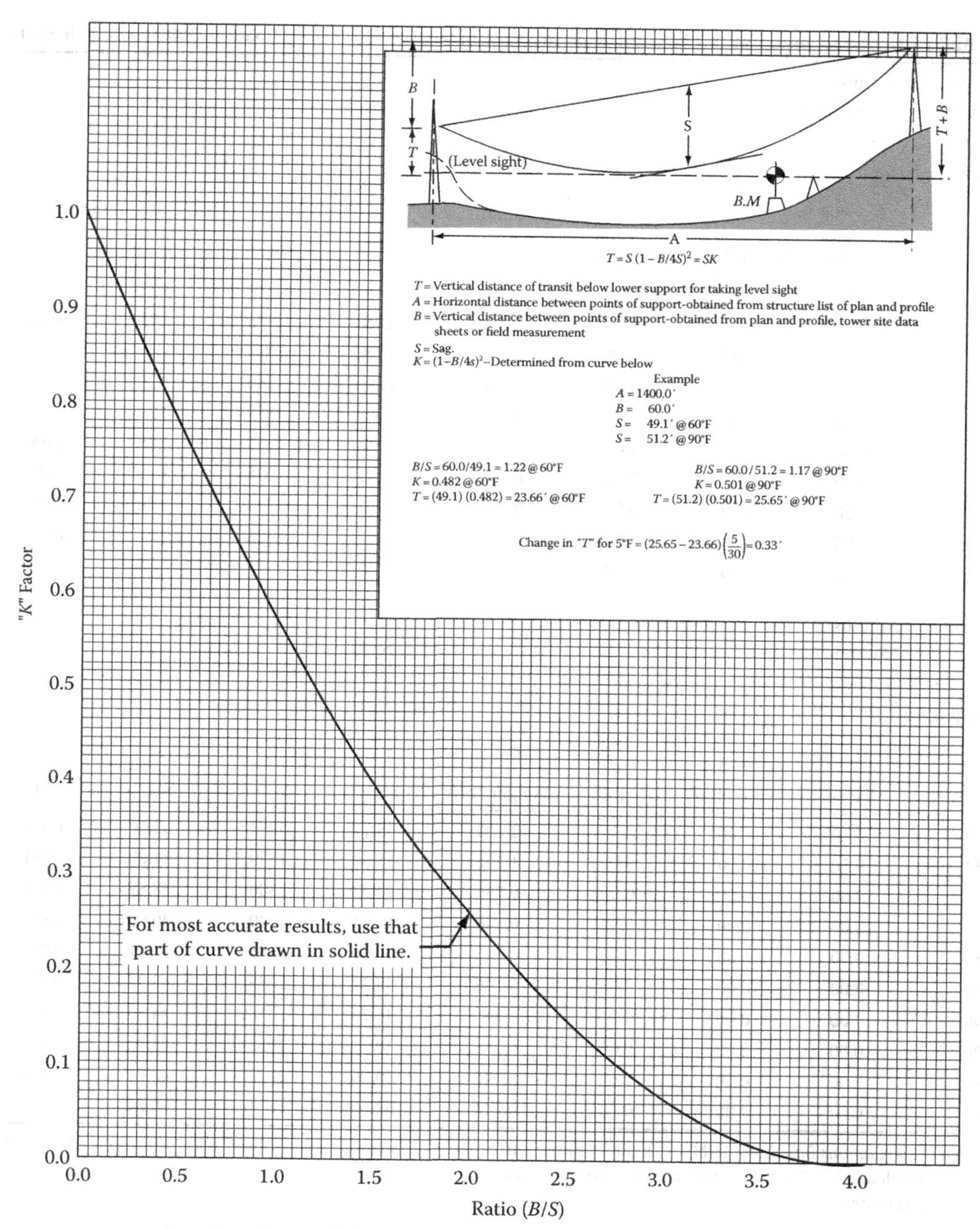

Sag is based on parabolic functions. If sag exceeds 5% of span, do not use this chart.

FIGURE 15.18 Conductor sagging by horizontal line of sight.

15.6.3.5 Sagging Accuracy

Sagging a conductor during construction of a new line or in the reconductoring of an old line involves many variables that can lead to a small degree of error. IEEE Std. 524–1993 suggests that all sags be within 6 in. of the stringing sag values. However, aside from measurement errors during sagging, errors in terrain measurement and variations in conductor properties, loading conditions, and hardware installation have led some utilities to allow up to 3 ft of margin in addition to the required minimum ground clearance.

TABLE 15.14 Sag-Tension Data 795 kcmil-Type 16 ACSR/SD with NESC Heavy Loading 300 and 1000 ft Spans

Conductor: Arbutus
795 kcmil-37 Strands AAC
Area = 0.6245 in.^{2}
Creep *is* a factor

Span = 300 ft

					Final		Initial	
Temp, °F	Ice, in.	Wind, lb/ft^2	*K*, lb/ft	Weight, lb/ft	Sag, ft	Tension, lb	Sag, ft	Tension, lb
0	0.50	4.00	0.30	2.125	3.97	6033	3.75	6383
32	0.50	0.00	0.00	1.696	4.35	4386	3.78	5053
−20	0.00	0.00	0.00	0.746	1.58	5319	1.39	6055
0	0.00	0.00	0.00	0.746	2.00	4208	1.59	5268
30	0.00	0.00	0.00	0.746	2.91	2889	2.06	4075
60	0.00	0.00	0.00	0.746	4.03	2085[a]	2.80	2999
90	0.00	0.00	0.00	0.746	5.13	1638	3.79	2215
120	0.00	0.00	0.00	0.746	6.13	1372	4.86	1732
167	0.00	0.00	0.00	0.746	7.51	1122	6.38	1319
212	0.00	0.00	0.00	0.746	8.65	975	7.65	1101

Conductor: Arbutus
795 kcmil-37 Strands AAC
Area = 0.6245 in.^{2}
Creep *is* a factor

Span = 1000 ft

NESC Heavy Loading District

					Final		Initial	
Temp, °F	Ice, in.	Wind, lb/ft^2	*K*, lb/ft	Weight, lb/ft	Sag, ft	Tension, lb	Sag, ft	Tension, lb
0	0.50	4.00	0.30	2.125	45.11	5953	44.50	6033
32	0.50	0.00	0.00	1.696	45.80	4679	44.68	4794
−20	0.00	0.00	0.00	0.746	40.93	2300	38.89	2418
0	0.00	0.00	0.00	0.746	42.04	2240	40.03	2350
30	0.00	0.00	0.00	0.746	43.66	2158	41.71	2258
60	0.00	0.00	0.00	0.746	45.24	2085[a]	43.32	2175
90	0.00	0.00	0.00	0.746	46.76	2018	44.89	2101
120	0.00	0.00	0.00	0.746	48.24	1958	46.42	2033
167	0.00	0.00	0.00	0.746	50.49	1873	48.72	1939
212	0.00	0.00	0.00	0.746	52.55	1801	50.84	1860

Note: Calculations based on (1) NESC light loading district and (2) tension limits: (a) initial loaded—60% RBS @ 0°F; (b) initial unloaded—25% RBS @ 60°F; and (c) final unloaded—15% RBS @ 60°F.

[a] Design condition.

15.6.3.6 Clipping Offsets

If the conductor is to be sagged in a series of suspension spans where the span lengths are reasonably close and where the terrain is reasonably level, then the conductor is sagged using conventional stringing sag tables and the conductor is simply clipped into suspension clamps that replace the travelers. If the conductor is to be sagged in a series of suspension spans where span lengths vary widely or more commonly, where the terrain is steep, then clipping offsets may need to be employed in order to yield vertical suspension strings after installation.

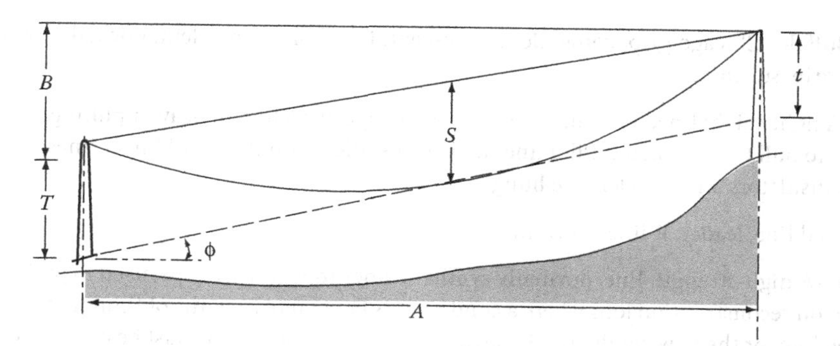

FIGURE 15.19 Conductor sagging for checking sag *S*.

Clipping offsets are illustrated in Figure 15.19, showing a series of steeply inclined spans terminated in a “snub” structure at the bottom and a “dead-end” structure at the top. The vector diagram illustrates a balance of total conductor tension in the travelers but an imbalance in the horizontal component of tension.

15.7 Defining Terms

Block—A device designed with one or more single sheaves, a wood or metal shell, and an attachment hook or shackle. When rope is reeved through two of these devices, the assembly is commonly referred to as a *block and tackle*. A *set of 4s* refers to a block and tackle arrangement utilizing two 4 in. double-sheave blocks to obtain four load-bearing lines. Similarly, a *set of 5s* or a *set of 6s* refers to the same number of load bearing lines obtained using two 5 in. or two 6 in. double-sheave blocks, respectively.

Synonyms: Set of 4s, set of 5s, set of 6s.

Bullwheel—A wheel incorporated as an integral part of a bullwheel puller or tensioner to generate pulling or braking tension on conductors or pulling lines, or both, through friction. A puller or tensioner normally has one or more pairs arranged in tandem incorporated in its design. The physical size of the wheels will vary for different designs, but 17 in. (43 cm) face widths and diameters of 5 ft (150 cm) are common. The wheels are power driven or retarded and lined with single- or multiple-groove neoprene or urethane linings. Friction is accomplished by reeving the pulling line or conductor around the groove of each pair.

Clipping-in—The transferring of sagged conductors from the traveler to their permanent suspension positions and the installing of the permanent suspension clamps.

Synonyms: Clamping, clipping.

Clipping offset—A calculated distance, measured along the conductor from the plum mark to a point on the conductor at which the center of the suspension clamp is to be placed. When stringing in rough terrain, clipping offset may be required to balance the horizontal forces on each suspension structure.

Grip, conductor—A device designed to permit the pulling of conductor without splicing on fittings, eyes, etc. It permits the pulling of a *continuous* conductor where threading is not possible. The designs of these grips vary considerably. Grips such as the Klein (Chicago) and Crescent utilize an open-sided rigid body with opposing jaws and swing latch. In addition to pulling conductors, this type is commonly used to tension guys and, in some cases, pull wire rope. The design of the come-along (pocketbook, suitcase, four bolt, etc.) incorporates a bail attached to the body of a clamp which folds to completely surround and envelope the conductor. Bolts are then used to close the clamp and obtain a grip.

Synonyms: Buffalo, Chicago grip, come-along, Crescent, four bolt, grip, Klein, pocketbook, seven bolt, six bolt, slip-grip, suitcase.

Line, pilot—A lightweight line, normally synthetic fiber rope, used to pull heavier pulling lines which in turn are used to pull the conductor. Pilot lines may be installed with the aid of finger lines or by helicopter when the insulators and travelers are hung.

Synonyms: Lead line, leader, P-line, straw line.

Line, pulling—A high-strength line, normally synthetic fiber rope or wire rope, used to pull the conductor. However, on reconstruction jobs where a conductor is being replaced, the old conductor often serves as the pulling line for the new conductor. In such cases, the old conductor must be closely examined for any damage prior to the pulling operations.

Synonyms: Bull line, hard line, light line, sock line.

Puller, bullwheel—A device designed to pull pulling lines and conductors during stringing operations. It normally incorporates one or more pairs of urethane- or neoprene-lined, power-driven, single- or multiple-groove bullwheels where each pair is arranged in tandem. Pulling is accomplished by friction generated against the pulling line which is reeved around the grooves of a pair of the bullwheels. The puller is usually equipped with its own engine which drives the bullwheels mechanically, hydraulically, or through a combination of both. Some of these devices function as either a puller or tensioner.

Synonym: Puller.

Puller, drum—A device designed to pull a conductor during stringing operations. It is normally equipped with its own engine which drives the drum mechanically, hydraulically, or through a combination of both. It may be equipped with synthetic fiber rope or wire rope to be used as the pulling line. The pulling line is paid out from the unit, pulled through the travelers in the sag section and attached to the conductor. The conductor is then pulled in by winding the pulling line back onto the drum. This unit is sometimes used with synthetic fiber rope acting as a pilot line to pull heavier pulling lines across canyons, rivers, etc.

Synonyms: Hoist, single drum hoist, single drum winch, tugger.

Puller, reel—A device designed to pull a conductor during stringing operations. It is normally equipped with its own engine which drives the supporting shaft for the reel mechanically, hydraulically, or through a combination of both. The shaft, in turn, drives the reel. The application of this unit is essentially the same as that for the drum puller previously described. Some of these devices function as either a puller or tensioner.

Reel stand—A device designed to support one or more reels and having the possibility of being skid, trailer, or truck mounted. These devices may accommodate rope or conductor reels of varying sizes and are usually equipped with reel brakes to prevent the reels from turning when pulling is stopped. They are used for either slack or tension stringing. The designation of reel trailer or reel truck implies that the trailer or truck has been equipped with a reel stand (jacks) and may serve as a reel transport or *payout* unit, or both, for stringing operations. Depending upon the sizes of the reels to be carried, the transporting vehicles may range from single-axle trailers to semi-trucks with trailers having multiple axles.

Synonyms: Reel trailer, reel transporter, reel truck.

Running board—A pulling device designed to permit stringing more than one conductor simultaneously with a single pulling line. For distribution stringing, it is usually made of lightweight tubing with the forward end curved gently upward to provide smooth transition over pole cross-arm rollers. For transmission stringing, the device is either made of sections hinged transversely to the direction of pull or of a hard-nose rigid design, both having a flexible pendulum tail suspended from the rear. This configuration stops the conductors from twisting together and permits smooth transition over the sheaves of bundle travelers.

Synonyms: Alligator, bird, birdie, monkey tail, sled.

Sag section—The section of line between snub structures. More than one sag section may be required in order to properly sag the actual length of conductor which has been strung.

Synonyms: Pull, setting, stringing section.

Site, pull—The location on the line where the puller, reel winder, and anchors (snubs) are located. This site may also serve as the pull or tension site for the next sag section.

Synonyms: Reel setup, tugger setup.

Site, tension—The location on the line where the tensioner, reel stands and anchors (snubs) are located. This site may also serve as the pull or tension site for the next sag section.

Synonyms: Conductor payout station, payout site, reel setup.

Snub structure—A structure located at one end of a sag section and considered as a *zero* point for sagging and clipping offset calculations. The section of line between two such structures is the sag section, but more than one sag section may be required in order to sag properly the actual length of conductor which has been strung.

Synonyms: 0 structure, zero structure.

Tensioner, bullwheel—A device designed to hold tension against a pulling line or conductor during the stringing phase. Normally, it consists of one or more pairs of urethane- or neoprene-lined, power braked, single- or multiple-groove bullwheels where each pair is arranged in tandem. Tension is accomplished by friction generated against the conductor which is reeved around the grooves of a pair of the bullwheels. Some tensioners are equipped with their own engines which retard the bullwheels mechanically, hydraulically, or through a combination of both. Some of these devices function as either a puller or tensioner. Other tensioners are only equipped with friction-type retardation.

Synonyms: Retarder, tensioner.

Tensioner, reel—A device designed to generate tension against a pulling line or conductor during the stringing phase. Some are equipped with their own engines which retard the supporting shaft for the reel mechanically, hydraulically, or through a combination of both. The shaft, in turn, retards the reel. Some of these devices function as either a puller or tensioner. Other tensioners are only equipped with friction type retardation.

Synonyms: Retarder, tensioner.

Traveler—A sheave complete with suspension arm or frame used separately or in groups and suspended from structures to permit the stringing of conductors. These devices are sometimes bundled with a center drum or sheave, and another traveler, and used to string more than one conductor simultaneously. For protection of conductors that should not be nicked or scratched, the sheaves are often lined with nonconductive or semiconductive neoprene or with nonconductive urethane. Any one of these materials acts as a padding or cushion for the conductor as it passes over the sheave. Traveler grounds must be used with lined travelers in order to establish an electrical ground.

Synonyms: Block, dolly, sheave, stringing block, stringing sheave, stringing traveler.

Winder reel—A device designed to serve as a recovery unit for a pulling line. It is normally equipped with its own engine which drives a supporting shaft for a reel mechanically, hydraulically, or through a combination of both. The shaft, in turn, drives the reel. It is normally used to rewind a pulling line as it leaves the bullwheel puller during stringing operations. This unit is not intended to serve as a puller, but sometimes serves this function where only low tensions are involved.

Synonyms: Take-up reel.

References

American Society of Civil Engineers Standard, *Minimum Design Loads for Buildings and Other Structures*, ASCE 7–88, New York, 1995.

Cahill, T., Development of low-creep ACSR conductor, *Wire Journal*, July 1973.

Ehrenburg, D.O., Transmission line catenary calculations, AIEE Paper, Committee on power transmission and distribution, July 1935.

Fink, D.G. and Beaty, H.W., *Standard Handbook for Electrical Engineers*, 13th edn., McGraw-Hill, New York, 1993.

Graphic Method for Sag Tension Calculations for ACSR and Other Conductors, Aluminum Company of America, Pittsburgh, PA, 1961.

IEEE Standard 524–1993, *IEEE Guide to the Installation of Overhead Transmission Line Conductors*, IEEE, New York, 1993.

National Electrical Safety Code, 2007 edition.

Overend, P.R. and Smith, S., *Impulse Time Method of Sag Measurement*, American Society of Civil Engineers, Reston, VA, 1986.

Stress-Strain-Creep Curves for Aluminum Overhead Electrical Conductors, Aluminum Association, Washington, DC, 1974.

Winkelman, P.F., Sag-tension computations and field measurements of Bonneville power administration, AIEE Paper 59-900, June 1959.

16

Corona and Noise

Giao N. Trinh
(retired)
Hydro-Québec Institute of Research

Modern electric power systems are often characterized by generating stations located far away from the consumption centers, with long overhead transmission lines to transmit the energy from the generating sites to the load centers. From the few tens of kilovolts in the early years of the twentieth century, the line voltage has reached the extra-high voltage (EHV) levels of 800 kV AC (Lacroix and Charbonneau, 1968) and 500 kV DC (Bateman et al., 1969) in the 1970s, and touched the ultrahigh voltage (UHV) levels of 1200 kV AC (Bortnik et al., 1988) and 600 kV DC (Krishnayya et al., 1988). Although overhead lines operating at high voltages are the most economical means of transmitting large amounts of energy over long distances, their exposure to atmospheric conditions constantly alters the surface conditions of the conductors and causes large variations in the corona activities on the line conductors.

Corona discharges follow an electron avalanche process whereby neutral molecules are ionized by electron impacts under the effect of the applied field (Raether, 1964). Since air is a particular mixture of nitrogen (79%), oxygen (20%), and various impurities, the discharge is significantly conditioned by the electronegative nature of oxygen molecules, which can readily capture free electrons to form negative ions and thus hamper the electron avalanche process (Loeb, 1965). Several modes of corona discharge can be distinguished; and while all corona modes produce energy losses, the streamer discharges also generate electromagnetic interference (RI), and audible noise (AN) in the immediate vicinity of high-voltage (HV) lines (Trinh and Jordan, 1968; Trinh, 1995a,b). These parameters are currently used to evaluate the corona performance of conductor bundles and to predict the energy losses and environmental impact of HV lines before their installation.

Adequate control of line corona is obtained by controlling the surface gradient at the line conductors. The introduction of bundled conductors by Whitehead in 1910 has greatly influenced the development of HV lines to today's EHVs (Whitehead, 1910). In effect, HV lines as we know them today would not exist without the bundled conductors. This chapter reviews the physical processes leading to the development of corona discharges on the line conductors and presents the current practices in selecting the line conductors.

16.1 Corona Modes (Trinh and Jordan, 1968; Trinh, 1995a)

In a nonuniform field gap in atmospheric air, corona discharges can develop over a whole range of voltages in a small region near the highly stressed electrode before the gap breaks down. Several criteria have been developed for the onset of corona discharge, the most familiar being the streamer criterion. They are all related to the development of an electron avalanche in the gas gap and can be expressed as

$$1-\gamma\exp\left[\int(\alpha-\eta)\mathrm{d}x\right]=0 \tag{16.1}$$

where

α and η are, respectively, the ionization and attachment coefficients in air
$\alpha - \eta$ is the net coefficient of ionization by electron impact of the gas
γ is a coefficient representing the efficiency of secondary processes in maintaining the ionization activities in the gap

The net coefficient of ionization varies with the distance x from the highly stressed electrode and the integral is evaluated for values of x where α' is positive.

A physical meaning may be given to the corona onset criteria given earlier. The onset conditions can be rewritten as

$$\exp\left[\int(\alpha-\eta)\mathrm{d}x\right]=\frac{1}{\gamma} \tag{16.2}$$

The left-hand side represents the avalanche development from a single electron and $1/\gamma$ the critical size of the avalanche to assure the stable development of the discharge.

The nonuniform field necessary for the development of corona discharges and the electronegative nature of air favor the formation of negative ions during the discharge development. Due to their relatively slow mobility, ions of both polarities from several consecutive electron avalanches accumulate in the low-field region of the gap and form ion space charges. To properly interpret the development of corona discharges, account must be taken of the active role of these ion space charges, which continuously modify the local field intensity and, hence, the development of corona discharges according to their relative buildup and removal from the region around the highly stressed electrode.

16.1.1 Negative Corona Modes

When the highly stressed electrode is at a negative potential, electron avalanches are initiated at the cathode and develop toward the anode in a continuously decreasing field. Referring to Figure 16.1, the nonuniformity of the field distribution causes the electron avalanche to stop at the boundary surface S_0, where the net ionization coefficient is zero, that is, $\alpha = \eta$. Since free electrons can move much faster than ions under the influence of the applied field, they concentrate at the avalanche head during its progression. A concentration of positive ions thus forms in the region of the gap between the cathode and the boundary surface, while free electrons continue to migrate across the gap. In air, free electrons rapidly attach themselves to oxygen molecules to form negative ions, which, because of the slow drift velocity, start to accumulate in the region of the gap beyond S_0. Thus, as soon as the first electron avalanche has developed, there are two ion space charges in the gap.

The presence of these space charges increases the field near the cathode, but it reduces the field intensity at the anode end of the gap. The boundary surface of zero ionization activity is therefore displaced toward the cathode. The subsequent electron avalanche develops in a region of slightly higher field intensity but covers a shorter distance than its predecessor. The influence of the ion space charge is such that it actually conditions the development of the discharge at the highly stressed electrode, producing three

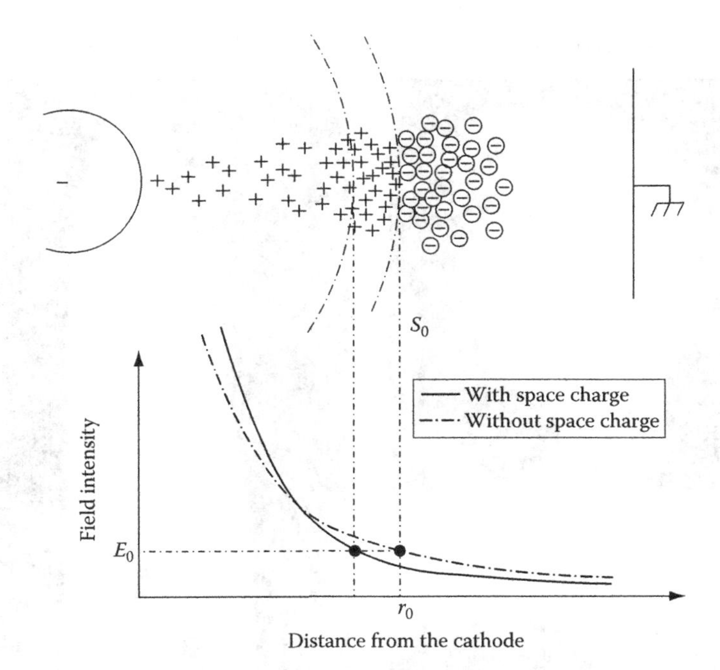

FIGURE 16.1 Development of an electron avalanche from the cathode. (From Trinh, N.G., *IEEE Electr. Insul. Mag.*, 11, 23, 1995a.)

modes of corona discharge with distinct electrical, physical, and visual characteristics (Figure 16.2). These are, respectively, with increasing field intensity: Trichel streamer, negative pulseless glow, and negative streamer. An interpretation of the physical mechanism of different corona modes is given in the following.

16.1.1.1 Trichel Streamer

Figure 16.2a shows the visual aspect of the discharge; its current and light characteristics are shown in Figure 16.3. The discharge develops along a narrow channel from the cathode and follows a regular pattern in which the streamer is initiated, developed, and suppressed; a short dead time follows before the cycle is repeated. The duration of an individual streamer is very short, a few tens of nanoseconds, while the dead time varies from a few microseconds to a few milliseconds, or even longer. The resulting discharge current consists of regular negative pulses of small amplitude and short duration, succeeding one another at the rate of a few thousand pulses per second. A typical Trichel current pulse is shown in Figure 16.3a where, it should be noted, the wave shape is somewhat influenced by the time constant of the measuring circuit. The discharge duration may be significantly shorter, as depicted by the light pulse shown in Figure 16.3c.

The development of Trichel streamers cannot be explained without taking account of the active roles of the ion space charges and the applied field. The streamer is initiated from the cathode by a free electron. If the corona onset conditions are met, the secondary emissions are sufficient to trigger new electron avalanches from the cathode and maintain the discharge activity. During the streamer development, several generations of electron avalanches are initiated from the cathode and propagate along the streamer channel. The avalanche process also produces two ion space charges in the gap, which gradually moves the boundary surface S_0 closer to the cathode. The positive ion cloud thus finds itself compressed at the cathode and, in addition, is partially neutralized at the cathode and by the negative ions produced in subsequent avalanches. This results in a net negative ion space charge, which eventually reduces the local field intensity at the cathode below the onset field and suppresses the discharge. The dead time is a period during which the remaining ion space charges are dispersed by the applied field. A new streamer will develop when the space charges in the immediate surrounding of the cathode have been cleared to a sufficient extent.

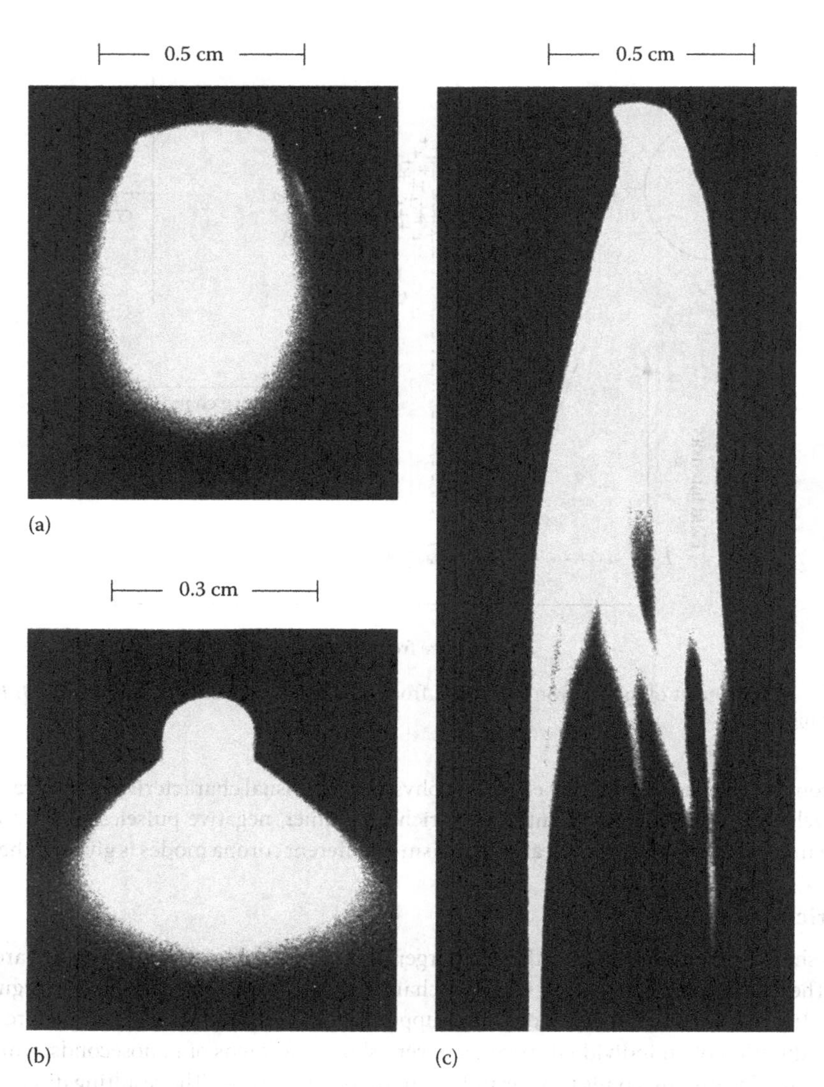

FIGURE 16.2 Corona modes at cathode: (a) Trichel streamers; (b) negative pulseless glow; (c) negative streamers. Cathode: spherical protrusion ($d = 0.8$ cm) on a sphere ($D = 7$ cm); gap 19 cm; time exposure 1/4 s. (From Trinh, N.G. and Jordan, I.B., *IEEE Trans.*, PAS-87, 1207, 1968; From Trinh, N.G., *IEEE Electr. Insul. Mag.*, 11, 23, 1995a. With permission.)

This mechanism depends on a very active electron attachment process to suppress the ionization activity within a few tens of nanoseconds following the beginning of the discharge. The streamer repetition rate is essentially a function of the removal rate of ion space charges by the applied field, and generally shows a linear dependence on the applied voltage. However, at high fields a reduction in the pulse repetition rate may be observed, which corresponds to the transition to a new corona mode.

16.1.1.2 Negative Pulseless Glow

The negative pulseless glow mode is characterized by a pulseless discharge current. As indicated by the well-defined visual aspect of the discharge (Figure 16.2b), the discharge itself is particularly stable, which shows the basic characteristics of a miniature glow discharge. Starting from the cathode, a cathode dark space can be distinguished, followed by a negative glow region, a Faraday dark space, and, finally, a positive column of conical shape. As with low-pressure glow discharges, these features of the pulseless glow discharge result from very stable conditions of electron emission from the cathode by

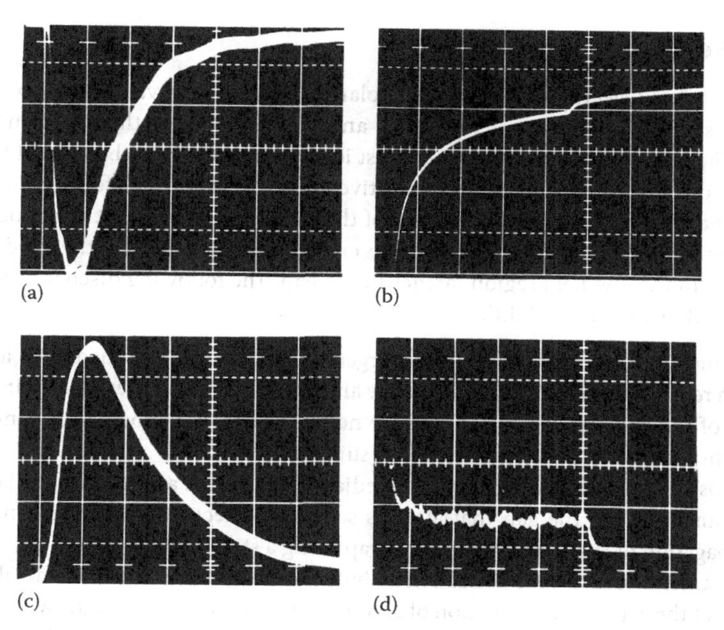

FIGURE 16.3 Current and light characteristics of Trichel streamer. Cathode: spherical protrusion ($d = 0.8$ cm) on a sphere ($D = 7$ cm); gap 19 cm. Scales: (a) current 350 μA/div., 50 ns/div. (b) 50 μA/div., 2 μs/div. Light: (c) 0.5 V/div., 20 ns/div. (d) 0.2 V/div., 2 μs/div. (From Trinh, N.G. and Jordan, I.B., *IEEE Trans.*, PAS-87, 1207, 1968; From Trinh, N.G., *IEEE Electr. Insul. Mag.*, 11, 23, 1995a.)

ionic bombardment. The electrons, emitted with very low kinetic energy, are first propelled through the cathode dark space, where they acquire sufficient energy to ionize the gas, and intensive ionization occurs at the negative glow region. At the end of the negative glow region, the electrons lose most of their kinetic energy and are again accelerated across the Faraday dark space before they can ionize the gas atoms in the positive column. The conical shape of the positive column is attributed to the diffusion of the free electrons in the low-field region.

These stable discharge conditions may be explained by the greater efficiency of the applied field in removing the ion space charges at higher field intensities. Negative ion space charges cannot build up sufficiently close to the cathode to effectively reduce the cathode field and suppress the ionization activities there. This interpretation of the discharge mechanism is further supported by the existence of a plateau in the Trichel streamer current and light pulses (Figure 16.3), which indicates that an equilibrium state exists for a short time between the removal and the creation of the negative ion space charge. It has been shown (Trinh and Jordan, 1970) that the transition from the Trichel streamer mode to the negative pulseless glow corresponds to an indefinite prolongation in time of one such current plateau.

16.1.1.3 Negative Streamer

If the applied voltage is increased still further, negative streamers may be observed, as illustrated in Figure 16.2c. The discharge possesses essentially the same characteristics observed in the negative pulseless glow discharge but here the positive column of the glow discharge is constricted to form the streamer channel, which extends farther into the gap. The glow discharge characteristics observed at the cathode imply that this corona mode also depends largely on electron emissions from the cathode by ionic bombardment, while the formation of a streamer channel characterized by intensive ionization denotes an even more effective space charge removal action by the applied field. The streamer channel is fairly stable. It projects from the cathode into the gap and back again, giving rise to a pulsating fluctuation of relatively low frequency in the discharge current.

16.1.2 Positive Corona Modes

When the highly stressed electrode is of positive polarity, the electron avalanche is initiated at a point on the boundary surface S_0 of zero net ionization and develops toward the anode in a continuously increasing field (Figure 16.4). As a result, the highest ionization activity is observed at the anode. Here again, due to the lower mobility of the ions, a positive ion space charge is left behind along the development path of the avalanche. However, because of the high field intensity at the anode, few electron attachments occur and the majority of free electrons created are neutralized at the anode. Negative ions are formed mainly in the low-field region farther in the gap. The following discharge behavior may be observed (Trinh and Jordan, 1968; Trinh, 1995a):

- The incoming free electrons are highly energetic and cannot be immediately absorbed by the anode. As a result, they tend to spread over the anode surface where they lose their energy through ionization of the gas particles, until they are neutralized at the anode, thus contributing to the development of the discharge over the anode surface.
- Since the positive ions are concentrated immediately next to the anode surface, they may produce a field enhancement in the gap that attracts secondary electron avalanches and promotes the radial propagation of the discharge into the gap along a streamer channel.
- During streamer discharge, the ionization activity is observed to extend considerably into the low-field region of the gap via the formation of corona globules, which propagate owing to the action of the electric field generated by their own positive ion space charge. Dawson (1965) has shown that if a corona globule is produced containing 10^8 positive ions within a spherical volume of 3×10^{-3} cm in radius, the ion space charge field is such that it attracts sufficient new electron avalanches to create a new corona globule a short distance away. In the meantime, the initial corona globule is neutralized, causing the corona globule to effectively move ahead toward the cathode.

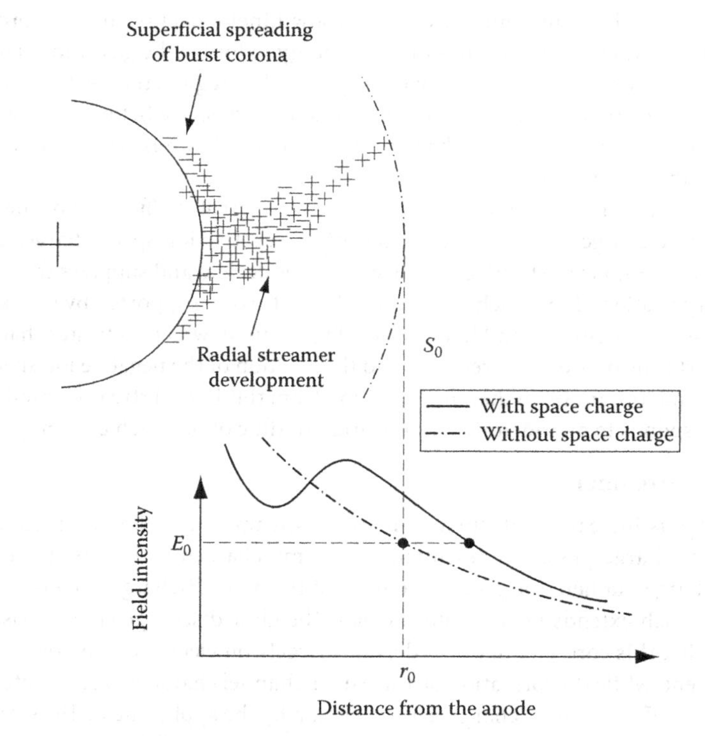

FIGURE 16.4 Development of an electron avalanche toward the anode. (From Trinh, N.G., *IEEE Electr. Insul. Mag.*, 11, 23, 1995a. With permission.)

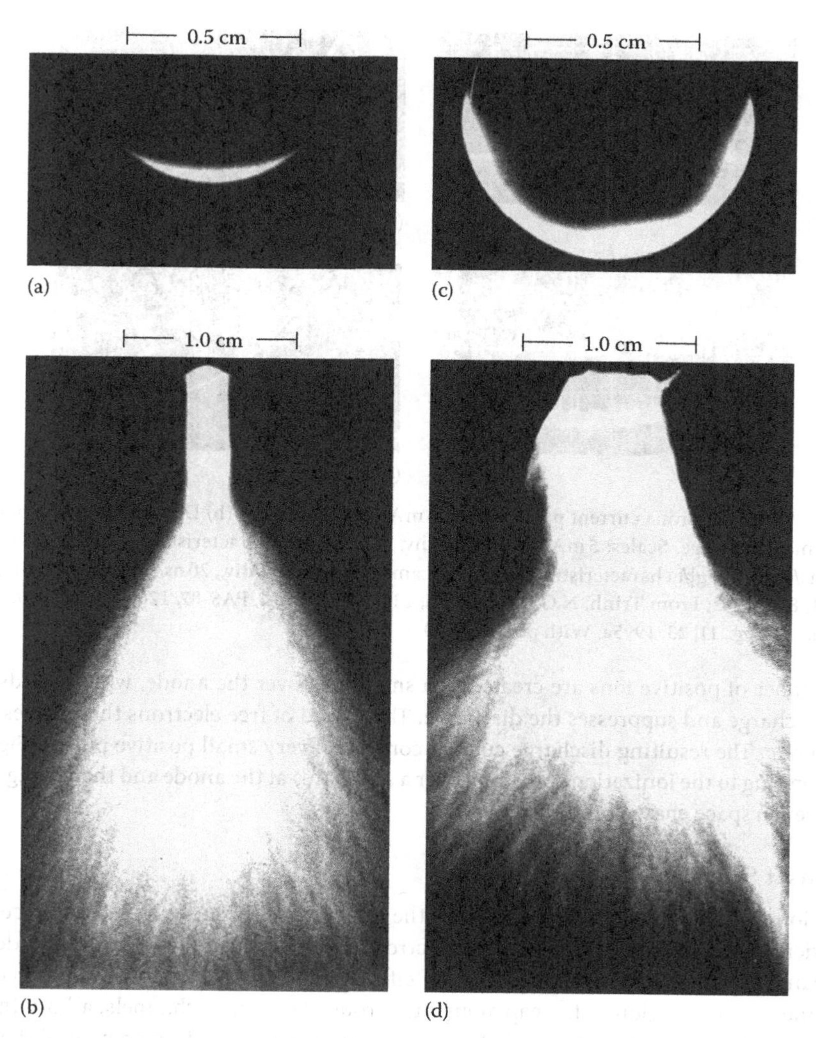

FIGURE 16.5 Corona modes at anode: (a) burst corona; (b) onset streamers; (c) positive glow corona; (d) breakdown streamers. Anode spherical protrusion (d = 0.8 cm) on a sphere (D = 7 cm); gap 35 cm; time exposure 1/4 s. (From Trinh, N.G. and Jordan, I.B., *IEEE Trans.*, PAS-87, 1207, 1968; From Trinh, N.G., *IEEE Electr. Insul. Mag.*, 11, 23, 1995a.)

The presence of ion space charges of both polarities in the anode region greatly affects the local distribution of the field, and, consequently, the development of corona discharge at the anode. Four different corona discharge modes having distinct electrical, physical, and visual characteristics can be observed at a highly stressed anode, prior to flashover of the gap. These are, respectively, with increasing field intensity (Figure 16.5): burst corona, onset streamers, positive glow, and breakdown streamers. An interpretation of the physical mechanisms leading to the development of these corona modes is given in the following.

16.1.2.1 Burst Corona

The burst corona appears as a thin luminous sheath adhering closely to the anode surface (Figure 16.5a). The discharge results from the spread of ionization activities at the anode surface, which allows the high-energy incoming electrons to lose their energy before neutralization at the anode. During this

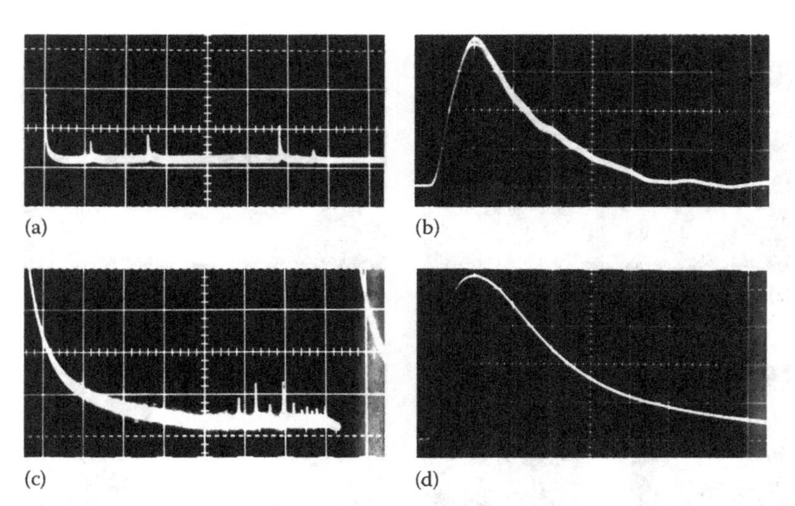

FIGURE 16.6 (a) Burst corona current pulse. Scales: 5 mA/div., 0.2 ms/div. (b) Development of burst corona following a streamer discharge. Scales: 5 mA/div., 0.2 ms/div. (c) Current characteristics of onset streamers. Scales: 7 mA/div., 50 ns/div. (d) Light characteristics of onset streamers. Scales: 1 V/div., 20 ns/div. (From Juette, G.W., *IEEE Trans.*, PAS-91, 865, 1972; From Trinh, N.G. and Jordan, I.B., *IEEE Trans.*, PAS-87, 1207, 1968; From Trinh, N.G., *IEEE Electr. Insul. Mag.*, 11, 23, 1995a. With permission.)

process, a number of positive ions are created in a small area over the anode, which builds up a local positive space charge and suppresses the discharge. The spread of free electrons then moves to another part of the anode. The resulting discharge current consists of very small positive pulses (Figure 16.6a), each corresponding to the ionization spreading over a small area at the anode and then being suppressed by the positive ion space charge produced.

16.1.2.2 Onset Streamer

The positive ion space charge formed adjacent to the anode surface causes a field enhancement in its immediate vicinity, which attracts subsequent electron avalanches and favors the radial development of onset streamers. This discharge mode is highly effective and the streamers are observed to extend farther into the low-field region of the gap along numerous filamentary channels, all originating from a common stem projecting from the anode (Figure 16.5b). During this development of the streamers, a considerable number of positive ions are formed in the low-field region. As a result of the cumulative effect of the successive electron avalanches and the absorption at the anode of the free electrons created in the discharge, a net residual positive ion space charge forms in front of the anode. The local gradient at the anode then drops below the critical value for ionization and suppresses the streamer discharge. A dead time is consequently required for the applied field to remove the ion space charge and restore the proper conditions for the development of a new streamer. The discharge develops in a pulsating mode, producing a positive current pulse of short duration, high amplitude, and relatively low repetition rate due to the large number of ions created in a single streamer (Figure 16.6c and d).

It has been observed that these first two discharge modes develop in parallel over a small range of voltages following corona onset. As the voltage is increased, the applied field rapidly becomes more effective in removing the ion space charge in the immediate vicinity of the electrode surface, thus promoting the lateral spread of burst corona at the anode. In fact, burst corona can be triggered just a few microseconds after suppression of the streamer (Figure 16.6b). This behavior can be explained by the rapid clearing of the positive ion space charge at the anode region, while the incoming negative ions encounter a high-enough gradient to shed their electrons, thus providing the seeding free electrons to initiate new avalanches and sustain the ionization activity over the anode surface in the form of burst corona. The latter will continue to develop until it is again suppressed by its own positive space charge.

As the voltage is raised even higher, the burst corona is further enhanced by a more effective space charge removal action of the field at the anode. During the development of the burst corona, positive ions are created and rapidly pushed away from the anode. The accumulation of positive ions in front of the anode results in the formation of a stable positive ion space charge that prevents the radial development of the discharge into the gap. Consequently, the burst corona develops more readily, at the expense of the onset streamer, until the latter is completely suppressed. A new mode, the positive glow discharge, is then established at the anode.

16.1.2.3 Positive Glow

A photograph of a positive glow discharge developing at a spherical protrusion is presented in Figure 16.5. This discharge is due to the development of the ionization activity over the anode surface, which forms a thin luminous layer immediately adjacent to the anode surface, where intense ionization activity takes place. The discharge current consists of a direct current (DC) superimposed by a small pulsating component with a high repetition rate, in the hundreds of kilohertz range. By analyzing the light signals obtained with photomultipliers pointing to different regions of the anode, it may be found that the luminous sheath is composed of a stable central region, from there, bursts of ionization activity may develop and project the ionizing sheath outward and back again, continuously, giving rise to the pulsating current component.

The development of the positive glow discharge may be interpreted as resulting from a particular combination of removal and creation of positive ions in the gap. The field is high enough for the positive ion space charge to be rapidly removed from the anode, thus promoting surface ionization activity. Meanwhile, the field intensity is not sufficient to allow radial development of the discharge and the formation of streamers. The main contribution of the negative ions is to supply the necessary triggering electrons to sustain ionization activity at the anode.

16.1.2.4 Breakdown Streamer

If the applied voltage is further increased, streamers are again observed and they eventually lead to breakdown of the gap. The development of breakdown streamers is preceded by local streamer spots of intense ionization activity, which may be seen moving slowly over the anode surface. The development of streamer spots is not accompanied by any marked change in the current or the light signal. Only when the applied field becomes sufficiently high to rapidly clear the positive ion space charges from the anode region does radial development of the discharge become possible, resulting in breakdown streamers.

Positive breakdown streamers develop more and more intensively with higher applied voltage and eventually cause the gap to break down. The discharge is essentially the same as the onset streamer type but can extend much farther into the gap. The streamer current is more intense and may occur at a higher repetition rate. A streamer crossing the gap does not necessarily result in gap breakdown, which proves that the filamentary region of the streamer is not fully conducting.

16.1.3 AC Corona

When alternating voltage is used, the gradient at the highly stressed electrode varies continuously, both in intensity and polarity. Different corona modes can be observed in the same cycle of the applied voltage. Figure 16.7 illustrates the development of different corona modes at a spherical protrusion as a function of the applied voltage. The corona modes can be readily identified by the discharge current. The following observations can be made:

- For short gaps, the ion space charges created in one half-cycle are absorbed by the electrodes in the same half-cycle. The same corona modes that develop near onset voltages can be observed, namely, negative Trichel streamers, positive onset streamers, and burst corona.

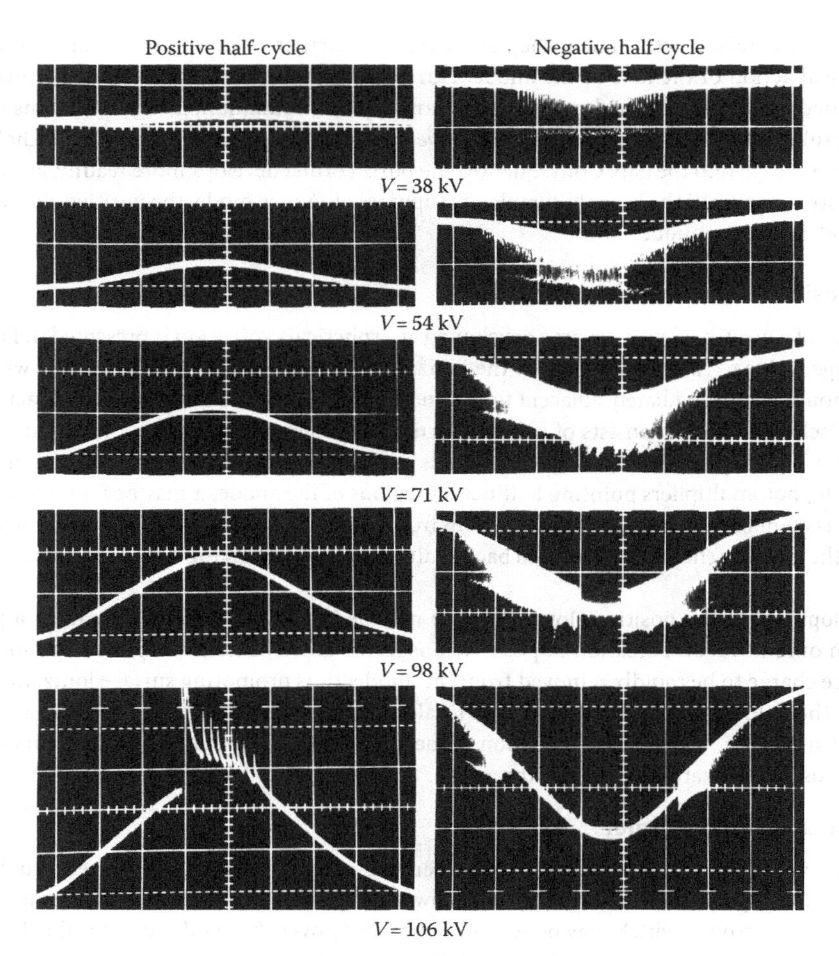

FIGURE 16.7 Corona modes under AC voltage. Electrode: conical protrusion (θ = 30°) on a sphere (D = 7 cm); gap 25 cm; R = 10 kΩ. Scales: 50 μA/div., 1.0 ms/div. (From Trinh, N.G. and Jordan, I.B., *IEEE Trans.*, PAS-87, 1207, 1968; From Trinh, N.G., *IEEE Electr. Insul. Mag.*, 11, 23, 1995a.)

- For long gaps, the ion space charges created in one half-cycle are not completely absorbed by the electrodes, leaving residual space charges in the gap. These residual space charges are drawn back to the region of high field intensity in the following half-cycle and can influence discharge development. Onset streamers are suppressed in favor of the positive glow discharge. The following corona modes can be distinguished: negative Trichel streamers, negative glow discharge, positive glow discharge, and positive breakdown streamers.
- Negative streamers are not observed under AC voltage, owing to the fact that their onset gradient is higher than the breakdown voltage that occurs during the positive half-cycle.

16.2 Main Effects of Corona Discharges on Overhead Lines (Trinh, 1995b)

The impact of corona discharges on the design of HV lines has been recognized since the early days of electric power transmission when the corona losses (CLs) were the limiting factor. Even today, CLs remain critical for HV lines below 300 kV. With the development of EHV lines operating at voltages between 300 and 800 kV, RIs become the designing parameters. For UHV lines operating at voltages above 800 kV, the AN appears to gain in importance over the other two parameters. The physical mechanisms of these effects—CLs, RI, and AN—and their current evaluation methods are discussed later.

16.2.1 Corona Losses

The movement of ions of both polarities generated by corona discharges, and subjected to the applied field around the line conductors, is the main source of energy loss. For AC lines, the movement of the ion space charges is limited to the immediate vicinity of the line conductors, corresponding to their maximum displacement during one half-cycle, typically a few tens of centimeters, before the voltage changes polarity and reverses the ionic movement. For DC lines, the ion displacement covers the whole distance separating the line conductors, and between the conductors and the ground.

CLs are generally described in terms of the energy losses per kilometer of the line. They are generally negligible under fair-weather conditions but can reach values of several hundreds of kilowatts per kilometer of line during foul weather. Direct measurement of CLs is relatively complex, but foul-weather losses can be readily evaluated in test cages under artificial rain conditions, which yield the highest energy loss. The results are expressed in terms of the generated loss W, a characteristic of the conductor to produce CLs under given operating conditions.

16.2.2 Electromagnetic Interference

Electromagnetic interference is associated with streamer discharges that inject current pulses into the conductor. These pulses of steep front and short duration have a high harmonic content, reaching the tens of megahertz range, as illustrated in Figure 16.8, which shows the typical frequency spectra associated with various streamer modes (Juette, 1972). A tremendous research effort was devoted to the subject during the years 1950–1980 in an effort to evaluate the RI from HV lines. The most comprehensive contributions were made by Moreau and Gary (1972a,b) of Électricité de France, who introduced the concept of the excitation function, $\Gamma(\omega)$, which characterizes the ability of a line conductor to generate RI under the given operating conditions.

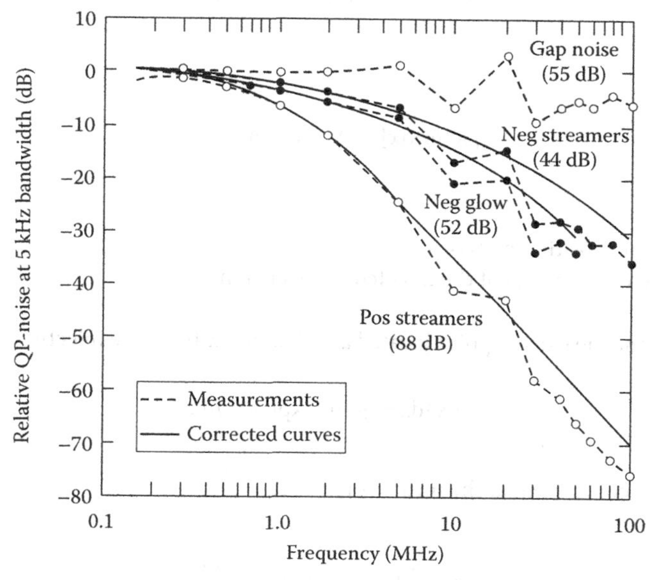

FIGURE 16.8 Relative frequency spectra for different noise types. (From Trinh, N.G., *IEEE Electr. Insul. Mag.*, 11, 5, 1995b; From Juette, G.W., *IEEE Trans.*, PAS-91, 865, 1972.)

Consider first the case of a single-phase line, where the contribution to the RI at the measuring frequency, ω, from corona discharges developing at a section dx of the conductor is

$$j0(\omega) = \frac{C}{2\pi\varepsilon 0}\Gamma(\omega)\,\mathrm{d}x \tag{16.3}$$

where C is the capacitance per unit length of the line conductor to ground.

Upon injection, the discharge current pulse splits itself into two identical current pulses of half amplitude propagating in opposite directions away from the discharge site. At a point of observation located at a distance x along the line from the discharge site, the noise current is distorted according to

$$i(\omega,x)\,\mathrm{d}x = i0(\omega)\exp(-\gamma x)\,\mathrm{d}x = i0(\omega)\exp(-\alpha x)\,\mathrm{d}x \tag{16.4}$$

where γ represents the propagation constant, which can be approximated by its real component α.

The total noise current circulating in the line conductor is the sum of all contributions from the corona discharges along the conductor and is given by

$$I(\omega) = \sqrt{\int_{-\infty}^{\infty}\left[i(\omega,x)\right]^2\mathrm{d}x} = \frac{i0(\omega)}{\sqrt{\alpha}} \tag{16.5}$$

For a multiphase line, because of the high-frequency nature of the noise current, the calculation of the interference field must take account of the mutual coupling among the conductors, which further complicates the process (Gary, 1972; Moreau and Gary, 1972a,b). Modal analysis provides a convenient means of evaluating the noise currents on the line conductors. In this approach, the noise currents are first transposed into their modal components, which propagate without distortion along the line conductors at their own velocity according to the relation

$$[i0(\omega)\,\mathrm{d}x] = [M][j0(\omega)\,\mathrm{d}x] \tag{16.6}$$

Consequently,

$$[j0(\omega)\,\mathrm{d}x] = [M]^{-1}[i0(\omega)\,\mathrm{d}x] \tag{16.7}$$

where

[M] is the modal transposition matrix
$j0(\omega)$ are the modal components of the injected noise current

The modal current at the measuring point located at a distance x from the injection point is

$$j(\omega,x)\,\mathrm{d}x = j0(\omega)\exp(-\alpha x)\,\mathrm{d}x \tag{16.8}$$

and the modal current component at the measuring point is

$$J(\omega) = \sqrt{\int_{-\infty}^{\infty}[j(\omega,x)]^2\,\mathrm{d}x} = \frac{j0(\omega)}{\sqrt{\alpha}} \tag{16.9}$$

or, in a general way

$$[J(\omega)] = \frac{1}{[\sqrt{\alpha}]}[j0(\omega)] = \frac{1}{[\sqrt{\alpha}]}[M]^{-1}[i0(\omega)] \tag{16.10}$$

Finally, the line current can be obtained from

$$[I(\omega)] = [M][J(\omega)] \tag{16.11}$$

The magnetic and electric fields produced by the noise currents in the line conductors can then be evaluated for assessment of the RIs. Moreau and Gary (1972a,b) obtained good agreement between calculated and experimental results with the symmetrical modes of Clarke for the modal transposition:

$$[M] = \begin{bmatrix} 1/\sqrt{6} & 1/2 & 1/\sqrt{3} \\ -2/\sqrt{6} & 0 & 1/\sqrt{3} \\ 1/\sqrt{6} & -1/2 & 1/\sqrt{3} \end{bmatrix} \tag{16.12}$$

The attenuation coefficients at 0.5 MHz are 11.1, 54, and 342 Np/m for the modal currents: and the magnetic ground was assumed to be located at a depth equal to the penetration depth of the magnetic field, dp, as defined by

$$dp = 2\sqrt{\frac{p}{\mu\omega}} \tag{16.13}$$

For a typical soil resistivity of 100 Ωm and a measuring frequency of 0.5 MHz, the depth of the magnetic ground is equal to 7.11 m. It is equal to 5.03 m at a measuring frequency of 1.0 MHz.

Circulation of the noise current in the line conductor effectively generates an RI field around the conductors, which is readily picked up by any radio or television receiver located in the vicinity of the HV line. The current practices characterize the interference field in terms of its electric component, $E(\omega)$, expressed in decibels (dB) above a reference level of 1 μV/m. Evaluation of the RI is usually made by first calculating the magnetic interference field $H(\omega)$ at the measuring point

$$H(\omega) = \sum_j \frac{1}{2\pi r_j} I_j(\omega) ar \tag{16.14}$$

The summation was made with respect to the number of phase conductors of the lines and their images with respect to the magnetic ground. The electric interference field can next be related to the magnetic interference field according to

$$E(\omega) = \sqrt{\frac{\mu 0}{\varepsilon 0}} H(\omega) \tag{16.15}$$

16.2.2.1 Television Interference

The frequency spectrum of corona discharges has cut-off frequencies around a few tens of megahertz. As a result, the interference levels at the television frequencies are very much attenuated. In fact, gap discharges, which generate sharp current pulse with nanosecond rise times, are the principal discharges that effectively interfere with the television reception. These discharges are produced by loose connections, a problem common on low-voltage distribution lines but rarely observed on HV transmission lines. Another source of interference is related to reflections of television signals at HV line towers, producing ghost images. However, the problem is not related in any way to corona activities on the line conductors (Juette, 1972).

16.2.3 Audible Noise

The high temperature in the discharge channel produced by the streamer creates a corresponding increase in the local air pressure. Consequently, a pulsating sound wave is generated from the discharge site, propagates through the surrounding ambient air, and is perfectly audible in the immediate vicinity of the HV lines. The typical octave-band frequency spectra of line corona in Figure 16.9 contain discrete components corresponding to the second and higher harmonics of the line voltage superimposed on a relatively broadband noise, extending well into the ultrasonic range (Ianna et al., 1974). The octave-band measurements in this figure show a sharp drop at frequencies over 20 kHz, due principally to the limited frequency response of the microphone and associated sound-level meter.

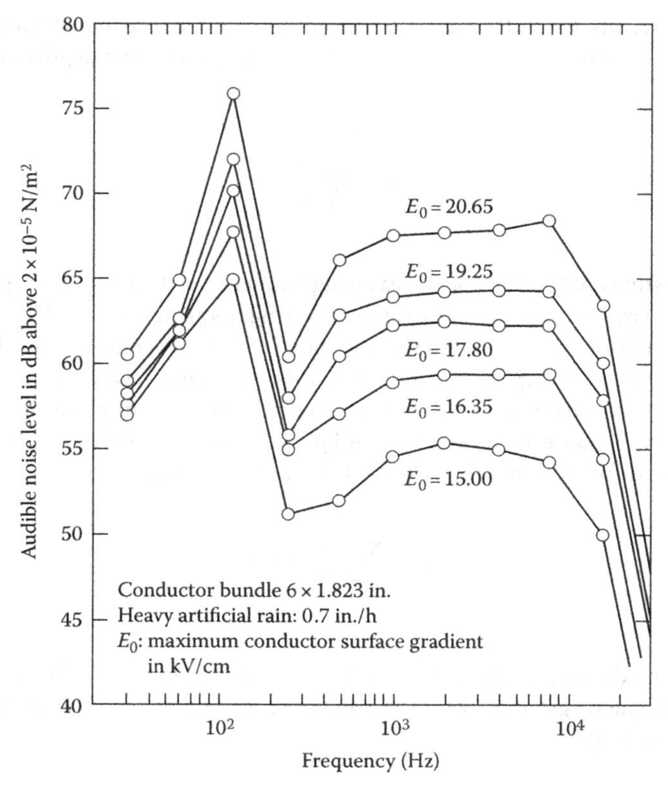

FIGURE 16.9 Octave-band frequency spectra of line corona audible noise at 10 m from the conductor. (From Trinh, N.G., *IEEE Electr. Insul. Mag.*, 11, 5, 1995b; From Trinh, N.G. and Maruvada, P.S., *IEEE Trans.*, PAS-96, 312, 1977. With permission.)

Similar to the case of RI, the ability of the line conductors to produce AN is characterized by the generated acoustic power density A, defined as the acoustic power produced per unit length of the line conductor under specific operating conditions. The acoustic power generated by corona discharges developing in a portion dx of the conductor is then

$$dA = A\,dx \tag{16.16}$$

Its contribution to the acoustic intensity at a measuring point located at a distance r from the discharge site is

$$dI = \frac{A}{4\pi r^2}dx \tag{16.17}$$

The acoustic intensity at the measuring point is the sum of all contributions from corona discharge distributed along the conductor:

$$I(R) = 2A\int_{-\infty}^{\infty} \frac{1}{4\pi(R^2 + x^2)}dx = \frac{A}{2R} \tag{16.18}$$

where
 R is the distance from the measuring point to the conductor
 the integral is evaluated in terms of the longitudinal distance x along the conductor

Finally, the acoustic intensity at the measuring point is the sum of the contributions from the different phase conductors of the line

$$I(R) = \sum_j I_j(R_j) \tag{16.19}$$

The sound pressure, usually expressed in terms of decibel (dBA) above a reference level of 2×10^{-5} N/m^2 is

$$p(r) = \sqrt{\rho 0 CI} \tag{16.20}$$

16.2.4 Example of Calculation

It is obvious from the preceding sections that the effects of corona discharges on HV lines—the CLs, the RIs, and AN—can be readily evaluated from the generated loss W, the excitation function $\Gamma(\omega)$, and the generated acoustic power density A of the conductor. The latter parameters are characteristics of the bundle conductor and are usually derived from tests in a test cage or on experimental line. An example calculation of the corona performance of an HV line is given in the following for the case of the Hydro-Québec's 735 kV lines under conditions of heavy rain. The line parameters are given in Table 16.1, together with the various corona-generated parameters taken from Trinh and Maruvada (1977). The calculation of the radio interference and AN levels will be made for a lateral distance of 15 m from the outer phase, i.e., at the limit of the right-of-way of the line.

Corona losses: The CLs are the sum of the losses generated at the three phases of the line, which amount to 127.63 kW/km.

TABLE 16.1 Hydro-Québec 735 kV Line

Distance between phase (m)	13.7	
Height of conductors (m)	19.8	
Number of subconductors	4	
Diameter of subconductor (cm)	3.05	
	Center phase	Outer phase
Electric field at the conductor surface (kVrms/cm)	19.79	18.46
Capacitance per unit length (pF/m)	10.57	
Generated loss W (W/m)	59.77	33.92
RI excitation function Γ (dB above $1\,\mu A/\sqrt{m}$)	43.52	39.59
Subconductor-generated acoustic power density A (dBA above 1 μW/m)	3.28	−0.24

Radio interference: The calculation of the radio interference requires that the noise current be first transformed into its modal components. Consider a noise current of unit excitation function $\Gamma a(\omega) = 1.0\,\mu A/\sqrt{m}$ circulating in phase A of the line. Because of the capacitive coupling, it induces currents to the other two phases of the line as well. For Hydro-Québec's 735 kV line, the capacitance matrix is

$$C = \begin{bmatrix} 11.204 & -2.241 & -0.73 \\ -2.241 & 11.605 & -2.241 \\ -0.73 & -2.241 & 11.204 \end{bmatrix}$$

and the noise current in phase A and its induced currents to phases B and C are

$$ia(\omega) = \begin{bmatrix} 11.204 \\ -2.241 \\ -0.73 \end{bmatrix}$$

The modal transformation using Equations 16.9 through 16.12 gives the following modal noise currents at the measuring point, taking into account the different attenuations of the modal currents:

$$Ja(\omega) = \begin{bmatrix} 16.472 & 10.321 & 2.31 \\ -30.497 & 0 & 1.998 \\ 16.472 & -10.321 & 2.31 \end{bmatrix}$$

These modal currents, once transformed back to the current mode, Equation 16.13, give the modal components of the noise currents flowing in the line conductors at the measuring point as related to the noise current injected to phase A:

$$Ia(\omega) = \begin{bmatrix} 6.725 & 7.298 & 1.333 \\ -13.449 & 0 & 1.333 \\ 6.725 & -7.298 & 1.333 \end{bmatrix}$$

These currents can then be used to calculate the magnetic and electric interference field using Equations 16.14 and 16.15:

$$Ha(\omega) = [0.0124 \quad 0.0449 \quad 0.0239]$$

$$Ea(\omega) = [4.674 \quad 16.938 \quad 9.017]$$

The corresponding electric interference level is 25.911 dB above 1 μV/m.

The aforementioned electric interference field and interference level are obtained assuming a noise excitation function of $1.0\mu A/\sqrt{m}$. For the case of interest, the excitation function at phase A is 39.59 dB and the corresponding interference level is 64.98 dB. By repeating the same process for the noise currents injected in phases B and C, one obtains effectively three sets of magnetic and electric field components generated by the circulation of the noise currents on the line conductors:

$$Eb(\omega) = [-8.653 \quad 0 \quad 7.80]$$

$$Ec(\omega) = [4.674 \quad -16.938 \quad 9.017]$$

Their contributions to the noise level are, respectively, 64.26 and 64.98 dB, resulting in a total noise level of 69.53 dB at the measuring point. The measuring frequency is 0.5 MHz.

Audible noise: Calculation of the AN is straightforward, since each phase of the line can be considered as an independent noise source. Consider the AN generated from phase A. The subconductor-generated acoustic power density is −0.24 dBA or 1.58×10^{-5} μW/m for the bundle conductor. The acoustic intensity at 15 m from the outer phase of the line as given by Equation 16.18 is 3.19×10^{-7} W/m^2 and the noise level is 55.14 dBA above 2×10^{-5} N/m^2.

By repeating the process for the other two phases of the line, the contributions to the acoustic intensity at the measuring point from the phases B and C of the line are 2.64×10^{-7} and 1.69×10^{-7} W/m^2, respectively, and the corresponding noise levels are 54.33 and 52.38 dBA. The total noise level is 58.87 dBA.

16.3 Impact on the Selection of Line Conductors

16.3.1 Corona Performance of HV Lines

Corona performance is a general term used to characterize the three main effects of corona discharges developing on the line conductors and their related hardware, namely, CLs, RI, and AN. All are sensitive to weather conditions, which dictate the corona activities. CLs can be described by a lump figure, which is equal to the total energy losses per kilometer of the line. Both the RI and the AN levels vary with the distance from the line and are best described by lateral profiles, which show the variations in the RI and AN level with the lateral distance from the line. Typical lateral profiles are presented in Figures 16.10 and 16.11 for a number of HV lines under foul-weather conditions. For convenience, the interference and noise levels at the edge of the right-of-way, typically 15 m from the outside phases of the line, are generally used to quantify the interference and noise level.

The time variations in the corona performance of HV lines is best described in terms of a statistical distribution, which shows the proportion of time that the energy losses, the RI, and AN exceed their specified levels. Figure 16.12 illustrates typical corona performances of Hydro-Québec's 735 kV lines as measured at the edge of the right-of-way. It can be seen that the RI and AN levels vary over wide ranges. In addition, the cumulative distribution curves show a typical inverted-S shape, indicating that the recorded data actually result from the combination of more than one population, usually associated with fair- and foul-weather conditions.

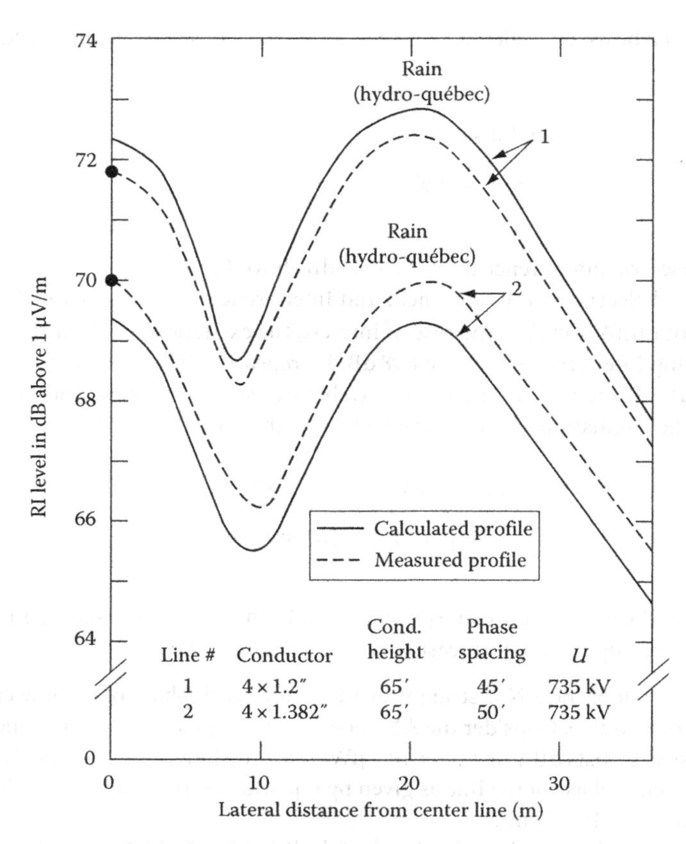

FIGURE 16.10 Comparison of calculated and measured RI performances of Hydro-Québec 735 kV lines at 1 MHz and using natural modes. (From Trinh, N.G., *IEEE Electr. Insul. Mag.*, 11, 5, 1995b; From Trinh, N.G. and Maruvada, P.S., *IEEE Trans.*, PAS-96, 312, 1977. With permission.)

DC coronas are less noisy than AC coronas. In effect, although DC lines can become very lossy during foul weather, the radio interference and ANs are significantly reduced. This behavior is related to the fact that water drops become elongated, remain stable, and produce glow corona modes rather than streamers in a DC field (Ianna et al., 1974).

16.3.2 Approach to Control the Corona Performance

The occurrence of corona discharges on line conductors is dictated essentially by the local field intensity, which, in turn, is greatly affected by the surface conditions, e.g., rugosity, water drops, snow, and ice particles, etc. For a smooth cylindrical conductor, the corona onset field is well described by Peek's experimental law

$$E_c = 30m\delta\left(1 + \frac{0.301}{\sqrt{\delta a}}\right) \tag{16.21}$$

where

E_c is the corona onset field
a is the radius of the conductor
m is an experimental factor to take account of the surface conditions

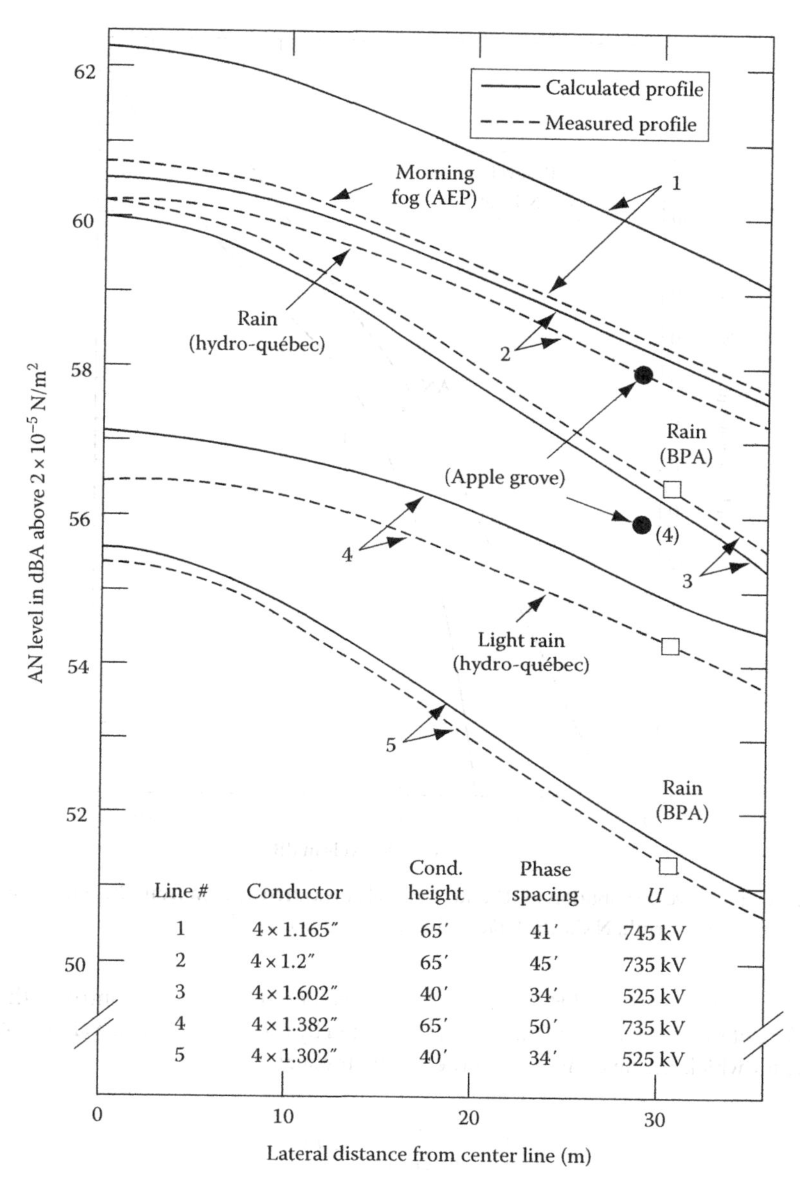

FIGURE 16.11 Comparison of calculated and measured AN performances of HV lines. (From Trinh, N.G., *IEEE Electr. Insul. Mag.*, 11, 5, 1995b; From Trinh, N.G. and Maruvada, P.S., *IEEE Trans.*, PAS-96, 312, 1977.)

Typical values of m are 0.8–0.9 for a dry-aged conductor, 0.5–0.7 for a conductor under foul-weather conditions, and δ is the relative air density factor.

The aforementioned corona onset condition emphasizes the great sensitivity of corona activities to the conductor surface condition and, hence, to changes in weather conditions. In effect, although the line voltage and the nominal conductor surface gradient remain constant, the surface condition factor varies continuously due to the exposure of line conductors to atmospheric conditions. The changes are particularly pronounced during foul weather as a result of the numerous discharge sites associated with water drops, snow, and ice particles deposited on the conductor surface.

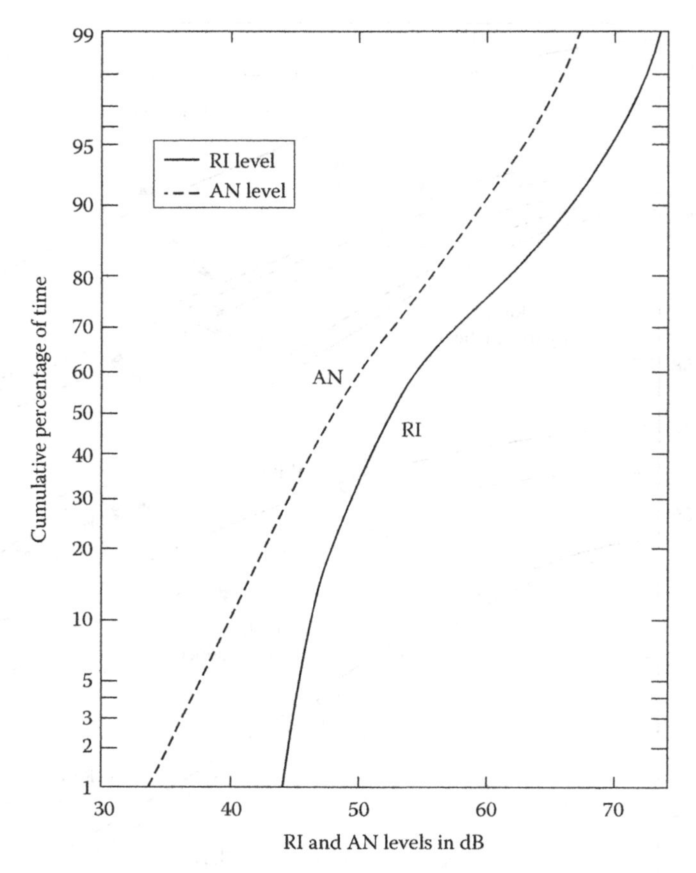

FIGURE 16.12 Cumulative distribution of RI and AN levels measured at 15 m from the outer phases of Hydro-Québec 735 kV lines. (From Trinh, N.G., *IEEE Electr. Insul. Mag.*, 11, 5, 1995b.)

Adequate corona performance of HV lines is generally achieved by a proper control of the field intensity at the surface of the conductor. It can be well illustrated by the simple case of a single-phase, single-conductor line for which the field intensity at the conductor surface is

$$E_0 = \frac{1}{\ln(2h/a)}\frac{U}{a} \leq Ec \tag{16.22}$$

It can be seen that the field intensity at the conductor surface is inversely proportional to its radius and, to a lesser extent, to the height of the conductor aboveground. By properly dimensioning the conductor, the field intensity at its surface can be kept below the fair-weather corona onset field for an adequate control of the corona activities and their undesirable effects.

With the single-conductor configuration, the size required for the conductor to be corona-free under fair-weather conditions is roughly proportional to the line voltage, and consequently will reach unrealistic values when the latter exceeds some 400 kV. Introduced in 1910 by Whitehead to increase the transmission capability of overhead lines (Whitehead, 1910), the concept of bundled conductors quickly revealed itself as an effective means of controlling the field intensity at the conductor surface, and hence, the line corona activities. This is well illustrated by the results in Table 16.2, which compare the single conductor design required to match the bundle performances in terms of power transmission capabilities, and the maximum conductor surface gradient for different line voltages. Bundled conductors are

TABLE 16.2 Comparison of Single and Bundled Conductors' Performances

Line voltage (kV)	400	735	1100
Distance between phases (m)	12	13.7	17
Number of subconductors	2	4	8
Bundle diameter (cm)	45	65	84
Conductor diameter (cm)	3.2	3.05	3.2
Corona onset gradient, $m = 0.85$, (kVrms/cm)	22.32	22.04	22.32
Maximum surface gradient (kVrms/cm)	16.3	19.79	17.3
Single conductor diameter of the same gradient (cm)	4.7	8.5	13.8
Transmission capability (GW)	0.5	2.0	4.9
Single conductor diameter of the same transmission capability (cm)	8.5	22	64

now used extensively in EHV lines rated 315 kV and higher; as a matter of fact, HV lines as we know them today would not exist without the introduction of conductor bundles.

16.3.3 Selection of Line Conductors

Even with the use of bundled conductors, it is not economically justifiable to design line conductors that would be corona-free under all weather conditions. The selection of line conductors is therefore made in terms of them being relatively corona-free under fair weather. While corona activities are tolerated under foul weather, their effects are controlled to acceptable levels at the edge of the right-of-way of the line. For AC lines, the design levels of 70 dB for the radio interference and 60 dBA for the AN at the edge of the right-of-way are often used (Trinh et al., 1974). These levels may be reached during periods of foul weather, and for a specified annual proportion of time, typically 15%–20%, depending on the local distribution of the weather pattern. The design process involves extensive field calculations and experimental testing to determine the number and size of the line conductors required to minimize the undesirable effects of corona discharges. Current practices in dimensioning HV-line conductors usually involve two stages of selection according to their worst-case and long-term corona performances.

16.3.3.1 Worst-Case Performance

Several conductor configurations (number, spacing, and diameter of the subconductors) are selected with respect to their worst-case performances which, for AC lines, correspond to foul-weather conditions, in particular heavy rain. Evaluation of the conductor worst-case performance is best done in test cages under artificial heavy rain conditions (Trinh and Maruvada, 1977). Test cages of square section, typically 3 m × 3 m, and a few tens of meters long, are adequate for evaluating full-size conductor bundles located along its central axis, for lines up to the 1500 kV class. The advantages of this experimental setup are the relatively modest test voltage required to reproduce the same field distribution on real-size bundled conductors, and the possibility of artificially producing the heavy rain conditions. The worst-case performance of various bundled conductors as defined by their generating quantities, namely, the corona-generated losses W, the RI excitation function Γ, and the AN-generated acoustic power density A, can then be determined over a wide range of surface gradients.

Under DC voltage, the worst-case corona performance is not directly related to foul-weather conditions. Although heavy rain was found to produce the highest losses, both the RI and the AN levels decrease under rain conditions. This behavior is related to the fact that under DC field conditions, the water droplets have an optimum shape, favorable to the development of stable glow-corona modes (Ianna et al., 1974). For this reason, test cage is less effective in evaluating the worst-case DC performance of bundled conductors.

A significant amount of data were gathered in cage tests at IREQ during the 1970s and provided the database for the development of a method to predict the worst-case performance of bundled conductors for AC voltage (Trinh and Maruvada, 1977). The latter was based on the following considerations:

- The corona performances of single conductors under artificial heavy rain conditions, as expressed by the generating functions, are a function of the conductor diameter and of the field intensity at its surface and can be expressed by the following empirical formula:

  $\Gamma_s(\text{dB}) = -93.35 + 92.42\log(E) + 43.03\log(\text{d})$ for the excitation function, Γ_s,
  $A_s(\text{dB}) = -123.94 + 82.84\log(E) + 48.28\log(\text{d})$ for the generated acoustic power density, A_s, and
  $W_s = e^{-17.41}E^{5.80}d^{2.46}$ for the corona-generated losses W_s.
  where E is expressed in kVrms/cm and d is in cm.

- The field intensity at the surface of a subconductor in a bundle is a function of its position and can be expressed as

$$E(\theta) = E_a + \frac{\Delta E}{2}\cos(\varphi - \alpha)$$

where

θ is the angle defining a point at the subconductor surface with respect to the point of maximum field intensity
φ is the angle defining a point at the subconductor surface with respect to the horizontal axis
α is the angle of the point of maximum field intensity at the subconductor surface, with respect to the horizontal axis
E_a is the average field intensity
ΔE is the difference between the extreme field intensities at the subconductor surface

Noting that for a given bundle conductor, the subconductor diameter is known, and the corona performances of the single conductor become a function of the field intensity at its surface only. As a result, the corona generating quantities of a bundle conductor under the actual operating conditions can be derived from that of the single conductor of the same size as the subconductor of the bundle, and taking into account of the actual distribution of the field intensity at the surface of the subconductors. The following expressions are obtained for the various generating quantities:

For the RI excitation function, Γ_b:

$$\Gamma_b = \frac{C_s}{C_b}\left[\sqrt{\sum_1^n \frac{1}{2\pi}\int_0^{2\pi} \Gamma s^2(E,d)\,d\phi}\right] \tag{16.23}$$

For the generated acoustic power density, A_b:

$$A_b = \left[\sum_1^n \sqrt{\frac{1}{2\pi}\int_0^{2\pi} A_s(E,d)\,d\phi}\right]^2 \tag{16.24}$$

For the generated losses, W_b:

$$W_b = \frac{C_b}{C_s}\left[\sum_{1}^{n} \frac{1}{2\pi}\int_{0}^{2\pi} W_s(E,d)\,d\phi\right] \quad (16.25)$$

where C_s and C_b are, respectively, the capacitances of the single conductor and the bundle conductor in the test cage.

Once the generating quantities are determined, the worst-case corona performances of the line can be evaluated as described in previous sections. The results presented in Figures 16.10 and 16.11, which compare the calculated and measured lateral RI and AN profiles of a number of HV lines, illustrate the good concordance of this approach. Commercial software exist that evaluate the worst-case performance of HV-line conductors using available experimental data obtained in cage tests under conditions of artificial heavy rain, making it possible to avoid undergoing tedious and expensive tests to help select the best configurations for line conductors for a given rating of the line.

16.3.3.2 Long-Term Corona Performance

Because of their wide range of variation in different weather conditions, representative corona performances of HV line are best evaluated in their natural environment. Test lines are generally used in this study that involves energizing the conductors for a sufficiently long period, usually 1 year to cover most of the weather conditions, and recording their corona performances together with the weather conditions. The higher cost of the long-term corona performance study usually limits its application to a small number of conductor configurations selected from their worst-case performance.

It should be noted that best results for the long-term corona performance evaluated on test lines are obtained when the weather pattern at the test site is similar to that existing along the actual HV line. A direct transposition of the results is then possible. If this condition is not met, some interpretation of the experimental data is needed. This is done by first decomposing the recorded long-term data into two groups, corresponding to the fair- and foul-weather conditions, then recombining these data according to the local weather pattern to predict the long-term corona performance along the line.

16.4 Conclusions

This chapter on transmission systems has reviewed the physics of corona discharges and discussed their impact on the design of HV lines, specifically in the selection of the line conductors. The following conclusions can be drawn.

Corona discharges can develop in different modes, depending on the equilibrium state existing under a given test condition, between the buildup and removal of ion space charges from the immediate vicinity of the highly stressed electrode. Three different corona modes—Trichel streamer, negative glow, and negative streamer—can be observed at the cathode with increasing applied field intensities. With positive polarity, four different corona modes are observed, namely, burst corona, onset streamers, positive glow, and breakdown streamers.

While all corona modes produce energy losses, the streamer discharges also generate RI and AN in the immediate vicinity of HV lines. These parameters are currently used to evaluate the corona performance of conductor bundles and to predict the energy losses and environmental impact of HV lines before their installation.

Adequate control of line corona is obtained by controlling the surface gradient at the line conductors. The introduction of bundled conductors in 1910 has greatly influenced the development of HV lines to today's EHVs.

Commercial software is available to select the bundle configuration: number and size of the subconductors, with respect to corona performances, which can be verified in test cages and lines in the early stage of new HV-line projects.

References

Bateman, L.A., Haywood, R.W., and Brooks, R.F., Nelson River DC transmission project, *IEEE Trans.*, PAS-88, 688, 1969.

Bortnik, I.M., Belyakov, N.N., Djakov, A.F., Horoshev, M.I., Ilynichin, V.V., Kartashev, I.I., Nikitin, O.A., Rashkes, V.S., Tikhodeyev, N.N., and Volkova, O.V., 1200 kV Transmission Line in the USSR: The First Results of Operation, in CIGRE Report No. 38-09, Paris, France, August 1988.

Dawson, G.A., A model for streamer propagation, *Zeitchrift fur Physik*, 183, 159, 1965.

Gary, C.H., The theory of the excitation function: A demonstration of its physical meaning, *IEEE Trans.*, PAS-91, 305, 1972.

Ianna, F., Wilson, G.L., and Bosak, D.J., Spectral characteristics of acoustic noise from metallic protrusion and water droplets in high electric fields, *IEEE Trans.*, PAS-93, 1787, 1974.

Juette, G.W., Evaluation of television interference from high-voltage transmission lines, *IEEE Trans.*, PAS-91, 865, 1972.

Krishnayya, P.C.S., Lambeth, P.J., Maruvada, P.S., Trinh, N.G., and Desilets, G., An Evaluation of the R&D Requirements for Developing HVDC Converter Stations for Voltages above ±600 kV, in CIGRE Report No. 14–07, Paris, France, August 1988.

Lacroix, R. and Charbonneau, H., Radio interference from the first 735-kV line of Hydro-Quebec, *IEEE Trans.*, PAS-87, 932, 1968.

Loeb, L.B., *Electrical Corona*, University of California Press, Berkeley, LA, 1965.

Moreau, M.R. and Gary, C.H., Predetermination of the radio-interference level of high voltage transmission lines—I: Predetermination of the excitation function, *IEEE Trans.*, PAS-91, 284, 1972a.

Moreau, M.R. and Gary, C.H., Predetermination of the radio-interference level of high voltage transmission lines—II: Field calculating method, *IEEE Trans.*, PAS-91, 292, 1972b.

Raether, H., *Electron Avalanches and Breakdown in Gases*, Butterworth Co., London, U.K., 1964.

Trinh, N.G., Partial discharge XIX: Discharge in air—Part I: Physical mechanisms, *IEEE Electr. Insul. Mag.*, 11, 23, 1995a.

Trinh, N.G., Partial discharge XX: Partial discharges in air—Part II: Selection of line conductors, *IEEE Electr. Insul. Mag.*, 11, 5, 1995b.

Trinh, N.G. and Jordan, I.B., Modes of corona discharges in air, *IEEE Trans.*, PAS-87, 1207, 1968.

Trinh, N.G. and Jordan, I.B., Trichel streamers and their transition into the pulseless glow discharge, *J. Appl. Phys.*, 41, 3991, 1970.

Trinh, N.G. and Maruvada, P.S., A method of predicting the corona performance of conductor bundles based on cage test results, *IEEE Trans.*, PAS-96, 312, 1977.

Trinh, N.G., Maruvada, P.S., and Poirier, B., A comparative study of the corona performance of conductor bundles for 1200-kV transmission lines, *IEEE Trans.*, PAS-93, 940, 1974.

Whitehead, J.B., Systems of Electrical Transmission, U.S. Patent No. 1,078,711, 1910.

17

Geomagnetic Disturbances and Impacts upon Power System Operation

John G. Kappenman
Metatech Corporation

17.1 Introduction

Recent analysis carried out for the EMP Commission, Federal Emergency Management Agency (FEMA), Federal Energy Regulatory Commission (FERC), North American Electric Reliability Corporation (NERC), and the U.S. National Academy of Sciences has determined that severe geomagnetic storms (i.e., space weather caused by solar activity) has the potential to cause crippling and long-duration damage to the North American electric power grid or any exposed power grid throughout the world (NRC 2008, Kappenman 2010, NERC/US DOE 2010). The primary impact to the power grid is the risk of widespread permanent damage to high-voltage transformers and other power delivery and production assets, which are key, scarce, and difficult to replace, of the high-voltage power network.

These storm events can have a continental and even planetary footprint causing widespread disruption, loss, and damage to the electric power supply for the United States or other similarly developed countries around the world. It is also estimated to be plausible on a 1 in 30 to 1 in 100 year time frame (Kappenman 2005). In short, this is potentially the largest and most plausible natural disaster that the United States could face, as the loss of electricity for extended durations would mean the collapse of nearly all other critical infrastructures, causing wide-scale loss of potable water, loss of perishable foods and medications, and many other disruptions to vital services necessary to sustain a nation's population. The severity of the threat geomagnetic storm impacts to present-day electric power grid infrastructures around the world have grown as the size of grids themselves have expanded by nearly a factor of 10 over the past 50 years, while at the same time they have become much more sensitive as higher EHV voltages and designs of transformers have evolved that react proportionately more to GIC exposure. These aspects of current design practices of electric grids have unknowingly and greatly escalated the risks and potential impacts from these threat environments. There has been no power grid design code that has ever taken into consideration these threat concerns.

Reliance of society on electricity for meeting essential needs has steadily increased over the years. This unique energy service requires coordination of electrical supply, demand, and delivery—all occurring at the same instant. Geomagnetic disturbances that arise from phenomena driven by solar activity commonly called space weather can cause correlated and geographically widespread disruption to these complex power grids. The disturbances to the earth's magnetic field causes geomagnetically induced currents (GICs, a near DC current typically with $f < 0.01$ Hz) to flow through the power system, entering and exiting the many grounding points on a transmission network. Geomagnetically induced currents are produced when shocks resulting from sudden and severe magnetic storms subject portions of the earth's surface to fluctuations in the planet's normally quiescent magnetic field. These fluctuations induce electric fields across the earth's surface—which causes GICs to flow through transformers, power system lines, and grounding points. Only a few amperes (amps) are needed to disrupt transformer operation, but over 300 A have been measured in the grounding connections of transformers in affected areas. Unlike threats due to ordinary weather, space weather can readily create large-scale problems because the footprint of a storm can extend across a continent. As a result, simultaneous widespread stress occurs across a power grid to the point where correlated widespread failures and even regional blackouts may occur.

Large impulsive geomagnetic field disturbances pose the greatest concern for power grids in close proximity to these disturbance regions. Large GICs are most closely associated with geomagnetic field disturbances that have high rate-of-change; hence a high-cadence and region-specific analysis of dB/dt of the geomagnetic field provides a generally scalable means of quantifying the relative level of GIC threat. These threats have traditionally been understood as associated with auroral electrojet intensifications at an altitude of ~100 km, which tend to locate at mid- and high-latitude locations during geomagnetic storms. However, both research and observational evidence has determined that the geomagnetic storm and associated GIC risks are broader and more complex than this traditional view (Kappenman 2005). Large GIC and associated power system impacts have been observed for differing geomagnetic disturbance source regions and propagation processes and in power girds at low geomagnetic latitudes (Erinmez et al. 2002). This includes the traditionally perceived impulsive disturbances originating from ionospheric electrojet intensifications. However, large GICs have also been associated with impulsive geomagnetic field disturbances such as those during an arrival shock of a large solar wind structure called coronal mass ejection (CME) that will cause brief impulsive disturbances even at very low latitudes. As a result, large GICs can be observed even at low- and mid-latitude locations for brief periods of time during these events (Kappenman 2003). Recent observations also confirm that geomagnetic field disturbances usually associated with an equatorial current system intensifications can be a source of large-magnitude and long-duration GIC in power grids at low and equatorial regions (Erinmez et al. 2002). High solar wind speed can also be the source of sustained pulsation of the geomagnetic field, (Kelvin–Helmholtz shearing) that has caused large GICs. The wide geographic extent of these disturbances implies GIC risks to power grids that have never considered the risk of GIC previously, largely because they were not at high-latitude locations.

Geomagnetic disturbances will cause the simultaneous flow of GICs over large portions of the interconnected high-voltage transmission network, which now span most developed regions of the world. As the GIC enters and exits the thousands of ground points on the high-voltage network, the flow path takes this current through the windings of large high-voltage transformers. GICs, when present in transformers on the system, will produce half-cycle saturation of these transformers, the root cause of all related power system problems. Since this GIC flow is driven by large geographic scale magnetic field disturbances, the impacts to power system operation of these transformers will be occurring simultaneously throughout large portions of the interconnected network. Half-cycle saturation produces voltage regulation and harmonic distortion effects in each transformer in quantities that build cumulatively over the network. The result can be sufficient to overwhelm the voltage regulation capability and the protection margins of equipment over large regions of the network. The widespread but correlated impacts can rapidly lead to systemic failures of the network. Power system designers and operators expect networks to be challenged by the terrestrial weather, and where those challenges were fully understood in the past, the system design has worked extraordinarily well. Most of these terrestrial weather challenges have largely been confined to much smaller regions than those encountered due to space weather. The primary design approach undertaken by the industry for decades has been to weave together a tight network, which pools resources and provides redundancy to reduce failures. In essence, an unaffected neighbor helps out the temporarily weakened neighbor. Ironically, the reliability approaches that have worked to make the electric power industry strong for ordinary weather, introduce key vulnerabilities to the electromagnetic coupling phenomena of space weather. As will be explained, the large continental grids have become in effect a large antenna to these storms. Further, space weather has a planetary footprint, such that the concept of unaffected neighboring system and sharing the burden is not always realizable. To add to the degree of difficulty, the evolution of threatening space weather conditions are amazingly fast. Unlike ordinary weather patterns, the electromagnetic interactions of space weather are inherently instantaneous. Therefore large geomagnetic field disturbances can erupt on a planetary scale within the span of a few minutes.

17.2 Power Grid Damage and Restoration Concerns

The onset of important power system problems can be assessed in part by experience from contemporary geomagnetic storms. At geomagnetic field disturbance levels as low as 60–100 nT/min (a measure of the rate of change in the magnetic field flux density over the earth's surface), power system operators have noted system upset events such as relay misoperation, the offline tripping of key assets, and even high levels of transformer internal heating due to stray flux in the transformer from GIC-caused half-cycle saturation of the transformer magnetic core. Reports of equipment damage have also included large electric generators and capacitor banks.

Power networks are operated using what is termed an "N–1" operation criterion. That is, the system must always be operated to withstand the next credible disturbance contingency without causing a cascading collapse of the system as a whole. This criterion normally works very well for the well-understood terrestrial environment challenges, which usually propagate more slowly and are more geographically confined. When a routine weather-related single-point failure occurs, the system needs to be rapidly adjusted (requirements typically allow a 10–30 min response time after the first incident) and positioned to survive the next possible contingency. Geomagnetic field disturbances during a severe storm can have a sudden onset and cover large geographic regions. They therefore cause near-simultaneous, correlated, multipoint failures in power system infrastructures, allowing little or no time for meaningful human interventions that are intended within the framework of the N–1 criterion. This is the situation that triggered the collapse of the Hydro Quebec power grid on March 13, 1989, when their system went from normal conditions to a situation where they sustained seven contingencies (i.e., N–7) in an elapsed time of 57 s; the province-wide blackout rapidly followed with a total elapsed time of 92 s from normal conditions to a complete collapse of the grid. For perspective, this occurred at a disturbance intensity of approximately 480 nT/min over the region (Figure 17.1). A recent examination by Metatech of historically large disturbance intensities indicated that disturbance levels greater than 2000 nT/min have been observed even in contemporary storms on at least three occasions over the past 30 years at geomagnetic latitudes of concern for the North American power grid infrastructure and most other similar world locations: August 1972, July 1982, and March 1989. Anecdotal information from older storms suggests that disturbance levels may have reached nearly 5000 nT/min, a level ~10 times greater than the environment which triggered the Hydro Quebec collapse (Kappenman 2005). Both observations and simulations indicate that as the intensity of the disturbance increases, the relative levels of GICs and related power system impacts will also proportionately increase. Under these scenarios, the scale and speed of problems that could occur on exposed power grids has the potential to cause wide spread and severe disruption of bulk power system operations. Therefore, as storm environments reach higher intensity levels, it becomes more likely that these events will precipitate widespread blackouts to exposed power grid infrastructures.

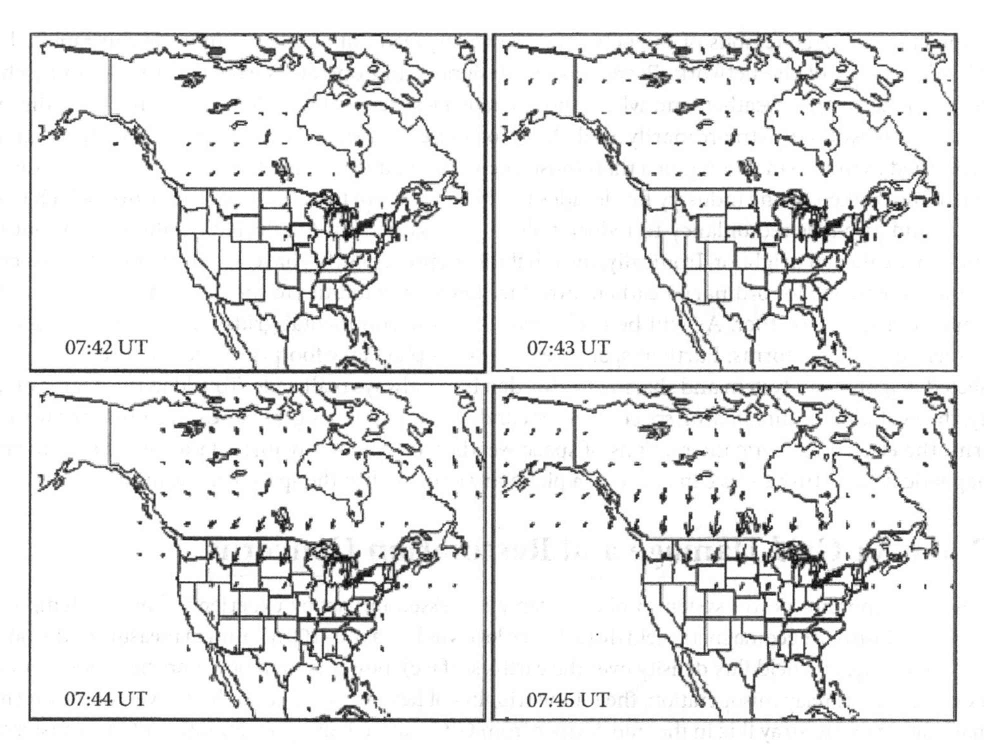

FIGURE 17.1 Four minutes of geomagnetic field disturbance on the March 13, 1989, Superstorm that triggered the Quebec grid collapse, from 7:42UT to 7:745UT.

17.3 Weak Link in the Grid: Transformers

The primary concern with GIC is the effect that they have on the operation of a large power transformer. Under normal conditions, the large power transformer is a very efficient device for converting one voltage level into another. Decades of design engineering and refinement have increased efficiencies and capabilities of these complex apparatus to the extent that only a few amperes of AC exciting current are necessary to provide the magnetic flux for the voltage transformation in even the largest modern power transformer. As GIC levels increase, the level of saturation of the transformer core and its impact on the operation of the power grid as a whole also increases.

However, in the presence of GIC, the near-direct current essentially biases the magnetic circuit of the transformer with resulting disruptions in performance. The three major effects produced by GIC in transformers are (1) the increased reactive power consumption of the affected transformer, (2) the increased even and odd harmonics generated by the half-cycle saturation, and (3) the possibilities of equipment damaging stray flux heating. These distortions can cascade problems by disrupting the performance of other network apparatus, causing them to trip off-line just when they are most needed to protect network integrity. For large storms, the spatial coverage of the disturbance is large and hundreds of transformers can be simultaneously saturated, a situation that can rapidly escalate into a network-wide voltage collapse. In addition, individual transformers may be damaged from overheating due to this unusual mode of operation, which can result in long-term outages to key transformers in the network. Damage of these assets can slow the full restoration of power grid operations.

Transformers use steel in their cores to enhance their transformation capability and efficiency, but this core steel introduces nonlinearities into their performance. Common design practice minimizes the effect of the nonlinearity while also minimizing the amount of core steel. Therefore, the transformers are usually designed to operate over a predominantly linear range of the core steel characteristics

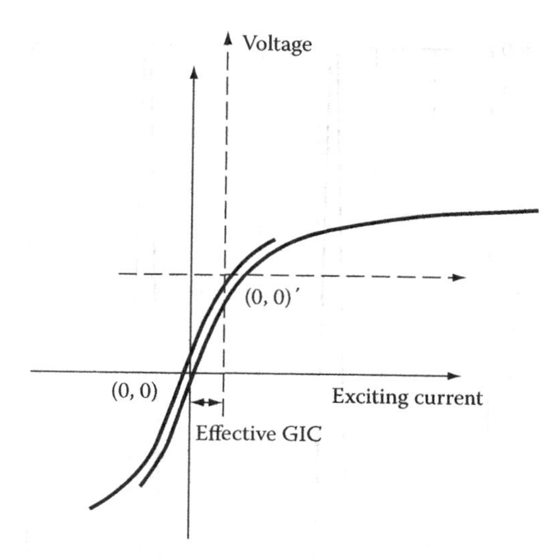

FIGURE 17.2 Transformer saturation characteristics for normal operation and for half-cycle saturation due to the presence of GIC.

(as shown in Figure 17.2), with only slightly nonlinear conditions occurring at the voltage peaks. This produces a relatively small exciting current (Figure 17.3). With GIC present, the normal operating point on the core steel saturation curve is offset and the system voltage variation that is still impressed on the transformer causes operation in an extremely nonlinear portion of the core steel characteristic for half of the AC cycle (Figure 17.2), hence the term half-cycle saturation.

Because of the extreme saturation that occurs on half of the AC cycle, the transformer now draws an extremely large asymmetrical exciting current. The waveform in Figure 17.3 depicts a typical example from field tests of the exciting current from a three-phase 600 MVA power transformer that has 75 A of GIC in the neutral (25 A per phase). Spectrum analysis reveals this distorted exciting current to be rich in even, as well as odd harmonics. As is well documented, the presence of even a small amount of GIC (3–4 A per phase or less) will cause half-cycle saturation in a large transformer.

Since the exciting current lags the system voltage by 90°, it creates reactive-power loss in the transformer and the impacted power system. Under normal conditions, transformer reactive power loss is very small. However, the several orders of magnitude increase in exciting current under half-cycle saturation also results in extreme reactive-power losses in the transformer. For example, the three-phase reactive power loss associated with the abnormal exciting current of Figure 17.3 produces a reactive power loss of over 40 MVars for this transformer alone. The same transformer would draw less than 1 MVar under normal conditions. Figure 17.4 provides a comparison of reactive power loss for two core types of transformers as a function of the amount of GIC flow.

Under a geomagnetic storm condition in which a large number of transformers are experiencing a simultaneous flow of GIC and undergoing half-cycle saturation, the cumulative increase in reactive power demand can be significant enough to impact voltage regulation across the network, and in extreme situations, lead to network voltage collapse.

The large and distorted exciting current drawn by the transformer under half-cycle saturation also poses a hazard to operation of the network because of the rich source of even and odd harmonic currents this injects into the network and the undesired interactions that these harmonics may cause with relay and protective systems or other power system apparatus. Figure 17.5 summarizes the spectrum analysis of the asymmetrical exciting current from Figure 17.3. Even and odd harmonics are present typically in the first 10 orders and the variation of harmonic current production varies somewhat with the level of GIC, the degree of half-cycle saturation, and the type of transformer core.

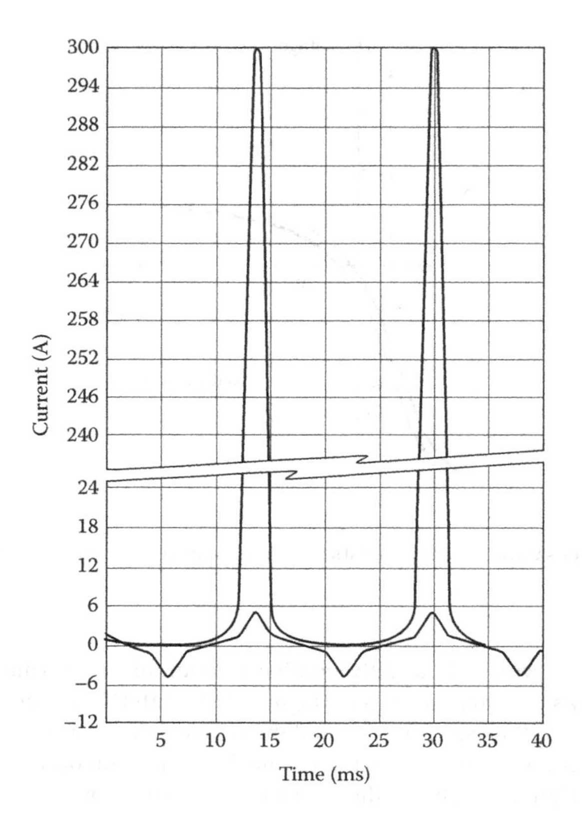

FIGURE 17.3 Transformer excitation current characteristics for normal operation and for half-cycle saturation due to the presence of GIC.

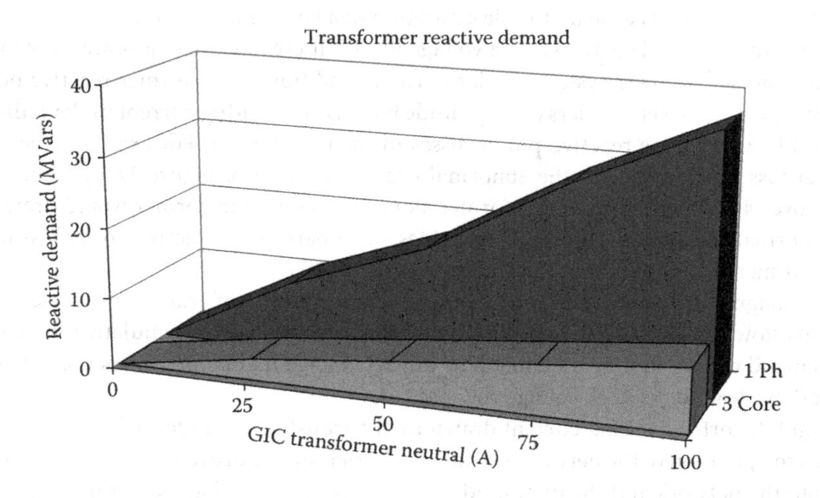

FIGURE 17.4 Transformer increased reactive power demands (MVARs) due to GIC for a typical 500 kV transformer for single phase and three-phase three-legged core type.

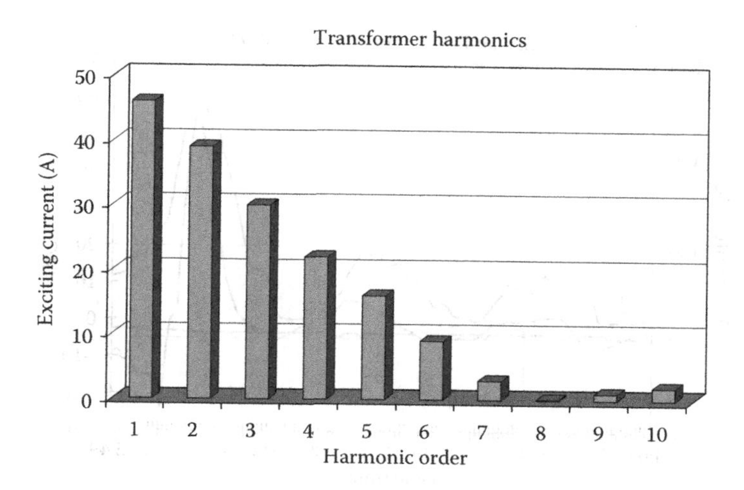

FIGURE 17.5 Example of even and odd harmonic spectrums of half-cycle saturated excitation current for the excitation current waveform shown in Figure 17.3.

With the magnetic circuit of the core steel saturated, the magnetic core will no longer contain the flow of flux within the transformer. This stray flux will impinge upon or flow through adjacent paths such as the transformer tank or core clamping structures. The flux in these alternate paths can concentrate to the densities found in the heating elements of a kitchen stove. This abnormal operating regime can persist for extended periods as GIC flows from storm events can last for hours. The hot spots that may then form can severely damage the paper-winding insulation, produce gassing and combustion of the transformer oil, or lead to other serious internal and or catastrophic failures of the transformer. Such saturation and the unusual flux patterns that result are not typically considered in the design process and, therefore, a risk of damage or loss of life is introduced.

One of the more thoroughly investigated incidents of transformer stray flux heating occurred in the Allegheny Power System on a 350 MVA 500/138 kV autotransformer at their Meadow Brook Substation near Winchester, Virginia. The transformer was first removed from service on March 14, 1989, because of high gas levels in the transformer oil which were a by-product of internal heating. The gas-in-oil analysis showed large increases in the amounts of hydrogen, methane, and acetylene, indicating core and tank heating. External inspection of the transformer indicated four areas of blistering or discolored paint due to tank surface heating. In the case of the Meadow Brook transformer, calculations estimate the flux densities were high enough in proximity to the tank to create hot spots approaching 400°C. Reviews made by Allegheny Power indicated that similar heating events (though less severe) occurred in several other large power transformers in their system due to the March 13 disturbance. Figure 17.6 is a recording that Allegheny Power made on their Meadow Brook transformer during a storm in 1992. This measurement shows an immediate transformer tank hot-spot developing in response to a surge in GIC entering the neutral of the transformer, while virtually no change is evident in the top oil readings. Because the hot spot is confined to a relatively small area, standard bulk top oil or other over-temperature sensors would not be effective deterrents to use to alarm or limit exposures for the transformer to these conditions.

Designing a large transformer that would be immune to GIC would be technically difficult and prohibitively costly. The ampere-turns of excitation (the product of the normal exciting current and the number of winding turns) generally determine the core steel volume requirements of a transformer. Therefore, designing for unsaturated operation with the high level of GIC present would require a core of excessive size. The ability to even assess existing transformer vulnerability is a difficult undertaking and can only be confidently achieved in extensive case-by-case investigations. Each transformer design (even from the same manufacturer) can contain numerous subtle design variations. These variations complicate the calculation of how and at what density the stray flux can impinge on internal structures

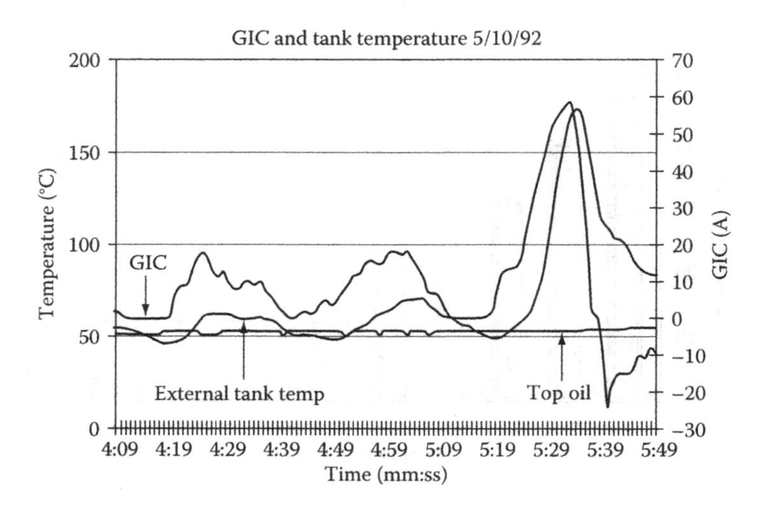

FIGURE 17.6 Observed Meadow Brook transformer hot-spot temperature for a minor storm on May 10, 1992.

in the transformer. However, the experience from contemporary space weather events is revealing and potentially paints an ominous outcome for historically large storms that are yet to occur on today's infrastructure. As a case in point, during a September 2004 Electric Power Research Industry workshop on transformer damage due to GIC, Eskom, the power utility that operates the power grid in South Africa (geomagnetic latitudes −27° to −34°), reported damage and loss of 15 large, high-voltage transformers (400 kV operating voltage) due to the geomagnetic storms of late October 2003. This damage occurred at peak disturbance levels of less than 100 nT/min in the region (Kappenman 2005).

17.4 Overview of Power System Reliability and Related Space Weather Climatology

The maintenance of the functional integrity of the bulk electric systems (i.e., power system reliability) at all times is a very high priority for the planning and operation of power systems worldwide. Power systems are too large and critical in their operation to easily perform physical tests of their reliability performance for various contingencies. The ability of power systems to meet these requirements is commonly measured by deterministic study methods to test the system's ability to withstand probable disturbances through computer simulations. Traditionally, the design criterion consists of multiple outage and disturbance contingencies typical of what may be created from relatively localized terrestrial weather impacts. These stress tests are then applied against the network model under critical load or system transfer conditions to define important system design and operating constraints in the network.

System impact studies for geomagnetic storm scenarios can now be readily performed on large complex power systems. For cases in which utilities have performed such analysis, the impact results indicate that a severe geomagnetic storm event may pose an equal or greater stress on the network than most of the classic deterministic design criteria now in use. Further, by the very nature that these storms impact simultaneously over large regions of the network, they arguably pose a greater degree of threat for precipitating a system-wide collapse than more traditional threat scenarios.

The evaluation of power system vulnerability to geomagnetic storms is, of necessity, a two-stage process. The first stage assesses the exposure to the network posed by the climatology. In other words, how large and how frequent can the storm driver be in a particular region? The second stage assesses the stress that probable and extreme climatology events may pose to reliable operation of the impacted network. This is measured through estimates of levels of GIC flow across the network and the manifestation of impacts such as sudden and dramatic increases in reactive power demands and implications on voltage regulation in the network. The essential aspects of risk management become the weighing of probabilities of storm events against the potential consequential impacts produced by a storm. From this analysis effort, meaningful operational procedures can be further identified and refined to better manage the risks resulting from storms of various intensities (Kappenman et al. 2000).

Successive advances have been made in the ability to undertake detailed modeling of geomagnetic storm impacts upon terrestrial infrastructures. The scale of the problem is enormous, the physical processes entail vast volumes of the magnetosphere, ionosphere, and the interplanetary magnetic field conditions that trigger and sustain storm conditions. In addition, it is recognized that important aspects and uncertainties of the solid-earth geophysics need to be fully addressed in solving these modeling problems. Further, the effects to ground-based systems are essentially contiguous to the dynamics of the space environment. Therefore, the electromagnetic coupling and resulting impacts of the environment on ground-based systems require models of the complex network topologies overlaid on a complex geological base that can exhibit variation of conductivities that can span 5 orders of magnitude.

These subtle variations in the ground conductivity play an important role in determining the efficiency of coupling between disturbances of the local geomagnetic field caused by space environment influences and the resulting impact on ground-based systems that can be vulnerable to GIC. Lacking full understanding of this important coupling parameter hinders the ability to better classify the climatology of space weather on ground-based infrastructures.

17.5 Geological Risk Factors and Geo-Electric Field Response

Considerable prior work has been done to model the geomagnetic induction effects in ground-based systems. As an extension to this fundamental work, numerical modeling of ground conductivity conditions have been demonstrated to provide accurate replication of observed geoelectric field conditions over a very broad frequency spectrum (Kappenman et al. 1997). Past experience has indicated that 1D earth conductivity models are sufficient to compute the local electric fields. Lateral heterogeneity of ground conductivity conditions can be significant over meso-scale distances (Kappenman 2001). In these cases, multiple 1D models can be used in cases where the conductivity variations are sufficiently large.

Ground conductivity models need to accurately reproduce geo-electric field variations that are caused by the considerable frequency ranges of geomagnetic disturbance events from the large-magnitude/low-frequency electrojet-driven disturbances to the low-amplitude but relatively high-frequency impulsive disturbances commonly associated with magnetospheric shock events. This variation of electromagnetic disturbances therefore require models accurate over a frequency range from 0.3 to as low as 0.00001 Hz. At these low frequencies of the disturbance environments, diffusion aspects of ground conductivities must be considered to appropriate depths. Therefore, skin depth theory can be used in the frequency domain to determine the range of depths that are of importance. For constant earth conductivities, the depths required are more than several hundred kilometers, although the exact depth is a function of the layers of conductivities present at a specific location of interest.

It is generally understood that the earth's mantle conductivity increases with depth. In most locations, ground conductivity laterally varies substantially at the surface over mesoscale distances; these conductivity variations with depth can range from 3 to 5 orders of magnitude. While surface conductivity can exhibit considerable lateral heterogeneity, conductivity at depth is more uniform, with conductivities ranging from values of 0.1 to 10 S/m at depths from 600 to 1000 km. If sufficient low-frequency measurements are available to characterize ground conductivity profiles, models of ground conductivity can be successfully applied over mesoscale distances and can be accurately represented by use of layered conductivity profiles or models.

For illustration of the importance of ground models on the response of geo-electric fields, a set of four example ground models have been developed that illustrate the probable lower to upper quartile response characteristics of most known ground conditions, considering there is a high degree of uncertainty in the plausible diversity of upper layer conductivities. Figure 17.7 provides a plot of the layered ground conductivity conditions for these four ground models to depths of 700 km. As shown, there can

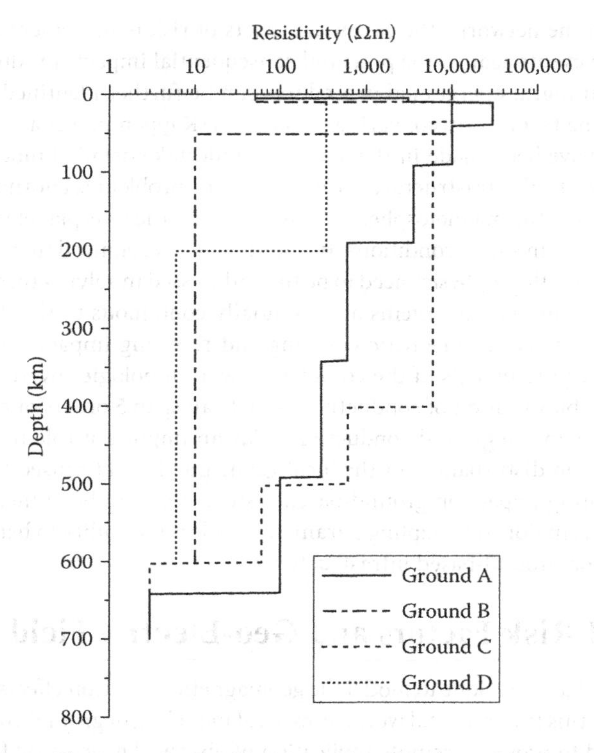

FIGURE 17.7 Resistivity profiles versus depth of four examples of the layered earth ground models.

be as much as four orders of magnitude variation in ground resistivity at various depths in the upper layers. Models A and B have very thin surface layers of relatively low resistivity. Models A and C are characterized by levels of relatively high resistivity until reaching depths exceeding 400 km, while models B and D have high variability of resistivity in only the upper 50–200 km of depth.

Figure 17.8 provides the frequency response characteristics for these same four-layered earth ground models of Figure 17.7. Each line plot represents the geo-electric field response for a corresponding incident magnetic field disturbance at each frequency. While each ground model has unique response characteristics at each frequency, in general, all ground models produce higher geo-electric field responses as the frequency of the incident disturbance increases. Also shown on this plot are the relative differences in geo-electric field response for the lowest and highest responding ground model at each decade of frequency. This illustrates that the response between the lowest and highest responding ground model can vary at discrete frequencies by more than a factor of 10. Also because the frequency content of an impulsive disturbance event can have higher frequency content (for instance due to a shock), the disturbance is acting upon the more responsive portion of the frequency range of the ground models (Kappenman 2003). Therefore, the same disturbance energy input at these higher frequencies produces a proportionately larger response in geo-electric field. For example, in most of the ground models, the geo-electric field response is a factor of 50 higher at 0.1 Hz compared to the response at 0.0001 Hz.

From the frequency response plots of the ground models as provided in Figure 17.8, some of the expected geo-electric field response due to geomagnetic field characteristics can be inferred. For example, Ground C provides the highest geo-electric field response across the entire spectral range, therefore, it would be expected that the time-domain response of the geo-electric field would be the highest for nearly all B field disturbances. At low frequencies, Ground B has the lowest geo-electric field response while at frequencies above 0.02 Hz, Ground A produces the lowest geo-electric field response. Because each of these ground models have both frequency-dependent and nonlinear variations in response, the resulting form of the geo-electric field waveforms would be expected to differ in form for the same B field input

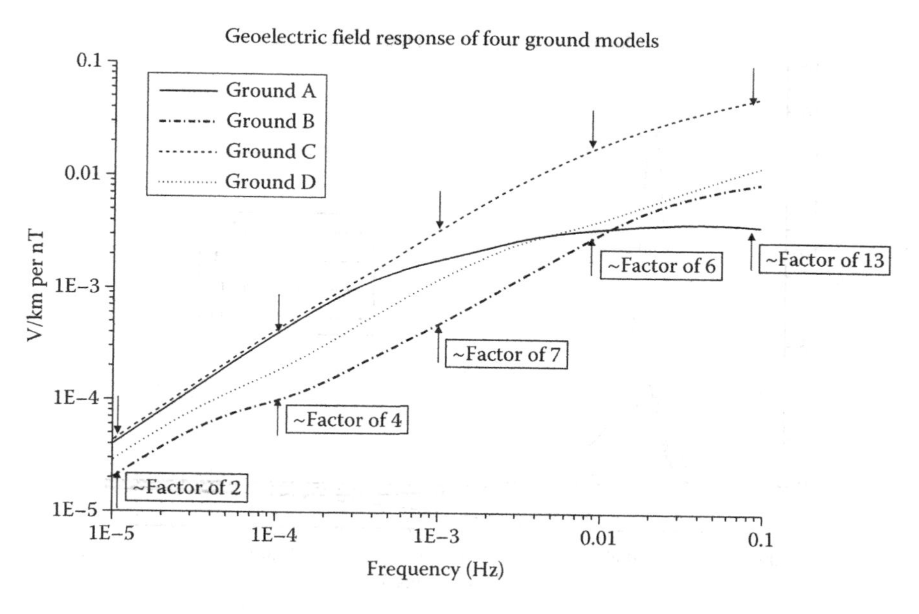

FIGURE 17.8 Frequency response of four examples of the ground models of Figure 17.7; max/min geo-electric field response characteristics shown at various discrete frequencies.

disturbance. In all cases, each of the ground models produces higher relative increasing geo-electric field response as the frequency of the incident B field disturbance increases. Therefore it should be expected that a higher peak geo-electric field should result for a higher spectral content disturbance condition.

A large electrojet-driven disturbance is capable of producing an impulsive disturbance as shown in Figure 17.9, which reaches a peak delta B magnitude of ~2000 nT with a rate of change (dB/dt) of 2400 nT/min. This disturbance scenario can be used to simulate the estimated geo-electric field response of the four example ground models. Figure 17.10 provides the geo-electric field responses for each of the four ground models for this 2400 nT/min B field disturbance. As expected, the Ground C model produces the largest geo-electric field reaching a peak of ~15 V/km, while Ground A is next largest and the Ground B model produces the smallest geo-electric field response. The Ground C geo-electric field

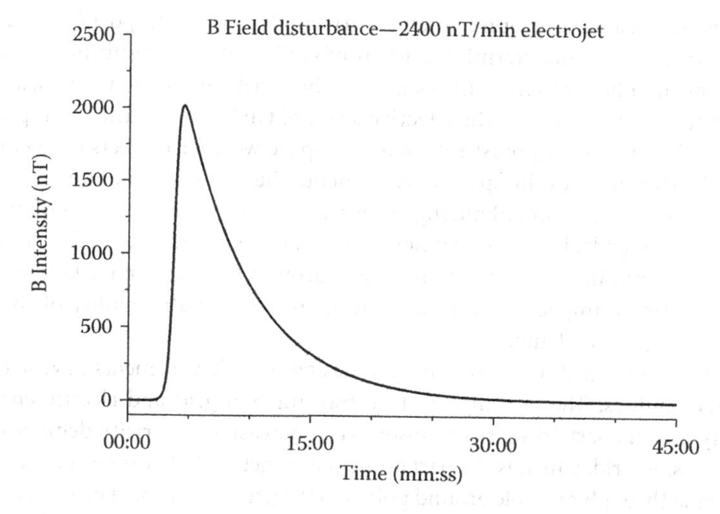

FIGURE 17.9 Waveform of an electrojet-driven geomagnetic field disturbance with 2400 nT/min rate of change intensity.

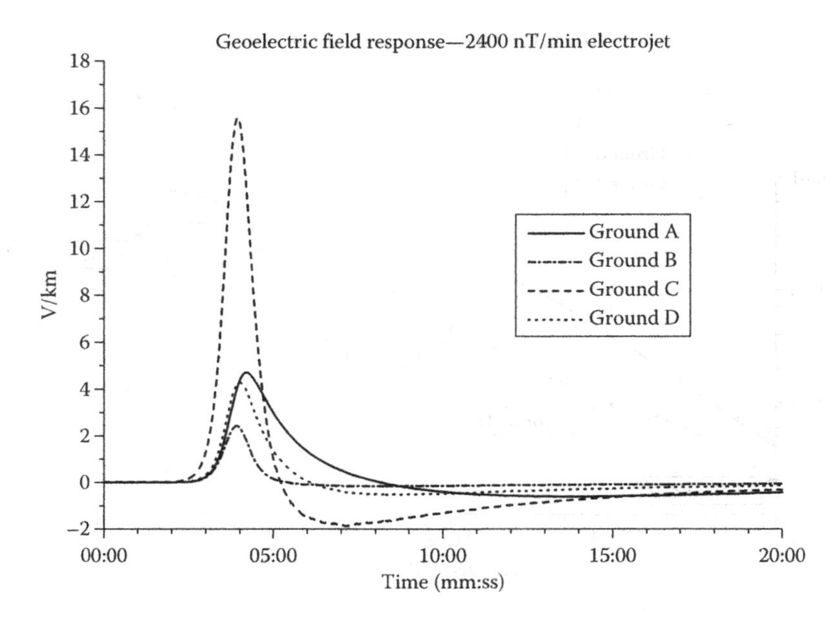

FIGURE 17.10 Geo-electric field response of the four examples of ground models to the 2400 nT/min disturbance conditions of Figure 17.9.

peak is more than six times larger than the peak geo-electric field for the Ground B model. It is also evident that significant differences result in the overall shape and form of the geo-electric field response. For example, the peak geo-electric field for the Ground A model occurs 17 s later than the time of the peak geo-electric field for the Ground B model. In addition to the differences in the time of peak, the waveforms also exhibit differences in decay rates. As is implied from this example, both the magnitudes of the geo-electric field responses and the relative differences in responses between models will change dependent on the source disturbance characteristics.

17.6 Power Grid Design and Network Topology Risk Factors

While the previous discussion on ground conductivity conditions are important in determining the geo-electric field response, and in determining levels of GICs and their resulting impacts, power grid design is also an important factor in the vulnerability of these critical infrastructures, a factor in particular that over time has greatly escalated the effective levels of GIC and operational impacts due to these increased GIC flows. Unfortunately, most research into space weather impacts on technology systems has focused upon the dynamics of the space environment. The role of the design and operation of the technology system in introducing or enhancing vulnerabilities to space weather is often overlooked. In the case of electric power grids, both the manner in which systems are operated and the accumulated design decisions engineered into present-day networks around the world have tended to significantly enhance geomagnetic storm impacts. The result is to increase the vulnerability of this critical infrastructure to space weather disturbances.

Both growth of the power grid infrastructure and design of its key elements have acted to introduce space weather vulnerabilities. The U.S. high-voltage transmission grid and electric energy usage have grown dramatically over the last 50 years in unison with increasing electricity demands of society. The high-voltage transmission grid, which is the part of the power network that spans long distances, couples almost like an antenna through multiple ground points to the geo-electric field produced by disturbances in the geomagnetic field. From Solar Cycle 19 in the late 1950s through Solar Cycle 22 in the early 1980s, the high-voltage transmission grid and annual energy usage have grown nearly tenfold (Figure 17.11).

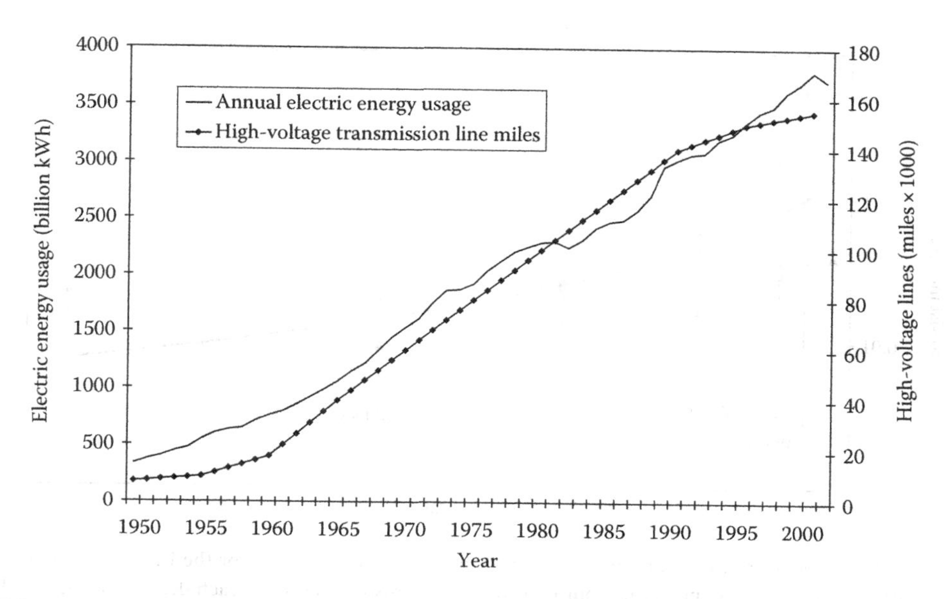

FIGURE 17.11 Growth of the high-voltage transmission network and annual electric energy usage in the United States over the past 50 years. In addition to increasing the total network size, the network has grown in complexity with the introduction of higher kV-rated lines that tend to carry larger GIC flows.

In short, the antenna that is sensitive to space weather disturbances is now very large. Similar development rates of transmission infrastructure have occurred simultaneously in other developed regions of the world.

As this network has grown in size, it has also grown in complexity and sets in place a compounding of risks that are posed to the power grid infrastructures for GIC events. Some of the more important changes in technology base that can increase impacts from GIC events include higher design voltages, changes in transformer design, and other related apparatus. The operating levels of high-voltage networks have increased from the 100–200kV thresholds of the 1950s to 400–765kV levels of present-day networks. With this increase in operating voltages, the average per unit length circuit resistance has decreased while the average length of the grid circuit increases. In addition, power grids are designed to be tightly interconnected networks, which present a complex circuit that is continental in size. These interrelated design factors have acted to substantially increase the levels of GIC that are possible in modern power networks.

In addition to circuit topology, GIC levels are determined by the size and the resistive impedance of the power grid circuit itself when coupled with the level of geo-electric field that results from the geomagnetic disturbance event. Given a geo-electric field imposed over the extent of a power grid, a current will be produced entering the neutral ground point at one location and exiting through other ground points elsewhere in the network. This can be best illustrated by examining the typical range of resistance per unit length for each kV class of transmission lines and transformers.

As shown in Figure 17.12, the average resistance per transmission line across the range of major kV-rating classes used in the current U.S. power grid decreases by a factor of more than 10. Therefore 115 and 765kV transmission lines of equal length can have a factor of ~10 difference in total circuit resistance. Ohm's law indicates that the higher-voltage circuits when coupled to the same geo-electric field would result in as much as ~10 times larger GIC flows in the higher voltage portions of the power grid. The resistive impedance of large power system transformers follows a very similar pattern: the larger the power capacity and kV-rating, the lower the resistance of the transformer. In combination, these design attributes will tend to collect and concentrate GIC flows in the higher kV-rated equipment. More important, the higher kV-rated lines and transformers are key network elements, as they are the long-distance heavy haulers of the power grid. The upset or loss of these key assets due to large GIC flows can rapidly cascade into geographically widespread disturbances to the power grid.

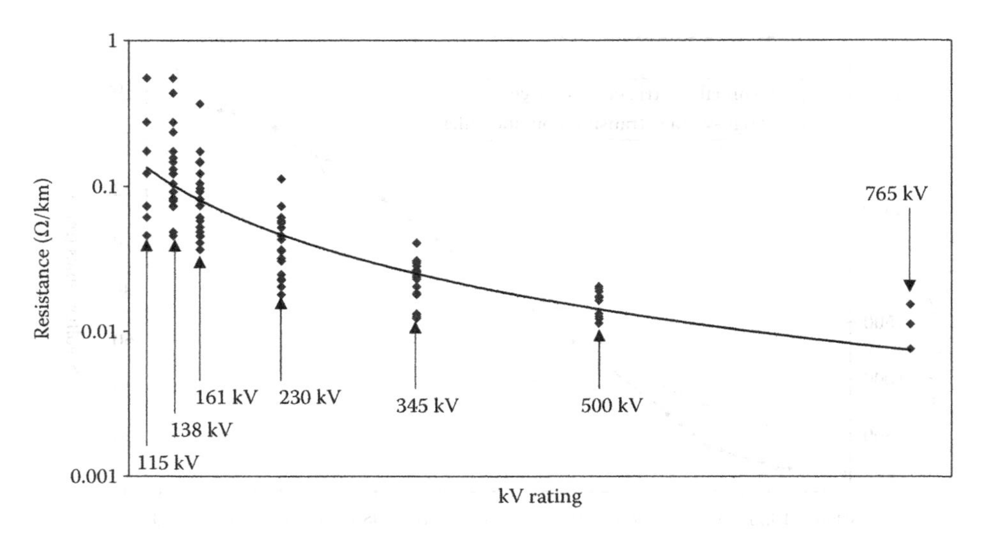

FIGURE 17.12 Range of transmission line resistance in major kV-rating classes for the U.S. electric power grid infrastructure, with a trend line indicating common conductor resistances used at each design voltage. The lower resistance for the higher voltage lines will also cause proportionately larger GIC flows.

Most power grids are highly complex networks with numerous circuits or paths and transformers for GIC to flow through. This requires the application of highly sophisticated network and electromagnetic coupling models to determine the magnitude and path of GIC throughout the complex power grid. However, for the purposes of illustrating the impact of power system design, a review will be provided using a single transmission line terminated at each end with a single transformer to ground connection. To illustrate the differences that can occur in levels of GIC flow at higher voltage levels, the simple demonstration circuit have also been developed at 138, 230, 345, 500, and 765 kV, which are common grid voltages used in the United States and Canada. In Europe, voltages of 130, 275, and 400 kV are commonly used for the bulk power grid infrastructures. For these calculations, a uniform 1.0 V/km geo-electric field disturbance conditions are used, which means that the change in GIC levels will result from changes in the power grid resistances alone. Also for uniform comparison purposes, a 100 km long line is used in all kV rating cases.

Figure 17.13 illustrates the comparison of GIC flows that would result for various U.S. infrastructure power grid kV ratings using the simple circuit and a uniform 1.0 V/km geo-electric field disturbance. In complex networks, such as those in the United States, some scatter from this trend line is possible

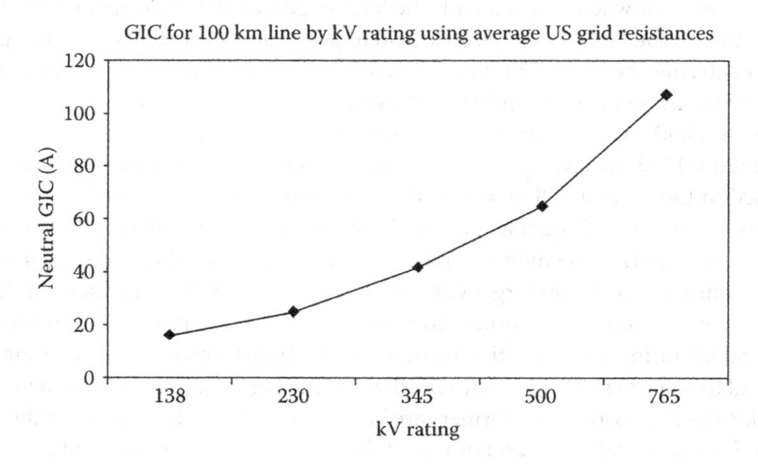

FIGURE 17.13 Average neutral GIC flows versus kV rating for a 100 km demonstration transmission circuit.

due to normal variations in circuit parameters such as line resistances that can occur in the overall population of infrastructure assets. Further, this was an analysis of simple "one-line" topology network, whereas real power grid networks have highly complex topologies, span large geographic regions, and present numerous paths for GIC flow, all of which tend to increase total GIC flows. Even this limited demonstration tends to illustrate that the power grid infrastructures of large grids in the United States and other locations of the world are increasingly exposed to higher GIC flows due to design changes that have resulted in reduced circuit resistance. Compounding this risk further, the higher kV portions of the network handle the largest bulk power flows and form the backbone of the grid. Therefore the increased GIC risk is being placed at the most vital portions of this critical infrastructure. In the United States, 345, 500, and 765 kV transmission systems are widely spread throughout the United States and especially concentrated in areas of the United States with high population densities.

One of the best ways to illustrate the operational impacts of large GIC flows is to review the way in which the GIC can distort the AC output of a large power transformer due to half-cycle saturation. Under severe geomagnetic storm conditions, the levels of geo-electric field can be many times larger than the uniform 1.0 V/km used in the prior calculations. Under these conditions, even larger GIC flows are possible. For example, in Figure 17.14, the normal AC current waveform in the high voltage winding of a 500 kV transformer under normal load conditions is shown (~300 A-rms, ~400 A-peak). With a large GIC flow in the transformer, the transformer experiences extreme saturation of the magnetic core for one-half of the AC cycle (half-cycle saturation). During this half-cycle of saturation, the magnetic core of the transformer draws an extremely large and distorted AC current from the power grid. This combines with the normal AC load current producing the highly distorted asymmetrically peaky waveform that now flows in the transformer. As shown, AC current peaks that are present are nearly twice as large compared to normal current for the transformer under this mode of operation. This highly distorted waveform is rich in both even and odd harmonics, which are injected into the system and can cause misoperations of sensors and protective relays throughout the network (Kappenman et al. 1981, 1989).

The design of transformers also acts to further compound the impacts of GIC flows in the high-voltage portion of the power grid. While proportionately larger GIC flows occur in these large high-voltage transformers, the larger high-voltage transformers are driven into saturation at the same few amperes of GIC

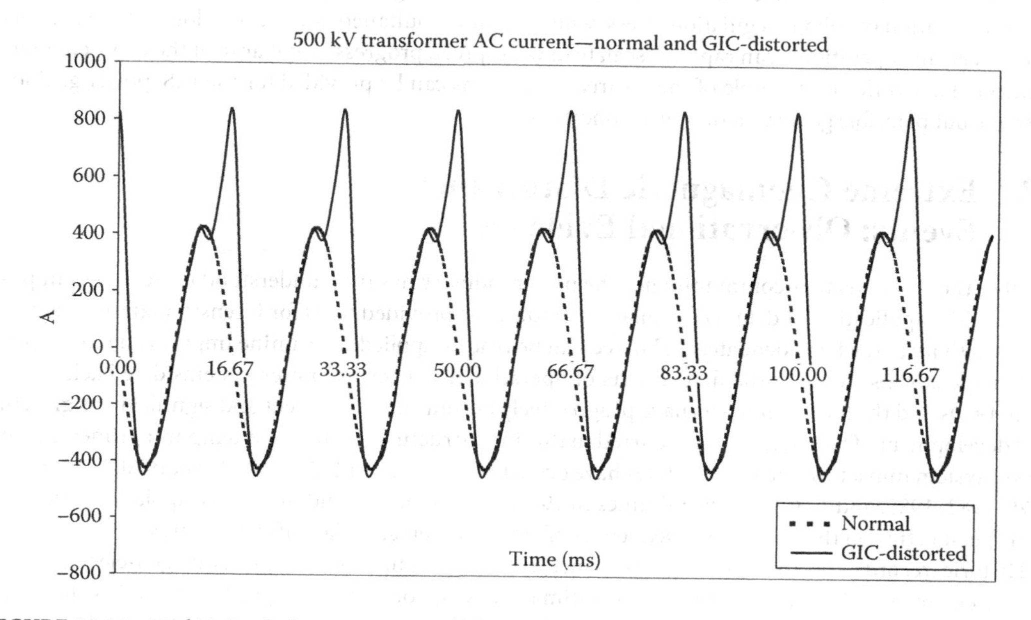

FIGURE 17.14 500 kV simple demonstration circuit simulation results—transformer AC currents and distortion due to GIC.

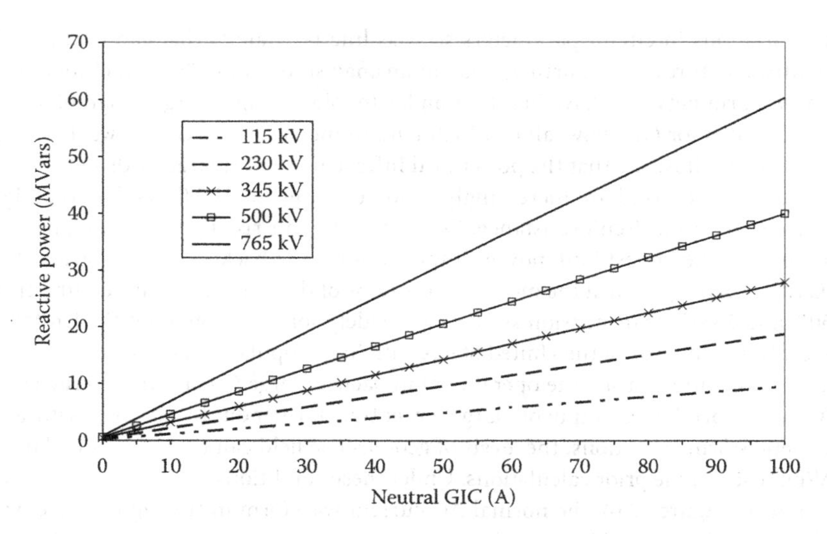

FIGURE 17.15 Comparison of the reactive power losses through transformers of increasing kV rating versus increasing levels of GIC flow. Higher kV-rated transformers will produce proportionately larger reactive power consumption on the grid compared to the same level of GIC flow in lower kV transformers.

exposure as those of lower-voltage transformers. More ominously, another compounding of risk occurs as these higher kV-rated transformers produce proportionately higher power system impacts than comparable lower-voltage transformers. As shown in Figure 17.15, because reactive power loss in a transformer is a function of the operating voltage, the higher kV-rated transformers will also exhibit proportionately higher reactive power losses due to GIC. For example, a 765 kV transformer will have approximately six times larger reactive power losses for the same magnitude of GIC flow as that of a 115 kV transformer.

All transformers on the network can be exposed to similar conditions simultaneously due to the wide geographic extent of most disturbances. This means that the network needs to supply an extremely large amount of reactive power to each of these transformers or voltage collapse of the network could occur. The combination of voltage regulation stress, which occurs simultaneously with the loss of key elements due to relay misoperations, can rapidly escalate to widespread progressive collapse of the exposed interconnected network. An example of these threat conditions can be provided for the U.S. power grid for extreme but plausible geomagnetic storm conditions.

17.7 Extreme Geomagnetic Disturbance Events: Observational Evidence

Neither the space weather community nor the power industry has fully understood these design implications. The application of detailed simulation models has provided tools for forensic analysis of recent storm activity and when adequately validated can be readily applied to examine impacts due to historically large storms. Some of the first reports of operational impacts to power systems date back to the early 1940s and the level of impacts have progressively become more frequent and significant as growth and development of technology has occurred in this infrastructure. In more contemporary times, major power system impacts in the United States have occurred in storms in 1957, 1958, 1968, 1970, 1972, 1974, 1979, 1982, 1983, and 1989 and several times in 1991. Both empirical and model extrapolations provide some perspective on the possible consequences of storms on present-day infrastructures.

Historic records of geomagnetic disturbance conditions and, more important, geo-electric field measurements provide a perspective on the ultimate driving force that can produce large GIC flows in power grids. Because geo-electric fields and resulting GIC are caused by the rate of change of the geomagnetic field, one of the most meaningful methods to measure the severity of impulsive geomagnetic field disturbances is by the magnitude of the geomagnetic field change per minute, measured in nanoteslas per minute (nT/min). For example, the regional disturbance intensity that triggered the Hydro Quebec collapse during the March 13, 1989 storm only reached an intensity of 479 nT/min. Large numbers of power system impacts in the United States were also observed for intensities that ranged from 300 to 600 nT/min during this storm. However, the most severe rate of change in the geomagnetic field observed during this storm reached a level of ~2000 nT/min over the lower Baltic. The last such disturbance with an intensity of ~2000 nT/min over North America was observed during a storm on August 4, 1972 when the power grid infrastructure was less than 40% of its current size.

Data assimilation models provide further perspectives on the intensity and geographic extent of the intense dB/dt of the March 1989 Superstorm. Figure 17.16 provides a synoptic map of the ground-level geomagnetic field disturbance regions observed at time 22:00UT. The previously mentioned lower Baltic region observations are embedded in an enormous westward electrojet complex during this period of time. Simultaneously with this intensification of the westward electrojet, an intensification of the eastward electrojet occupies a region across mid-latitude portions of the western United States. The features of the westward electrojet extend longitudinally ~120° and have a north–south cross-section ranging as much as 5°–10° in latitude.

Older storms provide even further guidance on the possible extremes of these specific electrojet-driven disturbance processes. A remarkable set of observations was conducted on rail communication circuits in Sweden that extend back nearly 80 years. These observations provide key evidence that allow for estimation of the geomagnetic disturbance intensity of historically important storms in an era where geomagnetic observatory data is unavailable. During a similarly intense westward electrojet disturbance on July 13–14, 1982, a ~100 km length communication circuit from Stockholm to Torreboda measured a peak geo-potential of 9.1 V/km (Lindahl). Simultaneous measurements at nearby Lovo observatory in central Sweden measured a dB/dt intensity of ~2600 nT/min at 24:00 UT on July 13. Figure 17.17 shows the delta Bx

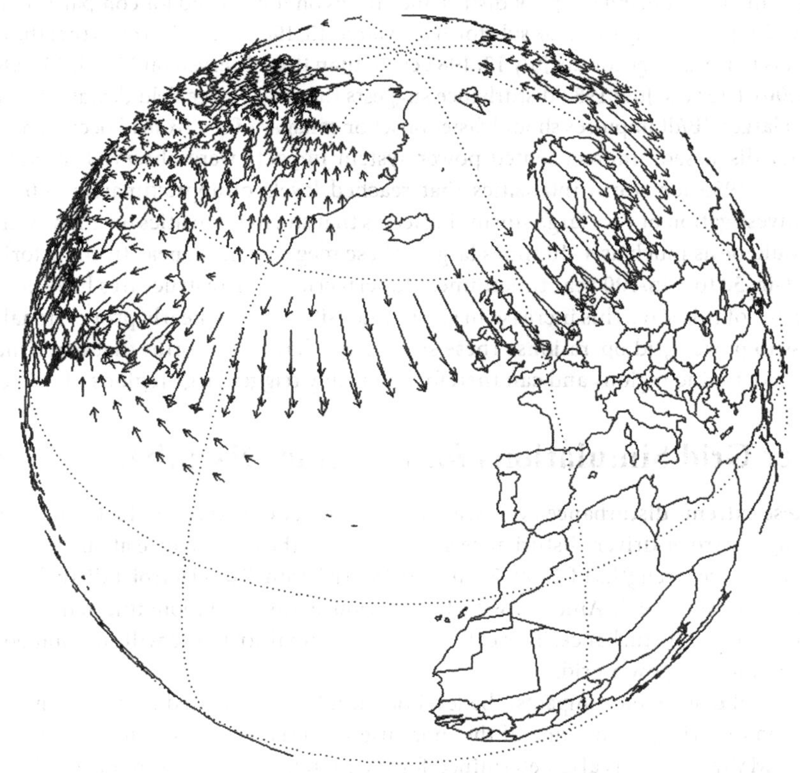

FIGURE 17.16 Extensive westward electrojet-driven geomagnetic field disturbances at time 22:00UT on March 13, 1989.

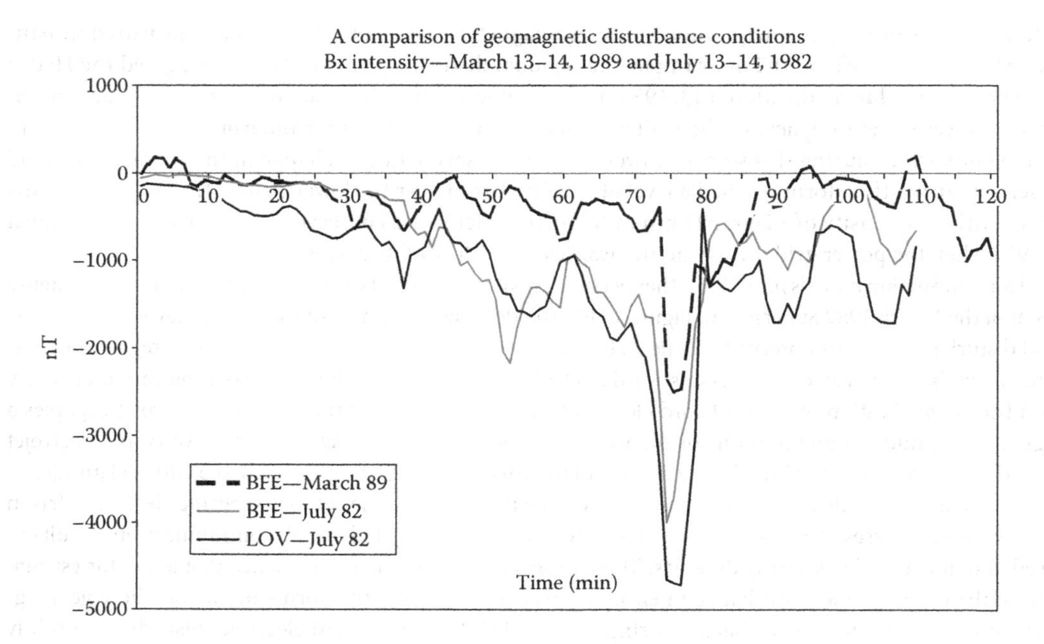

FIGURE 17.17 Comparison of observed delta Bx at Lovo and BFE on July 13–14, 1982, and March 13, 1989, electrojet intensification events.

observed at BFE and Lovo during the peak disturbance times on July 13 and for comparison purposes the delta Bx observed at BFE during the large substorm on March 13, 1989. This illustrates that the comparative level of delta Bx is twice as large for the July 13, 1982 event than that observed on March 13, 1989. The large delta Bx of >4000 nT for the July 1982 disturbance suggests that these large field deviations are capable of producing even larger dB/dt impulses should faster onset or collapse of the Bx field occur over the region.

As previously discussed, unprecedented power system impacts were observed in North America on March 13–14, 1989 for storm intensities that reached levels of approximately 300–600 nT/min. However, the investigation of very large storms indicates that storm intensities over many of these same U.S. regions could be as much as 4–10 times larger. These megastorms appear from historic data to be probable on a 1-in-50 to 1-in-100 year timeframe. Modern critical infrastructures have not as-yet been exposed to storms of this size. This increase in storm intensity causes a nearly proportional increase in resulting stress to power grid operations. These storms also have a footprint that can simultaneously threaten large geographic regions and can therefore plausibly trigger large regions of grid collapse.

17.8 Power Grid Simulations for Extreme Disturbance Events

Based upon these extreme disturbance events, a series of simulations were conducted for the entire U.S. power grid using electrojet-driven disturbance scenarios with the disturbance at 50° geomagnetic latitude and at disturbance strengths of 2400, 3600, and 4800 nT/min. The electrojet disturbance footprint was also positioned over North America with the previously discussed longitudinal dimensions of a large westward electrojet disturbance. This extensive longitudinal structure will simultaneously expose a large portion of the U.S. power grid.

In this analysis of disturbance impacts, the level of cumulative increased reactive demands (MVars) across the U.S. power grid provides one of the more useful measures of overall stress on the network. This cumulative MVar stress was also determined for the March 13, 1989 storm for the U.S. power grid, which was estimated using the current system model as reaching levels of ~7000–8000 MVars at times 21:44–21:57 UT. At these times, corresponding dB/dt levels in mid-latitude portions of the United States reached 350–545 nT/min as measured at various U.S. observatories. This provides a comparison benchmark that can be used to either compare absolute MVar levels or, the relative MVar level increases for the more severe disturbance scenarios. The higher intensity disturbances of 2400–4800 nT/min will have a proportionate effect on levels of GIC in the exposed network. GIC levels more than five times larger than those observed during the aforementioned periods in the March 1989 storm would be a probable. With the increase in GIC, a linear and proportionate increase in other power system impacts is likely. For example, transformer MVar demands increase with increases in transformer GIC. As larger GICs cause greater degrees of transformer saturation, the harmonic order and magnitude of distortion currents increase in a more complex manner with higher GIC exposures. In addition, greater numbers of transformers would experience sufficient GIC exposure to be driven into saturation, as generally higher and more widely experienced GIC levels would occur throughout the extensive exposed power grid infrastructure.

Figure 17.18 provides a comparison summary of the peak cumulative MVar demands that are estimated for the U.S. power grid for the March 89 storm, and for the 2400, 3600, and 4800 nT/min disturbances at the different geomagnetic latitudes. As shown, all of these disturbance scenarios are far larger in magnitude than the levels experienced on the U.S. grid during the March 1989 Superstorm. All reactive demands for the 2400–4800 nT/min disturbance scenarios would produce unprecedented in size reactive demand increases for the U.S. grid. The comparison with the MVar demand from the March 1989 Superstorm further indicates that even the 2400 nT/min disturbance scenarios would produce reactive demand levels at all of the latitudes that would be approximately six times larger than those estimated in March 1989. At the 4800 nT/min disturbance levels, the reactive demand is estimated, in total, to exceed 100,000 MVars. While these large reactive demand increases are calculated for illustration purposes, impacts on voltage regulation and probable large-scale voltage collapse across the network could conceivably occur at much lower levels.

This disturbance environment was further adapted to produce a footprint and onset progression that would be more geo-spatially typical of an electrojet-driven disturbance, using both the March 13, 1989 and July 13, 1982 storms as a template for the electrojet pattern. For this scenario, the intensity of the disturbance is decreased as it progresses from the eastern to western United States. The eastern portions of the United States are exposed to a 4800 nT/min disturbance intensity, while, west of the Mississippi, the disturbance intensity decreases to only 2400 nT/min. The extensive reactive power increase and extensive geographic boundaries of impact would be expected to trigger large-scale progressive collapse conditions, similar to the mode in which the Hydro Quebec collapse occurred. The most probable regions of expected

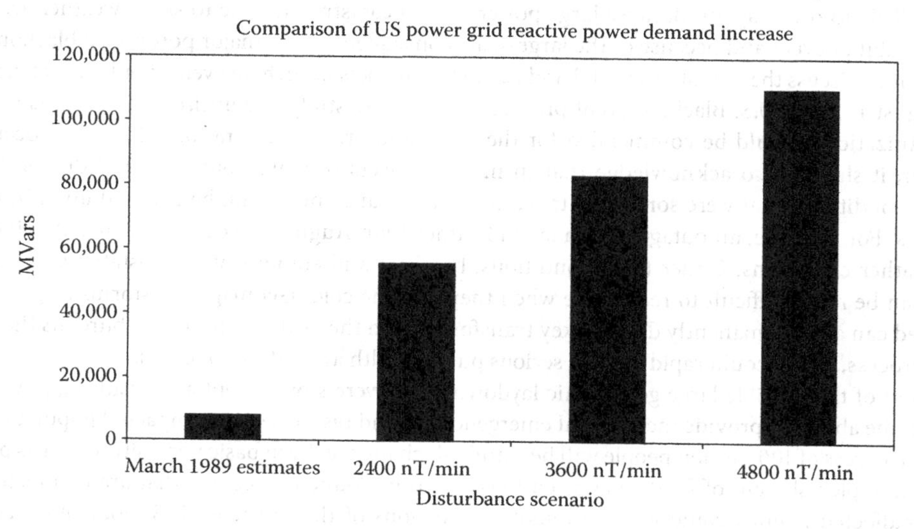

FIGURE 17.18 Comparison of estimated U.S. power grid reactive demands for the March 13, 1989, Superstorm and 2400, 3600, and 4800 nT/min disturbance scenarios at 50° geomagnetic latitude position over the United States.

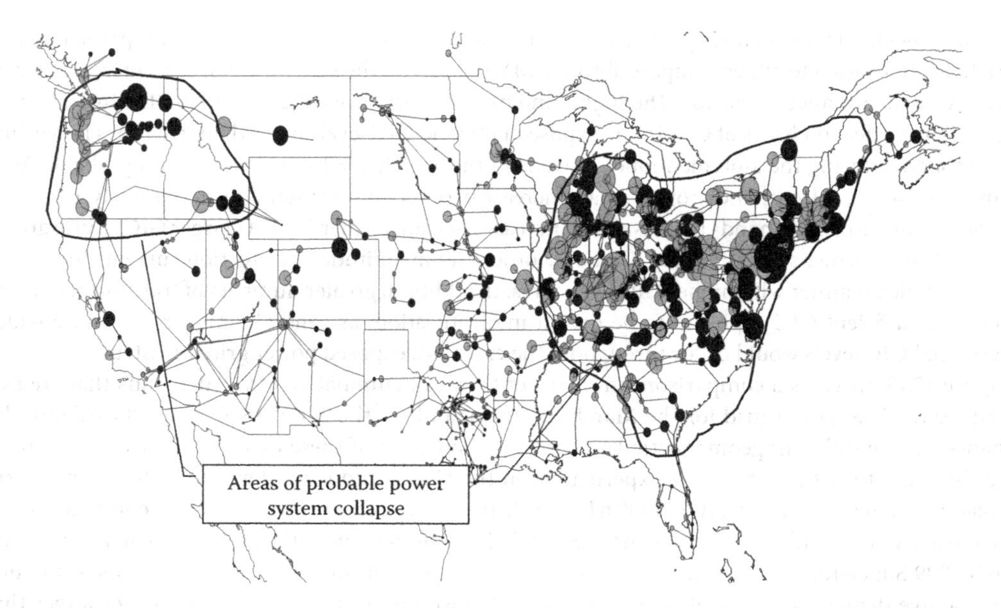

FIGURE 17.19 Regions of large GIC flows and possible power system collapse due to a 4800 nT/min disturbance scenario.

power system collapse can be estimated based upon the GIC levels and reactive demand increases in combination with the disturbance criteria as it applies to the U.S. power pools. Figure 17.19 provides a map of the peak GIC flows in the U.S. power grid (size of circle at each node indicates relative GIC intensity) and estimated boundaries of regions that likely could experience system collapse due to this disturbance scenario. This example shows one of many possible scenarios for how a large storm could unfold.

While these complex models have been rigorously tested and validated, this is an exceedingly complex task with uncertainties that can easily be as much as a factor of 2. However, just empirical evidence alone suggests that power grids in North America that were challenged to collapse for storms of 400–600 nT/min over a decade ago are not likely to survive the plausible but rare disturbances of 2000–5000 nT/min that long-term observational evidence indicates have occurred before and therefore may be likely to occur again. Because large power system catastrophes due to space weather are not a zero probability event and because of the large-scale consequences of a major power grid blackout, it is important to discuss the potential societal and economic impacts of such an event should it ever reoccur. The August 14, 2003 U.S. Blackout event provides a good case study, the utilities and various municipal organizations should be commended for the rapid and orderly restoration efforts that occurred. However, it should also acknowledge that in many respects this blackout occurred during highly optimal conditions, that were somewhat taken for granted and should not be counted upon in future blackouts. For example, an outage on January 14 rather than August 14 could have meant coincident cold weather conditions. Under these conditions, breakers and equipment at substations and power plants can be more difficult to reenergize when they become cold. Geomagnetic storms as previously discussed can also permanently damage key transformers on the grid which further burdens the restoration process, delays could rapidly cause serious public health and safety concerns.

Because of the possible large geographic laydown of a severe storm event and resulting power grid collapse, the ability to provide meaningful emergency aid and response to an impacted population that may be in excess of 100 million people will be a difficult challenge. Even basic necessities such as potable water and replenishment of foods may need to come from boundary regions that are unaffected and these unaffected regions could be very remote to portions of the impacted U.S. population centers. As previously suggested, adverse terrestrial weather conditions could cause further complications in restoration and re-supply logistics.

17.9 Conclusions

Contemporary models of large power grids and the electromagnetic coupling to these infrastructures by the geomagnetic disturbance environment have matured to a level in which it is possible to achieve very accurate benchmarking of storm geomagnetic observations and the resulting GIC. As abilities advance to model the complex interactions of the space environment with the electric power grid infrastructures, the ability to more rigorously quantify the impacts of storms on these critical systems also advances. This quantification of impacts due to extreme space weather events is leading to the recognition that geomagnetic storms are an important threat that has not been well recognized in the past.

References

Erinmez, I.A., S. Majithia, C. Rogers, T. Yasuhiro, S. Ogawa, H. Swahn, and J.G. Kappenman, Application of modelling techniques to assess geomagnetically induced current risks on the NGC transmission system, CIGRE Paper 39-304, Session 2002.

Kappenman, J.G., Chapter 13—An introduction to power grid impacts and vulnerabilities from space weather, NATO-ASI Book on *Space Storms and Space Weather Hazards*, I.A. Daglis, ed., NATO Science Series, Kluwer Academic Publishers, Dordrecht, the Netherlands, Vol. 38, pp. 335–361, 2001.

Kappenman, J.G., SSC events and the associated GIC risks to ground-based systems at low and mid-latitude locations, *AGU International Journal of Space Weather*, 1(3), 1016, 2003, doi: 10.1029/2003SW000009.

Kappenman, J.G., An overview of the impulsive geomagnetic field disturbances and power grid impacts associated with the violent Sun-Earth connection events of 29–31 October 2003 and a comparative evaluation with other contemporary storms, *Space Weather*, 3, S08C01, 2005, doi:10.1029/2004SW000128.

Kappenman, J.G., Great geomagnetic storms and extreme impulsive geomagnetic field disturbance events—An analysis of observational evidence including the Great Storm of May 1921, Paper for 35th COSPAR Assembly publication in Advances in Space Research, August 2005 Published by Elsevier Ltd on behalf of COSPAR. doi:10.1016/j.asr.2005.08.055.

Kappenman, J.G., Geomagnetic storms and their impacts on the U.S. power grid (Meta-R-319), ORNL-FERC Report, Weblink http://www.ornl.gov/sci/ees/etsd/pes/ferc_emp_gic.shtml, January 2010.

Kappenman, J.G., V.D. Albertson, and N. Mohan, Current transformer and relay performance in the presence of geomagnetically-induced currents, *IEEE PAS Transactions*, PAS-100, 1078–1088, March 1981.

Kappenman, J.G., D.L. Carlson, and G.A. Sweezy, GIC effects on relay and CT performance, Paper presented at the *EPRI Conference on Geomagnetically-Induced Currents*, San Francisco, CA, November 8–10, 1989.

Kappenman, J.G., W.A. Radasky, J.L. Gilbert, and I.A. Erinmez, Advanced geomagnetic storm forecasting: A risk management tool for electric power operations, *IEEE Plasma Society Special Issue on Space Plasmas*, 28(#6), 2114–2121, December 2000.

Kappenman, J.G., L.J. Zanetti, and W.A. Radasky, Space weather from a user's perspective: Geomagnetic storm forecasts and the power industry, *EOS Transactions of the American Geophysics Union*, 78(4), 37–45, January 28, 1997.

National Research Council, Severe space weather events—Understanding societal and economic impacts workshop report, Committee on the Societal and Economic Impacts of Severe Space Weather Events: A Workshop, ISBN: 0-309-12770-X, 131 pp., 2008.

NERC/US DOE Joint Report, High-impact, low-frequency event risk to the North American bulk power system, June 2010.

18

Lightning Protection

William A. Chisholm
Kinectrics/Université du Québec à Chicoutimi

The study of lightning predates electric power systems by many centuries. Observations of thunder were maintained in some areas for more than a millennium. Franklin and others established the electrical nature of lightning, and introduced the concepts of shielding and grounding to protect structures. Early power transmission lines used as many as six overhead shield wires, strung above the phase conductors and grounded at the towers for effective lightning protection. Later in the twentieth century, repeated strikes to tall towers, buildings, and power lines, contradicting the adage that "it never strikes twice," allowed systematic study of stroke current parameters. Improvements in electronics, computers, telecommunications, rocketry, and satellite technologies have all extended our knowledge about lightning, while at the same time exposing us to ever-increasing risks of economic damage from its consequences.

18.1 Ground Flash Density

The first return stroke from the direct termination of a negative, downward cloud-to-ground lightning flash is the dominant risk to power system components. Positive first strokes, negative subsequent strokes, and continuing currents can also cause specific problems. A traditional indicator of cloud-to-ground lightning activity is given by thunder observations, collected to World

Meteorological Organization standards and converted to Ground Flash Density (Anderson et al., 1984; MacGorman et al., 1984):

$$GFD = 0.04 \cdot TD^{1.25} \tag{18.1}$$

$$GFD = 0.054 \cdot TH^{1.1} \tag{18.2}$$

where

TD is the number of days with thunder per year
TH is the number of hours with thunder per year
GFD is the number of first cloud-to-ground strokes per square kilometer per year

Long-term thunder data suggest that GFD has a relative standard deviation of 30%.

Observations of optical transient density have been performed using satellites starting in 1995. These data have some of the same defects as thunder observations: cloud flash and ground flash activity is equally weighted and the observations are sporadic. However, statistical considerations as well as richly detailed observations of orographic terrain features now favor the use of optical transient density, reported by (Christian et al., 2003; NASA, 2006) over thunder observations to estimate ground flash density.

A good estimate of ground flash density can be obtained by dividing the optical transient density values in Figure 18.1 by a factor of 3.0. This average factor is valid in four different continents but may vary across regions, calling for a lower factor in some limited areas where storms have a higher ratio of positive to negative flashes.

Electromagnetic signals from individual lightning strokes are unique and have high signal-to-noise ratio at large distances. Many single-station lightning flash counters have been developed and calibrated, each with good discrimination between cloud flash and ground flash activity using simple electronic circuits (Heydt, 1982). It has also been feasible for more than 30 years (Krider et al., 1976) to observe these signals with two or more stations, and to triangulate lightning stroke locations on a continent-wide basis. Lightning location networks have improved continuously to the point where multiple ground strikes from a single flash can be resolved with high spatial and temporal accuracy and high probability of detection (CIGRE, 2009). A GFD value from these data should be based on approximately 400 counts

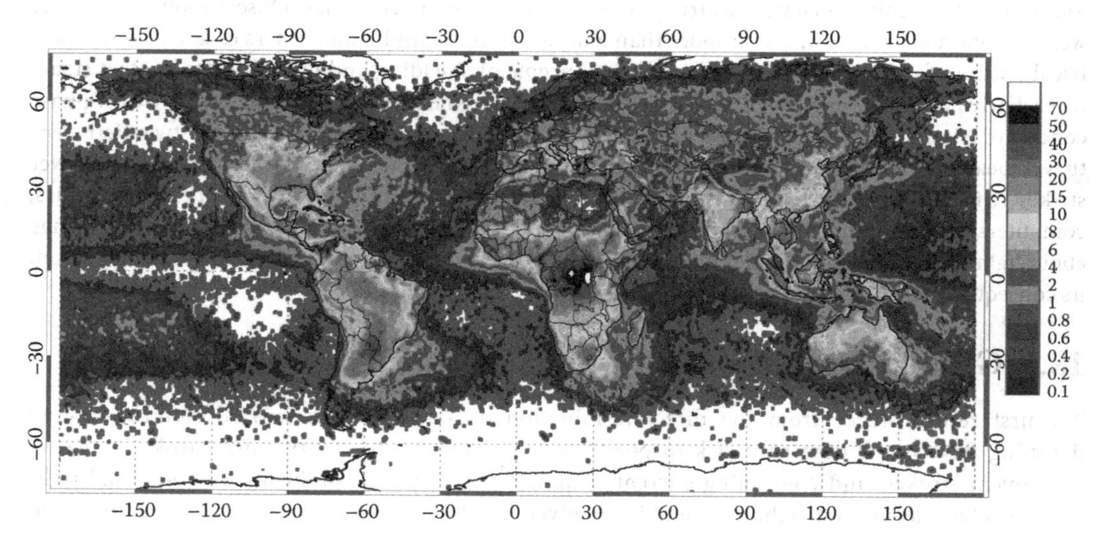

FIGURE 18.1 Observed optical transient density per km² per year from NASA (2006). The optical transient density (OTD) can be used to estimate lightning ground flash density N_g (flashes per km² per year) by dividing the observed values by 3.0.

TABLE 18.1 Typical Design Approaches for Overhead Transmission Lines

Optical Transient Density from Figure 18.1 (Transients per km² per Year)	Ground Flash Density Range (Flashes per km² per Year)	Typical Design Approaches
0.3–1	0.1–0.3	Unshielded, one or three-pole reclosing
1–3	0.3–1	Single overhead shield wire or unshielded with line arresters, upper phases, all towers
3–10	1–3	Two overhead shield wires
10–30	3–10	Two overhead shield wires with good grounding or line surge arresters
30–100	10–30	Three or more overhead and underbuilt shield wires with good grounding; line surge arresters; underground transmission cables

in each cell to reduce relative standard deviation of the observation process below 5%. In areas with moderate flash density, a minimum cell size of 20×20 km is appropriate.

18.2 Mitigation Methods

Lightning mitigation methods for transmission lines need to be appropriate for the expected long-term ground flash density and power system reliability requirements. Table 18.1 summarizes typical practices at five different levels of lightning activity to achieve a reliability of 1 outage per 100 km of line per year on an HV line.

Power system insulation is designed to withstand overvoltages that are generated within the power system, under steady state and also when components are switched. Unfortunately, even the weakest direct lightning stroke from a shielding failure to a phase conductor will cause an overvoltage that will flash over across an insulator that is not protected by a surge arrester nearby. Once an arc appears across an insulator, the power system fault current keeps this arc alive until voltage is removed by protective relay action. If the flash incidence is low, Table 18.1 shows that some utilities can simply accept a high tripout rate, up to 6 interruptions per 100 km per year, and can protect against the consequences using automatic reclosing and redundant paths.

Effective overhead shielding, with wires placed above the phase conductors to intercept flashes and divert them to ground, is the most common form of lightning protection on transmission lines in areas with moderate to high ground flash density.

When the overhead shield wire is struck, the potential difference on insulators is the sum of the resistive and inductive voltage rises on the tower, minus the coupled voltage on the phase conductors. The potential difference can lead to a "backflashover" from the tower to the phase conductor.

Backflashover is probable when peak stroke current is large, when footing resistance is high and when insulation strength is low. Simplified models (CIGRE, 1991; IEEE, 1997; EPRI, 2005) are available to carry out the lightning overvoltage calculations and coordinate the results with insulator strength, giving lightning outage rates. A schematic of this process is given as follows.

18.3 Stroke Incidence to Power Lines

The lightning leader, a thin column of electrically-charged plasma, develops from cloud down to the ground in a series of step breakdowns (Rakov and Uman, 2007). Near the ground, electric fields are high enough to satisfy the conditions for continuous positive leader inception upward from tall objects or conductors. Analysis of a single overhead conductor with this approach (Rizk, 1990) leads to

$$N_S = 3.8 \cdot \text{GFD} \cdot h^{0.45} \tag{18.3}$$

where

N_S is the number of strikes to the conductor per 100 km of line length per year
h is the average height of the conductor above ground in meters

In areas of moderate to high ground flash density, one or more overhead shield wires are usually installed above the phase conductors. This shielding usually has a success rate of greater than 95%, but adds nearly 10% to the cost of line construction and also wastes energy from induced currents. The leader inspection model (Rizk, 1990) has been developed to analyze shielding failures more accurately. The goal was to reduce the failure rate below the IEEE set reliability target of 0.05 per 100 km per year (IEEE, 1997).

18.4 Stroke Current Parameters

Once the downward leader contacts a power system component through an upward-connecting leader, the stored charge will be swept from the channel into a grounded object through a plasma channel with high internal impedance of 600–4000 Ω. With this high source impedance relative to the impedance of grounded structures, an impulse current source model is suitable.

Berger (1977) made the most reliable direct measurements of current and charge flow from negative downward cloud-to-ground lightning parameters on an instrumented tower from 1947 to 1977. Additional observations have been provided by many researchers and then summarized (Anderson and Eriksson, 1980; CIGRE, 1991; Takami and Okabe, 2007). The overall stroke current distribution can be approximated as lognormal with a mean of 31 kA and a log standard deviation of $\sigma_{\ln(I)} = 0.48$. The probability of exceeding a first return stroke peak current magnitude I can also be estimated from (CIGRE, 1991; IEEE, 1997; EPRI, 2005; IEEE, 2010):

$$P(I) = \frac{1}{1 + \left(\dfrac{I}{31\,\text{kA}}\right)^{2.6}} \tag{18.4}$$

The peak stroke current associated with a given probability level P can be obtained by inverting Equation 18.4 to obtain

$$I = (31\,\text{kA}) \cdot \left(\frac{1-P}{P}\right)^{(1/2.6)} \tag{18.5}$$

This leads to the following probability table.

Table 18.2 suggests that there will be a 15% chance that the first negative return stroke peak current will exceed 60 kA, and an 85% chance that it will exceed 16 kA.

The waveshape of the first return stroke current rises with a concave front, giving the maximum steepness near the crest of the wave, then decays with a time to half value of 50 μs or more. The median value of maximum steepness (CIGRE, 1991) is 24 kA/μs, with a log standard deviation of 0.60. Steepness has a strong correlation to the peak amplitude (CIGRE, 1991; Takami and Okabe, 2007) that allows simplified modeling using a single equivalent front time (peak current divided by peak rate of rise).

TABLE 18.2 Probability of Exceeding First Return Stroke Current in Lightning Flash

Probability	0.05	0.15	0.25	0.35	0.45	0.55	0.65	0.75	0.85	0.95
Peak current (kA)	96.2	60.4	47.3	39.3	33.5	28.7	24.4	20.3	15.9	10.0

The mean equivalent front is 1.4 μs for the median 31 kA current, and increases to 2.7 μs as peak stroke current increases to the 5% level of 100 kA (Takami and Okabe, 2007). An equivalent front time of 2 μs is recommended for simplified analysis of lightning performance (CIGRE, 1991; IEEE, 1997) with peak currents in the range of 50–150 kA.

18.5 Calculation of Lightning Overvoltage on Grounded Object

The peak voltage resulting from a lightning flash can be estimated from the sum of two components, the resistive voltage rise of the nearest ground electrode V_R and the inductive voltage rise V_L. The voltage rise V_L associated with conductor and tower series inductance L and the equivalent front time ($\Delta t = 2\,\mu s$) is $V_L = LI/\Delta t$. The V_L term will add to, and sometimes dominate, V_R.

18.6 Calculation of Resistive Voltage Rise V_R

The voltage rise V_R of the ground resistance R_f at each tower will be proportional to peak stroke current: $V_R = R_f I$. The resistance R_f of a tower base consisting of foundations, buried wires and anchor systems in close proximity, can be estimated closely using

$$R_f = \frac{\rho}{2\pi}\left[\frac{1}{g}\ln\left(\frac{11.8g^2}{A_{Total}}\right) + \frac{1}{l}\ln\left(\frac{A_{Total}}{2A_{Wire}}\right)\right] \tag{18.6}$$

where

- ρ is the soil resistivity (Ω-m)
- g is the geometric radius, given by the square root of the sum of the squares of the electrode extent in each direction (m)
- A_{Total} is the surface area (sides + base) of the hole needed to excavate the electrode (m²)
- l is the total length (m) of wire and foundations in the wire frame approximation to the electrode (infinite for solid electrodes)
- A_{Wire} is the surface area ($2\pi rl$) of the wire and concrete in the wire frame, with wire radius r, requiring $A_{Total} \geq 2A_{Wire}$
- ln is the natural logarithm function

For large surge currents, local ionization will reduce the second contact resistance term (varying as $1/l$) but not the first geometric resistance term varying as $1/g$ inside the square braces of Equation 18.6.

The soil resistivity ρ along a transmission line has a rather wide statistical distribution, typically with log standard deviation $\sigma_{\ln(\rho)}$ of 0.9. The variation of resistivity from tower to tower can be expressed as a probability function of the form

$$P(\rho) = \frac{1}{1 + \left(\dfrac{\rho}{\rho_{Median}}\right)^{1.85}} \tag{18.7}$$

The local soil resistivity associated with a given probability level P can be obtained by inverting Equation 18.7 to obtain

$$\rho = \rho_{Median} \cdot \left(\frac{1-P}{P}\right)^{(1/1.85)} \tag{18.8}$$

This leads to the following probability table.

TABLE 18.3 Probability of Exceeding Soil Resistivity ρ

Probability	0.05	0.15	0.25	0.35	0.45	0.55	0.65	0.75	0.85	0.95
ρ/ρ_{Median}	4.9	2.6	1.8	1.4	1.1	0.9	0.7	0.6	0.4	0.2

Table 18.3 suggests that there will be a 5% chance that the soil resistivity at any randomly selected tower will be 4.9 times higher than the median value over the entire line length.

18.7 Calculation of Inductive Voltage Rise V_L

Lumped inductance of a structure can be approximated from the expression

$$L = Z \cdot \tau = 60 \ln\left(\frac{2h}{r}\right) \cdot \frac{l}{c} \quad (18.9)$$

where

L is the inductance in Henries
Z is the element antenna impedance in ohms
h is the wire height above conducting ground (m)
r is the wire or overall structure radius (m)
l is the length of the wire or structure (m)
c is the speed of light (3×10^8 m/s)

In numerical analyses, series and shunt impedance elements can be populated using the same procedure. Tall transmission towers have longer travel times τ and thus higher inductance, which further exacerbates the increase of stroke incidence with line height. Thin steel pole structures, and wooden poles with bond wires of small radius r, will also have higher inductance than lattice towers with multiple paths to ground, giving a larger overall radius. The inductance of structures with guy wires is given by the parallel combination of the inductance of the central structure and the inductances of the individual guy wires, ignoring mutual coupling (CIGRE, 1991).

18.8 Calculation of Voltage Rise on Phase Conductor

The high electromagnetic fields surrounding any lighting flash illuminate nearby conductors and cause the flow of current, leading to induced voltages across insulators.

Fields from vertical lightning strokes to ground near overhead lines can induce overvoltages with 100–300 kV peak magnitude in nearby overhead lines without a direct flash termination. This is a particular concern only for MV and LV systems (IEEE, 2010).

In the case of a lightning flash directly to an overhead groundwire (OHGW), a small fraction of the overall current flows in horizontal directions, away from the flash location into every interconnected groundwire, shield wire beneath the phases and any phase conductor protected by a parallel line surge arrester. The voltage rise on each participant in this current flow increases common-mode voltage and reduces differential voltage across insulators through transverse electromagnetic (TEM) or surge-impedance coupling to insulated phases. Bundle configurations and corona can improve this desirable surge-impedance coupling to mitigate half of the total tower potential rise ($V_R + V_L$), but increasing separation between the phases and groundwires will reduce the effect.

Calculation of the coupling coefficients C_n on the undriven, unprotected phase conductors calls for registering the self and mutual surge impedances of each phase and groundwire, setting the voltage on the stricken conductors to unity and calculating the potential rise on undriven phases from the inverse of the resulting surge impedance matrix. Simplified methods for systems with one or two overhead groundwires (CIGRE, 1991; IEEE, 1997) consider voltage dependent corona effects as well as bundle conductor impedance.

The combined peak stress on an insulator under lightning surge conditions, V_{Pk} (kV), with a linear front time Δt of 2 μs, can be approximated by

$$V_{Pk} \approx \frac{I_{Pk} \cdot (1 - C_n)}{\dfrac{1}{R_f + \dfrac{L}{2\,\mu s}} + \dfrac{2n}{Z_{GW}}} \tag{18.10}$$

where

I_{Pk} is the peak first return stroke current (kA)
C_n is the surge impedance coupling coefficient from n groundwires, modified for corona effects
n is the number of groundwires, including OHGWs, underbuilt OPGW, and neutral wires and phases protected with line surge arresters
Z_{GW} is an average value of surge impedance of the groundwires (Ω)
R_f is the resistance of the stricken tower to ground from Equation 18.6 (Ω)
L is the inductance of the stricken tower from insulator location to ground (H) from Equation 18.7

18.9 Joint Distribution of Peak Voltage on Insulators

Since the peak stroke current and the resistivity at the base of a tower are statistically independent, the joint distribution of their voltage stress levels can be obtained by summing over the probability of all possible events.

Table 18.4 shows that the voltage stress on a transmission line insulator varies by a 50:1 range as a result of statistical variations in lightning peak current magnitude and tower-to-tower changes in soil resistivity.

Electrical utilities will often install additional buried grounding electrodes, such as vertical rods or radial counterpoise wires a meter below grade, at towers that have high soil resistivity. Construction specifications

TABLE 18.4 Probability Table for Insulator Voltage Rise on Untreated Line Median R_f of 15 Ω, L of 15 μH, Z_{GW} 500 Ω, n = 2, C_n = 0.3

V_{pk} (I_{pk}, R_f) (kV)		$P(I_{pk})$	0.05	0.15	0.25	0.35	0.45	0.55	0.65	0.75	0.85	0.95
$P(R_f)$	R_f (Ω)	I_{pk} (kA)	96.2	60.4	47.3	39.3	33.5	28.7	24.4	20.3	15.9	10.0
0.05	73.7		3314	2081	1630	1355	1154	989	842	700	548	344
0.15	38.3		2258	1418	1110	923	786	673	573	477	373	234
0.25	27.2		1828	1148	899	747	636	545	464	386	302	190
0.35	21.0		1561	980	768	638	543	466	396	330	258	162
0.45	16.7		1366	858	672	559	476	408	347	289	226	142
0.55	13.5		1209	759	594	494	421	361	307	255	200	126
0.65	10.7		1072	673	527	438	373	320	272	226	177	111
0.75	8.3		944	593	464	386	328	282	240	199	156	98
0.85	5.9		814	511	400	333	283	243	207	172	135	84
0.95	3.1		655	412	322	268	228	196	166	138	108	68

may call for achieving "20 Ω resistance where practical." Thus, the distribution of footing resistance R_f in Table 18.4 is modified by a "treatment rule" that follows this general model:

- If $R_f < 20\,\Omega$, do nothing.
- If $20\,\Omega < R_f < 40\,\Omega$, install enough grounding to reduce to 20 Ω.
- If $R_f > 40\,\Omega$, install enough grounding to reduce R_f by factor of 2.

This treatment strategy will improve the line outage rate as shown hereafter.

18.10 Insulation Strength

The lightning impulse flashover gradient (kV CFO per meter of dry arc distance) of typical transmission line insulator strings is linear over a wide range from 1 to 6 m. The critical flashover level (CFO) is the median voltage at which flashover occurs when tested with a standard lightning impulse voltage wave with 1.2 μs rise time and 50 μs time to half value, and is normally distributed with a relative standard deviation of about 5%. The CFO for full lightning impulse voltage waves scales linearly with insulator string dry arc distance as shown in Table 18.6.

The probability of flashover with 5% relative standard deviation can be approximated conveniently by Equation 18.11. For example, with V_{Pk} = 1136 kV applied to insulation having CFO = 1080 kV, the probability of flashover is 85%. However, accurate normal distribution functions are readily accessible in spreadsheets such as Excel and should be used where available:

$$P(flashover) \approx 1 - \frac{1}{1 + \left(\frac{V_{Pk}}{\mathrm{CFO}(t)}\right)^{35}} \tag{18.11}$$

The lightning impulse flashover voltage has a pronounced nonlinear volt–time characteristic, giving an increasing ability to withstand short-duration impulses at times t less than 10 μs compared to the full-wave CFO strength of 540 kV/m. The lightning surge itself peaks in an equivalent front time of about 2 μs. A simplified method may evaluate the possibility of flashover at this time t, resulting in a fixed strength of 822 kV per meter of dry arc distance $D_{Dry\,Arc}$ based on a volt–time characteristic as follows (Darveniza et al., 1975):

$$\mathrm{CFO} = D_{Dry\,Arc} \cdot \left(400 + \frac{710}{t^{0.75}}\right) \tag{18.12}$$

where

$D_{Dry\,Arc}$ is the dry arc distance of the insulator (m), in the range of 1–6 m
t is the time of flashover (μs), in the range of 0.3–14 μs
CFO is the peak of the applied standard lightning impulse voltage wave (kV) that causes a flashover 50% of the time

A volt–time curve approach such as Equation 18.12 remains valid up to the point in time when the applied voltage wave deviates significantly from the standard test wave. In the case of transmission lines, this point is well defined as the time at which cancelling reflections from the ground electrodes of nearby towers arrive, after a propagation time t_{Span} associated with 90% of the speed of light, c. Table 18.7 shows that the span length can thus change the critical flashover voltage by ±10%, leading to about ±30% changes in the predicted line outage rate.

18.11 Calculation of Transmission Line Outage Rate

The lightning outage rate of a transmission line is given by the number of flashes to the line, Equation 18.3, multiplied by the probability of flashover of each flash. Tables 18.4 and 18.5 have shown how the distribution of peak backflashover voltage stress across insulation varies for the probability distribution of peak first return stroke current I_{pk} and footing resistance R_f, considering that other factors such as tower inductance L_{twr}, number of OHGWs n, and the related coupling coefficient C_n in Equation 18.10 are all fixed. Tables 18.6 and 18.7 give the insulation characteristics as a function of insulator dry arc distance $D_{Dry\ Arc}$ and span length, which can also be calculated with Equation 18.12 for a particular line design or section. Thus, the calculation of a line outage rate simplifies into a calculation of the probability of flashover for each element in Table 18.4 or 18.5, summed over the entire range of probability as illustrated in Figure 18.2.

Computer programs and methods for calculating lightning outage rates (CIGRE, 1991; IEEE, 1997; Hileman, 1999) make use of the simplified concepts illustrated in Figure 18.2, but adding in calculation details related to

- Automatic calculation of individual conductor surge impedances Z_{GW} and coupling coefficients C_n at each phase conductor, incorporating nonlinear increase in C_n with increasing tower top voltage
- Automated analysis of the risk of a shielding failure and consequent flashover from a direct lightning flash to a phase conductor
- Integration of line voltage bias for every degree of phase (0°–360°) to establish the proportion of backflashover failures among phases

TABLE 18.5 Probability Table for Insulator Voltage Rise on Treated Line Median R_f of 15 Ω, Treatment to 20 Ω, L of 15 μH, Z_{GW} 500 Ω, n = 2, C_n = 0.3

V_{pk} (I_{pk}, R_f) (kV)		$P(I_{pk})$	0.05	0.15	0.25	0.35	0.45	0.55	0.65	0.75	0.85	0.95
$P(R_f)$	R_f (Ω)	I_{pk} (kA)	96.2	60.4	47.3	39.3	33.5	28.7	24.4	20.3	15.9	10.0
0.05	36.8		2204	1384	1084	901	767	657	560	465	364	229
0.15	20.0		1518	953	746	621	528	453	386	321	251	158
0.25	20.0		1518	953	746	621	528	453	386	321	251	158
0.35	20.0		1518	953	746	621	528	453	386	321	251	158
0.45	16.7		1366	858	672	559	476	408	347	289	226	142
0.55	13.5		1209	759	594	494	421	361	307	255	200	126
0.65	10.7		1072	673	527	438	373	320	272	226	177	111
0.75	8.3		944	593	464	386	328	282	240	199	156	98
0.85	5.9		814	511	400	333	283	243	207	172	135	84
0.95	3.1		655	412	322	268	228	196	166	138	108	68

TABLE 18.6 Peak Flashover Voltage (kV) versus Probability of Flashover for Insulator Strings Based on Critical Impulse Flashover Gradient of 540 kV/m and 5% Relative Standard Deviation

Dry Arc Distance (m)	Critical Impulse Flashover Level (kV)	Standard Deviation of CFO (kV)	Probability Level (Normal Distribution)									
			0.05	0.15	0.25	0.35	0.45	0.55	0.65	0.75	0.85	0.95
1	540	27	496	512	522	530	537	543	550	558	568	584
2	1080	54	991	1024	1044	1059	1073	1087	1101	1116	1136	1169
3	1620	81	1487	1536	1565	1589	1610	1630	1651	1675	1704	1753
4	2160	108	1982	2048	2087	2118	2146	2174	2202	2233	2272	2338

TABLE 18.7 Peak Flashover Voltage (kV) for 1-m (7 Standard Disk) Insulator String as Function of Span Length

Span (m)	Travel Time (μs)	CFO at t_{Span} (kV)	Standard Deviation of CFO (kV)	Probability Level (Normal Distribution) 0.05	0.15	0.25	0.35	0.45	0.55	0.65	0.75	0.85	0.95
200	1.5	929	46	852	881	897	911	923	935	947	960	977	1005
250	1.9	847	42	778	803	819	831	842	853	864	876	891	917
300	2.2	790	40	725	749	763	775	785	795	805	817	831	855
350	2.6	748	37	686	709	722	733	743	752	762	773	786	809

Insulator voltage stress factors

Median peak current, kA	31 kA	exponent (sigma ln I = 0.48)	2.6
Single OHGW surge impedance	500 Ω	Equiv linear front	2 μs

$$P(I) = \frac{1}{1 + \left(\frac{I}{31\ \text{kA}}\right)^{2.6}}$$

# of parallel groundwires	2	Including OHGW, UBGW and arrester protected phases
Coupling coefficient C_n	0.3	Considering number and proximity of parallel groundwires; bundle and corona effects
Tower inductance	15	μH effective from insulator position to ground
Median footing resistance	15	Ω exponent (sigma ln R_f = 0.9) 1.85
Span length	300	m
Power system voltage bias	81	kV, at least one phase will be positive relative to tower

$$P(R_f) = \frac{1}{1 + \left(\frac{R_f}{15\ \Omega}\right)^{1.85}}$$

Insulator voltage strength factors

Insulator distance $D_{Dry\,Arc}$	1	m
Span travel time t_{Span}	2.22	μs
CFO of dry arc distance	790	kV CFO at span travel time
Exponent for 5% Std Dev	35	

$$P(flashover) = 1 - \frac{1}{1 + \left(\frac{V_{Pk}}{CFO\,(t_{Span})}\right)^{35}}$$

Peak insulator voltage stress, modified by coupling

P	Raw R_f	Treated	0.05	0.15	0.25	0.35	0.45	0.55	0.65	0.75	0.85	0.95	Probability, P
			96.2	60.4	47.3	39.3	33.5	28.7	24.4	20.3	15.9	10.0	Peak current, kA
0.05	73.7	36.8	2285	1465	1164	982	848	738	640	546	445	310	
0.15	38.3	20.0	1599	1034	827	701	609	534	466	401	332	238	
0.25	27.2	20.0	1599	1034	827	701	609	534	466	401	332	238	
0.35	21.0	20.0	1599	1034	827	701	609	534	466	401	332	238	
0.45	16.7	16.7	1447	936	752	639	556	488	428	369	307	223	
0.55	13.5	13.5	1289	840	675	575	501	441	388	336	281	206	
0.65	10.7	10.7	1152	754	608	519	454	400	353	307	258	192	
0.75	8.3	8.3	1024	673	545	467	409	362	320	280	237	179	
0.85	5.9	5.9	894	592	481	413	364	323	287	253	215	165	
0.95	3.1	3.1	736	492	403	349	309	276	247	219	189	149	

P Footing resistance R_f, Ω

Probability of flashover based on volt–time curve of insulator string

1.000	1.000	1.000	1.000	0.922	0.085	0.001	0.000	0.000	0.000
1.000	1.000	0.832	0.015	0.000	0.000	0.000	0.000	0.000	0.000
1.000	1.000	0.832	0.015	0.000	0.000	0.000	0.000	0.000	0.000
1.000	1.000	0.832	0.015	0.000	0.000	0.000	0.000	0.000	0.000
1.000	0.998	0.154	0.001	0.000	0.000	0.000	0.000	0.000	0.000
1.000	0.894	0.004	0.000	0.000	0.000	0.000	0.000	0.000	0.000
1.000	0.161	0.000	0.000	0.000	0.000	0.000	0.000	0.000	0.000
1.000	0.004	0.000	0.000	0.000	0.000	0.000	0.000	0.000	0.000
0.987	0.000	0.000	0.000	0.000	0.000	0.000	0.000	0.000	0.000
0.078	0.000	0.000	0.000	0.000	0.000	0.000	0.000	0.000	0.000

Backflashover protection efficiency	79.2%	With 100 cells and empirical probability
	78.7%	With 10,000 cells and normal distribution function

FIGURE 18.2 Simplified spreadsheet calculation of backflashover protection efficiency based on Table 18.5 and Equation 18.11.

Advanced computer models are available to compute the possibility of multiple-phase or multi-circuit backflashover, and also to investigate the effects of applying transmission line surge arresters across selected insulators to limit their overvoltage stress and increase coupling coefficients on unprotected phases as suggested in (CIGRE, 2010).

18.12 Improving the Transmission Line Lightning Outage Rate

There are a number of options that affect the transmission line outage rate. A design with adequate shielding performance will use OHGWs to provide an estimated 0.05 shielding failures per 100 km year on new designs. It is difficult to reposition existing OHGWs on existing lines. If a study shows that time-correlated lightning outages on a line are the result of surges with low peak amplitudes (<20 kA), estimated from a lightning location system, then the application of transmission line surge arresters of suitable energy rating should be considered.

18.12.1 Increasing the Insulator Dry Arc Distance

Insulator dry arc distance, or the number of disks selected for insulator strings, has a remarkable effect on the lightning performance of transmission lines. At the 115 and 138 kV levels, it is common to use 7 or 8 standard (146 × 254 mm) disks or the equivalent polymer insulator length, giving $D_{Dry\,Arc}$ = 1–1.2 m. At 230 kV, 14 disks are common and EHV lines may use 23–26 disks at the 500 kV level for $D_{Dry\,Arc}$ = 3.4–3.8 m. This range of dry arc dimensions can change the lightning performance of a typical transmission line by a factor of 10 or more, as shown in Tables 18.8 through 18.10.

18.12.2 Modifying the Distribution of Footing Resistance

The simplified spreadsheet example of Figure 18.2 shows the relative outage rate in Tables 18.8 and 18.9 obtained when a utility makes an effort during construction to reduce most footing resistance values to less than 20 Ω "where feasible" using the modification schedule in Table 18.5. If no effort is made to improve grounding, leading to the untreated resistance and voltage stress values in Table 18.4, the efficiency of double OHGW protection decreases from 78.7% to 73.0%, meaning that the lightning fault

TABLE 18.8 Effect of Insulator Dry Arc Distance on Efficiency of Lightning Protection for Line with Median Footing Resistance R_f = 15 Ω, Treated to Reduce below 20 Ω Where Feasible[a]

Dry Arc Distance $D_{Dry\,Arc}$ (m)	Typical System Voltage (kV)	n = 2 (OHGW), C_n = 0.3		n = 3 (2 OHGW, UBGW), C_n = 0.4		n = 4 (2 OHGW + 2 Phases with Arresters), C_n = 0.5	
		Efficiency (%)	Line Fault Rate[a]	Efficiency (%)	Line Fault Rate[a]	Efficiency (%)	Line Fault Rate[a]
1	115	78.7	***1.00***	86.5	0.63	92.3	0.36
1.5	161	90.3	0.46	94.4	0.26	97.1	0.14
2	230	94.8	0.24	97.2	0.13	98.6	0.07
2.5	275	96.9	0.15	98.4	0.08	99.2	0.04
3	345	98.0	0.09	99.0	0.05	99.5	0.02
3.5	500	98.7	0.06	99.3	0.03	99.8	0.01

[a] Normalized to reference case with $D_{Dry\,Arc}$ = 1 m, Median R_f = 15 Ω, treated to 20 Ω or factor of 2 reduction (whichever is greater), n = 2, C_n = 0.3, 300 m span, L_{twr} = 15 μH. *Note:* The value of 1.00 is the reference case described in superscript a.

TABLE 18.9 Effect of Insulator Dry Arc Distance on Efficiency of Lightning Protection for Line with Median Footing Resistance $R_f = 15\,\Omega$, Untreated[a]

		$n = 2$ (OHGW), $C_n = 0.3$		$n = 3$ (2 OHGW, UBGW), $C_n = 0.4$		$n = 4$ (2 OHGW + 2 Phases with Arresters), $C_n = 0.5$	
Dry Arc Distance $D_{Dry\,Arc}$ (m)	Typical System Voltage (kV)	Efficiency (%)	Line Fault Rate[a]	Efficiency (%)	Line Fault Rate[a]	Efficiency (%)	Line Fault Rate[a]
1	115	73.0	1.27	82.0	0.84	89.5	0.49
1.5	161	85.9	0.66	91.7	0.39	95.7	0.20
2	230	91.8	0.39	95.5	0.21	97.8	0.10
2.5	275	94.8	0.24	97.4	0.12	98.8	0.06
3	345	96.5	0.16	98.3	0.08	99.3	0.03
3.5	500	97.6	0.11	98.9	0.05	99.6	0.02

[a] Normalized to reference case with $D_{Dry\,Arc} = 1$ m, Median $R_f = 15\,\Omega$, treated to $20\,\Omega$ or factor of 2 reduction (whichever is greater), $n = 2$, $C_n = 0.3$, 300 m span, $L_{twr} = 15\,\mu$H.

TABLE 18.10 Effect of Insulator Dry Arc Distance on Efficiency of Lightning Protection for Line with Median Footing Resistance $R_f = 30\,\Omega$, Treated to Reduce below $20\,\Omega$ Where Feasible[a]

		$n = 2$ (OHGW), $C_n = 0.3$		$n = 3$ (2 OHGW, UBGW), $C_n = 0.4$		$n = 4$ (2 OHGW + 2 Phases with Arresters), $C_n = 0.5$	
Dry Arc Distance $D_{Dry\,Arc}$ (m)	Typical System Voltage (kV)	Efficiency (%)	Line Fault Rate[a]	Efficiency (%)	Line Fault Rate[a]	Efficiency (%)	Line Fault Rate[a]
1	115	67.7	1.52	78.7	1.00	87.7	0.58
1.5	161	83.5	0.78	90.3	0.45	95.0	0.24
2	230	90.5	0.44	94.9	0.24	97.5	0.12
2.5	275	94.1	0.28	96.9	0.14	98.6	0.07
3	345	96.1	0.18	98.1	0.09	99.0	0.05
3.5	500	97.3	0.13	98.6	0.06	99.6	0.02

[a] Normalized to reference case with $D_{Dry\,Arc} = 1$ m, Median $R_f = 15\,\Omega$, treated to $20\,\Omega$ or factor of 2 reduction (whichever is greater), $n = 2$, $C_n = 0.3$, 300 m span, $L_{twr} = 15\,\mu$H.

rate would be 27% higher without treatment for a dry arc distance of 1 m with other line characteristics fixed. As dry arc distance increases, improved grounding makes a larger fractional reduction in a decreasing outage rate.

18.12.3 Increasing the Effective Number of Groundwires Using UBGW

The number of shield wires in parallel, n, has a direct role in Equation 18.10 as well as an indirect influence on the value of C_n, the electromagnetic coupling coefficient from all n driven shield wires (those carrying a small fraction of lightning current) and the insulated phase. The groundwires consist of the traditional overhead groundwires (OHGWs) as well as underbuilt groundwires (UBGWs) and any phases protected with line surge arresters, including circuits at lower distribution voltages. UBGWs have been used on transmission lines to provide convenient access to optical fibers, giving a protected location of the metal-sheathed OPGW that is not exposed to direct flashes, as well as to manage ac fault currents and to reduce electric and magnetic fields in urban areas. UBGWs are preferred to buried continuous counterpoise for the safety and lightning protection functions as they have reduced installation cost, less environmental impact, easier inspection, and greater physical security. The improved coupling coefficient C_n associated with a single UBGW is seen in Tables 18.8 and 18.9 to be roughly as effective as grounding improvements to maintain $20\,\Omega$ resistance "where feasible."

18.12.4 Increasing the Effective Number of Groundwires Using Line Surge Arresters

When selected properly, line surge arresters clip the transient overvoltages across insulators in lightning surge conditions to prevent flashovers across nearby insulators. The limit distance related to lightning equivalent front time typically means that arresters on one transmission tower are typically too far away to provide protection of insulators on the same phase of adjacent towers. Tables 18.8 through 18.10 suggest that, in addition to eliminating flashovers on the protected phases, the flow of current through the arresters and the resulting rise in potential on the protected phases make important improvements in the lightning performance of unprotected phases on the same tower. More detailed examples, including the use of arresters on a lower voltage circuit to protect a higher voltage circuit, are found in (CIGRE, 2010).

18.13 Conclusion

Direct lightning strokes to any overhead transmission line are likely to cause impulse flashover of supporting insulation, leading to a circuit interruption. The use of overhead shield wires, located above the phase conductors and grounded adequately at each tower, can reduce the risk of flashover by 70%–99.8% depending on insulation dry arc distance and soil conditions. Underbuilt groundwires and phases protected with line surge arresters both improve electromagnetic coupling and can further reduce the risk of backflashover to achieve protection efficiency that exceeds 90%, even for systems with 1 m dry arc distance.

References

Anderson, R. B. and Eriksson, A. J., Lightning parameters for engineering applications, *Electra* 69, 65–102, 1980.

Anderson, R. B., Eriksson, A. J., Kroninger, H., and Meal, D. V., Lightning and thunderstorm parameters, in *Lightning and Power Systems*, IEE Conference Publication 236, London, U.K., June 1984.

Berger, K., The earth flash, in *Lightning*, ed. Golde, R., Academic Press, London, U.K., 1977, pp. 119–190.

Christian, H. J., Blakeslee, R. J., Boccippio, D., Boeck, W., Buechler, D., Driscoll, K., Goodman, S., Hall. J., Koshak, W., Mach, D., and Stewart, M., Global frequency and distribution of lighting as observed from space by the optical transient detector, *J. Geophys. Res.* 108(D1), 4005, 2003.

CIGRE Working Group 01 (Lightning) of Study Committee 33, Guide to procedures for estimating the lightning performance of transmission lines, Technical Brochure 63, Paris, France, 1991.

CIGRE WG C4.404, Cloud-to-ground lightning parameters derived from lightning location systems—The effects of system performance, Technical Brochure 376, CIGRE, Paris, France, 2009.

CIGRE Working Group C4.301, Use of surge arresters for lightning protection of transmission lines, Technical Brochure 440, CIGRE, Paris, France, 2010.

Darveniza, M., Popolansky, F., and Whitehead, E. R., Lightning protection of UHV transmission lines, *Electra* 41, 39–69, 1975.

EPRI, *EPRI Transmission Line Reference Book, 200 kV and Above*, 3rd edn., Chapter 6. Lightning and grounding, Chisholm, W. A. and Anderson, J. G., EPRI, Palo Alto, CA, 2005, 1011974.

Heydt, G., Instrumentation, in *Handbook of Atmospherics*, vol. II, ed. Volland, H., CRC Press, Boca Raton, FL, 1982, pp. 203–256.

Hileman, A. R., *Insulation Coordination for Power Systems*, CRC Press, Boca Raton, FL, 1999.

IEEE Standard 1243–1997™, *IEEE Guide for Improving the Lightning Performance of Transmission Lines*, IEEE, Piscataway, NJ, December 1997.

IEEE Standard 1410–2010™, *IEEE Guide for Improving the Lightning Performance of Electric Power Overhead Distribution Lines*, IEEE, Piscataway, NJ, January 2011.

Krider, E. P., Noggle, R. C., and Uman, M. A., A gated, wideband direction finder for lightning return strokes, *J. Appl. Meteorol.* 15, 301, 1976.

MacGorman, D. R., Maier, M. W., and Rust, W. D., Lightning strike density for the contiguous United States from thunderstorm duration records, Report to U.S. Nuclear Regulatory Commission, NUREG/CR-3759, 1984.

NASA High Resolution Full Climatology version 2.2, LIS/OTD 0.5 Degree High Resolution Full Climatology (HRFC) http://gcmd.nasa.gov/records/GCMD_lohrfc.html, 2006 (accessed on December 16, 2011).

Rakov, V. and Uman, M. A., *Lightning Physics and Effects*, Cambridge University Press, Cambridge, U.K., 2007.

Rizk, F. A. M., Modeling of transmission line exposure to direct lightning strokes, *IEEE Trans. PWRD* 5(4), 1983, 1990.

Takami, J. and Okabe, S., Observational results of lightning current on transmission towers, *IEEE Trans. PWRD* 22(1), 547, 2007.

19

Reactive Power Compensation

Rao S. Thallam
Salt River Project

Géza Joós
McGill University

19.1 Need for Reactive Power Compensation

Except in a very few special situations, electrical energy is generated, transmitted, distributed, and utilized as alternating current (AC). However, AC has several distinct disadvantages. One of these is the necessity of reactive power that needs to be supplied along with active power. Reactive power can be leading or lagging. While it is the active power that contributes to the energy consumed, or transmitted, reactive power does not contribute to the energy. Reactive power is an inherent part of the "total power." Reactive power is either generated or consumed in almost every component of the system, generation, transmission, and distribution and eventually by the loads. The impedance of a branch of a circuit in an AC system consists of two components, resistance and reactance. Reactance can be either inductive or capacitive, which contributes to reactive power in the circuit. Most of the loads are inductive, and must be supplied with lagging reactive power. It is economical to supply this reactive power closer to the load in the distribution system.

In this chapter, reactive power compensation, mainly in transmission systems installed at substations, is discussed. Reactive power compensation in power systems can be either shunt or series. Both will be discussed.

19.1.1 Shunt Reactive Power Compensation

Since most loads are inductive and consume lagging reactive power, the compensation required is usually supplied by leading reactive power. Shunt compensation of reactive power can be employed either at load level, substation level, or at transmission level. It can be capacitive (leading) or inductive (lagging) reactive power, although in most cases as explained before, compensation is capacitive. The most common form of leading reactive power compensation is by connecting shunt capacitors to the line.

19.1.2 Shunt Capacitors

Shunt capacitors are employed at substation level for the following reasons:

1. Voltage regulation: The main reason that shunt capacitors are installed at substations is to control the voltage within required levels. Load varies over the day, with very low load from midnight to early morning and peak values occurring in the evening between 4 and 7 PM. Shape of the load curve also varies from weekday to weekend, with weekend load typically low. As the load varies, voltage at the substation bus and at the load bus varies. Since the load power factor is always lagging, a shunt-connected capacitor bank at the substation can raise voltage when the load is high. The shunt capacitor banks can be permanently connected to the bus (fixed capacitor bank) or can be switched as needed. Switching can be based on time, if load variation is predictable, or can be based on voltage, power factor, or line current.
2. Reducing power losses: Compensating the load lagging power factor with the bus-connected shunt capacitor bank improves the power factor and reduces current flow through the transmission lines, transformers, generators, etc. This will reduce power losses (I^2R losses) in this equipment.
3. Increased utilization of equipment: Shunt compensation with capacitor banks reduces kVA loading of lines, transformers, and generators, which means with compensation they can be used for delivering more power without overloading the equipment.

Reactive power compensation in a power system is of two types—shunt and series. Shunt compensation can be installed near the load, in a distribution substation, along the distribution feeder, or in a transmission substation. Each application has different purposes. Shunt reactive compensation can be inductive or capacitive. At load level, at the distribution substation, and along the distribution feeder, compensation is usually capacitive. In a transmission substation, both inductive and capacitive reactive compensations are installed.

19.2 Application of Shunt Capacitor Banks in Distribution Systems: A Utility Perspective

The Salt River Project (SRP) is a public power utility serving more than 720,000 (April 2000) customers in central Arizona. Thousands of capacitor banks are installed in the entire distribution system. The primary usage for capacitor banks in the distribution system is to maintain a certain power factor at peak loading conditions. The target power factor is 0.98 leading at system peak. This figure was set as an attempt to have a unity power factor on the 69 kV side of the substation transformer. The leading power factor compensates for the industrial substations that have no capacitors. The unity power factor maintains a balance with ties to other utilities.

The main purpose of the capacitors is not for voltage support, as the case may be at utilities with long distribution feeders. Most of the feeders in the SRP service area do not have long runs (substations are about 2 miles apart) and load tap changers on the substation transformers are used for voltage regulation.

The SRP system is a summer peaking system. After each summer peak, a capacitor study is performed to determine the capacitor requirements for the next summer. The input to the computer program for evaluating capacitor additions consists of three major components:

- Megawatts and megavars for each substation transformer at peak
- A listing of the capacitor banks with size and operating status at time of peak
- The next summer's projected loads

By looking at the present peak MW and Mvars and comparing the results to the projected MW loads, Mvar deficiencies can be determined. The output of the program is reviewed and a listing of potential needs is developed. The system operations personnel also review the study results and their input is included in making final decisions about capacitor bank additions.

Once the list of additional reactive power requirements is finalized, determinations are made about the placement of each bank. The capacitor requirement is developed on a per-transformer basis. The ratio of the kvar connected to kVA per feeder, the position on the feeder of existing capacitor banks, and any concentration of present or future load are all considered in determining the position of the new capacitor banks. All new capacitor banks are 1200 kvar. The feeder type at the location of the capacitor bank determines if the capacitor will be pole-mounted (overhead) or pad-mounted (underground).

Capacitor banks are also requested when new feeders are being proposed for master plan communities, large housing developments, or heavy commercial developments.

Table 19.1 shows the number and size of capacitor banks in the SRP system in 1998. Table 19.2 shows the number of line capacitors by type of control.

Substation capacitor banks (three or four per transformer) are usually staged to come on and go off at specific load levels.

TABLE 19.1 Number and Size of Capacitor Banks in the SRP System

	Number of Banks	
Kvar	Line	Station
150	1	
300	140	
450	4	
600	758	2
900	519	
1200	835	581
Total	2257	583

TABLE 19.2 SRP Line Capacitors by Type of Control

Type of Control	Number of Banks
Current	4
Fixed	450
Time	1760
Temperature	38 (used as fixed)
Voltage	5

19.3 Static VAR Control

Static VAR compensators, commonly known as SVCs, are shunt-connected devices, vary the reactive power output by controlling or switching the reactive impedance components by means of power electronics. This category includes the following equipment:

Thyristor-controlled reactors (TCR) with fixed capacitors (FC)
Thyristor switched capacitors (TSC)
Thyristor-controlled reactors in combination with mechanically or thyristor switched capacitors

SVCs are installed to solve a variety of power system problems:

1. Voltage regulation
2. Reduce voltage flicker caused by varying loads like arc furnace, etc.
3. Increase power transfer capacity of transmission systems
4. Increase transient stability limits of a power system
5. Increase damping of power oscillations
6. Reduce temporary overvoltages
7. Damp subsynchronous oscillations

A view of an SVC installation is shown in Figure 19.1.

19.3.1 Description of SVC

Figure 19.2 shows three basic versions of SVC. Figure 19.2a shows configuration of TCR with fixed capacitor banks. The main components of an SVC are thyristor valves, reactors, the control system, and the step-down transformer.

19.3.2 How Does SVC Work?

As the load varies in a distribution system, a variable voltage drop will occur in the system impedance, which is mainly reactive. Assuming the generator voltage remains constant, the voltage at the load bus will vary. The voltage drop is a function of the reactive component of the load current, and system and

FIGURE 19.1 View of SVC installation. (Photo courtesy of ABB Inc., Auburn Hills, MI.)

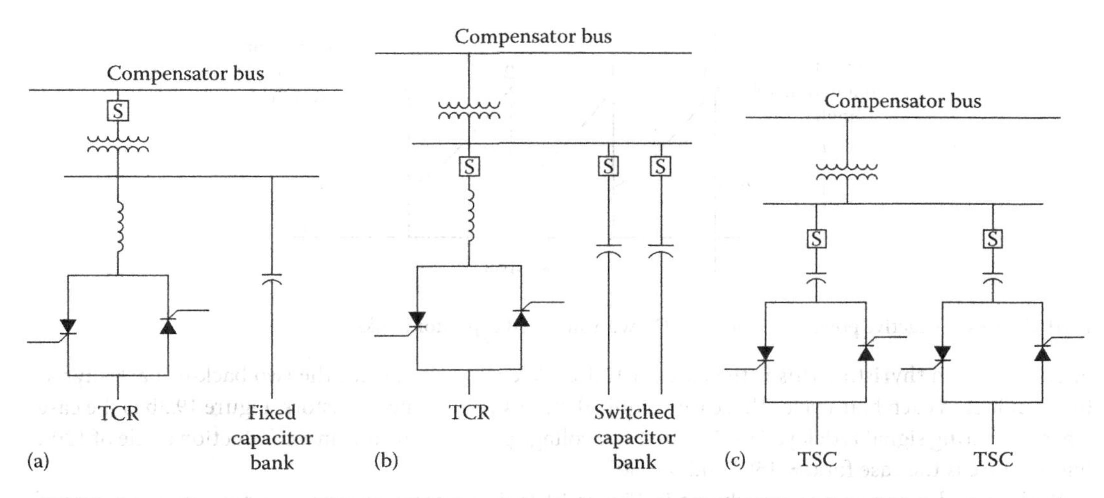

FIGURE 19.2 Three versions of SVC. (a) TCR with fixed capacitor bank; (b) TCR with switched capacitor banks; and (c) TSC compensator.

transformer reactance. When the loads change very rapidly, or fluctuate frequently, it may cause "voltage flicker" at the customers' loads. Voltage flicker can be annoying and irritating to customers because of the "lamp flicker" it causes. Some loads can also be sensitive to these rapid voltage fluctuations.

An SVC can compensate voltage drop for load variations and maintain constant voltage by controlling the duration of current flow in each cycle through the reactor. Current flow in the reactor can be controlled by controlling the gating of thyristors that control the conduction period of the thyristor in each cycle, from zero conduction (gate signal off) to full-cycle conduction. In Figure 19.2a, for example, assume the MVA of the fixed capacitor bank is equal to the MVA of the reactor when the reactor branch is conducting for full cycle. Hence, when the reactor branch is conducting full cycle, the net reactive power drawn by the SVC (combination of capacitor bank and TCR) will be zero. When the load reactive power (which is usually inductive) varies, the SVC reactive power will be varied to match the load reactive power by controlling the duration of the conduction of current in the thyristor-controlled reactive power branch. Figure 19.3 shows current waveforms for three conduction levels, 60°, 120°, and 180°. Figure 19.3a shows waveforms for thyristor gating angle (α) of 90°, which gives a conduction angle (σ)

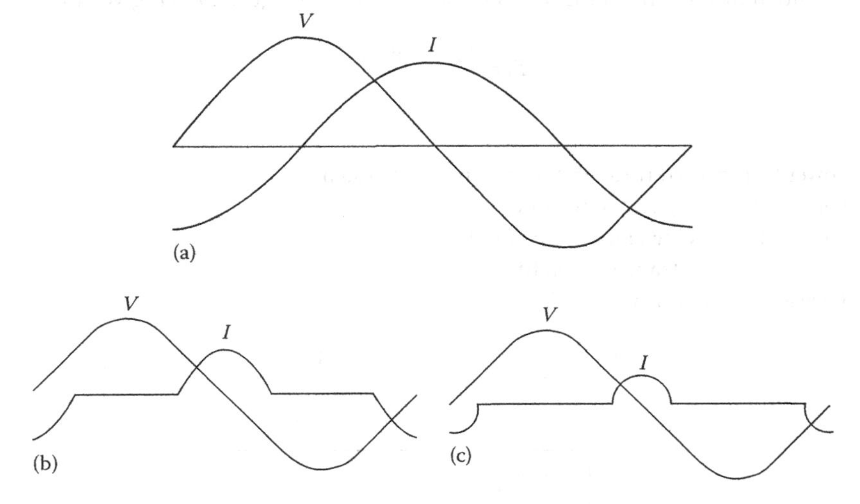

FIGURE 19.3 TCR voltage (V) and current (I) waveforms for three conduction levels. Thyristor gating angle = α; conduction angle = σ. (a) $\alpha = 90°$ and $\sigma = 180°$; (b) $\alpha = 120°$ and $\sigma = 120°$; and (c) $\alpha = 150°$ and $\sigma = 60°$.

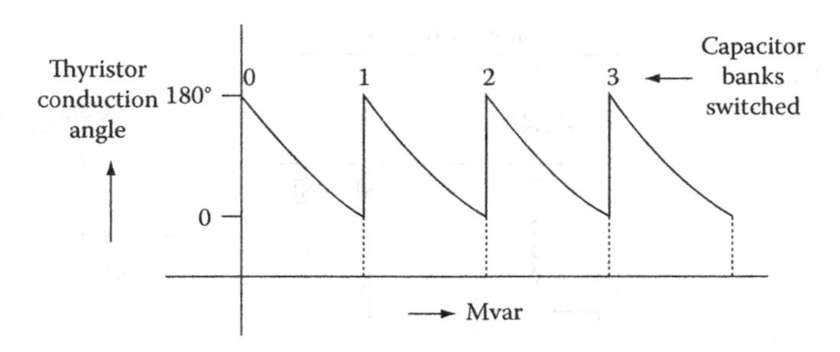

FIGURE 19.4 Reactive power variation of TCR with switched capacitor banks.

of 180° for each thyristor. This is the case for full-cycle conduction, since the two back-to-back thyristors conduct in each half-cycle. This case is equivalent to shorting the thyristors. Figure 19.3b is the case when the gating signal is delayed for 30° after the voltage peak, and results in a conduction angle of 120°. Figure 19.3c is the case for $\alpha = 150°$ and $\sigma = 60°$.

With a fixed capacitor bank as shown in Figure 19.2a, it is possible to vary the net reactive power of the SVC from 0 to the full capacitive VAR only. This is sufficient for most applications of voltage regulation, as in most cases only capacitive VARs are required to compensate the inductive VARs of the load. If the capacitor can be switched on and off, the Mvar can be varied from full inductive to full capacitive, up to the rating of the inductive and capacitive branches. The capacitor bank can be switched by mechanical breakers (see Figure 19.2b) if time delay (usually 5–10 cycles) is not a consideration, or they can be switched fast (less than 1 cycle) by thyristor switches (see Figure 19.2c).

Reactive power variation with switched capacitor banks for an SVC is shown in Figure 19.4.

19.4 Series Compensation

Series compensation is commonly used in high-voltage AC transmission systems. They were first installed in the late 1940s. Series compensation increases power transmission capability, both steady state and transient, of a transmission line. Since there is an increasing opposition from the public to construction of EHV transmission lines, series capacitors are attractive for increasing the capabilities of transmission lines. Series capacitors also introduce some additional problems for the power system. These will be discussed later.

Power transmitted through the transmission system (shown in Figure 19.5) is given by

$$P_2 = \frac{V_1 \cdot V_2 \cdot \sin\delta}{X_L} \tag{19.1}$$

where

P_2 is the power transmitted through the transmission system
V_1 is the voltage at sending end of the line
V_2 is the voltage at receiving end of transmission line
X_L is the reactance of the transmission line
δ is the phase angle between V_1 and V_2

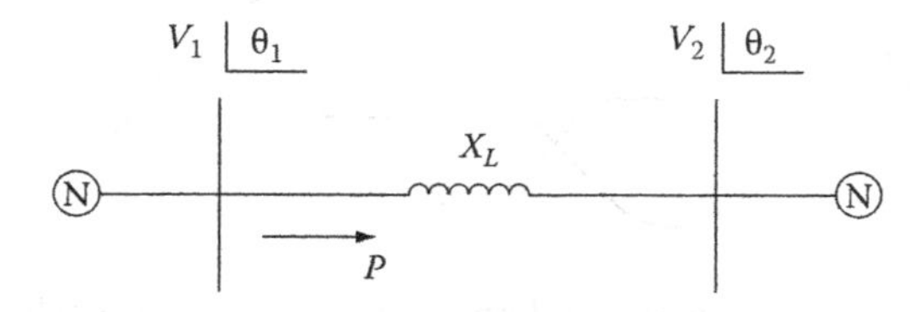

FIGURE 19.5 Power flow through transmission line.

Equation 19.1 shows that if the total reactance of a transmission system is reduced by installing capacitance in series with the line, the power transmitted through the line can be increased.

With a series capacitor installed in the line, Equation 19.1 can be written as

$$P_2 = \frac{V_1 \cdot V_2 \cdot \sin\delta}{X_L - X_C} \quad (19.2)$$

$$= \frac{V_1 \cdot V_2 \cdot \sin\delta}{X_L(1-K)} \quad (19.3)$$

where $K = X_C/X_L$ is the degree of compensation, usually expressed in percent. A 70% series compensation means the value of the series capacitor in ohms is 70% of the line reactance.

19.5 Series Capacitor Bank

A series capacitor bank consists of a capacitor bank, overvoltage protection system, and a bypass breaker, all elevated on a platform, which is insulated for the line voltage. See Figure 19.6. The overvoltage protection is comprised of a zinc oxide varistor and a triggered spark gap, which are connected in parallel to the capacitor bank, and a damping reactor. Prior to the development of the high-energy zinc

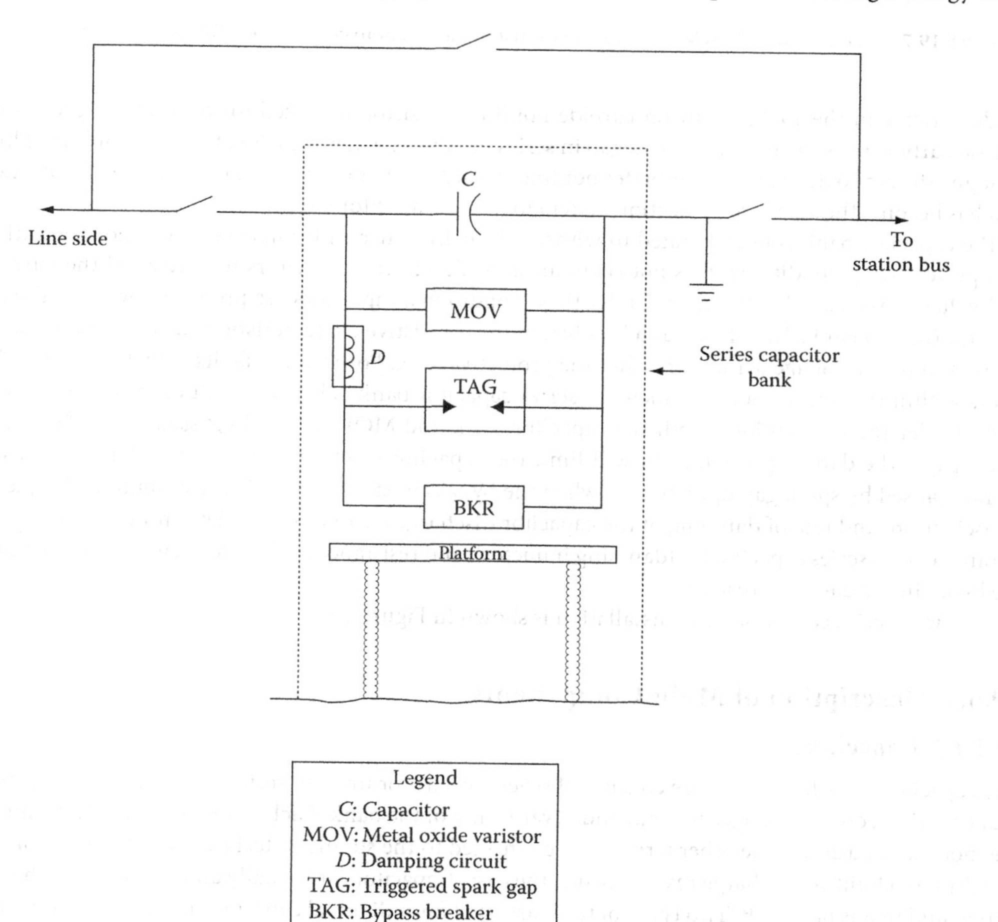

FIGURE 19.6 Schematic one-line diagram of series capacitor bank.

FIGURE 19.7 Aerial view of 500 kV series capacitor installation. (Photo courtesy of ABB Inc., Auburn Hills, MI.)

oxide varistor in the 1970s, a silicon carbide nonlinear resistor was used for overvoltage protection. Silicon carbide resistors require a spark gap in series because the nonlinearity of the resistors is not high enough. The zinc oxide varistor has better nonlinear resistive characteristics, provides better protection, and has become the standard protection system for series capacitor banks.

The capacitor bank is usually rated to withstand the line current for normal power flow conditions and power swing conditions. It is not economical to design the capacitors to withstand the currents and voltages associated with faults. Under these conditions, capacitors are protected by a metal oxide varistor (MOV) bank. The MOV has a highly nonlinear resistive characteristic and conducts negligible current until the voltage across it reaches the protective level. For internal faults, which are defined as faults within the line section in which the series capacitor bank is located, fault currents can be very high. Under these conditions, both the capacitor bank and MOV will be bypassed by the "triggered spark gap." The damping reactor (*D*) will limit the capacitor discharge current and damps the oscillations caused by spark gap operation or when the bypass breaker is closed. The amplitude, frequency of oscillation, and rate of damping of the capacitor discharge current will be determined by the circuit parameters, *C* (series capacitor), *L* (damping inductor), and resistance in the circuit, which in most cases are losses in the damping reactor.

A view of series capacitor bank installation is shown in Figure 19.7.

19.5.1 Description of Main Components

19.5.1.1 Capacitors

The capacitor bank for each phase consists of several capacitor units in series–parallel arrangement, to make up the required voltage, current, and Mvar rating of the bank. Each individual capacitor unit has one porcelain bushing. The other terminal is connected to the stainless steel casing. The capacitor unit usually has a built-in discharge resistor inside the case. Capacitors are usually all film design with insulating fluid that is non-PCB. Two types of fuses are used for individual capacitor units—internally fused or externally fused. Externally fused units are more commonly used in the United States. Internally fused capacitors are prevalent in European installations.

19.5.1.2 Metal Oxide Varistor

A metal oxide varistor (MOV) is built from zinc oxide disks in series and parallel arrangement to achieve the required protective level and energy requirement. One to four columns of zinc oxide disks are installed in each sealed porcelain container, similar to a high-voltage surge arrester. A typical MOV protection system contains several porcelain containers, all connected in parallel. The number of parallel zinc oxide disk columns required depends on the amount of energy to be discharged through the MOV during the worst-case design scenario. Typical MOV protection system specifications are as follows.

The MOV protection system for the series capacitor bank is usually rated to withstand energy discharged for all faults in the system external to the line section in which the series capacitor bank is located. Faults include single-phase, phase-to-phase, and three-phase faults. The user should also specify the fault duration. Most of the faults in EHV systems will be cleared by the primary protection system in three to four cycles. Backup fault clearing can be from 12 to 16 cycles duration. The user should specify whether the MOV should be designed to withstand energy for backup fault clearing times. Sometimes it is specified that the MOV be rated for all faults with primary protection clearing time, but for only single-phase faults for backup fault clearing time. Statistically, most of the faults are single-phase faults.

The energy discharged through the MOV is continuously monitored and if it exceeds the rated value, the MOV will be protected by the firing of a triggered air gap, which will bypass the MOV.

19.5.1.3 Triggered Air Gap

The triggered air gap provides a fast means of bypassing the series capacitor bank and the MOV system when the trigger signal is issued under certain fault conditions (e.g., internal faults) or when the energy discharged through the MOV exceeds the rated value. It typically consists of a gap assembly of two large electrodes with an air gap between them. Sometimes two or more air gaps in series can also be employed. The gap between the electrodes is set such that the gap assembly sparkover voltage without trigger signal will be substantially higher than the protective level of the MOV, even under the most unfavorable atmospheric conditions.

19.5.1.4 Damping Reactor

A damping reactor is usually an air-core design with parameters of resistance and inductance to meet the design goal of achieving the specified amplitude, frequency, and rate of damping. The capacitor discharge current when bypassed by a triggered air gap or a bypass breaker will be damped oscillation with amplitude, rate of damping, and frequency determined by circuit parameters.

19.5.1.5 Bypass Breaker

The bypass breaker is usually a standard line circuit breaker with a rated voltage based on voltage across the capacitor bank. In most of the installations, the bypass breaker is located separate from the capacitor bank platform and outside the safety fence. This makes maintenance easy. Both terminals of the breaker standing on insulator columns are insulated for the line voltage. It is usually an SF_6 puffer-type breaker, with controls at ground level.

19.5.1.6 Relay and Protection System

The relay and protection system for the capacitor bank is located at ground level, in the station control room, with information from and to the platform transmitted via fiber-optic cables. The present practice involves all measured quantities on the platform being transmitted to ground level, with all signal processing done at ground level.

19.5.2 Subsynchronous Resonance

Series capacitors, when radially connected to the transmission lines from the generation near by, can create a subsynchronous resonance (SSR) condition in the system under some circumstances. SSR can

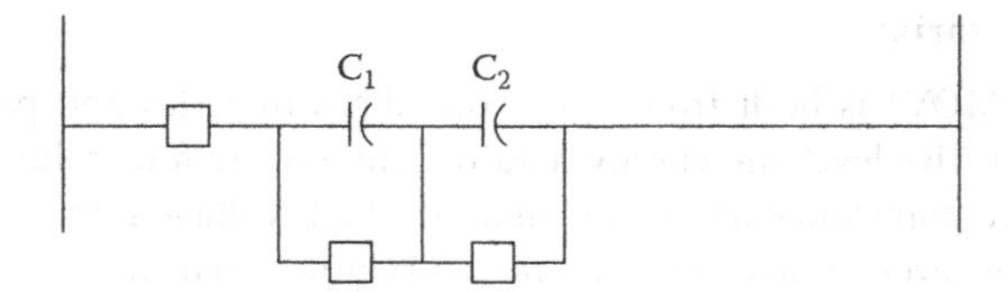

FIGURE 19.8 Breaker controlled variable series compensation.

cause damage to the generator shaft and insulation failure of the windings of the generator. This phenomenon is well described in several textbooks, given in the reference list at the end of this chapter.

19.5.3 Adjustable Series Compensation

The ability to vary the series compensation will give more control of power flow through the line, and can improve the dynamic stability limit of the power system. If the series capacitor bank is installed in steps, bypassing one or more steps with bypass breakers can change the amount of series compensation of the line. For example, as shown in Figure 19.8, if the bank consists of 33% and 67% of the total compensation, four steps, 0%, 33%, 67%, and 100%, can be obtained by bypassing both banks, smaller bank (33%), larger bank (67%), and not bypassing both banks, respectively.

Varying the series compensation by switching with mechanical breakers is slow, which is acceptable for the control of steady-state power flow. However, for improving the dynamic stability of the system, series compensation has to be varied quickly. This can be accomplished by thyristor-controlled series compensation (TCSC).

19.5.4 Thyristor-Controlled Series Compensation

Thyristor-controlled series compensation (TCSC) provides fast control and variation of the impedance of the series capacitor bank. To date (1999), three prototype installations, one each by ABB, Siemens, and the General Electric Company (GE), have been installed in the United States. TCSC is part of the flexible AC transmission system (FACTS), which is an application of power electronics for control of the AC system to improve the power flow, operation, and control of the AC system. TCSC improves the system performance for SSR damping, power swing damping, transient stability, and power flow control.

The latest of the three prototype installations is the one at the Slatt 500 kV substation in the Slatt–Buckley 500 kV line near the Oregon–Washington border in the United States. This is jointly funded by the Electric Power Research Institute (EPRI), the Bonneville Power Administration (BPA), and the General Electric Company (GE). A one-line diagram of the Slatt TCSC is shown in Figure 19.9. The capacitor bank (8 Ω) is divided into six identical TCSC modules. Each module consists of a capacitor (1.33 Ω), back-to-back thyristor valves controlling power flow in both directions, a reactor (0.2 Ω), and a varistor. The reactors in each module, in series with thyristor valves, limit the rate of change of current through the thyristors. The control of current flow through the reactor also varies the impedance of the combined capacitor–reactor combination, giving the variable impedance. When thyristor gating is blocked, complete line current flows through the capacitance only, and the impedance is 1.33 Ω capacitive (see Figure 19.10a). When the thyristors are gated for full conduction (Figure 19.10b), most of the line current flows through the reactor-thyristor branch (a small current flows through the capacitor) and the resulting impedance is 0.12 Ω inductive. If thyristors are gated for partial conduction only (Figure 19.10c), circulating current will flow between capacitor and inductor, and the impedance can be varied from 1.33 to 4.0 Ω, depending on the angle of conduction of the thyristor valves. The latter is called the vernier operating mode.

The complete capacitor bank with all six modules can be bypassed by the bypass breaker. This bypass breaker is located outside the main capacitor bank platform, similar to the case for the conventional

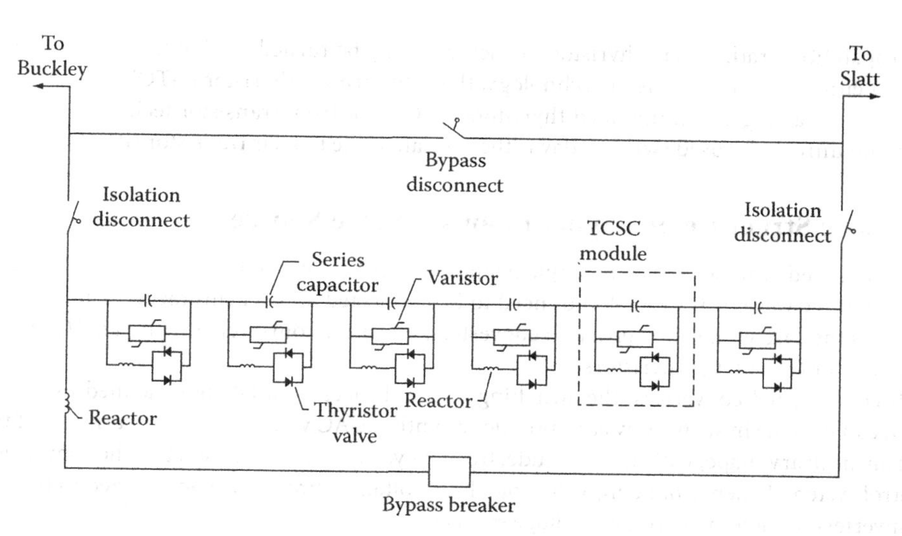

FIGURE 19.9 One-line diagram of TCSC installed at Slatt substation.

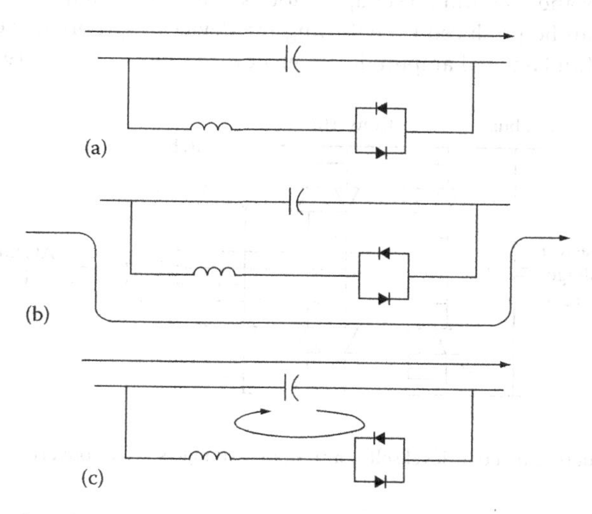

FIGURE 19.10 Current flow during various operating modes of TCSC. (a) No thyristor value current (gating blocked). (b) Bypassed with thyristor. (c) Inserted with vernier control, circulating some current through thyristor value.

series capacitor bank. There is also a reactor connected in series with the bypass breaker to limit the magnitude of capacitor discharge current through the breaker. All reactors are of air-core dry-type design and rated for the full line current rating. MOVs connected in parallel with the capacitors in each module provide overvoltage protection. The MOV for a TCSC requires significantly less energy absorption capability than is the case for a conventional series capacitor of comparable size, because gating of thyristor valves provides quick protection for faulted conditions.

19.6 Voltage Source Converter–Based Topologies

The alternative to thyristor-based compensators, in either a shunt configuration (SVC) or a series configuration (TCSC), is to make use of synchronous voltage source controllers based on voltage source converter topologies. These converters employ force-commutated switching devices, with turn-on and turn-off capabilities, rather than thyristors, which can only be turned on. Force-commutated devices are derived either from the thyristor technology, the gate turn-off thyristor (GTO), and the improved switch, the integrated gate-commutated thyristors (IGCT), or from transistor technologies. The more common and universally used switch today is the insulated gate bipolar transistor (IGBT).

19.6.1 Basic Structure of a Synchronous Voltage Source

Force-commutated switches allow the implementation of self-commutated converters, which can generate an arbitrary waveform. They do not need to be connected and synchronized to an AC source to operate, as is the case with thyristor-based converters. Thyristor converters require an AC grid voltage to commutate or turn off the thyristors.

In self-commutated converters, the switching of the devices on a DC bus, defined by means of a capacitor, can be done in such a way as to produce a synthetic AC voltage at the AC terminal of the converter of an arbitrary shape, with an amplitude, frequency, and phase that are set by the converter gating and control system. When producing a sinusoidal AC voltage source and synchronized to the AC grid, these converters become synchronous voltage sources.

The basic topology for the voltage source converter is the six-switch full-bridge topology, most commonly used in many low and medium power applications, but also in high power applications (Figure 19.11). A square wave can be produced by switching the devices appropriately once per AC cycle. It has a fully controllable fundamental amplitude, frequency, and phase (Figure 19.12). The amplitude is

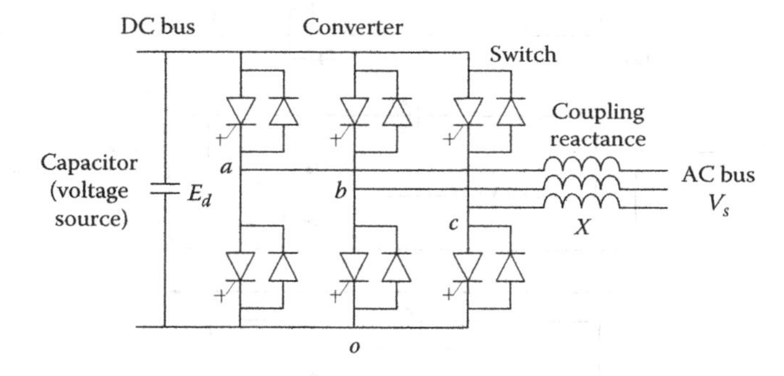

FIGURE 19.11 Basic structure of a two-level self-commutated voltage source converter.

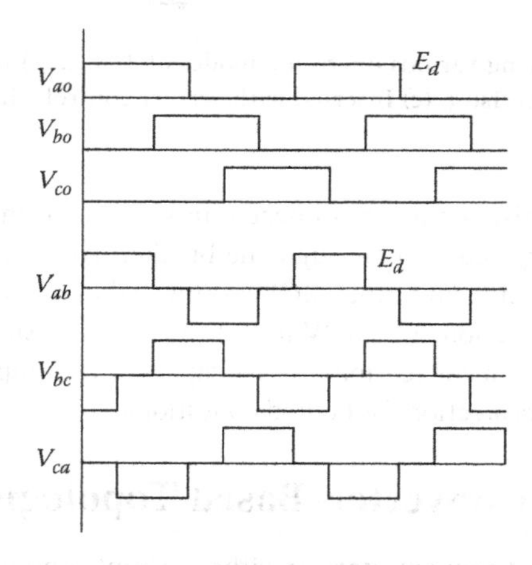

FIGURE 19.12 AC voltage waveforms of a self-commutated inverter with fundamental frequency gating and a constant DC bus voltage.

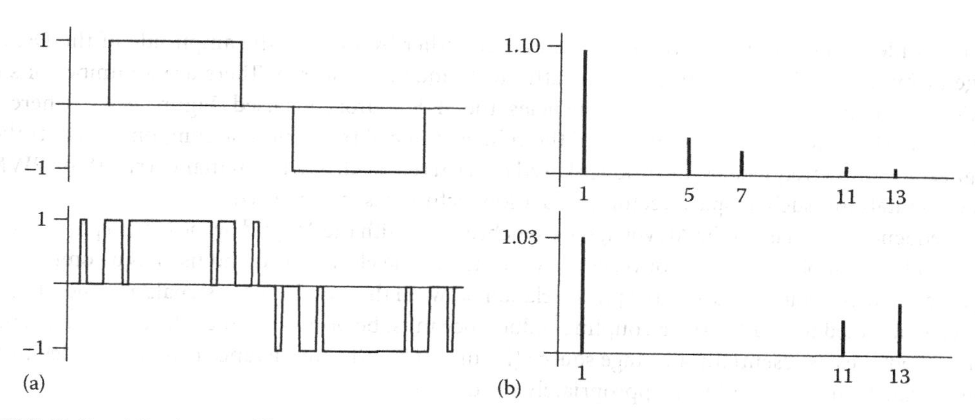

FIGURE 19.13 AC voltage and harmonic content of a self-commutated converter. (a) Waveforms for single pulse patterns and pulse width modulation (SHE) patterns (fifth and seventh harmonic elimination). (b) Harmonic spectrum for single pulse patterns and pulse width modulation (SHE) patterns.

dictated by the magnitude of the DC bus voltage, the frequency by a clock, and the phase by the position of the gating signal with respect to an AC reference waveform (the AC grid in compensators).

The output voltage consists of a fundamental voltage component, and low-order harmonics of frequency ($6n \pm 1$) and amplitude $1/n$, where n varies from 1 to infinity. The dominant harmonics for a three-phase waveform using fundamental switching frequency gating are the 5th, 7th, 11th, and 13th harmonics (Figure 19.13).

Harmonics can be reduced or eliminated using any one of the following methods or combinations of methods:

- Harmonic filters: These typically are passive LC tuned filters or low pass LC filters.
- Phase shifting transformers: By connecting n converters in series, for example, and feeding them with voltages that are phase shifted 60°/n, lower-order harmonics can be eliminated. For two converters fed from transformer secondary voltages shifted by 30°, the harmonics of order 5 and 7 are eliminated, leaving the 11th and 13th as the dominant harmonics.
- Pulse width modulation (PWM) techniques: The principle consists of introducing notches in the output AC voltage, such that a number of harmonic components are eliminated (or reduced) in the output waveform. There are a number of such techniques available. At low switching frequencies, the selective harmonic elimination (SHE) approach is used (Figure 19.13). At higher switching frequencies, techniques based on a carrier, such as a triangular carrier (Sine PWM), or on computations, such as space vector modulation techniques, naturally eliminate low-frequency harmonic components in the AC voltage.
- Multi-level and multi-module structures: Multiple levels can be created in the output voltage of a converter by using a number of topologies; the better known is the diode clamped capacitor multi-level inverter, usually in a three-level configuration. An alternative is to connect a number of basic converter modules in series, such as the two-level converter, appropriately phase shifted to eliminate a given number of harmonics at the combined output of the converter.
- Modular multi-level structures: In high-voltage and high-power applications, the use of a large number of low voltage devices in series is required to obtain the required AC output voltage magnitude. This is the case with any line-connected high-voltage compensator. Switching devices can then be configured in modules, each capable of a controlled low voltage output. A large number of these modules are then connected in series and gated appropriately to achieve a stepped AC waveform. With a large number of steps, an output voltage waveform that is close to sinusoidal is synthesized. This practically eliminates the need of filtering the AC waveform.

The amplitude of the output AC voltage is controlled either by varying the amplitude of the DC bus voltage or by using a PWM technique. The latter technique is preferred. There are a number of such techniques available. At low switching frequencies, the SHE approach is used (Figure 19.13), where the amplitude of the fundamental component of the voltage is one of the harmonic components controlled. At higher switching frequencies, techniques based on a carrier, such as a triangular carrier (Sine PWM), or on computations, such as space vector modulation techniques, can be used.

The frequency and phase of the AC voltage are synchronized with the AC grid by means of a phase-locked loop (PLL) for example, when the converter is connected to the electric grid and used as a compensator. The phase of the AC voltage is set by the phase relation between the switch gating signals and the AC grid.

When connected to an AC grid, a coupling inductance must be provided, since the converter, being a matrix of switches, is essentially a voltage source (Figure 19.11). This inductance can be provided by the coupling transformer and/or by an appropriately sized inductor.

19.6.2 Operation of Synchronous Voltage Sources

When connected to the AC grid, and assuming the general case of a power source on the DC side, the angle δ between the AC supply V_s and the fundamental component of the converter AC output voltage V_i can be set to any desired value (Figure 19.14). This angle defines the amount of power flowing between the converter DC bus and the AC supply. This power is controlled by the phase shift between the AC supply and the gating pattern of the converter AC voltage, and by its amplitude. The converter is reversible and the power transfer, neglecting losses, is given by the equation dictating the exchange of power between two synchronous AC sources:

$$P = 3\frac{V_i V_s}{X}\sin\delta \tag{19.4}$$

where X is the coupling reactance and voltages are expressed on a per-phase basis.

In addition, the converter can be operated at a leading or lagging power factor. The reactive power is given by

$$Q = 3\frac{V_i(V_i - V_s\cos\delta)}{X} \tag{19.5}$$

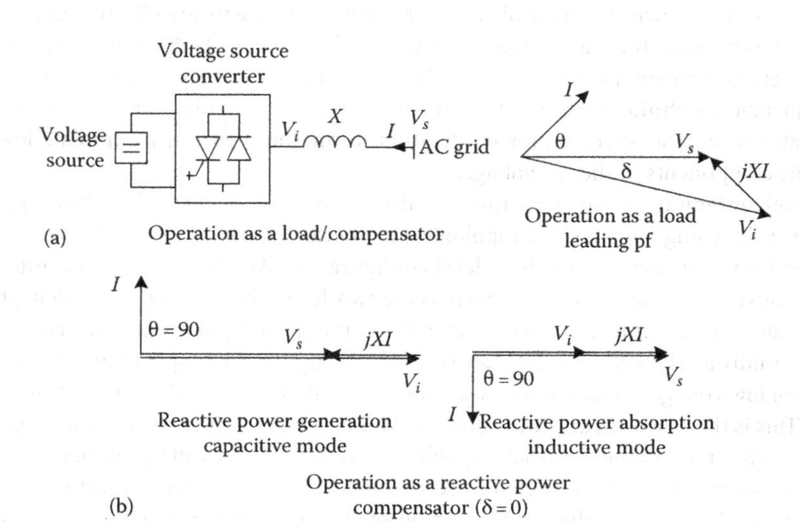

FIGURE 19.14 Operation of a synchronous voltage source. (a) Equivalent circuit (per phase) and phasor diagram—general case. (b) Operation as an SVC.

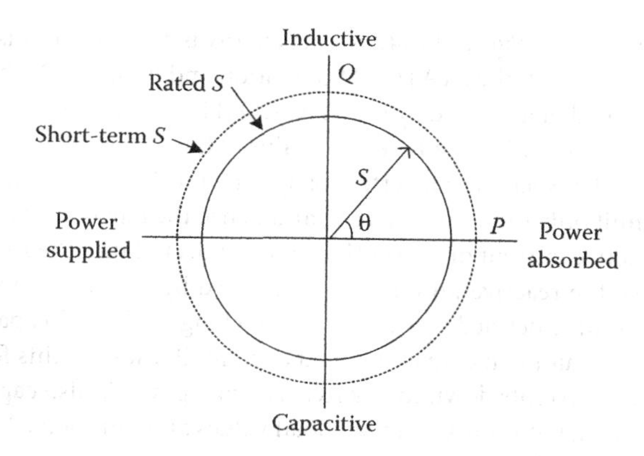

FIGURE 19.15 Power diagram of a self-commutated AC/DC converter with power factor control and constant AC rms current operation.

Operation is possible in all four quadrants of the Q–P plane (Figure 19.15). However, this requires that energy sources or storage devices be connected to the DC bus to supply or absorb real power. The more common storage devices in power systems are electric storage batteries and superconducting magnet energy storage (SMES) devices. This gives additional flexibility to the compensator in supporting the electric grid voltage.

The basic synchronous voltage source discussed previously can be connected to the grid in a shunt configuration. This is the basic structure of a static compensator (STATCOM). The same synchronous source can also be connected in series with the AC grid through a series transformer. It then implements a static synchronous series compensator (SSSC). Finally, because of the presence of the DC bus, a shunt and a series unit can share the same DC bus. The shunt-series structure is known as the unified power flow controller (UPFC).

It should be noted that most of the structures using synchronous voltage sources and presented next can be applied to, and have found applications, in transmission as well as distribution systems.

19.6.3 Static Compensator

A static compensator (STATCOM) provides variable reactive power from lagging to leading (Figure 19.16), but with no inductors or capacitors required for var generation, as demonstrated in the preceding section. Reactive power generation is achieved by regulating the terminal voltage of the converter. If the terminal voltage V_i of the voltage source converter is higher than the AC bus voltage, the STATCOM produces leading reactive power, that is, it operates as a capacitor. If V_i is lower than the bus voltage,

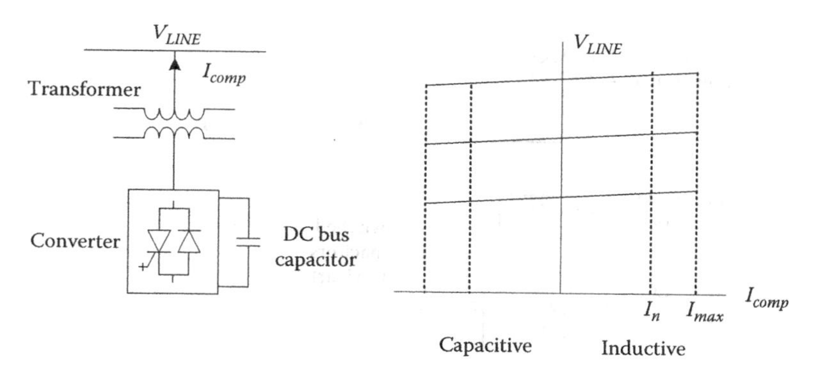

FIGURE 19.16 Static compensator (STATCOM)—operating characteristics.

it produces lagging reactive power (Figure 19.14). No net energy is required in this operation, other than supplying the losses associated with the STATCOM operation and with the DC bus capacitor.

The reactive power generated or absorbed by the STATCOM is not a function of the size of the capacitor on the DC bus of the converter, which only sets the value of the DC bus voltage. The capacitor voltage is regulated by the load angle between the converter voltage and the AC grid voltage, as per Equation 19.4. The capacitor is rated to limit only the ripple current, and hence the harmonics in the output voltage.

Unlike the SVC, the reactive current or power that can be injected into the AC grid does not depend upon the AC voltage, since the reactive power is not produced by means of capacitors and inductors, but by a synthetic voltage source defined by the DC bus. So long as the DC capacitor remains charged to its rated value, the compensator can deliver its rated current. Because of this feature, the STATCOM is capable of supplying rated currents down to low AC grid voltages. It is also capable of producing currents that are larger than rated, with short-term overload values and durations dependent on the design of the converter switches and cooling system.

The performance of the STATCOM is similar to that of a synchronous condenser (unloaded synchronous motor with varying excitation). However, the dynamic response is faster than that of a synchronous condenser and of an SVC, if switched at a frequency higher than fundamental frequency, as with PWM, for example. A STATCOM is more effective than an SVC for arc furnace flicker control because of its dynamic range.

The first demonstration STATCOM of ±100 Mvar rating was installed at the Tennessee Valley Authority's Sullivan substation in 1994 (Figure 19.17). To extend the range of the STATCOM in the capacitive region (with the inductive limit dictated by the STATCOM to 100 Mvar), mechanically switched capacitor banks are added. The control system (Figure 19.18) is similar to that of the typical SVC, with slope control as required. The controller also incorporates a reactive current inner loop and a synchronizing block (PLL).

The advantage of using the STATCOM in this installation has been the fast and coordinated response that it enables, particularly under contingencies. This has allowed deferral of the construction of an additional line or the use of a second transformer bank, resulting in significant cost savings.

STATCOM devices with storage have been successfully implemented, mostly at the distribution level and using the following storage devices: (a) battery storage on the DC bus, to achieve better ride-through

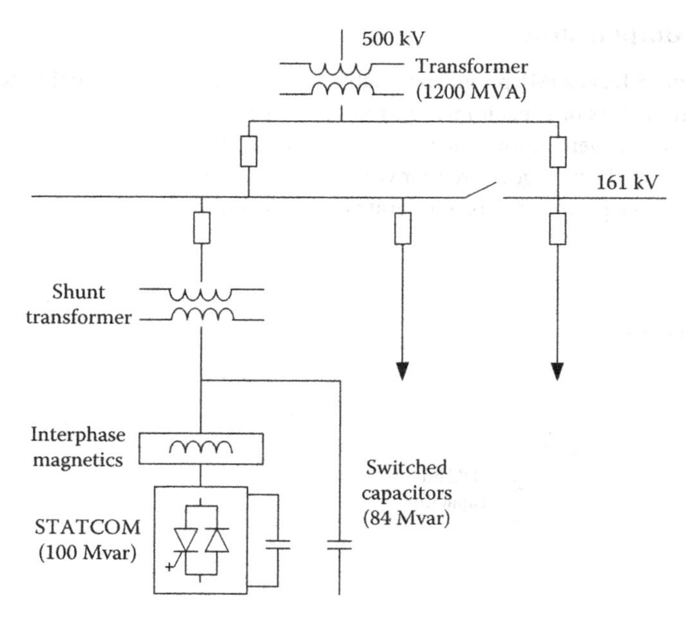

FIGURE 19.17 STATCOM configuration and associated equipment—TVA (United States).

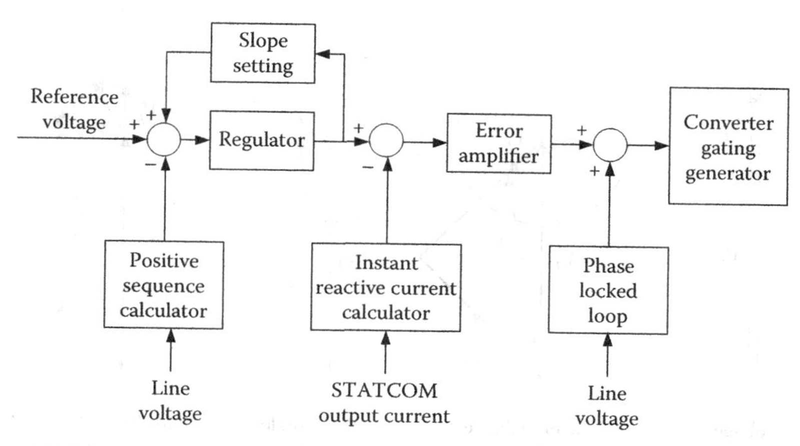

FIGURE 19.18 STATCOM control system configuration—TVA (United States).

and voltage regulation for critical loads in the case of faults and momentary loss of mains; (b) SMES, for applications in transmission and distribution, to support the system voltage in the case of faults on the system; recovery from faults has been found to be faster with the real power injection capability provided by the use of the stored energy.

19.6.4 Static Series Synchronous Compensator

The converter used in a STATCOM can also be used in the SSSC. This device is coupled to the AC grid by means of a series transformer (Figure 19.19), in place of a shunt transformer. It injects a voltage in quadrature with the AC line current, either leading or lagging, rather than injecting a controlled amount of leading or lagging reactive current, as in a STATCOM.

Since the equivalent reactance is the ratio of injected voltage over the line current, the SSSC can be used to emulate a variable inductor or capacitor in series with the line. As a variable series capacitor, it can be used in place of TCSC, with a more controllable characteristic and a faster dynamic response.

19.6.5 Unified Power Flow Controller

The DC bus of a STATCOM and of an SSSC can be connected together (Figure 19.20) to produce a device named the UPFC. It can exhibit the characteristics of both the STATCOM with shunt current injection, and the SSSC with series voltage injection, with added features. The device has 3 degrees

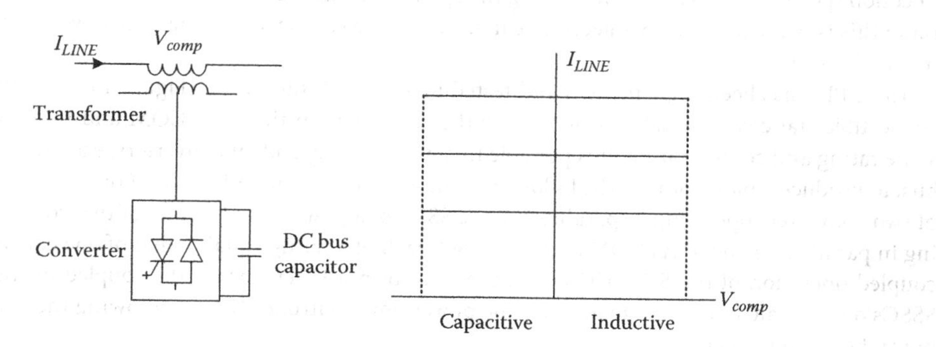

FIGURE 19.19 Static synchronous series compensator (SSSC)—characteristics.

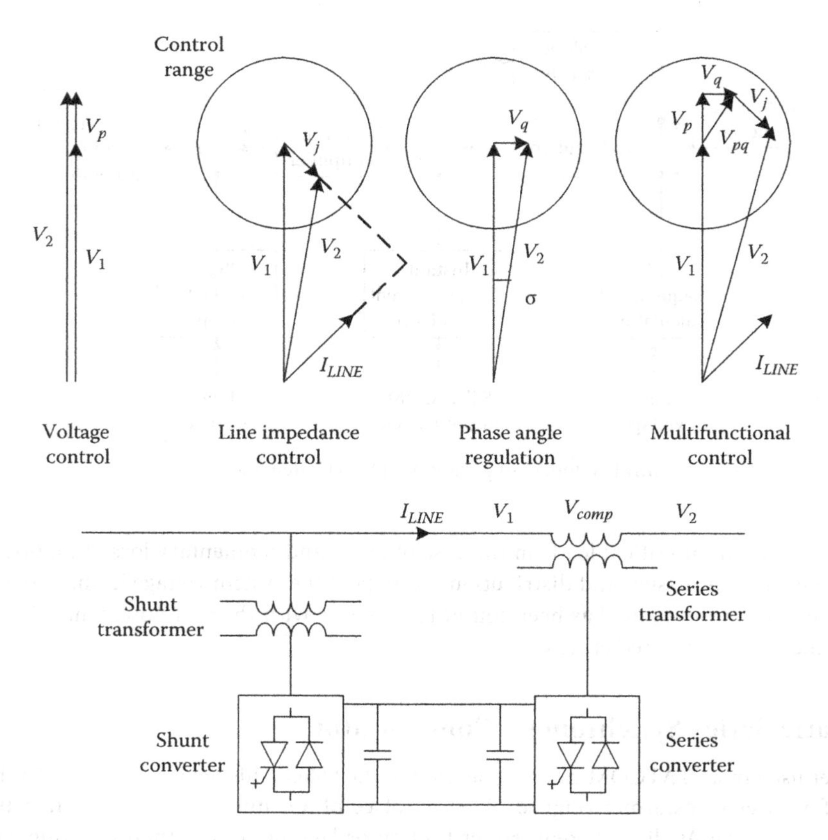

FIGURE 19.20 Unified power flow controller (UPFC)—operating modes.

of freedom, control of the reactive powers on the shunt and series connections, and of real power flowing through the common DC bus. This in turn allows real power injection on the shunt and series connections. The DC bus voltage is usually regulated from the shunt side, in a manner similar to a STATCOM.

Power injection on the shunt and series connections increases the flexibility and the number of modes of operation of the UPFC (Figure 19.20). In particular, the STATCOM and the SSSC can be operated the same way as the basic device, and if desired, with the added features provided such as real power injection. Among these are (a) the operation of the series side as a phase shifter, with real power being provided by the DC bus through the shunt side; (b) the operation as a variable line resistance injection (positive or negative), modifying the apparent resistance of the line and therefore the line X/R ratio; this is made possible by injecting real power on the series transformer, the power being provided by the shunt side.

The UPFC has been implemented and tested in an NYPA substation (Figure 19.21) in the form of a convertible static compensator (CSC). Since the converters in the STATCOM and the SSSC have the same rating and configuration, it is possible to have two independent converters, each with its own DC bus, to produce, among others, the following combinations: (1) one STATCOM of a power equal to that of two converters operating in parallel; (2) one SSSC of a power equal to that of two converters operating in parallel; (3) one STATCOM and one SSSC, each of a rating equal to that of one converter; (4) the coupled operation of the STATCOM and SSSC to form a UPFC; or (5) the coupled operation of two SSSCs on separate lines to form an interline power flow controller (IPFC), allowing the interchange of power between the two lines.

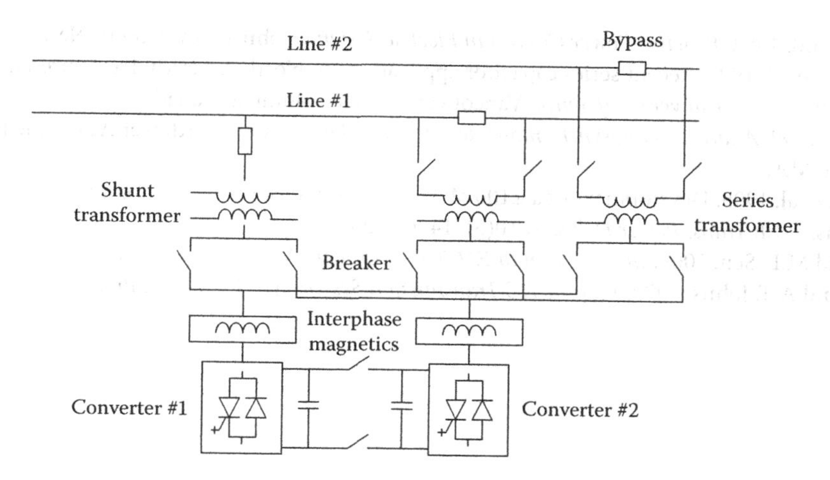

FIGURE 19.21 Convertible static compensator (CSC)—UPFC structure—NYPA (United States).

The potential benefits include (a) improving the voltage support and control; (b) increasing the power transfer capability through the substation; (c) relieving power transfer bottlenecks; (d) maximizing utilization of the transmission system; and (e) reducing power losses on the lines.

19.7 Defining Terms

Shunt capacitor bank: A large number of capacitor units connected in series and parallel arrangement to make up the required voltage and current rating, and connected between the high-voltage line and ground, between line and neutral, or between line and line.

Voltage flicker: Commonly known as "flicker" and "lamp flicker" is a rapid and frequent fluctuation of supply voltage that causes lamps to flicker. Lamp flicker can be annoying, and some loads are sensitive to these frequent voltage fluctuations.

Subsynchronous resonance: Per IEEE, SSR is an electric power system condition where the electric network exchanges energy with a turbine generator at one or more of the natural frequencies of the combined system below the synchronous frequency of the system.

References

Anderson, P.M., B.L. Agrawal, and J.E. Van Ness. 1990. *Subsynchronous Resonance in Power Systems*. IEEE Press, New York.

Anderson, P.M. and R.G. Farmer. 1996. *Series Compensation in Power Systems*. PBLSH! Inc., Encinitas, CA.

Gyugyi, L., R.A. Otto, and T.H. Putman. 1978. Principles and application of thyristor-controlled shunt compensators. *IEEE Trans. Power Appar. Syst.*, 97, 1935–1945.

Gyugyi, L. and E.R. Taylor. 1980. Characteristics of static thyristor-controlled shunt compensators for power transmission applications. *IEEE Trans. Power Appar. Syst.*, PAS-99, 1795–1804.

Hammad, A.E. 1986. Analysis of power system stability enhancement by static VAR compensators. *IEEE Trans. Power Syst.*, 1, 222–227.

Hingorani, N.G.L. and A. Gyugyi. 1999. *Understanding FACTS: Concepts and Technology of Flexible AC Transmission Systems*. Wiley-IEEE Press, New York.

Mathur, M. and R. Varma. 2002. *Thyristor-Based Facts Controllers for Electrical Transmission Systems*. Wiley-IEEE, New York.

Miller, T.J.E., Ed. 1982. *Reactive Power Control in Electric Systems*. John Wiley & Sons, New York.

Miske, S.A. Jr. et al. 1995. Recent series capacitor applications in North America. Paper presented at *CEA Electricity '95 Vancouver Conference*, Vancouver, British Columbia, Canada.

Padiyar, K.R. 1999. *Analysis of Subsynchronous Resonance in Power Systems*. Kluwer Academic Publishers, Boston, MA.

Schauder, C. et al. 1995. Development of a ±100 Mvar static condenser for voltage control of transmission systems. *IEEE Trans. Power Delivery*, 10(3), 1486–1496.

Sen, K.K. and M.L. Sen. 2009. *Introduction to FACTS Controllers*. Wiley-IEEE, New York.

Song, Y.H. and A.T. Johns. 1999. *Flexible AC Transmission Systems (FACTS)*. IEE Press, London, U.K.

20

Environmental Impact of Transmission Lines

George G. Karady
Arizona State University

20.1 Introduction

The appearance of the first transmission lines more than 100 years ago immediately started discussion and public concerns. When the first transmission line was built, more electrocutions occurred because of people climbing up the towers, flying kites, and touching wet conducting ropes. As the public became aware of the danger of electrocution, the aesthetic effect of the transmission lines generated public discussion. In fact, there is a story of Frank Lloyd Wright, the famous architect, calling President Roosevelt and demanding the removal of high-voltage lines obstructing his view in Scottsdale, Arizona. Undoubtedly, a transmission line corridor with several lines would disturb the appearance of a quite green valley.

The rapid increase in radio and television transmission has produced the occurrence of electromagnetic interference (EMI) problems. The high voltage on the transmission line produces corona discharge that generates electromagnetic waves. These waves disturb the radio and television reception, which resulted in public protests and opposition to build lines too near towns.

In the 1960s, the electrical field surrounding the high-voltage lines became subject to public concerns. The electrical field can produce minor sparks and small electric shocks under a high-voltage line. An example of this would be if a woman were to walk under a line holding an umbrella, the woman would feel the electric shocks produced by these small discharges.

In the 1970s, the transmission line current produced magnetic fields and became a public issue. Several newspaper articles discussed the adverse health effects of magnetic fields. This generated intensive research all over the world. The major concern is that exposure to magnetic fields caused cancer, mostly leukemia. The U.S. government report concluded that there was no evidence that moderate 60 Hz magnetic field caused cancer. However, this opinion is not shared by all.

This chapter will discuss the listed environmental effects of transmission lines.

20.2 Aesthetic Effects of Lines

The first transmission towers were small wooden poles that were tempting for children to climb but had no environmental impact. However, the increase of voltage resulted in large steel structures over 100 ft high and 50 ft wide.

In North America, the large wooden structures were common until the Second World War. The typical voltage of transmission lines with wooden poles is less than 132 kV, although 220 kV lines with H-frame wooden towers are also built in the Midwest.

Figure 20.1 shows a transmission line with H-frame wooden towers. This construction fits well in the rural environment and does not produce environmental concerns.

The increasing voltage and need for crossing large valleys and rivers resulted in the appearance of steel towers. These towers are welded or riveted lattice structures. Several different conductor arrangements are used. Figure 20.2a shows a lattice tower with conductors arranged horizontally. The horizontal arrangement increases the widths of the tower, which produces a more visible effect. Figure 20.2b shows a double circuit line with vertically arranged conductors. This results in a taller and more compact appearance.

The presented pictures demonstrate that the transmission lines with large steel towers are not very aesthetically pleasing. They do not blend in with the environment and can interrupt a beautiful landscape.

The increasing demand of electricity and the public objection to build new transmission lines resulted in the development of transmission line corridors. The utilities started to build lines in parallel on right-of-ways land that they already owned. Figure 20.3 shows a typical transmission line corridor. The appearance of the maze of conductors and large steel structures are not an aesthetically pleasing sight.

The public displeasure with the lattice tower triggered research work on the development of aesthetically more pleasing structures. Several attempts were made to develop nonmetallic transmission line structure using fiberglass rods, where the insulators are replaced by the tower itself. Although the development of nonmetallic structures was unsuccessful, the development of tubular steel towers led to a more pleasing appearance. Figure 20.4 shows tubular steel tower used in Arizona at the 220 kV high-voltage lines.

FIGURE 20.1 220 kV line with H-frame wooden towers.

FIGURE 20.2 High-voltage transmission lines. (a) Single circuit line with horizontally arranged conductors. (b) Double circuit line with vertically arranged conductors.

FIGURE 20.3 Transmission line corridor.

Figure 20.4 demonstrates that the slender tubular structure is less disturbing and aesthetically more pleasing. These towers blend in better with the desert environment and cause less visual interruptions.

The presented examples prove that the aesthetic appearance of the transmission lines is improving although even the best tower structures disturb the environment. The ultimate solution is the replacement of the lines by an underground cable system. Unfortunately, both technical and economic problems are preventing the use of underground energy transmission systems.

FIGURE 20.4 A 220 kV suspension tower.

20.3 Magnetic Field Generated by HV Lines

Several newspaper articles presented survey results showing that the exposure to magnetic fields increases the cancer occurrence. Studies linked the childhood leukemia to transmission line–generated magnetic field exposure. This triggered research in both biological and electrical engineering fields. The biological research studied the magnetic field effect on cells and performed statistical studies to determine the correlation between field exposure and cancer occurrence. The electrical engineering research aimed at the determination of magnetic field strength near transmission lines, electric equipment, motors, and appliances [15]. A related engineering problem is the reduction of magnetic field generated by lines and other devices.

In this chapter we will present a calculation method to determine a transmission line–generated magnetic field and summarize the major results of biological research.

20.3.1 Magnetic Field Calculation

The electric current in a cylindrical transmission line conductor generates magnetic field surrounding the conductor. The magnetic field lines are concentric circles. At each point around the conductor, the magnetic field strength or intensity is described by a field vector that is perpendicular to the radius drawn from the center of the conductor.

Figure 20.5 shows the current-carrying conductor, a circular magnetic field line, and the magnetic field vector H in a selected observation point. The magnetic field vector is perpendicular to the radius of the circular magnetic field line. The H field vector is divided into horizontal and vertical components. The location of both the observation point and the conductor is described by the x, y coordinates.

The magnetic field intensity is calculated by using the ampere law. The field intensity is

$$H = \frac{I}{2\pi r} = \frac{I}{2\pi\sqrt{(x_i - X)^2 + (y_i - Y)^2}}$$

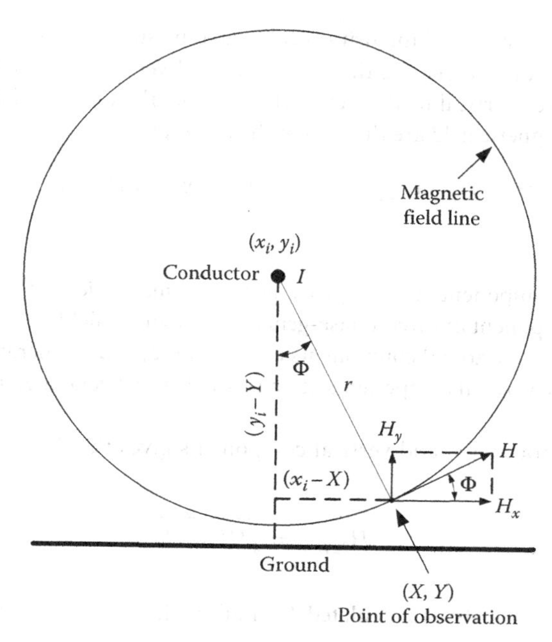

FIGURE 20.5 Magnetic field generation.

where

H is the field intensity in A/m
I is the current in the conductor
r is the distance from the conductor
(X, Y) are the coordinates of the observation point
(x_i, y_i) are the coordinates of the conductor

The horizontal and vertical components of the field are calculated from the triangle formed by the field vectors. The angle is calculated from the triangle formed with the coordinate's differences as shown in Figure 20.5:

$$\cos(\phi) = \frac{y_i - Y}{\sqrt{(x_i - X)^2 + (y_i - Y)^2}} \qquad \sin(\phi) = \frac{x_i - X}{\sqrt{(x_i - X)^2 + (y_i - Y)^2}}$$

$$\cos(\Phi) = \frac{x_i - X}{\sqrt{(x_i - X)^2 + (y_i - Y)^2}} \qquad \sin(\Phi) = \frac{y_i - Y}{\sqrt{(x_i - X)^2 + (y_i - Y)^2}}$$

The vertical and horizontal field components are

$$H_x = H\cos(\Phi) = \frac{I}{2\pi} \frac{x_i - X}{(x_i - X)^2 + (y_i - Y)^2}$$

$$H_y = H\sin(\Phi) = \frac{I}{2\pi} \frac{y_i - Y}{(x_i - X)^2 + (y_i - Y)^2}$$

In a three-phase system, each of the three-phase currents generates magnetic fields. The phase currents and corresponding field vectors are shifted by 120°. The three-phase currents are

$$I_1 = I \quad I_2 = Ie^{-120^\circ} \quad I_3 = Ie^{-240^\circ}$$

The three-phase-line-generated field intensity is calculated by substituting the conductor currents and coordinates in the equations describing the horizontal and vertical field components. This produces three horizontal and three vertical field vectors. The horizontal and vertical components of the three-phase-line-generated magnetic field are the sum of the three-phase components:

$$H_x = H_{x\_1} + H_{x\_2} + H_{x\_3} \quad H_y = X_{y\_1} + H_{y\_2} + H_{y\_3}$$

where

H_x is the horizontal component of three-phase-generated magnetic field
H_y is the vertical component of three-phase-generated magnetic field
$H_{x\_1}, H_{x\_2}, H_{x\_3}$ are the horizontal components of phases 1, 2, and 3 generated magnetic field
$H_{y\_1}, H_{y\_2}, H_{y\_3}$ are the vertical components of phases 1, 2, and 3 generated magnetic field

The vector sum of the horizontal and vertical components gives the three-phase-line-generated total magnetic field intensity:

$$H_{3\_phase} = \sqrt{H_x^2 + H_y^2}$$

The magnetic field flux density is calculated by multiplying the field intensity by the free space permeability:

$$\mu_o = 4\pi \times 10^{-7} \frac{Henry}{meter} \quad B_{3\_phase} = \mu_o H_{3\_phase}$$

For the demonstration of the expected results, we calculated a 500 kV transmission line–generated magnetic flux density under the line in 1 m distance from the ground. The conductors are arranged horizontally. The average conductor height is 24.38 m (80 ft); the distance between the conductors is 10.66 m (35 ft). The line current is 2000 A. Figure 20.6 shows the magnetic flux density distribution under the line in 1 m from the ground. The locations of the line conductors are marked on the figure. It can be seen that the maximum flux density is under the middle conductor and it decreases rapidly with distance.

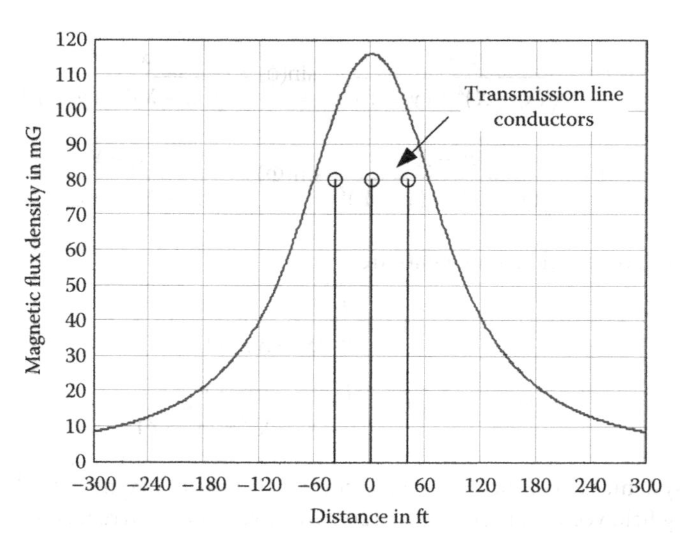

FIGURE 20.6 Magnetic field density under a 500 kV line when the load current is 2000 A.

The right-of-way is around 200 ft in this transmission line. The maximum flux density is around 116 mG (milligauss) or 11.6 μT and around 18 mG (1.8 μT) at the edge of the right-of-way.

Although the acceptable level of magnetic flux density is not specified by national or international standards, the utilities maintain less than 100 mG (10 μT) at the edge of the right-of-way and less than 10 mG (1 μT) at the neighboring residential area.

20.3.2 Health Effect of Magnetic Field

The health effects of magnetic fields are a controversial subject, which generated an emotional discussion. The first study that linked the occurrence of childhood leukemia to electrical current–generated magnetic fields was published in 1979 by Wertheimer and Leeper [1]. This was a statistical study where the electric wiring configuration near the house of the victim was related to the occurrence of childhood cancer. The researchers compared the wiring of the configuration including transmission lines close to the childhood leukemia victim's house and close to the house of a controlled population group. The study found a correlation between the occurrence of cancer and the power lines carrying high current. The study was dismissed because of inconsistencies and repeated in 1988 by Savitz et al. [2]. They measured the magnetic field in the victim's house and used the electric wiring configuration. The study found a modest statistical correlation between the cancer and wiring code but not between the cancer and the measured magnetic field. These findings initiated worldwide research on magnetic field health effects. The studies can be divided into three major categories:

- Epidemiological studies
- Laboratory studies
- Exposure assessment studies

20.3.2.1 Epidemiological Studies

These statistical studies connect the exposure to magnetic and electric fields to health effects, particularly to occurrence of cancer. The early studies investigated the childhood cancer occurrence and residential wiring [1–3]. This was followed by studies relating the occupation (electrical worker) to cancer occurrence. In this category, the most famous one is a Swedish study [4], which found elevated risk for lymphoma among electric workers. However, other studies found no elevated cancer risk [5]. The uncertainty in all of these studies is the assessment of actual exposure to electromagnetic fields. As an example, some of the studies estimated the exposure to magnetic field using the job title of the worker or the postal code where the worker lived. The results of these studies are inconclusive, some of the studies showed elevated risk to cancer, most of them not.

20.3.2.2 Laboratory Studies

These studies are divided into two categories—tissue studies and live animal studies. The tissue studies investigated the effect of electric and magnetic field on animal tissues. The studies showed that the electromagnetic field could cause chromosomal changes, single strand breaks, or alteration of ornithine decarboxylase, etc. [6,7]. Some of the studies speculate that the electromagnetic exposure can be a promoter of cancer together with other carcinogen material. The general conclusion is that the listed effects do not prove that the EMF can be linked to cancer or other health effects.

The study on live animals showed behavioral changes in rats and mice. Human studies observed changes of heart rates and melatonin production as a result of EMF exposure [8,9]. The problem with the laboratory studies are that they use a much higher field than what occurs in residential areas. None of these studies showed that the EMF produces toxicity that is typical for carcinogens. An overall conclusion is that laboratory studies cannot prove that magnetic fields are related to cancer in humans.

TABLE 20.1 Magnetic Maximum Permissible Exposure (MPE) Levels: Exposure of Head and Torso

	General Public		Controlled Environment	
Frequency Range (Hz)	*B*-Rms (mT)	*H*-Rms (A/m)	*B*-Rms (mT)	*H*-Rms (A/m)
<0.153	118	9.39×10^4	353	2.81×10^5
0.153–20	18.1/*f*	$1.44 \times 10^4/f$	54.3/*f*	$4.32 \times 10^4/f$
20–759	0.904	719	2.71	2.16×10^3
759–3000	687/*f*	$5.47 \times 10^5/f$	2060/*f*	$1.64 \times 10^6/f$

Source: IEEE Std C95.6-2002, IEEE Standard for Safety Levels with Respect to Human Exposure to Electromagnetic Fields, 0–3 kHz, Table 4.

f is frequency in Hz; MPEs refer to spatial maximum.

20.3.2.3 Exposure Assessment Studies

In the United States, the Electrical Power Research Institute led the research effort to assess the exposure to magnetic fields [10]. One of the interesting conclusions is the effect of ground current flowing through main water pipes. This current can generate a significant portion of magnetic fields in a residential area. Typically in 1 m distance from a TV, the magnetic field can be 0.01–0.2 μT; an electric razor and a fluorescent table lamp can produce a maximum of 0.3 μT. The worst is the microwave oven that can produce magnetic field around 0.3–0.8 μT in 1 m distance. The electric field produced by appliances varies between 30 and 130 V/m in a distance of 30 cm. The worst is the electric blanket that may generate 250 V/m [11].

The measurement of magnetic fields also created problems. EPRI developed a movable magnetic field measuring instrument. IEEE developed a standard ANSI/IEEE Std. 644 that presents a procedure to measure electric and magnetic field emitted by power lines. The conclusion is that both measuring techniques and instruments provide accurate exposure measurement.

The permitted maximum exposure to magnetic fields depends on the flux density and frequency.

International Commission on Non-Ionizing Radiation Protection (ICNIRP) guidelines of the reference levels for exposure to time varying 60 Hz magnetic fields are

- 833 mG for the general public
- 4167 G for the work environment

Institute of Electrical and Electronics Engineers (IEEE) Standard C95.6 provides the maximum EMF permissible exposure levels for the human body in the general public and in controlled environments. The permissible magnetic flux density and magnetic field values are listed in Table 20.1.

The IEEE standards are approximately an order of magnitude higher than the ICNIRP reference levels. A consensus has not been reached about health risks, and some individual scientists have proposed more restrictive exposure guidelines.

20.3.2.4 Summary

The health effect of magnetic field remains a controversial topic in spite of the U.S. Environmental Protection Agency report [12,13] that concluded that the low frequency, low level electric, and magnetic fields are not producing any health risks.

Many people believe that the prudent approach is the "prudent avoidance" to long-term exposure.

20.4 Electrical Field Generated by HV Lines

The energized transmission line produces electric field around the line. The high voltage on a transmission line drives capacitive current through the line. Typically, the capacitive current is maximum at the supply and linearly reduced to zero at the end of a no-loaded line, because of the evenly distributed line capacitance. The capacitive current generates sinusoidal variable charges on the conductors. The rms

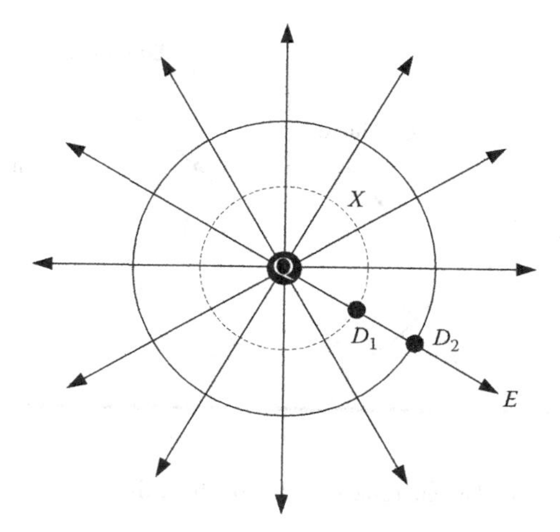

FIGURE 20.7 A charge-generated electric field.

value of the sinusoidal charge is calculated and expressed as coulomb per meter. The equations describing the relation between the voltage and charge were derived in Chapter 21. For a better understanding, we summarize the derivation of equations for field calculation.

Figure 20.7 shows a long energized cylindrical conductor. This conductor generates an electrical field. The emitted electrical field lines are radial and the field inside the conductor is zero. The electric field intensity is

$$E = \frac{D}{\varepsilon_o} = \frac{Q}{2\pi\varepsilon_o}\frac{1}{x} \quad \varepsilon_o = \frac{10^{-9}}{36\pi}\frac{Farad}{meter}$$

where

- D is the electric field flux density
- ε_o is the free place permeability
- Q is the charge on the conductor
- x is the radial distance
- E is the electric field intensity

The integral of the electric field between two points gives the voltage differences:

$$V_{D_1\_D_2} = \int_{D_1}^{D_2} \frac{Q}{2\pi\varepsilon_{ox}} dx = \frac{Q}{2\pi\varepsilon_o} \ln\left(\frac{D_2}{D_1}\right)$$

Typically, the three-phase transmission line is built with three conductors placed above the ground. The voltage between the conductors is the line-to-line voltage and between the conductor and ground is the line-to-ground voltage. As we described before, the line energization generates charges on the conductors. The conductor charges produce an electric field around the conductors. The electric field lines are radial close to the conductors. In addition to the electrical field, the conductor is surrounded by equipotential lines. The equipotential lines are circles in case of one conductor above the ground. The voltage difference between the conductor and the equipotential line is constant.

From a practical point of view, the voltage difference between a point in the space and the ground is important. This voltage difference is called space potential. Figure 20.8 shows the electric field lines and equipotential lines for a charged conductor aboveground.

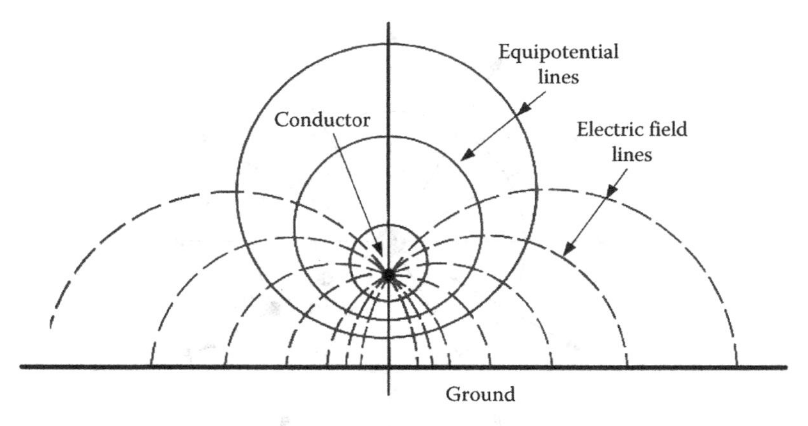

FIGURE 20.8 Electric field around an energized conductor above the ground.

20.4.1 Electric Charge Calculation

Figure 20.9 shows a three-phase, horizontally arranged transmission line. The ground in this figure is represented by the negatively charged image conductors. This means that each conductor of the line is represented by a positively charged line and a negatively charged image conductor. The voltage difference between the phase conductor and the corresponding image conductor is $2V_{ln}$. The electric charge on an energized conductor is calculated by repetitive use of the voltage difference equation presented before.

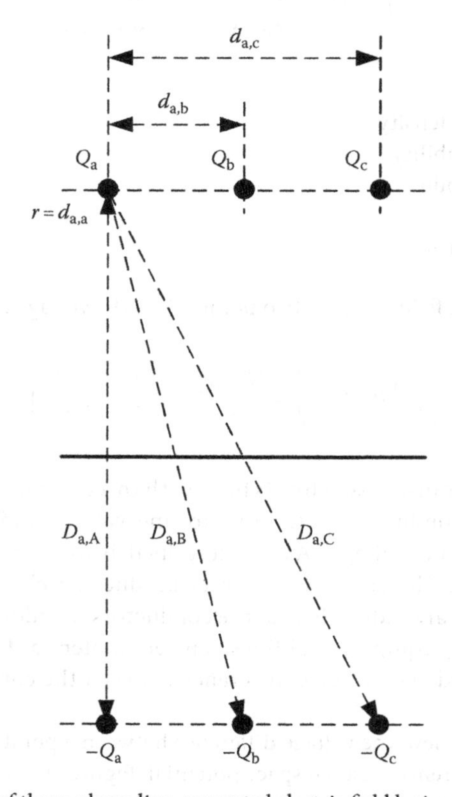

FIGURE 20.9 Representation of three-phase-line-generated electric field by image conductors.

The voltage difference between phase conductor "A" and its image conductor is generated by all charges (Q_a, Q_b, Q_c, and $-Q_a$, $-Q_b$, $-Q_c$) in the system. Using the voltage difference equations, we obtained the voltage difference between conductor A and its image:

$$V_{a,A} = \frac{Q_A}{2\pi\varepsilon_o}\ln\left(\frac{D_{a,A}}{r_{cond}}\right) + \frac{-Q_A}{2\pi\varepsilon_o}\ln\left(\frac{r_{cond}}{D_{a,A}}\right) + \frac{Q_B}{2\pi\varepsilon_o}\ln\left(\frac{D_{a,B}}{d_{a,b}}\right)$$

$$+\frac{-Q_B}{2\pi\varepsilon_o}\ln\left(\frac{d_{a,b}}{D_{a,B}}\right) + \frac{Q_C}{2\pi\varepsilon_o}\ln\left(\frac{D_{a,C}}{d_{a,c}}\right) + \frac{-Q_C}{2\pi\varepsilon_o}\ln\left(\frac{d_{a,c}}{D_{a,C}}\right)$$

This equation can be simplified by combining the $+Q$ and $-Q$ terms. The result is

$$V_{a,A} = 2V_{a\_ln} = \frac{2Q_A}{2\pi\varepsilon_o}\ln\left(\frac{D_{a,A}}{r_{cond}}\right) + \frac{2Q_B}{2\pi\varepsilon_o}\ln\left(\frac{D_{a,B}}{d_{a,b}}\right) + \cdots + \frac{2Q_C}{2\pi\varepsilon_o}\ln\left(\frac{D_{a,C}}{d_{a,c}}\right)$$

Further simplification is the division of both sides of the equation by two, which results in an equation for the line to neutral voltage. Similar equations can be derived for phases B and C. The results are

$$V_{a\_ln} = \frac{Q_A}{2\pi\varepsilon_o}\ln\left(\frac{D_{a,A}}{r_{cond}}\right) + \frac{Q_B}{2\pi\varepsilon_o}\ln\left(\frac{D_{a,B}}{d_{a,b}}\right) + \frac{Q_C}{2\pi\varepsilon_o}\ln\left(\frac{D_{a,C}}{d_{a,c}}\right)$$

$$V_{b\_ln} = \frac{Q_A}{2\pi\varepsilon_o}\ln\left(\frac{D_{b,A}}{d_{a,b}}\right) + \frac{Q_B}{2\pi\varepsilon_o}\ln\left(\frac{D_{b,B}}{r_{cond}}\right) + \frac{Q_C}{2\pi\varepsilon_o}\ln\left(\frac{D_{b,C}}{d_{b,c}}\right)$$

$$V_{c\_ln} = \frac{Q_A}{2\pi\varepsilon_o}\ln\left(\frac{D_{c,A}}{d_{c,b}}\right) + \frac{Q_B}{2\pi\varepsilon_o}\ln\left(\frac{D_{c,B}}{d_{b,c}}\right) + \frac{Q_C}{2\pi\varepsilon_o}\ln\left(\frac{D_{c,C}}{r_{cond}}\right)$$

In these equations, the line to neutral voltages and dimensions are given. The equations can be solved for the charges (Q_a, Q_b, Q_c).

20.4.2 Electric Field Calculation

The horizontal and vertical components of the electric field generated by the six charges (Q_a, Q_b, Q_c, and $-Q_a$, $-Q_b$, $-Q_c$) are calculated. The sum of the horizontal components and vertical components gives the X and Y components of the total electric field. The vector sum of the X and Y components gives the magnitude of the total field.

Figure 20.10 shows a Q charge–generated electric field. The field lines are radial to the charge.

The absolute value of electric field generated by a charge Q is described by the Gauss equation. The observation point coordinates are X and Y. The conductor coordinates are x_i and y_i.

The electric field magnitude is

$$E_i = \frac{Q_i}{2\pi r} = \frac{Q_i}{2\pi\sqrt{(x_i - X)^2 + (y_i - Y)^2}}$$

The Φ angle between the E vector and its vertical components is

$$\Phi = \text{atn}\left(\frac{x_i - X}{y_i - Y}\right)$$

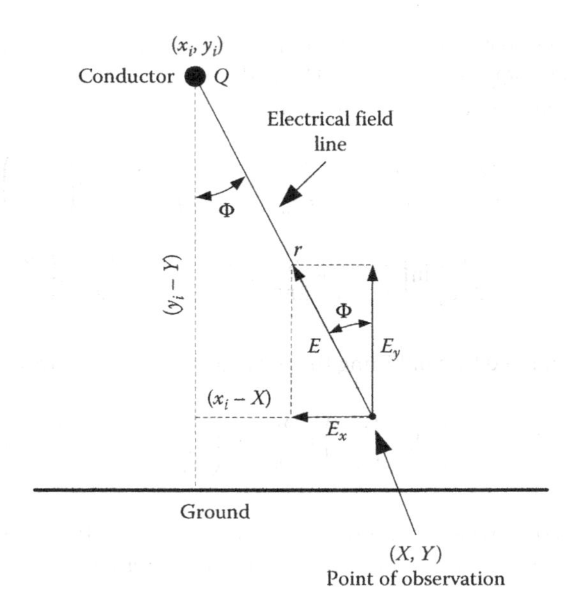

FIGURE 20.10 Electric field generated by a charge in an observation point (X, Y).

The horizontal and vertical components of the electric field are

$$E_{i\_x} = \frac{Q_i}{2\pi\sqrt{(x_i - X)^2 + (y_i - Y)^2}} \sin(\Phi_i) = \frac{Q_i}{2\pi} \frac{x_i - X}{[(x_i - X)^2 + (y_i - Y)^2]}$$

$$E_{i\_y} = \frac{Q_i}{2\pi\sqrt{(x_i - X)^2 + (y_i - Y)^2}} \cos(\Phi_i) = \frac{Q_i}{2\pi} \frac{y_i - Y}{[(x_i - X)^2 + (y_i - Y)^2]}$$

The x and y components generated by all six charges are calculated using the previous equations.

The magnitude of the total electric field is calculated by the summation of the components. The magnitude of the total field is

$$E = \sqrt{\left(\sum_i E_{i\_x}\right)^2 + \left(\sum_i E_{i\_y}\right)^2}$$

For the demonstration of the expected results, we calculated a 500 kV transmission line–generated electric field magnitude under the line in 1 m distance from the ground. The conductors are arranged horizontally. The average conductor height is 24.38 m (80 ft); the distance between the conductors is 10.66 m (35 ft). The line-to-ground voltage is

$$V_{\text{ln}} = \frac{500\,\text{kV}}{\sqrt{3}} = 288.7\,\text{kV}$$

Figure 20.11 shows the electric field distribution under the line in 1 m from the ground. The locations of the line conductors are marked on the figure. It can be seen that the maximum electric field is nearly under the side conductors. The electric field under the middle conductor is less than the side conductors because of the field cancellation caused by the 120° phase shift of the line voltages. The electric field decreases rapidly with the distance.

Typically, the electric field under high-voltage transmission lines varies between 2 and 10 kV/m.

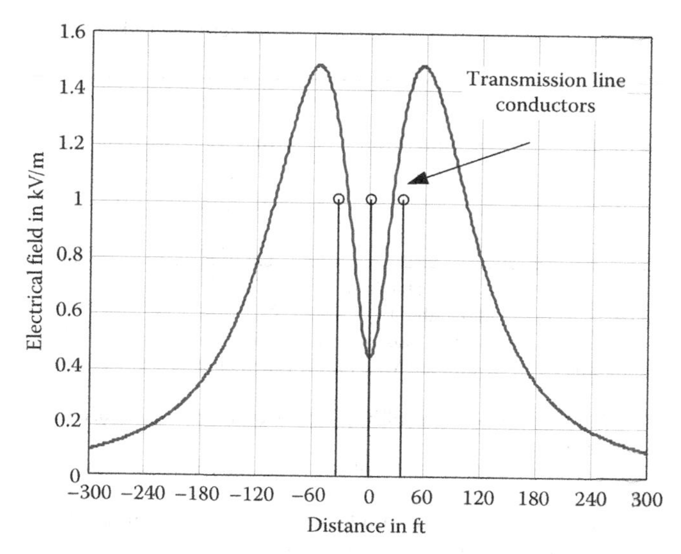

FIGURE 20.11 Electric field under a 500 kV line.

20.4.3 Environmental Effect of Electric Field

In general, the electric field generated by a transmission line has no harmful health effects. Large number of studies investigated the biological effect of small 1–20 kV/m, 60 Hz electrical fields. None of the studies has shown any harmful effects. However, the electrical field can produce annoying disturbances.

The electrical field surrounding a transmission line can charge ungrounded objects close to the space potential. If the object is large, like a truck, parking under the line affects the field distribution and space potential.

The simplest visualization of the problem is a truck parking under a transmission line; the rubber tires insulate the truck from the ground. The voltage difference between the truck and the ground is determined by the capacitance between the truck and the line, and the capacitance between the truck and ground. The two capacitances form a capacitive voltage divider. The truck potential to ground can be few kilovolts. A person standing on the ground and touching the truck will discharge the capacitor between the truck and ground. This produces a small spark discharge. The person touching the truck suffers minor electric shock, which is not dangerous but uncomfortable.

After the discharge the person touching the truck grounds it, which results in a constant current through the person. This current is determined by the capacitance between the object, in this case the truck and the line. EPRI-published *Redbook* (*Transmission Line Reference Book—345 kV and Above*) [14] gives an approximate formula for the expected current:

$$I_{\text{cap}} = 2\pi f \varepsilon_{\text{o}} E_y \,\text{surface}$$

where

E_y is the vertical component of the electric field
f is the frequency (60 Hz)
"surface" is the equivalent charge collecting area of the object
I_{cap} is the capacitive current flowing through the person grounding the object

Another potentially dangerous accident scenario is when a worker climbs on a wooden ladder to repair something close to a transmission line. A grounded coworker hands him a tool. This produces

TABLE 20.2 Environmental Electric Field Maximum Permissible Exposure (MPE), Whole Body Exposure

General Public		Controlled Environment	
Frequency Range (Hz)	*E*-Rms (V/m)	Frequency Range (Hz)	*E*-Rms (V/m)
1–368[a]	5000[b,c]	1–272[a]	20,000[d,e]
368–3000	$1.84 \times 10^6/f$	272–3000	$5.44 \times 10^6/f$
3000	614	3000	1,813

Source: IEEE Std C95.6-2002, IEEE Standard for Safety Levels with Respect to Human Exposure to Electromagnetic Fields, 0–3 kHz, Table 4.

[a] Limits below 1 Hz are not less than those specified at 1 Hz.

[b] Within power line rights-of-way, the MPE for the general public is 10 kV/m under normal load conditions.

[c] At 5 kV/m induced spark discharges will be painful to approximately 7% of adults (well-insulated individual touching ground).

[d] Painful discharges are readily encountered at 20 kV/m and are possible at 5–10 kV/m without protective measures.

[e] The limit of 20,000 V/m may be exceeded in the controlled environment when a worker is not within reach of a grounded conducting object. A specific limit is not provided in this standard.

a discharge and a minor spark, which is harmless. However, the shock may cause dropping the tool or falling off the ladder.

People walking under the line may experience a tingling sensation on their skin and hair stimulation if the electrical field is larger than 6–7 kV/m. This is an annoying but harmless effect.

IEEE Std C95.6 -2002 gives the maximum permissible exposure (MPE) levels for electric field. Table 20.2 shows the permissible whole body exposure levels to electric field.

The final conclusion is that the electric field should be less than 10 kV/m for the general public within the transmission line right-of-way.

20.5 Audible Noise

The corona discharge on the high-voltage transmission line generates audible noise. The corona discharge produced by a well-designed transmission line is very low in fair weather. Consequently, the transmission line produced audible noise in fair weather conditions is negligible.

Fog and light rain produce droplets on the surface of line conductors. The droplets increase the local electric field and generate corona discharge. The corona discharge produced air movement or pressure wave generates the audible noise. The light-rain- and fog-generated noise intensity varies, fluctuating depending on the level of wetting.

Heavy rain produces more or less constant noise. The corona discharge bursts the water droplets and disperses the water. However, the heavy rain replacing the dispersed water drops immediately.

Snowflakes also can increase corona level and audible noise. The dry, low temperature snow generally does not produce audible noise. The audible noise generated by wet melting snow can be significant and the noise level will be similar to the heavy rain–generated noise.

Typically the line-generated noise has two components:

- Broadband noise, which is mainly generated by the discharge on water droplets. This is a hissing, crackling noise with significant high-frequency components.
- Low-frequency humming noise with 120, 240 Hz, etc., components. This noise is generated by the oscillatory movement of the corona-generated ions around the conductors. The humming noise occurs mostly in good weather condition, if the line corona level is low.

From a practical point of view, the broadband noise is the most important. The utilities accept a noise level of 50–52 dB at the edge of right-of-way. The noise level is measured in dB. The base is 20 μPa.

The noise attenuates with the distance due to the divergence of the sound and the absorption of trees and other objects. Practical value is around 3 dB, when the distance is doubled.

A numerical example is presented to estimate the approximate sound level in a residential area if the sound level is 52 dB at the edge of the right-of-way. The distance between the line and the edge of right-of-way is 100 ft. The sound level in a distance of 200 ft is 52 − 3 dB = 49 dB and in a distance of 400 ft is 49 − 3 dB = 46 dB.

The transmission line–generated noise level can be reduced by reduction of corona discharge level. The most effective method is the use of bundle conductors. The rearrangement of the line conductors also can reduce corona discharge and audible noise.

The EPRI-published *Transmission Line Reference Book—345 kV and Above* [14] gives curves to estimate the expected audible noise level produced by a transmission line.

20.6 Electromagnetic Interference

The corona discharge produces radio noise and in lesser extent television (TV) disturbances around high-voltage transmission lines. This can be easily observed by all of us when we drive under a high-voltage line. The radio produces hissing, crackling noise close to the line or under the line, but disturbance disappearing fast as we drive away from the line crossing the highway. In a similar way, TV picture disturbance can be observed close to a transmission line. The disturbance varies from the snowy picture to the collapse of the picture.

The corona discharge causes short duration (few microseconds) repetitive current pulses. The repetition frequency can be in the MHz range. As was discussed before, the corona discharge is low in fair weather and increases rapidly in foul weather. The most severe EMI disturbance was observed during heavy rain, when the water droplets on the conductor caused corona discharge.

Additional sources of the EMI disturbance are discharge in faulty insulators or discharge generated by spikes, needles, and other sharp objects subjected to electric field. The sharp object produces an increase in the local electric field, which can lead to surface discharge. This discharge can produce EMI and unacceptable disturbances of local TV or radio reception.

The generated EMI disturbance decreases with the distance from the line. Typically, a 100 MHz signal decreases about 20 dB if we move 100 m from the line; simultaneously, a 1 MHz components attenuation is around 35–40 dB in the same distance. The radio and TV noise is measured in dB; the base is 1 μV/m.

The actual disturbance depends on the signal-to-noise ratio. As an example, the same level of EMI disturbance can produce an unacceptable radio or TV reception if the broadcasted signal is weak, and no disturbance in case of strong signal.

The EPRI-published *Transmission Line Reference Book—345 kV and Above* [14] gives curves to determine the expected radio or TV disturbance level produced by a transmission line.

References

1. Wertheimer, N. and Leeper, E., Electric wiring configuration and childhood cancer, *American Journal of Epidemiology*, 109(3), 273–284, March 1979.
2. Savitz, D.A., Wachtel, H., Barnes, F.A., John, E.M., and Tvrdik, R.G., Case control study of childhood cancer and residential exposure to electric and magnetic fields, *American Journal of Epidemiology*, 128(1), 21–38, January 1988.
3. London, S.J., Thomas, D.C., and Bowman, J.D., Exposure to residential electric and magnetic fields and risk of childhood leukemia, *American Journal of Epidemiology*, 131(9), 923–937, November 1992.
4. Floderus, B., Persson, T., Stenlund, C., Wennberg, A., Ost, A., and Knave, A., Occupational exposure to electromagnetic fields in relation to leukemia and brain tumor, *Cancer Causes Control*, 4(5), 465–476, 1993.

5. Tynes, T., Hanevik, M., and Vistnes, A.I., A nested case-control study of leukemia and brain tumors in Norwegian railway workers, *Conference Proceedings Fifteenth Annual Meeting of the Bioelectromagnetic Society*, Los Angeles, CA, 46 pp, June 1993.
6. Scarfi, M.R., Bersani, F., Cossarizza, A., Monti, D., Zeni, O., Lioi, M.B., Franceschetti, G., Capri, M., and Franceschi, C., 50 Hz, sinusoidal electric fields do not exert genotoxic effects (micronucleus formations) in human lymphocytes, *Radiation Research*, 235, 64–68, 1993.
7. Byus, C.V., Piper, S.A., and Adey, W.R., The effect of low energy 60 Hz environmental electromagnetic fields upon the growth related enzyme ornithine decarboxylase, *Carcinogenesis*, 8, 1385–1389, 1987.
8. Korpinen, L., Influence of 50 Hz electric and magnetic fields on the human heart, *Bioelectromagnetics*, 14(4), 329–340, 1993.
9. Graham, C., Cook, M.R., and Riffle, D.W., Human melatonin during continuous magnetic field exposure. *Bioelectromagnetics*, 18, 166–171, 1996.
10. EPRI Report, TR-100194, Survey of residential power magnetic field sources, Phase 1, RP2942, Electric Power Research Institute, Palo Alto, CA, 1990.
11. U.S. Environmental Protection Agency, Evaluation of the potential carcinogenicity of electromagnetic fields, EAP/600/6-90/005B, October 1990.
12. *Electric and Magnetic Field Fundamentals*, Electric Power Research Institute, Palo Alto, CA, March 1994.
13. *Electric and Magentic Fields: Prospective on Research Needs and Priorities for Improving Health Risk Assessment*, DIANE Publishing Company Medical, 65 pp, 1994.
14. *Transmission Line Reference Book—345 kV and Above*, 2nd edn., Electric Power Research Institute, Palo Alto, CA, 1987.
15. Kaune, W.T. and Zaffranella, L.E., Analysis of magnetic fields produced far from electric power lines, *IEEE Transactions on Power Delivery*, 7(4), 2082–2091, October 1992.

21

Transmission Line Reliability Methods

Brian Keel
Salt River Project

Vishal C. Patel
Southern California Edison Company

Hugh Stewart Nunn II
Salt River Project

21.1 Introduction

As the demand for, and societal dependence upon, electricity has grown, so too have the complexities of maintaining ever-growing transmission and distribution systems. From its inception in the late 1800s to today, the electric grid has undergone a significant growth, both in its geographic reach and transfer capability.

With this increasing demand came the need of longer distances for that electricity to be transmitted, due to the siting of large power-generating facilities away from major load centers. This required the development of higher voltages and ampacities, achieved through technological advances in the field of electric power transmission equipment, construction, and design. Longer transmission line distances were also driven by the interconnection of neighboring utility systems, which provided the key benefit of reliability through reserve sharing.

All of this, over time, has led to a change in the concept or definition of "transmission." Voltages, which might have once been considered "transmission" level are now commonplace for local distribution and subtransmission systems.

Today, typical extra high voltage (EHV) transmission line voltages in the United States are 765, 500, and 345 kV. Other common voltage classes utilized as transmission are 230, 161, 138, 115, and 69 kV. Many utilities differ in their definitions of what transmission is defined as. Per FERC Order 743, the bulk electric system has been defined as "greater than 100 kV"; thus, many entities will now use that definition for transmission. Some companies also use the term "subtransmission" as an intermediate level between distribution and transmission. In the western United States, 500, 345, and 230 kV make up the larger interconnection backbone, whereas 69–161 kV are typical voltage levels for local-area transmission systems.

In general, transmission line reliability is a greater concern than for distribution lines, since transmission substations typically serve as the source for delivering power to the subtransmission and/or distribution system from a variety of resources. Also, as mentioned previously, transmission typically connects multiple utility systems together, allowing for a more efficient, cost-effective, and reliable system. Along with this is also the fact that greater amounts of power, and thus customers, are at risk for every transmission outage or contingency. The existing NERC Transmission Planning (TPL) standards reflect this concept, as it does not allow for the loss of load when a single transmission element encounters an outage.

However, the NERC standards are not the initiating force behind the need to study outage performance. Transmission reliability concerns have been integral to power system planning since the inception of the lines themselves. Factors such as design, weather performance, common mode outages, and outage performance of lines are key facets of the overall design and specification of electrical systems.

Empirical data gathering and analysis are a necessary component of ensuring ongoing reliability and safety of systems. Engineers inherently seek to have a rule-of-thumb for a variety of issues; compiling outage data over time aids that continuous effort, helping to learn from mistakes and institute improvements over time.

21.2 Common Terminology for Analyzing Transmission Outage Data

One of the key issues when discussing data of any sort is the adherence to a specific set of definitions which is commonly understood. In the electric power industry, the majority of statistical/probabilistic analysis is focused on distribution system related outages. However, IEEE Standard 859[1] has definitions for power transmission. This standard was recently reaffirmed in 2008. Some of the basic definitions and how they are calculated are included hereafter.

In-service state: The component or unit is energized and fully connected to the system.

Outage state: The component or unit is not in the in-service state; that is, it is partially or fully isolated from the system.

Reporting period time: The duration of the reporting period (equals service time plus outage time).

Service time: The accumulated time one or more components or units are in the in-service state during the reporting period.

Outage time: The accumulated time one or more components or units are in the outage state during the reporting period.

Outage duration: The period from the initiation of an outage occurrence until the component or unit is returned to the in-service state.

Outage rate: The number of outage occurrences per unit of service time = # of outage occurrences/service time.

Availability = Service time/reporting period time.
Unavailability = Outage time/reporting period time.

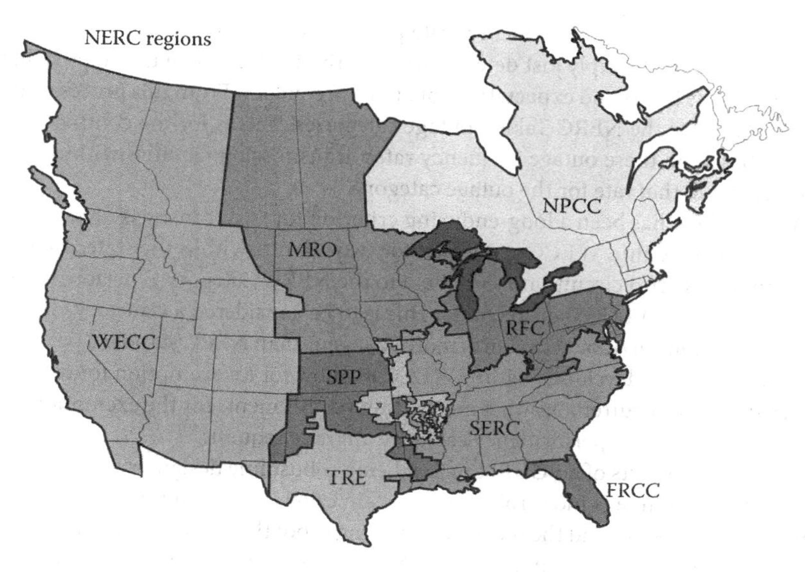

FIGURE 21.1 NERC regions.

21.3 Transmission Outage Data Sources and Current Data Gathering Efforts

The following list identifies transmission outage data gathering efforts throughout North America:

ECAR, the former regional reliability organization known as East Central Area Reliability, had a continuous data collection until after the 2004 data collection. ECAR gathered automatic and nonautomatic outage data. This data collection was halted in anticipation of the NERC TADS efforts.

The Canadian Electric Association or CEA has been gathering transmission circuit and transformer outage information since 1978. Canadian Electric Association gathers outage data for transmission lines but also for transformers, circuit breakers, cables, shunt reactors and capacitors, series capacitors, and static and synchronous compensators. The outage reports are available if a member of the CEA.

The MAPP, Mid-Continent Area Power Pool, did data collection in the 1970s and 1980s and helped in the development of common definitions and measures.

The WECC TRD, Western Electricity Coordinating Council, Transmission Reliability Database, development began with data collection in 2006. The WECC TRD data gathering is active and will be further explained in the next sections.

The NERC TADS, North American Electricity Corporation, Transmission Availability Database System, first year of data collection was 2008. The NERC TADS data gathering is active and will be further explained in the next sections Figure 21.1.[2]

21.4 Western Electricity Coordinating Council: Transmission Reliability Database

The WECC TRD was developed as a result of the PBRC, probabilistic-based reliability criteria—development in the late 1990s and creation of the workgroup called the RPEWG—reliability performance evaluation work group.

Data gathered for the WECC TRD are automatic outages for transmission elements greater than 200 kV. The data for nonautomatic/manual outages are not gathered, as these outages were deemed within WECC as having no benefit.

The members of WECC wanted to incorporate probabilistic planning into the WECC planning criteria and process instead of simply just deterministic methods. Because of this approach, the PBRC was developed to define the limits and expectations of outage frequency. From this process, outage frequency rates were developed for the NERC Table I outage categories. This is for the contingency Categories of A, B, C, and D. Based on these outage frequency rates; transmission facilities within WECC were then expected to perform at that rate for the outage category.

Within WECC, there has been a long-enduring criterion for two transmission circuits in a common corridor. This criterion within WECC states that for two circuits within the stated definition of a common corridor that these two circuits are to be held to the NERC Category C performance requirements as well as more stringent WECC requirements. This is to be considered a Category C contingency.

Because this common corridor criterion is more stringent than NERC standard performance requirements, there was a process developed within WECC to allow for an exemption for two corridor circuits from the more stringent requirements that met certain requirements for the exemption. This exemption process is called the PCUR or performance category upgrade request.

There are two main aspects of a PCUR; the first being robust line design or proving that the design of the potentially exempted line is more robust per certain criteria listed in the documents accompanying the robust line design process, and the second by showing proof that the circuits in the corridor will or do have an expected outage rate that is more reliable than the outage rate listed in the WECC criteria for this contingency. Therefore, the two circuits in the common corridor need to show outage MTBF greater than the listed rate of 0.033–0.33. The transmission owner that is requesting the PCUR needs to show that the outage frequency is greater than one in 30 years.

The TRD was developed to assist this process in two ways. The first to develop a database that potential PCUR transmission owners will have a database developed that can be used for typical or similar corridor outage rates for newly constructed or planned corridors. The second reason is to give the RPEWG the data it needs to follow or track the performance of existing corridors to monitor the performance relative to the outage rate. Also, to have a performance track of those exempted corridors, to ensure they are performing to expected values.

The assignment of developing the TRD began in the early 2000s, and the work on development of the database began in early 2004 and was formally accepted by the WECC in 2006. The data submittal of 2006 was the first year of TRD data submittal. As of this writing, there have been four TRD annual reports with data analyzed each year more and more to try and get to more meaningful results from the data submittal (Table 21.1).[3]

The following tables are compiled from data listed in the WECC TRD report. The current or latest year's data is listed as well as the average of the past four submittals.

The focus of the outage statistics in the TRD reports:

WECC transmission totals, momentary outage summary, sustained outage summary, outages per month, sustained transformer outage rate, transformer age summary, outage rate per bank by transformer age, outage rate of common corridor and common tower, outage rate per each physical attribute, voltage level outage cause code.

The focus of the TRD is to gather data of elements >200 kV and to focus on the physical design aspects of transmission circuits and transformers. This focus will allow a cooperation or coordination of all WECC members if particular equipment is seen to be failing at one TO or subregion and to give warning to other part of WECC of these failures. Each year outage rates are calculated for the number of overhead ground wires, conductors per phase, type of insulator, structure material, structure type, terrain type, elevation range. These results are shared in the WECC TRD report each year. Each year the RPEWG is developing different techniques to allow the witness of a systemic problem to convey to all of WECC.

The WECC TRD is a voluntary data submittal. But in 2009, 100% of WECC TOs submitted their data. This is due to the NERC TADS which is a mandatory data submittal and the WECC has designed the TRD data submittal to cover both data submittals of the WECC TRD and NERC TADS.

TABLE 21.1 WECC Disturban-Performance Table of Allow Able Effects on Other Systems

NERC and WECC Categories	Outage Frequency Associated with the Performance Category (Outage/Year)	Transient Voltage Dip Standard	Minimum Transient Frequency Standard	Post Transient Voltage Deviation Standard (See Note 3)
A	Not applicable	Nothing in addition to NERC		
B	≥0.33	Not to exceed 25% at load buses or 30% at non-load buses	Not below 59.6 Hz for 6 cycles or more at a load bus	Not to exceed 5% at any bus
C	0.033–0.33	Not to exceed 30% at any bus Not to exceed 20% for more than 40 cycles at load buses	Not below 59.0 Hz for 6 cycles or more at a load bus	Not to exceed 10% at any bus
D	<0.033	Nothing in addition to NERC		

Notes: (1) The WECC Disturbance-Performance Table applies equally to either a system with all elements in service, or a system with one element removed and the system adjusted. (2) As an example in applying the WECC Disturbance-Performance Table, a Category B disturbance in one system shall not cause a transient voltage dip in another system that is greater than 20% for more than 20 cycles at load buses, or exceed 25% at load buses or 30% at non-load buses at any time other than during the fault.

TABLE 21.2 2009 WECC TRD Data

Voltage	Total Number of Outages	Mileage
500 kV	309	19,343
345 kV	307	10,889
230 kV	910	44,393

TABLE 21.3 2009 WECC Momentary Outages

	Average		2009	
	Per Circuit	Per 100 miles	Per Circuit	Per 100 miles
500 kV	0.60	0.85	0.45	0.61
345 kV	1.00	1.21	0.83	1.07
230 kV	0.26	0.90	0.18	0.66

TABLE 21.4 2009 WECC Sustained Outages

	Average		2009	
	Per Circuit	Per 100 miles	Per Circuit	Per 100 miles
500 kV	0.94	1.32	0.73	0.99
345 kV	1.61	1.96	1.36	1.75
230 KV	0.42	1.49	0.38	1.39

From Table 21.2, we see that WECC is predominantly a 230 and 500 kV system with 1500–1600 outages per year across the entire WECC system.

In Table 21.3, the momentary outages are calculated for the WECC system and compared to the trailing year's data shown as average values. The values show that the system indices improved in 2009.

Table 21.4 shows the sustained outage rates for the WECC system. The values show that the system indices improved in 2009 versus the average values.

21.5 North American Electricity Reliability Corporation: Transmission Availability Database System

NERC TADS was started on the premise that transmission availability data will help to quantify NERC transmission system performance and reliability. NERC has taken the role of being an independent source of reliability performance information, a recommendation of the April 2004 United States—Canada power system outage task force report on the August 14, 2003 Blackout. The TADS task force was initiated in October 2006 to write the approach for transmission outage data reporting and measuring availability and performance. This became the NERC TADS.

NERC TADS collects outage data on specific lines while supplying utility aggregate transmission population information. NERC TADS will be implemented in two phases: Phase I is the collection of automatic outages began with the year 2008 data, Phase II adds the requirement of supplying the automatic outages as well as the nonautomatic outages for the calendar year. The first year of the nonautomatic outage data collection will be for the data collection of calendar year 2010.

The focus of the reporting in the TADS Reports:

There are a total of nine reports; there are NERC-wide and eight regional reports for each of the regions listing summary of NERC-wide results, all AC circuit metrics, all DC circuit metrics, all transformer metrics, and AC/DC back-to-back converter metrics.

The data structure for TADS will gather aggregate transmission population from the TOs while gathering specific data for each outage.

The purpose of TADS is to provide outage cause analysis and outage Event analysis. Event analysis will aid the in the determination of credible contingencies and will results in better understanding, and this understanding should be used to improve planning and operations.

In addition, trending each Regional Entity's performance against its own history will show that region's performance is changing over time. Given the vast physical differences between regions and TOs (weather, load density, geography, growth rate, system age, customer mix, impact of significant events, average circuit mileage, etc.), we believe that comparisons for the purposes of identifying relative performance between regions are not appropriate. Taken from the NERC TADS revised final report, September 26, 2007.

21.5.1 Data in Annual Reports

The data listed in Tables 21.5 and 21.6 shows the results for all of NERC for 2009. The numbers for the columns are total circuit outage frequency (TCOF), sustained circuit outage frequency (SCOF), and momentary circuit outage frequency (MCOF). More information can be found at www.nerc.com.[4]

TABLE 21.5 2009 NERC Wide Outage Summary

	Total Circuit Outage Frequency TCOF	Sustained Circuit Outage Frequency SCOF	Momentary Circuit Outage Frequency MCOF
500 kV	1.32	0.82	0.51
345 kV	2.18	1.13	1.05
230 kV	2.18	1.33	0.85

TABLE 21.6 NERC 2009 Outage Totals

	Momentary	Sustained
500 kV	163	262
345 kV	591	637
230 kV	878	137

21.6 Salt River Project Transmission Outage Data

What is Salt River Project (SRP)?

SRP is a municipal electric power and water utility that serves the Phoenix, AZ metropolitan area. The SRP system has

Line Parameters—Mileage, Rating, and Other Data

500 kV (Overhead pole miles)	357
230 kV (Overhead pole miles)	411
115 kV (Overhead pole miles)	264
69 kV (Overhead pole miles)	903
69 kV (Underground miles)	8

The SRP bulk transmission and subtransmission system consist of facilities and electrical equipment in the 500, 230, 115, and 69 kV voltage classes.

21.6.1 SRP Operating Environment

In order to reliably serve its customer load, SRP addresses specific design challenges associated with its operating environment. For example, temperatures in the Phoenix metro area can range from 20°F to 122°F. Due to high cooling needs, SRP's system peak load occurs annually during the summer after several hot days in a row with temperatures above 115°F. In order to address the extreme weather operating conditions, SRP designs its transmission in a manner that prevents thermal overloading of equipment during peak demand periods. Therefore, SRP's transmission system is highly reliable due to the development of an overall system design architecture, equipment specification, maintenance, and planning and operational standards. In order to present evidence that SRP's transmission system designs, operating and maintenance practices are adequate to meet the environmental demands, SRP measures its transmission reliability performance by collecting and analyzing transmission disturbance event data.

21.6.2 Transmission Event Data Capture

A transmission event or disturbance is defined as an abnormal system condition that may include electric system faults, equipment outages, frequency deviations, voltage sags, etc. Because SRP captures the data in the context of the event, the data quality is excellent and operations personnel are made aware of system events that may influence current and near-term system operations.

21.6.3 Transmission Event Data Characteristics

Transmission event data can be broadly classified into two distinct categories. One is a set of "random" events that represent environmental impacts regardless of the system design, maintenance, and operations. These random events contribute approximately 75% of the total annual transmission event count. The other category is classified as "nonrandom" events and is directly related to the system design, maintenance, and operation practices. Annual event counts of transmission system are very low compared to distribution system event counts. Current reliability reporting practices comingle random and nonrandom events in such a manner to often provide management with misleading information.

21.6.4 Nonrandom Event Performance Analysis of Actionable Transmission System Events

NREP represents a new reliability reporting and analysis concept that seeks to separate actionable information from random transmission event data. The underlying aspect of NREP is that electric utilities design their systems to perform in the environment in which they are expected to operate. NREP analysis differentiates transmission event data that reflects the fidelity of a transmission design, maintenance, and operating practices to the data that does not.

21.6.5 Potential Uses of the Nonrandom Event Performance, NREP, Feedback

Each utility uses a list of outage cause codes that the outages of the system are charged into for collection and, if necessary, more intense data scrutiny. SRP uses the following outage cause codes in that all 69 kV through 500 kV outage will be categorized.

21.6.6 Category Random

Animals, vehicle caused, bird contact, contamination, customer caused, debris in equipment, environmental condition, fire, foreign system, unknown with fault, inadvertent by public, lightning, rain, storm with unknown cause, unknown with no fault, vandalism, and wind.

21.6.7 Category Nonrandom

AC circuit equipment, breaker failure, communications, control, relay, EMS failure, inadvertent by SRP, unmitigated distribution fault, under-built line, shunt capacitor or reactor failure, power system condition, pole failure, series capacitor failure, AC substation equipment failure, vegetation or transformer failure.

There are three things that can be done from the feedback, do nothing, change maintenance spending, or change design. The feedback from NREP can give a better idea of how to proceed, if desired.

If the system shows a downward trending NREP, then the maintenance dollars could be potentially decreased to save money and to allow the NREP to raise enough to be within acceptable limits. How to define the acceptable limits?

If the system shows an upward trending NREP, then the time for maintenance spending increases is now or soon depending on how close to acceptable limits.

Maintenance spending is a quicker feedback mechanism for the NREP because design changes may take many, many years to see the influence because the existing system is very large given that the design changes will have impact over time as system is added.

In Figure 21.2, the SRP system NREP is shown for the past 6 years. The vertical bars represent all outages and the horizontal lines are the NREP values. The NREP value is shown to be decreasing.

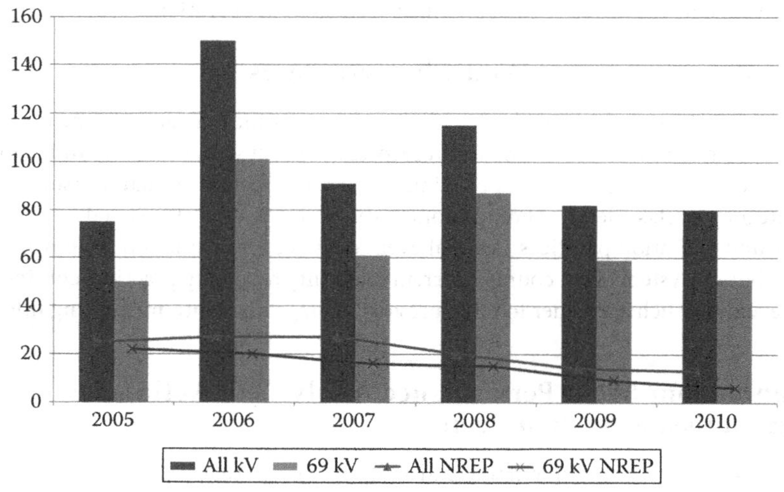

FIGURE 21.2 The SRP system NREP.

Also, only a small percentage of the total customer outages are transmission caused but there is always the possibility of transmission outage cascading, leading to multiple outages that may cause large scale customer outages or system-wide outages. Therefore, transmission outages may not be a great influence on customer outages normally, but the possibility always exists for this to turn large scale and regional.

Transmission outage data can serve the utility in the following ways:

1. Demonstrating the design, maintenance, and operational practices that meet or exceed the challenges posed by the environment
2. If the practices need to be changed to increase reliability
3. If improvements are needed operationally

Typically, SRP experiences four to six 69 kV line outages each year caused by an inadvertent action by SRP personnel. In 2008, SRP experienced 16 of these outages. There were two main causes that can be attributed to the increase. In 2008, relay maintenance revised their maintenance practices to include exercising inputs and outputs on relays. Also, at this time, three times the normal number of relays was upgraded in the first half of the year that contributed to this high number of outages. These practices were changed after 6 months and the outage rates returned to normal. The number of inadvertently caused outages returned to a typical value, 2 total, in 2009.

The number of transmission initiated outages versus the number of distribution caused customer outages:

2009	38/7343 = 0.5%
2008	86/8152 = 1.05%
2007	85/7908 = 1.08%
2006	114/7763 = 1.46%
2005	54/8133 = 0.66%

21.6.8 NREP Conclusion Section

NREP transmission event data analysis conveys the system's reliability of how well the system reacts to its environment. NREP provides an easily understood model to demonstrate to the stakeholders whether the system is performing adequately or not. Additionally, NREP analysis can provide a framework that enables a true utility to utility reliability performance comparison.

21.7 Southern California Edison Transmission Outage Data

Southern California Edison (SCE) is one of the largest electric utilities in the United States, providing electric service to nearly 14 million people throughout a 50,000 square mile service territory. The service territory is quite diverse geographically, exposing it to a variety of different climates. SCE has been gathering transmission outage data for 50-plus years. A summary of SCE's transmission and subtransmission line mileages as of 2010 is provided hereafter.

Line Mileages (approx)

500 kV (Overhead lines)	1031
230 kV (Overhead lines)	3379
230 kV (Underground lines)	1
115 kV (Overhead lines)	1874
115 kV (Underground lines)	16
66 kV (Overhead lines)	4838
66 kV (Underground lines)	254

SCE's system design philosophy differs in two key ways from other utilities. First, SCE utilizes a radial design. All loads at any "A-station" (see squares in the following diagram) serve load on lower voltages (i.e., subtransmission systems) radially, meaning the lower voltage levels (i.e., 66 and 115 kV) are not networked to other transmission stations during normal operation. Another way to think of this is that the subtransmission system is not operated electrically parallel to the transmission system. SCE is a participating transmission owner within the California Independent System Operator (CAISO), and thus the CAISO has operational control of the SCE transmission system. The majority of subtransmission and distribution systems SCE owns are still under the operational control of SCE. SCE follows this design philosophy since the original system development. This is due to the simplicity it brings in operating, planning, and designing subtransmission systems. Second, SCE is a large importer of electricity from outside the service territory, significantly depending on the EHV transmission system to import energy throughout the western United States (Figure 21.3).

Due to this slightly different philosophy and being a large importer of power with a vast and geographically diverse service territory, SCE requires a higher level of reliability from its transmission system. Rather than solely looking at the aggregate statistics, SCE's planning engineers focus their attention on ascertaining which outages are more frequent(or credible), by annually evaluating the outage data. This is in addition to availability reports and other regulatory obligations that SCE submits data to NERC, WECC, CAISO, and the state of California.

SCE Number of transmission circuit initiated customer outages versus the number of distribution circuit initiated customer outages:

2010	2/15,900 = 0.01%
2009	1/14,042 = 0.01%
2008	12/14,780 = 0.08%
2007	0/15,801 = 0.00%
2006	0/17,580 = 0.00%

SCE experiences a significant amount of 500 kV outages every year due to wildfires. As this is a common occurrence throughout its service territory annually, SCE criteria specify a separation of a third EHV circuit from any other two circuits within the same right-of-way. These criteria were developed due to the historical experience SCE had in the development of the Pacific AC Intertie lines.

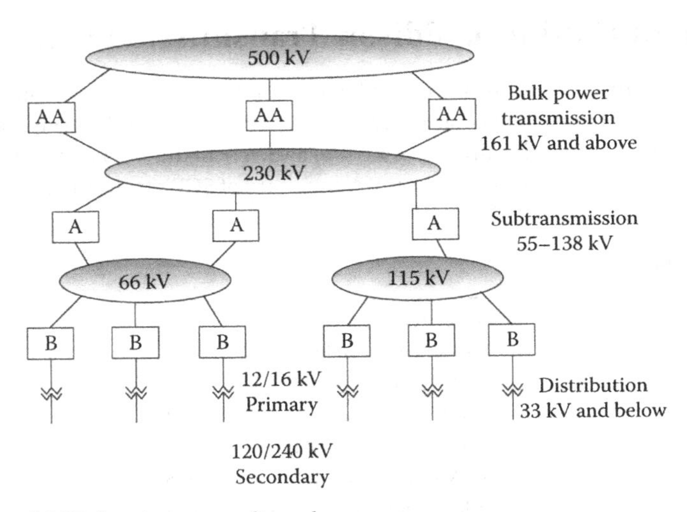

FIGURE 21.3 Typical SCE electric system voltage classes.

Due to SCE's dependence on the EHV system, it is essential not to put more than two EHV circuits in corridors exposed to extreme conditions, such as fire. Allowing a separation of 2000 ft or more between circuits can address this issue. These criteria serve as an example of how long-term data gathering efforts can guide utilities in prudently planning for the nuances of their unique service territories.

21.8 Conclusion

While there is a significant amount of work that has been performed regarding transmission outage data, the authors foresee that this work will only increase in the future, due to the even greater importance of reliability on transmission systems. The computational abilities of today, compared to that of a few decades ago, will allow for even more variations of data analysis and interpretation.

By the use of outage data over time and the vigilant investigation of outages and properly written outage cause codes can lead to potential feedback to the power system design and potential changes to the design over time. There may be potential feedback to maintenance practices as well.

Through experience, one can glean that the statistics and probabilities alone will not tell the whole story; specific details on certain lines can typically be an important piece of the puzzle during event analysis. However, in order to evaluate any large sample of data, it is understood that one must use the methods used in the examples mentioned earlier in the chapter. However, the authors urge readers to be cognizant of the fact that data can be presented in ways that could mislead; thus, it is always beneficial to obtain more specific information when making significant decisions.

The authors would like to thank all those involved in the compilation of this chapter, specifically those at both SCE and SRP, and their families.

References

1. IEEE Standard 859, IEEE Standard Terms for Reporting and Analyzing Outage Occurrences and Outage States of Electrical Transmission Facilities, 2008.
2. www.nerc.com, Key players, regional maps, accessed on March 11, 2011. http://www.nerc.com/fileUploads/File/AboutNERC/maps/NERC_Regions_color.jpg
3. www.wecc.biz, Standards, approved criteria, accessed March 11, 2011, http://www.wecc.biz/Standards/WECC%20Criteria/TPL-001%20thru%20004-WECC-1-CR%20-%20System%20Performance%20Criteria.pdf
4. www.nerc.com, Assessments and trends, transmission availability data systems(TADS), accessed March 11, 2011, http://www.nerc.com/docs/pc/tadstf/TADS_PC_Revised_Final_Report_09_26_07.pdf

22

High-Voltage Direct Current Transmission System

George G. Karady
Arizona State University

Géza Joós
McGill University

22.1 Introduction

High-voltage direct current (HVDC) energy transmission was developed in the late 1920s, and the first commercial HVDC submarine cable started operation in 1954. This system used mercury-arc valves, which required constant maintenance and rebuilding. With the development of the high-voltage and high-power thyristors in the early 1970s, thyristor valves gradually replaced the mercury-arc valves. Thyristor valves increased system reliability significantly with reduced maintenance. This resulted in fast development of the technology and the building of several new HVDC systems worldwide. Today, HVDC is a mature, well-developed technology.

The thyristor-based HVDC system uses current-commutated converters (valves), because the thyristors cannot be switched off. They switch off naturally at current zero, when the polarity of current changes.

The current source inverter–based system is designed to transfer large amount of power for long distance or bridged long water ways, sea crossing where the AC cable cannot be used because of the large capacitive current. The thyristor converter–based HVDC system is economical for transmission lines longer than 300 miles and transports 300–500 MW power. The typical application of this system is interconnection between two points. The taping of an HVDC line is difficult with thyristor-based converters. Another application is the back-to-back ties, which provides asynchronous interconnection between two AC systems without transmission line. The asynchronous interconnection permits the regulation of power transfer, blocks cascading failures, and prevents the increase of short-circuit current.

The classical HVDC system is based on a well-established and mature technology. Table 22.1 presents the list of most operating HVDC systems together with reliability data. The energy availability of most systems is close to 90%. The energy utilization is very low for some systems because they are mostly used for standby capacity. The utilization of other systems is more than 90%.

The forced energy unavailability is the amount of energy that cannot be transported due to forced outages. The scheduled energy unavailability is the amount of energy that cannot be transported due to scheduled maintenance outage. These data give information about the system maintenance requirements.

TABLE 22.1 System Energy Availability, Energy Utilization, and Converter Station Energy Unavailability

System	Year Commissioned	Maximum Continuous Capacity (MW)	Energy Availability Percent		Energy Utilization Percent[a]		Forced Energy Unavailability Percent[b]		Scheduled Energy Unavailability Percent	
			1991	1992	1991	1992	1991	1992	1991	1992
Thyristor Valves										
Eel River	1972	350	98.7	97.6	42.1	16.7	0.19	1.21	1.16	1.16
Skagarrek	1976/77	510	93.0	92.4	30.3	72.5	1.80	6.13	4.97	1.49
Vancouver Pole 2	1977/79	476	NA	82.9	NA	NA	NA	3.84	NA	13.30
Hamil	1977	110	99.9	99.5	31.1	45.7	0.05	0.02	0.05	0.46
Square Butte	1977	550	86.4	93.1	62.2	65.7	0.19	0.14	13.41	6.64
Shin-Shinano 1	1977	300	96.3	97.6	14.7	20.4	0.00	0.04	3.66	2.38
Shin-Shinano 2[c]	1992	300	—	99.2	—	53.1	—	0.00	—	0.79
Nelson River BP2	1978/85	2000	73.0	87.6	37.6	58.6	0.40	1.00	26.60	11.40
Hokkaido-Honshu	1979/80	300	98.2	97.3	27.1	21.3	0.00	0.00	1.82	2.72
CU	1979	1138	96.9	97.0	70.5	70.2	0.07	0.14	3.00	2.82
Vyborg	1981/84	1065	94.7	96.1	49.2	42.1	0.04	1.52	5.30	2.36
Dumrohr	1983	550	94.7	94.8	57.8	64.1	0.17	0.04	5.18	5.16
Gotland II	1983/87	320	99.0	95.9	33.6	31.7	0.04	0.02	0.92	2.10
Chateauguay	1984	1000	NA	84.1	NA	NA	NA	6.51	NA	9.36
Oklaunion	1985	220	98.9	91.5	64.4	67.8	0.00	8.11	1.06	0.34
Madawaska	1985	435	NA	92.4	NA	36.0	NA	0.52	NA	7.06
Itaipu BP1	1985/86	3150	87.1	91.0	60.7	50.2	0.41	0.25	12.29	8.80
Itaipu BP2	1989	3150	86.7	92.9	60.7	50.2	0.35	0.66	12.98	6.40
Miles City	1985	200[d]	95.0	95.8	53.2	77.1	1.60	0.24	3.43	3.98
Highgate	1985	200	97.0	95.9	57.3	66.4	0.11	0.03	2.93	4.06
Cross Channel 1	1985/86	1000	94.6	97.2	92.8	96.7	0.39	0.06	5.04	2.68
Cross Channel 2	1986	1000	97.1	96.6	95.2	96.8	0.13	0.72	2.51	2.70
Des Cantons[e]	1986	725	NA	92.1	NA	7.3	NA	1.35	NA	6.58

IPP	1986	1920[f]	93.5	96.6	68.2	80.6	1.44	0.09	4.81	3.32
Virginia Smith	1988	200	99.7	98.9	7.2	28.3	0.20	0.01	0.15	1.10
Konti-Skari 2	1988	300	98.2	98.4	31.8	53.0	0.17	0.15	1.67	1.44
Vindhyachal	1989	500	94.0	91.9	16.7	12.1	4.20	3.33	1.80	4.71
Fennoskan	1990	500	84.6		22.5		0.59		1.95	
Rihand-Delhi[g]	1991	1650	82.3	84.9	58.3	41.0	5.80	4.30	11.44	10.72
Radisson[e,h]	1991	2250	90.1	89.2	25.0	25.6	1.55	2.10	7.94	8.74
Nicolet[e]	1991	2138	NA	84.1	NA	1.7	NA	5.05	NA	10.80
New Zealand Pole 2	1992	700	—	94.2	—	35.3	—	0.16	—	5.47
Mercury-Arc Valves										
Konti Skan 1	1965	275	55.5[i]	96.2	8.2	42.7	13.35	0.85	0.11	2.96
Sakuma	1965	300	96.4	95.3	32.3	33.3	0.26	0.34	3.35	4.32
New Zealand Pole 1[j]	1965	600	97.3	58.3[k]	79.4	31.6	0.27	0.10	2.41	41.55
Vancouver Pole 1	1968/1969	312	NA	67.4	NA	43.9	NA	13.50	NA	19.20
Pacific Intertie	1970/1989	3100[l]	91.0	89.4	33.7	17.0	1.88	1.00	7.10	9.60
Nelson River BP1[m]	1973/1977	1669	91.8	82.8	55.6	48.6	1.39	0.80	6.70	16.40

[a] Based on maximum continuous capacity.
[b] Converter station outages only.
[c] Data for 1992 are for 7.2 months.
[d] 150 MW in other direction.
[e] Data are for one terminal.
[f] 1200 MW when only one pole in service.
[g] Data for 1991 are for 7.4 months for one pole.
[h] Data for 1991 are for 6 months.
[i] Value would be 86.5% excluding submarine cable repair.
[j] Data for 1991 are for bipolar.
[k] Availability affected by commissioning of new thyristor pole.
[l] Includes 400 and 1100 MW capacity of thyristor valve groups.
[m] Pole 1 valves being replaced by thyristor valves. First and second valve groups commissioned in June and November 1992, respectively.

An important reliability index is the number of thyristor failures per year, which is given as a percentage of the total number of thyristors. A typical value is under 0.5%. The data in Table 22.1 prove that classical HVDC is a very reliable system.

The recent development of high-power insulated gate bipolar transistors (IGBT) has caused revolutionary changes in HVDC technology and has made the voltage source converter-based DC transmission system possible. The first system has been built 10 years ago by ASEA Brown Bowery (ABB). Today, few manufacturers like Siemens and Alston also offer this technology, and several systems are successfully in operation.

Table 22.2 shows the list of existing and future HVDC projects using voltage source converters. In addition to the DC links, the voltage source IGBT converter is used for static VAR compensation.

The voltage source-based HVDC system is designed for less power than the classical thyristor-based HVDC system. One of the major advantages is the multiterminal DC or DC network building. Typical applications are as follows:

- Small isolated remote load supply
- Power supply to an island or offshore oil and gas platforms

TABLE 22.2 Light HVDC Systems

Name	Location	km	kV	MW	Year	Type
Terranora interconnector (Direktlink)	Australia—Mullumbimby	59	80	180	2000	IGBT
Eagle Pass, Texas B2B	USA—Eagle Pass, TX		15.9	36	2000	IGBT
Tjæreborg	Denmark—Tjæreborg/Enge	4.3	9	7	2000	IGBT
Cross Sound Cable	USA—New Haven, CT	40	150	330	2002	IGBT
Murraylink	Australia—Red Cliffs	177	±150	220	2002	IGBT
HVDC Troll	Norway—Kollsnes	70	60	80	2004	IGBT
Estlink	Estonia—Harku	105	150	350	2006	IGBT
NordE.ON 1	Germany—Diele	203	150	400	2009	IGBT
HVDC Valhall	Norway—Lista	292	150	78	2009	IGBT
Trans Bay Cable	USA—East Bay, Oakland, CA	88	200	400	2010	IGBT
Caprivi Link	Namibia—Gerus	970	500	300	2010	IGBT
SydVästlänken	Sweden—Hallsberg Norway—Oslo		400	1200	2013/2015	IGBT

FIGURE 22.1 Examples for HVDC transmission lines. (From Wikipedia, the free encyclopedia: High-voltage direct current.)

- Interconnection of asynchronous grids
- Infeed to cities by land cables
- Interconnection of small-scale (low-head hydro) generation
- Connection of offshore wind power generation to power grids

An HVDC transmission line has only two conductors as shown in Figure 22.1.

22.2 Current Source Converter–Based Classical HVDC System

Most classical HVDCs in operation are point-to-point transmission systems where a large amount of energy is transported between two regions. A typical example is the Pacific Intertie that transports the energy generated by hydro plants in Oregon and Washington to the Los Angeles area in summer and feeds the surplus energy from Los Angeles to Oregon in winter. These systems use line-commutated current source converters with thyristor valves. The operation of this converter requires synchronous generators or synchronous condensers in the AC network at both ends. The rating of the Pacific Intertie is 3100 MW, ±500 kV with a length of 846 miles. In China the rating of Biswanath-Agra is 3600 MW, ±600 kV.

The HVDC system is expensive because of the need for AC filters, DC filters, and a large iron-core smoothing reactor. Another problem is the reactive power consumption of both DC terminals. A large capacitor bank, a synchronous condenser, or a static VAR compensator is needed. This results in a further increase in the cost.

The current-commutated HVDC system cannot supply power to an AC system which has no local generation. The control of this system is complicated. It also requires fast communication channels between the two stations. Figure 22.2 shows the concept of an HVDC system with thyristor valves. The major components of the system are converter transformers, converters with thyristor valves, AC and DC filters, and a smoothing reactor.

The operation records of the more than 30 operating HVDC systems are very good. They are reliable with relatively low maintenance requirements. The first HVDC system between Gotland island and Swedish mainland operates at 150 kV and transports 30 MW power through a 60 mile submarine cable. This system was commissioned with mercury-arc valves in 1954 and was modernized in 1970.

The HVDC systems can be "monopolar" or "bipolar." Figure 22.3 shows the concept of these two arrangements.

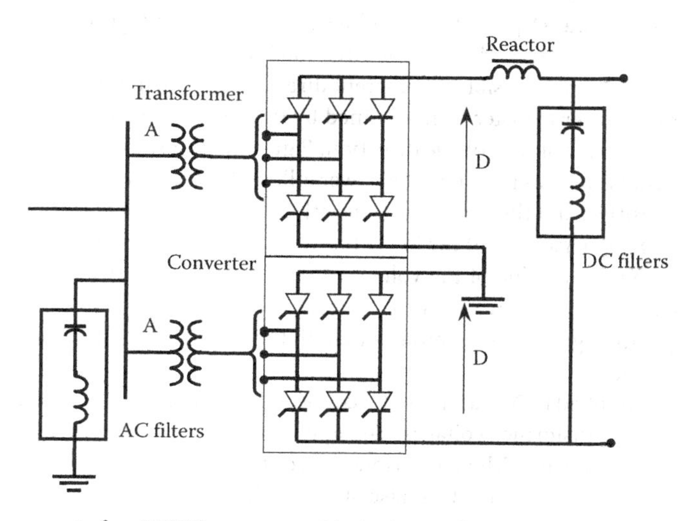

FIGURE 22.2 The concept of an HVDC converter with thyristor valves.

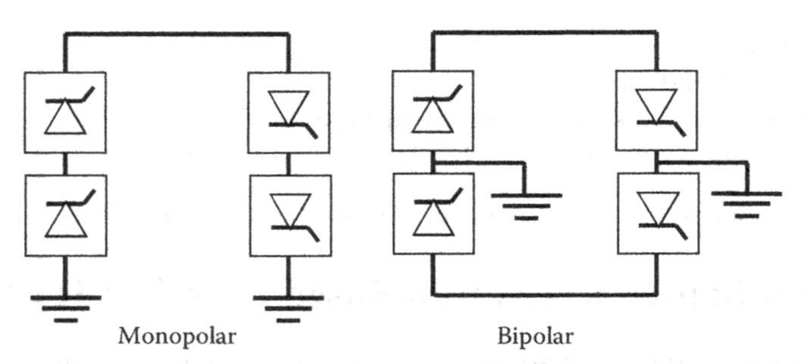

FIGURE 22.3 Monopolar and bipolar arrangements of the HVDC system.

In a monopolar arrangement, the system has only one conductor or cable, and the current returns through the ground or sea. The National Safety Electrical Code prohibits this type of operation for long time because the ground current produces corrosion and may endanger life. The monopolar operation is permitted in case of emergency for limited time only. The monopolar operation is definitely unacceptable in the heavily populated continental United States. However, more than six large HVDC systems operate in this mode worldwide. It is believed that the sea return is acceptable in underdeveloped areas. The obvious advantage of this system is the reduced number of conductors or cables.

The bipolar system has two conductors, and the midpoint of the system is grounded. This eliminates ground currents during normal operation. In case of faults on one cable, the system can operate in monopolar mode for short period of time.

22.2.1 Description of Classical HVDC

The typical architecture of an HVDC link with thyristor converters is shown in Figure 22.4. The link must be supplied both side by an AC system with short-circuit ratio larger than 10, to assure proper operation. The thyristor converter requires reactive power and sinusoidal supply voltage both side. The thyristor converter–based HVDC cannot supply a network without generation. It is not suitable for black start of an AC system. This type of HVDC line can provide emergency power only if the AC system can generate sinusoidal voltage and provide reactive power.

The system has two terminals interconnected by a DC transmission line. Each terminal has two converters connected in series. The midpoint of the converters is grounded. This is a bipolar system, where the converters generate $+V_{DC}$ and $-V_{DC}$ voltages. The higher voltage at converter 1 drives current through the DC line to converter 2. The current direction is determined by the orientation of the thyristors in the converters. The only possible DC current direction is shown in Figure 22.4.

The direction of the power transfer is determined by the voltage direction. The reverse of the power direction requires the reverse of the voltage direction. Figure 22.4 shows the case when converter 1 operates as rectifier and converter 2 as inverter. The power follows from converter 1 to converter 2. The upper terminal voltage is positive, and the lower one is negative. The power reversal requires the change of the upper terminal voltage to negative and the lower terminal voltage to positive.

In both cases, the absolute value of the voltage of converter 1 is slightly higher than converter 2 to maintain the current flow from converter 1 to converter 2.

The converter operation generates harmonics in both the AC and DC sides. The harmonics are eliminated by AC and DC filters.

The converters are supplied by Y-Y and Y-delta connected transformers. This generates 30° phase shift between the secondary transformer voltages. The upper converter is supplied by Y-Y transformer, and the lower by Y-delta transformer. The converters are connected in series; consequently, the generated DC voltages are added together. The 30° phase shift eliminates the 3rd–11th harmonics. This system requires only 13th and 15th harmonic filters.

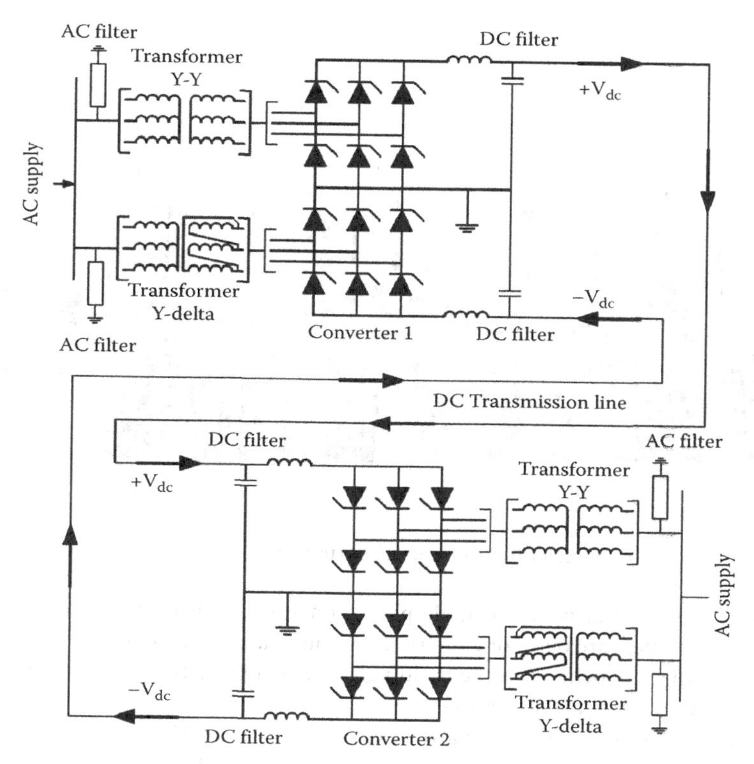

FIGURE 22.4 HVDC transmission system.

FIGURE 22.5 AC filters on a converter station. (Courtesy of ABB, Raleigh, North Carolina.)

The filters are tuned LC circuits containing inductance and capacitance in series. Figure 22.5 shows the filters arranged in a converter station yard. The figure shows the capacitor banks and inductances placed on racks. It can be seen that the space requirements for the filters are significant. The filters occupy close to the same area as the converter building.

The converter transformers can be single-phase units or three-phase transformers. The converter current is built up to square-shape pulses; consequently, the transformer must be designed to

FIGURE 22.6 Single-phase converter transformers. (Courtesy of ABB, Raleigh, North Carolina.)

tolerate harmonics. Figure 22.6 shows single-phase converter transformers built along the converter building. The transformers are outdoor constructions, but placed in front of the building, and the converter site bushing directly penetrates the building through a hole. The transformers are connected to the AC bus and connected in Y-delta and Y-Y.

Each converter contains six thyristor valves as shown in Figure 22.4. The thyristor valves are built with a large number of thyristors connected in series as shown in Figure 22.7. The high-power thyristor rating is 8–10 kV and 4–5 kA. A 600 kV valve requires over hundred thyristors connected in series. The thyristors are arranged into modules containing 5–10 thyristors connected in series. The water-cooled heat sinks are placed between the thyristors, and an inductance is connected in series with each module. This inductance reduces the derivative of current change during turn on.

The individual thyristors are protected by parallel connected CR snubber circuits (R, C), which assure equal voltage distribution during turn on and turn off of the thyristors. Figure 22.7 shows the concept of a valve module assembly.

Four modules with inductance are assembled in a rectangular insulated tray, which is shielded by rounded metal sheets. Ten to twelve series connected trays are assembled to form the valve. The trays or

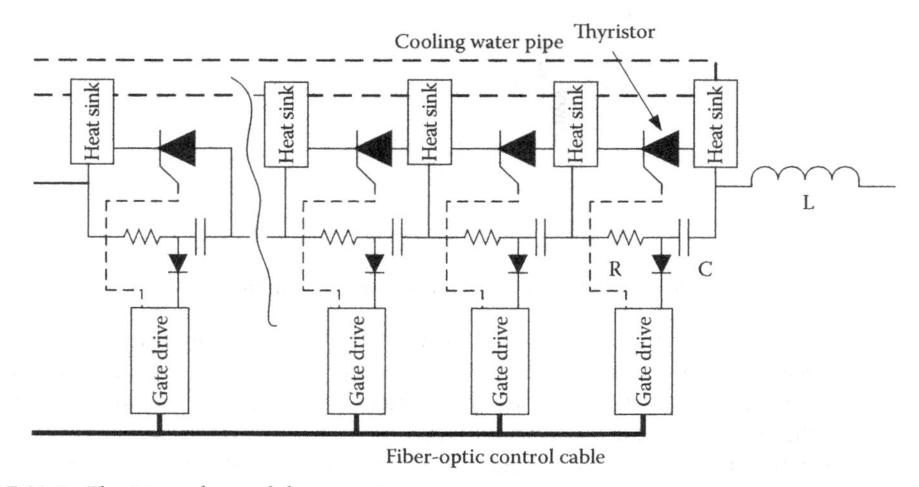

FIGURE 22.7 Thyristor valve module concept.

FIGURE 22.8 Valve hall with valves suspended from the ceiling. (Courtesy of ABB, Raleigh, North Carolina.)

layers are insulated by corrugated epoxy rods. Figure 22.8 shows the valves suspended from the ceiling. The figure also shows the end of the transformer bushings.

The thyristors are fired by direct light firing, or a driving circuit is used to generate the gate pulse. The driving circuit is controlled by a light signal through fiber-optic links. The driving circuits are powered by the voltage across the snubber circuit. The direct light firing requires very bright powerful light sources. The control of the firing through fiber-optic links requires less-power full-light sources. Figure 22.9 shows a high-power thyristor designed for direct light firing.

The thyristors are liquid cooled with deionized water, which is continuously circulated through a single-circuit cooling system with redundant pumps.

22.2.2 Operation of the HVDC System

The major building block of the classical HVDC system is the three-phase bridge converter shown in Figure 22.10. The converter is supplied by a three-phase voltage source and contains six thyristors. The thyristors are fired in sequence 1-2, 2-3, 3-4, 4-5, 5-6, 6-1. Always two thyristors are conducting simultaneously, one in the top row and the other in the bottom row. The transfer of current from one thyristor to the other is called commutation.

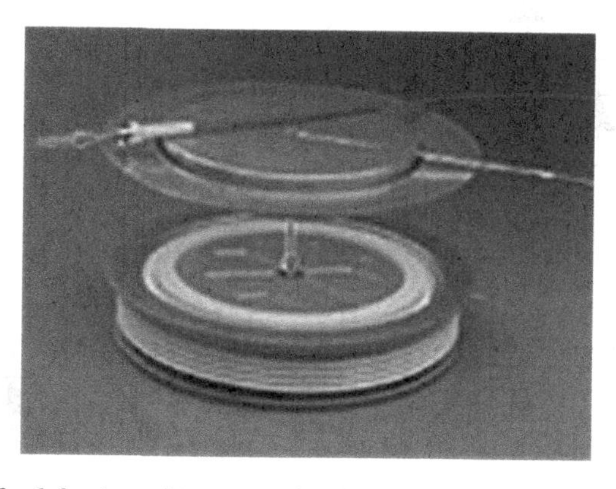

FIGURE 22.9 Direct light-fired thyristor. (Courtesy of Siemens, Charlotte, North Carolina.)

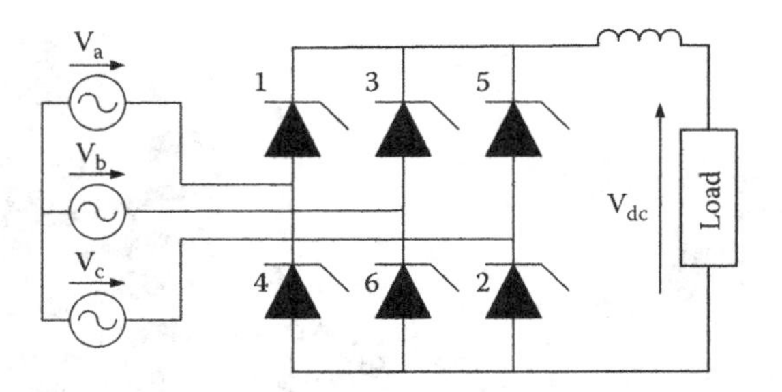

FIGURE 22.10 Simplified bridge converter for operation analysis.

Assuming that the supply inductance is negligible, the firing and follow-up conduction of the two thyristors connects the supply line-to-line voltages to the DC load. Consequently, the DC voltage will be equal with the line-to-line voltages.

Accordingly when thyristors

1-2 conduct $V_{dc} = V_{ac}$
2-3 conduct $V_{dc} = V_{bc}$
3-4 conduct $V_{dc} = V_{ba}$
4-5 conduct $V_{dc} = V_{ca}$
5-6 conduct $V_{dc} = V_{cb}$
6-1 conduct $V_{dc} = V_{ab}$

Figure 22.11 visualizes the variation of the output DC voltage production.

We also assume that the reactance on the DC side is large, which assures that the DC current is constant. This applies that the thyristor currents are square-shape pulses with 120° duration.

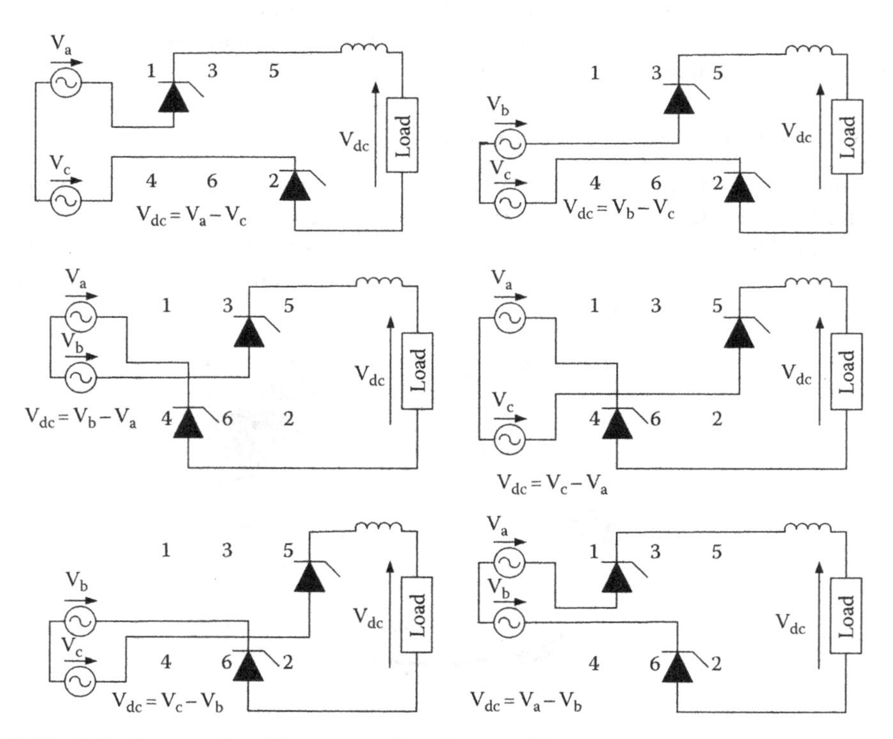

FIGURE 22.11 DC voltage produced by sequential conduction of thyristors.

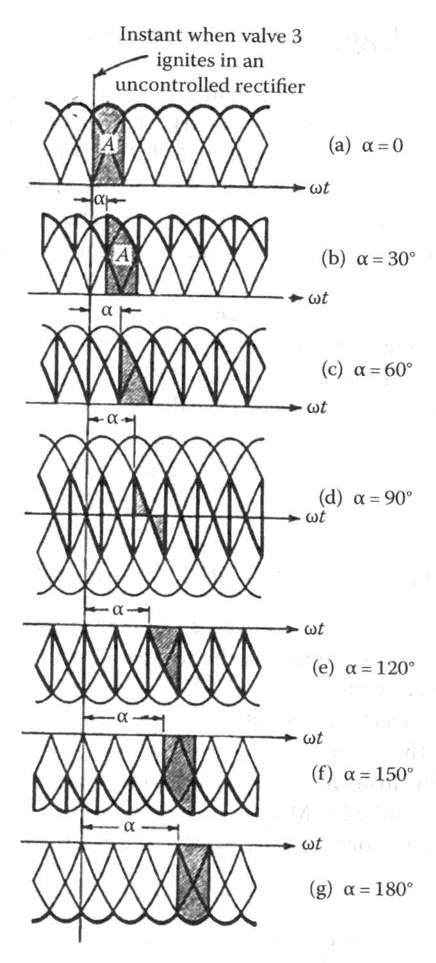

FIGURE 22.12 DC voltage wave shape without overlap. (From Kimbark, E.W., *Direct Current Transmission*, vol. 1, Wiley-Interscience, New York, 1971, Technology & Engineering, 508pp, Figure 18.)

In the real life, the supply inductance is not negligible, which results in overlapping during commutation. The overlapping angle is around 10°–15°. In this simplified analysis, we assume zero overlapping angle or instantaneous commutation.

The output DC voltage can be regulated by delaying the firing angle, which cuts out a section of the voltage as shown in Figure 22.11.

Figure 22.12 shows that the increase of delay angle reduces the average DC voltage. At 90° delay, the DC output voltage is zero, and beyond 90°, the voltage is negative. This implies that in the HVDC system of Figure 22.4 when converter 1 works as a rectifier, the typical delay angle is 5°–15°.

Simultaneously, converter 2 works as an inverter. In converter 1, the thyristors are oriented upward, and in converter 2, the thyristors are oriented downward. In converter 2, the thyristors are rotated by 180°, which reverse the output voltage. The inverter voltage must be less than the rectifier voltage to assure current flow. The voltage difference drives the DC current from the rectifier to the inverter. The difference is the voltage drop on the DC line.

This requires that converter 2 operates with a firing angle of 160°–170° delay angles. This produces negative voltage, but because converter 2 conduction direction is reversed, the inverter and rectifier voltage is in the same direction. The inverter typically operates constant firing angle of 165°–170°, and the current flow is controlled by the rectifier.

22.3 HVDC with Voltage Source Converters

The voltage source converter-based HVDC system is a fast developing technology, which uses IGBT switches and pulse-width modulation (PWM). The capacity of a voltage source converter-based HVDC system is limited to 1200 MW and ±320 kV in 2010. The semiconductor manufacturers increase the capacity of the IGBT switches, and with this, the capacity of the HVDC system is increasing. Presently, ABB and Siemens are selling IGBTs rated 6500 V, 600 A for HVDC systems. The IGBT has voltage-controlled capacitive gate, and it is shunted by a parallelly connected diode in reverse direction.

The first voltage source converter-based HVDC was introduced by ABB in 1997. Presently, more than 16 systems are in successful operation, and few new systems are under construction. However, the technical parameters of the voltage converter-based HVDC system indicate that this technology produces a nearly ideal transmission component that has the potential to change the conventional methods of electric power transmission and distribution. The market survey shows that ABB offers the HVDC light, Siemens the HVDC Plus system and ALSTON prepares a demonstration site with voltage source converter.

22.3.1 Description of HVDC with Voltage Source Converter

The basic module of the voltage source converter is the three-phase bridge built with insulated gate bipolar transistor (IGBT), shunted by diodes in the reverse direction. Figure 22.13 shows basic circuit diagram of the voltage source converter.

The system supplying the converter must not have high short-circuit capacity or at the receiving end not need to have generation. The converter can supply an island without generation. It has black start capacity. Typically a standard transformer and a series converter reactor (inductance) connect the converter to the network. Because of the PWM system, only small AC filter is needed.

At the DC side, two capacitors, connected in series, serve as a filter. The midpoint of the capacitors is grounded.

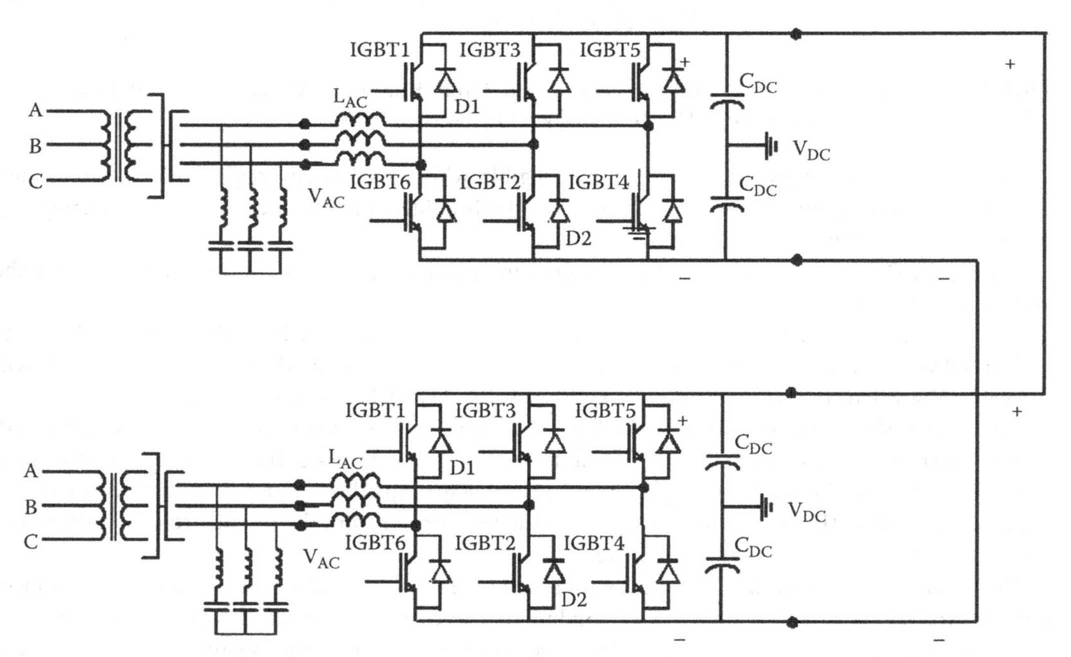

FIGURE 22.13 HVDC system with voltage source converter.

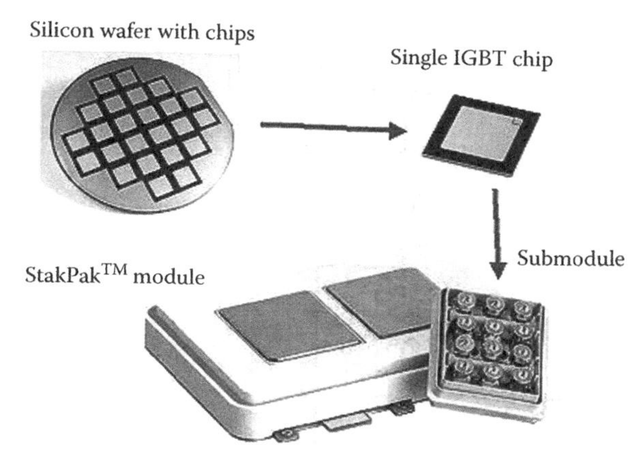

FIGURE 22.14 StakPak module of IGBTs. (Courtesy of ABB, Raleigh, North Carolina.)

The converters can operate automatically without communication between the stations. The system can regulate both the amplitude and phase angle of the AC voltage. This means the independent regulation of the active and reactive power. The direction of power transfer depends on the voltage. The current flows from the converter operating higher voltage than the other. The reversal of the power flow requires the reversal of current direction and not the reversal on the voltage. The system is suitable for multiterminal operation.

Each valve consists of several hundred IGBT connected in series. The valve is divided into series-connected modules called by ABB StakPak. Figure 22.14 shows the StakPak which can have up to 30 IGBTs connected in series. The IGBTs are cooled by deionized water. The even voltage distribution is assured by a parallelly connected voltage divider.

The voltage across each IGBT is rectified and provides power for the gate drive of the IGBT. An optical link from the ground controls the gate drive of the IGBTs. Figure 22.15 shows a section of an IGBT valve used for Light HVDC systems.

Siemens introduced the HVDC Plus system and built few projects. The major contribution is the development of a new type of multilevel converter. Figure 22.16b shows the basic circuit, which contains

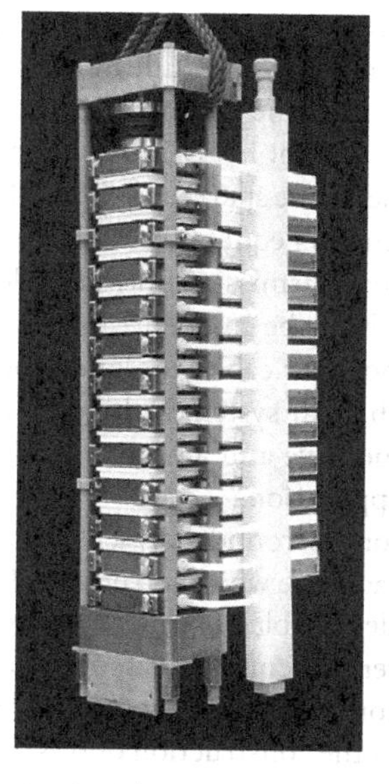

FIGURE 22.15 IGBT valve used for Light HVDC. (Courtesy of ABB, Raleigh, North Carolina.)

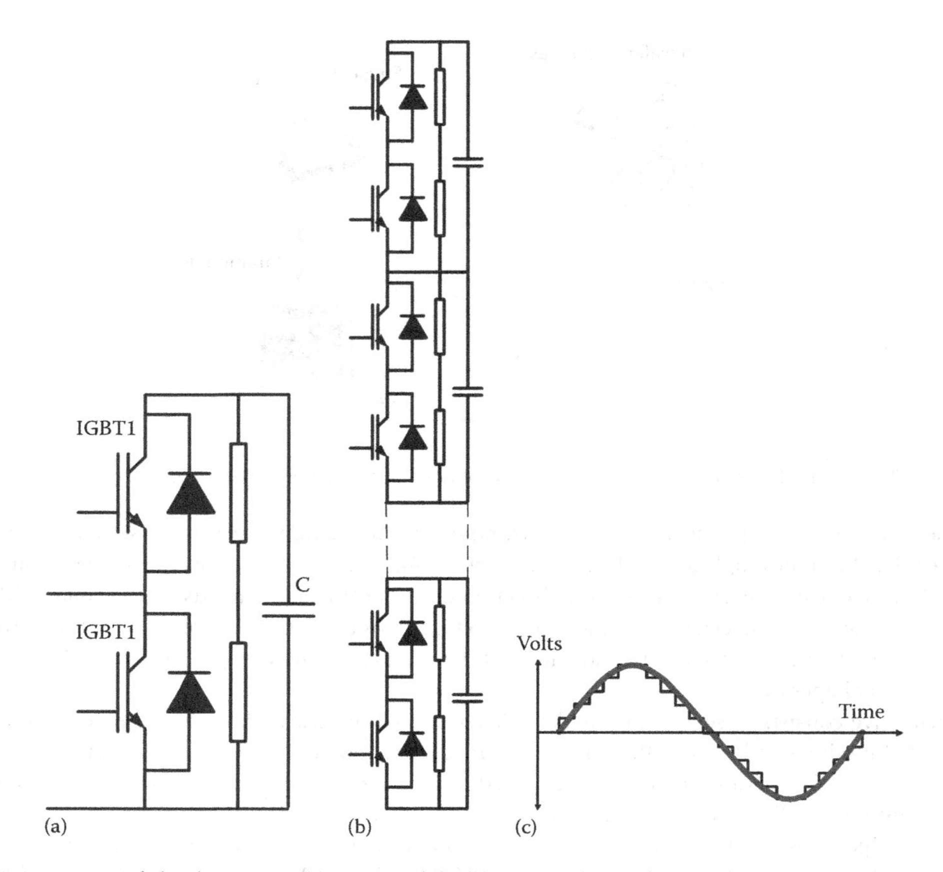

FIGURE 22.16 Multilevel converter. (a) Basic module, (b) generated wave-form, and (c) series connected modules form the valve.

two IGBTs and a capacitor. Several of these units are connected in series to form the valve as shown in Figure 22.16b. The converter produced voltage wave shown in Figure 22.16c.

The HVDC with voltage source converters is built indoor except the transformers. The station is built with modular equipment, in which the components are installed in enclosures in the factory. Figure 22.17 shows the photograph of a Light HVDC substation.

Figure 22.17 shows four transformers (one spare) and filters. The converters are in the building. In the foreground of the buildings are the cooling equipment. The fans are visible on the picture.

The Light HVDC system can supply transmission lines or cables. The advancement of cable technology promotes the use of land cables, which can be competitive with the overhead lines in difficult terrains.

The HVDC Light uses extruded polymer cables for land cable use and also for submarine cable use. The HVDC Light is an ungrounded bipolar system with a positive and a negative cable. As an overhead line, it needs also two conductors (positive and a negative). Figure 22.18 shows typical polymer cable pair designed for the HVDC Light application.

The cable has aluminum conductors surrounded by a black semiconduction layer, which reduces the electrical field on the conductor surface. The white extruded polymer is the main insulation of the cable. The polymer insulation is surrounded by black semiconducting layer and a grounded woven copper shield conductor. The next white layer prevents water penetration. The outer jacket of the cable is PVC.

Figure 22.19 shows the construction of a typical submarine cable. The coaxial single-conductor cable is XLPE insulated. It can be seen that the construction of this cable is more ragged than the cables used on land. This 400 kV 1000 A cable has two layers of shield conductors.

FIGURE 22.17 Light HVDC station. (Courtesy of ABB, Raleigh, North Carolina.)

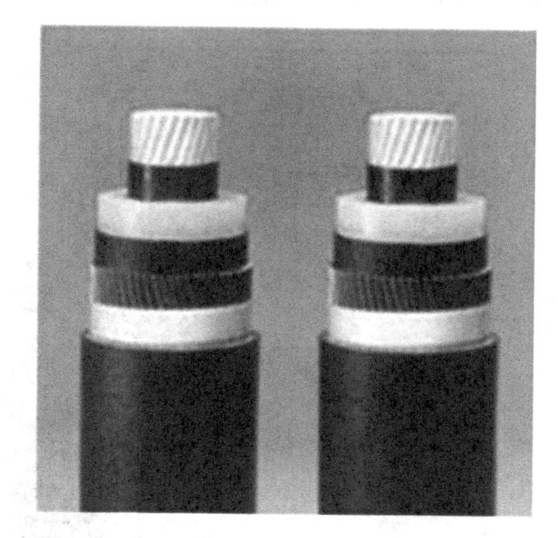

FIGURE 22.18 Pair of extruded polymer HVDC cables. (Courtesy of ABB, Raleigh, North Carolina.)

Submarine cable installation requires a cable ship (Figure 22.20). Pirelli operates the Giulio Verne cable ship, which is capable of laying cables in all weather conditions. The ship is equipped with a 7000 ton rotating turntable.

22.3.2 PWM Technology

The modern manufacturing industry requires AC motor drives that are accurately regulated. These drives should operate close to unity power factor and not generate significant current harmonics in the AC supply. These requirements led to the development of the PWM technique. Figure 22.21 shows the concept of PWM, and Figure 22.22 presents the generated voltage waveform.

The converter in Figure 22.21 can operate as a rectifier or as an inverter. It is built with six semiconductor switches, which are shunted by diodes. The semiconductor switch can be a transistor, IGBT, or a MOSFET. The six switches form a six-pulse bridge. A capacitor shunts the DC side, and three inductors are connected in series with each phase on the AC side. The switches are turned on in sequence, e.g., 12,

FIGURE 22.19 DC submarine cable, 500 MW, 400 kV, and 1250 A.

FIGURE 22.20 Giulio Verne cable laying ship. (Courtesy of Siemens.)

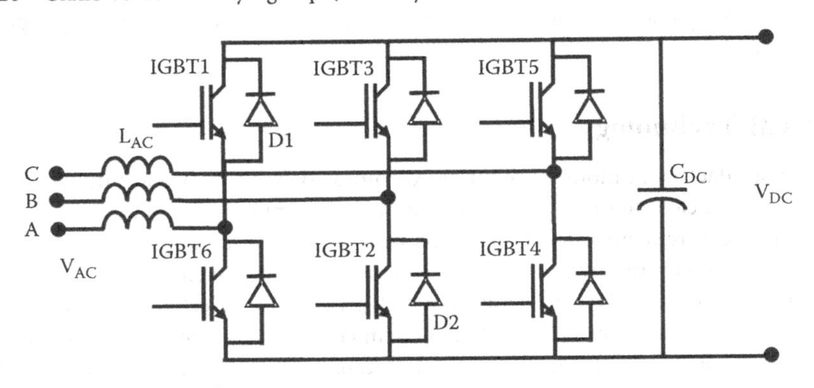

FIGURE 22.21 Circuit diagram of a PWM converter.

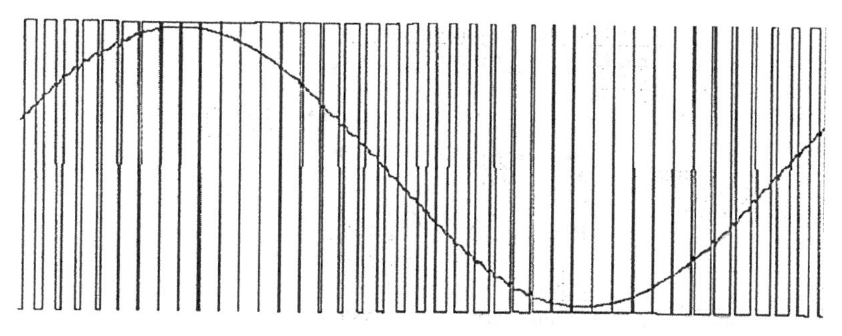

FIGURE 22.22 PWM voltage waveform.

34, 56, etc. The turn on of switch 1 connects the positive DC terminals to phase A, and the turn on of switch 2 connects the negative DC terminals to phase B. The switches are turned on for a short period of time, which generates a pulse train at the AC terminals.

Figure 22.22 shows the generated pulse train. It can be seen that the width of the pulses is modulated and hence the name PWM. The filtering of the output voltage produces a sinusoidal waveform.

The generation of the PWM waveform is illustrated in Figure 22.23. As shown in Figure 22.23a, a high-frequency triangular wave is compared with a 60 Hz reference sine wave to generate the control pulses shown in Figure 22.23b. The intersection of the triangular "carrier" and the sinusoidal "reference" determines the pulse width as demonstrated in Figure 22.23. If the sinusoidal reference voltage is higher than the triangular carrier wave, the upper semiconductor switch (1, 3, or 5) connects the phase terminal to the positive DC terminal (Figure 22.23a); if it is lower, the bottom switch (2, 4, or 6) connects the phase terminal to the negative DC terminal (Figure 22.23b).

The frequency spectrum of the generated pulse train contains the base 60 Hz component and other high-frequency components. The latter are multiples of the triangular carrier frequency. The AC inductance blocks the current harmonics at the AC side, which are further reduced by a high-pass filter.

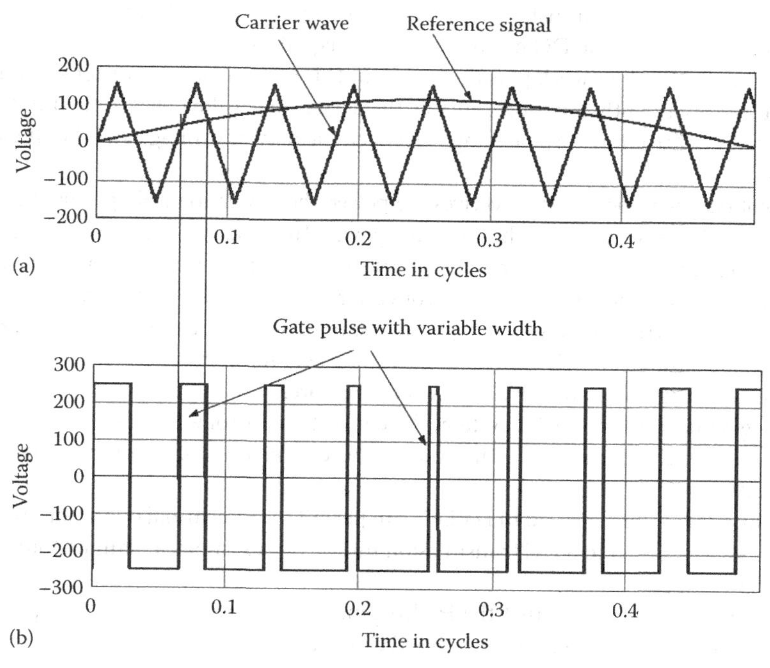

FIGURE 22.23 Generation of PWM voltage waveform. (a) Control and reference signal for PWM gate pulse generation, and (b) PWM gate pulse.

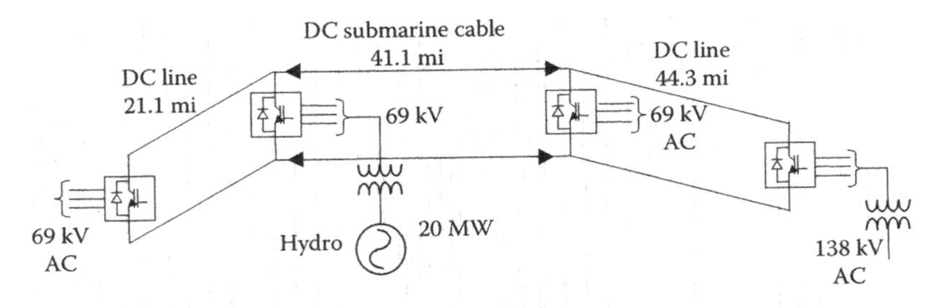

FIGURE 22.24 Example for a planned multiterminal DC system.

The DC capacitor reduces the harmonics at the DC side. The DC capacitor also controls the turn-off overvoltages, generated by the switches, by providing a low-impedance path. The output voltage can be controlled by the pulse pattern and by the DC voltage.

This converter can supply a passive AC system. It can be used to start an AC system after a fault. If the AC system has a voltage source, controlling the converter voltage phase angle can independently regulate the real and reactive power transfer. The active power depends mostly on the converter voltage phase angle, and the reactive power is dependent on the voltage magnitude. The converter can act as a motor or generator without mass and can provide either capacitive or inductive reactive power. The converter controls the AC and consequently does not contribute to the AC short-circuit current.

The PWM converter is an ideal device for energy transmission. It was proposed and developed more than a decade ago for low-power applications. The lack of high-power high-frequency switches has prevented application to HVDC.

For start up, the AC breaker is closed at one side. The diodes in the converter produce a DC voltage and energize the DC line. This charges the power supplies of the gate drive units that permits a start of the converter operation. The first converter that starts will control the DC voltage. The second converter that starts controls the power transfer. The reactive power is controlled independently at each station. The active power flowing in the DC network has to be equal to the active power transmitted from the first network to the second network plus the losses. In this system, one converter station maintains the DC voltage constant. The other station controls the active power flow within the limits of the system. This is achieved by controlling the phase angle between the network voltage and the sinusoidal reference control voltage.

If an AC fault occurs at the side that receives the power, the power-controlling converter is blocked. This interrupts the outgoing power, but not the incoming power. This results in a fast rise of DC voltage. The DC voltage-controlling converter will reduce or even reverse the incoming power to maintain the DC voltage level.

If the fault occurs on the AC side of the converter that controls the DC voltage, the converter is blocked, and a sudden drop in the DC voltage occurs. In this case, the remaining converter will control the DC voltage and simultaneously control its reactive power flow. The operation mode of this converter will be similar to the operation of a dynamic voltage restorer.

In case of a ground fault in the AC system, the converter control will reduce the DC voltage to limit the current flow to the prefault value. The voltage source converter will not increase the short-circuit current in the AC system.

Voltage source converters, applied to HVDC, can permit multiterminal operation. Several converters can be connected in parallel to a DC transmission line. As an example, a planned four-terminal circuit is shown in Figure 22.24.

One possible operation mode is that the Hydro generator maintains the DC voltage constant. This is achieved by modulation of the PWM pattern. The DC voltage is measured and compared with the reference DC voltage. The difference signal changes the amplitude or phase angle of the reference sine wave. With this control method, the Takatz Hydro maintains the power balance by keeping the DC voltage constant.

References

1. E. W. Kimbark, *Direct Current Transmission*, vol. 1, Wiley-Interscience, New York, 1971, Technology & Engineering, 289pp.
2. K. Padiyar, *HVDC Power Transmission System*, New Age International, New Delhi, 1990, Technology & Engineering, 508pp.
3. J. Arrillaga, *High Voltage Direct Current Transmission*, Institute of Engineering and Technology, 1998, Technology & Engineering, 299pp.
4. V. K. Sood, *HVDC and FACTS Controllers*, Springer, Berlin, Germany, 2004, 205pp.
5. SIEMENS, *HV Direct Current Transmission System (HVDC)*, http://www.energy.siemens.com/us/en/power-transmission/hvdc/
6. SIEMENS, *The Smart Way. HVDC PLUS*, http://www.energy.siemens.com/us/pool/hq/power-transmission/HVDC/HVDC_PLUS_The%20Smart%20Way.pdf
7. ABB, *The Classic HVDC Transmission*, http://www.abb.com/industries/us/9AAC30300393.aspx
8. ABB, *Brochures HVDC Light*, http://www.abb.com/industries/ap/db0003db004333/690e3da7796e1dcbc12574a900339e3d.aspx
9. ABB, *HVDC Light*, http://www.abb.com/industries/us/9AAC30300394.aspx
10. ALSTOM GRI, *HVDC*, http://www.alstom.com/grid/solutions/high-voltage-power-products/hvdc-transmission-systems/

23

Transmission Line Structures

Robert E. Nickerson
Consulting Engineer

Peter M. Kandaris
Salt River Project

Anthony M. DiGioia, Jr.
DiGioia, Gray and Associates, LLC

An overhead transmission line (OHTL) is a complex electric/mechanical system designed to transfer electricity between power substations. Structural elements must safely and reliably support current-carrying conductors while providing the necessary separation between individual line phases and the ground. Transmission lines are composed of many individual elements consisting of a variety of materials with a wide range of mechanical properties. The difference in performance of these components is evident in their mechanical characteristics, such as the following:

- Flexible vs. rigid
- Ductile vs. brittle
- Variant dispersions of strength
- Wear and deterioration occurring at different rates

Transmission lines consist of two separate structure systems: the structural support system comprised of towers/poles and foundations, and the wire system comprised of conductor, shield wire, insulators, and hardware (Aichinger et al., 2002). The structural support system is required to provide support for the wire system while accommodating ice and wind acting on both the structural support system and the wire system.

This discussion addresses the types of design practices normally required for the structural support system.

23.1 Transmission Line Design Practice

Traditional and modern transmission line design follows several common practices. These include providing overhead conductor and shield (ground) wire configurations that meet minimum clearance requirements based on the voltage level of the line as required by the National Electrical Safety Code

(NESC) (IEEE, 2006), as well as other applicable codes. The NESC and similar codes provide specified requirements for the separation of the following:

- Energized parts from other energized parts
- Energized parts from the support structure and other objects located along the right-of-way
- Energized parts aboveground

Transmission line loads can be classified as weather-related, accidental, or caused by construction or maintenance activities (Aichinger et al., 2002). Analysis of weather-related events is typically governed by national or regional codes (such as NESC), while accidental, construction, and maintenance events are primarily developed by utilities for the specific needs and conditions of their service territory, with many based on commonly accepted practices. Traditional design includes some probabilistic evaluations of weather-related events but relies mostly on a deterministic process from successful experience. Load cases for accidental events such as component breakage, wear or fatigue, structure failure from natural disasters or terrorism, and other unforeseen events are analyzed to provide designs that minimize their consequences to the OHTL as a whole and prevent an uncontrolled cascade-type system loss. These cases are considered special security requirements. Evaluations for safety requirements are applied to more predictable operational, construction, and maintenance events where calculated loads are analyzed using both regulations and standard codes of practice.

OHTLs are suspended and spaced to produce code-required conductor and shield wire clearances. Resultant structure loads are influenced by not only the sag/tensions but also variable meteorological conditions that act in both transverse and longitudinal directions. NESC provides requirements to evaluate these conditions that include combined ice and wind loading criteria divided into three distinct loading districts: heavy, medium, and light. Each district defines regional climate variations with different combinations of radial ice thickness, wind pressure, and ambient temperature. NESC requirements also give extreme wind loading, providing basic wind speeds that must be adjusted (with consideration for height and gust response) separately for both the wire and the support structure systems. The final NESC load criterion that must be considered is extreme ice with concurrent wind. The NESC provides loading maps for each of these requirements.

In recent years, NESC has adopted reliability-based design (RBD) using the load resistance factor design (LRFD) approach and now includes load factors and strength factors. Load factors consider the uncertainty of the load event, the possibility the design loads will be exceeded, the grade of construction, and structure function. Load factors, though, do not consider the type of material. Strength factors at installation are provided by NESC and take into consideration the type of material and deterioration characteristics.

The design of a transmission line normally includes the following steps:

1. The utility prepares an agenda of loading events consisting of
   a. Mandatory regulations from the NESC and other codes
   b. Climatic events assumed representative of the line's specific location
   c. Contingency (security) loading events of interest, i.e., broken conductor
   d. Safety requirements and expectations, i.e., maintenance loads, stringing, etc.

   Each of these loading events includes load factors to cover associated uncertainties to produce a set of factored design loads.
2. A ruling span is identified based on the sag/tension requirements for the preselected conductor.
3. Structure type is selected based on past experience, utility standards, recommendations of potential structure suppliers, terrain and construction issues, economics, and long-term maintenance concerns.
4. Ultimate design loads resulting from the ruling span are applied statically on components in the longitudinal, transverse, and vertical directions, and the structure is either designed by deterministic or reliability-based methods.

5. Using the loads and structure configuration, ground line reactions are calculated and used to complete the foundation design. Foundations are also designed using either deterministic or RBD methods.
6. The ruling span line configuration is adjusted to fit the actual right-of-way profile.
7. Structure/foundation designs are modified to account for variation in actual span lengths, changes in elevation, and running angles.

Traditional line design views the support structure as an isolated element supporting half span of overhead conductors and shield wires on either side of the structure. Using ruling span assumptions with similar span lengths and suspension supports in the tension section yields somewhat accurate results. Ruling span assumptions become less accurate in conditions where span lengths vary in hilly terrain. Under these conditions, sag differences can be much different from the ruling span assumptions. It should also be noted that inaccuracies when using ruling span assumptions under high temperature, unbalanced ice, and broken wire conditions are also evident. Thus, modern line design practice using computer-based programs enables a more accurate development of loads at each structure location, clearance to ground, and clearance to structure.

Inasmuch as structure types are available in a wide variety of configurations, materials, and costs, several iterations would normally be attempted in search of the most cost-effective line design based on total installed costs.

While traditional deterministic design using static loads is a convenient mathematical approach, it is obviously not representative of the real-world exposure of the structural support system. OHTLs are tens of yards wide and miles long and usually extend over many widely variant topographical and climatic zones, each capable of delivering unique events consisting of magnitude of load at a probability of occurrence. That component along the right-of-way that has the highest probability of occurrence of failure from a loading event becomes the weak link in the structure design and establishes the reliability level for the total line section. Since different components are made from different materials that have different response characteristics and that wear, age, and deteriorate at different rates, it is to be expected that the weak link

- Will likely be different in different line designs
- Will likely be different in different site locations within the same line
- Can change from one component to another over time

23.1.1 Transmission Line Support Structures

Structures used for transmission lines come in a wide range of materials, shapes, and configurations. Typical materials and shapes include but are not limited to the following:

- Steel (hot-rolled angles, plates, formed plate polygonal tubular members)
- Aluminum (extruded shapes, plates)
- Concrete (static cast, spun cast with pre- or posttension strands)
- Wood (glue-laminated poles and crossarms, poles, crossarms)
- Fiber-reinforced polymer (FRP) (crossarms, poles)

There are a variety of configurations used for transmission structures. These include the following:

- Lattice towers (steel, aluminum) (ASCE, 2000)
- Single shaft poles (steel, wood, FRP, concrete) (IEEE, 1991; ASCE, 2003; Magee, 2006)
- H-frame structures (steel [latticed and tubular], wood)
- Guyed structures (steel [latticed and tubular], concrete) (ASCE, 1997)
- Framed structures (tubular)

(**Note:** References indicate recommended design guides for various structure types.)

Utility standards, operational and maintenance procedures, installed cost, lifetime cost, structure performance, right-of-way access and terrain, and aesthetics are just some of the issues that must be considered when selecting a structure type for a transmission line. Long-term considerations for future upgrades may also impact final selection of material and structure configuration.

23.1.2 Transmission Line Foundations

The function of a transmission foundation is to transfer applied steady-state and transient loads into the surrounding soil and rock while limiting structure movement. Loads are conveyed to the subsurface at the ground line interface, where either a separate foundation system is installed and connected to the structure or the above-grade structure is directly buried and backfilled. Foundation systems can be categorized in the following general groups:

- Spread foundations (steel grillages and reinforced concrete)
- Reinforced concrete drilled shafts
- Direct embedment poles (steel, wood, concrete)
- Driven piles (steel, wood, concrete)
- Anchors (various materials and configurations)

Detailed descriptions of transmission foundation systems are provided by the Institute of Electrical and Electronics Engineers (IEEE) Standards 691 and 977 (IEEE, 2001, 2010).

Spread foundations typically support lattice tower structures using either a reinforced concrete or a prefabricated steel grillage footing for each leg. Guyed lattice structures use a combined system of one or more spread footings in conjunction with guy anchors.

Because of their ability to resist uplift loads, compression loads, and lateral overturning forces, drilled shaft foundations are used to support a wide variety of transmission system structures, including lattice towers, single shaft steel poles, and framed pole structures. Drilled shafts are connected to lattice towers typically with bent or angled structural steel angles embedded into the reinforced concrete foundation. Most often, single pole and framed pole structures (and occasionally lattice towers) include steel base plates at the bottom of the structure which are fixed to the foundation via steel anchor bolts in either a circular or rectangular pattern around the inside perimeter of the shaft. These anchor bolts can be partially or fully extended within the drilled shaft and can be made part of the reinforcement cage or be contained within and separate from reinforcement. Drilled shaft foundations are typically uniformly cylindrical but can be drilled with tapers, uniformly variable shaft sections, or belled bottoms.

Lower voltage lines with lighter loading conditions can be supported with single pole or frame pole structures directly embedded within a drilled shaft hole, then backfilled with a variety of natural and man-made materials. The annulus space around the pole can be filled with concrete, compacted native soil, or slurried aggregates either with or without cement. Permanent steel casing is sometimes incorporated with direct embedment foundations where groundwater is present.

Piles (or more commonly piles in closely spaced groups) are either mechanically driven or vibrated into the ground to support all types of transmission structures. Generally this is done where ground conditions are soft enough to accept the concrete, steel, or wood piles. Structures are fixed to pile groups often with cast-in-place reinforced concrete caps. Lightly loaded wood poles can be fixed to single steel or concrete piles via mechanical connection.

Foundation anchors for transmission structures encompass a large variety of types and materials and are used to directly support guyed structures or are encased within concrete-filled sockets with bottom anchor segments embedded into subsurface rock to support single shaft poles. Guyed helical anchorages are directly screwed into the ground; plated anchors are buried in angled trenches then backfilled; and grouted anchors are placed in predrilled holes then grouted to fill the surrounding space with high-strength cement or resin.

23.1.3 Factors Influencing Structure and Foundation Selection

There are a number of factors that can impact the selection of the structure and foundation type used in a transmission line. In some cases, the foundation requirements may dictate the structure selection; in others, it may dictate the structure that is critical. Some of the more significant issues are briefly identified in the following text.

Wire orientation: Flat, vertical, delta, single, or multicircuit configurations will all influence selection of structure type.

Right-of-way: Width of the right-of-way, blow out concerns, adjacent lines in the same right-of-way and terrain, all will influence structure selection and configuration.

Erection requirements: Clearly different structure types require different erection requirements. Latticed structures require an assembly yard or a flat area on the right-of-way to lie out and assemble the sections of the structure. Tapered steel poles require less assembly area but normally need a larger crane for installation of the structure. This may involve more extensive road work for crane access. Concrete poles require yet larger cranes (and often mats) for support when lifting the structure. As expected cost of assembly, erection, and installation of the foundations, all must be considered in developing a total installed cost.

Public concerns: Probably the most difficult factors to deal with arise as a result of the concerns of the general public living, working, or coming in proximity to the line. It is common practice to hold public hearings as part of the approval process for a new line. Such public hearings offer a platform for neighbors to express concerns about structure appearance and location that generally must be satisfactorily addressed before the required permit will be issued.

Often the public perceives transmission structures as "eyesores" and distractions in the local landscape. However, with the advent of sophisticated software packages, line models using different structure types and configurations can be used at public hearings. In some cases, these new tools have been helpful in mitigating public concerns. Other concerns include electromagnetic field effects (EMF) from the line, possible climbing access to the structures by the public and audible noise.

Inspection, assessment, and maintenance: During the design process, it is valuable to interact with line maintenance and inspection groups. Input from these specialists will provide not only a positive relationship but will also reduce the chance of fabrication or field changes to meet climbing requirements. Oftentimes, the owner's inspection and maintenance practices will influence the selection of the structure type for use in a specific line location. Inspections and assessment are normally made by humans who use diagnostic technologies to augment their personal observations. Inspectors must work from the most advantageous location when making inspections, and this can require climbing the structure. Methods can include observations from ground, fly-by patrol, climbing, bucket trucks, or helicopters. Likewise, there are certain maintenance activities that are required for particular structure types. The equipment necessary to maintain the structure should be taken into consideration during the structure type selection process to assure there will be no unexpected conflict between maintenance needs and right-of-way restrictions.

Future upgrading or uprating: Due to the difficulty of procuring rights-of-way and obtaining the necessary permits to build new lines, some utilities may select structure types for current line projects that more easily permit future upgrading and/or uprating initiatives or may be designed with additional capacity for future use.

Subsurface conditions: Site-specific ground conditions will control the ease or difficulty in the construction of various transmission foundations. Spread footings generally work best where shallow excavation is favored over deep drilling. Drilled shafts must maintain their shape without caving through the foundation construction process, so soils that are susceptible to caving must either be cased or held open with specialized drilling fluids. Alternately, stiff or strong soils and weak rock are ideal for drilled shaft and direct embedment construction techniques. The subsurface variables involved with foundation selection include site geology, soil/rock type, soil strength properties, and groundwater conditions.

Access: Foundation type is sometime dictated by the ability to bring equipment and material to the job site. Installation of drilled shafts or driven piles requires track or truck-mounted rigs that typically need substantial road access, while small spread footings and anchor systems can be carried by hand or small all-terrain vehicles to more difficult locations. Long distances to concrete ready-mix plants can increase the cost and possibly reduce the quality of cast-in-place foundations such as concrete spread footings and drilled shafts.

23.2 Current Design Practices

23.2.1 Deterministic Design Approach

The deterministic design approach, often referred to as the allowable stress design (ASD) approach, is shown schematically in Figure 23.1.

The maximum component design load Q_D, shown in Figure 23.1, is generally determined from one of the following load cases:

- Extreme wind
- Extreme ice with concurrent wind
- Broken conductor and/or overhead ground wires
- Construction and maintenance
- Legislated loads

Figure 23.2 exemplifies representative load cases for a typical steel lattice tower where different members of the structure are controlled by different load conditions.

Design load cases used in practice in North America are normally based on the NESC (IEEE, 2006), American Society of Civil Engineers (ASCE) Manual 74 (Wong and Miller, 2010), and/or the Canadian Standards Association (CSA) Standard C22.3 (CEI/IEC, 2006). The NESC divides the United States into three large global loading zones—heavy, medium, and light—and specifies radial ice thickness, wind pressure, and temperature relationships to define the minimum load levels that must be used within each loading zone. In addition, the NESC introduces the concept of

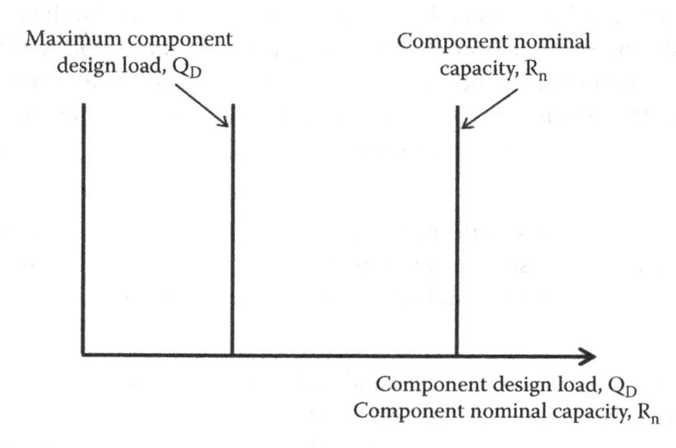

FIGURE 23.1 Deterministic design approach.

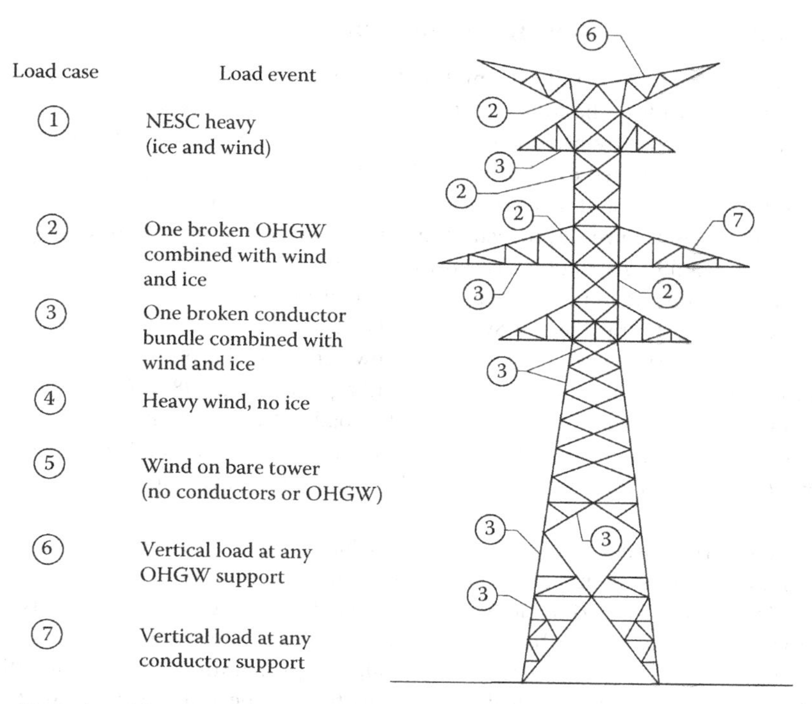

FIGURE 23.2 Typical steel lattice tower loading—load cases for structural elements.

safety factor in terms of an overload capacity factor (OCF) to cover uncertainties stemming from the following:

- Likelihood of occurrence of the specified load
- Dispersion of predicted strengths
- Grade of construction
- Deterioration of strength during service life
- Structure function (suspension, dead-end, angle)

Both ASCE Manual 74 and CSA Standard C22.3 include loads associated with a 50 year RP as the basis of design.

Nominal component capacities (R_ns) are based on using nominal material strength and deformation properties as needed by the component element design model being used by the designer. Component reliability is established by using the following design equation:

$$Q_D < \frac{R_n}{SF} \tag{23.1}$$

where

Q_D is the maximum component design load
R_n is the nominal capacity of the component
SF is the factor of safety

Above-grade structure component safety factors are typically established in various design codes. For foundations, the value of the safety factor adopted by the designer is based on the background and experience of the designer. However, surveys of the practice have shown that safety factors used in practice are quite variable, resulting in a wide variation in the level of reliability and cost for foundations.

23.2.2 Reliability-Based Design Approach

The RBD approach is founded on the assumption that component design loads are not unique but can vary significant over the life of the component—from relatively low loads (such as everyday wind events) to very high loads (such as extreme wind events). Reliability level is defined as the ability of a line (component) to perform its expected capability. In addition, the nominal strength of each component can vary due to differences in as-built material properties, dimensions, construction techniques, and design models. Thus, in an RBD approach, the variability and uncertainty in loads and component strengths are modeled by probability distribution as shown in Figure 23.3.

RBD applied to transmission line design is presented in detail by ASCE (Dagher, 2006) and the International Council on Large Electric Systems (CIGRE SC-22, 1995).

Consideration of these uncertainties to achieve a low but acceptable probability of failure is presented in the technical literature on structural reliability (Ang and Tang, 1984; Nowak and Collins, 2000). Reliability-based resistance factors, which separate the load (Q) and resistance (R) density functions, are developed for each component element design model so that failure will rarely occur. The probability of Q exceeding R is the probability of failure and is determined by convolution of the Q and R probability functions.

Using advanced first-order reliability methods, Ghannoum (1983) and Dagher et al. (1993) demonstrated that if the components of a transmission line system were designed for loads associated with a specific time return period (RP) and component capacities have low exclusion limits (e.g., 5%–10%), then the annual probability of failure (P_f) for these components is approximately equal to ½ · RP. For an RP of 50 years, P_f equals ½ times 50, or a probability of failure of 0.01. In addition, P_f was shown to be relatively unchanged with respect to the type and coefficient of variations of the Q and R density functions. Thus, the RBD of transmission line elements can be performed using the following equation:

$$R_5 \geq \text{Effect of } [\text{Dead Load} + \gamma Q_{50}] \tag{23.2}$$

where

R_5 is the 5% lower exclusion limit (LEL) capacity
γ is the load factor used to modify line reliability for return periods higher than 50 years
Q_{50} is the 50 year return period design load event

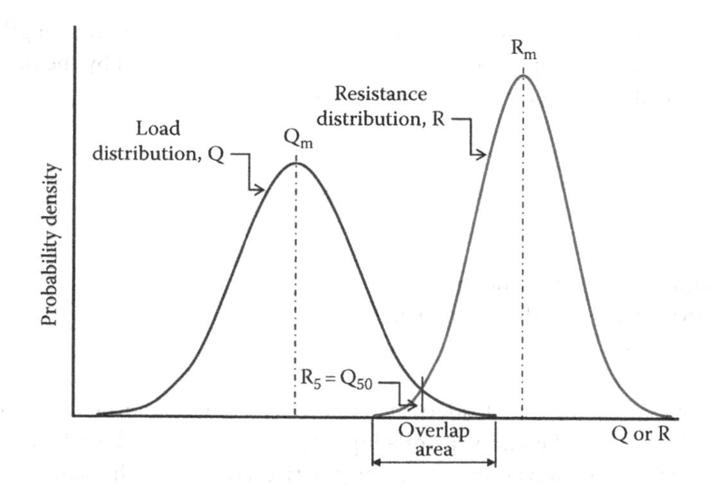

FIGURE 23.3 Combined load and resistance probability distributions.

TABLE 23.1 Approximate Load Factors to Convert "Extreme Wind Loads" from a 50 Year Event to Other Return Periods—ASCE Manual 74

Return Period (Years)	Extreme Wind Load Factor, γ
25	0.85
50	1.00
100	1.15
200	1.30
400	1.45

Source: Wong, C.J. and Miller, M.D., Guidelines for electrical transmission line structural loading, ASCE Manuals and Reports on Engineering Practice No. 74, American Society of Civil Engineers, Reston, VA, 2010.

TABLE 23.2 Approximate Load Factors to Convert "Extreme Ice with Concurrent Wind Loads" from a 50 Year Event to Other Return Periods—ASCE Manual 74

Return Period (Years)	Ice Thickness Factor	Concurrent Wind Load Factor, γ
25	0.80	1.00
50	1.00	1.00
100	1.25	1.00
200	1.50	1.00
400	1.85	1.00

Source: Wong, C.J. and Miller, M.D., Guidelines for electrical transmission line structural loading, ASCE Manuals and Reports on Engineering Practice No. 74, American Society of Civil Engineers, Reston, VA, 2010.

TABLE 23.3 Default Load Factors for Adjustment of Climatic Loads in Relation to Return Period vs. 50 Year Event—CSA C22.3 No. 60826

Return Period (Years)	Wind Speed Factor	Ice Thickness Factor	Ice Weight Factor
50	1.00	1.00	1.00
150	1.10	1.15	1.20
500	1.20	1.30	1.45

Source: Commission Électrotechnique Internationale/International Electrotechnical Commission, Design criteria of overhead transmission lines with Canadian deviations of the CEI/IEC, CSA Standard C22.3 No. 60826, Canadian Standards Association, Mississauga, Ontario, Canada, 2006.

Tables 23.1 through 23.3 provide load factor values (γ) for extreme wind loads, extreme ice and concurrent wind loads, and CSA C22.3 No. 60826 (CEI/IEC, 2006) loads, respectively.

The relationship between R_5 and R_n is given by the following equation:

$$R_5 = \Phi_5 R_n \tag{23.3}$$

where

R_5 and R_n are as defined previously

Φ_5 is the 5% LEL resistance (strength) factor

Thus the RBD equation is given as follows:

$$\Phi_5 R_n \geq \text{Effect of } [\text{Dead Load } + \gamma Q_{50}] \tag{23.4}$$

23.2.3 Security Level

It should be remembered, however, that the failure of every component does not necessarily progress into extensive damage. A comparison of the total risk that would result from the initial failure of components of interest can be accomplished by making a security-level check of the line design (Ostendorp, 1998). Security level can be described as the ability of a line to restrict progressive damage after the failure of the first component.

Since the OHTL is a contiguous mechanical system, the forces from the overhead conductors and shield wires (wire system) on one side of each tangent structure are balanced and restrained by those on the other side. When a critical component in the wire system fails, energy stored within tensioned elements is released suddenly and sets up unbalanced transients that can cause failure of critical components at the next structure. This can set off a cascading effect that will continue to travel down line until encountering a point in the line strong enough to withstand the unbalance. Unfortunately, a security check of the total line cannot be accomplished from the information describing the single structure in Figure 23.2, but perhaps some generalized observations can be drawn for demonstration purposes.

A structure designed for broken conductor bundle and broken shield wire contingencies would not appear to be subjected to a cascade from a broken bare conductor. But what if the conductor was coated with ice at the time? Since ice increases the energy trapped within the conductor prior to release, it might be of interest to determine how much ice would be enough to overcome the contingencies. Modern computer modeling would be employed to simulate ice coating of increasing thickness until the critical amount is defined. A proper micrometeorological study could then identify the probability of occurrence of a storm system capable of delivering that amount of ice at that specific location.

A security-level check can predict the amount of resulting losses and damages that would be expected from an initiating event compared to the other contingencies.

23.3 Foundation Design

Most often, foundation design is controlled by steady-state and transient loads (one or a combination of both). Construction and maintenance loads must be examined, but rarely influence foundation dimensions. Foundation performance criteria (such as rotation and displacement) result from deterministic evaluations of structure performance needs and can control final foundation size. Failure is not necessarily a catastrophic event, but the point where preestablished movement performance criteria are exceeded.

The IEEE/ASCE Transmission Structure Foundation Design Guide (Tedesco and DiGioia, 1983; DiGioia, 2002) gives a thorough presentation of transmission line foundation design. For either RBD or traditional ASD, the steps involved with transmission line foundation design remain the same:

- Perform subsurface investigation/obtain subsurface information
- Select subsurface foundation geotechnical design parameters
- Select appropriate foundation design model
- Apply loads to design model consistent with RBD or ASB

23.3.1 Subsurface Investigation

By virtue of their intended purpose, electrical transmission projects traverse large distances across widely varying geologic and geotechnical settings over many miles. Transmission foundation designers must balance cost-effective field investigation with the production of sufficient data for design of foundations that are economical and reliable. An initial qualitative preliminary assessment of subsurface variation should be performed by professionals skilled at geology or geologic engineering to optimize and apportion investigation sites according to the geologic strata. This process includes gathering of prior information in the form of soil reports, geologic maps, aerial photographs, hydrologic reports, etc., combined with new field observations and mapping (Kulhawy et al., 1983; Trautman and Kulhawy, 1983; Hunt, 1984; Grigoriu et al., 1987). A field subsurface investigation based on a thorough review of prior data follows and includes borings, test pits, in situ probes, and geophysical measures. The purpose is to establish one or more idealized profiles of the subsurface and gather samples for laboratory testing.

23.3.2 Foundation Geotechnical Design Parameters

One of the most difficult aspects of transmission line foundation design is the selection of representative soil/rock design properties for idealized subsurface profiles at each foundation location. The process is generally iterative: the properties are refined as more data are obtained from field in situ work and laboratory testing. Quite commonly, index parameters from standard penetration tests (SPT) and cone penetrometer tests (CPT) are correlated to the design properties as these values tend to be abundant from the investigations. Laboratory testing is used to refine the values along with high-quality in situ methods such as pressuremeter testing. Empirical correlations to subsurface properties are extensively used for estimating design parameters. Manuals and guides on this subject have been prepared by the Electric Power Research Institute (EPRI) focusing on the selection of geotechnical design parameters and investigation methods relative to transmission lines (Kulhawy and Mayne, 1990; DiGioia, 2010, Chapter 6). The selection of design parameters must always consider the design model and approach. It is important to understand how models apply soil properties to select reasonable low bound values for use in traditional allowable strength design or to determine nominal values for RBD.

23.3.3 Foundation Design Models

It is of great importance that the foundation designer selects a model that accurately reflects subsurface conditions and reactions to the applied loads. With the advent of modern computer analysis programs, there is a tendency to use the software at hand and make the design fit the program. Additionally, some utilities require specific models or software be used in their specifications. In any case, the designer needs to fully evaluate foundation models and understand the applications and limits.

Spread footings (cast-in-place concrete, precast concrete, grillages, pressed plates) provide basic resistance to axial forces (uplift and compression), while considering the load orientation (inclination and eccentricity) of the applied loads. Foundations must be designed to prevent shear failure and excess settlement in compression and have adequate size and depth to prevent uplift failure or excessive lift of foundation legs. The IEEE/ASCE foundation design guide (Tedesco and DiGioia, 1983; DiGioia, 2002) offers a number of models for spread footings using traditional design methods. AASHTO LRFD Bridge Design Specifications (AASHTO, 2010, Section 10.6) give detailed design methods for RBD of spread footings. Spread footings are most commonly used with lattice tower structures and as central bearing foundations for guyed lattice tower structures.

Drilled foundations (including reinforced concrete drilled shafts and direct embedment poles) support vertical compressive loads through a combination of side shear and end bearing, vertical uplift loads via side shear with foundation weight, and lateral loads with overturning moments using the lateral resistance of soil/rock within the embedded section. These foundations rely on a complex soil-structure interaction where movement mobilizes soil strength, transferring load in a nonuniform manner. With axial loads, this commonly referred to as a "t-z" effect and with lateral loads, a "p-delta" or "p-y" effect. EPRI has developed software (FADTools) to model electric system drilled shafts and direct embedment foundations. The lateral load program, MFAD, accurately models the rigid nature of short electric system shafts and is calibrated for both traditional strength design and RBD.

Pile foundations transmit axial compressive loads through soft soils to denser underlying soils or rock. Although pile foundations can provide substantial uplift and lateral resistance, these foundations are most often used for lattice towers which have low shear, low moment, and high axial load. Pile capacity determination is thoroughly described in the AASHTO LRFD Bridge Design Specifications (AASHTO, 2010, Section 10.7) for both ASD and RBD approaches.

Anchors offer resistance to upward loads transferred from either steel structure guy cables or the structure itself (tower leg or overturning moment within a shallow foundation). The anchor may be a buried steel plate or concrete slab or may be a grouted bar or cable within a drill hole. Anchorages may be prestressed to limit deformation of the supporting structure. Anchor capacity is usually designed based on the pull-out capacity of either a wedge of soil (dead-man anchor types), the side resistance of anchor rods, grouts, and surrounding soil/rock (grouted anchors), or both. The IEEE/ASCE foundation design guide (Tedesco and DiGioia, 1983; DiGioia, 2002) provides a more in-depth description of anchor types and general design methods. Many manufactured anchor systems provide proprietary design processes to be used with their products.

23.3.4 Foundation Reliability-Based Design

The resistance factor, Φ_5 (discussed in Section 23.2.2), can be determined for a specific foundation design model by using a calibration process (Bazán-Zurita et al., 2010). The process involves predicting the ultimate capacities of a given number of full-scale foundation load tests using the design model to be calibrated. Figure 23.4 is an example of calibrating the cylindrical shear uplift load design model using

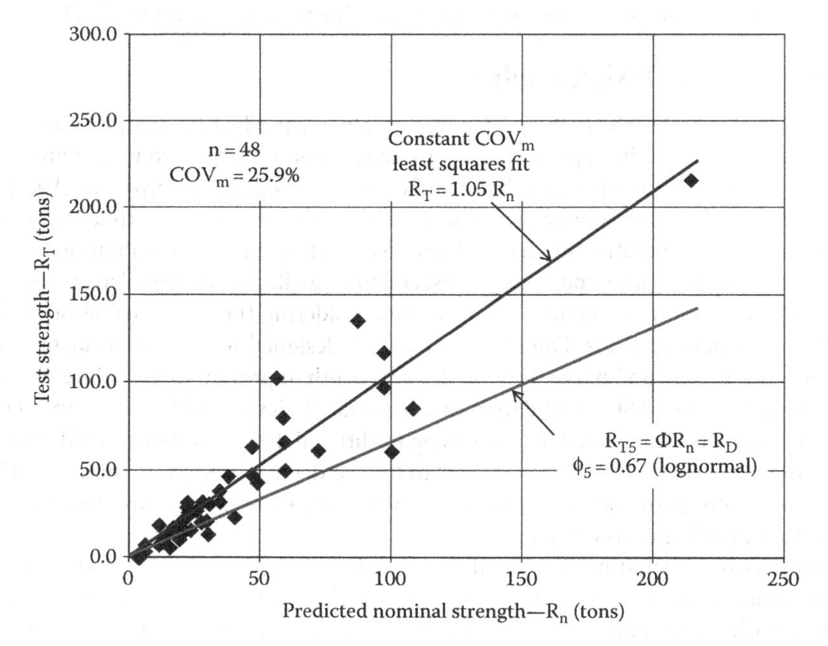

FIGURE 23.4 Cylindrical side shear design model predicted nominal ultimate uplift capacity, R_n, vs. interpreted test uplift capacity, R_T, for drilled shafts embedded in cohesive soils ($D/B \leq 10$).

48 full-scale drilled shaft uplift load tests, where each drilled shaft is embedded in a cohesive soil (Stas and Kulhawy, 1984).

The data presented in Figure 23.4 show that the cylindrical shear uplift load design model for drilled shafts embedded in cohesive soils has a 5% LEL resistance factor, Φ_5, of 0.67. This resistance factor is computed using the following equation:

$$\Phi_5 = m_m(1 - k_5 V_m) \tag{23.5}$$

where

m_m is the mean of the m-values for each test, wherein each m-value equals test resistance divided by predicted nominal capacity

V_m is the coefficient of variation of the m-values

As shown in Figure 23.4, $m_m = 1.05$ and $V_m = 25.9\%$ for the design model.

For a lognormal distribution of m, k_5 is given by the following equation:

$$k_5 = 0.01\ (1.64 - 0.00925 V_m), \tag{23.6}$$

where V_m is in percent.

Substituting with the preceding equations gives

$$\Phi_5 = m_m(1 - 1.64 \times 10^{-2} V_m + 9.25 \times 10^{-5} V_m^2) \tag{23.7}$$

Figure 23.5 provides plots of Φ_5 vs. V_m for values of m_m from 0.75 to 1.75.

The use of statistical data coupled with reliability theory meets the objective of providing a consistent level of safety in design. This approach, however, relies on a sufficient quantity and quality of test data

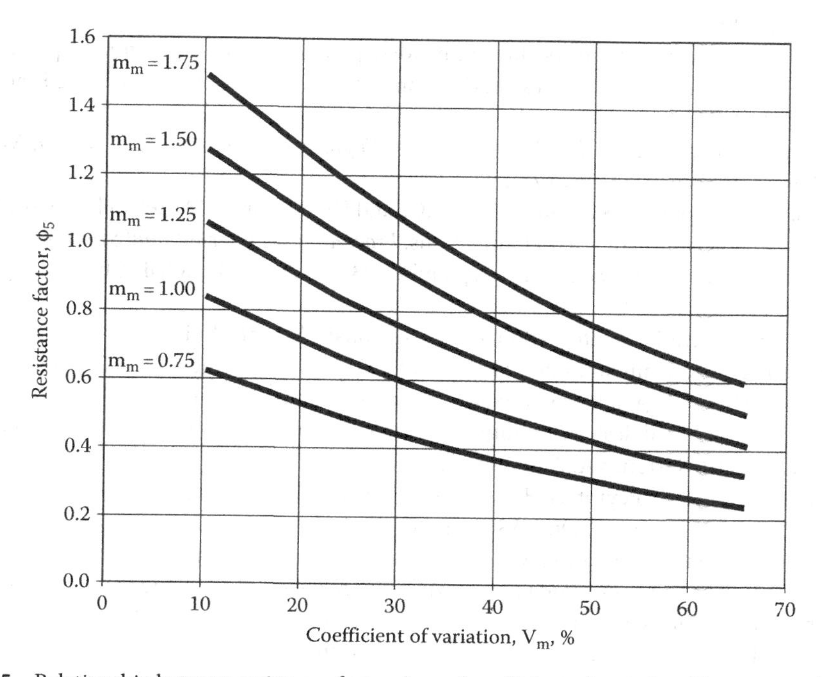

FIGURE 23.5 Relationship between resistance factor, ϕ_5, and coefficient of variation, V_m.

that, in many cases, is simply not available. When information is insufficient, many agencies and organizations tasked with development of resistance factors calibrate the results by curve-fitting of ASD safety factors. Calibration by fitting to ASD also offers an opportunity to adjust resistance factors developed from reliability theory to insure design results similar to ASD where justified (Allen, 2005). Monte Carlo simulation has also been used successfully for performing reliability analyses and developing resistance factors (Nowak and Collins, 2000; Allen, 2005).

No matter the method, estimation of foundation resistance factors must involve the use of engineering judgment. Limited or low-quality data should be taken into account in determining RBD methods. Poor or insufficient data can result in overly conservative resistance factors that produce designs in excess of ASD methods, thus discouraging designers in their use. The fact that transmission foundations have historically performed very well with relatively few failures indicates both success in the application of electric system foundation design and the opportunity to provide more economical and reliable designs in the future.

References

Aichinger, R., Bingle, N., Bowles, G.E., Dagher, H.J., Davidson, J.W., Fouad, F., Ishac, M., Lacoursiere, B., Oliphant, W.J., Randle, R.E., Rollins, M., Rubeiz, C.G., Vandergriend, L. Voda, M., West, D., Wolfe, R. Wong, E.J., and Zolotoochin, C. New ASCE/SEI manual of engineering practice: Structural reliability-based design of utility poles. *Electrical Transmission in a New Age: Proceedings of the Conference*, September 9–12, 2002, Omaha, NE. Reston, VA: American Society of Civil Engineers, pp. 64–135, 2002.

Allen, T.M. *Development of Geotechnical Resistance Factors and Downdrag Load Factors for LRFD Foundation Strength Limit State Design*. Washington, DC: FHWA-NHI-05-052, February 2005.

American Association of State Highway and Transportation Officials. *LRFD Highway Bridge Design Specifications*, 5th edn. Washington, DC: AASHTO, 2010.

American Society of Civil Engineers. Design of guyed electrical transmission structures. ASCE Manuals and Reports on Engineering Practice No. 91. New York: ASCE, 1997.

American Society of Civil Engineers. Design of latticed steel transmission structures. ASCE 10-97. Reston, VA: ASCE, 2000.

American Society of Civil Engineers. Recommended practice for fiber-reinforced polymer products for overhead utility line structures. ASCE Manuals and Reports on Engineering Practice No. 104. Reston, VA: ASCE, 2003.

Ang, A.H.S. and Tang, W.H. *Probability Concept in Engineering Planning and Design*. Vols. 1 and 2. New York: John Wiley & Sons, 1984.

Bazán-Zurita, E., Jarernprasert, S., Bazán-Arias, N.C., and DiGioia, Jr., A.M. Reliability-based strength factors for the geotechnical design of foundations. *Proceedings of International Symposium on Reliability Engineering and Risk Management*. September 23–26, 2010. Shanghai, China: Tongji University Press, pp. 822–827, 2010.

Commission Électrotechnique Internationale/International Electrotechnical Commission. Design criteria of overhead transmission lines with Canadian deviations of the CEI/IEC. CSA Standard C22.3 No. 60826. Mississauga, Ontario, Canada: Canadian Standards Association, 2006.

Dagher, H. Reliability-based design of utility pole structures. ASCE Manuals and Reports on Engineering Practice No. 111. Reston, VA: ASCE, 2006.

Dagher, H.J., Kulendran, S., Peyrot, A.H., Maamouri, M., and Lu, Q. System reliability in design of transmission structures. *J. Struct. Div.*, ASCE, 119(1): 323–340, 1993.

DiGioia, Jr., A.M. Overview of IEEE/ASCE foundation design guide and CIGRÉ WG 07: Foundation activities. *Electrical Transmission in a New Age: Proceedings of the Conference*, September 9–12, 2002, Omaha, NE. Reston, VA: American Society of Civil Engineers, pp. 309–317, 2002.

DiGioia, Jr., A.M. Guide for reliability-based design of transmission line structure foundations, Report 1019959. Palo Alto, CA: Electric Power Research Institute, 246 pp., 2010.

Ghannoum, E. Probabilistic design of transmission lines part I: Probability calculations and structural reliability, and part II: Design criteria corresponding to a target reliability. *IEEE Trans. Power Appar. Syst.*, PAS-102.9, 3057–3079, 1983.

Grigoriu, M.D., Kulhawy, F.H., Spry, M.J., and Filippas, O.B. Probabilistic site strategy for transmission lines. *Foundations for Transmission Line Towers: Proceedings of a Session*, GSP 8, April 27, 1987, Atlantic City, New Jersey. New York: ASCE, pp. 1–14, 1987.

Hunt, R.E. *Geotechnical Engineering Investigation Manual*. New York: McGraw-Hill, 1984.

Institute of Electrical and Electronics Engineers. IEEE Trial-use design guide for wood transmission structures. Standard 751. New York: IEEE, 1991.

Institute of Electrical and Electronics Engineers. IEEE guide for transmission structure foundation design and testing. Standard 691. Piscataway, NJ: IEEE, 2001.

Institute of Electrical and Electronics Engineers. National Electrical Safety Code. ANSI/IEEE Standard C2-2007. New York: IEEE, 336 pp., 2006.

Institute of Electrical and Electronics Engineers. Guide to installation of foundations for transmission line structure. Standard 977. New York: IEEE, 2010.

International Council on Large Electric Systems. Improved design criteria of overhead transmission lines based on reliability concepts, CIGRE SC-22 Report, October 1995.

Kulhawy, F.H. and Mayne, P.W. Manual on estimating soil properties for foundation design, Report EL-6800. Palo Alto, CA: Electric Power Research Institute, 306 pp., 1990.

Kulhawy, F.H., Trautmann, C.H., Beech, J.F., O'Rourke, T.D., McGuire, W., Wood, W.A., and Capano, C. Transmission line structure foundations for uplift-compression loading. Report EL-2870. Palo Alto, CA: Electric Power Research Institute, 412 pp., 1983.

Magee, W.L. Design of steel transmission pole structures. ASCE Standard ASCE 48-05. Reston, VA: ASCE, 2006.

Nowak, A.S. and Collins, K.R. *Reliability of Structures*. Boston, MA: McGraw-Hill, 2000.

Ostendorp, M. Longitudinal loading and cascading failure assessment for transmission line upgrades. *ESMO Conference '98*, Orlando, FL, April 26–30, pp. 324–359, 1998.

Stas, C.V. and F.H. Kulhawy. Critical evaluation of design methods for foundations under axial uplift and compression loading. Report EL-3771. Palo Alto, CA: Electric Power Research Institute, 198 pp., 1984.

Tedesco, P.A. and DiGioia, Jr., A.M. IEEE/ASCE transmission structure foundation design guide. *Proceedings of a Symposium, ASCE National Spring Convention*, May 16–20, 1983, Philadelphia, PA. New York: ASCE, 1983.

Trautmann, C.H. and Kulhawy, F.H. Data sources for engineering geologic studies. *Bull. Assoc. Eng. Geol.*, 20(4): 439–454, 1983.

Wong, C.J. and Miller, M.D. Guidelines for electrical transmission line structural loading. ASCE Manuals and Reports on Engineering Practice No. 74. Reston, VA: American Society of Civil Engineers, 2010.

24

Advanced Technology High-Temperature Conductors

James R. Hunt
Salt River Project

24.1 Introduction

Since the days of Nikola Tesla and Thomas Edison, electrical transmission and distribution (T&D) engineers have evaluated the merits of various bare overhead conductors. Copper was employed early on, due to its excellent conductivity. Soon, however, usage trended toward aluminum, due to its lighter weight and lower cost. These advantages overcame the roughly 40% lower conductivity of aluminum compared to copper. Another factor was that copper was considered a strategic material in high demand during World War I.

As long-distance transmission applications emerged, it was clear that line construction economics could be optimized with longer spans, and the strength of aluminum was insufficient. This deficiency was remedied in two ways—either using stronger aluminum alloy conductors (AAC) or supporting outer aluminum strands with a steel core. For more than 100 years, composite conductors made of aluminum for good conductivity and steel for high strength have been the first choice of T&D engineers. In many cases, the venerable aluminum conductor steel reinforced (ACSR) conductor is still the optimal economic choice. Its place as an industry standard has been secure for over 100 years.

Higher conductor operating temperatures offer an opportunity to reduce the initial cost of a transmission line, at the expense of higher resistive losses, especially when the line loading is high.

Nowadays, many utility companies are using a short time horizon in assessing total lifetime cost of construction and operations, and are focused on cutting the initial capital cost. ACSR operating temperatures have generally been limited to 93°C–100°C; otherwise annealing of the aluminum strands will cause the conductor to lose a significant portion of its initial strength and render it unable to meet the design requirements.

To add flexibility in conductor choices, manufacturers began offering aluminum conductor steel supported (ACSS), with the aluminum fully annealed, so there was full reliance on the steel for strength.

This new composition allowed a typical maximum operating temperature of 200°C. A key weakness of this product is that the thermal coefficient of expansion of steel is high, and taller structures are required to meet clearance requirements at maximum sag. Starting with the decade of the 1990s, a trend emerged in which this "high-temperature high-sag high-loss" conductor was and continues to be widely used in new construction.

The demand for electrical energy continues to increase, either through new residential, commercial, and industrial development, or through new electrically powered devices at existing locations. The need for electrical transmission capability also increases as generation sources are added, and the mix of generation changes, as is evidenced in the recent trend toward renewable resources. Transmission planners are often faced with a choice between adding new lines and increasing the capacity of existing ones. Expanding the throughput in a given corridor is sometimes the only choice available, especially in a fully developed urban setting.

Power line ratings are generally limited by the thermal capability of the conductor, or by clearance to the ground, other wires, or nearby structures. Increased capacity can be obtained by installing an additional conductor per phase, or by reconductoring with a higher-capacity conductor. However, when expanding the number of conductors in a bundle, structure strength will likely be insufficient unless the upgrade was envisioned in the original design. When replacing an ACSR conductor with ACSS, the maximum sag is significantly greater, and clearance is normally an issue. As a result, the cost of these options can be exorbitant since most or all of the existing poles or towers may have to be replaced.

Even when the benefits yield a good value proposition, it may be impossible to find a construction window in which the outage of the existing facility is acceptable.

These considerations led to the search for high-temperature low-sag (HTLS) conductors that can provide a sizeable rating increase, when a direct replacement occurs, with HTLS wire substituted for AAC or ACSR (CIGRE 2009, Douglass 2002, Jones 2006, Kavanaugh 2010, and Zheng 2010). Several products have been launched to fill this need. The list includes the following:

- Aluminum conductor composite core (ACCC)
- Aluminum conductor composite reinforced (ACCR)
- Gap-type ACSR conductor (GTACSR)
- INVAR-supported conductor

While conductors with a conventional steel core are still the predominant players in the marketplace, the HTLS offerings have established a toehold and have been a valuable addition to the engineers' menu of choices. For example, over 2000 circuit miles of ACCC have been installed. Engineers should be familiar with the options, the pros and cons of each, and special situations that are most likely to call for a high-capacity specialty conductor.

24.2 General Considerations

T&D conductors are generally composed of multiple layers of wire strands for flexibility. When trying to achieve a high ampacity, one of the possibilities is to specify trapezoidal shaped strands. For a given overall conductor diameter, this adds weight but significantly increases conductivity and reduces ohmic resistance to electrical current flow. "Trap wire" is available in any conductor type. However, most ACSR and ACSS are still sold with round strands, because the cost of strand shaping exceeds the expense to beef up the structures to accommodate the added loading due to wind on a larger diameter wire.

Some of the materials used in HTLS conductors come at a cost premium. An example is the added expense of using a high-temperature annealed aluminum alloy. There is a compensation, however; designing lines with HTLS conductors will open the possibility of using longer spans or shorter structures.

The nomenclature used for an HTLS conductor of a given outside diameter has sometimes been selected to match the names used for standard ACSR or ACSS conductors. A common conductor size such as Drake may also exist in ACCC or ACCR. Of course, this is only a shorthand way to refer to relative size, and the specific characteristics of the conductor will govern the design.

It is not certain that any conductor would survive a direct hit from a high-powered rifle. However, the brittleness of ACCR makes it a particular concern, and galvanic corrosion could occur after damage to the outer fiberglass strands of the ACCC core. Even rare events should receive due consideration of the probability and repercussions. Gunfire damage testing at Western Area Power Administration (Clark 2010) evaluated five conductor types: AAC, ACSR, ACSS, ACCC, and ACCR. A 12-gauge shotgun (3.5 in. magnum shell with #2 steel shot) caused only superficial damage. In additional testing, conductors were tensioned to 6000 lb, and a 30-06 caliber hunting rifle (180 grain bullet with ballistic tip) was fired directly at the core of each conductor type. No conductor fared well. Convention steel core ACSR was destroyed, and loss of tension was 100%. The 30-06 rifle test caused total core failures in ACCC and ACCR also.

24.3 Aluminum Conductor Composite Core

Development of ACCC began in the early 2000s. It consists of a hybrid carbon and glass fiber composite core which utilizes a high-temperature epoxy resin matrix to bind thousands of individual fibers into a unified load-bearing tensile member. The central carbon fiber core is surrounded by high-grade glass fibers to improve flexibility and toughness, and prevent galvanic corrosion. The composite core exhibits a high strength to weight ratio and has a very low coefficient of thermal expansion which reduces conductor sag under high electrical load/high-temperature conditions. The composite core is surrounded by aluminum strands to carry electrical current. The conductive strands are generally trapezoidal in shape to provide the greatest conductivity and lowest electrical resistance for any given conductor diameter (Figure 24.1).

The maximum recommended conductor temperature is given as 180°C for continuous operation, and 200°C for a short-term emergency. One design element used to allow operation at high temperatures is the use of fully annealed Type 1350-0 aluminum. For an EHV application, where lower conductor temperature operation is assured, nonannealed alloys are appropriate and can be used to increase the overall conductor strength.

The ratio and type of carbon and glass fibers used in standard ACCC core provide a tensile strength of 320 KSI and a modulus of elasticity of 16 MSI. Strength and modulus are design variables that can be optimized to a particular set of requirements, such as a long span or a heavy ice or wind region. Thus, the choices include ultralow sag and ultrahigh strength. A twisted pair variation is also being offered.

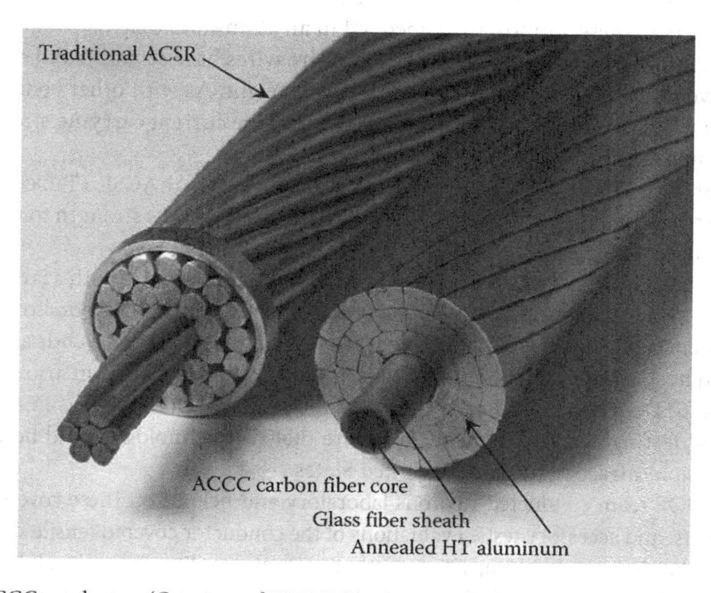

FIGURE 24.1 ACCC conductor. (Courtesy of CTC Cable Corporation, Irvine, CA.)

ACCC was originally developed as a reconductoring option, typically to replace ACSR. Several of the ACCC conductor designs were created to match an ACSR counterpart. As an example, Drake ACCC/TW has the same outside diameter and approximately the same weight as Drake ACSR. The lighter core is offset by additional aluminum. In Europe, ACCC designs were introduced in mm^2 sizes and given city names. Standards for the general category of polymer matrix composite conductors (PMCC) are being developed by ASTM and IEC, and will help users in understanding and selecting conductors such as ACCC.

Applications for ACCC have primarily been geared to upgrading the capacity of existing lines. However, some of the attributes of ACCC are cited as reasons for consideration in other circumstances. These include the following:

- Very high strength designs with good self-damping, for long spans
- Better elasticity, for lower probability of failure in a heavy ice load condition
- Corrosion resistance

One concern about ACCC is that it may be susceptible to breakage during installation. Strict attention must be paid to guidance from the manufacturer regarding the diameter of sheaves (pulleys), pulling tensions, and rate of change of the tension. If too much tension is applied around a tight radius, or if tension is applied in a sudden jerky fashion, the core can be damaged. There have been numerous successful installations where construction crews stayed within the pulling guides. There have also been a few instances where the wire broke because crews failed to comply with the instructions.

Other concerns expressed about ACCC include how stable the organic compounds are at high temperature, and whether there might be a failure mechanism similar to brittle fracture in polymer composite insulators. As noted earlier, there is an extensive body of research, as well as about a decade of field experience, to draw upon when evaluating if and how to apply a PMCC (Alawar et al. 2005, Bosze et al. 2006a, b, c, Burks et al. 2010, Engdahl et al. 2009).

24.4 Aluminum Conductor Composite Reinforced

Published articles on ACCR technology date back to the mid-1990s (Deve and McCullough 1995). The core of ACCR gets its strength from tens of thousands of strands of an advanced material—a ceramic aluminum oxide fiber. These strands are embedded in an aluminum matrix, so the resulting core wire has the look and feel of a common metal. While the core wires have strength and stiffness comparable to steel, they have much higher conductivity and lower weight. As with other commercial HTLS conductors, in order to allow elevated operating temperatures, the current carrying strands are made of an aluminum–zirconium alloy.

ACCR can carry twice the current of the same size conventional ACSR (Table 24.1). Its core has a coefficient of thermal expansion that is roughly half that of steel, and its strength to weight ratio is better than double (Figure 24.2).

The team that developed this technology was headed by a manufacturer with a large research staff and expertise in high-tech materials. It also included independent scientists, a conductor manufacturer with stranding experience, producers of different styles of line hardware for dead ends and suspensions, test laboratories, and host utilities with facilities that served as test beds. The team also received encouragement from the U.S. Department of Energy, because the DOE concluded it was in the national interest to accelerate field testing and deployment, to ensure that the technology could be used in upgrading electric transmission infrastructure in the United States.

The industry/DOE team conducted rigorous laboratory and field tests. These covered material properties, conductor tests, and accessory tests. Evaluations of the conductor covered tensile strength, stress-strain behavior, electrical resistance, axial impact strength, torsional ductility, short-circuit behavior, crush resistance, lightning resistance, and so on (Johnson et al. 2001, McCullough 2006, 3M Company 2003).

TABLE 24.1 Comparison of Core Properties: ACCR vs. Steel

Core Material	Strength (KSI)	Density (lbs/in^3)	Strength/Density	Coefficient of Thermal Expansion (10^{-6}/°F)
ACSR/ACSS steel	185	0.282	656	6.7
ACCR aluminum matrix	200	0.122	1681	3.5

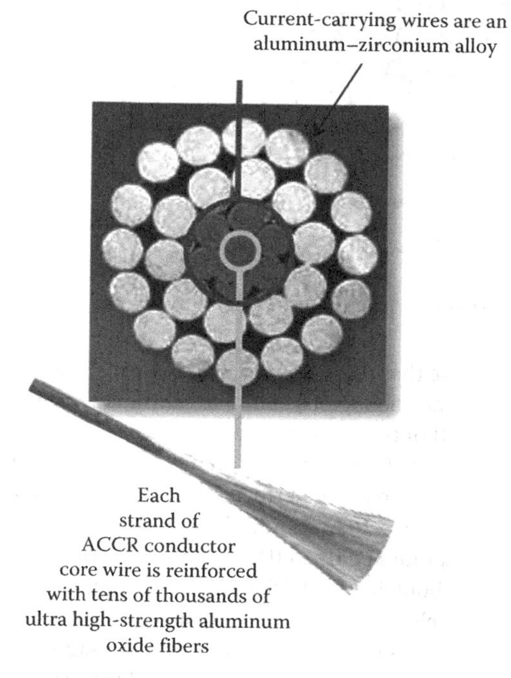

FIGURE 24.2 ACCR conductor. (Courtesy of 3M Corporation, Maplewood, MN.)

The first 230 kV installation took place on a Western Area Power Administration line near Fargo, North Dakota, in 2003. That locale is subject to extreme weather conditions, including low temperatures, ice, and high wind speeds. Testing was also conducted at the opposite end of the spectrum, in metro Phoenix, Arizona. ACCR was installed to deliver the entire output of a Salt River Project generating unit. The high load factor and extreme desert southwest heat created an interesting test environment. The tests included both compression and formed wire-type accessories, as well as vibration dampers.

Advantages touted for ACCR include being well suited for long-span crossings, regions with heavy ice loads, and installations in corrosive environments.

In general, installation techniques are similar to those for ACSR, AAC, or ACSS. However, despite the ACCR core wires being similar in appearance to aluminum, they are quite brittle and will snap if bent to a tight radius. Early on, the manufacturer conducted extensive tests and developed guidelines for pulling tension and diameter of sheaves and bull wheels. Because installations have been carefully supervised, and crews have followed the guidelines, no conductor damage during installation has occurred to date.

24.5 Gap-Type ACSR Conductor

This conductor has a strikingly different geometrical configuration than ACSR. It is made up of heat-resistant aluminum strands surrounding a steel core. The outer aluminum layer(s) can be made of either round or trapezoidal strands. However, the furthest inside layer of aluminum must be comprised of trapezoidal strands, configured to form a gap between the steel core and the aluminum layers.

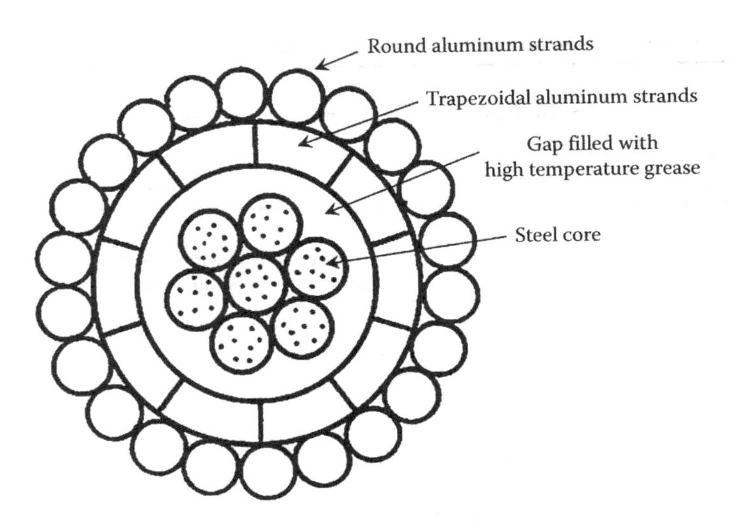

FIGURE 24.3 Gap-type conductor.

The gap can be filled with a grease that is resistant to high temperatures. Among other advantages, this arrangement of wires enhances the conductor's ability to self-dampen aeolian vibration (Figure 24.3).

When GTACSR is installed, all of the tension is applied to the steel core. The aluminum layers hang loosely. As a result, at temperatures higher than at stringing, all of the mechanical strength is supplied by the steel. At lower temperatures, sag performance is similar to conventional ACSR. Installing GTACSR is considerably more complex than standard conductor.

As an application example (Zamora et al. 2001), in Spain, an existing line was limited by clearance at an operating temperature of about 50°C. ACSS and INVAR conductors had similar performance to ACSR in the temperature range below 50°C and were not able to significantly increase the line capacity. However, assuming that GTACSR was installed at 20°C, the flatter sag-temperature characteristic above that temperature allowed for a significant benefit and increased the maximum temperature to 80°C.

24.6 INVAR-Supported Conductor

The key to reduced conductor sag with the INVAR approach is the use of an iron–nickel alloy for the core wires. This specialty alloy includes about 36%–38% nickel, and it has a very favorable coefficient of thermal expansion. For a given increase in temperature, INVAR's elongation is approximately one-third that of steel. The low-sag core is paired with specially formulated aluminum conductor strands, which are alloyed with zirconium to allow a continuous operating temperature up to 210°C. The core wires can be galvanized (zinc-coated) or aluminum-clad in areas where there is a particular concern about corrosion.

At low temperatures, some of the INVAR conductor's mechanical strength comes from the aluminum alloy current carrying strands, and in that mode, the sag-temperature characteristic is similar to ACSR. Above a transition temperature in the range of 85°C–100°C, all of the mechanical strength is provided by the INVAR core. Thus, in the higher-temperature range, the sag-temperature relationship is very flat.

INVAR conductors have the advantage of being similar to the familiar ACSR: metallic, ductile, and unquestioned longevity. There are no special precautions required during installation. As with other HTLS conductors, it can be used as a direct replacement for ACSR, yielding a twofold increase in line capacity without replacing any poles or towers.

Another similarity in benefits with other HTLS wires is that ecological or environmental benefits can result, since a right-of-way can transmit the same power with fewer or shorter structures, and fewer wires. Many stakeholders feel that the ultimate plus for the environment is to avoid building new lines at all, and meet transmission capacity needs by uprating existing lines. Substantial cost savings are possible when new right-of-way acquisition is avoided and tower modifications are not required.

24.7 Testing: The Sequential Mechanical Test

In view of the novel make-up of the advanced conductors that are available in the marketplace, certain companies have pressed for more exhaustive tests than have traditionally been conducted on conventional conductors. An example of this is the sequential mechanical test proposed by Eric Engdahl and Bruce Freimark of American Electric Power. The purpose of this test is to simulate the multiple mechanical challenges that a conductor experiences over its installed life. Rather than performing various independent tests, in this case, the same conductor is subjected to a series of stresses, one after the other.

The sequential mechanical test consists of the following:

- Sheave test
- Aeolian vibration test
- Galloping test
- Load cycling test
- Tensile test to failure

The test protocol includes demanding requirements in each of the five segments of the evaluation. The criteria for success consist of no visible damage during the first four tests, and withstanding at least 100% of rated breaking strength (RBS) before failure in the tensile test. The following paragraphs describe the ACCR test in particular.

In the sheave test, a 125 ft length of the conductor is subjected to various tensions and break-over angles, to simulate forces that would occur during line construction. Some adjustments in the setup are allowed, in order to comply with manufacturer recommendations and standard practices. When testing a particular size of ACCR (1033 kcmil TW-T13), the elements of this test were the following:

- Twenty passes around a 28 in. sheave with a break-over angle of 10°, at 20% RBS tension.
- Seven passes over the 28 in. sheave with a break-over angle of 20°, at 20% RBS tension.
- Finally, three passes with a break-over angle of 30°, at 20% RBS tension. Per a normal field approach for ACCR, this step was conducted with an array of seven 7 in. roller, with an effective bending radius of 60 in.

The test setup consisted of a triangular loop. In addition to the test sheave, there were a 55 in. drive sheave and a 55 in. idle sheave at the other corners. Tension was applied using a hydraulic cylinder attached to the test sheave. The break-over angle was adjusted by inserting appropriate length wire rope slings, connected using Kellum grips. Line tensions were determined using a dynamometer in the test loop and a load cell between the test sheave and the hydraulic cylinder. With this setup, the conductor actually experienced two passes around 55 in. sheaves for every pass over the test sheave.

The setup for the aeolian vibration test calls for an active span that is about 65% of the overall span. This ensures that vibration activity in the back span is less than in the test span. About 80 ft of conductor is stretched between dead ends and tensioned to 20% RBS. A formed wire-type suspension assembly separated the active and back spans, and a shaker was positioned about 9 ft from the dead end of the active span. The test was conducted for 100,000,000 cycles, at a frequency of slightly over 30 Hz and an amplitude of one-half conductor diameter.

To prepare for the galloping test, the aeolian vibration test setup was adjusted, with the back span being increased to the same length as the test span. A galloper was positioned about 5 ft from one dead end. Conductor tension was set around 4% of RBS, and the test was conducted for 100,000 cycles, at a frequency of 1.88 Hz and a peak-to-peak amplitude of 26 in. (1/25 of the span length). Another detail of the test setup was that guides were used to ensure that the conductor movement was all vertical, with no horizontal component.

Once the sheave, aeolian vibration, and galloping tests were completed, and no visible damage was observed in a careful examination of the conductor (particularly focusing on the wire under the suspension assembly, where damage was most likely to have occurred), setup for a load cycling test commenced. The test sample consisted of the approximately 10 ft length of conductor under the suspension, plus an additional 5 ft on either end, for a total length of 20 ft. The wire was gripped by a compression dead end at one end and a resin socket at the other, and instrumentation was provided for direct strain measurement during the test.

The load cycling sequence consisted of increasing tensions, with a load hold of varying lengths:

- 10%, 20%, 30%, 40%, 50%, 60%, and 70% RBS—5 min hold
- 85% RBS—30 min hold
- Decrease tension back to 10% RBS

This was repeated for four cycles. On the fifth cycle, the hold at 85% RBS was extended to 3 hours.

To prepare for the final tensile test to failure, the sample from the load cycling test was cut in half and fitted with a resin socket on each end, to ensure even loading of each conductor strand. The sample was loaded until it failed in tension, with the loading steadily increased at a rate of 1% strain per minute. In the case of the ACCR test, the sample failed at 109% of the RBS.

For more details of sequential mechanical test results, see references by Engdahl et al. (2009) for ACCC and McCullough (2009) for ACCR.

24.8 Conclusion

Research and development activity in the last decade has culminated in new choices. The list of T&D conductors in the marketplace has expanded. Particularly when there is a system need for added throughput in an existing corridor, the new HTLS conductors can be viable economic choices. Each product has unique advantages and concerns. It is important to make an informed decision when considering

- ACCC
- ACCR
- GTACSR
- INVAR-supported conductor

Whether the requirement is a capacity upgrade, a long span river crossing, or a region subject to heavy ice loading, HTLS conductor may be the preferred choice.

References

Alawar, A., E. J. Bosze, and S. R. Nutt, A composite core conductor for low sag at high temperatures, *IEEE Transactions on Power Delivery*, 20: 2193–2199, 2005.

Albizu, I., A. J. Mazon, V. Valverde, and G. Buigues, Aspects to take into account in the application of mechanical calculation to high-temperature low-sag conductors, *IET Generation, Transmission & Distribution*, 4(5): 631–640, 2009.

Baldick, R. and R. P. O'Neill, Estimates of comparative costs for up-rating transmission capacity, *IEEE Transactions on Power Delivery*, 24(2): 961–969, 2009.

Bosze, E. J., A. Alawar, A. Lim, J. Randy, Y. I. Tsai, and S. Nutt, Comparison of ACCC/TW with ACCR, ACSS and ACSR, University of Southern California M. C. Gill Foundation Composites Center, 2006a.

Bosze, E. J., A. Alawar, Y. I. Tsai, S. R. Nutt, D. Bryant, and G. Bowles, Performance of a new overhead conductor design using a carbon/glass fiber composite core, *International Conference on Overhead Lines*, March 2006, Fort Collins, CO, pp. 121–135, 2006b.

Bosze, E. J., Y. I. Tsai, E. Barjasteh, S. R. Nutt, and D. Bryant, Long-term durability of the composite core conductor ACCC/TW, *International Conference on Overhead Lines*, March 2006, Fort Collins, CO, pp. 516–525, 2006c.

Burks, B., D. L. Armentrout, and M. Kumosa, Failure prediction analysis of an ACCC conductor subjected to thermal and mechanical stresses, *IEEE Transactions on Dielectrics and Electrical Insulation*, 17: 588–596, 2010.

CIGRE Working Group B2.08, Large overhead line crossings, Technical Brochure 396, October 2009.

Clark, R., Ballistics test comparing HTLS vs. conventional conductor designs, *IEEE PES Conductors Work Group Meeting*, July 26, 2010.

Deve, H. and C. McCullough, Continuous-fiber reinforced aluminum composites—A new generation, *The Journal of Minerals, Metals & Materials Society*, 47(7): 33–37, July 1995.

Douglass, D., High-temperature, low-sag transmission conductors, EPRI report 1001811, June 2002.

Engdahl, E. K., C. Pon, D. Witt, E. J. Bosze, and D. C. Bryant, ACCC conductor combined cyclic load test report, American Electric Power & CRTC Cable Corp., 2009.

Johnson, D. J., T. L. Anderson, and H. E. Deve, A new generation of high performance conductors, *IEEE Power Engineering Society Summer Meeting*, July 2001, Vancouver, British Columbia, Canada, Vol. 1, pp. 175–179, 2001.

Jones, W. D., More heat, less sag [power cable upgrades], *IEEE Spectrum*, 43(6): 16–18, 2006.

Kavanagh, T. and O. Armstrong, An evaluation of high temperature low sag conductors for uprating the 220 kV transmission network in Ireland, *Universities Power Engineering Conference*, August 31–September 3, 2010, Cardiff, U.K., pp. 1–5, 2010.

3M Company, Aluminum conductor composite reinforced technical notebook (795 kcmil family), 2003. http://multimedia.3 m.com/mws/mediawebserver?mwsId=66666UuZjcFSLXTtN8Tt4xfEEVuQEcuZgVs6EVs6E666666–&fn=ACCR1.pdf

McCullough, C., Update on ACCR conductor, *Proceedings of the IEEE TP&C Line Design Meeting*, Albuquerque, NM, 2006.

McCullough, C., AEP 1033TW-T13 ACCR sequential mechanical test report summary, 3 M Company, October 26, 2009.

Zamora, I., A. J. Mazon, R. Criado, C. Alonzo, and J. R. Saenz, Uprating using high-temperature electrical conductors, CRIED2001 Conference Publication No. 482, 2001.

Zheng, Y.-X., J.-H. Zhang, J.-Y. Yang, H.-F. Su, and Y.-J. Li, Research of new technologies to anti-icing disaster and anti-earthquake in transmission and distribution engineering, *2010 International Conference on Power System Technology (POWERCON)*, October 24–28, 2010, Hangzhou, China, pp. 1–6, 2010.

IV

Distribution Systems

William H. Kersting

William H. Kersting was born in Santa Fe, New Mexico. He received his BSEE from New Mexico State University (NMSU), Las Cruces, New Mexico, and his MSEE from Illinois Institute of Technology, Chicago, Illinois. He joined the faculty at New Mexico State University in 1962 and served as professor of electrical engineering and director of the Electric Utility Management Program until his retirement in 2002. He is currently a consultant for Milsoft Utility Solutions. He is also a partner in WH Power Consultants, Las Cruces, New Mexico.

Professor Kersting is a life fellow of the Institute of Electrical and Electronics Engineers. He received the Edison Electric Institutes' Power Engineering Educator Award in 1979 and the NMSU Westhafter Award for Excellence in Teaching in 1977. Prior to joining NMSU, he was employed as a distribution engineer at El Paso Electric Company. Professor Kersting has been an active member of the IEEE Power Engineering Education Committee and the Distribution System Analysis Subcommittee.

25

Power System Loads

Raymond R. Shoults
University of Texas at Arlington

Larry D. Swift
University of Texas at Arlington

The physical structure of most power systems consists of generation facilities feeding bulk power into a high-voltage bulk transmission network that in turn serves any number of distribution substations. A typical distribution substation will serve from 1 to as many as 10 feeder circuits. A typical feeder circuit may serve numerous loads of all types. A light to medium industrial customer may take service from the distribution feeder circuit primary, while a large industrial load complex may take service directly from the bulk transmission system. All other customers, including residential and commercial, are typically served from the secondary of distribution transformers that are in turn connected to a distribution feeder circuit. Figure 25.1 illustrates a representative portion of a typical configuration.

25.1 Load Classification

The most common classification of electrical loads follows the billing categories used by the utility companies. This classification includes residential, commercial, industrial, and other. Residential customers are domestic users, whereas commercial and industrial customers are obviously business and industrial users. Other customer classifications include municipalities, state and federal government agencies, electric cooperatives, educational institutions, etc.

Although these load classes are commonly used, they are often inadequately defined for certain types of power system studies. For example, some utilities meter apartments as individual residential customers, while others meter the entire apartment complex as a commercial customer. Thus, the common classifications overlap in the sense that characteristics of customers in one class are not unique to that class. For this reason some utilities define further subdivisions of the common classes.

A useful approach to classification of loads is by breaking down the broader classes into individual load components. This process may altogether eliminate the distinction of certain of the broader classes, but it is a tried and proven technique for many applications. The components of a particular load, be it residential, commercial, or industrial, are individually defined and modeled. These load components as a whole constitute the composite load and can be defined as a "load window."

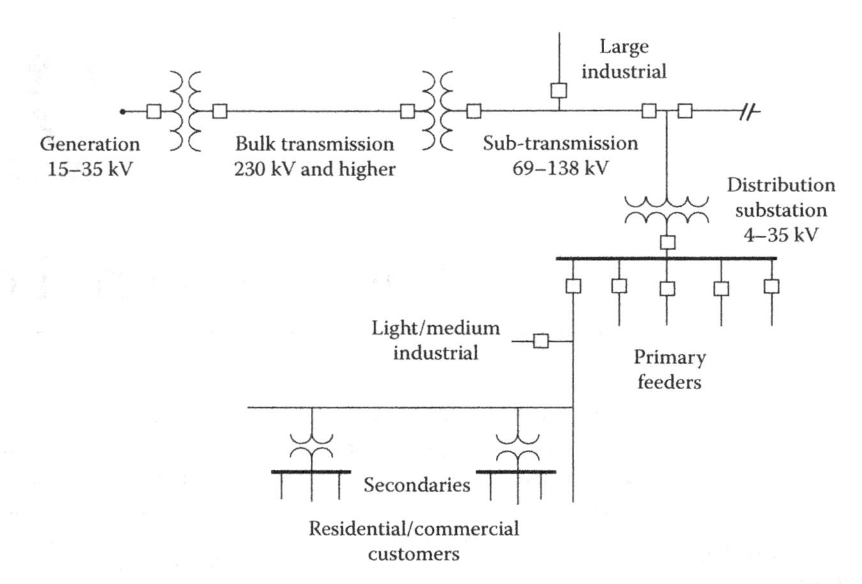

FIGURE 25.1 Representative portion of a typical power system configuration.

25.2 Modeling Applications

It is helpful to understand the applications of load modeling before discussing particular load characteristics. The applications are divided into two broad categories: static ("snap-shot" with respect to time) and dynamic (time varying). Static models are based on the steady-state method of representation in power flow networks. Thus, static load models represent load as a function of voltage magnitude. Dynamic models, on the other hand, involve an alternating solution sequence between a time-domain solution of the differential equations describing electromechanical behavior and a steady-state power flow solution based on the method of phasors. One of the important outcomes from the solution of dynamic models is the time variation of frequency. Therefore, it is altogether appropriate to include a component in the static load model that represents variation of load with frequency. The lists below include applications outside of Distribution Systems but are included because load modeling at the distribution level is the fundamental starting point.

Static applications: Models that incorporate only the voltage-dependent characteristic include the following.

1. Power flow (PF)
   a. Distribution power flow (DPF)
   b. Harmonic power flow (HPF)
   c. Transmission power flow (TPF)
2. Voltage stability (VS)

Dynamic applications: Models that incorporate both the voltage- and frequency-dependent characteristics include the following.

- Transient stability (TS)
- Dynamic stability (DS)
- Operator training simulators (OTS)

Strictly power-flow based solutions utilize load models that include only voltage dependency characteristics. Both voltage and frequency dependency characteristics can be incorporated in load modeling for those hybrid methods that alternate between a time-domain solution and a power flow solution, such as found in Transient Stability and Dynamic Stability Analysis Programs, and Operator Training Simulators (EPRI User's Manual, 1992; EPRI Final Report EL-5003, 1987; Kundur, 1994).

Load modeling in this section is confined to static representation of voltage and frequency dependencies. The effects of rotational inertia (electromechanical dynamics) for large rotating machines are discussed in Chapters 13 and 17 of *Power System Stability and Control*. Static models are justified on the basis that the transient time response of most composite loads to voltage and frequency changes is fast enough so that a steady-state response is reached very quickly.

25.3 Load Modeling Concepts and Approaches

There are essentially two approaches to load modeling: component based and measurement based. Load modeling research over the years has included both approaches (EPRI, 1981, 1984, 1985). Of the two, the component-based approach lends itself more readily to model generalization. It is generally easier to control test procedures and apply wide variations in test voltage and frequency on individual components.

The component-based approach is a "bottom-up" approach in that the different load component types comprising load are identified. Each load component type is tested to determine the relationship between real and reactive power requirements versus applied voltage and frequency. A load model, typically in polynomial or exponential form, is then developed from the respective test data. The range of validity of each model is directly related to the range over which the component was tested. For convenience, the load model is expressed on a per-unit basis (i.e., normalized with respect to rated power, rated voltage, rated frequency, rated torque if applicable, and base temperature if applicable). A composite load is approximated by combining appropriate load model types in certain proportions based on load survey information. The resulting composition is referred to as a "load window."

The measurement approach is a "top-down" approach in that measurements are taken at either a substation level, feeder level, some load aggregation point along a feeder, or at some individual load point. Variation of frequency for this type of measurement is not usually performed unless special test arrangements can be made. Voltage is varied using a suitable means and the measured real and reactive power consumption recorded. Statistical methods are then used to determine load models. A load survey may be necessary to classify the models derived in this manner. The range of validity for this approach is directly related to the realistic range over which the tests can be conducted without damage to customers' equipment. Both the component and measurement methods were used in the EPRI research projects EL-2036 (1981) and EL-3591 (1984, 1985). The component test method was used to characterize a number of individual load components that were in turn used in simulation studies. The measurement method was applied to an aggregate of actual loads along a portion of a feeder to verify and validate the component method.

25.4 Load Characteristics and Models

Static load models for a number of typical load components appear in Tables 25.1 and 25.2 (EPRI, 1984, 1985). The models for each component category were derived by computing a weighted composite from test results of two or more units per category. These component models express per-unit real power and reactive power as a function of per-unit incremental voltage and/or incremental temperature and/or per-unit incremental torque. The incremental form used and the corresponding definition of variables are outlined below:

$\Delta V = V_{act} - 1.0$ (incremental voltage in per unit)

$\Delta T = T_{act} - 95°F$ (incremental temperature for Air Conditioner model), or

$= T_{act} - 47°F$ (incremental temperature for Heat Pump model)

$\Delta\tau = \tau_{act} - \tau_{rated}$ (incremental motor torque, per unit)

TABLE 25.1 Static Models of Typical Load Components—AC, Heat Pump, and Appliances

Load Component	Static Component Model
1-ϕ central air conditioner	$P = 1.0 + 0.4311 * \Delta V + 0.9507 * \Delta T + 2.070 * \Delta V^2 + 2.388 * \Delta T^2 - 0.900 * \Delta V * \Delta T$
	$Q = 0.3152 + 0.6636 * \Delta V + 0.543 * \Delta V^2 + 5.422 * \Delta V^3 + 0.839 * \Delta T^2 - 1.455 * \Delta V * \Delta T$
3-ϕ central air conditioner	$P = 1.0 + 0.2693 * \Delta V + 0.4879 * \Delta T + 1.005 * \Delta V^2 - 0.188 * \Delta T^2 - 0.154 * \Delta V * \Delta T$
	$Q = 0.6957 + 2.3717 * \Delta V + 0.0585 * \Delta T + 5.81 * \Delta V^2 + 0.199 * \Delta T^2 - 0.597 * \Delta V * \Delta T$
Room air conditioner (115 V rating)	$P = 1.0 + 0.2876 * \Delta V + 0.6876 * \Delta T + 1.241 * \Delta V^2 + 0.089 * \Delta T^2 - 0.558 * \Delta V * \Delta T$
	$Q = 0.1485 + 0.3709 * \Delta V + 1.5773 * \Delta T + 1.286 * \Delta V^2 + 0.266 * \Delta T^2 - 0.438 * \Delta V * \Delta T$
Room air conditioner (208/230 V rating)	$P = 1.0 + 0.5953 * \Delta V + 0.5601 * \Delta T + 2.021 * \Delta V^2 + 0.145 * \Delta T^2 - 0.491 * \Delta V * \Delta T$
	$Q = 0.4968 + 2.4456 * \Delta V + 0.0737 * \Delta T + 8.604 * \Delta V^2 - 0.125 * \Delta T^2 - 1.293 * \Delta V * \Delta T$
3-ϕ heat pump (heating mode)	$P = 1.0 + 0.4539 * \Delta V + 0.2860 * \Delta T + 1.314 * \Delta V^2 - 0.024 * \Delta V * \Delta T$
	$Q = 0.9399 + 3.013 * \Delta V - 0.1501 * \Delta T + 7.460 * \Delta V^2 - 0.312 * \Delta T^2 - 0.216 * \Delta V * \Delta T$
3-ϕ heat pump (cooling mode)	$P = 1.0 + 0.2333 * \Delta V + 0.5915 * \Delta T + 1.362 * \Delta V^2 + 0.075 * \Delta T^2 - 0.093 * \Delta V * \Delta T$
	$Q = 0.8456 + 2.3404 * \Delta V - 0.1806 * \Delta T + 6.896 * \Delta V^2 + 0.029 * \Delta T^2 - 0.836 * \Delta V * \Delta T$
1-ϕ heat pump (heating mode)	$P = 1.0 + 0.3953 * \Delta V + 0.3563 * \Delta T + 1.679 * \Delta V^2 + 0.083 * \Delta V * \Delta T$
	$Q = 0.3427 + 1.9522 * \Delta V - 0.0958 * \Delta T + 6.458 * \Delta V^2 - 0.225 * \Delta T^2 - 0.246 * \Delta V * \Delta T$
1-ϕ heat pump (cooling mode)	$P = 1.0 + 0.3630 * \Delta V + 0.7673 * \Delta T + 2.101 * \Delta V^2 + 0.122 * \Delta T^2 - 0.759 * \Delta V * \Delta T$
	$Q = 0.3605 + 1.6873 * \Delta V + 0.2175 * \Delta T + 10.055 * \Delta V^2 - 0.170 * \Delta T^2 - 1.642 * \Delta V * \Delta T$
Refrigerator	$P = 1.0 + 1.3958 * \Delta V + 9.881 * \Delta V^2 + 84.72 * \Delta V^3 + 293 * \Delta V^4$
	$Q = 1.2507 + 4.387 * \Delta V + 23.801 * \Delta V^2 + 1540 * \Delta V^3 + 555 * \Delta V^4$
Freezer	$P = 1.0 + 1.3286 * \Delta V + 12.616 * \Delta V^2 + 133.6 * \Delta V^3 + 380 * \Delta V^4$
	$Q = 1.3810 + 4.6702 * \Delta V + 27.276 * \Delta V^2 + 293.0 * \Delta V^3 + 995 * \Delta V^4$
Washing machine	$P = 1.0 + 1.2786 * \Delta V + 3.099 * \Delta V^2 + 5.939 * \Delta V^3$
	$Q = 1.6388 + 4.5733 * \Delta V + 12.948 * \Delta V^2 + 55.677 * \Delta V^3$
Clothes dryer	$P = 1.0 - 0.1968 * \Delta V - 3.6372 * \Delta V^2 - 28.32 * \Delta V^3$
	$Q = 0.209 + 0.5180 * \Delta V + 0.363 * \Delta V^2 - 4.7574 * \Delta V^3$
Television	$P = 1.0 + 1.2471 * \Delta V + 0.562 * \Delta V^2$
	$Q = 0.2431 + 0.9830 * \Delta V + 1.647 * \Delta V^2$
Fluorescent lamp	$P = 1.0 + 0.6534 * \Delta V - 1.65 * \Delta V^2$
	$Q = -0.1535 - 0.0403 * \Delta V + 2.734 * \Delta V^2$
Mercury vapor lamp	$P = 1.0 + 0.1309 * \Delta V + 0.504 * \Delta V^2$
	$Q = -0.2524 + 2.3329 * \Delta V + 7.811 * \Delta V^2$
Sodium vapor lamp	$P = 1.0 + 0.3409 * \Delta V -2.389 * \Delta V^2$
	$Q = 0.060 + 2.2173 * \Delta V + 7.620 * \Delta V^2$
Incandescent	$P = 1.0 + 1.5209 * \Delta V + 0.223 * \Delta V^2$
	$Q = 0.0$
Range with oven	$P = 1.0 + 2.1018 * \Delta V + 5.876 * \Delta V^2 + 1.236 * \Delta V^3$
	$Q = 0.0$
Microwave oven	$P = 1.0 + 0.0974 * \Delta V + 2.071 * \Delta V^2$
	$Q = 0.2039 + 1.3130 * \Delta V + 8.738 * \Delta V^2$
Water heater	$P = 1.0 + 0.3769 * \Delta V + 2.003 * \Delta V^2$
	$Q = 0.0$
Resistance heating	$P = 1.0 + 2 * \Delta V + \Delta V^2$
	$Q = 0.0$

TABLE 25.2 Static Models of Typical Load Components—Transformers and Induction Motors

Load Component	Static Component Model
Transformer	
Core loss model	$P = \frac{\text{KVA (rating)}}{\text{KVA (system base)}}[0.00267V^2 + 0.73 \times 10^{-9} \times e^{13.5V^2}]$
	$Q = \frac{\text{KVA (rating)}}{\text{KVA (system base)}}[0.00167V^2 + 0.268 \times 10^{-13} \times e^{22.76V^2}]$
	Where V is voltage magnitude in per unit
1-ϕ motor	$P = 1.0 + 0.5179 * \Delta V + 0.9122 * \Delta\tau + 3.721 * \Delta V^2 + 0.350 * \Delta\tau^2 - 1.326 * \Delta V * \Delta\tau$
Constant torque	$Q = 0.9853 + 2.7796 * \Delta V + 0.0859 * \Delta\tau + 7.368 * \Delta V^2 + 0.218 * \Delta\tau^2 - 1.799 * \Delta V * \Delta\tau$
3-ϕ motor (1–10 HP)	$P = 1.0 + 0.2250 * \Delta V + 0.9281 * \Delta\tau + 0.970 * \Delta V^2 + 0.086 * \Delta\tau^2 - 0.329 * \Delta V * \Delta\tau$
Const. torque	$Q = 0.7810 + 2.3532 * \Delta V + 0.1023 * \Delta\tau - 5.951 * \Delta V^2 + 0.446 * \Delta\tau^2 - 1.48 * \Delta V * \Delta\tau$
3-ϕ motor (10 HP/above)	$P = 1.0 + 0.0199 * \Delta V + 1.0463 * \Delta\tau + 0.341 * \Delta V^2 + 0.116 * \Delta\tau^2 - 0.457 * \Delta V * \Delta\tau$
Const. torque	$Q = 0.6577 + 1.2078 * \Delta V + 0.3391 * \Delta\tau + 4.097 * \Delta V^2 + 0.289\Delta\tau^2 - 1.477 * \Delta V * \Delta\tau$
1-ϕ motor	$P = 1.0 + 0.7101 * \Delta V + 0.9073 * \Delta\tau + 2.13 * \Delta V^2 + 0.245 * \Delta\tau^2 - 0.310 * \Delta V * \Delta\tau$
Variable torque	$Q = 0.9727 + 2.7621 * \Delta V + 0.077 * \Delta\tau + 6.432 * \Delta V^2 + 0.174 * \Delta\tau^2 - 1.412 * \Delta V * \Delta\tau$
3-ϕ motor (1–10 HP)	$P = 1.0 + 0.3122 * \Delta V + 0.9286 * \Delta\tau + 0.489 * \Delta V^2 + 0.081 * \Delta\tau^2 - 0.079 * \Delta V * \Delta\tau$
Variable torque	$Q = 0.7785 + 2.3648 * \Delta V + 0.1025 * \Delta\tau + 5.706 * \Delta V^2 + 0.13 * \Delta\tau^2 - 1.00 * \Delta V * \Delta\tau$
3-ϕ motor (10 HP and above)	$P = 1.0 + 0.1628 * \Delta V + 1.0514 * \Delta\tau \angle 0.099 * \Delta V^2 + 0.107 * \Delta\tau^2 + 0.061 * \Delta V * \Delta\tau$
Variable torque	$Q = 0.6569 + 1.2467 * \Delta V + 0.3354 * \Delta\tau + 3.685 * \Delta V^2 + 0.258 * \Delta\tau^2 - 1.235 * \Delta V * \Delta\tau$

If ambient temperature is known, it can be used in the applicable models. If it is not known, the temperature difference, ΔT, can be set to zero. Likewise, if motor load torque is known, it can be used in the applicable models. If it is not known, the torque difference, Δτ, can be set to zero.

Based on the test results of load components and the developed real and reactive power models as presented in these tables, the following comments on the reactive power models are important.

- The reactive power models vary significantly from manufacturer to manufacturer for the same component. For instance, four load models of single-phase central air-conditioners show a Q/P ratio that varies between 0 and 0.5 at 1.0 p.u. voltage. When the voltage changes, the ΔQ/ΔV of each unit is quite different. This situation is also true for all other components, such as refrigerators, freezers, fluorescent lights, etc.
- It has been observed that the reactive power characteristic of fluorescent lights not only varies from manufacturer to manufacturer, from old to new, from long tube to short tube, but also varies from capacitive to inductive depending upon applied voltage and frequency. This variation makes it difficult to obtain a good representation of the reactive power of a composite system and also makes it difficult to estimate the ΔQ/ΔV characteristic of a composite system.
- The relationship between reactive power and voltage is more non-linear than the relationship between real power and voltage, making Q more difficult to estimate than P.
- For some of the equipment or appliances, the amount of Q required at the nominal operating voltage is very small; but when the voltage changes, the change in Q with respect to the base Q can be very large.
- Many distribution systems have switchable capacitor banks either at the substations or along feeders. The composite Q characteristic of a distribution feeder is affected by the switching strategy used in these banks.

25.5 Static Load Characteristics

The component models appearing in Tables 25.1 and 25.2 can be combined and synthesized to create other more convenient models. These convenient models fall into two basic forms: exponential and polynomial.

25.5.1 Exponential Models

The exponential form for both real and reactive power is expressed in Equations 25.1 and 25.2 below as a function of voltage and frequency, relative to initial conditions or base values. Note that neither temperature nor torque appear in these forms. Assumptions must be made about temperature and/or torque values when synthesizing from component models to these exponential model forms.

$$P = P_o\left[\frac{V}{V_o}\right]^{\alpha_v}\left[\frac{f}{f_o}\right]^{\alpha_f} \tag{25.1}$$

$$Q = Q_o\left[\frac{V}{V_o}\right]^{\beta_v}\left[\frac{f}{f_o}\right]^{\beta_f} \tag{25.2}$$

The per-unit models of Equations 25.1 and 25.2 are as follows.

$$P_u = \frac{P}{P_o}\left[\frac{V}{V_o}\right]^{\alpha_v}\left[\frac{f}{f_o}\right]^{\alpha_f} \tag{25.3}$$

$$Q_u = \frac{Q}{P_o} = \frac{Q_o}{P_o}\left[\frac{V}{V_o}\right]^{\beta_v}\left[\frac{f}{f_o}\right]^{\beta_f} \tag{25.4}$$

The ratio Q_o/P_o can be expressed as a function of power factor (pf) where ± indicates a lagging/leading power factor, respectively.

$$R = \frac{Q_o}{P_o} = \pm\sqrt{\frac{1}{pf^2} - 1}$$

After substituting R for Q_o/P_o, Equation 25.4 becomes the following.

$$Q_u = R\left[\frac{V}{V_o}\right]^{\beta_v}\left[\frac{f}{f_o}\right]^{\beta_f} \tag{25.5}$$

Equations 25.1 and 25.2 (or 25.3 and 25.5) are valid over the voltage and frequency ranges associated with tests conducted on the individual components from which these exponential models are derived. These ranges are typically ±10% for voltage and ±2.5% for frequency. The accuracy of these models outside the test range is uncertain. However, one important factor to note is that in the extreme case of voltage approaching zero, both P and Q approach zero.

TABLE 25.3 Parameters for Voltage and Frequency Dependencies of Static Loads

Component/Parameters	pf	α_v	α_f	β_v	β_f	N_m	pf_{nm}	α_{vnm}	α_{fnm}	β_{vnm}	β_{fnm}
Resistance space heater	1.0	2.0	0.0	0.0	0.0	0.0	—	—	—	—	—
Heat pump space heater	0.84	0.2	0.9	2.5	−1.3	0.9	1.0	2.0	0.0	0.0	0.0
Heat pump/central AC	0.81	0.2	0.9	2.5	−2.7	1.0	—	—	—	—	—
Room air conditioner	0.75	0.5	0.6	2.5	−2.8	1.0	—	—	—	—	—
Water heater and range	1.0	2.0	0.0	0.0	0.0	0.0	—	—	—	—	—
Refrigerator and freezer	0.84	0.8	0.5	2.5	−1.4	0.8	1.0	2.0	0.0	0.0	0.0
Dish washer	0.99	1.8	0.0	3.5	−1.4	0.8	1.0	2.0	0.0	0.0	0.0
Clothes washer	0.65	0.08	2.9	1.6	1.8	1.0	—	—	—	—	—
Incandescent lighting	1.0	1.54	0.0	0.0	0.0	0.0	—	—	—	—	—
Clothes dryer	0.99	2.0	0.0	3.3	−2.6	0.2	1.0	2.0	0.0	0.0	0.0
Colored television	0.77	2.0	0.0	5.2	−4.6	0.0	—	—	—	—	—
Furnace fan	0.73	0.08	2.9	1.6	1.8	1.0	—	—	—	—	—
Commercial heat pump	0.84	0.1	1.0	2.5	−1.3	0.9	1.0	2.0	0.0	0.0	0.0
Heat pump comm. AC	0.81	0.1	1.0	2.5	−1.3	1.0	—	—	—	—	—
Commercial central AC	0.75	0.1	1.0	2.5	−1.3	1.0	—	—	—	—	—
Commercial room AC	0.75	0.5	0.6	2.5	−2.8	1.0	—	—	—	—	—
Fluorescent lighting	0.90	0.08	1.0	3.0	−2.8	0.0	—	—	—	—	—
Pump, fan, (motors)	0.87	0.08	2.9	1.6	1.8	1.0	—	—	—	—	—
Electrolysis	0.90	1.8	−0.3	2.2	0.6	0.0	—	—	—	—	—
Arc furnace	0.72	2.3	−1.0	1.61	−1.0	0.0	—	—	—	—	—
Small industrial motors	0.83	0.1	2.9	0.6	−1.8	1.0	—	—	—	—	—
Industrial motors large	0.89	0.05	1.9	0.5	1.2	1.0	—	—	—	—	—
Agricultural H_2O pumps	0.85	1.4	5.6	1.4	4.2	1.0	—	—	—	—	—
Power plant auxiliaries	0.80	0.08	2.9	1.6	1.8	1.0	—	—	—	—	—

EPRI-sponsored research resulted in model parameters such as found in Table 25.3 (EPRI, 1987; Price et al., 1988). Eleven model parameters appear in this table, of which the exponents α and β and the power factor (pf) relate directly to Equations 25.3 and 25.5. The first six parameters relate to general load models, some of which include motors, and the remaining five parameters relate to nonmotor loads—typically resistive type loads. The first is load power factor (pf). Next in order (from left to right) are the exponents for the voltage (α_v, α_f) and frequency (β_v, β_f) dependencies associated with real and reactive power, respectively. N_m is the motor-load portion of the load. For example, both a refrigerator and a freezer are 80% motor load. Next in order are the power factor (pf_{nm}) and voltage (α_{vnm}, α_{fnm}) and frequency (β_{vnm}, β_{fnm}) parameters for the nonmotor portion of the load. Since the refrigerator and freezer are 80% motor loads (i.e., $N_m = 0.8$), the nonmotor portion of the load must be 20%.

25.5.2 Polynomial Models

A polynomial form is often used in a Transient Stability program. The voltage dependency portion of the model is typically second order. If the nonlinear nature with respect to voltage is significant, the order can be increased. The frequency portion is assumed to be first order. This model is expressed as follows.

$$P = P_o\left[a_o + a_1\left(\frac{V}{V_o}\right) + a_2\left(\frac{V}{V_o}\right)^2\right][1 + D_p\Delta f] \tag{25.6}$$

$$Q = Q_o\left[b_o + b_1\left(\frac{V}{V_o}\right) + b_2\left(\frac{V}{V_o}\right)^2\right][1 + D_q\Delta f] \tag{25.7}$$

where

$a_o + a_1 + a_2 = 1$
$b_o + b_1 + b_2 = 1$
$D_p \equiv$ real power frequency damping coefficient, per unit
$D_q \equiv$ reactive power frequency damping coefficient, per unit
$\Delta f \equiv$ frequency deviation from scheduled value, per unit

The per-unit form of Equations 25.6 and 25.7 is the following.

$$P_u = \frac{P}{P_o}\left[a_o + a_1\left(\frac{V}{V_o}\right) + a_2\left(\frac{V}{V_o}\right)^2\right][1 + D_p\Delta f] \tag{25.8}$$

$$Q = \frac{Q}{P_o} = \frac{Q_o}{P_o}\left[b_o + b_1\left(\frac{V}{V_o}\right) + b_2\left(\frac{V}{V_o}\right)^2\right][1 + D_q\Delta f] \tag{25.9}$$

25.5.3 Combined Exponential and Polynomial Models

The two previous kinds of models may be combined to form a synthesized static model that offers greater flexibility in representing various load characteristics (EPRI, 1987; Price et al., 1988). The mathematical expressions for these per-unit models are the following.

$$P_u = \frac{P_{poly} + P_{exp1} + P_{exp2}}{P_o} \tag{25.10}$$

$$Q_u = \frac{Q_{poly} + Q_{exp1} + Q_{exp2}}{P_o} \tag{25.11}$$

where

$$P_{poly} = a_0 + a_1\left(\frac{V}{V_o}\right) + a_3\left(\frac{V}{V_o}\right)^2 \tag{25.12}$$

$$P_{exp1} = a_4\left(\frac{V}{V_o}\right)^{\alpha_1}[1 + D_{p1}\Delta f] \tag{25.13}$$

$$P_{exp2} = a_5\left(\frac{V}{V_o}\right)^{\alpha_2}[1 + D_{p2}\Delta f] \tag{25.14}$$

The expressions for the reactive components have similar structures. Devices used for reactive power compensation are modeled separately.

TABLE 25.4 Static Load Frequency Damping Characteristics

Component	Frequency Parameters	
	D_p	D_q
Three-phase central AC	1.09818	−0.663828
Single-phase central AC	0.994208	−0.307989
Window AC	0.702912	−1.89188
Duct heater w/blowers	0.528878	−0.140006
Water heater, electric cooking	0.0	0.0
Clothes dryer	0.0	−0.311885
Refrigerator, ice machine	0.664158	−1.10252
Incandescent lights	0.0	0.0
Florescent lights	0.887964	−1.16844
Induction motor loads	1.6	−0.6

The flexibility of the component models given here is sufficient to cover most modeling needs. Whenever possible, it is prudent to compare the computer model to measured data for the load.

Table 25.4 provides typical values for the frequency damping characteristic, D, that appears in Equations 25.6 through 25.9, 25.13, and 25.14 (EPRI, 1979; Warnock and Kilpatrick, 1986). Note that nearly all of the damping coefficients for reactive power are negative. This means that as frequency declines, more reactive power is required which can cause an exacerbating effect for low-voltage conditions.

25.5.4 Comparison of Exponential and Polynomial Models

Both models provide good representation around rated or nominal voltage. The accuracy of the exponential form deteriorates when voltage significantly exceeds its nominal value, particularly with exponents (α) greater than 1.0. The accuracy of the polynomial form deteriorates when the voltage falls significantly below its nominal value when the coefficient a_o is non zero. A nonzero a_o coefficient represents some portion of the load as constant power. A scheme often used in practice is to use the polynomial form, but switch to the exponential form when the voltage falls below a predetermined value.

25.5.5 Devices Contributing to Modeling Difficulties

Some load components have time-dependent characteristics that must be considered if a sequence of studies using static models is performed that represents load changing over time. Examples of such a study include Voltage Stability and Transient Stability. The devices that affect load modeling by contributing abrupt changes in load over periods of time are listed below.

Protective relays: Protective relays are notoriously difficult to model. The entire load of a substation can be tripped off line or the load on one of its distribution feeders can be tripped off line as a result of protective relay operations. At the utilization level, motors on air conditioner units and motors in many other residential, commercial, and industrial applications contain thermal and/or over-current relays whose operational behavior is difficult to predict.

Thermostatically controlled loads: Air conditioning units, space heaters, water heaters, refrigerators, and freezers are all controlled by thermostatic devices. The effects of such devices are especially troublesome to model when a distribution load is reenergized after an extended outage (cold-load pickup). The effect of such devices to cold-load pickup characteristics can be significant.

Voltage regulation devices: Voltage regulators, voltage controlled capacitor banks, and automatic LTCs on transformers exhibit time-dependent effects. These devices are present at both the bulk power and distribution system levels.

Discharge lamps (mercury vapor, sodium vapor, and fluorescent lamps): These devices exhibit time-dependent characteristics upon restart, after being extinguished by a low-voltage condition—usually about 70%–80% of rated voltage.

25.6 Load Window Modeling

The static load models found in Tables 25.1 and 25.2 can be used to define a composite load referred to as the "load window" mentioned earlier. In this scheme, a distribution substation load or one of its feeder loads is defined in as much detail as desired for the model. Using the load window scheme, any number of load windows can be defined representing various composite loads, each having as many load components as deemed necessary for accurate representation of the load. Figure 25.2 illustrates the load window concept. The width of each subwindow denotes the percentage of each load component to the total composite load.

Construction of a load window requires certain load data be available. For example, load saturation and load diversity data are needed for various classes of customers. These data allow one to (1) identify the appropriate load components to be included in a particular load window, (2) assign their relative percentage of the total load, and (3) specify the diversified total amount of load for that window. If load modeling is being used for Transient Stability or Operator Training Simulator programs, frequency dependency can be added. Let P(V) and Q(V) represent the composite load models for P and Q,

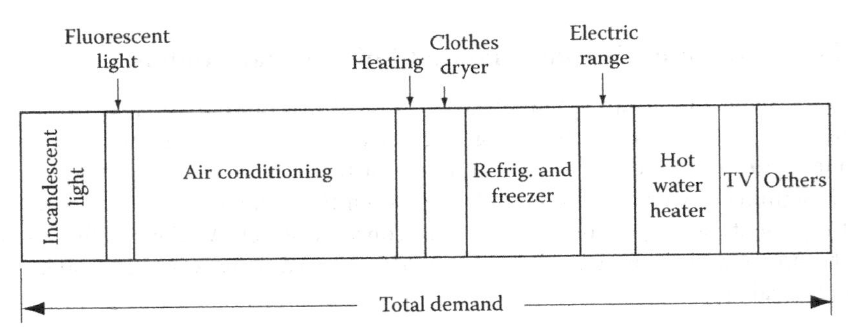

FIGURE 25.2 A typical load window with % composition of load components.

TABLE 25.5 Composition of Six Different Load Window Types

	LW 1	LW 2	LW 3	LW 4	LW 5	LW 6
Load Window Type	Res. 1	Res. 2	Res. 3	Com 1	Com 2	Indust
Load Component	(%)	(%)	(%)	(%)	(%)	(%)
3-phase central AC	25	30	10	35	40	20
Window type AC	5	0	20	0	0	0
Duct heater with blower	5	0	0	0	0	0
Water heater, range top	10	10	10	0	0	0
Clothes dryer	10	10	10	0	0	0
Refrigerator, ice machine	15	15	10	30	0	0
Incandescent lights	10	5	10	0	0	0
Fluorescent lights	20	30	30	25	30	10
Industrial (induct. motor)	0	0	0	10	30	70

respectively, with only voltage dependency (as developed using components taken from Tables 25.1 and 25.2). Frequency dependency is easily included as illustrated below.

$$P = P(V) \times (1 + D_p \Delta f)$$

$$Q = Q(V) \times (1 + D_q \Delta f)$$

Table 25.5 shows six different composite loads for a summer season in the southwestern portion of the United States. This "window" serves as an example to illustrate the modeling process. Note that each column must add to 100%. The entries across from each component load for a given window type represent the percentage of that load making up the composite load.

References

EPRI User's Manual—Extended Transient/Midterm Stability Program Package, version 3.0, June 1992.

General Electric Company, Load modeling for power flow and transient stability computer studies, EPRI Final Report EL-5003, January 1987 (four volumes describing LOADSYN computer program).

Kundur, P., *Power System Stability and Control*, EPRI Power System Engineering Series, McGraw-Hill, Inc., New York, pp. 271–314, 1994.

Price, W.W., Wirgau, K.A., Murdoch, A., Mitsche, J.V., Vaahedi, E., and El-Kady, M.A., Load modeling for power flow and transient stability computer studies, *IEEE Trans. Power Syst.*, 3(1): 180–187, February 1988.

Taylor, C.W., *Power System Voltage Stability*, EPRI Power System Engineering Series, McGraw-Hill, Inc., New York, pp. 67–107, 1994.

University of Texas at Arlington, Determining load characteristics for transient performances, EPRI Final Report EL-848, May 1979 (three volumes).

University of Texas at Arlington, Effect of reduced voltage on the operation and efficiency of electrical loads, EPRI Final Report EL-2036, September 1981 (two volumes).

University of Texas at Arlington, Effect of reduced voltage on the operation and efficiency of electrical loads, EPRI Final Report EL-3591, June 1984 and July 1985 (three volumes).

Warnock, V.J. and Kirkpatrick, T.L., Impact of voltage reduction on energy and demand: Phase II, *IEEE Trans. Power Syst.*, 3(2): 92–97, May 1986.

26

Distribution System Modeling and Analysis

William H. Kersting
New Mexico State University

26.1 Modeling

Radial distribution feeders are characterized by having only one path for power to flow from the source (distribution substation) to each customer. A typical distribution system will consist of one or more distribution substations consisting of one or more "feeders." Components of the feeder may consist of the following:

- Three-phase primary "main" feeder
- Three-phase, two-phase ("V" phase), and single-phase laterals
- Step-type voltage regulators or load tap changing transformer (LTC)
- In-line transformers
- Shunt capacitor banks
- Three-phase, two-phase, and single-phase loads
- Distribution transformers (step-down to customer's voltage)

The loading of a distribution feeder is inherently unbalanced because of the large number of unequal single-phase loads that must be served. An additional unbalance is introduced by the nonequilateral conductor spacings of the three-phase overhead and underground line segments.

Because of the nature of the distribution system, conventional power-flow and short-circuit programs used for transmission system studies are not adequate. Such programs display poor convergence characteristics for radial systems. The programs also assume a perfectly balanced system so that a single-phase equivalent system is used.

If a distribution engineer is to be able to perform accurate power-flow and short-circuit studies, it is imperative that the distribution feeder be modeled as accurately as possible. This means that three-phase models of the major components must be utilized. Three-phase models for the major components will be developed in the following sections. The models will be developed in the "phase frame" rather than applying the method of symmetrical components.

Figure 26.1 shows a simple one-line diagram of a three-phase feeder; it illustrates the major components of a distribution system. The connecting points of the components will be referred to as "nodes."

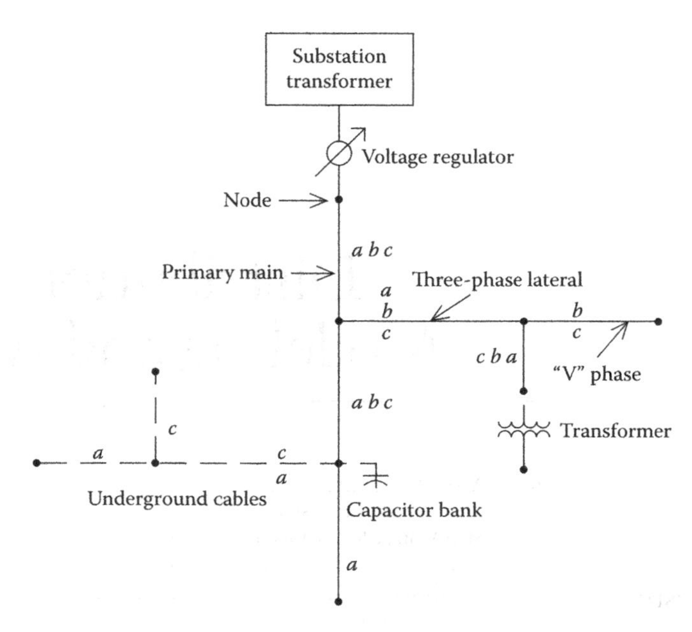

FIGURE 26.1 Distribution feeder.

Note in the figure that the phasing of the line segments is shown. This is important if the most accurate models are to be developed.

The following sections will present generalized three-phase models for the "series" components of a feeder (line segments, voltage regulators, transformer banks). Additionally, models are presented for the "shunt" components (loads, capacitor banks). Finally, the "ladder iterative technique" for power-flow studies using the models is presented along with a method for computing short-circuit currents for all types of faults.

26.1.1 Line Impedance

The determination of the impedances for overhead and underground lines is a critical step before analysis of the distribution feeder can begin. Depending upon the degree of accuracy required, impedances can be calculated using Carson's equations where no assumptions are made, or the impedances can be determined from tables where a wide variety of assumptions are made. Between these two limits are other techniques, each with their own set of assumptions.

26.1.1.1 Carson's Equations

Since a distribution feeder is inherently unbalanced, the most accurate analysis should not make any assumptions regarding the spacing between conductors, conductor sizes, or transposition. In a classic paper, John Carson developed a technique in 1926 whereby the self and mutual impedances for ncond overhead conductors can be determined. The equations can also be applied to underground cables. In 1926, this technique was not met with a lot of enthusiasm because of the tedious calculations that had to be done on the slide rule and by hand. With the advent of the digital computer, Carson's equations have now become widely used.

In his paper, Carson assumes the earth is an infinite, uniform solid, with a flat uniform upper surface and a constant resistivity. Any "end effects" introduced at the neutral grounding points are not large at power frequencies, and therefore are neglected. The original Carson equations are given in Equations 26.1 and 26.2.

Self-impedance:

$$\hat{z}_{ii} = r_i + 4\omega P_{ii} G + j\left(X_i + 2\omega G \cdot \ln \frac{S_{ii}}{RD_i} + 4\omega Q_{ii} G \right) \quad \Omega/\text{mile} \tag{26.1}$$

Mutual impedance:

$$\hat{z}_{ij} = 4\omega P_{ij} G + j\left(2\omega G \cdot \ln \frac{S_{ij}}{D_{ij}} + 4\omega Q_{ij} G \right) \quad \Omega/\text{mile} \tag{26.2}$$

$$X_i = 2\omega G \cdot \ln \frac{RD_i}{GMR_i} \quad \Omega/\text{mile} \tag{26.3}$$

$$P_{ij} = \frac{\pi}{8} - \frac{1}{3\sqrt{2}} k_{ij} \cos(\theta_{ij}) + \frac{k_{ij}^2}{16} \cos(2\theta_{ij}) \cdot \left(0.6728 + \ln \frac{2}{k_{ij}} \right) \tag{26.4}$$

$$Q_{ij} = -0.0386 + \frac{1}{2} \cdot \ln \frac{2}{k_{ij}} + \frac{1}{3\sqrt{2}} k_{ij} \cos(\theta_{ij}) \tag{26.5}$$

$$k_{ij} = 8.565 \times 10^{-4} \cdot S_{ij} \cdot \sqrt{\frac{f}{\rho}} \tag{26.6}$$

where

- $\hat{z}_{ii}$ is the self impedance of conductor i in Ω/mile
- $\hat{z}_{ij}$ is the mutual impedance between conductors i and j in Ω/mile
- r_i is the resistance of conductor i in Ω/mile
- $\omega = 2\pi f$ is the system angular frequency in radians per second
- G is the 0.1609347×10^{-3} Ω/mile
- RD_i is the radius of conductor i in feet
- GMR_i is the Geometric mean radius of conductor i in feet
- f is the system frequency in Hertz
- ρ is the resistivity of earth in Ω-meters
- D_{ij} is the distance between conductors i and j in ft. (see Figure 4.4)
- S_{ij} is the distance between conductor i and image j in ft. (see Figure 4.4)
- θ_{ij} is the angle between a pair of lines drawn from conductor i to its own image and to the image of conductor j (see Figure 4.4)

As indicated above, Carson made use of conductor images; that is, every conductor at a given distance above ground has an image conductor the same distance below ground. This is illustrated in Figure 26.2.

26.1.1.2 Modified Carson's Equations

Only two approximations are made in deriving the "modified Carson equations." These approximations involve the terms associated with P_{ij} and Q_{ij}. The approximations are shown below:

$$P_{ij} = \frac{\pi}{8} \tag{26.7}$$

$$Q_{ij} = -0.03860 + \frac{1}{2} \ln \frac{2}{k_{ij}} \tag{26.8}$$

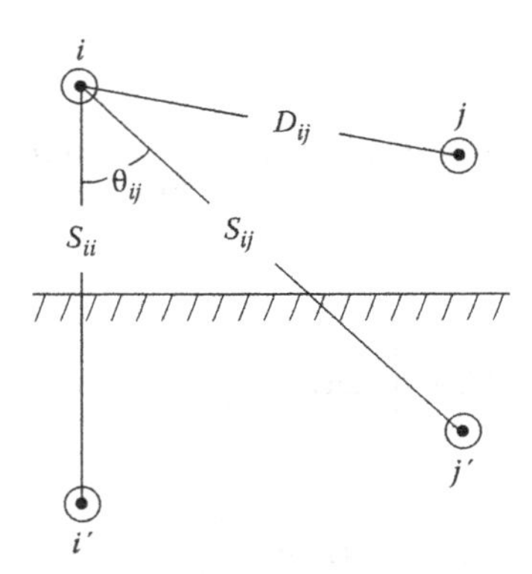

FIGURE 26.2 Conductors and images.

It is also assumed

f is the frequency = 60 Hz

r is the resistivity = 100 Ω m

Using these approximations and assumptions, Carson's equations reduce to:

$$\hat{z}_{ii} = r_i + 0.0953 + j0.12134\left(\ln\frac{1}{\text{GMR}_i} + 7.93402\right) \quad \Omega/\text{mile} \tag{26.9}$$

$$\hat{z}_{ij} = 0.0953 + j0.12134\left(\ln\frac{1}{D_{ij}} + 7.93402\right) \quad \Omega/\text{mile} \tag{26.10}$$

26.1.1.3 Overhead and Underground Lines

Equations 26.9 and 26.10 can be used to compute an ncond × ncond "primitive impedance" matrix. For an overhead four wire, grounded wye distribution line segment, this will result in a 4 × 4 matrix. For an underground grounded wye line segment consisting of three concentric neutral cables, the resulting matrix will be 6 × 6. The primitive impedance matrix for a three-phase line consisting of m neutrals will be of the form

$$[z_{\text{primitive}}] = \left[\begin{array}{ccc|ccc} \hat{z}_{aa} & \hat{z}_{ab} & \hat{z}_{ac} & \hat{z}_{anl} & \cdot & \hat{z}_{anm} \\ \hat{z}_{ba} & \hat{z}_{bb} & \hat{z}_{bc} & \hat{z}_{bnl} & \cdot & \hat{z}_{bnm} \\ \hat{z}_{ca} & \hat{z}_{cb} & \hat{z}_{cc} & \hat{z}_{cnl} & \cdot & \hat{z}_{cnm} \\ \hline \hat{z}_{nla} & \hat{z}_{nlb} & \hat{z}_{nlc} & \hat{z}_{nlnl} & \cdot & \hat{z}_{nlnm} \\ \cdot & \cdot & \cdot & \cdot & \cdot & \cdot \\ \hat{z}_{nma} & \hat{z}_{nmb} & \hat{z}_{nmc} & \hat{z}_{nmnl} & \cdot & \hat{z}_{nmnm} \end{array}\right] \tag{26.11}$$

In partitioned form Equation 20.11 becomes

$$[z_{\text{primitive}}] = \begin{bmatrix} [\hat{z}_{ij}] & [\hat{z}_{in}] \\ [\hat{z}_{nj}] & [\hat{z}_{nn}] \end{bmatrix} s \tag{26.12}$$

26.1.1.4 Phase Impedance Matrix

For most applications, the primitive impedance matrix needs to be reduced to a 3 × 3 phase frame matrix consisting of the self and mutual equivalent impedances for the three phases. One standard method of reduction is the "Kron" reduction (1952) where the assumption is made that the line has a multigrounded neutral. The Kron reduction results in the "phase impedances matrix" determined by using Equation 26.13 below:

$$[z_{abc}]=[\hat{z}_{ij}]-[\hat{z}_{in}][\hat{z}_{nn}]^{-1}[\hat{z}_{nj}] \tag{26.13}$$

It should be noted that the phase impedance matrix will always be of rotation a–b–c no matter how the phases appear on the pole. That means that always row and column 1 in the matrix will represent phase a, row and column 2 will represent phase b, row and column 3 will represent phase c.

For two-phase (V-phase) and single-phase lines in grounded wye systems, the modified Carson equations can be applied, which will lead to initial 3 × 3 and 2 × 2 primitive impedance matrices. Kron reduction will reduce the matrices to 2 × 2 and a single element. These matrices can be expanded to 3 × 3 phase frame matrices by the addition of rows and columns consisting of zero elements for the missing phases. The phase frame matrix for a three-wire delta line is determined by the application of Carson's equations without the Kron reduction step.

The phase frame matrix can be used to accurately determine the voltage drops on the feeder line segments once the currents flowing have been determined. Since no approximations (transposition, for example) have been made regarding the spacing between conductors, the effect of the mutual coupling between phases is accurately taken into account. The application of Carson's equations and the phase frame matrix leads to the most accurate model of a line segment. Figure 26.3 shows the equivalent circuit of a line segment.

The voltage equation in matrix form for the line segment is given by the following equation:

$$\begin{bmatrix} V_{ag} \\ V_{bg} \\ V_{cg} \end{bmatrix}_n = \begin{bmatrix} V_{ag} \\ V_{bg} \\ V_{cg} \end{bmatrix}_m + \begin{bmatrix} Z_{aa} & Z_{ab} & Z_{ac} \\ Z_{ba} & Z_{bb} & Z_{bc} \\ Z_{ca} & Z_{cb} & Z_{cc} \end{bmatrix} \begin{bmatrix} I_a \\ I_b \\ I_c \end{bmatrix} \tag{26.14}$$

where

$$Z_{ij} = z_{ij} \times \text{length}$$

The phase impedance matrix is defined in Equation 26.15. The phase impedance matrix for single-phase and V-phase lines will have a row and column of zeros for each missing phase

$$[Z_{abc}] = \begin{bmatrix} Z_{aa} & Z_{ab} & Z_{ac} \\ Z_{ba} & Z_{bb} & Z_{bc} \\ Z_{ca} & Z_{cb} & Z_{cc} \end{bmatrix} \tag{26.15}$$

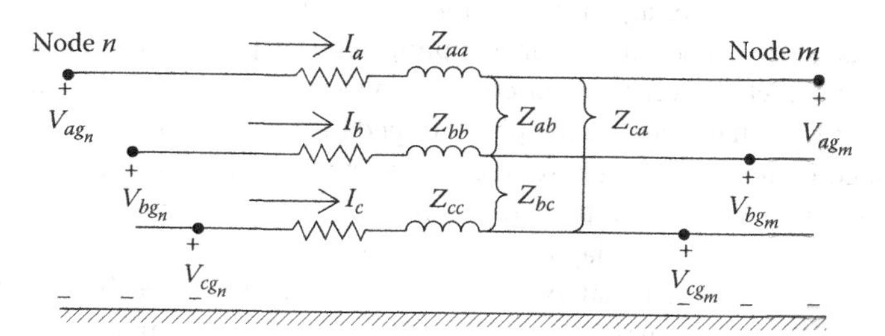

FIGURE 26.3 Three-phase line segment.

Equation 26.14 can be written in condensed form as

$$[VLG_{abc}]_n = [VLG_{abc}]_m + [Z_{abc}][I_{abc}] \tag{26.16}$$

This condensed notation will be used throughout the document.

26.1.1.5 Sequence Impedances

Many times the analysis of a feeder will use the positive and zero sequence impedances for the line segments. There are basically two methods for obtaining these impedances. The first method incorporates the application of Carson's equations and the Kron reduction to obtain the phase frame impedance matrix. The 3 × 3 "sequence impedance matrix" can be obtained by

$$[z_{012}] = [A_s]^{-1}[z_{abc}][A_s]\Omega/\text{mile} \tag{26.17}$$

where

$$[A_s] = \begin{bmatrix} 1 & 1 & 1 \\ 1 & a_s^2 & a_s \\ 1 & a_s & a_s^2 \end{bmatrix} \tag{26.18}$$

$$a_s = 1.0\angle 120 \quad a_s^2 = 1.0\angle 240$$

The resulting sequence impedance matrix is of the form:

$$[z_{012}] = \begin{bmatrix} z_{00} & z_{01} & z_{02} \\ z_{10} & z_{11} & z_{12} \\ z_{20} & z_{21} & z_{22} \end{bmatrix} \Omega/\text{mile} \tag{26.19}$$

where
- z_{00} is the zero sequence impedance
- z_{11} is the positive sequence impedance
- z_{22} is the negative sequence impedance

In the idealized state, the off-diagonal terms of Equation 26.19 would be zero. When the off-diagonal terms of the phase impedance matrix are all equal, the off-diagonal terms of the sequence impedance matrix will be zero. For high-voltage transmission lines, this will generally be the case because these lines are transposed, which causes the mutual coupling between phases (off-diagonal terms) to be equal. Distribution lines are rarely if ever transposed. This causes unequal mutual coupling between phases, which causes the off-diagonal terms of the phase impedance matrix to be unequal. For the nontransposed line, the diagonal terms of the phase impedance matrix will also be unequal. In most cases, the off-diagonal terms of the sequence impedance matrix are very small compared to the diagonal terms and errors made by ignoring the off-diagonal terms are small.

Sometimes the phase impedance matrix is modified such that the three diagonal terms are equal and all of the off-diagonal terms are equal. The usual procedure is to set the three diagonal terms of the phase impedance matrix equal to the average of the diagonal terms of Equation 26.15 and the off-diagonal terms equal to the average of the off-diagonal terms of Equation 26.15. When this is done, the self and mutual impedances are defined as

$$z_s = \frac{1}{3}(z_{aa} + z_{bb} + z_{cc}) \tag{26.20}$$

$$z_m = \frac{1}{3}(z_{ab} + z_{bc} + z_{ca}) \tag{26.21}$$

The phase impedance matrix is now defined as

$$[z_{abc}] = \begin{bmatrix} z_s & z_m & z_m \\ z_m & z_s & z_m \\ z_m & z_m & z_s \end{bmatrix} \tag{26.22}$$

When Equation 26.17 is used with this phase impedance matrix, the resulting sequence matrix is diagonal (off-diagonal terms are zero). The sequence impedances can be determined directly as

$$\begin{aligned} z_{00} &= z_s + 2z_m \\ z_{11} &= z_n = z_s - z_m \end{aligned} \tag{26.23}$$

A second method that is commonly used to determine the sequence impedances directly is to employ the concept of geometric mean distances (GMDs). The GMD between phases is defined as

$$D_{ij} = \text{GMD}_{ij} = \sqrt[3]{D_{ab}D_{bc}D_{ca}} \tag{26.24}$$

The GMD between phases and neutral is defined as

$$D_{in} = \text{GMD}_{in} = \sqrt[3]{D_{an}D_{bn}D_{cn}} \tag{26.25}$$

The GMDs as defined above are used in Equations 26.9 and 26.10 to determine the various self and mutual impedances of the line resulting in

$$\hat{z}_{ii} = r_i + 0.0953 + j0.12134\left[\ln\left(\frac{1}{\text{GMR}_i}\right) + 7.93402\right] \tag{26.26}$$

$$\hat{z}_{nn} = r_n + 0.0953 + j0.12134\left[\ln\left(\frac{1}{\text{GMR}_n}\right) + 7.93402\right] \tag{26.27}$$

$$\hat{z}_{ij} = 0.0953 + j0.12134\left[\ln\left(\frac{1}{D_{ij}}\right) + 7.93402\right] \tag{26.28}$$

$$\hat{z}_{in} = 0.0953 + j0.12134\left[\ln\left(\frac{1}{D_{in}}\right) + 7.93402\right] \tag{26.29}$$

Equations 26.26 through 26.29 will define a matrix of order ncond × ncond, where ncond is the number of conductors (phases plus neutrals) in the line segment. Application of the Kron reduction (Equation 26.13) and the sequence impedance transformation (Equation 26.23) lead to the following expressions for the zero, positive, and negative sequence impedances:

$$z_{00} = \hat{z}_{ii} + 2\hat{z}_{ij} - 3\left(\frac{\hat{z}_{in}^2}{\hat{z}_{nn}}\right)\Omega/\text{mile} \quad (26.30)$$

$$z_{11} = z_{22} = \hat{z}_{ii} - \hat{z}_{ij}$$

$$z_{11} = z_{22} = r_i + j0.12134 \times \ln\left(\frac{D_{ij}}{\text{GMR}_i}\right)\Omega/\text{mile} \quad (26.31)$$

Equation 26.31 is recognized as the standard equation for the calculation of the line impedances when a balanced three-phase system and transposition are assumed.

Example 26.1

The spacings for an overhead three-phase distribution line are constructed as shown in Figure 26.4. The phase conductors are 336,400 26/7 ACSR (Linnet) and the neutral conductor is 4/0 6/1 ACSR.

1. Determine the phase impedance matrix.
2. Determine the positive and zero sequence impedances.

Solution

From the table of standard conductor data, it is found that

$$33,400\,26/7\,\text{ACSR:} \quad \text{GMR} = 0.0244\,\text{ft}$$

$$\text{Resistance} = 0.306\,\Omega/\text{mile}$$

$$4/0\,6/1\,\text{ACSR:} \quad \text{GMR} = 0.00814\,\text{ft}$$

$$\text{Resistance} = 0.5920\,\Omega/\text{mile}$$

From Figure 26.4 the following distances between conductors can be determined:

$$D_{ab} = 2.5\,\text{ft} \quad D_{bc} = 4.5\,\text{ft} \quad D_{ca} = 7.0\,\text{ft}$$

$$D_{an} = 5.6569\,\text{ft} \quad D_{bn} = 4.272\,\text{ft} \quad D_{cn} = 5.0\,\text{ft}$$

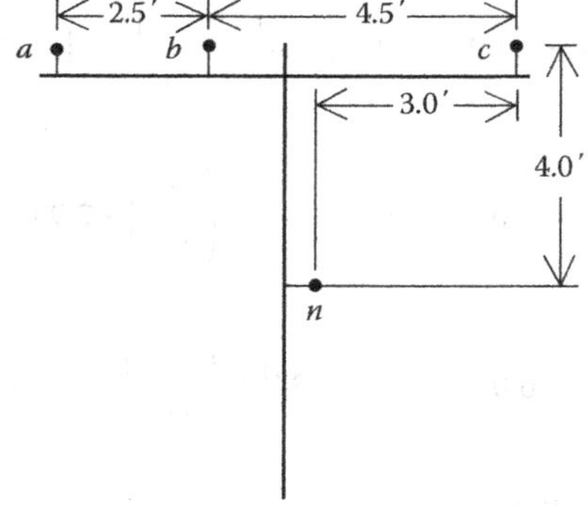

FIGURE 26.4 Three-phase distribution line spacings.

Applying Carson's modified equations (Equations 26.9 and 26.10) results in the primitive impedance matrix.

$$[\hat{z}] = \begin{bmatrix} 0.4013+j1.4133 & 0.0953+j0.8515 & 0.0953+j0.7266 & 0.0953+j0.7524 \\ 0.0953+j0.8515 & 0.4013+j1.4133 & 0.0953+j0.7802 & 0.0953+j0.7865 \\ 0.0953+j07266 & 0.0953+j0.7802 & 0.4013+j1.4133 & 0.0953+j0.7674 \\ 0.0953+j0.7524 & 0.0953+j0.7865 & 0.0953+j0.7674 & 0.6873+j1.5465 \end{bmatrix} \quad (26.32)$$

The Kron reduction of Equation 26.13 results in the phase impedance matrix

$$[z_{abc}] = \begin{bmatrix} 0.4576+j1.0780 & 0.1560+j0.5017 & 0.1535+j0.3849 \\ 0.1560+j0.5017 & 0.4666+j1.0482 & 0.1580+j0.4236 \\ 0.1535+j0.3849 & 0.1580+j0.4236 & 0.4615+j1.0651 \end{bmatrix} \Omega/\text{mile} \quad (26.33)$$

The phase impedance matrix of Equation 26.33 can be transformed into the sequence impedance matrix with the application of Equation 26.17

$$[z_{012}] = \begin{bmatrix} 0.7735+j1.9373 & 0.0256+j0.0115 & -0.0321+j0.0159 \\ -0.0321+j0.0159 & 0.3061+j0.6270 & -0.0723-j0.0060 \\ 0.0256+j0.0115 & 0.0723+j0.0059 & 0.3061+j0.6270 \end{bmatrix} \Omega/\text{mile} \quad (26.34)$$

In Equation 26.34, the 1,1 term is the zero sequence impedance, the 2,2 term is the positive sequence impedance, and the 3,3 term is the negative sequence impedance. Note that the off-diagonal terms are not zero, which implies that there is mutual coupling between sequences. This is a result of the nonsymmetrical spacing between phases. With the off-diagonal terms nonzero, the three sequence networks representing the line will not be independent. However, it is noted that the off-diagonal terms are small relative to the diagonal terms.

In high-voltage transmission lines, it is usually assumed that the lines are transposed and that the phase currents represent a balanced three-phase set. The transposition can be simulated in this example by replacing the diagonal terms of Equation 26.33 with the average value of the diagonal terms ($0.4619 + j1.0638$) and replacing each off-diagonal term with the average of the off-diagonal terms ($0.1558 + j0.4368$). This modified phase impedance matrix becomes

$$[zl_{abc}] = \begin{bmatrix} 0.3619+j1.0638 & 0.1558+j0.4368 & 0.1558+j0.4368 \\ 0.1558+j0.4368 & 0.3619+j1.0638 & 0.1558+j0.4368 \\ 0.1558+j0.4368 & 0.1558+j0.4368 & 0.3619+j1.0638 \end{bmatrix} \Omega/\text{mile} \quad (26.35)$$

Using this modified phase impedance matrix in the symmetrical component transformation, Equation 26.17 results in the modified sequence impedance matrix

$$[zl_{012}] = \begin{bmatrix} 0.7735+j1.9373 & 0 & 0 \\ 0 & 0.3061+j0.6270 & 0 \\ 0 & 0 & 0.3061+j0.6270 \end{bmatrix} \Omega/\text{mile} \quad (26.36)$$

Note now that the off-diagonal terms are all equal to zero, meaning that there is no mutual coupling between sequence networks. It should also be noted that the zero, positive, and negative sequence impedances of Equation 26.36 are exactly equal to the same sequence impedances of Equation 26.34.

The results of this example should not be interpreted to mean that a three-phase distribution line can be assumed to have been transposed. The original phase impedance matrix of Equation 26.33 must be used if the correct effect of the mutual coupling between phases is to be modeled.

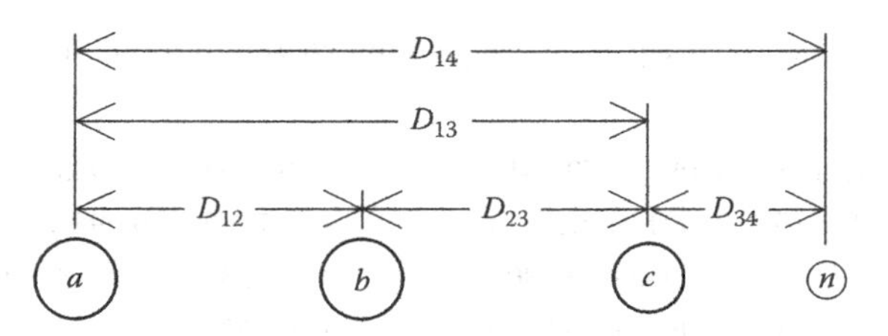

FIGURE 26.5 Three-phase underground with additional neutral.

26.1.1.6 Underground Lines

Figure 26.5 shows the general configuration of three underground cables (concentric neutral, or tape shielded) with an additional neutral conductor.

Carson's equations can be applied to underground cables in much the same manner as for overhead lines. The circuit of Figure 26.5 will result in a 7 × 7 primitive impedance matrix. For underground circuits that do not have the additional neutral conductor, the primitive impedance matrix will be 6 × 6.

Two popular types of underground cables in use today are the "concentric neutral cable" and the "tape shield cable." To apply Carson's equations, the resistance and GMR of the phase conductor and the equivalent neutral must be known.

26.1.1.7 Concentric Neutral Cable

Figure 26.6 shows a simple detail of a concentric neutral cable. The cable consists of a central phase conductor covered by a thin layer of nonmetallic semiconducting screen to which is bonded the insulating material. The insulation is then covered by a semiconducting insulation screen. The solid strands of concentric neutral are spiraled around the semiconducting screen with a uniform spacing between strands. Some cables will also have an insulating "jacket" encircling the neutral strands.

In order to apply Carson's equations to this cable, the following data needs to be extracted from a table of underground cables:

d_c is the phase conductor diameter (in.)
d_{od} is the nominal outside diameter of the cable (in.)
d_s is the diameter of a concentric neutral strand (in.)
GMR_c is the geometric mean radius of the phase conductor (ft)
GMR_s is the geometric mean radius of a neutral strand (ft)
r_c is the resistance of the phase conductor (Ω/mile)
r_s is the resistance of a solid neutral strand (Ω/mile)
k is the number of concentric neutral strands

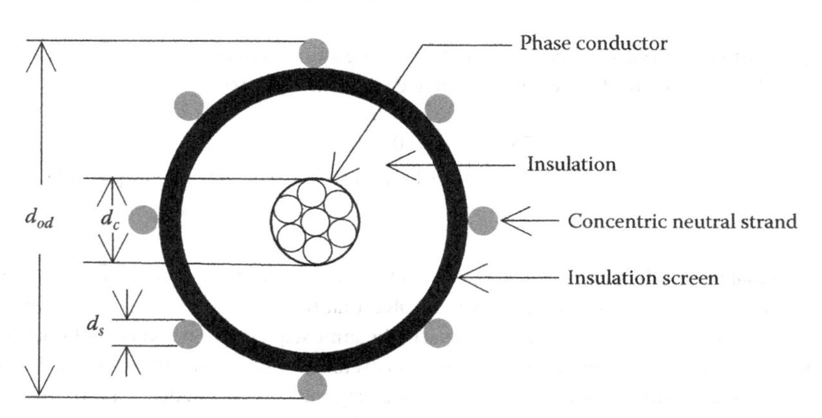

FIGURE 26.6 Concentric neutral cable.

The geometric mean radii of the phase conductor and a neutral strand are obtained from a standard table of conductor data. The equivalent geometric mean radius of the concentric neutral is given by

$$\text{GMR}_{cn} = \sqrt[k]{\text{GMR}_s \times kR^{k-1}} \text{ ft} \tag{26.37}$$

where R is the radius of a circle passing through the center of the concentric neutral strands

$$R = \frac{d_{od} - d_s}{24} \text{ ft} \tag{26.38}$$

The equivalent resistance of the concentric neutral is

$$r_{cn} = \frac{r_s}{k} \, \Omega/\text{mile} \tag{26.39}$$

The various spacings between a concentric neutral and the phase conductors and other concentric neutrals are as follows:

Concentric neutral to its own phase conductor

$$D_{ij} = R \quad \text{(Equation 21.38 above)}$$

Concentric neutral to an adjacent concentric neutral

$$D_{ij} = \text{center-to-center distance of the phase conductors}$$

Concentric neutral to an adjacent phase conductor

Figure 26.7 shows the relationship between the distance between centers of concentric neutral cables and the radius of a circle passing through the centers of the neutral strands.

The GMD between a concentric neutral and an adjacent phase conductor is given by the following equation:

$$D_{ij} = \sqrt[k]{D_{nm}^k - R^k} \text{ ft} \tag{26.40}$$

where D_{nm} is the center-to-center distance between phase conductors.

For cables buried in a trench, the distance between cables will be much greater than the radius R and therefore very little error is made if D_{ij} in Equation 26.40 is set equal to D_{nm}. For cables in conduit, that assumption is not valid.

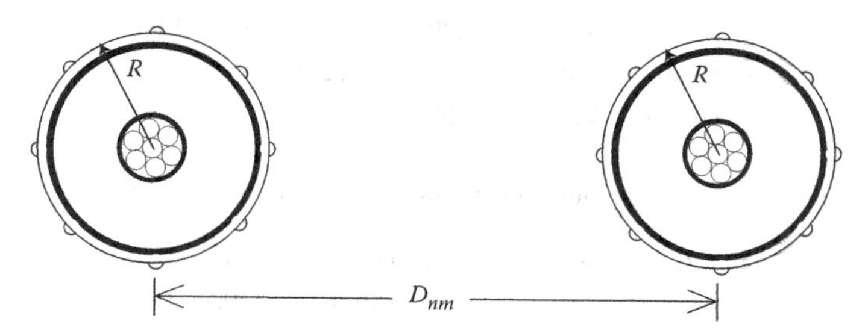

FIGURE 26.7 Distances between concentric neutral cables.

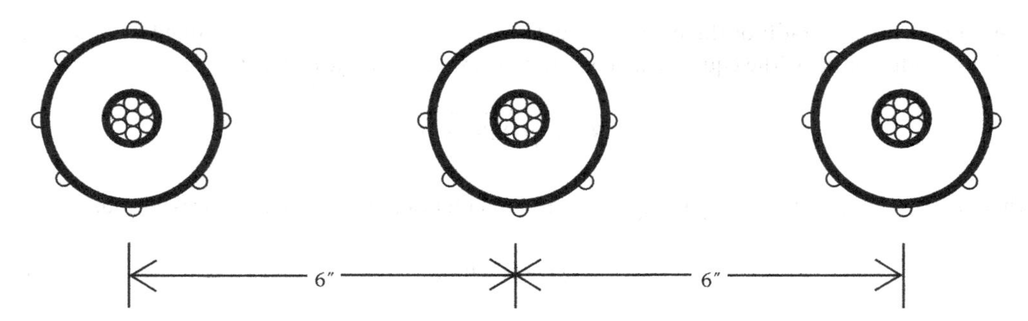

FIGURE 26.8 Three-phase concentric neutral cable spacing.

Example 26.2

Three concentric neutral cables are buried in a trench with spacings as shown in Figure 26.8. The cables are 15 kV, 250,000 CM stranded all aluminum with 13 strands of #14 annealed coated copper wires (1/3 neutral). The data for the phase conductor and neutral strands from a conductor data table are

250,000 AA phase conductor:	$GMR_p = 0.0171$ ft, resistance = 0.4100 Ω/mile
#14 copper neutral strands	$GMR_s = 0.00208$ ft, resistance = 14.87 Ω/mile
Diameter (d_s) = 0.0641 in.	

The equivalent GMR of the concentric neutral (Equation 26.37) = 0.04864 ft
The radius of the circle passing through strands (Equation 26.38) = 0.0511 ft
The equivalent resistance of the concentric neutral (Equation 26.39) = 1.1440 Ω/mile
Since R (0.0511 ft) is much less than D_{12} (0.5 ft) and D_{13} (1.0 ft), then the distances between concentric neutrals and adjacent phase conductors are the center-to-center distances of the cables.

Applying Carson's equations results in a 6 × 6 primitive impedance matrix. This matrix in partitioned form (Equation 26.12) is:

$$[z_{ij}] = \begin{bmatrix} 0.5053 + j1.4564 & 0.0953 + j1.0468 & 0.0953 + j0.9627 \\ 0.0953 + j1.0468 & 0.5053 + j1.4564 & 0.0953 + j1.0468 \\ 0.0953 + j0.9627 & 0.0953 + j1.0468 & 0.5053 + j1.4564 \end{bmatrix}$$

$$[z_{in}] = \begin{bmatrix} 0.0953 + j1.3236 & 0.0953 + j1.0468 & 0.0953 + j0.9627 \\ 0.0953 + j1.0468 & 0.0953 + j1.3236 & 0.0953 + j1.0468 \\ 0.0953 + j0.9627 & 0.0953 + j1.0468 & 0.0953 + j1.3236 \end{bmatrix}$$

$$[z_{nj}] = [z_{in}]$$

$$[z_{nn}] = \begin{bmatrix} 1.2393 + j1.3296 & 0.0953 + j1.0468 & 0.0953 + j0.9627 \\ 0.0953 + j1.0468 & 1.2393 + j1.3296 & 0.0953 + j1.0468 \\ 0.0953 + j0.9627 & 0.0953 + j1.0468 & 1.2393 + j1.3296 \end{bmatrix}$$

Using the Kron reduction (Equation 26.13) results in the phase impedance matrix

$$[z_{abc}] = \begin{bmatrix} 0.7982 + j0.4463 & 0.3192 + j0.0328 & 0.2849 - j0.0143 \\ 0.3192 + j0.0328 & 0.7891 + j0.4041 & 0.3192 + j0.0328 \\ 0.2849 - j0.0143 & 0.3192 + j0.0328 & 0.7982 + j0.4463 \end{bmatrix} \Omega/\text{mile}$$

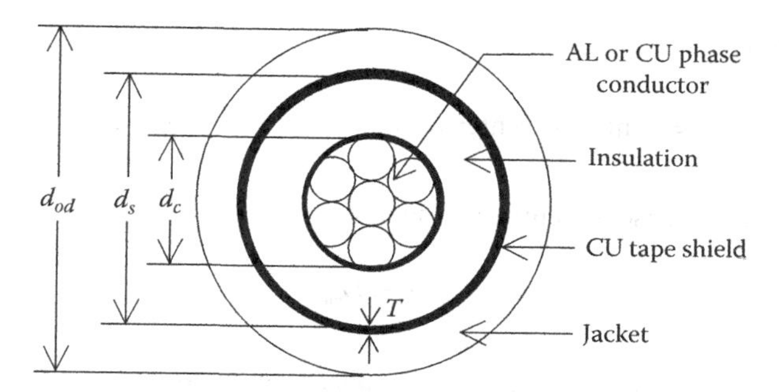

FIGURE 26.9 Taped shielded cable.

The sequence impedance matrix for the concentric neutral three-phase line is determined using Equation 26.3. The resulting sequence impedance matrix is

$$[z_{012}] = \begin{bmatrix} 1.4106 + j0.4665 & -0.0028 - j0.0081 & -0.0056 + j0.0065 \\ -0.0056 + j0.0065 & 0.4874 + j0.4151 & -0.0264 + j0.0451 \\ -0.0028 - j0.0081 & 0.0523 + j0.0003 & 0.4867 + j0.4151 \end{bmatrix} \Omega/\text{mile}$$

26.1.1.8 Tape Shielded Cables

Figure 26.9 shows a simple detail of a tape shielded cable.

Parameters of Figure 26.9 are

d_c is the diameter of phase conductor (in.)
d_s is the inside diameter of tape shield (in.)
d_{od} is the outside diameter over jacket (in.)
T is the thickness of copper tape shield in mils
= 5 mils (standard)

Once again, Carson's equations will be applied to calculate the self-impedances of the phase conductor and the tape shield as well as the mutual impedance between the phase conductor and the tape shield.

The resistance and GMR of the phase conductor are found in a standard table of conductor data.

The resistance of the tape shield is given by

$$r_{\text{shield}} = \frac{18.826}{d_s T} \Omega/\text{mile} \tag{26.41}$$

The resistance of the tape shield given in Equation 26.41 assumes a resistivity of 100 Ω m and a temperature of 50°C. The diameter of the tape shield d_s is given in inches and the thickness of the tape shield T is in mils.

The GMR of the tape shield is given by

$$\text{GMR}_{\text{shield}} = \frac{(d_s/2) - (T/2000)}{12} \text{ft} \tag{26.42}$$

The various spacings between a tape shield and the conductors and other tape shields are as follows:

Tape shield to its own phase conductor

$$D_{ij} = \text{GMR}_{\text{tape}} = \text{radius to midpoint of the shield} \tag{26.43}$$

Tape shield to an adjacent tape shield

$$D_{ij} = \text{center-to-center distance of the phase conductors} \tag{26.44}$$

Tape shield to an adjacent phase or neutral conductor

$$D_{ij} = D_{nm} \tag{26.45}$$

where D_{nm} is the center-to-center distance between phase conductors.

In applying Carson's equations for both concentric neutral and tape shielded cables, the numbering of conductors and neutrals is important. For example, a three-phase underground circuit with an additional neutral conductor must be numbered as

1 is the phase conductor #1
2 is the phase conductor #2
3 is the phase conductor #3
4 is the neutral of conductor #1
5 is the neutral of conductor #2
6 is the neutral of conductor #3
7 is the additional neutral conductor (if present).

Example 26.3

A single-phase circuit consists of a 1/0 AA tape shielded cable and a 1/0 CU neutral conductor as shown in Figure 26.10.

Cable Data: 1/0 AA
Inside diameter of tape shield = d_s = 1.084 in.
Resistance = 0.97 Ω/mile
GMR_p = 0.0111 ft
Tape shield thickness = T = 8 mils

Neutral Data: 1/0 Copper, 7 strand
Resistance = 0.607 Ω/mile
GMR_n = 0.01113 ft
Distance between cable and neutral = D_{nm} = 3 in.

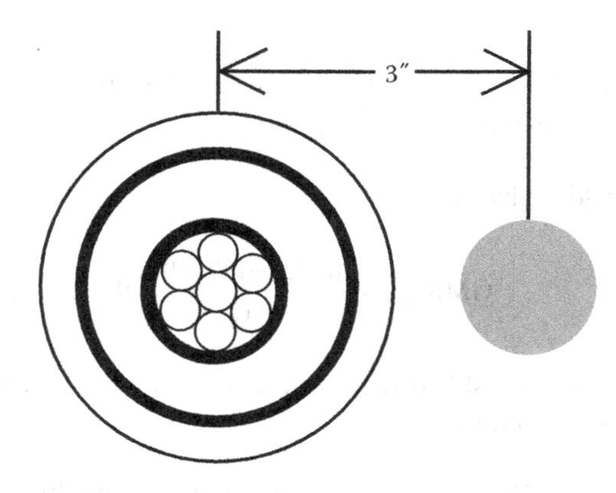

FIGURE 26.10 Single-phase tape shield with neutral.

The resistance of the tape shield is computed according to Equation 26.41:

$$r_{shield} = \frac{18.826}{d_s T} = \frac{18.826}{1.084 \times 8} = 2.1705\,\Omega/\text{mile}$$

The GMR of the tape shield is computed according to Equation 26.42:

$$GMR_{shield} = \frac{(d_s/2) - (T/2000)}{12} = \frac{(1.084/2) - (8/2000)}{12} = 0.0455\,\text{ft}$$

Using the relations defined in Equations 26.43 through 26.45 and Carson's equations results in a 3 × 3 primitive impedance matrix:

$$z_{primitive} = \begin{bmatrix} 1.0653 + j1.5088 & 0.0953 + j1.3377 & 0.0953 + j1.1309 \\ 0.0953 + j1.3377 & 2.2658 + j1.3377 & 0.0953 + j1.1309 \\ 0.0953 + j1.1309 & 0.0953 + j1.1309 & 0.7023 + j1.5085 \end{bmatrix} \Omega/\text{mile}$$

Applying Kron's reduction method will result in a single impedance that represents the equivalent single-phase impedance of the tape shield cable and the neutral conductor.

$$zl_p = 1.3368 + j0.6028\,\Omega/\text{mile}$$

26.1.2 Shunt Admittance

When a high-voltage transmission line is less than 50 miles in length, the shunt capacitance of the line is typically ignored. For lightly loaded distribution lines, particularly underground lines, the shunt capacitance should be modeled.

The basic equation for the relationship between the charge on a conductor to the voltage drop between the conductor and ground is given by

$$Q_n = C_{ng} V_{ng} \tag{26.46}$$

where

Q_n is the charge on the conductor
C_{ng} is the capacitance between the conductor and ground
V_{ng} is the voltage between the conductor and ground

For a line consisting of ncond (number of phase plus number of neutral) conductors, Equation 26.46 can be written in condensed matrix form as

$$[Q] = [C][V] \tag{26.47}$$

where

$[Q]$ is the column vector of order ncond
$[C]$ is the ncond × ncond matrix
$[V]$ is the column vector of order ncond

Equation 26.47 can be solved for the voltages

$$[V] = [C]^{-1}[Q] = [P][Q] \tag{26.48}$$

where

$$[P] = [C]^{-1} \tag{26.49}$$

26.1.2.1 Overhead Lines

The determination of the shunt admittance of overhead lines starts with the calculation of the "potential coefficient matrix" (Glover and Sarma, 1994). The elements of the matrix are determined by

$$P_{ii} = 11.17689 \times \ln \frac{S_{ii}}{RD_i} \tag{26.50}$$

$$P_{ij} = 11.17689 \times \ln \frac{S_{ij}}{D_{ij}} \tag{26.51}$$

See Figure 26.2 for the following definitions.

S_{ii} is the distance between a conductor and its image below ground (ft)
S_{ij} is the distance between conductor i and the image of conductor j below ground (ft)
D_{ij} is the overhead spacing between two conductors (ft)
RD_i is the radius of conductor i (ft)

The potential coefficient matrix will be an ncond × ncond matrix. If one or more of the conductors is a grounded neutral, then the matrix must be reduced using the Kron method to an $n$phase × $n$phase matrix $[P_{abc}]$.

The inverse of the potential coefficient matrix will give the $n$phase × $n$phase capacitance matrix $[C_{abc}]$. The shunt admittance matrix is given by

$$[y_{abc}] = j\omega[C_{abc}]\mu\text{S/mile} \tag{26.52}$$

where $\omega = 2\pi f = 376.9911$.

Example 26.4

Determine the shunt admittance matrix for the overhead line of Example 26.1. Assume that the neutral conductor is 25 ft above ground.

Solution

For this configuration, the image spacing matrix is computed to be

$$[S] = \begin{bmatrix} 58 & 58.0539 & 58.4209 & 54.1479 \\ 58.0539 & 58 & 58.1743 & 54.0208 \\ 58.4209 & 58.1743 & 58 & 54.0833 \\ 54.1479 & 54.0208 & 54.0835 & 50 \end{bmatrix} \text{ft}$$

The primitive potential coefficient matrix is computed to be

$$[P_{\text{primitive}}] = \begin{bmatrix} 84.56 & 35.1522 & 23.7147 & 25.2469 \\ 35.4522 & 84.56 & 28.6058 & 28.359 \\ 23.7147 & 28.6058 & 84.56 & 26.6131 \\ 25.2469 & 28.359 & 26.6131 & 85.6659 \end{bmatrix}$$

Kron reduce to a 3 × 3 matrix

$$[P]=\begin{bmatrix}77.1194 & 26.7944 & 15.8714\\ 26.7944 & 75.172 & 19.7957\\ 15.8714 & 19.7957 & 76.2923\end{bmatrix}$$

Invert [P] to determine the shunt capacitance matrix

$$[Y_{abc}]=j376.9911[C_{abc}]=\begin{bmatrix}j5.6711 & -j1.8362 & -j0.7033\\ -j1.8362 & j5.9774 & -j1.169\\ -j0.7033 & -j1.169 & j5.391\end{bmatrix}\mu\text{S/mile}$$

Multiply $[C_{abc}]$ by the radian frequency to determine the final three-phase shunt admittance matrix.

26.1.2.2 Underground Lines

Because the electric fields of underground cables are confined to the space between the phase conductor and its concentric neutral to tape shield, the calculation of the shunt admittance matrix requires only the determination of the "self" admittance terms.

26.1.2.3 Concentric Neutral

The self-admittance in μS/mile for a concentric neutral cable is given by

$$Y_{cn}=j\frac{77.582}{\ln(R_b/R_a)-(1/k)\ln(kR_n/R_b)} \tag{26.53}$$

where

R_b is the radius of a circle to center of concentric neutral strands (ft)
R_a is the radius of phase conductor (ft)
R_n is the radius of concentric neutral strand (ft)
k is the number of concentric neutral strands

Example 26.5

Determine the three-phase shunt admittance matrix for the concentric neutral line of Example 26.2.

Solution

$$R_b=R=0.0511\text{ ft}$$

Diameter of the 250,000 AA phase conductor = 0.567 in.

$$R_b=\frac{0.567}{24}=0.0236\text{ ft}$$

Diameter of the #14 CU concentric neutral strand = 0.0641 in.

$$R_n=\frac{0.0641}{24}=0.0027\text{ ft}$$

Substitute into Equation 26.53:

$$Y_{cn} = j\frac{77.582}{\ln(R_b/R_a)-(1/k)\ln(kR_n/R_b)} = j\frac{77.582}{\ln(0.0511/0.0236)-\frac{1}{13}\ln((13\times 0.0027)/0.0511)} = j96.8847$$

The three-phase shunt admittance matrix is:

$$[Y_{abc}] = \begin{bmatrix} j96.8847 & 0 & 0 \\ 0 & j96.8847 & 0 \\ 0 & 0 & j96.8847 \end{bmatrix} \mu\text{S/mile}$$

26.1.2.4 Tape Shield Cable

The shunt admittance in μS/mile for tape shielded cables is given by

$$Y_{ts} = j\frac{77.586}{\ln(R_b/R_a)}\ \mu\text{S/mile} \quad (26.54)$$

where
R_b is the inside radius of the tape shield
R_a is the radius of phase conductor

Example 26.6

Determine the shunt admittance of the single-phase tape shielded cable of Example 26.3 in Section 26.1.1.

Solution

$$R_b = \frac{d_s}{24} = \frac{1.084}{24} = 0.0452$$

The diameter of the 1/0 AA phase conductor = 0.368 in.

$$R_a = \frac{d_p}{24} = \frac{0.368}{24} = 0.0153$$

Substitute into Equation 26.54:

$$Y_{ts} = j\frac{77.586}{\ln(R_b/R_a)} = j\frac{77.586}{\ln(0.0452/0.0153)} = j71.8169\ \mu\text{S/mile}$$

26.1.3 Line Segment Models

26.1.3.1 Exact Line Segment Model

The exact model of a three-phase line segment is shown in Figure 26.11. For the line segment in Figure 26.11, the equations relating the input (node n) voltages and currents to the output (node m) voltages and currents are

$$[VLG_{abc}]_n = [a][VLG_{abc}]_m + [b][I_{abc}]_m \quad (26.55)$$

$$[I_{abc}]_n = [c][VLG_{abc}]_m + [d][I_{abc}]_m \quad (26.56)$$

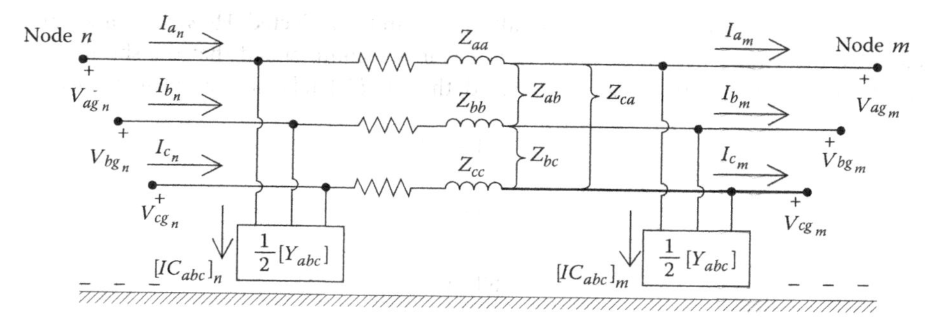

FIGURE 26.11 Three-phase line segment model.

where

$$[a] = [U] - \frac{1}{2}[Z_{abc}][Y_{abc}] \tag{26.57}$$

$$[b] = [Z_{abc}] \tag{26.58}$$

$$[c] = [Y_{abc}] - \frac{1}{4}[Z_{abc}][Y_{abc}]^2 \tag{26.59}$$

$$[d] = [U] - \frac{1}{2}[Z_{abc}][Y_{abc}] \tag{26.60}$$

In Equations 26.57 through 26.60, the impedance matrix $[Z_{abc}]$ and the admittance matrix $[Y_{abc}]$ are defined earlier in this document.

Sometimes it is necessary to determine the voltages at node m as a function of the voltages at node n and the output currents at node m. The necessary equation is

$$[VLG_{abc}]_m = [A][VLG_{abc}]_n - [B][I_{abc}]_m \tag{26.61}$$

where

$$[A] = \left([U] + \frac{1}{2}[Z_{abc}][Y_{abc}]\right)^{-1} \tag{26.62}$$

$$[B] = \left([U] + \frac{1}{2}[Z_{abc}][Y_{abc}]\right)^{-1}[Z_{abc}] \tag{26.63}$$

$$[U] = \begin{bmatrix} 1 & 0 & 0 \\ 0 & 1 & 0 \\ 0 & 0 & 1 \end{bmatrix} \tag{26.64}$$

In many cases the shunt admittance is so small that it can be neglected. However, for all underground cables and for overhead lines longer than 15 miles, it is recommended that the shunt admittance be included. When the shunt admittance is neglected, the $[a]$, $[b]$, $[c]$, $[d]$, $[A]$, and $[B]$ matrices become

$$[a]=[U] \tag{26.65}$$

$$[b]=[Z_{abc}] \tag{26.66}$$

$$[c]=[0] \tag{26.67}$$

$$[d]=[U] \tag{26.68}$$

$$[A]=[U] \tag{26.69}$$

$$[B]=[Z_{abc}] \tag{26.70}$$

When the shunt admittance is neglected, Equations 26.55, 26.56, and 26.61 become

$$[VLG_{abc}]_n=[VLG_{abc}]_m+[Z_{abc}][I_{abc}]_m \tag{26.71}$$

$$[I_{abc}]_n=[I_{abc}]_m \tag{26.72}$$

$$[VLG_{abc}]_m=[VLG_{abc}]_n-[Z_{abc}][I_{abc}]_m \tag{26.73}$$

If an accurate determination of the voltage drops down a line segment is to be made, it is essential that the phase impedance matrix $[Z_{abc}]$ be computed based upon the actual configuration and phasing of the overhead or underground lines. No assumptions should be made, such as transposition. The reason for this is best demonstrated by an example.

Example 26.7

The phase impedance matrix for the line configuration in Example 26.1 was computed to be

$$[z_{abc}]=\begin{bmatrix} 0.4576+j1.0780 & 0.1560+j0.5017 & 0.1535+j0.3849 \\ 0.1560+j0.5017 & 0.4666+j1.0482 & 0.1580+j0.4236 \\ 0.1535+j0.3849 & 0.1580+j0.4236 & 0.4615+j1.0651 \end{bmatrix}\Omega/\text{mile}$$

Assume that a 12.47 kV substation serves a load 1.5 miles from the substation. The metered output at the substation is balanced 10,000 kVA at 12.47 kV and 0.9 lagging power factor. Compute the three-phase line-to-ground voltages at the load end of the line and the voltage unbalance at the load.

Solution

The line-to-ground voltages and line currents at the substation are

$$[VLG_{abc}]=\begin{bmatrix} 7200\angle 0 \\ 7200\angle -120 \\ 7200\angle 120 \end{bmatrix} \quad [I_{abc}]_n=\begin{bmatrix} 463\angle -25.84 \\ 463\angle -145.84 \\ 463\angle 94.16 \end{bmatrix}$$

Solve Equation 26.71 for the load voltages:

$$[VLG_{abc}]_m = [VLG_{abc}]_n - 1.5[Z_{abc}][I_{abc}]_n = \begin{bmatrix} 6761.10\angle 2.32 \\ 6877.7\angle -122.43 \\ 6836.33\angle 117.21 \end{bmatrix}$$

The voltage unbalance at the load using the NEMA definition is

$$V_{\text{unbalance}} = \frac{\max(V_{\text{deviation}})}{V_{\text{avg}}} 100 = 0.937\%$$

The point of Example 26.7 is to demonstrate that even though the system is perfectly balanced at the substation, the unequal mutual coupling between the phases results in a significant voltage unbalance at the load; significant because NEMA requires that induction motors be derated when the voltage unbalance is 1% or greater.

26.1.3.2 Approximate Line Segment Model

Many times the only data available for a line segment will be the positive and zero sequence impedances. An approximate three-phase line segment model can be developed by applying the "reverse impedance transformation" from symmetrical component theory.

Using the known positive and zero sequence impedances, the "sequence impedance matrix" is given by

$$[Z_{\text{seq}}] = \begin{bmatrix} Z_0 & 0 & 0 \\ 0 & Z_+ & 0 \\ 0 & 0 & Z_+ \end{bmatrix} \tag{26.74}$$

The reverse impedance transformation results in the following "approximate phase impedance matrix:"

$$[Z_{\text{approx}}] = [A_s][Z_{\text{seq}}][A_s]^{-1} = \frac{1}{3}\begin{bmatrix} (2Z_+ - Z_0) & (Z_0 - Z_+) & (Z_0 - Z_+) \\ (Z_0 - Z_+) & (2Z_+ - Z_0) & (Z_0 - Z_+) \\ (Z_0 - Z_+) & (Z_0 - Z_+) & (2Z_+ - Z_0) \end{bmatrix} \tag{26.75}$$

Notice that the approximate phase impedance matrix is characterized by the three diagonal terms being equal and all mutual terms being equal. This is the same result that is achieved if the line is assumed to be transposed. Substituting the approximate phase impedance matrix into Equation 26.71 results in

$$\begin{bmatrix} V_{an} \\ V_{bn} \\ V_{cn} \end{bmatrix}_n = \begin{bmatrix} V_{an} \\ V_{bn} \\ V_{cn} \end{bmatrix}_m + \frac{1}{3}\begin{bmatrix} (2Z_+ - Z_0) & (Z_0 - Z_+) & (Z_0 - Z_+) \\ (Z_0 - Z_+) & (2Z_+ - Z_0) & (Z_0 - Z_+) \\ (Z_0 - Z_+) & (Z_0 - Z_+) & (2Z_+ - Z_0) \end{bmatrix}\begin{bmatrix} I_a \\ I_b \\ I_c \end{bmatrix}_n \tag{26.76}$$

Equation 26.76 can be expanded and an equivalent circuit for the approximate line segment model can be developed. This approximate model is shown in Figure 26.12.

The errors made by using this approximate line segment model are demonstrated in the following example.

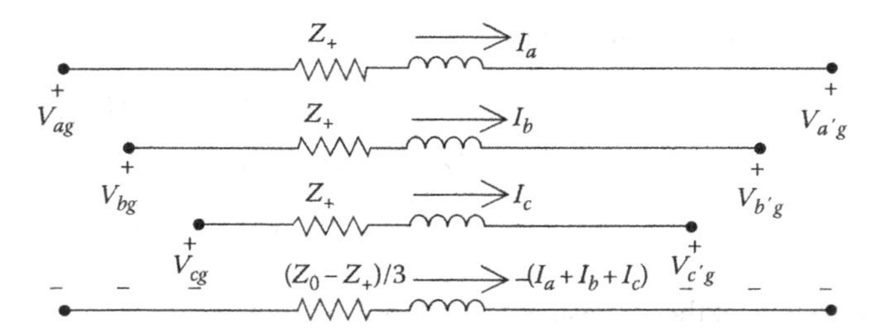

FIGURE 26.12 Approximate line segment model.

Example 26.8

For the line of Example 26.7, the positive and zero sequence impedances were determined to be

$$Z_+ = 0.3061 + j0.6270\,\Omega/\text{mile}$$

$$Z_0 = 0.7735 + j1.9373\,\Omega/\text{mile}$$

Solution

The sequence impedance matrix is

$$[z_{\text{seq}}] = \begin{bmatrix} 0.7735 + j1.9373 & 0 & 0 \\ 0 & 0.3061 + j0.6270 & 0 \\ 0 & 0 & 0.3061 + j0.6270 \end{bmatrix}$$

Performing the reverse impedance transformation results in the approximate phase impedance matrix.

$$[z_{\text{approx}}] = [A_s][z_{\text{seq}}][A_s]^{-1} = \begin{bmatrix} 0.4619 + j1.0638 & 0.1558 + j0.4368 & 0.1558 + j0.4368 \\ 0.1558 + j0.4368 & 0.4619 + j1.0638 & 0.1558 + j0.4368 \\ 0.1558 + j0.4368 & 0.1558 + j0.4368 & 0.4619 + j1.0638 \end{bmatrix}$$

Note in the approximate phase impedance matrix that the three diagonal terms are equal and all of the mutual terms are equal.

Use the approximate impedance matrix to compute the load voltage and voltage unbalance as specified in Example 26.1.

Note that the voltages are computed to be balanced. In the previous example it was shown that when the line is modeled accurately, there is a voltage unbalance of almost 1%.

26.1.4 Step-Voltage Regulators

A step-voltage regulator consists of an autotransformer and a LTC mechanism. The voltage change is obtained by changing the taps of the series winding of the autotransformer. The position of the tap is determined by a control circuit (line drop compensator). Standard step regulators contain a reversing switch enabling a ±10% regulator range, usually in 32 steps. This amounts to a 5/8% change per step or 0.75 V change per step on a 120 V base.

A type B step-voltage regulator is shown in Figure 26.13. There is also a type A step-voltage regulator where the load and source sides of the regulator are reversed from that shown in Figure 26.13. Since the type B regulator is more common, the remainder of this section will address the type B step-voltage regulator.

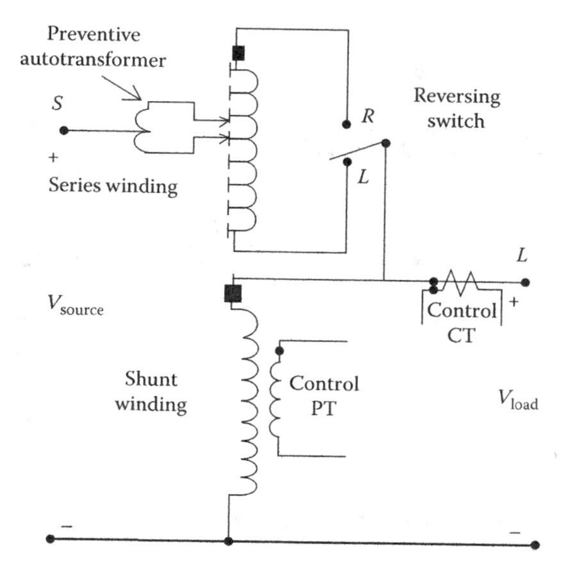

FIGURE 26.13 Type B step-voltage regulator.

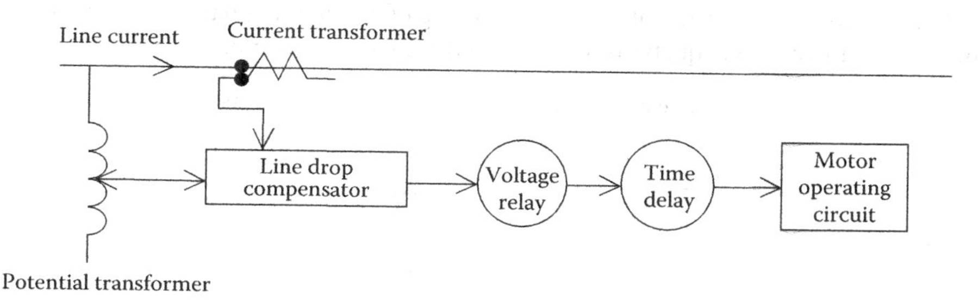

FIGURE 26.14 Regulator control circuit.

The tap changing is controlled by a control circuit shown in the block diagram of Figure 26.14. The control circuit requires the following settings:

1. Voltage level: The desired voltage (on 120 V base) to be held at the "load center." The load center may be the output terminal of the regulator or a remote node on the feeder.
2. Bandwidth: The allowed variance of the load center voltage from the set voltage level. The voltage held at the load center will be ±1/2 of the bandwidth. For example, if the voltage level is set to 122 V and the bandwidth set to 2 V, the regulator will change taps until the load center voltage lies between 121 and 123 V.
3. Time delay: Length of time that a raise or lower operation is called for before the actual execution of the command. This prevents taps changing during a transient or short time change in current.
4. Line drop compensator: Set to compensate for the voltage drop (line drop) between the regulator and the load center. The settings consist of *R* and *X* settings in volts corresponding to the equivalent impedance between the regulator and the load center. This setting may be zero if the regulator output terminals are the load center.

The rating of a regulator is based on the kVA transformed, not the kVA rating of the line. In general this will be 10% of the line rating since rated current flows through the series winding that represents the ±10% voltage change.

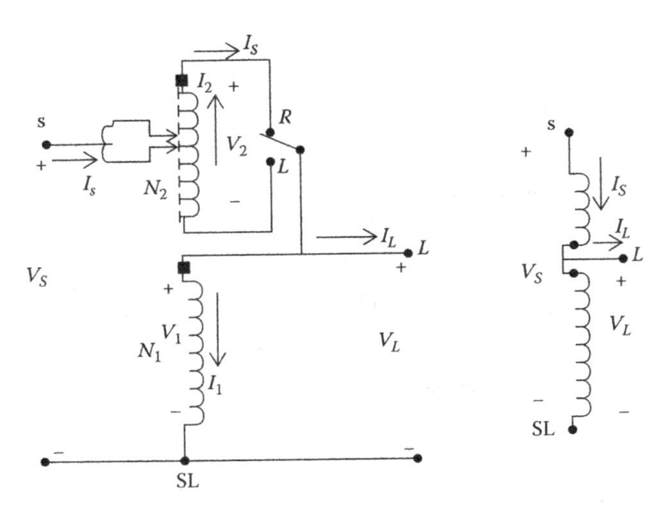

FIGURE 26.15 Type B voltage regulator in the raise position.

26.1.4.1 Voltage Regulator in the Raise Position

Figure 26.15 shows a detailed and abbreviated drawing of a type B regulator in the raise position. The defining voltage and current equations for the type B regulator in the raise position are as follows:

Voltage equations	Current equations	
$\frac{V_1}{N_1} = \frac{V_2}{N_2}$	$N_1 I_1 = N_2 I_2$	(26.77)
$V_S = V_1 - V_2$	$I_L = I_S - I_1$	(26.78)
$V_L = V_1$	$I_2 = -I_S$	(26.79)
$V_2 = \frac{N_2}{N_1} V_1 = \frac{N_2}{N_1} V_L$	$I_1 = \frac{N_2}{N_1} I_2 = \frac{N_2}{N_1} I_S$	(26.80)
$V_S = \left(1 - \frac{N_2}{N_1}\right) V_l$	$I_L = \left(1 - \frac{N_2}{N_1}\right) I_S$	(26.81)
$V_S = a_R V_L$	$I_L = a_R I_S$	(26.82)
$a_R = 1 - \frac{N_2}{N_1}$		(26.83)

Equations 26.82 and 26.83 are the necessary defining equations for modeling a regulator in the raise position.

26.1.4.2 Voltage Regulator in the Lower Position

Figure 26.16 shows the detailed and abbreviated drawings of a regulator in the lower position. Note in the figure that the only difference between the lower and the raise models is that the polarity of the series winding and how it is connected to the shunt winding is reversed.

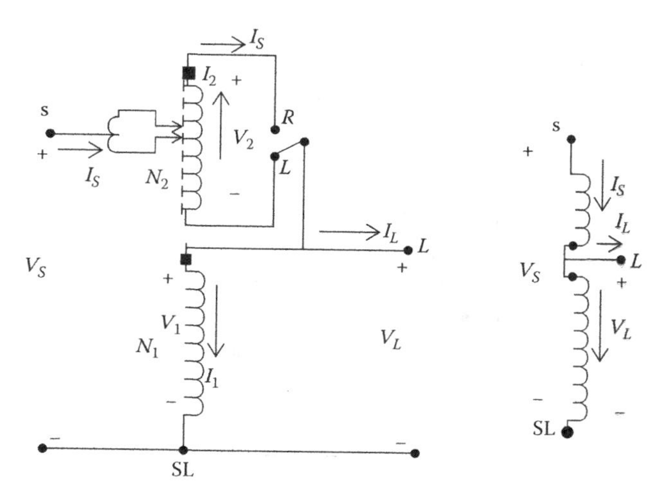

FIGURE 26.16 Type B regulator in the lower position.

The defining voltage and current equations for a regulator in the lower position are as follows:

Voltage equations / Current equations

$$\frac{V_1}{N_1} = \frac{V_2}{N_2} \qquad N_1 I_1 = N_2 I_2 \tag{26.84}$$

$$V_S = V_1 + V_2 \qquad I_L = I_S - I_1 \tag{26.85}$$

$$V_L = V_1 \qquad I_2 = -I_S \tag{26.86}$$

$$V_2 = \frac{N_2}{N_1} V_1 = \frac{N_2}{N_1} V_L \qquad I_1 = \frac{N_2}{N_1} I_2 = \frac{N_2}{N_1}(-I_S) \tag{26.87}$$

$$V_S = \left(1 + \frac{N_2}{N_1}\right) V_1 \qquad I_L = \left(1 - \frac{N_2}{N_1}\right) I_S \tag{26.88}$$

$$V_S = a_R V_L \qquad I_L = a_R I_S \tag{26.89}$$

$$a_R = 1 + \frac{N_2}{N_1} \tag{26.90}$$

Equations 26.83 and 26.90 give the value of the effective regulator ratio as a function of the ratio of the number of turns on the series winding (N_2) to the number of turns on the shunt winding (N_1). The actual turns ratio of the windings is not known. However, the particular position will be known. Equations 26.83 and 26.90 can be modified to give the effective regulator ratio as a function of the tap position. Each tap changes the voltage by 5/8% or 0.00625 per unit. On a 120 V base, each step change results in a change of voltage of 0.75 V. The effective regulator ratio can be given by

$$a_R = 1 \mp 0.00625 \cdot Tap \tag{26.91}$$

In Equation 26.91, the minus sign applies to the "raise" position and the positive sign for the "lower" position.

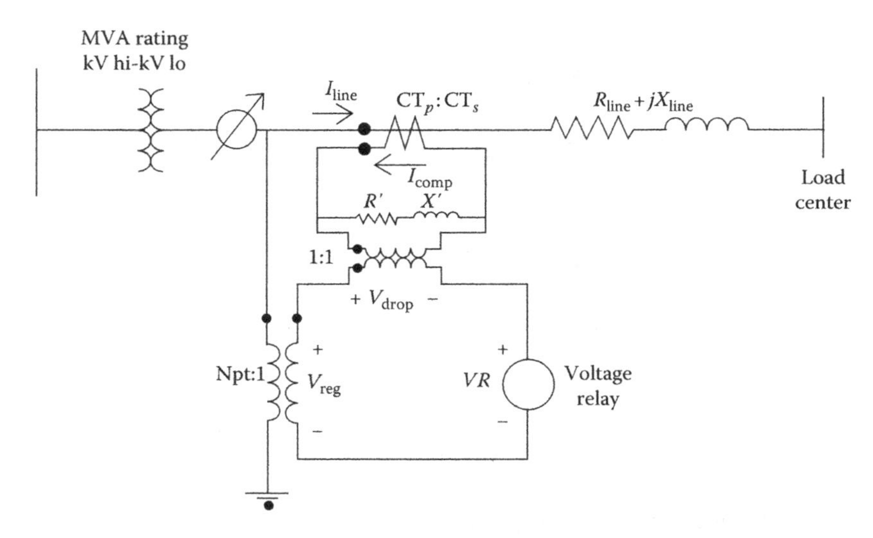

FIGURE 26.17 Line drop compensator circuit.

26.1.4.3 Line Drop Compensator

The changing of taps on a regulator is controlled by the "line drop compensator." Figure 26.17 shows a simplified sketch of the compensator circuit and how it is connected to the circuit through a potential transformer and a current transformer.

The purpose of the line drop compensator is to model the voltage drop of the distribution line from the regulator to the load center. Typically, the compensator circuit is modeled on a 120 V base. This requires the potential transformer to transform rated voltage (line-to-neutral or line-to-line) down to 120 V. The current transformer turns ratio ($CT_p:CT_s$) where the primary rating (CT_p) will typically be the rated current of the feeder. The setting that is most critical is that of R' and X'. These values must represent the equivalent impedance from the regulator to the load center. Knowing the equivalent impedance in Ohms from the regulator to the load center (R_{line_ohms} and X_{line_ohms}), the required value for the compensator settings are calibrated in volts and determined by

$$R'_{volts} + jX'_{volts} = (R_{line\_ohms} + jX_{line\_ohms}) \cdot \frac{Ct_p}{N_{pt}} \text{V} \tag{26.92}$$

The value of the compensator settings in ohms is determined by

$$R'_{ohms} + jX'_{ohms} = \frac{R'_{volts} + jX'_{volts}}{Ct_s} \Omega \tag{26.93}$$

It is important to understand that the value of $R_{line\_ohms} + jX_{line\_ohms}$ is not the impedance of the line between the regulator and the load center. Typically the load center is located down the primary main feeder after several laterals have been tapped. As a result, the current measured by the CT of the regulator is not the current that flows all the way from the regulator to the load center. The proper way to determine the line impedance values is to run a power-flow program of the feeder without the regulator operating. From the output of the program, the voltages at the regulator output and the load center are known. Now the "equivalent" line impedance can be computed as

$$R_{line} + jX_{line} = \frac{V_{regulator\_output} - V_{load\_center}}{I_{line}} \Omega \tag{26.94}$$

In Equation 26.94, the voltages must be specified in system volts and the current in system amps.

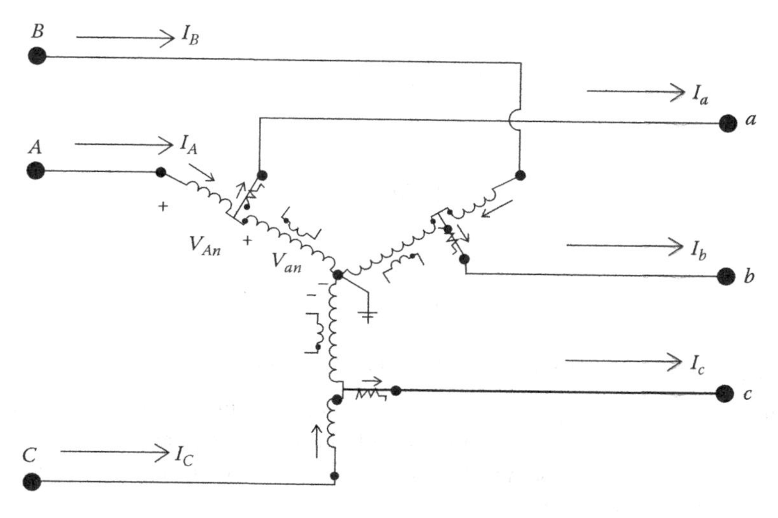

FIGURE 26.18 Wye connected type B regulators.

26.1.4.4 Wye Connected Regulators

Three single-phase regulators connected in wye are shown in Figure 26.18. In Figure 26.18 the polarities of the windings are shown in the raise position. When the regulator is in the lower position, a reversing switch will have reconnected the series winding so that the polarity on the series winding is now at the output terminal.

Regardless of whether the regulator is raising or lowering the voltage, the following equations apply.

26.1.4.5 Voltage Equations

$$\begin{bmatrix} V_{An} \\ V_{Bn} \\ V_{Cn} \end{bmatrix} = \begin{bmatrix} a_{R\_a} & 0 & 0 \\ 0 & a_{R\_b} & 0 \\ 0 & 0 & a_{R\_c} \end{bmatrix} \begin{bmatrix} V_{an} \\ V_{bn} \\ V_{cn} \end{bmatrix} \tag{26.95}$$

Equation 26.95 can be written in condensed form as

$$[VLN_{ABC}] = [aRV_{abc}][VLN_{abc}] \tag{26.96}$$

also

$$[VLN_{abc}] = [aRV_{ABC}][VLN_{ABC}] \tag{26.97}$$

where

$$[aRV_{ABC}] = [aRV_{abc}]^{-1} \tag{26.98}$$

26.1.4.6 Current Equations

$$\begin{bmatrix} I_A \\ I_B \\ I_C \end{bmatrix} = \begin{bmatrix} \frac{1}{a_{R\_a}} & 0 & 0 \\ 0 & \frac{1}{a_{R\_b}} & 0 \\ 0 & 0 & \frac{1}{a_{R\_c}} \end{bmatrix} \begin{bmatrix} I_a \\ I_b \\ I_c \end{bmatrix} \tag{26.99}$$

or

$$[I_{ABC}] = [aRI_{abc}][I_{abc}] \tag{26.100}$$

also

$$[I_{abc}] = [aRI_{ABC}][I_{ABC}] \tag{26.101}$$

where

$$[aRI_{ABC}] = [aRI_{abc}]^{-1} \tag{26.102}$$

where $0.9 \le a_{R\_abc} \le 1.1$ in 32 steps of 0.625% per step (0.75 V/step on 120 V base).

Note: The effective turn ratios (a_{R_a}, a_{R_b}, and a_{R_c}) can take on different values when three single-phase regulators are connected in wye. It is also possible to have a three-phase regulator connected in wye where the voltage and current are sampled on only one phase and then all three phases are changed by the same value of aR (number of taps).

26.1.4.7 Closed Delta Connected Regulators

Three single-phase regulators can be connected in a closed delta as shown in Figure 26.19. In the figure, the regulators are shown in the raise position. The closed delta connection is typically used in three-wire delta feeders. Note that the potential transformers for this connection are monitoring the load side line-to-line voltages and the current transformers are monitoring the load side line currents.

Applying the basic voltage and current Equations 26.77 through 26.83 of the regulator in the raise position, the following voltage and current relations are derived for the closed delta connection.

$$\begin{bmatrix} V_{AB} \\ V_{BC} \\ V_{CA} \end{bmatrix} = \begin{bmatrix} a_{R\_ab} & 1-a_{R\_bc} & 0 \\ 0 & a_{R\_bc} & 1-a_{R\_ca} \\ 1-a_{R\_ab} & 0 & a_{R\_ca} \end{bmatrix} \begin{bmatrix} V_{ab} \\ V_{bc} \\ V_{ca} \end{bmatrix} \tag{26.103}$$

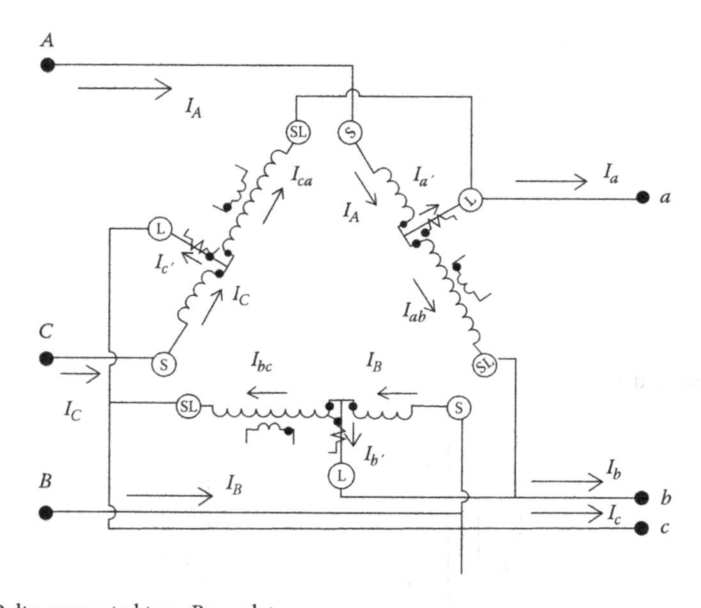

FIGURE 26.19 Delta connected type B regulators.

Equation 26.101 in abbreviated form can be written as

$$[VLL_{ABC}] = [aRVD_{abc}][VLL_{abc}] \tag{26.104}$$

When the load side voltages are known, the source side voltages can be determined by

$$[VLL_{abc}] = [aRVD_{ABC}][VLL_{ABC}] \tag{26.105}$$

where

$$[aRVD_{ABC}] = [aRVD_{abc}]^{-1} \tag{26.106}$$

In a similar manner, the relationships between the load side and source side line currents are given by

$$\begin{bmatrix} I_a \\ I_b \\ I_c \end{bmatrix} = \begin{bmatrix} a_{R\_ab} & 0 & 1-a_{R\_ca} \\ 1-a_{R\_ab} & a_{R\_bc} & 0 \\ 0 & 1-a_{R\_bc} & a_{R\_ca} \end{bmatrix} \begin{bmatrix} I_A \\ I_B \\ I_C \end{bmatrix} \tag{26.107}$$

or

$$[I_{abc}] = [AID_{ABC}][I_{ABC}] \tag{26.108}$$

also

$$[I_{ABC}] = [AID_{abc}][I_{abc}] \tag{26.109}$$

where

$$[IAD_{abc}] = [IAD_{ABC}]^{-1} \tag{26.110}$$

The closed delta connection can be difficult to apply. Note in both the voltage and current equations that a change of the tap position in one regulator will affect voltages and currents in two phases. As a result, increasing the tap in one regulator will affect the tap position of the second regulator. In most cases the bandwidth setting for the closed delta connection will have to be wider than that for wye connected regulators.

26.1.4.8 Open Delta Connection

Two single-phase regulators can be connected in the "open" delta connection. Shown in Figure 26.20 is an open delta connection where two single-phase regulators have been connected between phases *AB* and *CB*.

Two other open connections can also be made where the single-phase regulators are connected between phases *BC* and *AC* and also between phases *CA* and *BA*.

The open delta connection is typically applied to three-wire delta feeders. Note that the potential transformers monitor the line-to-line voltages and the current transformers monitor the line currents. Once again, the basic voltage and current relations of the individual regulators are used to determine the relationships between the source side and load side voltages and currents.

For all three open connections, the following general equations will apply:

$$[VLL_{ABC}] = [aRV_{abc}][VLL_{abc}] \tag{26.111}$$

$$[VLL_{abc}] = [aRV_{ABC}][VLL_{ABC}] \tag{26.112}$$

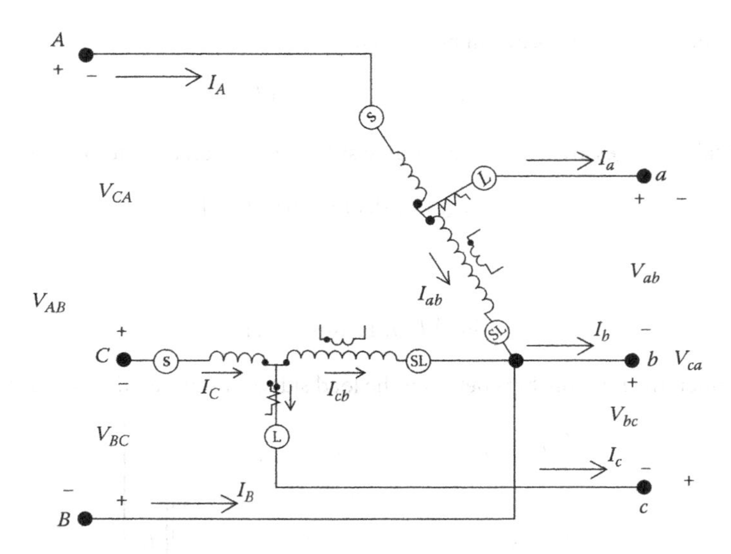

FIGURE 26.20 Open delta type B regulator connection.

$$[I_{ABC}] = [aRI_{abc}][I_{abc}] \tag{26.113}$$

$$[I_{abc}] = [aRI_{ABC}][I_{ABC}] \tag{26.114}$$

The matrices for the three open connections are defined as follows:

Phases *AB* and *CB*

$$[aRV_{abc}] = \begin{bmatrix} a_{R\_A} & 0 & 0 \\ 0 & a_{R\_C} & 0 \\ -a_{R\_A} & -a_{R\_C} & 0 \end{bmatrix} \tag{26.115}$$

$$[aRV_{ABC}] = \begin{bmatrix} \frac{1}{a_{R\_A}} & 0 & 0 \\ 0 & \frac{1}{a_{R\_C}} & 0 \\ -\frac{1}{a_{R\_A}} & -\frac{1}{a_{R\_C}} & 0 \end{bmatrix} \tag{26.116}$$

$$[aRI_{abc}] = \begin{bmatrix} \frac{1}{a_{R\_A}} & 0 & 0 \\ -\frac{1}{a_{R\_A}} & 0 & -\frac{1}{a_{R\_C}} \\ 0 & 0 & \frac{1}{a_{R\_C}} \end{bmatrix} \tag{26.117}$$

$$[aRI_{ABC}] = \begin{bmatrix} a_{R\_A} & 0 & 0 \\ -a_{R\_A} & 0 & a_{R\_C} \\ 0 & 0 & a_{R\_C} \end{bmatrix} \tag{26.118}$$

Phases *BC* and *AC*

$$[aRV_{abc}]=\begin{bmatrix}0 & -a_{R\_B} & -a_{R\_A}\\ 0 & a_{R\_B} & 0\\ 0 & 0 & a_{R\_A}\end{bmatrix} \quad (26.119)$$

$$[aRV_{ABC}]=\begin{bmatrix}0 & -\frac{1}{a_{R\_B}} & -\frac{1}{a_{R\_A}}\\ 0 & \frac{1}{a_{R\_B}} & 0\\ 0 & 0 & \frac{1}{a_{R\_A}}\end{bmatrix} \quad (26.120)$$

$$[aRI_{abc}]=\begin{bmatrix}\frac{1}{a_{R\_A}} & 0 & 0\\ 0 & \frac{1}{a_{R\_B}} & 0\\ -\frac{1}{a_{R\_A}} & -\frac{1}{a_{R\_B}} & 0\end{bmatrix} \quad (26.121)$$

$$[aRI_{ABC}]=\begin{bmatrix}a_{R\_A} & 0 & 0\\ 0 & a_{R\_B} & 0\\ -a_{R\_A} & -a_{R\_B} & 0\end{bmatrix} \quad (26.122)$$

Phases *CA* and *BA*

$$[aRV_{abc}]=\begin{bmatrix}a_{R\_B} & 0 & 0\\ -a_{R\_B} & 0 & -a_{R\_C}\\ 0 & 0 & a_{R\_C}\end{bmatrix} \quad (26.123)$$

$$[aRV_{ABC}]=\begin{bmatrix}\frac{1}{a_{R\_B}} & 0 & 0\\ -\frac{1}{a_{R\_B}} & 0 & -\frac{1}{a_{R\_C}}\\ 0 & 0 & \frac{1}{a_{R\_C}}\end{bmatrix} \quad (26.124)$$

$$[aRI_{abc}]=\begin{bmatrix}0 & -\frac{1}{a_{R\_B}} & -\frac{1}{a_{R\_C}}\\ 0 & \frac{1}{a_{R\_B}} & 0\\ 0 & 0 & \frac{1}{a_{R\_C}}\end{bmatrix} \quad (26.125)$$

$$[aRI_{ABC}]=\begin{bmatrix}0 & -a_{R\_B} & -a_{R\_C}\\ 0 & a_{R\_B} & 0\\ 0 & 0 & a_{R\_C}\end{bmatrix} \quad (26.126)$$

26.1.4.9 Generalized Equations

The voltage regulator models used in power-flow studies are generalized for the various connections in a form similar to the *ABCD* parameters that are used in transmission line analysis. The general form of the power-flow models in matrix form is

$$[V_{ABC}] = [a][V_{abc}] + [b][I_{abc}] \tag{26.127}$$

$$[I_{ABC}] = [c][V_{abc}] + [d][I_{abc}] \tag{26.128}$$

$$[V_{abc}] = [A][V_{ABC}] + [B][I_{abc}] \tag{26.129}$$

Depending upon the connection, the matrices $[V_{ABC}]$ and $[V_{abc}]$ can be either line-to-line or line-to-ground. The current matrices represent the line currents regardless of the regulator connection. For all voltage regulator connections, the generalized constants are defined as

$$[a] = [aRV_{abc}] \tag{26.130}$$

$$[b] = [0] \tag{26.131}$$

$$[c] = [0] \tag{26.132}$$

$$[d] = [aRI_{abc}] \tag{26.133}$$

$$[A] = [aRV_{ABC}] \tag{26.134}$$

$$[B] = [0] \tag{26.135}$$

26.1.5 Transformer Bank Connections

Unique models of three-phase transformer banks applicable to radial distribution feeders have been developed (Kersting, 1999). Models for the following three-phase connections are included in this document:

- Delta–grounded wye
- Grounded wye–delta
- Ungrounded wye–delta
- Grounded wye–grounded wye
- Delta–delta

Figure 26.21 defines the various voltages and currents for the transformer bank models. The models can represent a step-down (source side to load side) or a step-up (source side to load side) transformer bank. The notation is such that the capital letters *A*, *B*, *C*, *N* will always refer to the source side of the bank and the lower case letters *a*, *b*, *c*, *n* will always refer to the load side of the bank. It is assumed that

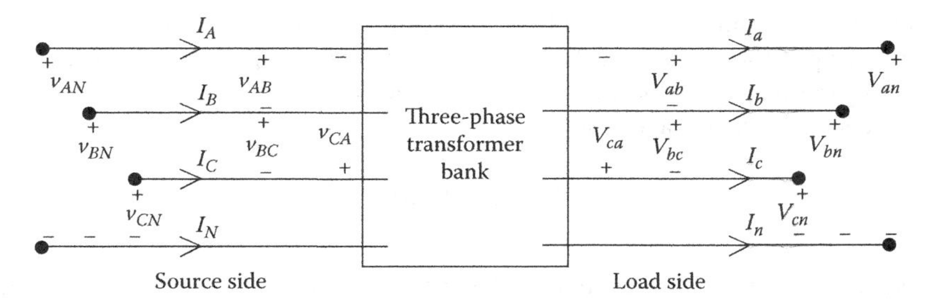

FIGURE 26.21 General transformer bank.

all variations of the wye–delta connections are connected in the "American Standard Thirty Degree" connection. The standard is such that:

Step-down connection
V_{AB} leads V_{ab} by 30°
I_A leads I_a by 30°

Step-up connection
V_{ab} leads V_{AB} by 30°
I_a leads I_A by 30°

26.1.5.1 Generalized Equations

The models to be used in power-flow studies are generalized for the connections in a form similar to the ABCD parameters that are used in transmission line analysis. The general form of the power-flow models in matrix form are

$$[VLN_{ABC}] = [a_t][VLN_{abc}] + [b_t][I_{abc}] \tag{26.136}$$

$$[I_{ABC}] = [c_t][V_{abc}] + [d_t][I_{abc}] \tag{26.137}$$

$$[VLN_{abc}] = [A_t][VLN_{ABC}] - [B_t][I_{abc}] \tag{26.138}$$

In Equations 26.136 through 26.138, the matrices $[VLN_{ABC}]$ and $[VLN_{abc}]$ will be the equivalent line-to-neutral voltages on delta and ungrounded wye connections and the line-to-ground voltages for grounded wye connections.

When the "ladder technique" or "sweep" iterative method is used, the "forward" sweep is assumed to be from the source working toward the remote nodes. The "backward" sweep will be working from the remote nodes toward the source node.

26.1.5.2 Common Variable and Matrices

All transformer models will use the following common variable and matrices:

- Transformer turns ratio

$$n_t = \frac{V_{\text{rated\_source}}}{V_{\text{rated\_load}}} \tag{26.139}$$

where

$V_{\text{rated\_source}}$ is the transformer winding rating on the source side. Line-to-line voltage for delta connections and line-to-neutral for wye connections

$V_{\text{rated\_load}}$ is the transformer winding rating on the load side. Line-to-line voltage for delta connections and line-to-neutral for wye connections

Note that the transformer "winding" ratings may be either line-to-line or line-to-neutral, depending upon the connection. The winding ratings can be specified in actual volts or per-unit volts using the appropriate base line-to-neutral voltages.

- Source to load matrix voltage relations:

$$[V_{ABC}] = [AV][V_{abc}] \tag{26.140}$$

The voltage matrices may be line-to-line or line-to-neutral voltages depending upon the connection.

- Load to source matrix current relations:

$$[I_{abc}] = [AI][I_{ABC}] \tag{26.141}$$

The current matrices may be line currents or delta currents depending upon the connection.

- Transformer impedance matrix:

$$[Zt_{abc}] = \begin{bmatrix} Zt_a & 0 & 0 \\ 0 & Zt_b & 0 \\ 0 & 0 & Zt_c \end{bmatrix} \tag{26.142}$$

The impedance elements in the matrix will be the per-unit impedance of the transformer windings on the load side of the transformer whether it is connected in wye or delta.

- Symmetrical component transformation matrix:

$$[A_s] = \begin{bmatrix} 1 & 1 & 1 \\ 1 & a_s^2 & a_s \\ 1 & a_s & a_s^2 \end{bmatrix} \tag{26.143}$$

where $a_s = 1\ \angle 120$

- Phaseshift matrix

$$[T_s] = \begin{bmatrix} 1 & 0 & 0 \\ 0 & t_s^* & 0 \\ 0 & 0 & t_s \end{bmatrix} \tag{26.144}$$

where $t_s = (1/\sqrt{3})\angle 30$

- Matrix to convert line-to-line voltages to equivalent line-to-neutral voltages:

$$[W] = [A_s][T_s][A_s]^{-1} = \frac{1}{3}\begin{bmatrix} 2 & 1 & 0 \\ 0 & 2 & 1 \\ 1 & 0 & 2 \end{bmatrix} \tag{26.145}$$

Example: $[VLN] = [W][VLL]$

- Matrix to convert delta currents into line currents:

$$[DI]=\begin{bmatrix}1 & 0 & -1\\ -1 & 1 & 0\\ 0 & -1 & 1\end{bmatrix} \tag{26.146}$$

Example: $[I_{abc}] = [DI][ID_{abc}]$

- Matrix to convert line-to-ground or line-to-neutral voltages to line-to-line voltages:

$$[D]=\begin{bmatrix}1 & -1 & 0\\ 0 & 1 & -1\\ -1 & 0 & 1\end{bmatrix} \tag{26.147}$$

Example: $[VLL_{abc}] = [D][VLN_{abc}]$

26.1.5.3 Per-Unit System

All transformer models were developed so that they can be applied using either "actual" or "per-unit" values of voltages, currents, and impedances. When the per-unit system is used, all per-unit voltages (line-to-line and line-to-neutral) use the line-to-neutral base as the base voltage. In other words, for a balanced set of three-phase voltages, the per-unit line-to-neutral voltage magnitude will be 1.0 at rated voltage and the per-unit line-to-line voltage magnitude will be the $\sqrt{3}$. In a similar fashion, all currents (line currents and delta currents) are based on the base line current. Again, $\sqrt{3}$ relationship will exist between the line and delta currents under balanced conditions. The base line impedance will be used for all line impedances and for wye and delta connected transformer impedances. There will be different base values on the two sides of the transformer bank.

Base values are computed following the steps listed below:

- Select a base three-phase kVA_{base} and the rated line-to-line voltage, $kVLL_{\text{source}}$, on the source side as the base line-to-line voltage.
- Based upon the voltage ratings of the transformer bank, determine the rated line-to-line voltage, $kVLL_{\text{load}}$, on the load side.
- Determine the transformer ratio, a_x, as

$$a_x = \frac{kVLL_{\text{source}}}{kVLL_{\text{load}}} \tag{26.148}$$

- The source side base values are computed as

$$kVLN_S = \frac{kVLL_S}{\sqrt{3}} \tag{26.149}$$

$$I_S = \frac{kVA_{\text{base}}}{\sqrt{3}kVLL_{\text{source}}} \tag{26.150}$$

$$Z_S = \frac{kVLL_{\text{source}}^2 1000}{kVA_B} \tag{26.151}$$

- The load side base values are computed by

$$kVLN_L = \frac{kVLN_S}{a_x} \tag{26.152}$$

$$I_L = a_x I_S \tag{26.153}$$

$$Z_L = \frac{Z_S}{a_x^2} \tag{26.154}$$

The matrices $[a_t]$, $[b_t]$, $[c_t]$, $[d_t]$, $[A_t]$, and $[B_t]$ (see Equations 26.136 through 26.138) for each connection are defined as follows:

26.1.5.4 Matrix Definitions

26.1.5.4.1 Delta–Grounded Wye

Backward sweep:

$$[VLN_{ABC}] = [a_t][VLG_{abc}] + [b_t][I_{abc}]$$

$$[I_{ABC}] = [c_t][VLG_{abc}] + [d_t][I_{abc}]$$

Forward sweep:

$$[VLG_{abc}] = [A_t][VLN_{ABC}] - [B_t][I_{abc}]$$

The matrices used for the step-down connection are

$$[a_t] = \frac{-n_t}{3}\begin{bmatrix} 0 & 2 & 1 \\ 1 & 0 & 2 \\ 2 & 1 & 0 \end{bmatrix}$$

$$[b_t] = \frac{-n_t}{3}\begin{bmatrix} 0 & 2Zt_b & Zt_c \\ Zt_a & 0 & 2Zt_c \\ 2Zt_a & Zt_b & 0 \end{bmatrix}$$

$$[c_t] = \begin{bmatrix} 0 & 0 & 0 \\ 0 & 0 & 0 \\ 0 & 0 & 0 \end{bmatrix}$$

$$[d_t] = \frac{1}{n_t}\begin{bmatrix} 1 & -1 & 0 \\ 0 & 1 & -1 \\ -1 & 0 & 1 \end{bmatrix}$$

$$[A_t] = \frac{1}{n_t}\begin{bmatrix} 1 & 0 & -1 \\ -1 & 1 & 0 \\ 0 & -1 & 1 \end{bmatrix}$$

$$[B_t] = \begin{bmatrix} Zt_a & 0 & 0 \\ 0 & Zt_b & 0 \\ 0 & 0 & Zt_c \end{bmatrix}$$

26.1.5.4.2 Ungrounded Wye–Delta

Power-flow equations:

Backward sweep:

$$[VLN_{ABC}] = [a_t][VLN_{abc}] + [b_t][I_{abc}]$$

$$[I_{ABC}] = [c_t][VLN_{abc}] + [d_t][I_{abc}]$$

Forward sweep:

$$[VLN_{abc}] = [A_t][VLN_{ABC}] - [B_t][I_{abc}]$$

Matrices used for the step-down connection are

$$[a_t] = n_t \begin{bmatrix} 1 & -1 & 0 \\ 0 & 1 & -1 \\ -1 & 0 & 1 \end{bmatrix}$$

$$[b_t] = \frac{n_t}{3} \begin{bmatrix} Zt_{ab} & -Zt_{ab} & 0 \\ Zt_{bc} & 2Zt_{bc} & 0 \\ -2Zt_{ca} & -Zt_{ca} & 0 \end{bmatrix}$$

$$[c_t] = \begin{bmatrix} 0 & 0 & 0 \\ 0 & 0 & 0 \\ 0 & 0 & 0 \end{bmatrix}$$

$$[d_t] = \frac{1}{3n_t} \begin{bmatrix} 1 & -1 & 0 \\ 1 & 2 & 0 \\ -2 & -1 & 0 \end{bmatrix}$$

$$[A_t] = \frac{1}{3n_t} \begin{bmatrix} 2 & 1 & 0 \\ 0 & 2 & 1 \\ 1 & 0 & 2 \end{bmatrix}$$

$$[B_t] = \frac{1}{9} \begin{bmatrix} 2Zt_{ab} + Zt_{bc} & 2Zt_{bc} - 2Zt_{ab} & 0 \\ 2Zt_{bc} - 2Zt_{ca} & 4Zt_{bc} - Zt_{ca} & 0 \\ Zt_{ab} - 4Zt_{ca} & -Zt_{ab} + 2Zt_{ca} & 0 \end{bmatrix}$$

where Zt_{ab}, Zt_{bc}, and Zt_{ca} are the transformer impedances inside the delta secondary connection.

26.1.5.4.3 Grounded Wye–Delta

Power-flow equations:

Backward sweep:

$$[VLG_{ABC}] = [a_t][VLN_{abc}] + [b_t][I_{abc}]$$

$$[I_{ABC}] = [c_t][VLN_{abc}] + [d_t][I_{abc}]$$

Forward sweep:

$$[VLG_{abc}]=[A_t][VLN_{ABC}]-[B_t][I_{abc}]$$

The matrices used for the step-down connection are

$$[a_t]=n_t\begin{bmatrix}1 & -1 & 0\\ 0 & 1 & -1\\ -1 & 0 & 1\end{bmatrix}$$

$$[b_t]=\frac{n_t}{Zt_{ab}+Zt_{bc}+Zt_{ca}}\begin{bmatrix}Zt_{ab}Zt_{ca} & -Zt_{ab}Zt_{bc} & 0\\ Zt_{bc}Zt_{ca} & Zt_{bc}(Zt_{ca}+Zt_{ab}) & 0\\ Zt_{ca}(-Zt_{ab}-Zt_{bc}) & -Zt_{bc}Zt_{ca} & 0\end{bmatrix}$$

$$[c_t]=\begin{bmatrix}0 & 0 & 0\\ 0 & 0 & 0\\ 0 & 0 & 0\end{bmatrix}$$

$$[d_t]=\frac{1}{n_t(Zt_{ab}+Zt_{bc}+Zt_{ca})}\begin{bmatrix}Zt_{ca} & -Zt_{bc} & 0\\ Zt_{ca} & Zt_{ab}+Zt_{ca} & 0\\ -Zt_{ab}-Zt_{bc} & -Zt_{ca} & 0\end{bmatrix}$$

$$[A_t]=\frac{1}{3n_t}\begin{bmatrix}2 & 1 & 0\\ 0 & 2 & 1\\ 1 & 0 & 2\end{bmatrix}$$

$$[B_t]=\frac{1}{3\sum Zt}\begin{bmatrix}2Zt_{ab}Zt_{ca}+Zt_{bc}Zt_{ca} & -2Zt_{ab}Zt_{bc}+Zt_{bc}(Zt_{ab}+Zt_{ca}) & 0\\ 2Zt_{bc}Zt_{ca}-Zt_{bc}(Zt_{ab}+Zt_{bc}) & 2Zt_{bc}(Zt_{ab}+Zt_{ca})-Zt_{bc}Zt_{ca} & 0\\ Zt_{ab}Zt_{ca}-2Zt_{ca}(Zt_{ab}+Zt_{bc}) & -Zt_{ab}Zt_{bc}-2Zt_{bc}Zt_{ca} & 0\end{bmatrix}$$

where

$$\sum Zt=Zt_{ab}+Zt_{bc}+Zt_{ca}$$

26.1.5.4.4 The Grounded Wye–Grounded Wye Connection

Power-flow equations:

Backward sweep:

$$[VLG_{ABC}]=[a_t][VLG_{abc}]+[b_t][I_{abc}]$$

$$[I_{ABC}]=[c_t][VLG_{abc}]+[d_t][I_{abc}]$$

Forward sweep:

$$[VLG_{abc}]=[A_t][VLG_{ABC}]-[B_t][I_{abc}]$$

The matrices used are

$$[a_t] = n_t \begin{bmatrix} 1 & 0 & 0 \\ 0 & 1 & 0 \\ 0 & 0 & 1 \end{bmatrix}$$

$$[b_t] = n_t \begin{bmatrix} Zt_a & 0 & 0 \\ 0 & Zt_b & 0 \\ 0 & 0 & Zt_c \end{bmatrix}$$

$$[c_t] = \begin{bmatrix} 0 & 0 & 0 \\ 0 & 0 & 0 \\ 0 & 0 & 0 \end{bmatrix}$$

$$[d_t] = \frac{1}{n_t} \begin{bmatrix} 1 & 0 & 0 \\ 0 & 1 & 0 \\ 0 & 0 & 1 \end{bmatrix}$$

$$[A_t] = \frac{1}{n_t} \begin{bmatrix} 1 & 0 & 0 \\ 0 & 1 & 0 \\ 0 & 0 & 1 \end{bmatrix}$$

$$[B_t] = \begin{bmatrix} Zt_a & 0 & 0 \\ 0 & Zt_b & 0 \\ 0 & 0 & Zt_c \end{bmatrix}$$

26.1.5.4.5 Delta–Delta

Power-flow equations:

Backward sweep:

$$[VLN_{ABC}] = [a_t][VLN_{abc}] + [b_t][I_{abc}]$$

$$[I_{ABC}] = [c_t][VLN_{abc}] + [d_t][I_{abc}]$$

Forward sweep:

$$[VLN_{abc}] = [A_t][VLN_{ABC}] - [B_t][I_{abc}]$$

The matrices used are

$$[a_t] = \frac{n_t}{3} \begin{bmatrix} 2 & -1 & -1 \\ -1 & 2 & -1 \\ -1 & -1 & 2 \end{bmatrix}$$

$$[b_t] = \frac{n_t}{3\sum Zt} \begin{bmatrix} 2Zt_{ab}Zt_{ca} + Zt_{bc}Zt_{ca} & -2Zt_{ab}Zt_{bc} + Zt_{bc}(Zt_{ab} + Zt_{ca}) & 0 \\ 2Zt_{bc}Zt_{ca} - Zt_{bc}(Zt_{ab} + Zt_{bc}) & 2Zt_{bc}(Zt_{ab} + Zt_{ca}) - Zt_{bc}Zt_{ca} & 0 \\ Zt_{ab}Zt_{ca} - 2Zt_{ca}(Zt_{ab} + Zt_{bc}) & -Zt_{ab}Zt_{bc} - 2Zt_{bc}Zt_{ca} & 0 \end{bmatrix}$$

where

$$\sum Zt = Zt_{ab} + Zt_{bc} + Zt_{ca}$$

$$[c_t] = \begin{bmatrix} 0 & 0 & 0 \\ 0 & 0 & 0 \\ 0 & 0 & 0 \end{bmatrix}$$

$$[d_t] = \frac{1}{n_t}\begin{bmatrix} 1 & 0 & 0 \\ 0 & 1 & 0 \\ 0 & 0 & 1 \end{bmatrix}$$

$$[A_t] = \frac{1}{3n_t}\begin{bmatrix} 2 & -1 & -1 \\ -1 & 2 & -1 \\ -1 & -1 & 2 \end{bmatrix}$$

$$[B_t] = \frac{1}{3\sum Zt}\begin{bmatrix} 2Zt_{ab}Zt_{ca} + Zt_{bc}Zt_{ca} & -2Zt_{ab}Zt_{bc} + Zt_{bc}(Zt_{ab} + Zt_{ca}) & 0 \\ 2Zt_{bc}Zt_{ca} - Zt_{bc}(Zt_{ab} + Zt_{bc}) & 2Zt_{bc}(Zt_{ab} + Zt_{ca}) - Zt_{bc}Zt_{ca} & 0 \\ Zt_{ab}Zt_{ca} - 2Zt_{ca}(Zt_{ab} + Zt_{bc}) & -Zt_{ab}Zt_{bc} - 2Zt_{bc}Zt_{ca} & 0 \end{bmatrix}$$

where

$$\sum Zt = Zt_{ab} + Zt_{bc} + Zt_{ca}$$

26.1.5.5 Thevenin Equivalent Circuit

The study of short-circuit studies that occur on the load side of a transformer bank requires the three-phase Thevenin equivalent circuit referenced to the load side terminals of the transformer. In order to determine this equivalent circuit, the Thevenin equivalent circuit up to the primary terminals of the "feeder" transformer must be determined. It is assumed that the transformer matrices as defined above are known for the transformer connection in question. A one-line diagram of the total system is shown in Figure 26.22.

The desired Thevenin equivalent circuit on the secondary side of the transformer is shown in Figure 26.23.

In Figure 26.22 the system voltage source $[ELN_{ABC}]$ will typically be a balanced set of per-unit voltages. The Thevenin equivalent voltage on the secondary side of the transformer will be:

$$[Eth_{abc}] = [A_t]\cdot[ELN_{ABC}] \tag{26.155}$$

The Thevenin equivalent impedance in Figure 26.23 from the source to the primary terminals of the feeder transformer is given by

$$[Zth_{abc}] = [A_t][Zsys_{ABC}][d_t] + [B_t] \tag{26.156}$$

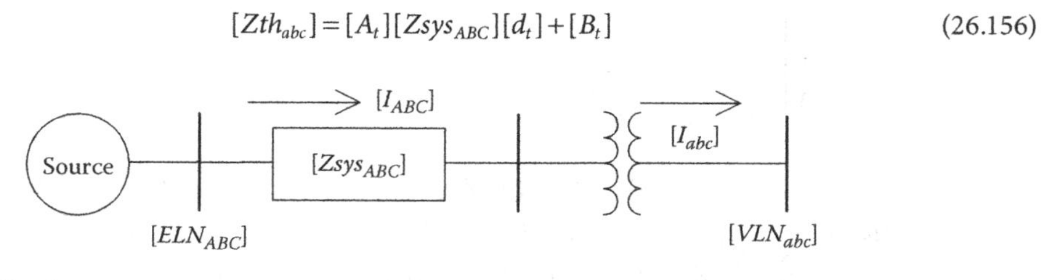

FIGURE 26.22 Total system.

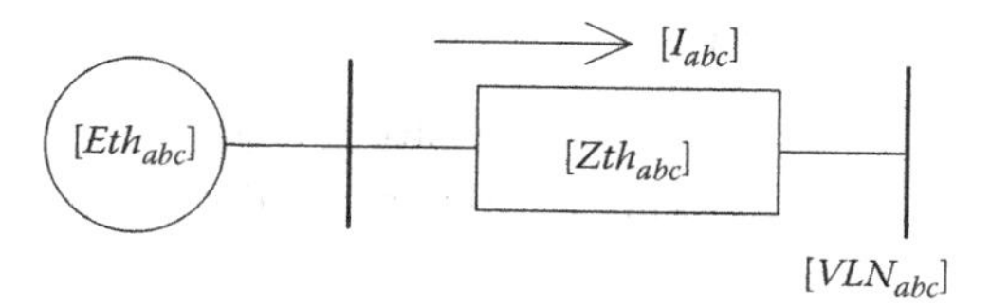

FIGURE 26.23 Three-phase Thevenin equivalent circuit.

The values of the source side Thevenin equivalent circuit will be the same regardless of the type of connection of the feeder transformer. For each three-phase transformer connection, unique values of the matrices [Eth_{abc}] and [Zth_{abc}] are defined as functions of the source side Thevenin equivalent circuit. These definitions are shown for each transformer connection below.

26.1.5.6 Center Tapped Transformers

The typical single-phase service to a customer is 120/240 V. This is provided from a center tapped single-phase transformer through the three-wire secondary and service drop to the customer's meter. The center tapped single-phase transformer with the three-wire secondary and 120/240 V loads can be modeled as shown in Figure 26.24.

Notice in Figure 26.24 that three impedance values are required for the center tapped transformer. These three impedances typically will not be known. In fact, usually only the magnitude of the transformer impedance will be known as found on the nameplate. In order to perform a reasonable analysis, some assumptions have to be made regarding the impedances. It is necessary to know both the per-unit R_A and the X_A components of the transformer impedance. References Gonen (1986) and Hopkinson (1977) are two sources for typical values. From Hopkinson (1977) the center tapped transformer impedances as a function of the transformer impedance are given. For interlaced transformers the three impedances are given by

$$Z_0 = 0.5R_A + j0.8X_A$$
$$Z_1 = R_A + j0.4X_A \tag{26.157}$$
$$Z_2 = R_A + j0.4X_A$$

The equations for the noninterlaced design are

$$Z_0 = 0.25R_A + j0.6X_A$$
$$Z_1 = 1.5R_A + j3.3X_A \tag{26.158}$$
$$Z_2 = 1.5R_A + j3.1X_A$$

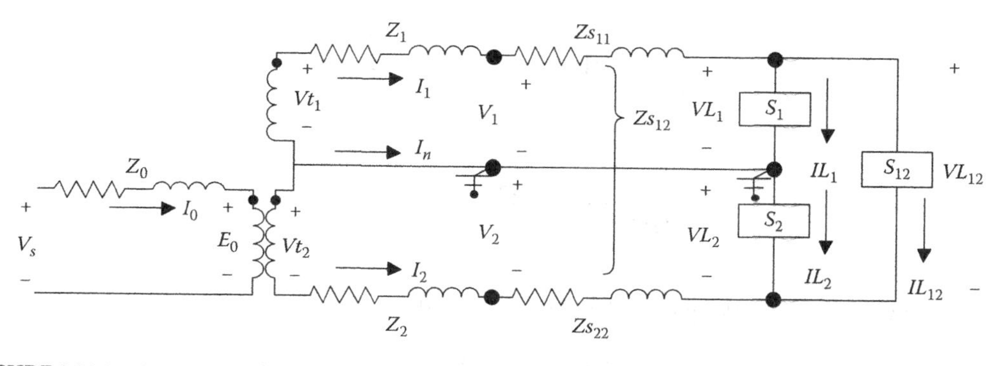

FIGURE 26.24 Center tapped single-phase transformer with secondary.

The transformer turns ratio is defined as

$$n_t = \frac{\text{high side rated voltage}}{\text{low side half winding rated voltage}} \tag{26.159}$$

$$\text{Example: } n_t = \frac{7200}{120} = 60$$

With reference to Figure 26.24, note that the secondary current I_1 flows out of the dot of the secondary half winding whereas the current I_2 flows out of the undotted terminal. This is done in order to simplify the voltage drop calculations down the secondary. The basic transformer equations that must apply at all times are

$$\begin{aligned} E_0 &= n_t Vt_1 = n_t Vt_2 \\ I_0 &= \frac{1}{n_t}(I_1 - I_2) \end{aligned} \tag{26.160}$$

General matrix equations similar to those of the three-phase transformer connections are used in the analysis. For the backward sweep (working from the load toward the source), the equations are

$$\begin{aligned} [V_{ss}] &= [a_{ct}][V_{12}] + [b_{ct}][I_{12}] \\ [I_{00}] &= [d_{ct}][I_{12}] \end{aligned} \tag{26.161}$$

where

$$[V_{ss}] = \begin{bmatrix} V_s \\ V_s \end{bmatrix}_{00}$$

$$[I_{00}] = \begin{bmatrix} I_0 \\ I_0 \end{bmatrix}$$

$$[I_{12}] = \begin{bmatrix} I_1 \\ I_2 \end{bmatrix}$$

$$[a_{ct}] = \begin{bmatrix} n_t & 0 \\ 0 & n_t \end{bmatrix}$$

$$[b_{ct}] = \begin{bmatrix} n_t\left(Z_1 + \dfrac{Z_0}{n_t^2}\right) & -\dfrac{Z_0}{n_t} \\ \dfrac{Z_0}{n_t} & -n_t\left(Z_2 + \dfrac{Z_0}{n_t^2}\right) \end{bmatrix} \tag{26.162}$$

$$[d_{ct}] = \frac{1}{n_t}\begin{bmatrix} 1 & -1 \\ 1 & -1 \end{bmatrix}$$

For the forward sweep (working from the source toward the loads) the equations are

$$[V_{12}] = [A_{ct}][V_{ss}] - [B_{ct}][I_{12}] \tag{26.163}$$

where

$$[A_{ct}] = \frac{1}{n_t}\begin{bmatrix} 1 & 0 \\ 0 & 1 \end{bmatrix}$$

$$[B_{ct}] = \begin{bmatrix} Z_1 + \frac{Z_0}{n_t^2} & -\frac{Z_0}{n_t^2} \\ \frac{Z_0}{n_t^2} & -\left(Z_2 + \frac{Z_0}{n_t^2}\right) \end{bmatrix} \tag{26.164}$$

The three-wire secondary is modeled by first applying the Carson's equations and Kron reduction method to determine the 2 × 2 phase impedance matrix:

$$[Z_s] = \begin{bmatrix} Z_{s11} & Z_{s12} \\ Z_{s21} & Z_{s22} \end{bmatrix} \tag{26.165}$$

The backward sweep equation becomes

$$[V_{12}] = [a_s][VL_{12}] + [b_s][I_{12}]$$

where

$$[a_s] = \begin{bmatrix} 1 & 0 \\ 0 & 1 \end{bmatrix}$$

$$[b_s] = [Z_s] \tag{26.166}$$

The forward sweep equation is:

$$[VL_{12}] = [A_s][V_{12}] - [B_s][I_{12}] \tag{26.167}$$

where

$$[A_s] = [a_s]^{-1}$$

$$[B_s] = [Z_s] \tag{26.168}$$

26.1.6 Load Models

Loads can be represented as being connected phase-to-phase or phase-to-neutral in a four-wire wye systems or phase-to-phase in a three-wire delta system. The loads can be three-phase, two-phase, or single-phase with any degree of unbalance and can be modeled as

- Constant real and reactive power (constant PQ)
- Constant current
- Constant impedance
- Any combination of the above

The load models developed in this document are used in the iterative process of a power-flow program. All models are initially defined by a complex power per phase and either a line-to-neutral (wye load) or a line-to-line voltage (delta load). The units of the complex power can be in volt-amperes and volts or per-unit volt-amperes and per-unit volts.

For both the wye and delta connected loads, the basic requirement is to determine the load component of the line currents coming into the loads. It is assumed that all loads are initially specified by their complex power ($S = P + jQ$) per phase and a line-to-neutral or line-to-line voltage.

26.1.6.1 Wye Connected Loads

Figure 26.25 shows the model of a wye connected load.

The notation for the specified complex powers and voltages is as follows:

$$\text{Phase } a\text{: } |S_a|\angle\theta_a = P_a + jQ_a \quad \text{and} \quad |V_{an}|\angle\delta_a \tag{26.169}$$

$$\text{Phase } b\text{: } |S_b|\angle\theta_b = P_b + jQ_b \quad \text{and} \quad |V_{bn}|\angle\delta_b \tag{26.170}$$

$$\text{Phase } c\text{: } |S_c|\angle\theta_c = P_c + jQ_c \quad \text{and} \quad |V_{cn}|\angle\delta_c \tag{26.171}$$

1. Constant real and reactive power loads

$$\begin{aligned}
IL_a &= \left(\frac{S_a}{V_{an}}\right)^* = \frac{|S_a|}{|V_{an}|}\angle\delta_a - \theta_a = |IL_a|\angle\alpha_a \\
IL_b &= \left(\frac{S_b}{V_{bn}}\right)^* = \frac{|S_b|}{|V_{bn}|}\angle\delta_b - \theta_b = |IL_b|\angle\alpha_b \\
IL_c &= \left(\frac{S_c}{V_{cn}}\right)^* = \frac{|S_c|}{|V_{cn}|}\angle\delta_c - \theta_c = |IL_c|\angle\alpha_c
\end{aligned} \tag{26.172}$$

In this model the line-to-neutral voltages will change during each iteration until convergence is achieved.

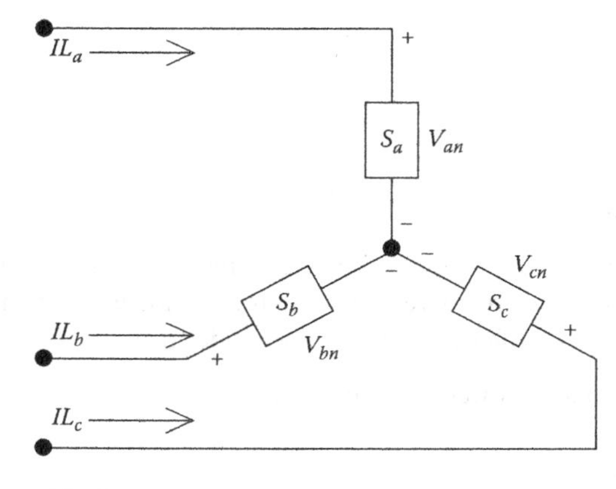

FIGURE 26.25 Wye connected load.

2. Constant impedance loads
   The "constant load impedance" is first determined from the specified complex power and line-to-neutral voltages according to the following equation:

$$Z_a = \frac{|V_{an}|^2}{S_a^*} = \frac{|V_{an}|^2}{|S_a|} \angle\theta_a = |Z_a| \angle\theta_a$$
$$Z_b = \frac{|V_{bn}|^2}{S_b^*} = \frac{|V_{bn}|^2}{|S_b|} \angle\theta_b = |Z_b| \angle\theta_b \tag{26.173}$$
$$Z_c = \frac{|V_{cn}|^2}{S_c^*} = \frac{|V_{cn}|^2}{|S_c|} \angle\theta_c = |Z_c| \angle\theta_c$$

   The load currents as a function of the constant load impedances are given by the following equation:

$$IL_a = \frac{V_{an}}{Z_a} = \frac{|V_{an}|}{|Z_a|} \angle\delta_a - \theta_a = |IL_a| \angle\alpha_a$$
$$IL_b = \frac{V_{bn}}{Z_b} = \frac{|V_{bn}|}{|Z_b|} \angle\delta_b - \theta_b = |IL_b| \angle\alpha_b \tag{26.174}$$
$$IL_c = \frac{V_{cn}}{Z_c} = \frac{|V_{cn}|}{|Z_c|} \angle\delta_c - \theta_c = |IL_c| \angle\alpha_c$$

   In this model the line-to-neutral voltages will change during each iteration until convergence is achieved.
3. Constant current loads
   In this model the magnitudes of the currents are computed according to Equation 26.172 and then held constant while the angle of the voltage (δ) changes during each iteration. In order to keep the power factor constant, the angles of the load currents are given by

$$IL_a = |IL_a| \angle\delta_a - \theta_a$$
$$IL_b = |IL_b| \angle\delta_b - \theta_b \tag{26.175}$$
$$IL_c = |IL_c| \angle\delta_c - \theta_c$$

4. Combination loads
   Combination loads can be modeled by assigning a percentage of the total load to each of the above three load models. The total line current entering the load is the sum of the three components.

26.1.6.2 Delta Connected Loads

Figure 26.26 shows the model of a delta connected load.

The notation for the specified complex powers and voltages is as follows:

$$\text{Phase } ab\text{: } |S_{ab}| \angle\theta_{ab} = P_{ab} + jQ_{ab} \quad \text{and} \quad |V_{ab}| \angle\delta_{ab} \tag{26.176}$$

$$\text{Phase } bc\text{: } |S_{bc}| \angle\theta_{bc} = P_{bc} + jQ_{bc} \quad \text{and} \quad |V_{bc}| \angle\delta_{bc} \tag{26.177}$$

$$\text{Phase } ca\text{: } |S_{ca}| \angle\theta_{ca} = P_{ca} + jQ_{ca} \quad \text{and} \quad |V_{ca}| \angle\delta_{ca} \tag{26.178}$$

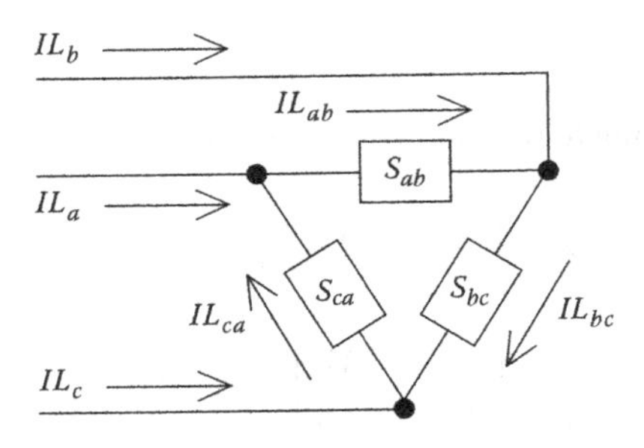

FIGURE 26.26 Delta connected load.

1. Constant real and reactive power loads

$$IL_{ab} = \left(\frac{S_{ab}}{V_{ab}}\right)^* = \frac{|S_{ab}|}{|V_{ab}|}\angle\delta_{ab} - \theta_{ab} = |IL_{ab}|\angle\alpha_{ab}$$

$$IL_{bc} = \left(\frac{S_{bc}}{V_{bc}}\right)^* = \frac{|S_{bc}|}{|V_{bc}|}\angle\delta_{bc} - \theta_{bc} = |IL_{bc}|\angle\alpha_{bc} \quad (26.179)$$

$$IL_{ca} = \left(\frac{S_{ca}}{V_{ca}}\right)^* = \frac{|S_{ca}|}{|V_{ca}|}\angle\delta_{ca} - \theta_{ca} = |IL_{ca}|\angle\alpha_{ca}$$

In this model the line-to-line voltages will change during each iteration until convergence is achieved.

2. Constant impedance loads
The constant load impedance is first determined from the specified complex power and line-to-neutral voltages according to the following equation:

$$Z_{ab} = \frac{|V_{ab}|^2}{S_{ab}^*} = \frac{|V_{ab}|^2}{|S_{ab}|}\angle\theta_{ab} = |Z_{ab}|\angle\theta_{ab}$$

$$Z_{bc} = \frac{|VL_{bc}|^2}{S_{bc}^*} = \frac{|V_{bc}|^2}{|S_{bc}|}\angle\theta_{bc} = |Z_{bc}|\angle\theta_{bc} \quad (26.180)$$

$$Z_{ca} = \frac{|V_{ca}|^2}{S_{ca}^*} = \frac{|V_{ca}|^2}{|S_{ca}|}\angle\theta_{ca} = |Z_{ca}|\angle\theta_{ca}$$

The load currents as a function of the constant load impedances are given by the following equation:

$$IL_{ab} = \frac{V_{ab}}{Z_{ab}} = \frac{|V_{anb}|}{|Z_{ab}|}\angle\delta_{ab} - \theta_{ab} = |IL_{ab}|\angle\alpha_{ab}$$

$$IL_{bc} = \frac{V_{bc}}{Z_{bc}} = \frac{|V_{bc}|}{|Z_{bc}|}\angle\delta_{bc} - \theta_{bc} = |IL_{bc}|\angle\alpha_{bc} \quad (26.181)$$

$$IL_{ca} = \frac{V_{ca}}{Z_{ca}} = \frac{|V_{ca}|}{|Z_{ca}|}\angle\delta_{ca} - \theta_{ca} = |IL_{ca}|\angle\alpha_{ca}$$

In this model the line-to-line voltages in Equation 26.181 will change during each iteration until convergence is achieved.

3. Constant current loads
   In this model the magnitudes of the currents are computed according to Equation 26.179 and then held constant while the angle of the voltage (δ) changes during each iteration. This keeps the power factor of the load constant.

$$\begin{aligned} IL_{ab} &= |IL_{ab}|\angle\delta_{ab} - \theta_{ab} \\ IL_{bc} &= |IL_{bc}|\angle\delta_{bc} - \theta_{bc} \\ IL_{ca} &= |IL_{ca}|\angle\delta_{ca} - \theta_{ca} \end{aligned} \tag{26.182}$$

4. Combination loads
   Combination loads can be modeled by assigning a percentage of the total load to each of the above three load models. The total delta current for each load is the sum of the three components.
   The line currents entering the delta connected load for all models are determined by

$$\begin{bmatrix} IL_a \\ IL_b \\ IL_c \end{bmatrix} = \begin{bmatrix} 1 & 0 & -1 \\ -1 & 1 & 0 \\ 0 & -1 & 1 \end{bmatrix} \begin{bmatrix} IL_{ab} \\ IL_{bc} \\ IL_{ca} \end{bmatrix} \tag{26.183}$$

In both the wye and delta connected loads, single-phase and two-phase loads are modeled by setting the complex powers of the missing phases to zero. In other words, all loads are modeled as three-phase loads and by setting the complex power of the missing phases to zero, the only load currents computed using the above equations will be for the nonzero loads.

26.1.7 Shunt Capacitor Models

Shunt capacitor banks are commonly used in a distribution system to help in voltage regulation and to provide reactive power support. The capacitor banks are modeled as constant susceptances connected in either wye or delta. Similar to the load model, all capacitor banks are modeled as three-phase banks with the kVAr of missing phases set to zero for single-phase and two-phase banks.

26.1.7.1 Wye Connected Capacitor Bank

A wye connected capacitor bank is shown in Figure 26.27. The individual phase capacitor units are specified in kVAr and kV. The constant susceptance for each unit can be computed in either Siemans or per unit. When per unit is desired, the specified kVAr of the capacitor must be divided by the base single-phase kVAr and the kV must be divided by the base line-to-neutral kV.

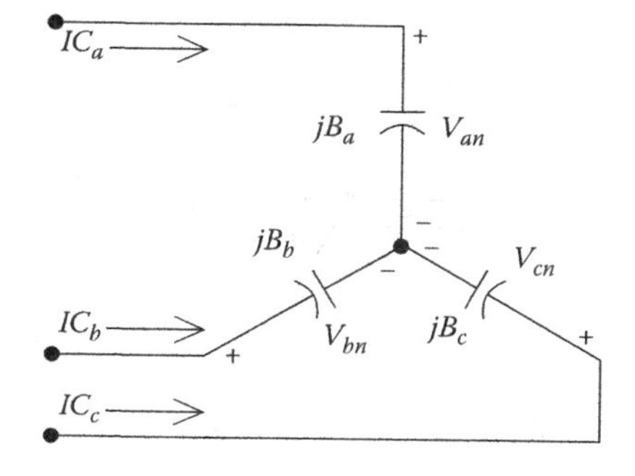

FIGURE 26.27 Wye connected capacitor bank.

The susceptance of a capacitor unit is computed by

$$B_{\text{actual}} = \frac{\text{kVAr}}{\text{kV}^2 1000} \text{ Siemans} \tag{26.184}$$

$$B_{pu} = \frac{\text{kVAr}_{pu}}{V_{pu}^2} \text{ per unit} \tag{26.185}$$

where

$$\text{kVAr}_{pu} = \frac{\text{kVAr}_{\text{actual}}}{\text{kVA}_{\text{single\_phase\_base}}} \tag{26.186}$$

$$V_{pu} = \frac{\text{kV}_{\text{actual}}}{\text{kV}_{\text{line\_to\_neutral\_base}}} \tag{26.187}$$

The per-unit value of the susceptance can also be determined by first computing the actual value (Equation 26.184) and then dividing by the base admittance of the system.

With the susceptance computed, the line currents serving the capacitor bank are given by

$$\begin{aligned} IC_a &= jB_a V_{an} \\ IC_b &= jB_b V_{bn} \\ IC_c &= jB_c V_{cn} \end{aligned} \tag{26.188}$$

26.1.7.2 Delta Connected Capacitor Bank

A delta connected capacitor bank is shown in Figure 26.28.

Equations 26.184 through 26.187 can be used to determine the value of the susceptance in actual Siemans or per unit. It should be pointed out that in this case, the kV will be a line-to-line value of the voltage. Also, it should be noted that in Equation 26.187, the base line-to-neutral voltage is used to compute the per-unit line-to-line voltage. This is a variation from the usual application of the per-unit system where the actual line-to-line voltage would be divided by a base line-to-line voltage in order to get the per-unit line-to-line voltage. That is not done here so that under normal conditions, the per-unit line-to-line voltage will have a magnitude of $\sqrt{3}$ rather than 1.0. This is done so that Kirchhoff's current law (KCL) at each node of the delta connection will apply for either the actual or per-unit delta currents.

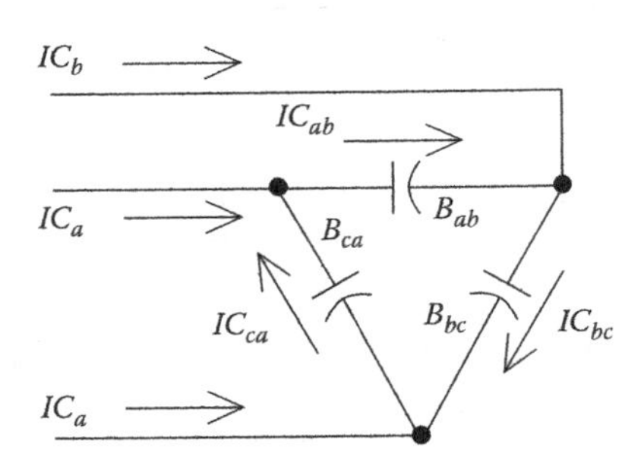

FIGURE 26.28 Delta connected capacitor bank.

The currents flowing in the delta connected capacitors are given by

$$\begin{aligned} IC_{ab} &= jB_{ab}V_{ab} \\ IC_{bc} &= jB_{bc}V_{bc} \\ IC_{ca} &= jB_{ca}V_{ca} \end{aligned} \tag{26.189}$$

The line currents feeding the delta connected capacitor bank are given by

$$\begin{bmatrix} IC_a \\ IC_b \\ IC_c \end{bmatrix} = \begin{bmatrix} 1 & 0 & -1 \\ -1 & 1 & 0 \\ 0 & -1 & 1 \end{bmatrix} \begin{bmatrix} IC_{ab} \\ IC_{bc} \\ IC_{ca} \end{bmatrix} \tag{26.190}$$

26.2 Analysis

26.2.1 Power-Flow Analysis

The power-flow analysis of a distribution feeder is similar to that of an interconnected transmission system. Typically what will be known prior to the analysis will be the three-phase voltages at the substation and the complex power of all the loads and the load model (constant complex power, constant impedance, constant current, or a combination). Sometimes, the input complex power supplied to the feeder from the substation is also known.

In Sections 26.1.3 through 26.1.5, phase frame models were presented for the series components of a distribution feeder. In Sections 26.1.6 and 26.1.7, models were presented for the shunt components (loads and capacitor banks). These models are used in the "power-flow" analysis of a distribution feeder.

A power-flow analysis of a feeder can determine the following by phase and total three-phase:

- Voltage magnitudes and angles at all nodes of the feeder
- Line flow in each line section specified in kW and kVAr, amps and degrees, or amps and power factor
- Power loss in each line section
- Total feeder input kW and kVAr
- Total feeder power losses
- Load kW and kVAr based upon the specified model for the load

Because the feeder is radial, iterative techniques commonly used in transmission network power-flow studies are not used because of poor convergence characteristics (Trevino, 1970). Instead, an iterative technique specifically designed for a radial system is used. The ladder iterative technique (Kersting and Mendive, 1976) will be presented here.

26.2.1.1 The Ladder Iterative Technique

26.2.1.1.1 Linear Network

A modification of the ladder network theory of linear systems provides a robust iterative technique for power-flow analysis. A distribution feeder is nonlinear because most loads are assumed to be constant kW and kVAr. However, the approach taken for the linear system can be modified to take into account the nonlinear characteristics of the distribution feeder.

For the ladder network in Figure 26.29, it is assumed that all of the line impedances and load impedances are known along with the voltage at the source (V_s).

The solution for this network is to assume a voltage at the most remote load (V_5). The load current I_5 is then determined as

$$I_5 = \frac{V_5}{ZL_5} \tag{26.191}$$

For this "end-node" case, the line current I_{45} is equal to the load current I_5. The voltage at node 4 (V_4) can be determined using Kirchhoff's voltage law (KVL):

$$V_4 = V_5 + Z_{45}I_{45} \tag{26.192}$$

The load current I_4 can be determined and then KCL applied to determine the line current I_{34}.

$$I_{34} = I_{45} + I_4 \tag{26.193}$$

KVL is applied to determine the node voltage V_3. This procedure is continued until a voltage (V_1) has been computed at the source. The computed voltage V_1 is compared to the specified voltage V_s. There will be a difference between these two voltages. The ratio of the specified voltage to the compute voltage can be determined as

$$\text{Ratio} = \frac{V_s}{V_1} \tag{26.194}$$

Since the network is linear, all of the line and load currents and node voltages in the network can be multiplied by the Ratio for the final solution to the network.

26.2.1.1.2 Nonlinear Network

The linear network of Figure 26.29 is modified to a nonlinear network by replacing all of the constant load impedances by constant complex power loads as shown in Figure 26.30.

The procedure outlined for the linear network is applied initially to the nonlinear network. The only difference being that the load current (assuming constant P and Q) at each node is computed by

$$I_n = \left(\frac{S_n}{V_n}\right)^* \tag{26.195}$$

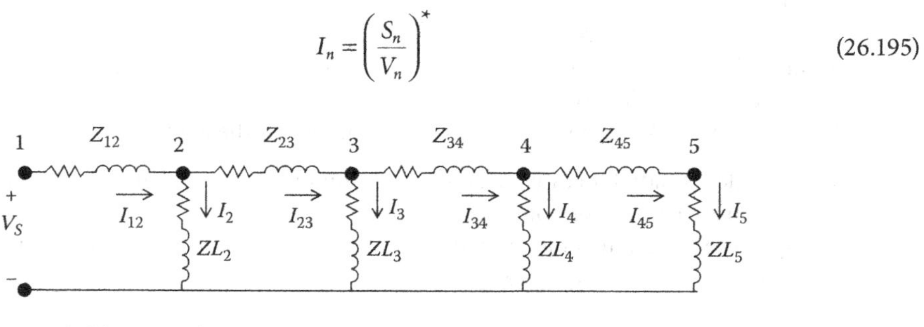

FIGURE 26.29 Linear ladder network.

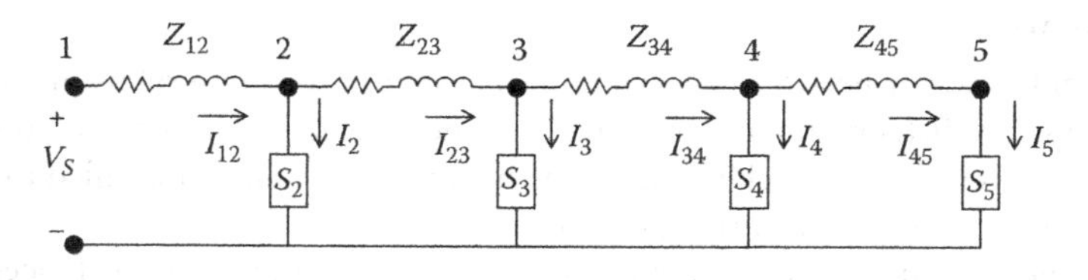

FIGURE 26.30 Non-linear ladder network.

The backward sweep will determine a computed source voltage V_1. As in the linear case, this first iteration will produce a voltage that is not equal to the specified source voltage V_s. Because the network is nonlinear, multiplying currents and voltages by the ratio of the specified voltage to the computed voltage will not give the solution. The most direct modification to the ladder network theory is to perform a forward sweep. The forward *d* sweep commences by using the specified source voltage and the line currents from the backward sweep. KVL is used to compute the voltage at node 2 by

$$V_2 = V_s - Z_{12} I_{12} \tag{26.196}$$

This procedure is repeated for each line segment until a "new" voltage is determined at node 5. Using the new voltage at node 5, a second backward sweep is started that will lead to a new computed voltage at the source. The backward and forward sweep process is continued until the difference between the computed and specified voltage at the source is within a given tolerance.

26.2.1.1.3 General Feeder

A typical distribution feeder will consist of the "primary main" with laterals tapped off the primary main, and sublaterals tapped off the laterals, etc., Figure 26.30 shows an example of a typical feeder.

The ladder iterative technique for the feeder of Figure 26.31 would proceed as follows:

1. Assume voltages (1.0 per unit) at the "end" nodes (6, 8, 9, 11, and 13).
2. Starting at node 13, compute the node current (load current plus capacitor current if present).
3. With this current, apply KVL to calculate the node voltages at 12 and 10.
4. Node 10 is referred to as a "junction" node since laterals branch in two directions from the node. This feeder goes to node 11 and computes the node current. Use that current to compute the voltage at node 10. This will be referred to as "the most recent voltage at node 10."
5. Using the most recent value of the voltage at node 10, the node current at node 10 (if any) is computed.
6. Apply KCL to determine the current flowing from node 4 toward node 10.
7. Compute the voltage at node 4.

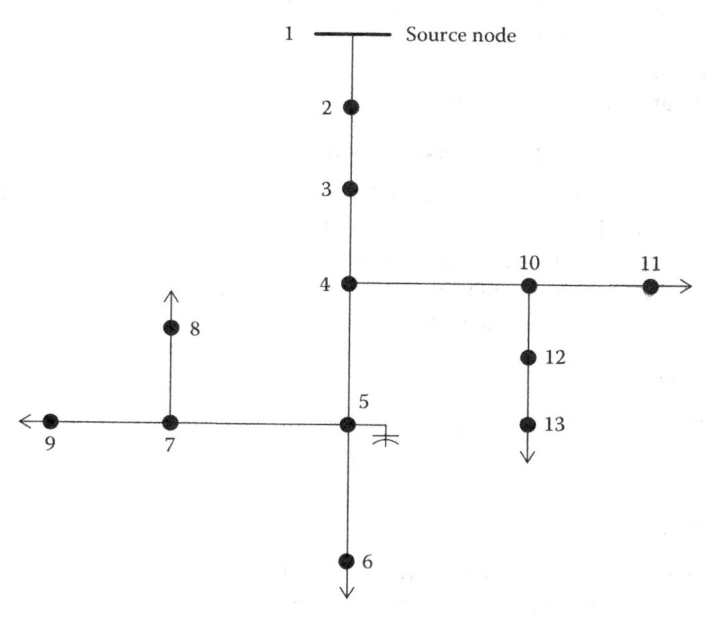

FIGURE 26.31 Typical distribution feeder.

8. Node 4 is a junction node. An end-node downstream from node 4 is selected to start the forward sweep toward node 4.
9. Select node 6, compute the node current, and then compute the voltage at junction-node 5.
10. Go to downstream end-node 8. Compute the node current and then the voltage at junction-node 7.
11. Go to downstream end-node 9. Compute the node current and then the voltage at junction-node 7.
12. Compute the node current at node 7 using the most recent value of node 7 voltage.
13. Apply KCL at node 7 to compute the current flowing on the line segment from node 5 to node 7.
14. Compute the voltage at node 5.
15. Compute the node current at node 5.
16. Apply KCL at node 5 to determine the current flowing from node 4 toward node 5.
17. Compute the voltage at node 4.
18. Compute the node current at node 4.
19. Apply KCL at node 4 to compute the current flowing from node 3 to node 4.
20. Calculate the voltage at node 3.
21. Compute the node current at node 3.
22. Apply KCL at node 3 to compute the current flowing from node 2 to node 3.
23. Calculate the voltage at node 2.
24. Compute the node current at node 2.
25. Apply KCL at node 2.
26. Calculate the voltage at node 1.
27. Compare the calculated voltage at node 1 to the specified source voltage.
28. If not within tolerance, use the specified source voltage and the backward sweep current flowing from node 1 to node 2 and compute the new voltage at node 2.
29. The forward sweep continues using the new upstream voltage and line segment current from the forward sweep to compute the new downstream voltage.
30. The forward sweep is completed when new voltages at all end nodes have been completed.
31. This completes the first iteration.
32. Now repeat the backward sweep using the new end voltages rather than the assumed voltages as was done in the first iteration.
33. Continue the backward and forward sweeps until the calculated voltage at the source is within a specified tolerance of the source voltage.
34. At this point, the voltages are known at all nodes and the currents flowing in all line segments are known. An output report can be produced giving all desired results.

26.2.1.2 The Unbalanced Three-Phase Distribution Feeder

The previous section outlined the general procedure for performing the ladder iterative technique. This section will address how that procedure can be used for an unbalanced three-phase feeder.

Figure 26.32 is the one-line diagram of an unbalanced three-phase feeder. The topology of the feeder in Figure 26.32 is the same as the feeder in Figure 26.31. Figure 26.32 shows more detail of the feeder however. The feeder in Figure 26.32 can be broken into the series components and the shunt components.

26.2.1.2.1 Series Components

The series components of a distribution feeder are

- Line segments
- Transformers
- Voltage regulators

Models for each of the series components have been developed in prior areas of this section. In all cases, models (three-phase, two-phase, and single-phase) were developed in such a manner that they can be generalized. Figure 26.33 shows the "general model" for each of the series components.

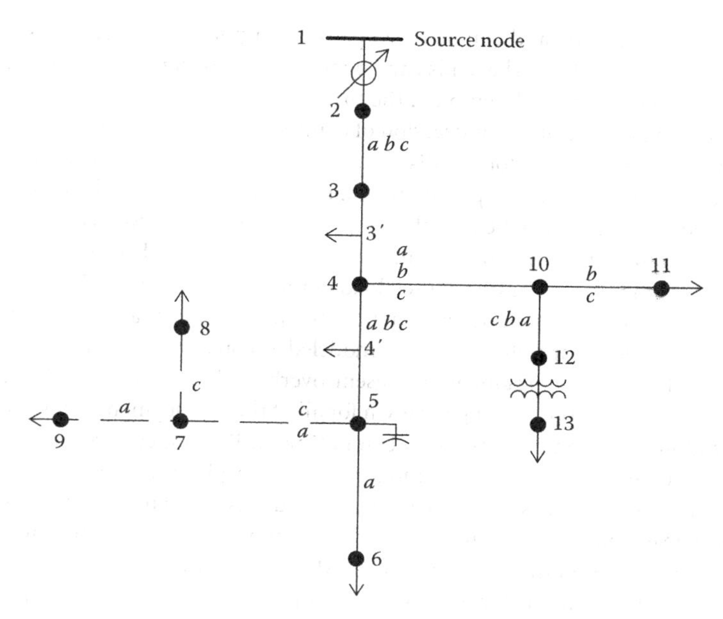

FIGURE 26.32 Unbalanced three-phase distribution feeder.

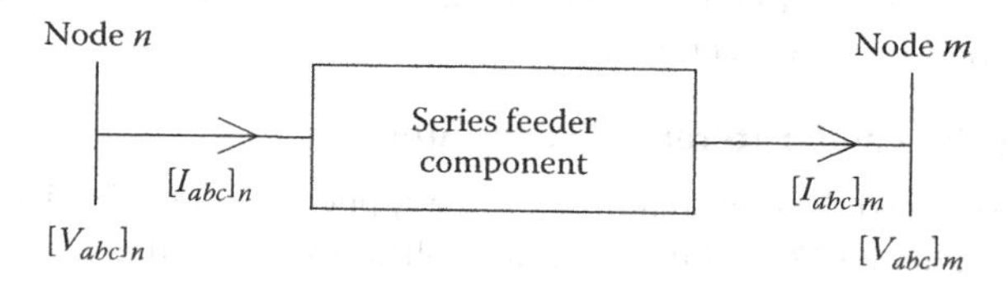

FIGURE 26.33 Series feeder component.

The general equations defining the "input" (node n) and "output" (node m) voltages and currents are given by

$$[V_{abc}]_n = [a][V_{abc}]_m + [b][I_{abc}]_m \tag{26.197}$$

$$[I_{abc}]_n = [c][V_{abc}]_m + [d][I_{abc}]_m \tag{26.198}$$

The general equation relating the output (node m) and input (node n) voltages is given by

$$[V_{abc}]_m = [A][V_{abc}]_n + [B][I_{abc}]_m \tag{26.199}$$

In Equations 26.197 through 26.199, the voltages are line-to-neutral for a four-wire wye feeder and equivalent line-to-neutral for a three-wire delta system. For transformers and voltage regulators, the voltages are line-to-neutral for terminals that are connected to a four-wire wye and line-to-line when connected to a three-wire delta.

26.2.1.2.2 *Shunt Components*

The shunt components of a distribution feeder are

- Spot loads
- Distributed loads
- Capacitor banks

Spot loads are located at a node and can be three-phase, two-phase, or single-phase and connected in either a wye or a delta connection. The loads can be modeled as constant complex power, constant current, constant impedance, or a combination of the three.

Distributed loads are located at the midsection of a line segment. A distributed load is modeled when the loads on a line segment are uniformly distributed along the length of the segment. As in the spot load, the distributed load can be three-phase, two-phase, or single-phase and connected in either a wye or a delta connection. The loads can be modeled as constant complex power, constant current, constant impedance, or a combination of the three. To model the distributed load, a "dummy" node is created in the center of a line segment with the distributed load of the line section modeled at this dummy node.

Capacitor banks are located at a node and can be three-phase, two-phase, or single-phase and can be connected in a wye or delta. Capacitor banks are modeled as constant admittances.

In Figure 26.32 the solid line segments represent overhead lines while the dashed lines represent underground lines. Note that the phasing is shown for all of the line segments. In the area of the Section 26.1.1, the application of Carson's equations for computing the line impedances for overhead and underground lines was presented. There it was pointed out that two-phase and single-phase lines are represented by a 3×3 matrix with zeros set in the rows and columns of the missing phases.

In the area of the Section 26.1.2, the method for the computation of the shunt capacitive susceptance for overhead and underground lines was presented. Most of the time the shunt capacitance of the line segment can be ignored; however, for long underground segments, the shunt capacitance should be included.

The "node" currents may be three-phase, two-phase, or single-phase and consist of the sum of the load current at the node plus the capacitor current (if any) at the node.

26.2.1.3 Applying the Ladder Iterative Technique

The previous section outlined the steps required for the application of the ladder iterative technique. For the general feeder of Figure 26.32 the same outline applies. The only difference is that Equations 26.197 and 26.198 are used for computing the node voltages on the backward sweep and Equation 26.199 is used for computing the downstream voltages on the forward sweep. The $[a]$, $[b]$, $[c]$, $[d]$, $[A]$, and $[B]$ matrices for the various series components are defined in the following areas of this section:

- Line segments: Line segment models
- Voltage regulators: Step-voltage regulators
- Transformer banks: Transformer bank connections

The node currents are defined in the following area:

- Loads: Load models
- Capacitors: Shunt capacitor models

26.2.1.4 Final Notes

26.2.1.4.1 Line Segment Impedances

It is extremely important that the impedances and admittances of the line segments be computed using the exact spacings and phasing. Because of the unbalanced loading and resulting unbalanced line currents, the voltage drops due to the mutual coupling of the lines become very important. It is not unusual to observe a voltage rise on a lightly loaded phase of a line segment that has an extreme current unbalance.

26.2.1.4.2 Power Loss

The real power losses of a line segment must be computed as the difference (by phase) of the input power to a line segment minus the output power of the line segment. It is possible to observe a negative power loss on a phase that is lightly loaded compared to the other two phases. Computing power loss as the phase current squared times the phase resistance does not give the actual real power loss in the phases.

26.2.1.4.3 Load Allocation

Many times the input complex power (kW and kVAr) to a feeder is known because of the metering at the substation. This information can be either total three-phase or for each individual phase. In some cases the metered data may be the current and power factor in each phase.

It is desirable to have the computed input to the feeder match the metered input. This can be accomplished (following a converged iterative solution) by computing the ratio of the metered input to the computed input. The phase loads can now be modified by multiplying the loads by this ratio. Because the losses of the feeder will change when the loads are changed, it is necessary to go through the ladder iterative process to determine a new computed input to the feeder. This new computed input will be closer to the metered input, but most likely not within a specified tolerance. Again, a ratio can be determined and the loads modified. This process is repeated until the computed input is within a specified tolerance of the metered input.

26.2.1.5 Short-Circuit Analysis

The computation of short-circuit currents for unbalanced faults in a normally balanced three-phase system has traditionally been accomplished by the application of symmetrical components. However, this method is not well-suited to a distribution feeder that is inherently unbalanced. The unequal mutual coupling between phases leads to mutual coupling between sequence networks. When this happens, there is no advantage to using symmetrical components. Another reason for not using symmetrical components is that the phases between which faults occur is limited. For example, using symmetrical components, line-to-ground faults are limited to phase a to ground. What happens if a single-phase lateral is connected to phase *b* or *c*? This section will present a method for short-circuit analysis of an unbalanced three-phase distribution feeder using the phase frame (Kersting, 1980).

26.2.1.5.1 General Theory

Figure 26.34 shows the unbalanced feeder as modeled for short-circuit calculations. In Figure 26.34, the voltage sources E_a, E_b, and E_c represent the Thevenin equivalent line-to-ground voltages at the faulted bus. The matrix $[ZTOT]$ represents the Thevenin equivalent impedance matrix at the faulted bus. The fault impedance is represented by Z_f in Figure 26.34.

Kirchhoff's voltage law in matrix form can be applied to the circuit of Figure 26.33.

$$\begin{bmatrix} E_a \\ E_b \\ E_c \end{bmatrix} = \begin{bmatrix} Z_{aa} & Z_{ab} & Z_{ac} \\ Z_{ba} & Z_{bb} & Z_{bc} \\ Z_{ca} & Z_{cb} & Z_{cc} \end{bmatrix} \begin{bmatrix} If_a \\ If_b \\ If_c \end{bmatrix} + \begin{bmatrix} Z_f & 0 & 0 \\ 0 & Z_f & 0 \\ 0 & 0 & Z_f \end{bmatrix} \begin{bmatrix} If_a \\ If_b \\ If_c \end{bmatrix} + \begin{bmatrix} V_{ax} \\ V_{bx} \\ V_{cx} \end{bmatrix} + \begin{bmatrix} V_{xg} \\ V_{xg} \\ V_{xg} \end{bmatrix} \quad (26.200)$$

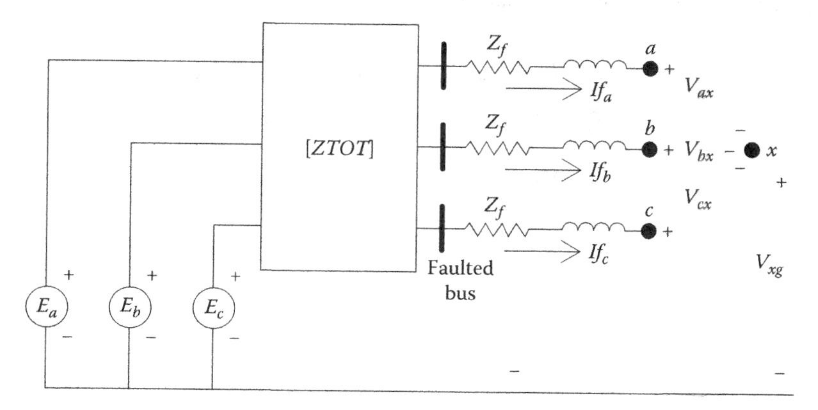

FIGURE 26.34 Unbalanced feeder short-circuit analysis model.

Equation 26.188 can be written in compressed form as

$$[E_{abc}] = [ZTOT][If_{abc}] + [ZF][If_{abc}] + [V_{abcx}] + [V_{xg}] \tag{26.201}$$

Combine terms in Equation 26.201.

$$[E_{abc}] = [ZEQ][If_{abc}] + [V_{abcx}] + [V_{xg}] \tag{26.202}$$

where

$$[ZEQ] = [ZTOT] + [ZF] \tag{26.203}$$

Solve Equation 26.202 for the fault currents:

$$[If_{abc}] = [YEQ][E_{abc}] - [YEQ][V_{abcx}] - [YEQ][V_{xg}] \tag{26.204}$$

where

$$[YEQ] = [ZEQ]^{-1} \tag{26.205}$$

Since the matrices $[YEQ]$ and $[E_{abc}]$ are known, define

$$[IP_{abc}] = [YEQ][E_{abc}] \tag{26.206}$$

Substituting Equation 26.206 into Equation 26.204 results in the following expanded equation:

$$\begin{bmatrix} If_a \\ If_b \\ If_c \end{bmatrix} = \begin{bmatrix} IP_a \\ IP_b \\ IP_c \end{bmatrix} - \begin{bmatrix} Y_{aa} & Y_{ab} & Y_{ac} \\ Y_{ba} & Y_{bb} & Y_{bc} \\ Y_{ca} & Y_{cb} & Y_{cc} \end{bmatrix} \begin{bmatrix} V_{ax} \\ V_{bx} \\ V_{cx} \end{bmatrix} - \begin{bmatrix} Y_{aa} & Y_{ab} & Y_{ac} \\ Y_{ba} & Y_{bb} & Y_{bc} \\ Y_{ca} & Y_{cb} & Y_{cc} \end{bmatrix} \begin{bmatrix} V_{xg} \\ V_{xg} \\ V_{xg} \end{bmatrix} \tag{26.207}$$

Performing the matrix operations in Equation 26.195:

$$\begin{aligned} If_a &= IP_a - (Y_{aa}V_{ax} + Y_{ab}V_{bx} + Y_{ac}V_{cx}) - Y_aV_{xg} \\ If_b &= IP_b - (Y_{ba}V_{ax} + Y_{bb}V_{bx} + Y_{bc}V_{cx}) - Y_bV_{xg} \\ If_c &= IP_c - (Y_{ca}V_{ax} + Y_{cb}V_{bx} + Y_{cc}V_{cx}) - Y_cV_{xg} \end{aligned} \tag{26.208}$$

where

$$\begin{aligned} Y_a &= Y_{aa} + Y_{ab} + Y_{ac} \\ Y_b &= Y_{ba} + Y_{bb} + Y_{bc} \\ Y_c &= Y_{ca} + Y_{cb} + Y_{cc} \end{aligned} \tag{26.209}$$

Equation 26.208 become the general equations that are used to simulate all types of short circuits. Basically there are three equations and seven unknowns (If_a, If_b, If_c, V_{ax}, V_{bx}, V_{cx}, and V_{xg}). The other three variables in the equations (IP_a, IP_b, and IP_c) are functions of the total impedance and the Thevenin voltages and are therefore known. In order to solve Equation 26.208, it will be necessary to specify four of the seven unknowns. These specifications are functions of the type of fault being simulated. The additional required four knowns for various types of faults are given below:

Three-phase faults

$$V_{ax} = V_{bx} = V_{cx} = 0$$
$$I_a + I_b + I_c = 0 \tag{26.210}$$

Three-phase-to-ground faults

$$V_{ax} = V_{bx} = V_{cx} = V_{xg} = 0 \tag{26.211}$$

Line-to-line faults (assume i–j fault with phase k unfaulted)

$$V_{ix} = V_{jx} = 0$$
$$If_k = 0 \tag{26.212}$$
$$If_i = If_j = 0$$

Line-to-line-to-ground faults (assume i–j to ground fault with k unfaulted)

$$V_{ix} = V_{jx} = V_{xg} = 0$$
$$V_{kx} = \frac{IP_k}{Y_{kk}} \tag{26.213}$$

Line-to-ground faults (assume phase k fault with phases i and j unfaulted)

$$V_{kx} = V_{xg} = 0$$
$$If_i = If_j = 0 \tag{26.214}$$

Notice that Equations 26.212 through 26.214 will allow the simulation of line-to-line, line-to-line-to-ground, and line-to-ground faults for all phases. There is no limitation to b–c faults for line-to-line and a–g for line-to-ground as is the case when the method of symmetrical components is employed.

References

Carson, J.R., Wave propagation in overhead wires with ground return, *Bell Syst. Tech. J.*, 5, 539–554, 1926.

Glover, J.D. and Sarma, M., *Power System Analysis and Design*, 2nd edn., PWS Publishing Company, Boston, MA, Chapter 5, 1994.

Gonen, T., *Electric Power Distribution System Engineering*, McGraw-Hill Book Company, New York, 1986.

Hopkinson, R.H., Approximate distribution transformer impedances, from an internal GE Memo dated August 30, 1977.

Kersting, W.H., Distribution system short circuit analysis, *25th Intersociety Energy Conversion Conference*, Reno, NV, August 1980.

Kersting, W.H., Milsoft transformer models—Theory, research Report, Milsoft Integrated Solutions, Inc., Abilene, TX, 1999.

Kersting, W.H. and Mendive, D.L., An application of ladder network theory to the solution of three-phase radial load-flow problems, *IEEE Conference Paper presented at the IEEE Winter Power Meeting*, New York, January 1976.

Kron, G., Tensorial analysis of integrated transmission systems, Part I: The six basic reference frames, *AIEE Trans.*, 71, 505–512, 1952.

Trevino, C., Cases of difficult convergence in load-flow problems, IEEE Paper no. 71-62-PWR, presented at the *IEEE Summer Power Meeting*, July 12–17, 1970, Los Angeles, CA, 1970.

27

Power System Operation and Control

George L. Clark
Alabama Power Company

Simon W. Bowen
Alabama Power Company

27.1 Implementation of Distribution Automation

The implementation of "distribution automation" (DA) within the continental United States is as diverse and numerous as the utilities themselves. Particular strategies of implementation utilized by various utilities have depended heavily on environmental variables such as size of the utility, urbanization, and available communication paths. The current level of interest in DA is the result of the following:

- The August 14, 2003, northeast blackout, which focused attention on infrastructure deficiencies and increased industry attention on sensor technology and digital control systems.
- Government and industry initiatives such as the DOE's GridWise Architecture Council, EPRI's IntelliGrid program, the Energy Independence and Security Act (EISA) of 2007, the American Recovery and Reinvestment Act (ARRA) of 2009, the publishing of the NIST Smart Grid Roadmap, and subsequent Smart Grid Interoperability Panel have led to significant investment in distribution research and development projects.
- The availability of low-cost, high-performance general purpose microprocessors, embedded processors, and digital signal processors, which have extended technology choices by blurring

the lines between traditional RTU (remote terminal unit), PLC (programmable logic controller), meter, and relay technologies, specifically capabilities that include meter accuracy measurements and calculations with power quality information including harmonic content.
- Continuous improvement in processor performance in host servers for the same or lower cost, lower cost of memory, and in particular the movement to Windows and Linux architectures.
- The threat of deregulation and competition as a catalyst to automate.
- Strategic benefits to be derived (e.g., potential of reduced labor costs, better planning from better information, optimizing of capital expenditures, reduced outage time, increased customer satisfaction).

While not meant to be all inclusive, this section on DA attempts to provide some dimension to the various alternatives available to the utility engineer. The focus will be on providing insight on the elements of automation that should be included in a scalable and extensible system. The approach will be to describe the elements of a "typical" DA system in a simple manner, offering practical observations as required.

27.2 Distribution SCADA History

SCADA (supervisory control and data acquisition) is the foundation for the DA system. The ability to remotely monitor and control electric power system facilities found its first application within the power generation and transmission sectors of the electric utility industry. The ability to significantly influence the utility bottomline through the effective dispatch of generation and the marketing of excess generating capacity provided economic incentive. The interconnection of large power grids in the Midwestern and the southern United States (1962) created the largest synchronized system in the world. The blackout of 1965 prompted the U.S. Federal Power Commission to recommend closer coordination between regional coordination groups (Electric Power Reliability Act of 1967) and gave impetus to the subsequent formation of the National Electric Reliability Council (1970). From that time (1970) forward, the priority of the electric utility has been to engineer and build a highly reliable and secure transmission infrastructure. The importance and urgency of closer coordination was reemphasized with the northeast blackout of 2003. Transmission SCADA became the base for the large energy management systems that were required to manage the transmission grid.

Distribution SCADA was not given equal consideration during this period. For electric utilities, justification for automating the distribution system, while being highly desirable, was not readily attainable based on a high cost/benefit ratio due to the size of the distribution infrastructure and cost of communication circuits. Still, there were tactical applications deployed on parts of distribution systems that were enough to keep the dream alive.

The first real deployments of distribution SCADA systems began in the late 1980s and early 1990s when SCADA vendors delivered reasonably priced "small" SCADA systems on low-cost hardware architectures to the small co-ops and municipality utilities. As the market expanded, SCADA vendors who had been providing transmission SCADA began to take notice of the distribution market. These vendors initially provided host architectures based on VAX/VMS and later on Alpha/OpenVMS platforms and on UNIX platforms. These systems were required for the large distribution utility (100,000–250,000 point ranges). These systems often resided on company-owned LANs with communication front-end (CFE) processors and user interface (UI) attached either locally on the same LAN or across a WAN.

In the mid-1980s, EPRI published definitions for DA and associated elements. The industry generally associates DA with the installation of automated distribution line devices such as switches, reclosers, sectionalizers, etc. The author's definition of DA encompasses the automation of the distribution substations and the distribution line devices. The automated distribution substations and the automated distribution line devices are then operated as a system to facilitate the operation of the electric distribution system.

The EISA of 2007 provided renewed impetus for the automation of distribution system. Title XIII of the EISA 2007 specifically states the "policy of the United States to support the modernization of the nation's electricity transmission and distribution system to maintain a reliable and secure electricity infrastructure." The emerging Smart Grid is described in Title XIII and characterized by the increased use of digital technology, dynamic optimization of grid operations, use of distributed resources to support the grid, and the deployment of "smart" technologies including DA. The EPRI Green Circuits initiative, which is in response to EISA 2007 Title XIII, utilizes DA telemetry to obtain the necessary feeder information to achieve the Smart Grid objectives. Thus, distribution SCADA and DA are viable and available technologies to advance the modernization of transmission and distribution system.

EISA 2007 Title XIII also provides for the development of standards for communication and interoperability of devices deployed in the Smart Grid. The National Institute of Standards and Technology (NIST) is the coordinating authority for protocol and model standards development to support smart grid device interoperability. In the development of protocol standards, NIST solicits input from the GridWise Architecture Council, the Institute of Electrical and Electronics Engineers (IEEE), and other interested parties. Interoperability is a key element of the standards development. The interoperability aspect of the standards is verified by industry-supported testing. NIST created priority actions plans (PAP) for the developing standards required for an interoperable Smart Grid.

27.2.1 SCADA System Elements

At a high level, the elements of a DA system can be divided into three main areas:

- SCADA application and servers
- DMS applications and servers
- Trouble management applications and servers

27.2.2 Distribution SCADA

As was stated in the introduction, the SCADA system is foundational to the distribution management system (DMS) architecture. A SCADA system should have all of the infrastructure elements to support the multifaceted nature of DA and the higher level applications of a DMS. A distribution SCADA system's primary functions, e.g., telemetry, alarming, event recording, and remote control of field equipment, all support distribution operations. Historically, SCADA systems have been notorious for their lack of support for the import and, more importantly, the export of power system data values. A modern SCADA system should support the engineering budgeting and planning functions by providing a well-defined and documented application programming interface for access to power system data without requiring possession of an operational workstation. The main elements of a SCADA system are

- Host equipment
- Communication infrastructure (network and serial communications)
- Field devices (in sufficient quantity to support operations and telemetry requirements of a DMS platform)

27.2.3 Host Equipment

The authors feel that the essential elements of a distribution SCADA host are

- Host servers (redundant servers with backup/failover capability)
- Communication front-end nodes (network based)
- FGUIs
- Relational database server (for archival of historical power system values) and data server/Web server (for access to near real-time values and events)

The elements and components of the typical DA system are illustrated in Figure 27.1.

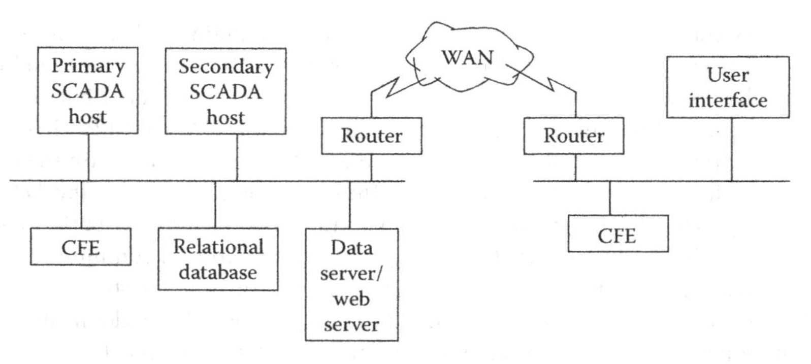

FIGURE 27.1 DA system architecture.

27.2.4 Host Computer System

27.2.4.1 SCADA Servers

As SCADA has proven its value in operation during inclement weather conditions, service restoration, and daily operations, the dependency on SCADA has created a requirement for highly available and high-performance systems. High-performance servers with abundant physical memory, RAID hard disk systems, and LAN connection are typical of today's SCADA high-performance systems. Redundant server hardware operating in a "live" backup/failover mode is required to meet the high availability criteria. In meeting the high availability criteria, electric utilities may also include a remote SCADA host configuration for disaster recovery.

27.2.5 Communication Front-End Processors

Most utilities will utilize more than one communication transport with the particular choice based on system requirements, license availability for licensed frequencies, coverage, loading, and economics. However, the preponderance of SCADA host to field device communications still depends heavily on serial communications. That is to say, no matter what the communication medium used, the electrical interface to the SCADA system (CFE) is still most often a serial interface, not a network interface. The host/RTU interface requirement is filled by the CFE. The CFE can come in several forms based on bus architecture (older CFE technologies were most often based on VME or PCI bus systems with custom serial controllers). Currently, CFE architectures are predominately Intel/Windows architectures with the serial controller function performed by the main processor instead of specialized serial controllers. Current trends include the use of terminal server architectures. Location of the CFE in relation to the SCADA server can vary based on requirement. In some configurations, the CFE is located on the LAN with the SCADA server. In other cases, existing communications hubs may dictate that the CFE resides remotely across a WAN at the communication hub. The incorporation of the WAN into the architecture requires a more robust CFE application to compensate for intermittent interruptions of network connectivity (relatively speaking—comparing WAN to LAN communication reliability).

The advent of new architectures for CFEs will offer new capabilities and opportunities for sharing data within the utility. The ability to serve data through a nonproprietary protocol such as ICCP offers the possibility for rethinking SCADA architectures within large utilities that may have more than one SCADA system or more than one audience for SCADA information.

In general, the CFE will include three functional devices: a network/CPU board, either dedicated serial cards or terminal servers, and possibly a time code receiver. Functionality should include the ability to download configuration and scan tables. The CFE should also support the ability to dead band values (i.e., report only those analog values that have changed by a user-defined amount). Even when exception scanning/reporting is used, the CFE, network, and SCADA servers should be capable of supporting worst-case conditions (i.e., all points changing outside of the dead band limits), which typically occur during severe system disturbances. Deterministic communications with known data solicitation rates facilitate the sizing of the SCADA database and the performance of the SCADA system during wide-area storm events. Deterministic serial communications with the RTU are required for secure predictable data acquisition and supervisory control.

27.2.6 Full Graphics User Interface

The current distribution SCADA UI is a full graphics (FG) UI. In the mid-1990s, the SCADA vendors implemented their FGUI on low-cost NT and XP workstations using third-party applications to emulate the X11 Windows system. Today, most UI is natively integrated into the Windows architecture or implemented as a "browser"-like application. FG displays provide the ability to display power system data along with the electric distribution facilities in a geographical (or semi geographical) perspective. The advantage of using a FG interface becomes evident (particularly for distribution utilities) as SCADA is deployed beyond the substation fence where feeder diagrams become critical to distribution operations.

27.2.7 Relational Databases, Data Servers, and Web Servers

The traditional SCADA systems were poor providers of data to anyone not connected to the SCADA system by an operational console. This occurred due to the proprietary nature of the performance (in memory) database and its design optimization for putting scanned data in and pushing display values out. Power system quantities such as bank and feeder loading (MW, MWH, MQH, and ampere loading) and bus volts provide valuable information to the distribution planning engineer. The maintenance engineer frequently uses the externalized SCADA data to identify trends and causality information to provide more effective and efficient equipment maintenance. The availability of event (log) data is important in postmortem analysis. The use of relational databases, data servers, and Web servers by the corporate and engineering functions provides access to power system information and data while isolating the SCADA server from nonoperations personnel.

27.2.8 Host to Field Communications

There are many communication media available to distribution SCADA for host/remote communications today. Some SCADA implementations utilize a network protocol over fiber to connect the SCADA hosts to substation automation systems; typically, this is more often found in a small co-op or PUD who may have a relatively small substation count. Communication technologies such as frame relay, multiple address system (MAS) radio, 900 MHz unlicensed, cell-based phone and radio, and even satellite find common usage today. Early in the twenty-first century, new technologies emerged that were expected to enter the mix of host/RTU communications (e.g., WiFi, WiMAX, and even broadband over power line [BPL] are possibilities at least for data acquisition). These technologies have not had broad adoption as of the writing of this chapter. (The authors do not recommend supervisory control over BPL.) The authors believe that a mixture of all of these technologies will be utilized for RTU communications. However, we feel that broadband radio networks supporting both serial and IP-based communications will predominate.

Radio technologies offer good communications value. One such technology is the MAS radio. The MAS operates in the 900 MHz range and is omnidirectional, providing radio coverage in an area with radius up to 20–25 miles depending on terrain. A single MAS master radio can communicate with many remote sites. The 900 MHz remote radio depends on a line-of-sight path to the MAS master radio. Protocol and bandwidth limit the number of RTUs that can be communicated with by a master radio. The protocol limit is simply the address range supported by the protocol. Bandwidth limitations can be offset by the use of efficient protocols or slowing down the scan rate to include more remote units. Spread-spectrum and point-to-point radio (in combination with MAS) offers an opportunity to address specific communication problems, e.g., terrain changes or buildings within the MAS radio line-of-sight. At the present time, MAS radio is preferred (authors' opinion) to packet radio (another new radio technology); MAS radio communications tend to be more deterministic, providing for smaller timeout values on communication no-responses and controls.

Wireless communications support the wide area deployment requirement for DA. The MAS radio infrastructure meets this requirement. However, the commercial cellular infrastructure is another option for consideration by the utility. While the coverage may not include the entire utility service territory, the commercial cellular option can potentially be complementary to the wide area deployment requirement. The guarantee of wireless service should be considered to ensure the operations of the electric system during the clear, blue-sky day and during a wide area system disturbance event.

27.3 Field Devices

DA field devices are multifeatured installations meeting a broad range of control, operations, planning, and system performance issues for the utility personnel. Each device provides specific functionality, supports system operations, includes fault detection, captures planning data, and records power quality information. These devices are found in the distribution substation and at selected locations along the distribution line. The multifeatured capability of the DA device increases its ability to be integrated into the electric distribution system. The functionality and operations capabilities complement each other with regard to the control and operation of the electric distribution system. The fault detection feature is the "eyes and ears" for the operating personnel. The fault detection capability becomes increasingly more useful with the penetration of DA devices on the distribution line.

The real-time data collected by the SCADA system are provided to the planning engineers for inclusion in the radial distribution line studies. As the distribution system continues to grow, the utility makes annual investments to improve the electric distribution system to maintain adequate facilities to meet the increasing load requirements. The use of the real-time data permits the planning engineers to optimize the annual capital expenditures required to meet the growing needs of the electric distribution system.

The power quality information includes capturing harmonic content to the 15th harmonic or greater and recording percent total harmonic distortion (%THD). This information is used to monitor the performance of the distribution electric system.

27.3.1 Modern RTU

Today's modern RTU is modular in construction with advanced capabilities to support functions that heretofore were not included in the RTU design. The modular design supports installation configurations ranging from the small point count required for the distribution line pole-mounted units to the very large point count required for large bulk-power substations and power plant switchyard installations. The modern RTU modules include analog units with 9 points, control units with 4 control pair points, status units with 16 points, and communication units with power supply. The RTU installation requirements are met by accumulating the necessary number of modern RTU modules to support the analog, control, status, and communication requirements for the site to be automated. Packaging of the minimum point count RTUs is available for the distribution line requirement. The substation automation requirement has the option of installing the traditional RTU in one cabinet with connections to the substation devices or distributing the RTU modules at the devices within the substation with fiber-optic communications between the modules. The distributed RTU modules are connected to a data concentrating unit which in turn communicates with the host SCADA computer system.

The modern RTU accepts direct AC inputs from a variety of measurement devices including line-post sensors, current transformers, potential transformers, station service transformers, and transducers. Direct AC inputs with the processing capability in the modern RTU support fault current detection and harmonic content measurements. The modern RTU has the capability to report the magnitude, direction, and duration of fault current with time tagging of the fault event to 1 ms resolution. Monitoring and reporting of harmonic content in the distribution electric circuit are capabilities that are included in the modern RTU. The digital signal processing capability of the modern RTU supports the necessary calculations to report %THD for each voltage and current measurement at the automated distribution line or substation site.

The modern RTU includes logic capability to support the creation of algorithms to meet specific operating needs. Automatic transfer schemes have been built using automated switches and modern RTUs with the logic capability. This capability provides another option to the distribution line engineer when developing the method of service and addressing critical load concerns. The logic capability in the modern RTU has been used to create the algorithm to control distribution line switched capacitors for operation on a per-phase basis. The capacitors are switched on at zero voltage crossing and switched off at zero current crossing. The algorithm can be designed to switch the capacitors for various system parameters such as voltage, reactive load, time, etc. The remote control capability of the modern RTU then allows the system operator to take control of the capacitors to meet system reactive load needs.

The modern RTU has become a dynamic device with increased capabilities. The new logic and input capabilities are being exploited to expand the uses and applications of the modern RTU.

27.3.2 PLCs and IEDs

PLCs and intelligent electronic devices (IEDs) are components of the DA system, which meet specific operating and data gathering requirements. While there is some overlap in capability with the modern RTU, the authors are familiar with the use of PLCs for automatic isolation of the faulted power transformer in a two-bank substation and automatic transfer of load to the unfaulted power transformer to maintain an increased degree of reliability. The PLC communicates with the modern RTU in the substation to facilitate the remote operation of the substation facility. The typical PLC can support serial communications to a SCADA server. The modern RTU has the capability to communicate via an RS-232 interface with the PLC.

IEDs include electronic meters, electronic relays, and controls on specific substation equipment such as breakers, regulators, LTC on power transformers, etc. The IEDs also have the capability to support serial communications to a SCADA server. The authors' experience indicates that substation IEDs are either connected to a substation automation master via a substation LAN or reporting to the modern RTU (and thus to the SCADA host) via a serial interface using ASCII or vendor-specific protocol. Recent improvement in measurement accuracy and inclusion of power quality (harmonic content) especially in the realm of electronic relays are making the IED an important part of the substation protection and automation strategy.

27.3.3 Substation

The installation of the SCADA technology in the DA substation provides for the full automation of the distribution substation functions and features. The modular RTU supports various substation sizes and configuration. The load on the power transformer is monitored and reported on a per-phase basis. The substation low-side bus voltage is monitored on a per-phase basis. The distribution feeder breaker is fully automated. Control of all breaker control points is provided, including the ability to remotely set up the distribution feeder breaker to support energized distribution line work. The switched capacitor banks and substation regulation are controlled from the typical modular RTU installation. The load on the distribution feeder breaker is monitored and reported on a per-phase basis as well as on a three-phase basis. This capability is used to support the normal operations of the electric distribution system and to respond to system disturbances. The installation of the SCADA technology in the DA substation eliminates the need to dispatch personnel to the substation except for periodic maintenance and equipment failure.

Substation automation solutions are being installed, utilizing dedicated processing nodes with FG interfaces for local operations at the substation with IED integration with data concentration. These solutions have the capability to provide a local area SCADA control system incorporating feeder automation along the connected feeders. Substation SCADA telemetry is also provided through either serial or network connections to the centralized SCADA host and possibly to other engineering applications, which are located on utility networks. The NIST Roadmap is promoting IEC 61850 as the substation network protocol. However, IEC 61850 has not been widely adopted by U.S. utilities. DNP3 is de facto protocol selection at least by the U.S. utilities. IEC 61850 has a broader acceptance and deployment in the European utility environment.

27.3.4 Line

The DA distribution line applications include line monitoring, pole-mounted reclosers, gang-operated switches equipped with motor operators, switched capacitor banks, pole-mounted regulators, and pad-mounted automatic transfer switchgear. The modular RTU facilitates the automation of the distribution line applications. The use of the line-post sensor facilitates the monitoring capability on a per-phase basis. The direct AC input from the sensors to the RTU supports monitoring of the normal load, voltage, and power factor measurements, and also the detection of fault current. The multifeatured distribution line DA device can be used effectively to identify the faulted sections of the distribution circuit during system disturbances, isolate the faulted sections, and restore service to the unfaulted sections of the distribution circuit. The direct AC inputs to the RTU also support the detection and reporting of harmonics and the %THD per phase for voltage and current. Fault detection (forward and reverse) per phase as well as fault detection on the residual current is supported in the RTU. Vendor offerings now include package solutions with controllers that provide functionality and telemetry accuracy with harmonic content equivalent to the traditional RTU, which communicate directly to the SCADA host.

27.3.5 Other Line Controller Schemes

Vendors are providing package schemes for automation of the distribution feeder. The scheme provides a self-contained solution for the operation of the portion of the distribution feeder controlled within the scheme. The solution typically supports fault isolation and restoration to the unfaulted portion of the distribution feeder within the scheme. Feeder telemetry is supported from devices within the schemes. Communications within the scheme are typically included in the vendor package scheme. The vendor will work with the host utility to establish communications between the package scheme and the utility's distribution SCADA system. The introduction of a subcommunication system in the system-wide communications for distribution SCADA may result in data latency. This latency of feeder telemetry data transmission between dissimilar communication systems needs to be considered when incorporating package scheme devices into an advanced DMS solution.

27.3.6 Tactical and Strategic Implementation Issues

As the threat of deregulation and competition emerges, retention of industrial and large commercial customers will become the priority for the electric utility. Every advantage will be sought by the electric utility to differentiate itself from other utilities. Reliable service, customer satisfaction, fast storm restorations, and power quality will be the goals of the utility. Differing strategies will be employed based on the customer in question and the particular mix of goals that the utility perceives will bring customer loyalty.

For large industrial and commercial customers, where the reliability of the electric service is important and outages of more than a few seconds can mean lost production runs or lost revenue, tactical automation solutions may be required. Tactical solutions are typically transfer schemes or switching schemes that can respond independently of operator action, reporting the actions that were initiated in response to loss of` preferred service and/or line faults. The requirement to transfer source power or reconfigure a section of the electric distribution system to isolate and reconnect in a matter of seconds is the primary criteria. Tactical automation based on local processing provides the solution.

In cases where there are particularly sensitive customer requirements, tactical solutions are appropriate. When the same requirements are applied to a large area and/or customer base, a strategic solution based on a distribution management platform is preferred. This solution requires a DMS with a system operational model that reflects the current configuration of the electric distribution system. Automatic fault isolation and restoration applications, which can reconfigure the electric distribution system, require a "whole and dynamic system" model in order to operate correctly and efficiently.

27.3.7 Distribution Management Platform

So, while tactical automation requirements exist and have significant impact and high profile, goals that target system issues require a strategic solution. A DMS is the capstone for automation of the distribution system and includes advance distribution applications, integrated SCADA, integrated trouble call and outage management, and distribution operations training simulator (DOTS) at a minimum.

27.3.8 Advanced Distribution Applications

Transmission EMS systems have had advanced applications for many years. The distribution management platform will include advanced applications for distribution operations. A true DMS should include advanced applications such as volt/VAR control, automatic fault isolation and service restoration, operational power flows, contingency analysis, loss minimization, switching management, etc.

27.4 Integrated SCADA System

A functional DMS platform should be fully integrated with the distribution SCADA system. The SCADA–DMS interface should be fully implemented with the capability of passing data [discrete indication (status) and values (analog)] bidirectionally. The SCADA interface should also support device control. Figure 27.2 details the components of a DMS.

27.4.1 Trouble Call and Outage Management System

In addition to the base SCADA functionality and high-level DMS applications, the complete DA system will include a trouble call and outage management system (TCOMS). TCOMS collects trouble calls received by human operators and interactive voice recorders (IVR). The trouble calls are fed to an analysis/prediction engine that has a model of the distribution system with customer to electrical address relationships. Outage prediction is presented on a FG display that overlays the distribution system on CAD base information. A TMS also provides for the dispatch and management of crews, customer callbacks, accounting, and reports. A SCADA interface to a TCOMS provides the means to provide confirmed (SCADA telemetry) outage information to the prediction engine. Figure 27.3 shows a typical TCOMS.

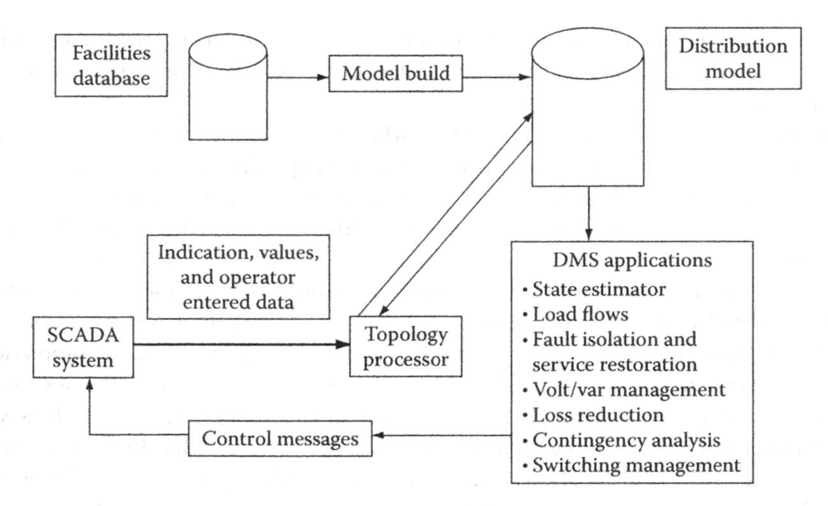

FIGURE 27.2 A DMS platform with SCADA interface.

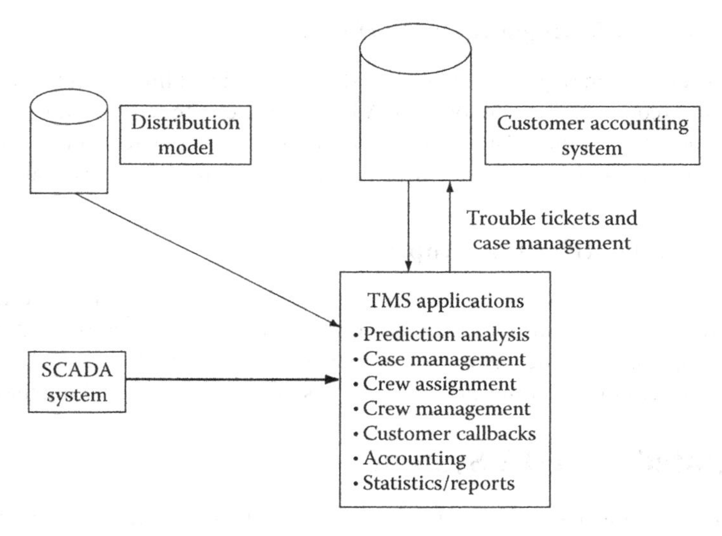

FIGURE 27.3 A TCOMS platform with SCADA interface.

27.4.2 Distribution Operations Training Simulator

With the graying of the American workforce and subsequent loss of expertise, there is a requirement to provide better training for the distribution operator. A DOTS will provide the ability to train and test the distribution operator with real-world scenarios captured (and replayed) through the DOTS. The DOTS instructor will be able to "tweak" the scenarios, varying complexity and speed of the simulation, providing the distribution operator with the opportunity to learn best practices and to test his skills in an operational simulation without consequences of making operational mistakes on the "real distribution system."

27.5 Security

In today's environment, security of control systems has become an important topic. The dependence by electric utilities on digital control systems for operations coupled with the threat of terrorist activity whether by governments or individuals is beyond the scope of this chapter. However, it should be noted that most distribution SCADA systems (unlike transmission SCADA and EMS systems which are often on their own separate network) often reside on the utilities corporate networks, elevating the risk of exposure to viruses, worms, and Trojan horses.

Every electric utility, no matter what size, should have the appropriate policy and procedures in place to secure their distribution "control system" from malicious or accidental harm. Securing administrator accounts, password aging policies, passwords with requirements on length and requirements on the mixture of character types, two factor authentication, virus protection, firewalls, intrusion detection, and securing the physical and electronic perimeter have all become a part of the vocabulary for SCADA system support staffs.

27.6 Practical Considerations

27.6.1 Choosing the Vendor

27.6.1.1 Choosing a Platform Vendor

In choosing a platform (SCADA, DMS, TCOMS) vendor, there are several characteristics that should be kept in mind (these should be considered as a rule of thumb based on experience of what works and what does not). Choosing the right vendor is as important as choosing the right software package.

Vendor characteristics that the authors consider important are the following:

- A strong "product" philosophy. Having a strong product philosophy is typically a chicken and egg proposition. Which came first, the product or the philosophy? Having a baseline SCADA application can be a sign of maturity and stability. Did the platform vendor get there by design or did they back into it? Evidence of a product philosophy includes a baseline system that is in production and enhancements that are integrated in a planned manner with thorough testing on the enhancement and regression testing on the product along with complete and comprehensive documentation.
- A documented development and release path projected 3–5 years into the future.
- By inference from the first two bullets, a vendor who funds planned product enhancements from internal funds.
- A strong and active user group that is representative of the industry and industry drivers.
- A platform vendor that actively encourages its user group by incentive (e.g., dedicating part of its enhancement funding to user group initiatives).
- A vendor that is generally conservative in moving its platform to a new technology; one that does not overextend its own resources.
- Other considerations.
- As much as possible, purchase the platform as an off-the-shelf product (i.e., resist the urge to ask for customs that drive your system away from the vendor's baseline).
- If possible, maintain/develop your own support staff.

All "customization" should be built around the inherent capabilities and flexibility of the system (i.e., do not generate excessive amounts of new code). Remember, you will have to reapply any code that you may have developed to every new release, or worse, you will have to pay the vendor to do it for you.

27.7 Standards

27.7.1 Internal Standards

The authors highly recommend the use of standards (internal to your organization) as a basis for ensuring a successful DA or SCADA program. Well-documented construction standards that specify installation of RTUs, switches, and line sensors with mechanical and electrical specifications will ensure consistent equipment installations from site to site. Standards that cover nontrivial but often overlooked issues can often spell the difference between acceptance and rejection by operational users and provide the additional benefit of having a system that is "maintainable" over the 10–20 years (or more) life of a system. Standards that fall in this category include standards that cover point-naming conventions, symbol standards, display standards, and the all-important operations manual.

27.7.2 Industry Standards

In general, standards fall into two categories: standards that are developed by organizations and commissions (e.g., EPRI, IEEE, ANSI, IEC, CCITT, ISO) and de facto standards that become standards by virtue of widespread acceptance. As an example of what can occur, the reader is invited to consider what has happened in network protocols over the recent past with regard to the OSI model and TCP/IP.

International Electrotechnical Commission (IEC) is developing the Common Information Model (CIM) for the utility business units. The CIM is a suite of protocol standards including IEC 61850 for smart substation devices, IEC 61970 for the transmission planning model, and IEC 61968 for the distribution business unit.

Past history of SCADA and automation has been dominated by the proprietary nature of the various system vendor offerings. Database schemas and RTU communication protocols are exemplary of proprietary design philosophies utilized by SCADA platform and RTU vendors. Electric utilities that operate as part of the interconnected power grid have been frustrated by the lack of ability to share power system data between dissimilar energy management systems. The same frustration exists at the device level; RTU vendors, PLC vendors, electronic relay vendors, and meter vendors each having their own product protocols have created a "tower of Babel" problem for utilities. Recently, several communications standards organizations and vendor consortiums have proposed standards to address these deficiencies in intersystem data exchange, intrasystem data exchange (corporate data exchange), and device level interconnectivity. Some of the more notable examples of network protocol communication standards are ICCP (intercontrol center protocol), UCA (utility communication architecture), CCAPI (control center applications interface), and UIB (utility integration bus). For database schemas, EPRI's CIM (common information model) is gaining supporters. In RTU, PLC, and IED communications, DNP 3.0 has also received much attention from the industry's press.

In 2010, IEEE announced the ratification of its IEEE 1815 Distributed Network Protocol (DNP3) standard for electric power systems communications. The IEEE announcement stated, in part, that "the new standard improves device interoperability and strengthens security protocols. IEEE 1815 is expected to play a significant role in the development and deployment of Smart Grid technologies." The NIST PAP 12 provides for DNP3 mapping to IEC 61850 objects. IEEE 1815 supports the achievement of NIST PAP 12. In light of the number of standards that have appeared (and then disappeared) and the number of possibly competing "standards" that are available today, the authors, while acknowledging the value of standards, prefer to take (and recommend) a cautious approach to standards. A wait-and-see posture may be an effective strategy. Standards by definition must have proven themselves over time. Difficulties in immediately embracing new standards are due in part to vendors having been allowed to implement only portions of a standard, thereby nullifying the hopefully "plug-and-play" aspect for adding new devices. Also, the trend in communication protocols has been to add functionality in an attempt to be all inclusive, which has resulted in an increased requirement on bandwidth. Practically speaking, utilities that have already existing infrastructure may find it economical to resist the deployment of new protocols. In the final analysis, as in any business decision, a "standard" should be accepted only if it adds value and benefit that exceeds the cost of implementation and deployment.

27.8 Deployment Considerations

The definition of the automation technology to be deployed should be clearly delineated. This definition includes the specification of the host systems, the communication infrastructure, the automated end-use devices, and the support infrastructure. This effort begins with the development of a detailed installation plan that takes into consideration the available resources. The pilot installation will never be any more than a pilot project until funding and manpower resources are identified and dedicated to the enterprise of implementing the technologies required to automate the electric distribution system. The implementation effort is best managed on an annual basis with stated incremental goals and objectives for the installation of automated devices. With the annual goals and objectives identified, then the budget process begins to ensure that adequate funding is available to support the implementation plan. To ensure adequate time to complete the initial project tasks, the planning should begin 18–24 months prior to the budget year. During this period, the identification of specific automation projects is completed. The initial design work is commenced with the specification of field automation equipment (e.g., substation RTU based on specific point count requirements and distribution line RTU). The verification of the communication to the selected automation site is an urgent early consideration in order to minimize the cost of achieving effective remote communications. As the installation year approaches, the associated automation equipment (e.g., switches, motor operators, sensors) must be verified to ensure that adequate supplies are stocked to support the implementation plan.

The creation of a SCADA database and display is on the critical path for new automated sites. The database and display are critical to the efficient completion of the installation and checkout tasks. Data must be provided to the database and display team with sufficient lead time to create the database and display for the automated site. The database and display are subsequently used to check out the completed automated field device. The point assignment (PA) sheet is a project activity that merits serious attention. The PA sheet is the basis for the creation of the site-specific database in the SCADA system. The PA sheet should be created in a consistent and standard fashion. The importance of an accurate database and display cannot be overemphasized. The database and display form the basis for the remote operational decisions for the electric distribution system using the SCADA capability. Careful coordination of these project tasks is essential to the successful completion of the annual automation plans.

Training is another important consideration during the deployment of the automation technology. The training topics are as varied as the multidisciplined nature of the DA project. Initial training requirements include the system support personnel in the use and deployment of the automation platform, the end user (operator) training, and installation teams. Many utilities now install new distribution facilities using energized line construction techniques. The automated field device adds a degree of complexity to the construction techniques to ensure adherence to safe practices and construction standards. These training issues should be addressed at the outset of the planning effort to ensure a successful DA project.

27.8.1 Support Organization

The support organization must be as multidisciplined as the DA system is multifeatured. The support to maintain a deployed DA system should not be underestimated. Functional teams should be formed to address each discipline represented within the DA system. The authors recommend forming a core team that is made up of representation from each area of discipline or area of responsibility within the DA project. These areas of discipline include the following:

- Host SCADA system
- UI
- Communication infrastructure

- Facilities design personnel for automated distribution substation and distribution line devices
- System software and interface developments
- Installation teams for automated distribution substation and distribution line devices
- End users (i.e., the operating personnel)

The remaining requirement for the core team is project leadership with responsibility for the project budget, scheduling, management reports, and overall direction of the DA project. The interaction of the various disciplines within the DA team will ensure that all project decisions are supporting the overall project goals. The close coordination of the various project teams through the core team is essential to minimizing decision conflict and maximizing the synergy of project decisions. The involvement of the end user at the very outset of the DA project planning cannot be overemphasized. The operating personnel are the primary users of the DA technology. The participation of the end user in all project decisions is essential to ensure that the DA product meets business needs and improves the operating environment in the operating centers. One measure of good project decisions is found in the response of the end user. When the end user says, "I like it," then the project decision is clearly targeting the end user's business requirements. With this goal achieved, the DA system is then in a position to begin meeting other corporate business needs for real-time data from the electric distribution system.

28

Hard to Find Information (on Distribution System Characteristics and Protection)*

Jim Burke
Quanta Technology

28.1 Overcurrent Protection

28.1.1 Introduction

The distribution system shown in Figure 28.1 illustrates many of the features of a distribution system making it unique. The voltage level of a distribution system can be anywhere from about 5 kV to as high as 35 kV with the most common voltages in the 15 kV class. Areas served by a given voltage are proportional to the voltage itself, indicating that, for the same load density, a 35 kV system can serve considerably longer lines. Lines can be as short as a mile or two and as long as 20 or 30 miles.

* This chapter is a collection of unrelated "hard to find" information pertinent to the distribution engineer. This material has been excerpted from Mr. Burke's guides entitled *Hard to Find Information About Distribution Systems*, Volumes I and II. These two documents are used in seminars, in conjunction with Mr. Burke's book, *Power Distribution Engineering: Fundamentals and Applications*, also published by Taylor & Francis.

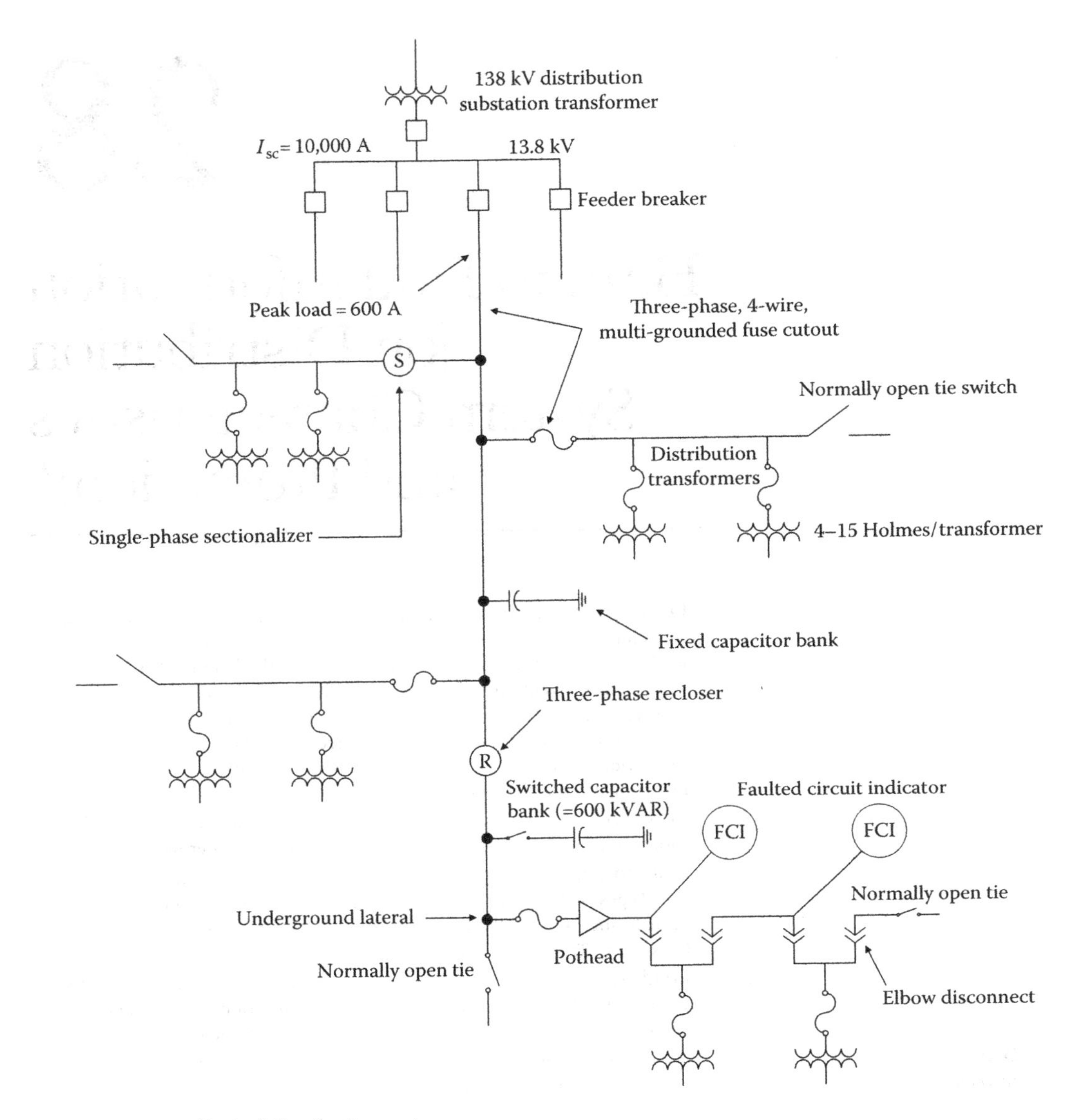

FIGURE 28.1 Typical distribution system.

Typically, however, lines are generally 10 miles or less. Short-circuit levels at the substation are dependent on voltage level and substation size. The average short-circuit level at a distribution substation has been shown, by survey, to be about 10,000 A. Feeder load current levels can be as high as 600 A but rarely exceed about 400 A with many never exceeding a couple of hundred amperes.

28.1.2 Fault Levels

There are two types of faults: low impedance and high impedance. A high-impedance fault is considered to be a fault that has a high Z due to the contact of the conductor to the earth, i.e., Z_f is high. By this definition, a bolted fault at the end of a feeder is still classified as a low-impedance fault. A summary of findings on faults and their effects is as follows.

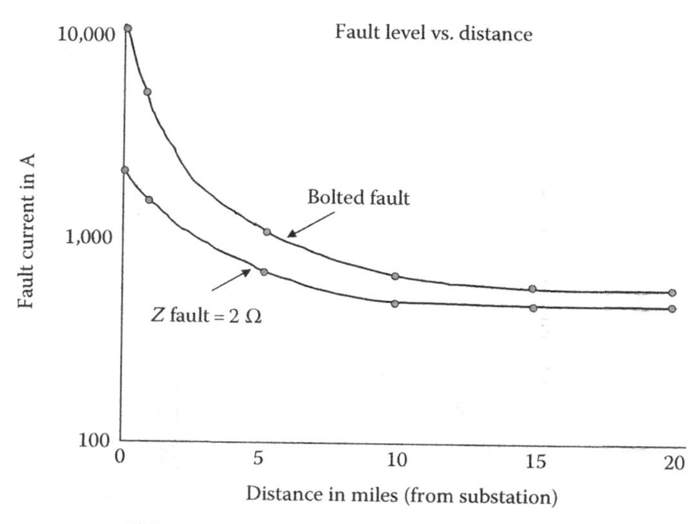

FIGURE 28.2 Low-impedance faults.

28.1.2.1 Low-Impedance Faults

Low-impedance faults or bolted faults can be either very high in current magnitude (10,000 A or above) or fairly low, for example, 300 A at the end of a long feeder. Faults that can be detected by normal protective devices are all low-impedance faults. These faults are such that the calculated value of fault current assuming a "bolted fault" and the actual are very similar. Most detectable faults, per study data, do indeed show that fault impedance is close to 0 Ω. This implies that the phase conductor either contacts the neutral wire or that the arc to the neutral conductor has a very low impedance. An EPRI study performed by the author over 10 years ago indicated that the maximum fault impedance for a detectable fault was 2 Ω or less. Figure 28.2 indicates that 2 Ω of fault impedance influences the level of fault current depending on location of the fault. As can be seen, 2 Ω of fault impedance considerably decreases the level of fault current for close-in faults but has little effect for faults some distance away. What can be concluded is that *fault impedance does not significantly affect faulted circuit indicator performance* since low level faults are not greatly altered.

28.1.2.2 High-Impedance Faults

High-impedance faults are faults that are low in value, i.e., generally less than 100 A due to the impedance between the phase conductor and the surface on which the conductor falls. Figure 28.3 illustrates that most surface areas, whether wet or dry, do not conduct well. If one considers the fact that an 8 ft ground rod sunk into the earth more often than not results in an impedance of 100 Ω or greater, then it is not hard to visualize the fact that a conductor simply lying on a surface cannot be expected to have a low impedance. These faults, called high-impedance faults, do not contact the neutral and do not arc to the neutral. They are not detectable by any conventional means and are not to be considered at all in the evaluation of fault current indicators (FCIs) and most other protective devices.

28.1.3 Surface Current Levels

See Figure 28.3.

28.1.4 Reclosing and Inrush

On most systems where most faults are temporary, the concept of reclosing and the resulting inrush currents are a fact of life. Typical reclosing cycles for breakers and reclosers are different and are shown in Figure 28.4.

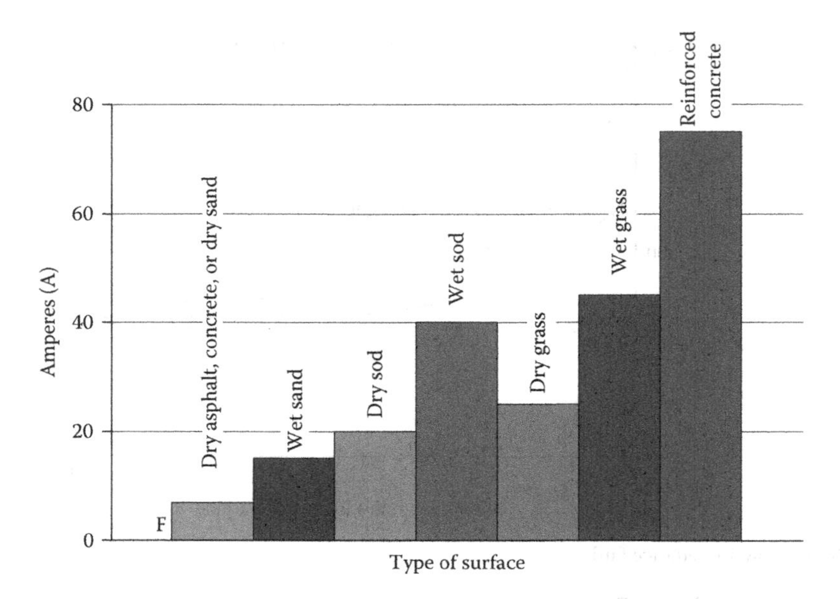

FIGURE 28.3 High-impedance fault current levels.

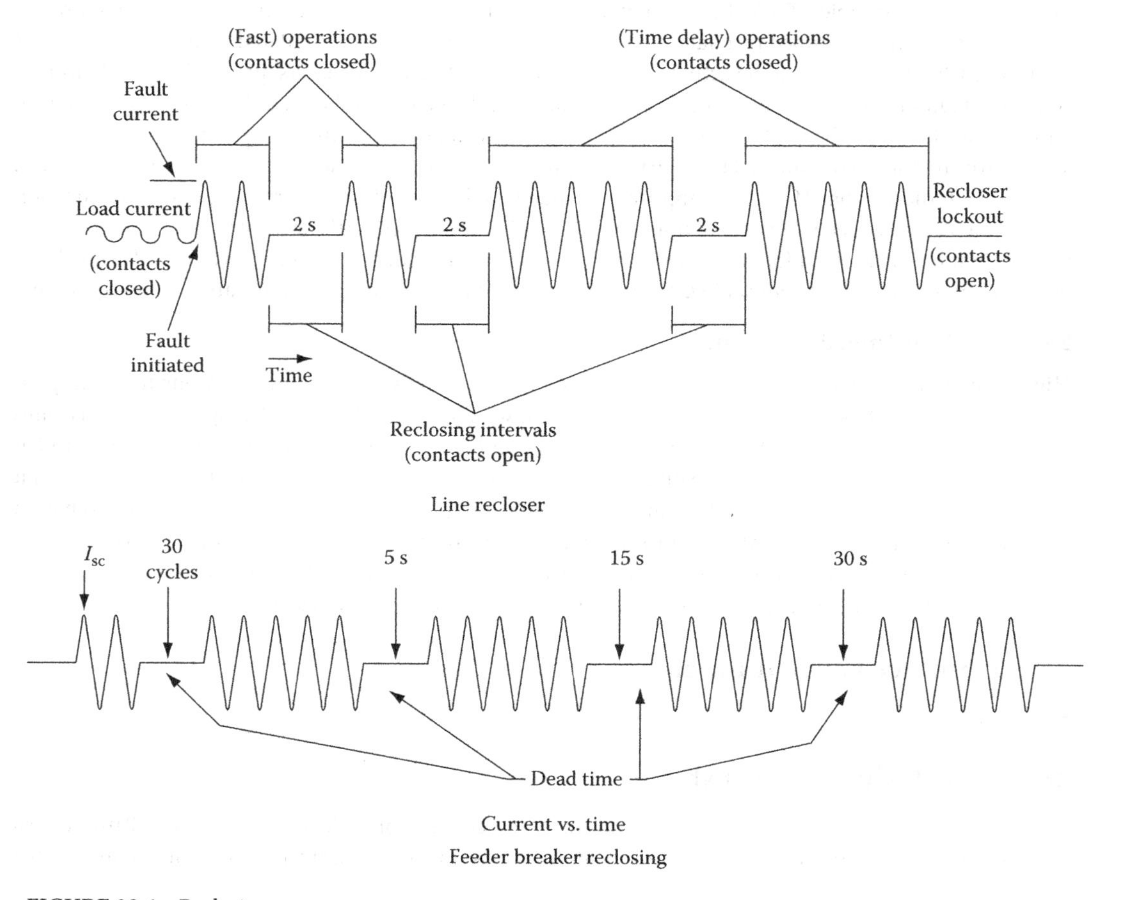

FIGURE 28.4 Reclosing sequences.

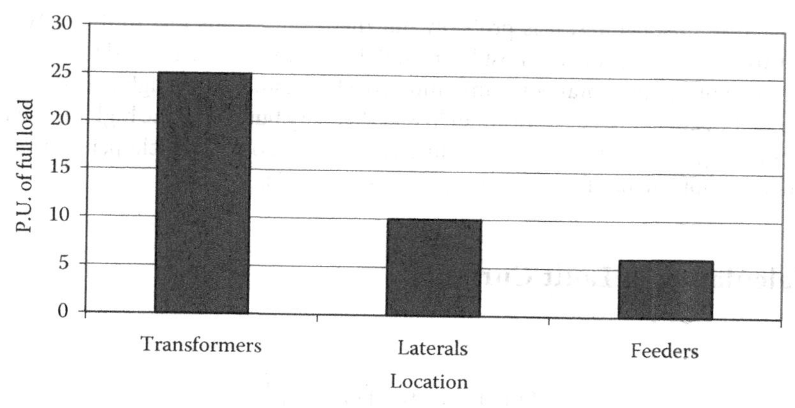

FIGURE 28.5 Magnitudes of inrush current.

These reclosing sequences produce inrush primarily resulting from the connected transformer kVA. This inrush current is high and can approach the actual fault current level in many instances. Figure 28.5 shows the relative magnitude of these currents. What keeps most protective devices from operating is that the duration of the inrush is generally short and as a consequence will not melt a fuse or operate a time delay relay.

28.1.5 Cold Load Pickup

Cold load pickup, occurring as the result of a permanent fault and long outage, is often maligned as the cause of many protective device misoperations. Figure 28.6 illustrates several cold load pickup curves developed by various sources. These curves are normally considered to be composed of the following three components:

1. Inrush—lasting a few cycles
2. Motor starting—lasting a few seconds
3. Loss of diversity—lasting many minutes

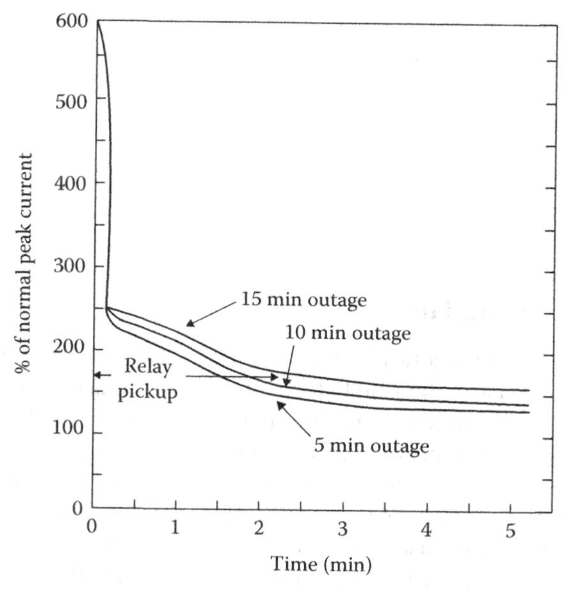

FIGURE 28.6 Cold load inrush current characteristics for distribution circuits.

When a lateral fuse misoperates, it is probably not the result of this loss of diversity, i.e., the fuse is overloaded. This condition is rare on most laterals. Relay operation during cold load pickup is generally the result of a trip of the instantaneous unit and probably results from high inrush. Likewise, an FCI operation would not appear to be the result of loss of diversity but rather the high inrush currents. Since inrush occurs during all energization and not just as a result of cold load pickup, it can be concluded that cold load pickup is not a major factor in the application of FCIs.

28.1.6 Calculation of Fault Current

Line Faults

$$\text{Line-to-neutral fault} = \frac{E}{\sqrt{3}2Z_\ell}$$

where

Z_ℓ is the line impedance

$2Z_\ell$ is the loop impedance assuming the impedance of the phase conductor and the neutral conductor are equal (some people use a 1.5 factor)

$$\text{Line-to-line fault} = \frac{E}{2Z_\ell}$$

Transformer Faults

$$\text{Line to neutral or three-phase} = \frac{E}{\sqrt{3}Z_T}$$

$$\text{Line to line} = \frac{E}{2(Z_T + Z_\ell)}$$

where

$$Z_\ell = \sqrt{R_L^2 + X_L^2}$$

$$Z_T = \frac{Z_{T\%}10E^2}{\text{kVA}}$$

28.1.7 Current Limiting Fuses

Current limiting fuses (CLFs) use a fusible element (usually silver) surrounded by sand (Figure 28.7). When the element melts, it causes the sand to turn into fulgurite (glass). Since glass is a good insulator, this results in a high resistance in series with the faults. This limits not only the magnitude of the fault but also the energy. All this can happen in less than a half cycle.

CLFs are very good at interrupting high currents (e.g., 50,000 A). They historically have had trouble (general purpose and backup) with low level fault currents and overloads, where the fuse tube melts before the element (i.e., these two fuses are not considered to be "full range," since they do not necessarily interrupt low currents that melt the element). There are now "full range" CLFs in the market (see Figure 28.7).

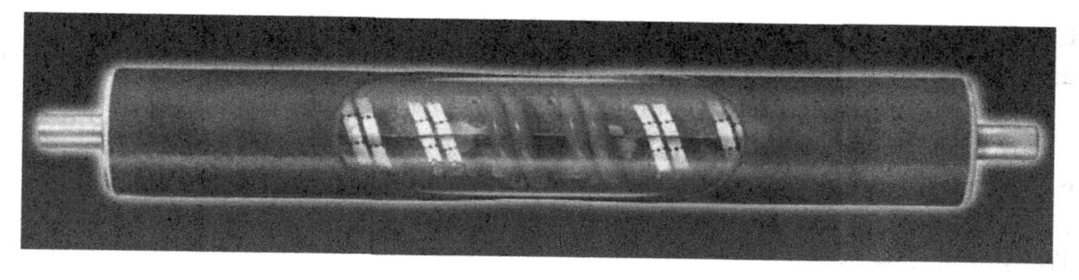

FIGURE 28.7 Full range CLF. (Courtesy of T&B, Memphis, TN. With permission.)

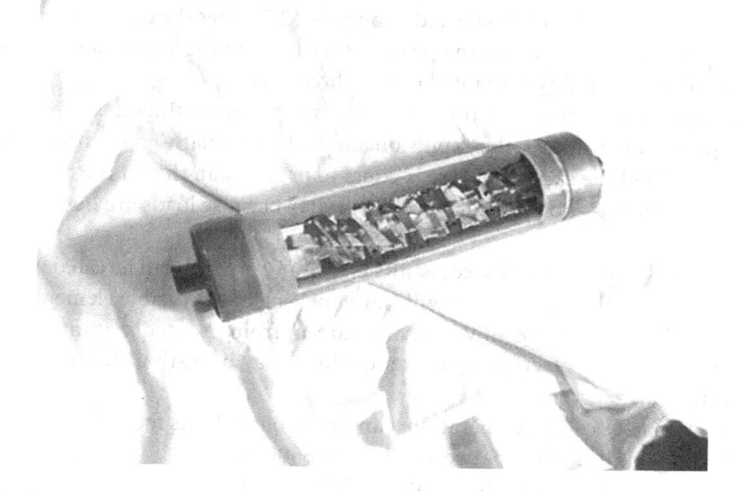

FIGURE 28.8 Backup CLF.

The three types of CLFs are defined as follows:

- General purpose—a fuse capable of interrupting all currents from the rated maximum interrupting current down to the current that causes melting of the fusible element in 1 h
- Backup—a fuse capable of interrupting all currents from the rated maximum interrupting current down to the rated minimum interrupting current (Figure 28.8)
- Full range—a fuse capable of interrupting all currents from the rated maximum current down to any current that melts the element

28.1.8 Rules for Application of Fuses

1. Cold load pickup.

After 15 min outage	200% for 0.5 s
	140% for 5 s
After 4 h, all electric	300% for 5 min

2. "Damage" curve—75% of minimum melt.
3. Two expulsion fuses cannot be coordinated if the available fault current is great enough to indicate an interruption of less than 0.8 cycles.
4. "T"-SLOW and "K"-FAST.
5. CLFs can be coordinated in the subcycle region.

6. Capacitor protection:
   a. The fuse should be rated for 135% of the normal capacitor current (per the industry standard). However, manufacturers recommend at least 165%. Use 165% to prevent nuisance operations. The fuse should also clear within 300 s for the minimum short-circuit current.
   b. If current exceeds the maximum case rupture point, a CLF must be used.
   c. CLFs should be used if a single parallel group exceeds 300 kVAR.
7. Transformer:
   a. Inrush—12 times for 0.1 s.
   b. Twenty-five times for 0.01 s.
   c. Self protected—primary fuse rating is 10–14 times continuous when secondary breaker is used.
   d. Self protected—weak link is selected to be about 2.5 times the continuous when no secondary breaker is used (which means that minimum melt is in the area of four to six times rating).
   e. Conventional—primary fuse rated two to three times.
   f. General-purpose current limiting—two to three times continuous.
   g. Backup current limiting—the expulsion and CLF are usually coordinated such that the minimum melt I^2t of the expulsion fuse is equal to or less than that of the backup CLF.
8. Conductor burndown—not as great a problem today because loads are higher and hence conductors are larger.
9. General purpose—one that will successfully clear any current from its rated maximum interrupting current down to the current that will cause melting of the fusible element in 1 h.
10. Backup—one that will successfully clear any current from its rated maximum interrupting down to the rated minimum interrupting current, which may be at the 10 s time period on the minimum melting time–current curve.
11. CLF—approximately 1/4 cycle operation; can limit energy by as much as 60 to 1.
12. Weak link—in oil is limited to between 1500 and 3500 A.
13. Weak link—in cutout is limited to 6,000–15,000 asymmetrical.
14. Lightning minimum fuse size to reduce nuisance operations (12 T-SLOW), (25 K-FAST).
15. Energy stored in inductance = $\frac{1}{2}\text{Li}^2$.
16. The maximum voltage produced by a CLF typically will not exceed 3.1 times the fuse rated maximum voltage.
17. The minimum sparkover allowed for a gapped arrester is 1.5 × 1.414 = 2.1 times arrester rating.
18. General practice is to keep the minimum sparkover of a gapped arrester at about 2.65 × arrester rating.
19. Metal oxide varistors (MOVs) do not have a problem with CLF "kick voltages."

28.1.9 More Overcurrent Rules

1. *Hydraulically controlled reclosers* are limited to about 10,000 A for the 560 A coil and 6,000 A for the 100 A coil.
2. Many companies set ground minimum trip at maximum load level and phase trip at two times load level.
3. A *K factor* of 1 (now used in the standards) means the interrupting current is constant for any operating voltage. A recloser is rated on the maximum current it can interrupt. This current generally remains constant throughout the operating voltage range.
4. A *recloser* is capable of its full interrupting rating for a complete four-operation sequence. The sequence is determined by the standard. A breaker is subject to derating.
5. A recloser can handle any degree of asymmetrical current. A breaker is subject to an *S* factor derating.
6. A *sectionalizer* is a self-contained circuit-opening device that automatically isolates a faulted portion of a distribution line from the source only after the line has been de-energized by an upline primary protective device.

7. A *power fuse* is applied close to the substation (2.8–169 kV and X/R between 15 and 25).
8. A *distribution fuse* is applied farther out on the system (5.2–38 kV and X/R between 8 and 15).
9. The *fuse tube* (in cutout) determines the interrupting capability of the fuse. There is an auxiliary tube that usually comes with the fuse that aids in low current interruption.
10. Some *expulsion fuses* can handle 100% continuous and some 150%.
11. Type "K" is a fast fuse link with a speed ratio of melting time–current characteristics from 6 to 8.1. (Speed is the ratio of the 0.1 s minimum melt current to the 300 s minimum melt current. Some of the larger fuses use the 600 s point.)
12. Type "T" is a slow fuse link with a speed ratio of melt time–current characteristics from 10 to 13.
13. After about 10 fuse link operations, the *fuse holder* should be replaced.
14. *Slant ratings* can be used on grounded wye, wye, or delta systems as long as the line-to-neutral voltage of the system is lower than the smaller number and the line-to-line voltage is lower than the higher number. A slant-rated cutout can withstand the full line-to-line voltage, whereas a cutout with a single voltage rating could not withstand the higher line-to-line voltage.
15. *Transformer fusing*—25 at 0.01, 12 at 0.1, and 3 at 10 s.
16. *Unsymmetrical transformer connections (delta/wye)*

Fault Type	Multiplying Factor
Three phase	N
Phase to phase	87 (N)
Phase to ground	1.73 (N)

where N is the ratio of $V_{primary}/V_{secondary}$.

17. Multiply the high side device *current points* by the appropriate factor.
18. *K factor for load side fuses*:
    a. Two fast operations and dead time 1–2 s = 1.35.
19. *K factor for source side fuses*:
    a. Two fast—two delayed and dead time of 2 s = 1.7.
    b. Two fast—two delayed and dead time of 10 s = 1.35.
    c. Sometimes these factors go as high as 3.5 so check.
20. *Sequence coordination*—Achievement of true "trip coordination" between an upline electronic recloser and a downline recloser is made possible through a feature known as "sequence" coordination. Operation of sequence coordination requires that the upline electronic recloser be programmed with "fast curves" whose control response time is slower than the clearing time of the downline recloser fast operation, through the range of fault currents within the reach of the upline recloser. Assume a fault beyond the downline recloser that exceeds the minimum trip setting of both reclosers. The downline recloser trips and clears before the upline recloser has a chance to trip. However, the upline control does see the fault and the subsequent cutoff of fault current. The sequence coordination feature then advances its control through its fast operation, such that both controls are at their second operation, even though only one of them has actually tripped. Should the fault persist, and a second fast trip occur, sequence coordination repeats the procedure. Sequence coordination is active only on the programmed fast operations of the upline recloser. In effect, sequence coordination maintains the downline recloser as the faster device.
21. *Recloser time–current characteristics*:
    a. Some curves are average. Maximum is 10% higher.
    b. Response curves are the responses of the sensing device and do not include arc extinction.
    c. Clearing time is measured from fault initiation to power arc extinction.
    d. The response time of the recloser is sometimes the only curve given. To obtain the interrupting time, you must add approximately 0.045 s to the curve (check…they are different).
    e. Some curves show maximum clearing time. On the new electronic reclosers, you usually get a control response curve and a clearing curve.

22. The "75% rule" considers TCC tolerances, ambient temperature, preloading, and predamage. Predamage only uses 90%.
23. *A backup CLF* with a designation like "12 K" means that the fuse will coordinate with a K link rated 12 A or less.
24. *Capacitor Fusing*:
    a. The 1.35 factor may result in nuisance fuse operations. Some utilities use 1.65.
    b. Case rupture is not as big a problem as years ago due to all film designs.
    c. Tank rupture curves may be probable or definite in nature. Probable means there is a probability chance of not achieving coordination. Definite indicates there is effectively no chance of capacitor tank rupture with the proper 0% probability curve.
    d. T links are generally used up to about 25 A and K link above that to reduce nuisance fuse operations from lightning.
25. *Line impedance*—typical values for line impedance (350 kcm) on a per mile basis are as follows:

	$Z_{positive}$	Z_0
Cable UG	$0.31 + j0.265$	$1.18 + j0.35$
Spacer	$0.3 + j0.41$	$1.25 + j2.87$
Tree wire	$0.3 + j0.41$	$1.25 + j2.87$
Armless	$0.3 + j0.61$	$0.98 + j2.5$
Open	$0.29 + j0.66$	$0.98 + j2.37$

26. 1A–3B is necessary when sectionalizers are used downstream from the recloser.
27. Vacuum reclosers have interrupting ratings as high as 10–20 kA.
28. Highest recloser continuous ratings are 800 and 1200 A.
29. Sectionalizer actuating current should be <80% of backup device trip current.
30. Interrupting ratings of cutouts are approximately 7–10 kA symmetrical.
31. *K factor* can mean a "voltage range" factor or a "shift factor" caused by the recloser heating up the fuse.
32. *Sectionalizer counts* should normally be one count less than the operations to lockout of the breaker or recloser.
33. *Sectionalizer memory* time must be greater than cumulative trip and recloser time.
34. Fuses melt at about 200% of rating.
35. Sectionalizers have momentary ratings for 1 and 10 s.
36. *Twenty-five percent rule* for fuses includes preload, ambient temperature, and predamage.

28.1.10 Capacitor Fusing

1. Purpose of fusing:
    a. To isolate faulted bank from system
    b. To protect against bursting
    c. To give indication
    d. To allow manual switching (fuse control)
    e. To isolate faulted capacitor from bank
2. Recommended rating:
    a. The continuous-current capability of the fuse should be at least 135%, according to the IEEE standards, and 165%, according to the manufacturers, of the normal capacitor bank (for delta and floating wye banks, the factor may be reduced to 150% if necessary). Use 165% to prevent nuisance fuse operation.
    b. The total clearing characteristics of the fuse link must be coordinated with the capacitor "case bursting" curves.

3. Tests have shown that expulsion fuse links will not satisfactorily protect against violent rupture where the fault current through the capacitor is greater than 5000 A.
4. The capacitor bank may be connected in a floating wye to limit short-circuit current to less than 5000 A.
5. *Inrush*—for a single bank, the inrush current is always less than the short-circuit value at the bank location.
6. *Inrush*—for parallel banks, the inrush current is always much greater than for a single bank.
7. *Expulsion fuses* offer the following advantages:
   a. They are inexpensive and easily replaced.
   b. They offer a positive indication of operation.
8. *CLFs* are used where
   a. A high available short circuit exceeds the expulsion or nonvented fuse rating.
   b. A CLF is needed to limit the high energy discharge from adjacent parallel capacitors effectively.
   c. A nonventing fuse is needed in an enclosure.
9. The fuse link rating should be such that the link will melt in 300 s at 240%–350% of normal load current.
10. The fuse link rating should be such that it melts in 1 s at not over 220 A and in 0.015 s at not over 1700 A.
11. The fuse rating must be chosen through the use of melting time–current characteristic curves, because fuse links of the same rating, but of different types and makes, have a wide variation in the melting time at 300 s and at high currents.
12. *Safe zone*—usually greater damage than a slight swelling.
   a. *Zone* 1—suitable for locations where case rupture or fluid leakage would present no hazard.
   b. *Zone* 2—suitable for locations that have been chosen after careful consideration of possible consequences associated with violent case ruptures.
   c. *Hazardous zone*—unsafe for most applications. The case will often rupture with sufficient violence to damage adjacent units.
13. Manufacturers normally recommend that the group fuse size be limited by the 50% probability curve or the upper boundary of Zone 1.
14. Short-circuit current in an open wye bank is limited to approximately three times the normal current.
15. CLFs can be used for delta or grounded wye banks, provided there is sufficient short-circuit current to melt the fuse within one-half cycle.

28.1.11 Conductor Burndown

Conductor burndown is a function of (1) conductor size, (2) whether the wire is bare or covered, (3) the magnitude of the fault current, (4) climatic conditions such as wind, and (5) the duration of the fault current.

If burndown is less of a problem today than in years past, it must be attributed to the trend of using heavier conductors and a lesser use of covered conductors. However, extensive outages and hazards to life and property still occur as the result of primary lines being burned down by flashover, tree branches falling on lines, etc. Insulated conductors, which are used less and less, anchor the arc at one point and thus are the most susceptible to being burned down. With bare conductors, except on multigrounded neutral circuits, the motoring action of the current flux of an arc always tends to propel the arc along the line away from the power source until the arc elongates sufficiently to automatically extinguish itself. However, if the arc encounters some insulated object, the arc will stop traveling and may cause line burndown.

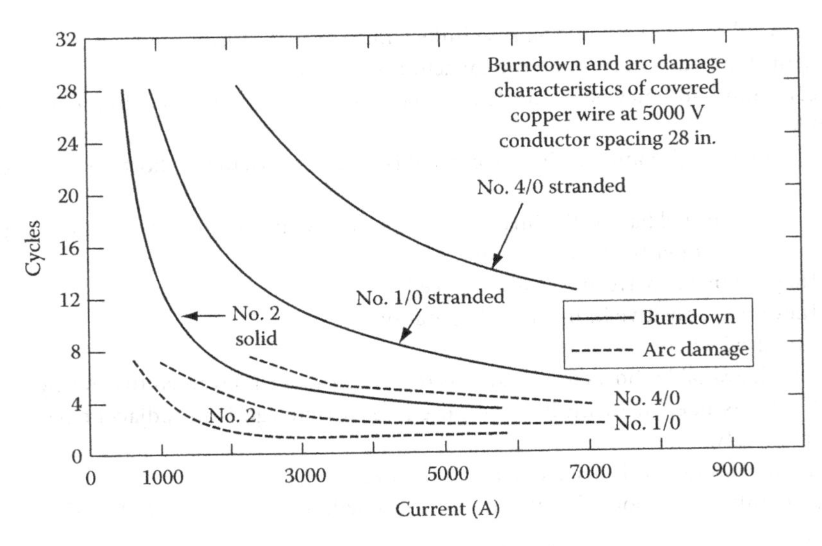

FIGURE 28.9 Burndown characteristics of several weatherproof conductors.

With tree branches falling on bare conductors, the arc may travel away and clear itself; however, the arc will generally reestablish itself at the original point and continue this procedure until the line burns down or the branch falls off the line. Limbs of soft spongy wood are more likely to burn clear than hard wood. However ½ in. diameter branches of any wood, which cause a flashover, are apt to burn the lines down unless the fault is cleared quickly enough.

Figure 28.9 shows the burndown characteristics of several weatherproof conductors. Arc damage curves are given as arc is extended by traveling along the phase wire; it is extinguished but may be reestablished across the original path. Generally, the neutral wire is burned down.

28.1.12 Protective Device Numbers

The devices in the switching equipment are referred to by numbers, with appropriate suffix letters (when necessary), according to the functions they perform. These numbers are based on a system that has been adopted as standard for automatic switchgear by the American Standards Association (Table 28.1).

28.1.13 Protection Abbreviations

CS—Control switch
X—Auxiliary relay
Y—Auxiliary relay
YY—Auxiliary relay
Z—Auxiliary relay

1. To denote the location of the main device in the circuit or the type of circuit in which the device is used or with which it is associated, or otherwise identify its application in the circuit or equipment, the following are used:

   N—Neutral
   SI—Seal-in

TABLE 28.1 Protective Device Numbers

Device No.	Function and Definition
11	*Control power transformer* is a transformer that serves as the source of AC control power for operating AC devices
24	*Bus-tie circuit breaker* serves to connect buses or bus sections together
27	*AC undervoltage relay* is one which functions on a given value of single-phase AC under voltage
43	*Transfer device* is a manually operated device that transfers the control circuit to modify the plan of operation of the switching equipment or of some of the devices
50	*Short-circuit selective relay* is one which functions instantaneously on an excessive value of current
51	*AC overcurrent relay* (inverse time) is one which functions when the current in an AC circuit exceeds a given value
52	*AC circuit breaker* is one whose principal function is usually to interrupt short-circuit or fault currents
64	*Ground protective relay* is one which functions on failure of the insulation of a machine, transformer, or other apparatus to ground; this function is, however, not applied to devices 51N and 67N connected in the residual or secondary neutral circuit of current transformers
67	*AC power directional or AC power directional overcurrent relay* is one which functions on a desired value of power flow in a given direction or on a desired value of overcurrent with AC power flow in a given direction
78	*Phase-angle measuring relay* is one which functions at a predetermined phase angle between voltage and current
87	*Differential current relay* is a fault-detecting relay that functions on a differential current of a given percentage or amount

2. To denote parts of the main device, the following are used:

   H—High set unit of relay
   L—Low set unit of relay
   OC—Operating coil
   RC—Restraining coil
   TC—Trip coil

3. To denote parts of the main device such as auxiliary contacts that move as part of the main device and are not actuated by external means. These auxiliary switches are designated as follows:

   "a"—closed when main device is in energized or operated position
   "b"—closed when main device is in de-energized or nonoperated position

4. To indicate special features, characteristics, and the conditions when the contacts operate, or are made operative or placed in the circuit, the following are used:

   A—Automatic
   ER—Electrically reset
   HR—Hand rest
   M—Manual
   TDC—Time-delay closing
   TDDO—Time-delay dropping out
   TDO—Time-delay opening

To prevent any possible conflict, one letter or combination of letters has only one meaning on individual equipment. Any other words beginning with the same letter are written out in full each time, or some other distinctive abbreviation is used.

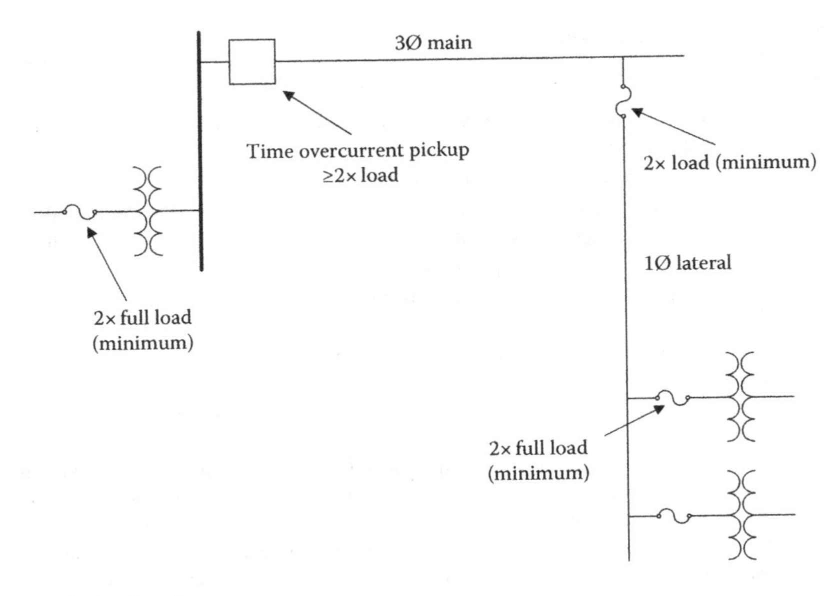

FIGURE 28.10 Burke 2× rule.

28.1.14 Simple Coordination Rules

There are few things more confusing in distribution engineering than trying to find out rules of overcurrent coordination, i.e., what size fuse to pick or where to set a relay, etc. The patented (*just kidding*) Burke 2× rule states that when in doubt, pick a device of twice the rating of what it is you are trying to protect, as shown in Figure 28.10. This rule picks the minimum value you should normally consider and is generally as good as any of the much more complicated approaches you might see. For various reasons, you might want to go higher than this, which is usually okay. To go lower, you will generally get into trouble. One exception to this rule is the fusing of capacitors where minimum size fusing is important to prevent case rupture.

28.1.15 Lightning Characteristics

1. Stroke currents
   a. Maximum—220,000 A
   b. Minimum—200 A
   c. Average—10,000–15,000 A
2. Rise times—1–100 μs
3. Lightning polarity—approximately 95% are negative
4. Annual variability (Empire State Building)
   a. Maximum number of hits—50
   b. Average—21
   c. Minimum—3
5. Direct strokes to T line—one per mile per year with keraunic levels between 30 and 65
6. Lightning discharge currents in distribution arresters on primary distribution lines (composite of urban and rural)
   a. Maximum measured to date—approximately 40,000 A
   b. 1% of records at least 22,000 A
   c. 5% of records at least 10,500 A
   d. 10% of records at least 6,000 A
   e. 50% of records at least 1,500 A

TABLE 28.2 Lightning Discharge Current vs. Location

Col. 1	Col. 2	Col. 3	Col. 4
Urban Circuits (%)	Semiurban Circuits (%)	Rural Circuits (%)	Discharge Currents (A)
20	35	45	1,000
1.6	7	12	5,000
0.55	3.5	6	10,000
0.12	0.9	2.4	20,000
		0.4	40,000

7. Percent of distribution arresters receiving lightning currents at least as high as in Col. 4 in Table 28.2
8. Number of distribution arrester operations per year (excluding repeated operations on multiple strokes):
   a. Average on different systems—0.5 to 1.1 per year
   b. Maximum recorded—6 per year
   c. Maximum number of successive operations of one arrester during one multiple lightning stroke—12 operations

28.1.16 Arc Impedance

Although arcs are quite variable, a commonly accepted value for currents between 70 and 20,000 A has been an arc drop of 440 V/ft, essentially independent of current magnitude:

$Z_{arc} = 440l/I \quad l = \text{length of arc (in feet)} \quad I = \text{current}$

Assume

$I_F = 5000 \quad A = I$

Arc length = 2 ft

$Z_{arc} = 440 \times (2/5000) = 0.176\,\Omega$ i.e., Arc impedance is pretty small.

Let us say you have a 120 V secondary fault and the distance between the phase and neutral is 1 ft. If the current level was 500 A, then the arc resistance would be $(440 \times 1)/500 = 0.88\,\Omega$, which is significant in its effect on fault levels.

28.2 Transformers

28.2.1 Saturation Curve

See Figure 28.11.

28.2.2 Insulation Levels

Table 28.3 gives the American standard test levels for insulation of distribution transformers.

28.2.3 Δ-Y Transformer Banks

Figure 28.12 is a review of fault current magnitudes for various secondary faults on a Δ-Y transformer bank connection.

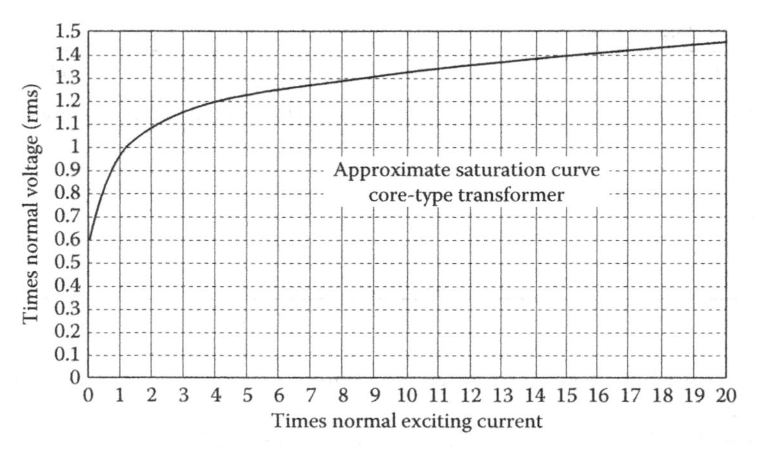

FIGURE 28.11 Transformer saturation curve.

TABLE 28.3 Insulation Levels for Transformer Windings and Bushings

		Windings			Bushings		
		Impulse Tests (1.2 × 50 Wave)			Bushing Withstand Voltages		
		Chopped Wave					
Insulation Class and Nominal Bushing Rating	Low-Frequency Dielectric Tests	Minimum Time to Flashover		Full Wave	60-Cycle 1 min Dry	60-Cycle 10 s Wet	Impulse 1.2 × 50 Wave
kV	kV	kV	μs	kV	kV (rms)	kV (rms)	kV (Crest)
1.2	10	36	1.0	10	10	6	30
5.0	19	69	1.5	60	21	20	60
8.66	26	88	1.6	75	27	24	75
15.0	34	110	1.8	95	35	30	95
25.0	40	145	1.9	125	70	60	150
34.5	70	175	3.0	150	95	95	200
46.0	95	290	3.0	250	120	120	250
69.0	140	400	3.0	350	175	175	350

28.2.3.1 Transformer Loading

When the transformer is overloaded, the high temperature decreases the mechanical strength and increases the brittleness of the fibrous insulation. Even though the insulation strength of the unit may not be seriously decreased, transformer failure rate increases due to this mechanical brittleness.

- Insulation life of the transformer is where it loses 50% of its tensile strength. A transformer may continue beyond its predicted life if it is not disturbed by short-circuit forces, etc.
- The temperature of top oil should never exceed 100°C for power transformers with a 55° average winding rise insulation system. Oil overflow or excessive pressure could result.
- The temperature of top oil should not exceed 110°C for those with a 65°C average winding rise.
- Hotspot should not exceed 150°C for 55°C systems and 180°C for 65°C systems. Exceeding these temperatures could result in free bubbles that could weaken dielectric strength.
- Peak short duration loading should never exceed 200%.
- Standards recommend that the transformer should be operated for normal life expectancy. In the event of an emergency, a 2.5% loss of life per day for a transformer may be acceptable.
- Percent daily load for normal life expectancy with 30°C cooling air (see Table 28.4).

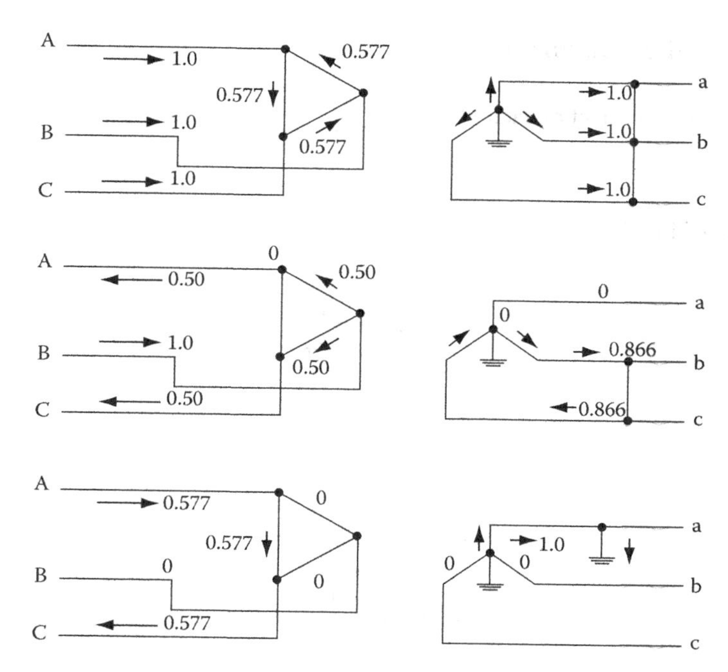

FIGURE 28.12 Δ-Y transformer banks.

TABLE 28.4 Distribution Transformer Overload with Normal Loss of Life

Duration of Peak Load (h)	Self-Cooled with % Load before Peak of		
	50	70	90
0.5	189	178	164
1	158	149	139
2	137	132	124
4	119	117	113
8	108	107	106

28.3 Instrument Transformers

28.3.1 Two Types

1. Potential (usually 120 V secondary)
2. Current (5 A secondary at rated primary current)

28.3.2 Accuracy

Three factors will influence accuracy:

1. Design and construction of transducer
2. Circuit conditions (V, I, and f)
3. Burden (in general, the higher the burden, the greater the error)

28.3.3 Potential Transformers

$$\text{Ratio correction factor (RCF)} = \frac{\text{True ratio}}{\text{Marked ratio}} \quad \text{(RCF generally} > 1)$$

Burden is measured in VA,

$$\therefore \text{VA} = \frac{E^2}{Z_b}$$

Assume

$$\text{True ratio} = \frac{10}{0.9} = 11.1$$

$$\Rightarrow \text{RCF} = \frac{11.1}{10} = 1.11$$

$$\text{Marked ratio} = \frac{10}{1} = 10$$

Voltage at secondary is low and must be compensated by 11% to get the actual primary voltage using the marked ratio.

28.3.4 Current Transformer

True ratio = marked ratio × RCF

$$\therefore \text{RCF} = \frac{\text{True ratio}}{\text{Marked ratio}}$$

28.3.5 H-Class

Burdens are in series,
for example, 10H200 ⇒ 10% error at 200 V

$$\therefore 20(5\,\text{As}) = 100\,\text{A} \Rightarrow Z_b = \frac{200}{100} = 2\,\Omega$$

$$\Rightarrow 5\,\text{A to } 100\,\text{A} \quad \text{has} \leq 10\% \text{ error if } Z_b = 4\,\Omega$$

or

If $Z_b = 4\,\Omega$
200 V/4 Ω = 50 A (10 times normal)
H-class—constant magnitude error (variable %)
L-class—constant % error (variable magnitude)

Example

True ratio = marked ratio × RCF

Assume marked is 600/5 or 120:1 at rated amps and 2 Ω

At 100% A true = 120 × 1.002 × 5 secondary

Primary = 600 × 1.002 = 601.2

At 20% A true = 600 × 0.2 × 1.003 = 120.36 (marked was 120)

28.3.6 Current Transformer Facts

1. Bushing current transformers (BCTs) tend to be accurate more on high currents (due to large core and less saturation) than other types.
2. At low currents, BCTs are less accurate due to their larger exciting currents.
3. Rarely, if ever, it is necessary to determine the phase–angle error.
4. Accuracy calculations need to be made only for three-phase and single-phase to ground faults.
5. CT burden decreases as secondary current increases, because of saturation in the magnetic circuits of relays and other devices. At high saturation, the impedance approaches the DC resistance.
6. It is usually sufficiently accurate to add series burden impedance arithmetically.
7. The reactance of a tapped coil varies as the square of the coil turns, and the resistance varies approximately as the turns.
8. Impedance varies as the square of the pickup current.
9. Burden impedances are always connected in wye.
10. "RCF" is defined as that factor by which the marked ratio of a current transformer must be multiplied to obtain the true ratio. These curves are considered standard application data.
11. The secondary-excitation-curve method of accuracy determination does not lend itself to general use except for bushing-type, or other, CTs with completely distributed secondary leakage, for which the secondary leakage reactance is so small that it may be assumed to be zero.
12. The curve of rms terminal voltage vs. rms secondary current is approximately the secondary-excitation curve for the test frequency.
13. ASA accuracy classification:
    a. This method assumes CT is supplying 20 times its rated secondary current to its burden.
    b. The CT is classified on the basis of the maximum rms value of voltage that it can maintain at its secondary terminals without its ratio error exceeding a specified amount.
    c. "H" stands for high internal secondary impedance.
    d. "L" stands for low internal secondary impedance (bushing type).
    e. 10H800 means the ratio error is 10% at 20 times rated voltage with a maximum secondary voltage of 800 and high internal secondary impedance.
    f. Burden (max)—maximum specified voltage/20 × rated second.
    g. The higher the number after the letter, the better the CT.
    h. A given 1200/5 bushing CT with 240 secondary turns is classified as 10L400: if a 120-turn completely distributed tap is used, then the applicable classification is 10L200.
    i. For the same voltage and error classifications, the H transformer is better than the L for currents up to 20 times rated.

28.3.7 Glossary of Transducer Terms

Voltage transformers—They are used whenever the line voltage exceeds 480 V or whatever lower voltage may be established by the user as a safe voltage limit. They are usually rated on a basis of 120 V secondary voltage and used to reduce primary voltage to usable levels for transformer-rated meters.

Current transformers—Current transformers are usually rated on a basis of 5 A secondary current and used to reduce primary current to usable levels for transformer-rated meters and to insulate and isolate meters from high-voltage circuits.

Current transformer ratio—Current transformer ratio is the ratio of primary to secondary current. For current transformer rated 200:5, the ratio is 200:5 or 40:1.

Voltage transformer ratio—Voltage transformer ratio is the ratio of primary to secondary voltage. For voltage transformer rated 480:120, the ratio is 4:1, 7200:120, or 60:1.

Transformer ratio (TR)—TR is the total ratio of current and voltage transformers. For 200:5 CT and 480:120 PT, TR = 40 × 4 = 160.

Weatherability—Transformers are rated as indoor or outdoor, depending on construction (including hardware).

Accuracy classification—Accuracy classification is the accuracy of an instrument transformer at specified burdens. The number used to indicate accuracy is the maximum allowable error of the transformer for specified burdens. For example, 0.3 accuracy class means the maximum error will not exceed 0.3% at stated burdens.

Rated burden—Rated burden is the load that may be imposed on the transformer secondaries by associated meter coils, leads, and other connected devices without causing an error greater than the stated accuracy classification.

Current transformer burdens—Current transformer burdens are normally expressed in ohms impedance such as B-0.1, B-0.2, B-0.5, B-0.9, or B-1.8. Corresponding volt–ampere values are 2.5, 5.0, 12.5, 22.5, and 45.

Voltage transformer burdens—Voltage transformer burdens are normally expressed as volt–amperes at a designated power factor (pf). It may be *W*, *X*, *M*, *Y*, or *Z*, where *W* is 12.5 VA at 0.10 pf, *X* is 25 VA at 0.70 pf, *M* is 35 VA at 0.20 pf, *Y* is 75 VA at 0.85 pf, and *Z* is 200 VA at 0.85 pf. The complete expression for a current transformer accuracy classification might be 0.3 at B-0.1, B-0.2, and B-0.5, while the potential transformer might be 0.3 at *W*, *X*, *M*, and *Y*.

Continuous thermal rating factor (TRF)—Continuous TRF is normally designated for current transformers and is the factor by which the rated primary current is multiplied to obtain the maximum allowable primary current without exceeding temperature rise standards and accuracy requirements. For example, if a 400:5 CT has a TRF of 4.0, the CT will continuously accept 400 × 4 or 1600 primary amperes with 5 × 4 or 20 A from the secondary. The thermal burden rating of a voltage transformer shall be specified in terms of the maximum burden in volt–amperes that the transformer can carry at rated secondary voltage without exceeding a given temperature rise.

Rated insulation class—Rated insulation class denotes the nominal (line-to-line) voltage of the circuit on which it should be used. Associated Engineering Company has transformers rated for 600 V through 138 kV.

Polarity—The relative polarity of the primary and secondary windings of a current transformer is indicated by polarity marks (usually white circles), associated with one end of each winding. When current enters at the polarity end of the primary winding, a current in phase with it leaves the polarity end of the secondary winding. Representation of primary marks on wiring diagrams is shown as black squares.

Hazardous open circulating—The operation of CTs with the secondary winding open can result in a high voltage across the secondary terminals, which may be dangerous to personnel or equipment. Therefore, the secondary terminals should always be short circuited before a meter is removed from service. This may be done automatically with a bypass in the socket or by a test switch for A-base meters.

28.4 Loading

Probably no area of distribution engineering causes more confusion than does loading. Reading the standards does not seem to help much since everyone appears to have their own interpretation. Manufacturers of equipment are very conservative since they really never know how the user will actually put the product to use so they must expect the worst. On the other hand, many users seem to take the approach that since it did not fail last year with traditional overloading values, it will not fail this year either. In fact, it will not fail until after retirement. Heck! "Save a buck and get a promotion." The author of this document is not a psychology major and frankly has no idea of what the thinking was when much of the following was produced. The material that follows, however, was taken from sources with excellent reputation. Use it with caution.

28.4.1 Transformer Loading Basics

1. All modern transformers have insulation systems designed for operation at 65°C average winding temperature and 80°C hottest-spot winding rise over ambient in an average ambient of 30°C. This means
   a. 65°C average winding rise + 30°C ambient = 95°C average winding temperature.
   b. 80°C hottest-spot rise + 30°C ambient = 110°C hottest spot.
      (Old system: 55°C winding rise + 30°C ambient = 85°C average winding temperature)
   c. 65°C hottest spot + 30°C ambient = 95°C hottest spot.
2. Notice that 95°C is the average winding temperature for the new insulation system and the hottest spot for the old—a source of immense confusion for many of us.
3. The temperature of the top oil should not exceed 100°C. Obviously, top-oil temperature is always less than hottest spot.
4. The maximum hotspot temperature should not exceed 150°C for a 55°C rise transformer or 180°C for a 65°C rise transformer.
5. Peak 0.5 h loading should not exceed 200%.
6. The conditions of 30°C ambient temperature and 100% load factor establish the basis of transformer ratings.
7. The ability of the transformer to carry more than nameplate rating under certain conditions without exceeding 95°C is basically due to the fact that top-oil temperature does not instantaneously follow changes in transformer load due to thermal storage.
8. An average loss of life of 1% per year (or 5% in any emergency) incurred during emergency operations is considered reasonable.
9. Most companies do not allow normal daily peaks to exceed the permissible load for normal life expectancy.
10. The firm capacity is usually the load that the substation can carry with one supply line or one transformer out of service.
11. "Emergency 24 h firm capacity" usually means a loss of life of 1% but is sometimes as much as 5% or 6%.
12. The following measures can be used for emergency conditions lasting more than 24 h:
    a. Portable fans.
    b. Water spray.
    c. Interconnect cooling equipment of FOA units.
    d. Use transformer thermal relays to drop certain loads.

28.4.2 Examples of Substation Transformer Loading Limits

The following is an example of maximum temperature limits via the IEEE for a 65°C rise transformer:

	IEEE Normal Life Expectancy
Top-oil temperature	105°C
Hotspot temperature	120°C

This next example shows the loading practice of various utilities for substation transformers:

	Utility A	Utility B	Utility C	Utility D	Utility E	Utility F	Utility G
Normal conditions							
Top oil	95	110	95	95	95	110	110
Hotspot	125	130	120	110	120	140	120
Emergency conditions							
Top oil	110	110	110	110	110	110	110
Hotspot	140	140	140	130	140	140	140

What happens when the hotspot is raised from 125°C to 130°C? This is shown as follows:

Maximum Hotspot (°C)	% Loss of Life, Annual
125	0.3366
130	0.5372

An example of the effect of load cycle (3 h peak with 70% preload for 13 h and 45% load for 8 h) and ambient on transformer capability via the ANSI guide is shown as follows:

	Peak Load for Normal Life Expectancy		Emergency Peak Load with 24 h Loss of Life	
Transformer Type	10°C Ambient	30°C Ambient	0.25%	1.0%
20,000—OA	30,000	24,200	28,400	32,000
15,000/2,000—OA/FA	28,700	23,800	27,500	30,700
12,000/16,000/ 20,000—OA/FA/FOA	27,500	23,200	26,800	29,700
20,000—FOA	27,500	23,200	26,800	29,700

The following is the effect on transformer ratings for various limits of top-oil temperature:

	MVA	Top-Oil Temperature (°C)
Normal rating	50	95
New rating	55	105
Emergency rating	59	110

28.4.3 Distribution Transformers

The loading of distribution transformers varies more widely than substation units. Some utilities try to never exceed the loading of the transformer nameplate. Others, particularly those using TLM, greatly overload smaller distribution transformers with no apparent increase in failure rates. An example of one utilities practice is as follows:

	Padmounted		Submersible	
kVA	Install Range	Removal Point	Install Range	Removal Point
25	0–40	55	0–34	42
50	41–69	88	35–64	79
75	70–105	122	65–112	112
100	106–139	139	113–141	141

28.4.4 Ampacity of Overhead Conductors

The following table shows the rating of conductors via a typical utility:

Conductor	ACSR		All Aluminum	
Size	Normal	Emergency	Normal	Emergency
1/0	319	331	318	334
2/0	365	379	369	388
3/0	420	435	528	450
4/0	479	496	497	523
267	612	641	576	606
336	711	745	671	705
397	791	830	747	786

28.4.5 Emergency Ratings of Equipment

The following are some typical 2 h overload ratings of various substation equipment. Use at your own risk:

Station transformer	140%
Current transformer	125%
Breakers	110%
Reactors	140%
Disconnects	110%
Regulators	150%

28.5 Miscellaneous Loading Information

The following are some miscellaneous loading information and thoughts from a number of actual utilities:

1. *Commercial and industrial transformer loading*

Load Factor (%)	Transformer Load Limit (%)
0–64	130
65–74	125
75–100	120

2. *Demand factor*:
   a. Lights—50%
   b. Air conditioning—70%
   c. Major appliances—40%
3. *Transformer loading*:
   a. Distribution transformer life is in excess of five times the present guide levels.
   b. Distribution guide shows that life expectancy is about 500,000 h for 100°C hottest-spot operation, compared to 200,000 h for a power transformer. Same insulation system.
   c. Using present loading guides, only 2.5% of power transformer thermal life is used up after 15 years.

d. Results of one analysis showed that the transition from acceptable to unacceptable risk (approximately an order of magnitude) was accompanied (by this utility) by only an 8.5% investment savings and a 12% increase in transformer loading.
e. Application of transformers in *excess of normal loading* can cause the following:
   i. Evolution of *free gas* from insulation of winding and lead conductors.
   ii. Evolution of free gas from insulation adjacent to metallic structural parts linked by magnetic flux produced by winding or lead currents may also reduce dielectric strength.
   iii. Operation at high temperatures will cause *reduced mechanical strength* of both conductor and structural insulation.
   iv. Thermal expansion of conductors, insulation materials, or structural parts at high temperature may result in permanent *deformations* that could contribute to mechanical or dielectric failures.
   v. *Pressure buildup in bushings* for currents above rating could result in leaking gaskets, loss of oil, and ultimate dielectric failure.
   vi. *Increased resistance in the contacts* of tap changers can result from a buildup of oil decomposition products in a very localized high temperature region.
   vii. Reactors and current transformers are also at risk.
   viii. *Oil expansion* could become greater than the holding capacity of the tank.
f. *Aging or deterioration of insulation* is a time function of temperature, moisture content, and oxygen content. With modern oil preservation systems, the moisture and oxygen contributions to insulation deterioration can be minimized, leaving insulation temperature as the controlling parameter.
g. Distribution and power transformer model tests indicate that the *normal life expectancy* at a continuous hottest-spot temperature of 110°C is 20.55 years.
h. Input into *a transformer loading program* should be as follows:
   i. Transformer characteristics (loss ratio, top-oil rise, hottest-spot rise, total loss, gallons of oil, and weight of tank and fittings).
   ii. Ambient temperatures.
   iii. Initial continuous load.
   iv. Peak load durations and the specified daily percent loss of life.
   v. Repetitive 24 h load cycle if desired.
i. Maximum permitted loading is 200% for a power transformer and 300% for a distribution transformer.
j. Suggested limits of loading for distribution transformers are as follows:
   i. Top oil—120°C.
   ii. Hottest spot—200°C.
   iii. Short time (0.5 h)—300%.
k. Suggested limits for power transformers are as follows:
   i. Top oil—100°C.
   ii. Hottest spot—180°C.
   iii. Maximum loading—200%.
l. Overload limits for *coordination of bushings* with transformers are as follows:
   i. Ambient air—40°C maximum.
   ii. Transformer top oil—110°C maximum.
   iii. Maximum current—two times bushing rating.
   iv. Bushing insulation hottest spot—150°C maximum.
m. Current ratings for the *load tap changer* (*LTC*) are the following:
   i. Temperature rise limit of 20°C for any current carrying contact in oil when carrying 1.2 times the maximum rated current of the LTC.
   ii. Capable of 40 breaking operations at twice the rate current and kVA.

n. Planned *loading beyond nameplate* rating defines a condition wherein a transformer is so loaded that its hottest-spot temperature is in the temperature range of 120°C–130°C.
o. Long-term *emergency loading* defines a condition wherein a power transformer is so loaded that its hottest-spot temperature is in the temperature range of 120°C–140°C.
p. The *principal gases* found dissolved in the mineral oil of a transformer are as follows:
   i. *Nitrogen*: from external atmosphere or from gas blanket over the free surface of the oil.
   ii. *Oxygen*: from external atmosphere.
   iii. *Water*: from moisture absorbed in cellulose insulation or from decomposition of the cellulose.
   iv. *Carbon dioxide*: from thermal decomposition of cellulose insulation.
   v. *Carbon monoxide*: from thermal decomposition of cellulose insulation.
   vi. *Other gases*: may be present in very small amounts (e.g., acetylene) as a result of oil or insulation decomposition by overheated metal, partial discharge, arcing, etc. These are very important in any analysis of transformers, which may be in the process of failing.
q. *Moisture* affects insulation strength, pf, aging, losses, and the mechanical strength of the insulation. Bubbles can form at 140°C, which enhance the chances of partial discharge and the eventual breakdown of the insulation as they rise to the top of the insulation. If a transformer is to be overloaded, it is important to know the moisture content of the insulation, especially if it is an older transformer. Bubbles evolve fast, so temperature is important to bubble formation but not the time at that temperature. Transformer insulation with 3.5% moisture content should not be operated above nameplate for a hottest spot of 120°C. Tests have shown that the use of circulated oil for the drying process takes some time. For a processing time of 70 h, the moisture content of the test transformers was reduced from 2% to 1.9% at a temperature of 50°C–75°C. Apparently only surface moisture was affected. A more effective method is to remove the oil and heat the insulation under vacuum.

29

Real-Time Control of Distributed Generation

Murat Dilek
Electrical Distribution Design, Inc.

Robert P. Broadwater
Virginia Polytechnic Institute and State University

Distributed generation (DG) can be operated to control voltages and power flows within the distribution system. Improvements in distribution system reliability and overall power system efficiency can be realized. For load growth with short-lived peaks that occur during extreme weather, DGs may provide lower-cost solutions than other approaches to system capacity upgrades.

DG provides a means for increasing the capacity of existing distribution facilities. When considering increasing distribution system capacity, DGs can be an alternative to new substation addition and replacing existing equipment with larger ones. A DG installed at the distribution level releases capacity throughout the system, from transmission through distribution. Transmission system losses are eliminated, and distribution system losses are reduced.

Some customer facilities have DGs that are installed for back-up power. These DGs are employed during grid-power outages or periods of high-cost grid power. They are operated for only a small fraction of time over the year. Moreover, back-up DGs are usually oversized, which means that they can provide more power than their facility loads need. These DGs can be equipped with a set of devices that will enable them to seamlessly interconnect to the grid and be dispatched if needed. The available capacity from such DGs can then be used for utility purposes.

DGs across many circuits in distribution areas can be controlled from a single control point. That is, such DGs can be aggregated into a block of generation and made available for transmission system use.

Although specifically intended for DGs, the aggregate control may also include other means of capacity release. When equipped with the necessary control and interconnection instrumentation, capacitors can be involved in aggregate control also. Some loads may also participate in the aggregation process in the form of curtailable or interruptible load. The aggregate control handles the collection of all of these participating entities.

The total power made available to the transmission system by the aggregate control is exhibited as a capacity release. That is, it is not the power injected into the transmission system from the distribution side, rather it is less power drawn by the distribution side. In the discussion to follow, the phrases DG power by aggregate control and capacity release by aggregate control are used interchangeably.

The aggregate control of DGs may serve a number of purposes. For instance, aggregated DGs can be activated if the transmission system or the distribution utility is having supply emergencies. Thus, DG aggregation provides a means to increase operating reserve. DGs can also help utilities manage energy purchases during times when the transmission grid electricity price is excessively high.

In the next section, local control for common DGs is discussed first. Next, controlling a group of DGs as an aggregate is addressed. Then, the DG as part of a hierarchical control system for controlling voltages and system power flows is investigated. Finally, load estimation for real-time *DG control* and also for forecasting aggregate control of DGs is presented.

29.1 Local Site *DG Control*

A DG operates basically in two modes in regard to being connected to the utility grid. In parallel mode, the DG remains connected to the grid. Hence, both the DG and the grid provide power for the local load in the customer facility (or DG site). In stand-alone (isolated or island) mode the DG is the sole power source to the local loads. In this section, consideration will be given only to DGs operating in parallel with the grid.

There are several forms of control for parallel DG. In one form of control, a local controller maintains a constant kW and kVar generation. In most cases, the local load is greater than the DG. Therefore, the power mismatch is supplied by the grid.

In another form of local control, the DG is controlled in order to maintain a constant power flow at the point of common coupling (PCC)—the point where the DG site interfaces with the grid, which is basically the metering point. The power flow maintained might be from the grid into the DG site (import) or from the site into the grid (export). As the local load varies, the local controller acts to change the kW and kVar generation at the DG in an attempt to keep the power flow constant at the PCC.

The most common DGs in service utilize synchronous machines. They prevail in grid-scale power exchanges between the utility and DG sites. Internal combustion (IC) engines and combustion turbines are the main prime movers for the synchronous generators. IC engines are much more common. Diesel fuel and natural gas are chosen for powering these engines.

The control of a synchronous machine is achieved by adjusting the fuel flow into the engine and the excitation of the generator. The fuel flow control by the governor determines the horsepower (kW) developed on the shaft of the engine. In a parallel DG, the shaft speed must be maintained very close to system frequency. The governor uses the kW set-point signal from the local controller and the speed signal from the DG output. The governor adjusts the fuel control to cause the kW output of the DG to match the kW set point that is set by the local controller.

The excitation control achieved by the voltage regulator determines terminal voltage and kVar output of the generator. Parallel DGs are required not to actively participate in regulating voltage at the PCC where the grid is supposed to set the voltage. Therefore, the excitation control is used to adjust kVar generation only. Rather than a kVar set point, a power factor (pf) set point is used for the excitation control. The local controller feeds the pf set point to the regulator. The regulator then adjusts the excitation to match the pf measured at the DG to the provided pf setting.

Basic functionality of the control system for parallel-connected DGs can be seen in Figure 29.1. For simplicity, it is assumed that the customer facility has only one DG. The local control receives the desired kW and kVar generation set points from an upper-level controller. The strategy can be a constant kW and kVar generation level for the DG or a constant kW and kVar flow at the PCC. Based on the control strategy, the local controller sends the required set points to the controller of the DG. An operator can supervise the control process and intervene as needed.

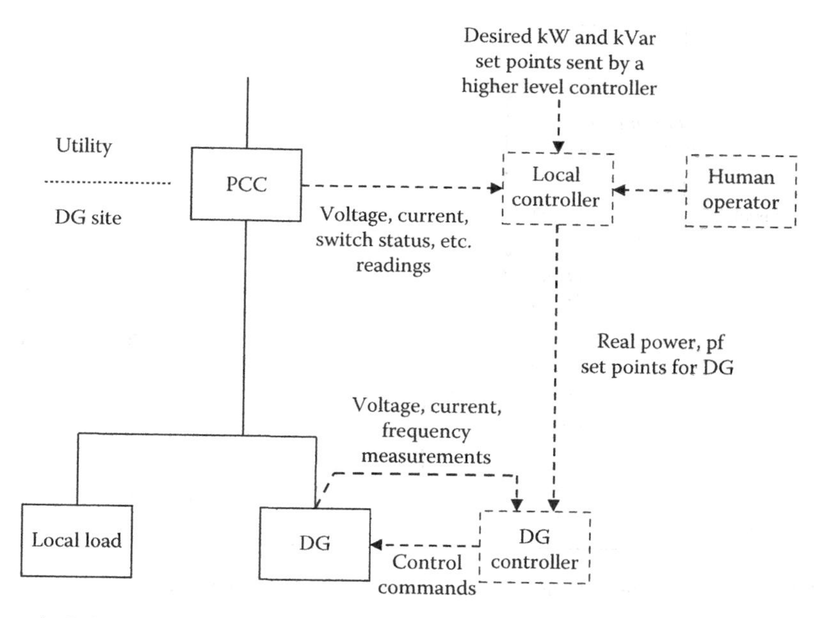

FIGURE 29.1 Block diagram for local control of a parallel DG at a customer site.

29.2 Hierarchical Control: Real-Time Control

The hierarchical *DG control* consists of three levels and is illustrated in Figure 29.2. The control functionality is used for two purposes: (1) for real-time *DG control* and (2) for forecasting future generation.

The aggregate control at level 3 shown in Figure 29.2 groups DGs together from many distribution circuits within a distribution service area. The aggregate control talks to both a transmission system entity (let us refer to this entity as the independent system operator, ISO) at a higher level and the circuit controls below at level 2. Each circuit might have a number of DG sites from which the circuit can import power. Each such DG site has a local controller (level 1) that can handle the import/export processes as explained in the previous section.

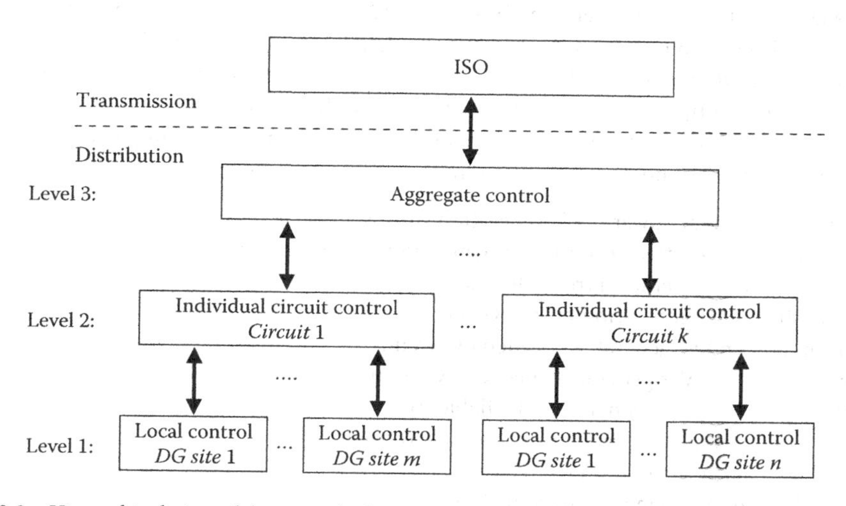

FIGURE 29.2 Hierarchical view of the control of aggregated DGs.

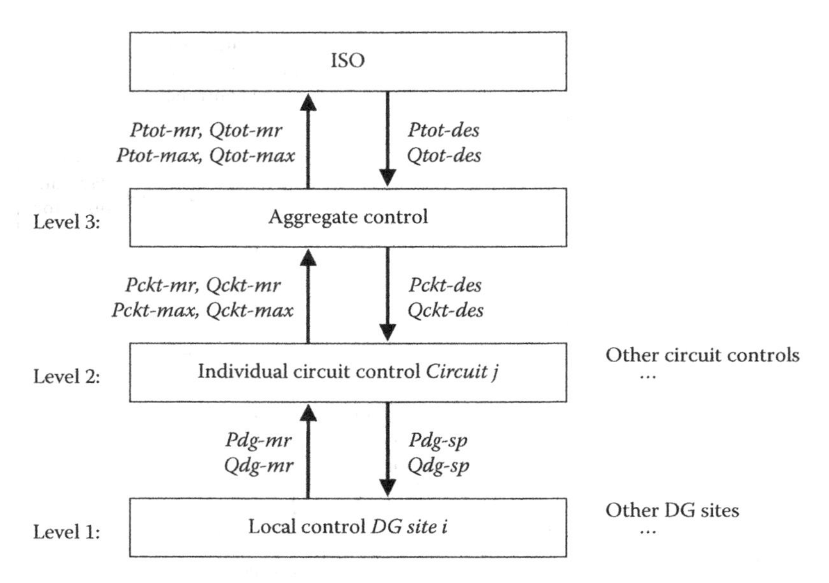

FIGURE 29.3 Data flow among ISO, aggregate controller, controller of a particular *Circuit j*, and controller of a particular *DG site i* in *Circuit j*.

The challenge of *DG control* is to implement the control without having to install measurement or monitoring equipment throughout the many miles of the distribution circuits. Each circuit control at level 2 has a model of the corresponding circuit, which includes such data as any existing circuit measurements and historical load measurements. Given weather and circuit conditions, the circuit control can make use of the available circuit model to estimate the power flows rather than measure the flows via instrumentation that would have to be installed throughout the circuit. This will be discussed further.

In essence, the aggregate control evaluates the DG power present at its lower levels and informs the ISO about the DG power that can be made available for transmission system use. After some negotiations, the ISO informs the aggregate control of the power it needs. The aggregate control then talks to the circuit controls in an attempt to provide the requested power in the best way possible. Data traffic among the layers of the control hierarchy in Figure 29.2 can be seen in Figure 29.3. Note that in order to simplify the discussion only a partial view of the data flow is presented. The view shown considers one circuit and one DG site in that circuit. One can extend this view to understand the data flow for the general case where multiple circuits with multiple DGs would be involved.

The data flow will be examined from two perspectives: flow from lower to higher layers and flow from higher to lower layers. The nomenclature used in Figure 29.3 is as follows:

Pdg-mr: must-run real power (kW) generation from DG site
Qdg-mr: must-run reactive power (kVar) generation from DG site
Pdg-sp: desired kW generation from DG site
Qdg-sp: desired kVar generation from DG site
Pckt-mr: must-run kW generation needed by circuit
Qckt-mr: must-run kVar generation needed by circuit
Pckt-max: maximum kW generation available from circuit
Qckt-max: maximum kVar generation available from circuit
Pckt-des: desired kW generation from circuit
Qckt-des: desired kVar generation from circuit

Ptot-mr: total must-run kW generation needed by all circuits
Qtot-mr: total must-run kVar generation needed by all circuits
Ptot-max: total kW generation available from all circuits
Qtot-max: total kVar generation available from all circuits
Ptot-des: total desired kW generation needed by ISO from aggregate *DG control*
Qtot-des: total desired kVar generation needed by ISO from aggregate *DG control*

29.2.1 Data Flow to Upper Layers

As mentioned earlier, level-2 circuit controllers have their corresponding circuit models, which are used to estimate power flows throughout the circuits. Given weather and circuit conditions such as voltage and current measurements taken at the start of circuit, the circuit controllers evaluate flows and voltages for the circuits. Consider for example *Circuit j* shown in level 2 in Figure 29.3. The circuit controller of *Circuit j* examines voltages and loadings in the circuit. If there exist any circuit problems such as undervoltage or overloaded locations in the circuit, then the circuit controller attempts to use the controllable DGs in the circuit to eliminate the problems. If employing the DGs helps to solve the circuit problems, then the DG kW and kVar generation levels at which the problems disappear are recorded. Such generation quantities are labeled as "must run," which means that the circuit itself needs that DG for solving its own problems.

Consider *DG site i* at level 1 in Figure 29.3. *Pdg-mr* and *Qdg-mr* represent the kW and kVar amounts that *DG site i* needs to produce in order to remove the problems that *Circuit j* will experience. *Pdg-mr* and *Qdg-mr* will be zero if no circuit problems occur when the *DG site i* produces no power.

Each circuit controller at level 2 sums up must-run generation. Each circuit controller also calculates the total available generation within the circuit. Must-run and maximum generation amounts are passed to the aggregate control at level 3. In Figure 29.3, *Pckt-mr*, *Qckt-mr*, *Pckt-max*, and *Qckt-max* indicate must-run and maximum generations from *Circuit j*. Note that *Pckt-max* and *Qckt-max* may also include curtailable load and reactive power available from capacitors installed in *Circuit j*. The *Circuit j* controller at level 2 may also know the type and operating characteristics of the DGs. Therefore, *Pckt-max* and *Qckt-max* may actually be further subdivided into available base-load generation and available load-following generation.

The aggregate control at level 3 sums both the totals of must-run generation and the maximum available generation across the individual circuit controllers at level 2. These sums are communicated to the ISO, as indicated by *Ptot-mr*, *Qtot-mr*, *Ptot-max*, and *Qtot-max* in Figure 29.3. Generation costs may also be communicated to the ISO, which is not considered here.

29.2.2 Data Flow to Lower Layers

The aggregate control negotiates with the ISO. When the negotiation is complete, the ISO informs the aggregate control of the total desired real and reactive generation. *Ptot-des* and *Qtot-des* in Figure 29.3 indicate the kW and kVar amounts requested by the ISO, respectively.

The aggregate control takes the total amount of desired generation and divides it among the DGs in the circuits under its control. *Pckt-des* and *Qckt-des*, for instance, represent kW and kVar generation that the aggregate control allocates for *Circuit j* to provide. A circuit controller at level 2 addresses control for all DG sites located in the corresponding circuit. Each circuit controller determines the generation sharing among the individual generators, based upon economic and reliability considerations. Thus, kW and kVar generation levels for all DGs under a circuit are calculated and communicated to the corresponding local controllers at DG sites. These kW and kVar values become the set points for the generator controllers. For instance, *Pdg-sp* and *Qdg-sp* in Figure 29.3 are the kW and kVar set points for the DG at *DG site i* in *Circuit j*.

29.3 Control of DGs at Circuit Level

Basic functions used in circuit-level control are depicted in Figure 29.4. The direction of arrows in the figure is interpreted such that what is at the tail-side of an arrow is available for use by what is at the head of the arrow. For instance, the arrow between *Power Flow* and *DG Control* indicates that *Power Flow* is used by the *DG Control* task. That is, *DG Control* can run *Power Flow* and obtain power flow results. Similarly, it can be seen that circuit measurements are made available for use in the load scaling.

All the functions shown in Figure 29.4 share the same circuit model and circuit data. Exchange of results among these functions takes place through the common circuit model. The circuit model and data include the following:

- Topology information of the circuit
- Type, status, rating, configuration, impedance, and/or admittance of the components present in the circuit
- Location and class of loads connected throughout the circuit
- Historical load measurements
- Load research data for the various classes of loads

Typically, measurements are taken at a very limited number of locations such as at the start of the circuit and DG sites. Therefore, the main task is to use the circuit model and the available measurements to estimate the flows in the circuit. That is, the majority of flows are determined by calculations instead of measurements obtained via data acquisition systems.

The most common scenario concerning control is as follows. Real-time current and voltage measurements taken at the start of the circuit are fed into the circuit model. Real-time kW and kVar measurements taken at the DGs are also fed into the model. *Power Flow* then calculates voltages and currents throughout the circuit. Since the load data (location, class, historical measurements, and load research data such as load curves, coincidence, and diversity factors) are already available, *Power Flow* uses *Load Scaling* for matching the calculated flows to the measurements. *Load Scaling* adjusts the circuit loads until the calculated flows match the measured flows. This is thus an estimation process for the loads that result in the measured flows.

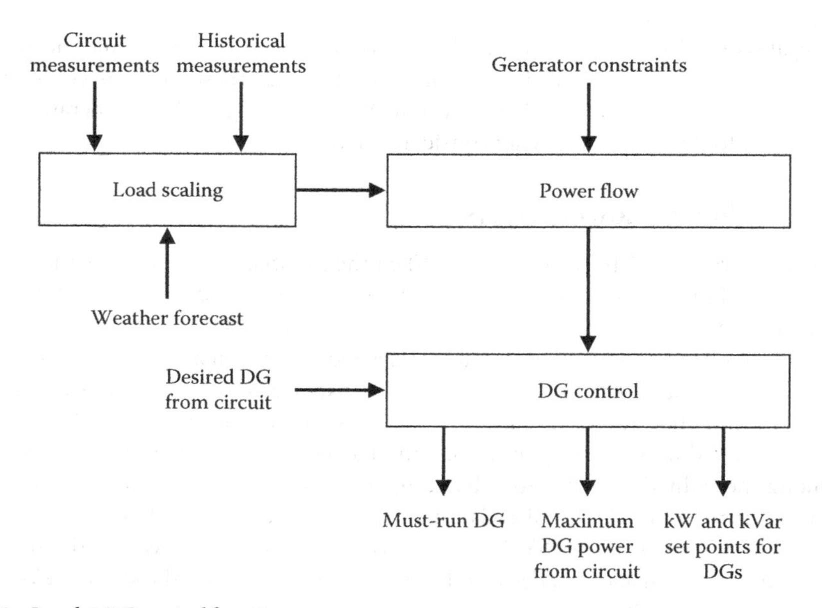

FIGURE 29.4 Level-2 *DG control* functions.

In case real-time circuit measurements are not available, historical measurements and weather data are used to estimate loading. From this information, the flows at the start of circuit can be estimated. Then the estimated flows are used as if they were measurements at the start of the circuit, and *Load Scaling* again adjusts load sizes so that the estimated and measured flows match.

Once the circuit flows are estimated, *DG Control* can check to see if there are any circuit problems such as overloaded equipment and/or locations with voltages below specified limits. If problems exist, *DG Control* runs power flow calculations and uses the controllable DGs to attempt to eliminate the problems. If the problems are removed, the generation levels required are referred to as must-run generation.

In another scenario, suppose that initially there are no problems in the circuit. However, the real-time kW and kVar DG measurements show that some DGs are running. In this case, *DG Control* tries reducing the generation to check if the no-problem condition can be obtained with less DG. If so, the reduced generation levels will be reported as must run.

Besides the must-run generation, *DG Control* also calculates the total power that can be dispatched by the circuit control. Circuit loading and generator constraints are used in this process. When DGs are dispatched, circuit losses and voltage profiles in the circuit are affected. Therefore, when looked at from the transmission side, the maximum power flow change that the DGs can achieve is greater than their rated capacities. The additional capacity achieved is due to reduced losses in the circuit and DG effects on circuit voltage profiles.

The explanation given in the preceding paragraphs is from the point of view of what happens in level 2 when data flows upward in the control hierarchy. The result of this flow is must-run generation levels and additional capacity release that can be provided for the transmission side. On the other hand, when the data flows downward from level 3, the aggregate control informs the circuit control of how much DG power is desired from the circuit. This desired power quantity is given as an input parameter to *DG Control* as shown in Figure 29.4. *DG Control* then evaluates how the desired power can be realized from the participating DGs in the circuit. This is basically an assignment problem: How much power should each generator produce so that the desired total power can be obtained in the most effective way possible? Generator constraints, fuel costs, generator operating characteristics, circuit-loss effects, reliability effects, and other parameters can be considered in this assignment process. At the end, the settings for kW and kVar generation that need to be supplied from individual DG sites are determined and sent to local controllers.

29.3.1 Estimating Loading throughout Circuit

The control of the DGs at the circuit level constitutes a major computational element in the control hierarchy. As stated earlier, the control primarily uses estimates of circuit conditions rather than measurements. Estimating the loading of customers throughout the circuit model plays a central role in the success of the control. Because system load is usually monitored at only a few points in the circuit, determining circuit loads accurately is a challenging process. In general, load is monitored at substations, major system equipment locations, and major customer (load) sites. Besides such load data, the only load information commonly available is billing-cycle customer kilowatt-hour (kWh) consumptions. The estimation of load has features described next.

Historical load measurements: Historical load measurements consist of monthly kWh measurements or periodic (such as every 15-min or hourly) kW/kVar measurements obtained at customer sites. These measurements are used in the estimation of loading at each customer site in the circuit model.

Load research statistics: With the help of electronic recorders, utilities can automatically gather hourly sample load data from diverse classes of customers. This raw data (load research data) is then analyzed to obtain load research statistics. The purpose of load research statistics is to convert kWh measurements to kW and kVar load estimates. Load research statistics consist of the following elements:

- *Kilowatt-hour parsing factors* are defined as a function of customer class. They represent the fractional annual energy use as a function of the day of the year. Thus, they vary from 0 to 1. They are used to split a kWh measurement made across monthly boundaries into estimates of how much of the measurement was used in each month.
- *kWh-to-peak-kW conversion coefficients* (referred to as C-factors) are used to convert kWh measurements for a customer to peak-kW estimates. The C-factor is calculated as a function of class of customer, type of month, type of day, and weather condition. C-factor curves are typically parameterized by the customer class, type of day, and weather condition, and plotted against the month of year.
- *Diversity factors* are used to find the aggregated demand of a group of customers. It is defined as the ratio of the sum of individual noncoincident customer peaks in the group to the coincident peak demand of the group itself. The diversity factors are greater than unity. They are defined as function of class of customer, type of month, type of day, weather conditions, and number of customers. Diversity factor curves are typically parameterized by the customer class, type of day, type of month, and weather condition, and plotted against number of customers.
- *Diversified load curves* are parameterized by class of customer, type of month, type of day, and weather conditions. They show the expected energy use for each hour of the day. Diversified load curves may be used to estimate loading as a function of the hour of day. Diversified load curves may be normalized by dividing each point on the diversified load curve by the peak of the diversified curve itself.
- *Temperature/humidity load sensitivity coefficients* are defined as a function of class of customer. They are used to scale loads to take into account temperature/humidity load sensitivities. They are calculated by correlating load research data with the weather conditions that existed at the time the load research measurements were made.

Start-of-circuit measurements: Start-of-circuit measurements generally consist of voltage magnitude, current magnitude, and/or power flows. They are used to affect scaling of estimated loads throughout the distribution circuit model such that the power flow solution matches the start-of-circuit measurements.

Examples of load research statistics for a residential class of customer are shown in Figures 29.5 through 29.9. Figure 29.5 illustrates a parsing-factor curve as a function of the day of the year. The parsing factor may be used together with monthly kWh measurements to estimate the energy usage between any 2 days of the year.

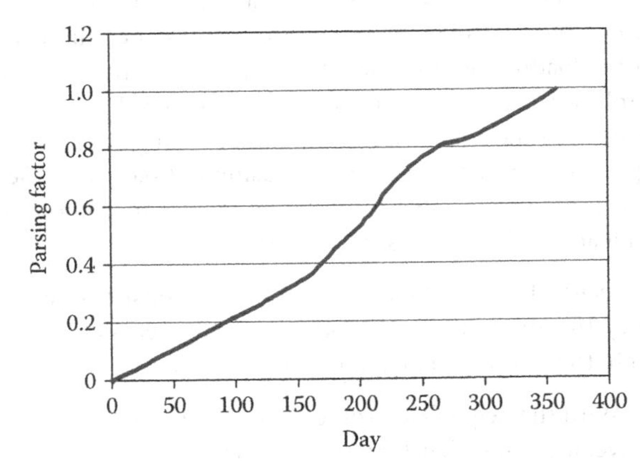

FIGURE 29.5 A representative parsing-factor curve for residential customer.

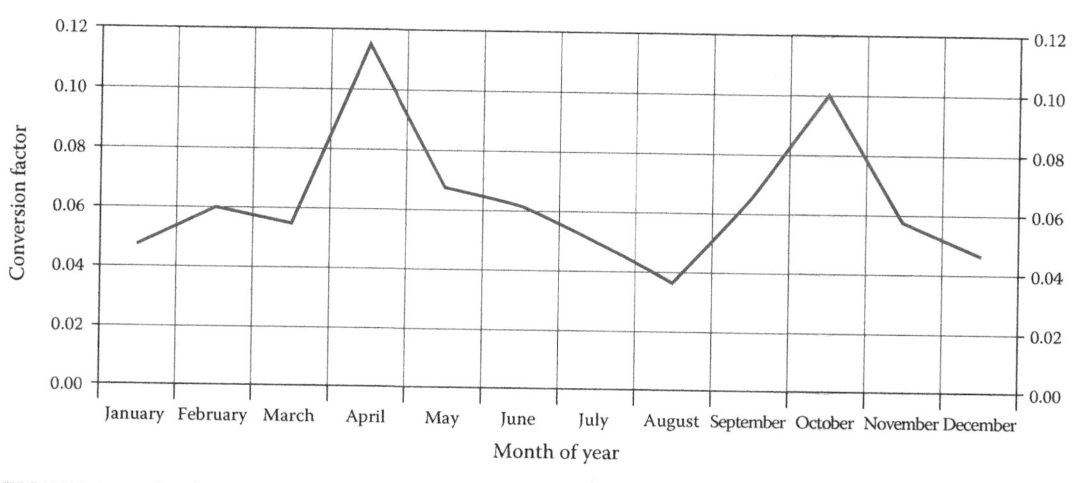

FIGURE 29.6 kWh-to-peak-kW conversion coefficients for residential class for weekdays at normal weather conditions.

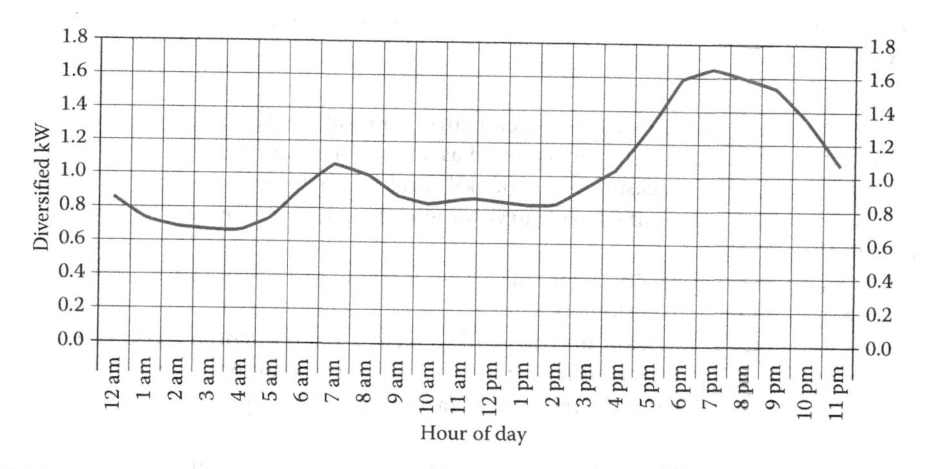

FIGURE 29.7 Diversified load curve for residential class for weekdays during February at normal weather conditions.

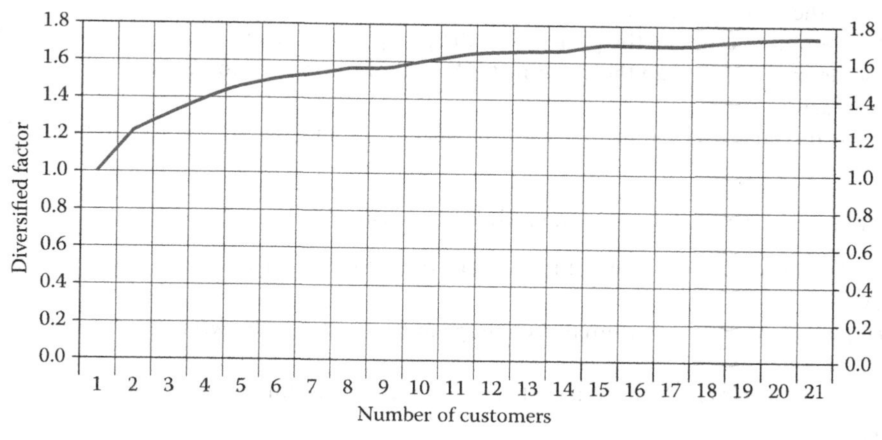

FIGURE 29.8 Diversity factor curve for residential class for weekdays during February at normal weather conditions.

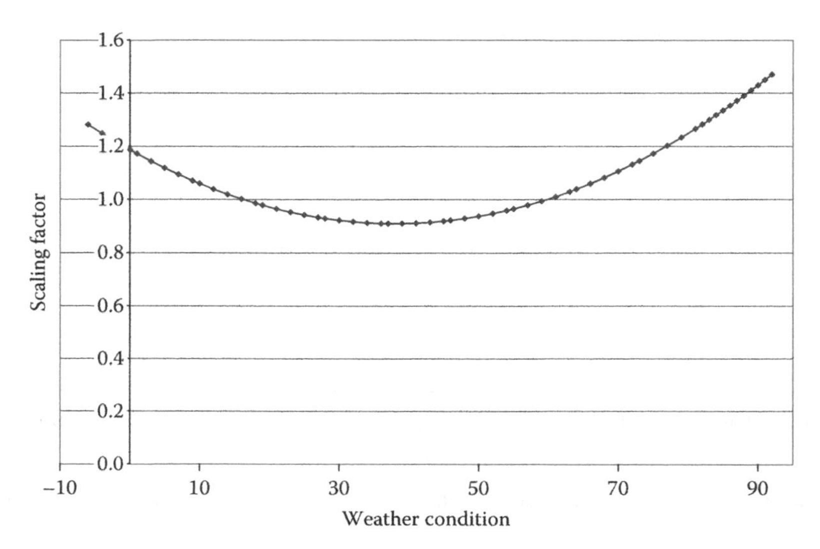

FIGURE 29.9 Representative variation of load scaling factors for residential customers as a function of weather condition.

Figure 29.6 illustrates a representative C-factor curve for residential customers for weekdays at typical weather conditions, where the C-factor is plotted as a function of month. Values read from this curve may be used to convert kWh measurements into kW-peak estimates for weekdays.

Figure 29.7 illustrates a diversified load curve for weekdays during February at normal temperatures as a function of hour of day.

Figure 29.8 illustrates a diversity factor curve for weekdays during February at normal temperatures as a function of the number of customers.

Figure 29.9 represents variation of load scaling factors for residential customers as a function of weather condition. Note that weather condition incorporates not only the temperature, but also other factors such as humidity and wind speed. Variations in these quantities are compounded into a single index.

As an example of calculating a load estimate at a point in a circuit, assume the following (where for simplicity, weather considerations have been neglected):

- Below the point selected, the circuit is radial.
- It is desired to estimate the peak-kW of the group of customers for a weekday in February. It is also desired to calculate the combined kW load of the two customers at 2 pm on a weekday in February.
- There are only two customers of the same load research class, say Class R, fed from the selected point.

Assume that each customer only has monthly kWh measurements as given by:

$KWH_{m1}(Jan18, Feb16)$ = Measured kWh usage of first customer between the dates January 18 and February 16.

$KWH_{m1}(Feb17, Mar17)$ = Measured kWh usage of first customer between the dates February 17 and March 17.

$KWH_{m2}(Jan20, Feb17)$ = Measured kWh usage of second customer between the dates January 20 and February 17.

$KWH_{m2}(Feb18, Mar19)$ = Measured kWh usage of second customer between the dates February 18 and March 19.

The first step is to estimate the energy usages of each customer during the month of February. Note that the recorded measurements do not directly reflect the February usages. Parsing factors provide the ability to estimate the February energy usage from the two measurements available.

Let $p(MonX)$ represent the parsing-factor value for customers of Class R for day X in month Mon. Then using the kWh measurements given above, the estimated kWh energy use of Customer 1 during February is calculated as follows:

$$KWH_{e1}(Feb1, Feb16) = KWH_{m1}(Jan18, Feb16) \times \frac{p(Feb16) - p(Feb1)}{p(Feb16) - p(Jan18)}$$

$$KWH_{e1}(Feb17, Feb28) = KWH_{m1}(Feb17, Mar17) \times \frac{p(Feb28) - p(Feb17)}{p(Mar17) - p(Feb17)}$$

$$KWH_{e1}(Feb) = KWH_{e1}(Feb1, Feb16) + KWH_{e1}(Feb17, Feb28)$$

where

$KWH_{e1}(Feb)$ is the estimated kWh usage of Customer 1 during February

$KWH_{e1}(Feb1, Feb16)$ is the estimated kWh usage of Customer 1 between February 1 and February 16

$KWH_{e1}(Feb17, Feb28)$ is the estimated kWh usage of Customer 1 between February 17 and February 28

A similar calculation can be performed to estimate the kWh usage of Customer 2 during February:

$$KWH_{e2}(Feb1, Feb17) = KWH_{m2}(Jan20, Feb17) \times \frac{p(Feb17) - p(Feb1)}{p(Feb17) - p(Jan20)}$$

$$KWH_{e2}(Feb18, Feb28) = KWH_{m2}(Feb18, Mar19) \times \frac{p(Feb28) - p(Feb18)}{p(Mar19) - p(Feb18)}$$

$$KWH_{e2}(Feb) = KWH_{e2}(Feb1, Feb17) + KWH_{e2}(Feb18, Feb28)$$

The next step is to estimate the peak kW demand by the two customers together. For this, we use the C-factors and diversity factors (d) from the load research statistics. Consider the following:

$C(weekday, Feb, R)$ = The kWh-to-peak-kW factor value for customers of Class R for weekdays during February.

$d(weekday, Feb, R, 2)$ = The diversity-factor value for two customers of Class R for weekdays during February.

$KW_{peak}(Sum)$ = Sum of the individual kW peaks (noncoincident peaks) for the two customers.

$KW_{peak}(Group)$ = The group peak (coincident peak) for the two customers.

Then,

$$KW_{peak}(Sum) = C(weekday, Feb, R) \times (KWH_{e1}(Feb) + KWH_{e2}(Feb))$$

$$KW_{peak}(Group) = KW_{peak}(Sum)/d(weekday, Feb, R, 2)$$

This is the first answer that was sought, which is the estimated peak of the group of customers on a weekday in February. The diversified load curve corresponding to the given conditions can be referred to for finding the time point (hour) of day at which the peak would occur. One can examine kW demands at any other time points, say at 2 pm, as well. The normalized diversified load curve is used for this purpose. The normalized curve has the maximum value of unity at its peak-kW time point. To estimate the load of the two customers at 2 pm on a weekday in February, let *k*(*2 pm, weekday, Feb*) be the normalized diversified load curve value for customers of Class R at 2 pm for weekdays during February. Then, the estimated load at 2 pm, $KW_e(Group,\ 2\ pm,\ weekday,\ Feb)$ is given by

$$KW_e(Group, 2pm, weekday, Feb) = k(2pm, weekday, Feb) \times KW_{peak}(Group)$$

29.3.2 Siting DGs for Improving Efficiency and Reliability

Along with voltage and power flow control, DGs can be placed within the distribution system for simultaneously improving efficiency and reliability. That is, there are many locations within a circuit from which a DG can implement some desired voltage or flow control, and of these many locations, the location that results in the optimum improvement in efficiency and/or reliability can be selected.

Within a system of circuits, the circuits can be reconfigured via switching operations and DG can be shifted from one circuit to another in order to implement some desired control. With such switching operations, the DG does not necessarily need to be operated as an island. That is, a DG that is connected to an unenergized circuit may be switched to an energized circuit, and then brought on line. Thus, a DG can be placed to serve a number of circuits, and can be looked at as increasing both efficiency and reliability for the system of circuits.

For a single circuit or a system of circuits, the DG site placement for best reliability is generally not the same as the placement for best efficiency. Percent changes in system reliability and efficiency can be used to determine desirable locations from a limited set of geographical locations where the DG may be placed.

To obtain good locations for efficiency and/or reliability improvements, exhaustive searches and/or optimization methods may be applied. The exhaustive search approach often works well because there are generally only a very limited number of physical sites for placing DGs. This is due to constraints placed on the siting from community impact and available land considerations.

The method that is used to site the DG should take into account the time-varying loading of the circuits involved. Placing a DG based upon just peak loading conditions will generally not result in the best reliability or efficiency when the entire time-varying load pattern is considered.

29.4 Hierarchical Control: Forecasting Generation

The load estimation discussed above is combined with a weather forecast and used to forecast system loading on an hourly basis. This load forecast is used to provide a generation schedule to the ISO, and is typically performed for the 24 h of the next day. Forecasting the next day's generation uses functionality found in levels 2 and 3 of the hierarchical control shown in Figure 29.2.

The load forecast is used to predict a schedule of must-run generation located in the distribution system. The forecast is also used to provide the ISO with the amount of base load and load following generation available at the distribution level. The amount of available generation is a function of the circuit loading. DGs provide the possibility of causing the power to flow from the distribution system to the transmission system. Since typical distribution systems are not designed to handle reverse power flows, including reverse fault currents, IEEE 1574 recommends that DGs be operated at a generation level that is 25% or less of existing circuit loading. This is taken into account when calculating the maximum amount of generation available from the distribution system.

In the forecast, DG that is just for standby must be treated specially. The load that the standby generation serves is what must be reported to the ISO as a capacity release, and not the capability of the standby generation itself. Load research statistics coupled with the weather forecast are used to estimate the hourly variation of the load that is served by the standby generation. It is this release of load estimate that is then reported to the ISO.

References

Broadwater, R.P., Sargent, A., Yarali, A., Shaalan, H.E., and Nazarko, J., Estimating substation peaks from load research data. *IEEE Transactions on Power Delivery*, 12(1): 451–456, 1997.

Daley, J.M. and Siciliano, R.L., Application of emergency and standby generation for distributed generation. I. Concepts and hypotheses. *IEEE Transactions on Industry Applications*, 39(4): 1214–1225, 2003.

IEEE Std. 1547-2003, *Standard for Interconnecting Distributed Resources with Electric Power Systems*. Institute of Electrical and Electronics Engineers, New York, 2003.

NREL SR-560-34779, *Aggregation of Distributed Generation Assets in New York State*. National Renewable Energy Laboratory (NREL), Golden, CO, 2004.

Sargent, A., Broadwater, R.P., Thompson, J.C., and Nazarko, J., Estimation of diversity and kWHR-to-peak-kW factors from load research data. *IEEE Transactions on Power Systems*, 9(3): 1450–1456, 1994.

Westinghouse Electric Cooperation, *Electric Utility Engineering Reference Book—Distribution Systems*, Vol. 3. Westinghouse Electric Cooperation, East Pittsburg, PA, 1965.

30

Distribution Short-Circuit Protection

Tom A. Short
Electric Power Research Institute

Overcurrent protection or short-circuit protection is very important on any electrical power system, and the distribution system is no exception. Circuit breakers and reclosers, expulsion fuses, and current-limiting fuses—these protective devices interrupt fault current, a vital function. Short-circuit protection is the selection of equipment, placement of equipment, selection of settings, and coordination of devices to efficiently isolate and clear faults with as little impact on customers as possible.

Of top priority, good fault protection clears faults quickly to prevent

- *Fires and explosions*
- *Further damage to utility equipment such as transformers and cables*

Secondary goals of protection include practices that help reduce the impact of faults on the following:

- *Reliability* (long-duration interruptions): In order to reduce the impact on customers, reclosing of circuit breakers and reclosers automatically restores service to customers. Having more protective devices that are properly coordinated assures that the fewest customers possible are interrupted and makes fault-finding easier.
- *Power quality* (voltage sags and momentary interruptions): Faster tripping reduces the duration of voltage sags. Coordination practices and reclosing practices impact the number and severity of momentary interruptions.

30.1 Basics of Distribution Protection

Circuit interrupters should only operate for faults, not for inrush, not for cold-load pickup, and not for transients. Additionally, protective devices should coordinate to interrupt as few customers as possible.

The philosophies of distribution protection differ from transmission-system protection and industrial protection. In distribution systems, protection is not normally designed to have backup. If a protective device fails to operate, the fault may burn and burn (until an upstream device is manually opened). Of course, protection coverage should overlap, so that if a protective device fails due to an internal short circuit (which is different from failing to open), an upstream device operates for the internal fault in the downstream protector. Backup is not a mandatory design constraint (and is impractical to achieve in all cases).

Most often, we base distribution protection on standardized settings, standardized equipment, and standardized procedures. Standardization makes operating a distribution company easier if designs are consistent. The engineering effort to do a coordination study on every circuit reduces considerably.

Another characteristic of distribution protection is that it is not always possible to fully coordinate all devices. Take fuses. With high fault currents, it is impossible to coordinate two fuses in series because the high current can melt and open both fuses in approximately the same time. Therefore, close to the substation, fuse coordination is nonexistent. There are several other situations where coordination is not possible. Some low-level faults are very difficult—some would say impossible—to detect. A conductor in contact with the ground may draw very little current. The "high-impedance" fault of most concern (because of danger to the public) is an energized, downed wire.

30.1.1 Reach

A protective device must clear all faults in its protective zone. This "zone" is defined by the following:

- *Reach*: The reach of a protective device is the maximum distance from a protective device to a fault for which the protective device will operate.

Lowering a relay pickup setting or using a smaller fuse increases the reach of the protective device (increasing the device's *sensitivity*). Sensitivity has limits; if the setting or size is too small, the device trips unnecessarily from overloads, from inrush, from cold-load pickup.

We have several generic or specific methods to determine the reach of a protective device. Commonly, we estimate the minimum fault current for faults along the line and choose the reach of the device as the point where the minimum fault current equals the magnitude where the device will operate. Some common methods of calculating the reach are as follows:

- *Percentage of a bolted line-to-ground fault*: The minimum ground fault current is some percentage (usually 25%–75%) of a bolted fault.
- *Fault resistance*: Assume a maximum value of fault resistance when calculating the current for a single line-to-ground fault. Common values of fault resistance used are 1–2, 20, 30, and 40 Ω. Rural Electrification Administration (REA) standards use 40 Ω.

Other options for determining the reach are as follows:

- *Point based on a maximum operating time of a device*: Define the reach as the point giving the current necessary to operate a protective device in a given time (with or without assuming any fault impedance). Example: The REA has recommended taking the reach of a fuse as the point where the fuse will just melt for a single line-to-ground fault in 20 s with a fault resistance of 30 or 40 Ω.
- *Point based on a multiplier of the device setting*: Choose the point where the fault current is some multiple of the device rating or setting. Example: The reach of a fuse is the point where the bolted fault current is six times the fuse rating.

None of these methods are exact. Some faults will always remain undetectable (high-impedance faults). The trick is to try to clear all high-current faults without being overly conservative.

Assuming a high value of fault resistance (20–40 Ω) is overly conservative, so avoid it. For a 12.47 kV system (7.2 kV line to ground), the fault current with a 40 Ω fault impedance is less than 180 A (this ignores the system impedance—additional system impedance reduces the calculated current even more). Using typical relay/recloser setting philosophies, which say that the rating of the recloser must be less than half of the minimum fault current, a recloser must be less than 90 A, which effectively limits the load current to an unreasonably low value. In many (even most, for some utilities) situations this is unworkable. Faults with arc impedances greater than 2 Ω are not common (see Short, 2004), so take the approach that the minimum fault is close to the bolted fault current. On the other hand, high-impedance faults (common during downed conductors) generally draw less than 50 A and have impedances of over 100 Ω. The 40 Ω rule does not guarantee that a protective device will clear high-impedance faults, and in most cases would not improve high-impedance fault detection.

30.1.2 Inrush and Cold-Load Pickup

When an electrical distribution system energizes, components draw a high, short-lived inrush; the largest component magnetizes the magnetic material in distribution transformers (in most cases, it is more accurate to say *remagnetizes* since the core is likely magnetized in a different polarity if the circuit is energized following a short-duration interruption). The transformer inrush characteristics important for protection are as follows:

- At a distribution transformer, inrush can reach peak magnitudes of 30 times the transformer's full-load rating.
- Relative to the transformer rating, inrush has higher peak magnitudes for smaller transformers, but the time constant is longer for larger transformers. Of course, on an absolute basis (amperes), a larger transformer draws more inrush.
- Sometimes inrush occurs, and sometimes it does not, depending on the point on the voltage waveform at which the reclosing occurs.
- System impedance limits the peak inrush.

The system impedance relative to the transformer size is an important concept since it limits the peak inrush for larger transformers and larger numbers of transformers. If one distribution transformer is energized by itself, the transformer is small relative to the source impedance, so the peak inrush maximizes. If a tap with several transformers is energized, the equivalent connected transformer is larger relative to the system impedance, so the peak inrush is lower (but the duration is extended). Several transformers energizing at once pull the system voltage down. This reduction in voltage causes less inrush current to be drawn from each transformer. For a whole feeder, the equivalent transformer is even larger, so less inrush is observed. Some guidelines for estimating inrush are as follows:

- *One distribution transformer*: 30 times the crest of the full-load current.
- *One lateral tap*: 12 times the crest of the full-load current of the total connected kVA.
- *Feeder*: 5 times the connected kVA up to about half of the crest of the system available fault current.

At the feeder level, inrush was only reported to cause tripping by 15% of responders to an IEEE survey (IEEE Working Group on Distribution Protection, 1995). When a three-phase circuit is reclosed, the ground relay is most likely to operate since the inrush seen by a ground relay can be as high as the peak inrush on the phases (and it is usually set lower than the phase settings). An instantaneous relay element is most sensitive to inrush, but the instantaneous element is almost always disabled for the first reclose attempt. The ground instantaneous element could operate if a significant single-phase lateral is reconnected.

Transformers are not the only elements that draw inrush; others include resistive lighting and heating elements and motors. Incandescent filaments can draw eight times normal load current. The time constant for the incandescent filaments is usually very short; the inrush is usually finished after a half cycle. Motor starting peak currents are on the order of six times the motor rating. The duration is longer than transformer inrush with durations typically from 3 to 10 s.

Cold-load pickup is the extra load following an extended interruption due to loss of the normal diversity between customers. Following an interruption, the water in water heaters cools down and refrigerators warm up. When the power is restored, all appliances that need to catch up energize at once. In cold weather, following an extended interruption, heaters all come on at once (so it is especially bad with high concentrations of resistance heating). In hot weather, houses warm up, so all air conditioners start following an interruption.

Cold-load pickup can be over three times the load prior to the interruption. As diversity is regained, the load slowly drops back to normal. This time constant varies depending on the types of loads and the duration of the interruption. Cold-load pickup is often divided into transformer inrush which last a few cycles, motor starting and accelerating currents which last a few seconds, and finally just the load due to loss of diversity which can last many minutes. Figure 30.1 shows the middle-range time frame with motor starting and accelerating currents.

It is important to select relay settings and fuse sizes high enough to avoid operations due to cold-load pickup. Even so, cold-load pickup problems are hard to avoid in some situations. A survey of utilities reported 75% having experienced cold-load pickup problems (IEEE Working Group on Distribution Protection, 1995). When a cold-load pickup problem occurs at the substation level, the most common way to reconnect is to sectionalize and pick the load up in smaller pieces. For this reason, cold-load pickup problems are not widespread—after a long interruption, utilities usually sectionalize anyway to get customers back on more quickly. Two other ways that are sometimes used to energize a circuit are to raise relay settings or even to block tripping (not recommended unless as a very last resort).

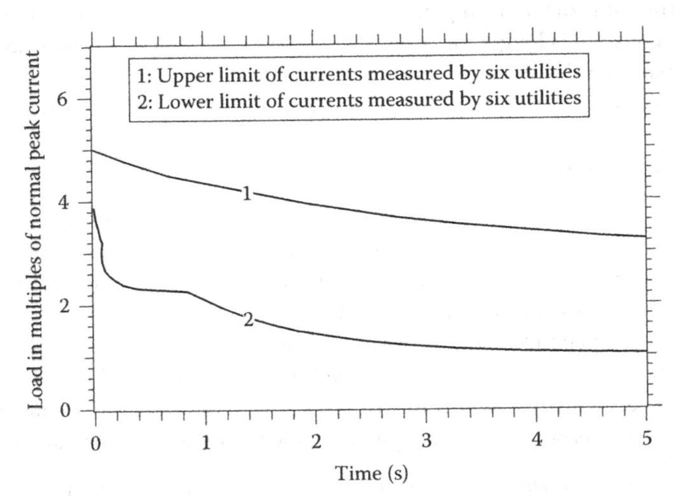

FIGURE 30.1 Ranges of cold-load pickup current from tests by six utilities. (Data from Smithley, R.S., Normal relay settings handle cold load, *Electrical World*, pp. 52–54, June 15, 1959.)

In order to pick relays, recloser settings, and fuses, we often plot a cold-load pickup curve on a time–current coordination graph along with the protection equipment characteristics. Points can be taken from the curves in Figure 30.1. It is also common to choose one or two points to represent cold-load pickup. Three-hundred percent of full-load current at 5 s is a common point.

Distribution protective devices tend to have steep time–overcurrent characteristics, meaning that they operate much faster for higher currents. K-link fuses and extremely inverse relays are most commonly used, and these happen to have the steepest characteristics. This is no coincidence; a distribution protective device must operate fast for high currents (most faults) and slow for lower currents. This characteristic gives a protective device a better chance to ride through inrush and cold-load pickup.

30.2 Protection Equipment

30.2.1 Circuit Interrupters

All circuit interrupters—including circuit breakers, fuses, and reclosers—operate on some basic principles. All devices interrupt fault current during a zero crossing. To do this, the interrupter creates an arc. In a fuse, an arc is created when the fuse element melts, and in a circuit breaker or recloser, an arc is created when the contacts mechanically separate. An arc conducts by ionizing gasses, which leads to a relatively low-impedance path.

After the arc is created, the trick is to increase the dielectric strength across the arc so that the arc clears at a current zero. Each half cycle, the ac current momentarily stops as the current is reversing directions. During this period when the current is reversing, the arc is not conducting and is starting to de-ionize, and in a sense, the circuit is interrupted (at least temporarily). Just after the arc is interrupted, the voltage across the now-interrupted arc path builds up. This is the *recovery voltage*. If the dielectric strength builds up faster than the recovery voltage, then the circuit stays interrupted. If the recovery voltage builds up faster than the dielectric strength, the arc breaks down again. Several methods used to increase the dielectric strength of the arc are discussed in the following paragraphs. The general methods are as follows:

- *Cooling the arc*: The ionization rate decreases with lower temperature.
- *Pressurizing the arc*: Dielectric strength increases with pressure.
- *Stretching the arc*: The ionized-particle density is reduced by stretching the arc stream.
- *Introducing fresh air*: Introducing de-ionized gas into the arc stream helps the dielectric strength to recover.

An air blast breaker blasts the arc stream into chutes that quickly lengthen and cool the arc. Blowout coils can move the arc by magnetically inducing motion. Compressed air blasts can blow the arc away from the contacts.

The arc in the interrupter has enough resistance to make it very hot. This can wear contact terminals, which have to be replaced after a given number of operations. If the interrupter fails to clear with the contacts open, the heat from the arc builds high pressure that can breach the enclosure, possibly in an explosive manner.

In an oil device, the heat of an arc decomposes the oil and creates gasses that are then ionized. This process takes heat and energy out of the arc. To enhance the chances of arc extinction in oil, fresh oil can be forced across the path of the arc. Lengthening the arc also helps improve the dielectric recovery. In an oil circuit breaker, the contact parting time is long enough that there may be several restrikes before the dielectric strength builds up enough to interrupt the circuit.

Vacuum devices work because the dielectric strength increases rapidly at very low pressures (because there are very few gas molecules to ionize). Normally, when approaching atmospheric pressures, the dielectric breakdown of air decreases as pressure decreases, but for very low pressures, the dielectric breakdown goes back up. The pressure in vacuum bottles is 10^{-6}–10^{-8} torr. A vacuum device only needs a very short separation distance (about 8–10 mm for a 15 kV circuit breaker). Interruption is quick since the mechanical travel time is small. The separating contacts draw an arc (it still takes a current zero to clear). Sometimes, vacuum circuit breakers chop the current, causing voltage spikes. The arc is a metal vapor consisting of particles melted from each side. Contact erosion is low, so vacuum devices are low maintenance and have a long life. Restrikes are uncommon.

SF_6 is a gas that is a very good electrical insulator, so it has rapid dielectric recovery. At atmospheric pressures, the dielectric strength is 2.5 times that of air, and at higher pressures, the performance is even better. SF_6 is very stable, does not react with other elements, and has good temperature characteristics. One type of device blows compressed SF_6 across the arc stream to increase the dielectric strength. Another type of SF_6 interrupter used in circuit breakers and reclosers has an arc spinner which is a setup that uses the magnetic field from a coil to cause the arc to spin rapidly (bringing it in contact with un-ionized gas). SF_6 can be used as the insulating medium as well as the interrupting medium. SF_6 devices are low maintenance, have short opening times, and most do not have restrikes.

Since interrupters work on the principle of the dielectric strength increasing faster than the recovery voltage, the X/R ratio can make a significant difference in the clearing capability of a device. In an inductive circuit, the recovery voltage rises very quickly since the system voltage is near its peak when the current crosses through zero. Asymmetry increases the peak magnitude of the fault current. For this reason, the capability of most interrupters decreases with higher X/R ratios. Some interrupting equipment is rated based on a symmetrical current basis while other equipment is based on asymmetrical current. Whether based on a symmetrical or asymmetrical basis, the interrupter has asymmetric interrupting capability.

30.2.2 Circuit Breakers

Circuit breakers are often used in the substation on the bus and on each feeder. Circuit breakers are available with very high interrupting and continuous current ratings. The interrupting medium in circuit breakers can be any of vacuum, oil, air, or SF_6. Oil and vacuum breakers are most common on distribution stations with newer units being mainly vacuum with some SF_6.

Circuit breakers are tripped with external relays. The relays provide the brains to control the opening of the circuit breaker, so the breaker coordinates with other devices. The relays also perform reclosing functions.

Circuit breakers are historically rated as constant MVA devices. A symmetrical short-circuit rating is specified at the maximum rated voltage (for more ratings information, see ANSI C37.06-1997; IEEE Std. C37.04-1999; IEEE Std. C37.010-1999). Below the maximum rated voltage (down to a specified minimum value), the circuit breaker has more interrupting capability. The minimum value where the circuit breaker is a constant-MVA device is specified by the constant K:

$$\text{Symmetrical interrupting capability} = \begin{cases} V_R \cdot I_R & \text{for } \dfrac{V_R}{K} < V \leq V_R \\ K \cdot I_R & \text{for } V \leq \dfrac{V_R}{K} \end{cases}$$

where

I_R is the rated symmetrical rms short-circuit current operating at V_R

V_R is the maximum rms line-to-line rated voltage

V is the operating voltage (also rms line-to-line)

K = voltage range factor = ratio of the maximum rated voltage to the lower limit in which the circuit breaker is a constant MVA device

Newer circuit breakers are rated as constant current devices ($K = 1$).

Consider a 15 kV class breaker application on a 12.47 kV system where the maximum voltage will be assumed to be 13.1 kV (105%). For an ANSI-rated 500-MVA class breaker with $V_R = 15$ kV, $K = 1.3$, and

TABLE 30.1 15 kV Class Circuit Breaker Short-Circuit Ratings

	500 MVA	750 MVA	1000 MVA
Rated voltage, kV	15	15	15
K, voltage range factor	1.3	1.3	1.3
Short circuit at max voltage rating	18	28	37
Maximum symmetrical interrupting, kA	23	36	48
Close and latch rating			
$1.6K \times$ rated short-circuit current, kA (asym)	37	58	77
$2.7K \times$ rated short-circuit current, kA (peak)	62	97	130

I_R = 18 kA, the symmetrical interrupting capability would be 20.6 kA (15/13.1 × 18). Circuit breakers are often referred to by their MVA class designation (1000 MVA class for example). Typical circuit breaker ratings are shown in Table 30.1.

Circuit breakers must also be derated if the reclose cycle could cause more than two operations and if the operations occur within less than 15 s. The percent reduction is given by

$$D = d_1(n-2) + \frac{d_1(15-t_1)}{15} + \frac{d_1(15-t_2)}{15} + \cdots$$

where

D is the total reduction factor, %

$$d_1 = \text{calculating factor, \%} = \begin{cases} 3\% & \text{for } I_R < 18\text{ kV} \\ \dfrac{I_R}{6} & \text{for } I_R > 18\text{ kA} \end{cases}$$

n is the total number of openings

t_n is the nth time interval less than 15 s

The interrupting rating is then $(100 - D)I_R$. The permissible tripping delay is also a standard. For the given delay period, the circuit breaker must withstand K times the rated short-circuit current between closing and interrupting. A typical delay is 2 s.

Continuous current ratings are independent of interrupting ratings (although higher continuous ratings usually go along with higher interrupting ratings). Standard continuous ratings include 600, 1200, 2000, and 3000 A (the 600 and 1200 A circuit breakers are most common for distribution substations).

A circuit breaker also has a *momentary* or *close and latch* short-circuit rating (also called the first-cycle capability). During the first cycle of fault current, a circuit breaker must be able to withstand any current up to a multiple of the short-circuit rating. The rms current should not exceed $1.6K \times I_R$ and the peak (crest) current should not exceed $2.7K \times I_R$.

The circuit breaker interrupting time is defined as the interval between energizing the trip circuit and the interruption of all phases. Most distribution circuit breakers are five-cycle breakers. Older breakers interrupt in eight cycles.

Distribution circuit breakers are three-phase devices. When the trip signal is received, all three phases are tripped. All three will not clear simultaneously because the phase current zero crossings are separated. The degree of separation between phases is usually one-half to one cycle.

30.2.3 Circuit Breaker Relays

Several types of relays are used to control distribution circuit breakers. Distribution circuits are almost always protected by overcurrent relays that use inverse time overcurrent characteristics. An inverse time–current characteristic means that the relay will operate faster with increased current.

The main types of relays are as follows:

- *Electromechanical relays*: The induction disk relay has long been the main relay used for distribution overcurrent protection. The relay is like an induction motor with contacts. Current through the CT leads induces flux in the relay magnetic circuit. These flux linkages cause the relay disk to turn. A larger current turns the disk faster. When the disk travels a certain distance, the contacts on the disk meet stationary contacts to complete the relay trip circuit. An instantaneous relay functionality can be provided: A plunger surrounded by a coil or a disk cup design operates quickly if the current is above the relay pickup. Most electromechanical relays are single phase.
- *Static relays*: Analog electronic circuitry (like op-amps) provide the means to perform a time–current characteristic that approximates that of the electromechanical relay.
- *Digital relays*: The most modern relay technology is fully digital based on microprocessor components.

Electromechanical relays have reliably served their function and will continue to be used for many years. The main characteristics that should be noted as it affects coordination are overtravel and reset time. *Overtravel* occurs because of the inertia in the disk. The disk will keep turning for a short distance even after the short circuit is interrupted. A typical overtravel of 0.1 s is assumed when applying induction relays. An induction disk cannot instantly turn back to the neutral position. This *reset time* should be considered when applying reclosing sequences. It is not desirable to reclose before the relay resets or *ratcheting* to a trip can occur.

Digital relays are slowly replacing electromechanical relays. The main advantages of digital relays are as follows:

- *More relay functions*: One relay performs the functions of several electromechanical relays. One relay can provide both instantaneous and time–overcurrent relay protection for three phases, plus the ground, and perform reclosing functions. This can result in considerable space and cost savings. Some backup is lost with this scheme if a relay fails. One option to provide relay backup is to use two digital relays, each with the same settings.
- *New protection schemes*: Advanced protection schemes are possible that provide more sensitive protection and better coordination with other devices. Two good examples for distribution protection are negative-sequence relaying and sequence coordination. Advanced algorithms for high-impedance fault detection are also possible.
- *Other auxiliary functions*: Fault location algorithms, fault recording, and power quality recording functions.

Digital relays have another advantage: internal diagnostics with ability to self-test. With digital technology, the relay is less prone to drift over time from mechanical movements or vibrations. Digital relays also avoid relay overtravel and ratcheting that are constraints with electromechanical relays (though some digital relays do reset like an electromagnetic relay).

Digital relays do have disadvantages. They are a relatively new technology. Computer technology has a poor reputation as far as reliability. If digital relays were as unreliable as a typical personal computer, we would have many more interruptions and many fires caused by uncleared faults. Given that, most digital relays have proven to be reliable and are gaining more and more acceptance by utilities.

Just as computer technology continues to advance rapidly, digital relays are also advancing. While it is nice to have new features, technical evolution can also mean that relay support becomes more difficult. Each relay within a certain family has to have its own supporting infrastructure for adjusting the relay settings, uploading and downloading data, and testing the relay. Each relay requires a certain amount of crew learning and training. As relays evolve, it becomes more difficult to maintain a variety of digital relays. The physical form and connections of digital relays are not standardized. As a contrast, electromechanical relays change very little and require a relatively stable support infrastructure. Equipment standardization helps minimize the support infrastructure required.

TABLE 30.2 Relay Designations

	Westinghouse/ ABB Designation	General Electric Designation
Moderately inverse	CO-7	
Inverse time	CO-8	IAC-51
Very inverse	CO-9	IAC-53
Extremely inverse	CO-11	IAC-77

The time–current characteristics are based on the historically dominant manufacturers of relays. Westinghouse relays have a CO family of relays, and the General Electric relays are IAC (see Table 30.2). Most relays (digital and electromechanical) follow the characteristics of the GE or Westinghouse relays. For distribution overcurrent protection, the extremely inverse relays are most often used (CO-11 or IAC-77).

The time–current curves for induction relays can be approximated by the following equation (Benmouyal and Zocholl, 1994):

$$t = TD\left(\frac{A}{M^p - 1} + B\right)$$

where

t is the trip time, s
M is the multiple of pickup current ($M > 1$)
TD is the time dial setting
A, B, p are curve shaping constants

With actual induction disk relays, the constants A and B change with the time dial setting, but with digital relays, they stay constant.

Standardized characteristics of relays have been defined by IEEE (IEEE Std. C37.112-1996). This is an attempt to make relay characteristics consistent (since the relay curve can be adjusted to almost anything in a digital relay). The equations for the standardized inverse relay characteristics are shown in Table 30.3. Figure 30.2 compares the shapes of these curves. The standard allows relays to have tripping times within 15% of the curves. The standard also specifies the relay reset time for $0 < M < 1$ as

$$t = TD\left(\frac{t_r}{M^2 - 1}\right)$$

where t_r is the reset time, s for $M = 0$

TABLE 30.3 IEEE Standardized Relay Curve Equation Constants

	A	B	p
Moderately inverse	0.0515	0.114	0.02
Very inverse	19.61	0.491	2.0
Extremely inverse	28.2	0.1217	2.0

Source: IEEE Std. C37.112-1996, IEEE standard inverse-time characteristic equations for overcurrent relays.

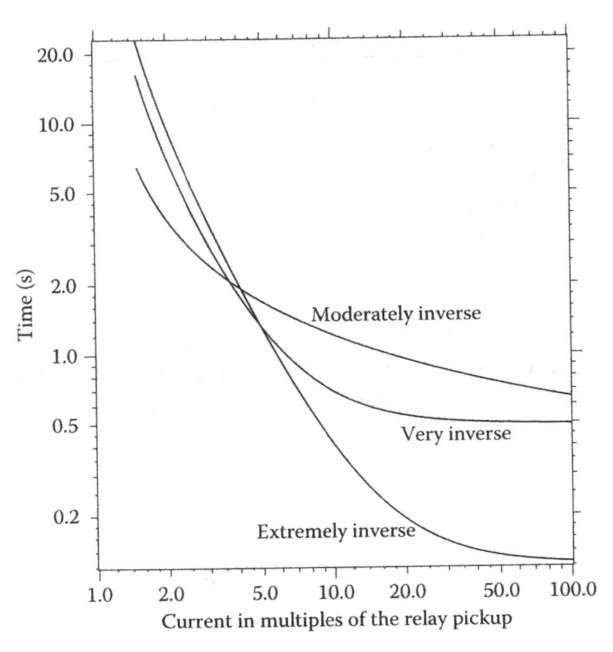

FIGURE 30.2 Relay curves following the IEEE standardized characteristics for a time dial = 5.

30.2.4 Reclosers

A recloser is a specialty distribution protective device capable of interrupting fault current and automatically reclosing. The official definition of a recloser is as follows:

Automatic circuit recloser: A self-controlled device for automatically interrupting and reclosing an alternating-current circuit, with a predetermined sequence of opening and reclosing followed by resetting, hold closed, or lockout.

Like a circuit breaker, interruption occurs at a natural current zero. The interrupting medium of a recloser is most commonly vacuum or oil. The insulating medium is generally oil, air, a solid dielectric, or SF_6. The recloser control can be electronic, electromechanical (the relay for tripping is electromechanical, and the reclosing control is electronic), or hydraulic. A hydraulic recloser uses springs and hydraulic systems for timing and actuation.

The interrupting rating of a recloser is based on a symmetrical current rating. The interrupting current rating does not change with voltage. There is an exception that some reclosers have a higher interrupting current if operated at a significantly lower voltage than the rating. Smaller reclosers with a 50–200 A continuous rating typically have interrupting ratings of 2–5 kA (these would normally be feeder reclosers). Larger reclosers that could be used in substations have continuous current ratings as high as 1120 A and interrupting ratings of 6–16 kA. Historically, reclosers with series coil types had coil ratings of 25, 35, 50, 70, 100, 140, 200, 280, 400, and 560 A (each rating is approximately 1.4 times higher than the next lower rating).

Reclosers are tested at a specified X/R ratio as specified in ANSI/IEEE C37.60-1981. A typical test value is $X/R = 16$. While a lower X/R ratio at the point of application does not mean you can increase the rating of a recloser, the recloser must be derated if the X/R ratio is larger than that specified.

There are some other differences with recloser ratings vs. circuit breaker ratings (Cooper Power Systems, 1994). Reclosers do not have to be derated for multiple operations. Reclosers do not have a separate closing and latching (or first-cycle) rating. The symmetrical current rating is sufficient to handle the asymmetry during the first cycle as long as the circuit X/R ratio is lower than the tested value.

Reclosers have many distribution applications. We find reclosers in the substation as feeder interrupters instead of circuit breakers. An IEEE survey found that 51% of station feeder interrupting devices were reclosers (IEEE Working Group on Distribution Protection, 1995). Reclosers are used more in smaller stations and circuit breakers more in larger stations. Three-phase reclosers can be used on the main feeder to provide necessary protection coverage on longer circuits, along with improved reliability. Overhead units and padmounted units are available. Reclosers are available as single-phase units, so they can be used on single-phase taps instead of fuses. Another common application is in autoloop automation schemes to automatically sectionalize customers after a fault.

Since reclosers are devices built for distribution circuits; some have features that are targeted to distribution circuit needs. Three-phase units are available that can operate each phase independently (so a single-phase fault will only open one phase). Some reclosers have a feature called sequence coordination to enhance coordination between multiple devices.

The time–current characteristics of hydraulic reclosers have letter designations: A, B, and C. The A is a fast curve that is used similarly to an instantaneous relay element, and the B and C curves have extra delay ("delayed" and "extra delayed"). For a hydraulically controlled recloser, the minimum trip threshold is twice the full-load rating of the trip coil of the recloser and is normally not adjustable. On electronically controlled reclosers, the minimum trip threshold is adjustable independently of the rating (analogous to setting the pickup of a time–overcurrent relay).

30.2.5 Expulsion Fuses

Expulsion fuses are the most common protective device on distribution circuits. Fuses are low-cost interrupters that are easily replaced (when in cutouts). Interruption is relatively fast and can occur in a half of a cycle for large currents. An expulsion fuse is a simple concept: A fusible element made of tin or silver melts under high current. Expulsion fuses are most often applied in a fuse cutout. In a fuse tube, after the fuse element melts, an arc remains. The arc, which has considerable energy, causes a rapid pressure buildup. This forces much of the ionized gas out of the bottom of the cutout (see Figure 30.3), which helps to prevent the arc from reigniting at a current zero. The extreme pressure, the stretching of the arc, and the turbulence help increase the dielectric strength of the air and clear the arc at a current zero. A fuse tube also has an organic fiber liner that melts under the heat of the

FIGURE 30.3 Example operation of an expulsion fuse during a fault. (Courtesy of the Long Island Power Authority, Uniondale, NY.)

arc and emits fresh, nonionized gases. At high currents, the expulsion action predominates, while at lower currents, the deionizing gases increase the dielectric strength the most.

The "expulsion" characteristics should be considered by crews when placing a cutout on a structure. Avoid placement where a blast of hot, ionized gas blown out the bottom of the cutout could cause a flashover on another phase or other energized equipment. Implement and enforce safety procedures whenever a cutout is switched in (because it could be switching into a fault), including eye protection, arc resistant clothing, and, of course, avoiding the bottom of the cutout.

The *speed ratio* of a fuse quantifies how steep the fuse curve is. The speed ratio is defined differently depending on the size of the fuse (IEEE Std. C37.40-1993):

$$\text{Speed ratio for fuse ratings of 100 A and under} = \frac{\text{melting current at 0.1 s}}{\text{melting current at 300 s}}$$

$$\text{Speed ratio for ratings above 100 A} = \frac{\text{melting current at 0.1 s}}{\text{melting current at 600 s}}$$

Industry standards specify two types of expulsion fuses, the most commonly used fuses. The "K" link is a relatively fast fuse, and the "T" is somewhat slower. K links have a speed ratio of 6–8. T links have a speed ratio of 10–13. The K link is the most commonly used fuse for transformers and for line taps. The K and T fuse links are standardized well enough that they are interchangeable among manufacturers for most applications.

Two time–current curves are published for expulsion fuses: the minimum melt curve and the maximum total clear curve. The minimum-melt time is 90% of the average melt time to account for manufacturing tolerances. The total clearing time is the average melting time plus the arcing time plus manufacturing tolerances. Figure 30.4 shows the two published curves for 50 A K and T fuse links. The manufacturer's minimum melt curves for fuses less than or equal to 100 A normally start at 300 s, and those over 100 A start at 600 s.

The time–current characteristics for K and T links are standardized at three points (ANSI C37.42-1989). The minimum and maximum allowed melting current is specified for durations of 0.1, 10, and either 300 s (for fuses rated 100 A or less) or 600 s (for larger fuses).

Published fuse curves are for no loading and an ambient temperature of 25°C. Both loading and ambient temperature change the fuse melting characteristic. Load current causes the most dramatic difference, especially when a fuse is overloaded. Figure 30.5 shows the effect of loading on fuse melting time. Figures 30.6 through 30.9 show time–current curves for K and T links.

For operation outside of this ambient range, the fuse melting time changes. The melting characteristic of tin fuse links changes 3.16% for each 10°C above or below 25°C, so a fuse operating in a 50°C ambient will operate in 92% of the published time (100% – (25/10)3.16%). Silver links are less sensitive to temperature (0.9% melting change for each 10°C above or below 25°C).

The I^2t of a fuse is often needed to coordinate between fuses. Table 30.4 shows the minimum melt I^2t of K and T links estimated from the time–current curves at 0.01 s. This number is also useful to estimate melting characteristics for high currents below the published time–current characteristics, which generally have a minimum time of 0.01 s.

The 6, 10, 15, 25, 40, 65, 100, 140, and 200 A fuses are standard ratings that are referred to as *preferred* fuses. The 8, 12, 20, 30, 50, and 80 A fuses are *intermediate* fuses. The designations are provided because two adjacent fuses (for those below 100 A) will not normally coordinate. A 40 and a 30 A fuse will not coordinate, but a 40 and a 25 A fuse will coordinate up to some maximum current. Most utilities pick a standard set of fuses to limit the number of fuses stocked.

K or T links with tin fuse elements can carry 150% of the nominal current rating indefinitely. It is slightly confusing that a 100 A fuse can operate continuously up to 150 A. Overloaded fuses, although

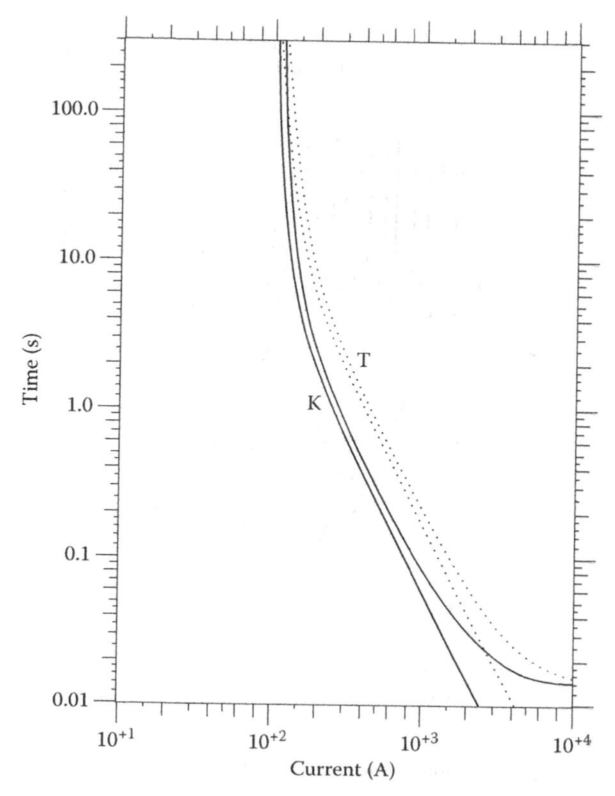

FIGURE 30.4 Minimum melt and total clearing curves for a T and K link (50 A).

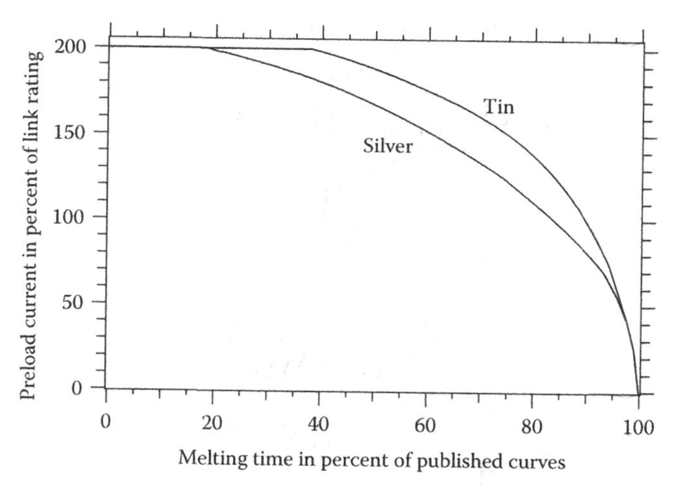

FIGURE 30.5 Effect of loading on fuse melting time. (Adapted from Cooper Power Systems, *Electrical Distribution—System Protection,* 3rd edn., 1990. With permission from Cooper Industries, Inc.)

they can be safely overloaded, operate significantly faster when overloaded, which could cause miscoordination. In contrast to tin links, silver links have no continuous overload capability.

Other nonstandard fuses are available from manufacturers for special purposes. One type of specialty fuse is a fuse even slower than a T link that is used to provide better coordination with upstream circuit breakers or reclosers in a fuse-saving scheme. Another notable type of specialty fuse is a *surge-resistant* fuse that responds slowly to fast currents (such as surges) but faster to lower currents.

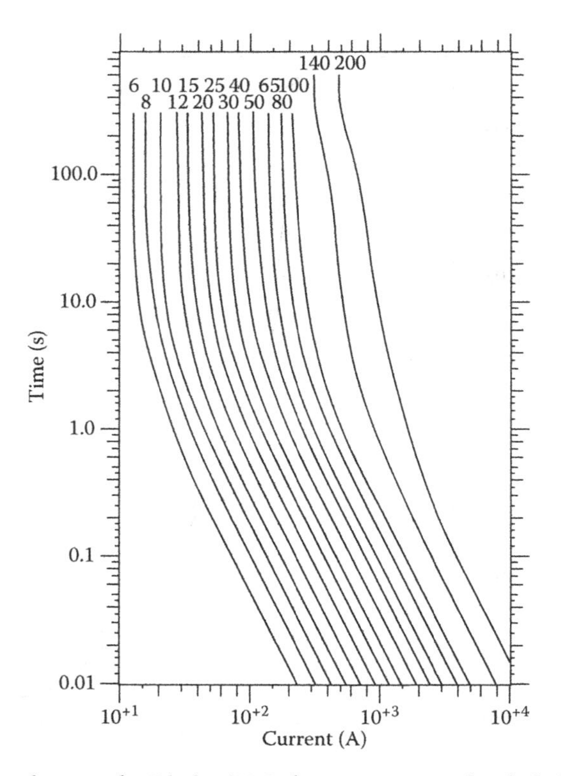

FIGURE 30.6 Minimum melt curves for K links. (S&C Electric Company silver links.)

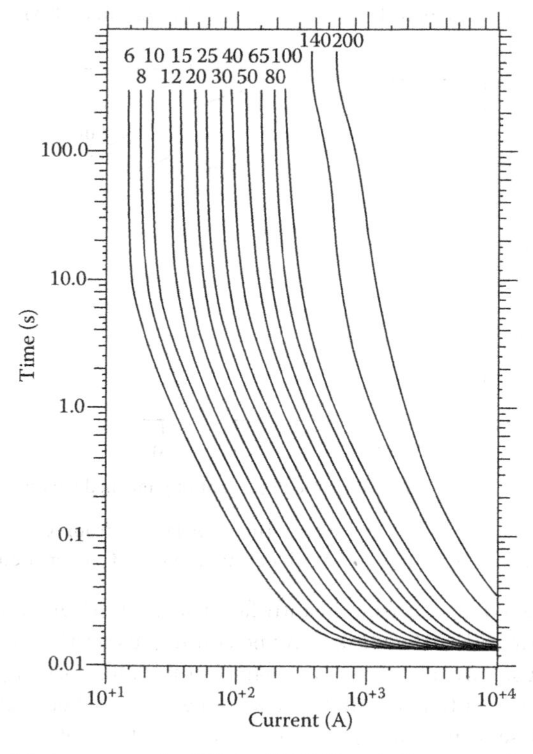

FIGURE 30.7 Maximum total clear curves for K links. (S&C Electric Company silver links.)

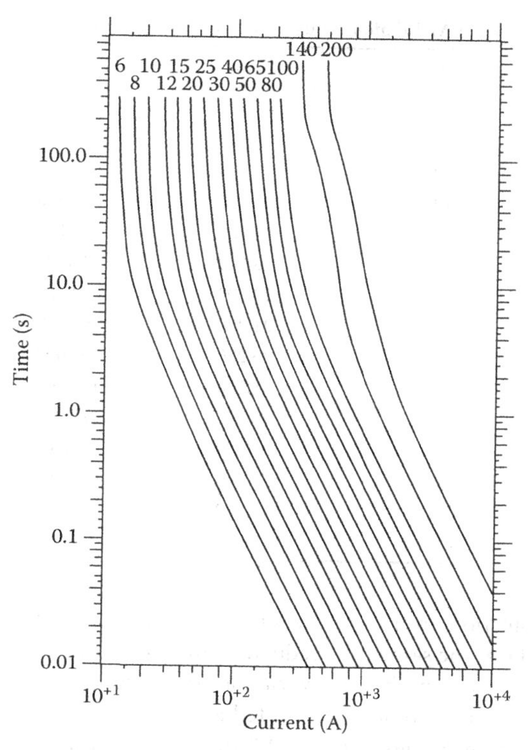

FIGURE 30.8 Minimum melt curves for T links. (S&C Electric Company silver links.)

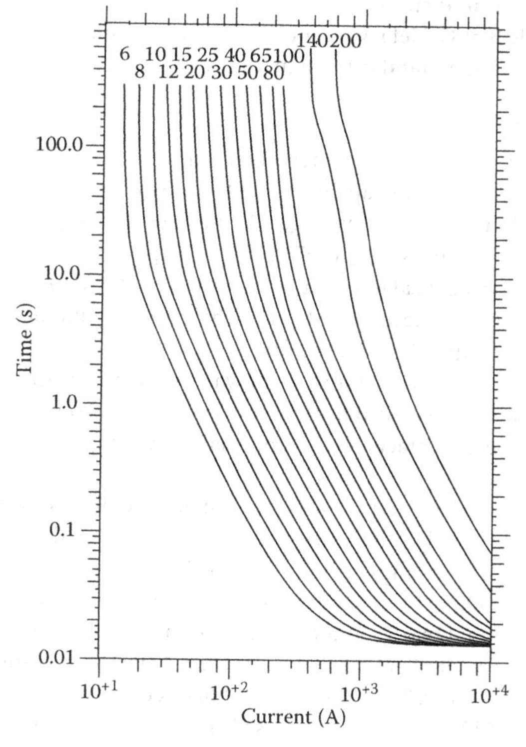

FIGURE 30.9 Maximum total clear curves for T links. (S&C Electric Company silver links.)

TABLE 30.4 Fuse Minimum-Melt I^2t (A^2-s)

Rating, A	K Links	T Links
6	534	1,490
8	1,030	2,770
10	1,790	5,190
12	3,000	8,810
15	5,020	15,100
20	8,500	24,500
25	13,800	40,200
30	21,200	65,500
40	36,200	107,000
50	58,700	173,000
65	90,000	271,000
80	155,000	425,000
100	243,000	699,000
140	614,000	1,570,000
200	1,490,000	3,960,000

These achieve better protection on transformers for secondary faults and faster operation for internal transformer failures while at the same time reducing nuisance fuse operations due to lightning.

Expulsion fuses under oil are another fuse variation. These "weak-links" are used on CSP (completely self-protected) transformers and some padmounted and vault transformers. Since they are not easily replaced, they have very high ratings—at least 2.5 times the transformer full load current and much higher if a secondary circuit breaker is used.

Transformers on underground circuits use a variety of fuses. For padmounted transformers, a common fuse is the replaceable Bay-O-Net style expulsion fuse. The time–current characteristics of this fuse do not follow one of the industry standard designations.

30.2.5.1 Fuse Cutouts

The cutout is an important part of the fuse interrupter. The cutout determines the maximum interrupting capability, the continuous current capability, the load-break capability, the basic lightning impulse insulation level (BIL), and the maximum voltage. Cutouts are typically available in 100, 200, and 300 A continuous ratings (ANSI standard sizes [ANSI C37.42-1989]).

Cutouts are rated on a symmetrical basis. Cutouts are tested at X/R ratios of 8 to 12, so if the X/R ratio at the application point is higher than the test value, the cutout should be derated. The fuse line holder determines the interrupting capability, not the fuse link.

Most cutouts are of the open variety with a removable fuse holder that is placed in a cutout with a porcelain-bushing-type support. We also find enclosed cutouts and open-link cutouts. Open links have a fuse link suspended between contacts. Open links have a much lower interrupting capability (1.2 kA symmetrical).

Many cutouts available are full-rated cutouts that can be used on any type of system where the maximum line-to-line voltage is less than the cutout rating. Cutouts are also available that have *slant* voltage ratings, which provide two ratings such as 7.8/15 kV that are meant for application on grounded circuits (IEEE Std. C37.48-1997). One cutout will interrupt any current up to its interrupting rating and up to the lower voltage rating. On a grounded distribution system, in most situations, any cutout can be applied that has the lower slant rating voltage greater than the maximum line-to-ground voltage. On a 12.47Y/7.2 kV grounded distribution system, a 7.8/15 kV cutout could be used. If the system were ungrounded, a full-rated 15 kV cutout must be used. For three-phase grounded circuits, the recovery voltage is the line-to-line voltage for a line-to-line fault rather than the line-to-ground voltage (requiring a higher voltage rating). In this case, the slant-rated cutouts are designed and tested so that two cutouts in series will interrupt a current up to its interrupting rating and up to the higher voltage rating. The two cutouts share the recovery voltage (even considering the differences in the melting times of the two fuses). On grounded systems, there are cases where slant-rated cutouts are "under-rated"—any time that a phase-to-phase fault could happen that would only be cleared by one cutout. This includes constructions where multiple circuits share a pole or cases where cutouts are applied on different poles.

Most cutouts used on distribution systems do not have load break capability. If the cutout is opened under load, it can draw an arc that will not clear. It is not an uncommon practice for crews to open cutouts under load (if it draws an arc, they slam it back in). Cutouts with load-break capability are available, usually capable of interrupting 100–300 A. Cutouts with load-break capability usually use an arc chute. A spring pulls the arc quickly through the arc chute where the arc is stretched, cooled, and interrupted. A load-break tool is available that can open standard cutouts (with no load break capability of its own) under load up to 600–900 A. Utilities also sometimes use solid blades in cutouts instead of a fuse holders; then crews can use the cutout as a switch.

30.2.6 Current-Limiting Fuses

Current-limiting fuses (CLFs) are another interrupter having the unique ability to reduce the magnitude of the fault current. CLFs consist of fusible elements in silicon sand (see Figure 30.10). When fault current melts the fusible elements, the sand melts and creates a narrow tube of glass called a *fulgerite*. The voltage across the arc in the fulgerite greatly increases. The fulgerite constricts the arc. The sand helps cool the arc (which means it takes energy from the arc). The sand does not give off ionizable gas when it melts, and it absorbs electrons, so the arc has very little ionizable air to use as a conductor. Without ionizable air, the arc is choked off, and the arc resistance becomes very high. This causes a back voltage that quickly reduces the current. The increase in resistance also lowers the X/R ratio of the circuit, causing a premature current zero. At the current zero, the arc extinguishes. Since the X/R ratio is low, the voltage zero and current zero occur very close together, so there will be very little transient recovery voltage (the high arc voltage comes just after the element melts). Because the current-limiting fuse forces an early current zero, the fuse can clear the short circuit in much less than one half of a cycle.

Current-limiting fuses are noted for their very high fault-clearing capability. CLFs have symmetrical maximum interrupt ratings up to 50 kA; contrast that to expulsion fuses which may have typical maximum interrupt ratings of 3.5 kA in oil and 13 kA in a cutout. Current-limiting fuses also completely contain the arc during operation and are noiseless with no pressure buildup.

Current-limiting fuses are widely used for protection of equipment in high fault current areas. Table 30.5 shows the percentages of utilities that use CLFs. The major reason given for the use of CLFs is safety, and the second most common reason is high fault currents in excess of expulsion fuse ratings.

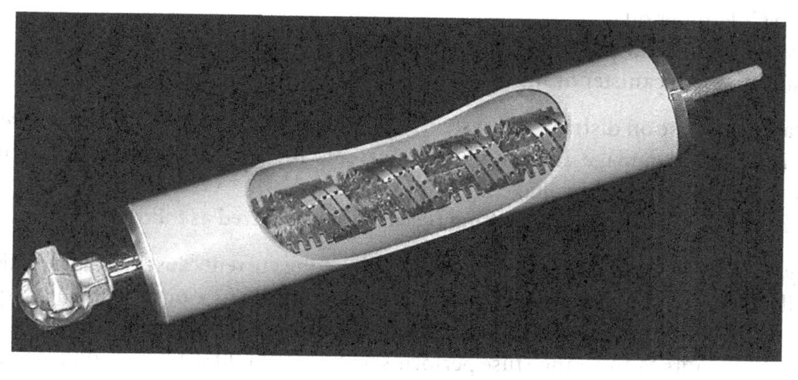

FIGURE 30.10 Example backup current-limiting fuse. (From Hi-Tech Fuses, Inc., Hickory, NC.)

TABLE 30.5 Use of Current-Limiting Fuses as Reported in a 1995 IEEE Survey

	5 kV	15 kV	25 kV	35 kV
General purpose	15%	29%	30%	18%
Backup	15%	38%	43%	30%
On OH line laterals	5%	6%	9%	3%
On UG line laterals	7%	18%	20%	18%

Source: IEEE Working Group on Distribution Protection, *IEEE Trans. Power Deliv.*, 10(1), 176–186, January 1995.

There are three types of current-limiting fuses (IEEE Std. C37.40-1993):

- *Backup*: A fuse capable of interrupting all currents from the maximum rated interrupting current down to the minimum rated interrupting current.
- *General purpose*: A fuse capable of interrupting all currents from the maximum rated interrupting current down to the current that causes melting of the fusible element in 1 h.
- *Full range*: A fuse capable of interrupting all currents from the rated interrupting current down to the minimum continuous current that causes melting of the fusible element(s), with the fuse applied at the maximum ambient temperature specified by the manufacturer.

Current-limiting fuses are very good at clearing high-current faults. They have a much harder time with low-current faults or overloads. For a low-level fault, the fusible element will not melt, but it will get very hot and can melt the fuse hardware resulting in failure. This is why the most common CLF application is as a backup in series with an expulsion fuse. The expulsion fuse clears low-level faults, and the CLF clears high-current faults. Current-limiting fuses have very steep melting and clearing curves, much steeper than expulsion links. Many current-limiting fuses have steeper characteristics than I^2t. At low currents, heat from the notches transfers to the un-notched portion; at high currents, the element melts faster because heat cannot escape from the notched areas fast enough to delay melting.

General-purpose fuses usually use two elements in series: one for the high-current faults and one for the low-current faults. General-purpose fuses could fail for overloads, so restrict their application to situations where overloads are not present or are protected by some other device (such as a secondary circuit breaker on a transformer).

Full-range fuses provide even better low-current capability and can handle overloads and low-level faults without failing (as long as the temperature is within rating).

Current-limiting fuses can be applied in several ways, including

- Backup current-limiting fuse in series with an expulsion fuse in a cutout
- Full-range current-limiting fuse in a cutout
- Backup CLF under oil
- Full-range (or general-purpose) fuse under oil
- CLF in a dry-well canister or insulator

The best locations for use on distribution systems are close to the substation. This is where they are most appropriate for limiting damage due to high fault currents and where they are most useful for reducing the magnitude and duration of a voltage sag.

Some of the drawbacks of current-limiting fuses are summarized as follows:

- *Voltage kick*: When a CLF operates, the rapidly changing current causes a voltage spike ($V = Ldi/dt$). Usually, this is not severe enough to cause problems for the fuse or for customer equipment.
- *Limited overload capability*: A backup or general-purpose fuse does not do well for overloads or low-current faults. A full-range fuse performs better but could still have problems with a transient overcurrent that partially melts the fuse.

- *Coordination issues*: A current-limiting fuse may be difficult to coordinate with expulsion fuses or reclosers or other distribution protective devices. CLFs are fast enough that they almost have to be used in a fuse-blowing scheme (fuse saving will not work because the fuse will be faster than the circuit breaker).
- *Cost*: High cost relative to an expulsion fuse.

Current-limiting fuses limit the energy at the location of the fault. This provides safety to workers and the public. Arc damage to life and property occurs in several ways:

- *Pressure wave*: The fault arc pressure wave damages equipment and personnel.
- *Heat*: The fault arc heat burns personnel and can start fires.
- *Pressure buildup in equipment*: An arc in oil causes pressure buildup that can rupture equipment.

All of these effects are related to the arc energy and all are greatly reduced with current-limiting fuses. Distribution transformers are a common application of current-limiting fuses to prevent them from failing violently due to internal failures.

30.3 Transformer Fusing

The primary purpose of a transformer fuse is to disconnect the transformer from the circuit if it fails. Some argue that the fuse should also protect for secondary faults. The fuse cannot effectively protect the transformer against overloads.

Engineers most commonly pick fuse sizes for distribution transformers from a fusing table developed by the utility, transformer manufacturer, or fuse manufacturer. These tables are developed based on criteria for applying a fuse such that the fuse should not have false operations from inrush and cold-load pickup.

One way to pick a fuse is to plot cold-load pickup and inrush points on a time–current coordination graph and pick a fuse with a minimum melt or damage curve that is above the cold-load and inrush points. Most fusing tables are developed this way. A fuse should withstand the cold-load and inrush points given in Table 30.6. The inrush points are almost universal, but the cold-load pickup points are more variable (and they should be since cold-load pickup characteristics change with predominant load types). An example application of the points given in Table 30.6 for a 50 kVA, 7.2 kV

TABLE 30.6 Inrush and Cold-Load Pickup Withstand Points for Transformer Fusing

	Full-Load Current Multiplier	Duration, s
Cold-load pickup	2	100
	3	10
	6	1
Inrush points	12	0.1
	25	0.01

Source: Amundson, R.H., High voltage fuse protection theory & considerations, in *IEEE Tutorial Course on Application and Coordination of Reclosers, Sectionalizers, and Fuses*, 1980, Publication 80 EHO157-8-PWR; Cook, C.J. and Niemira, J.K., Overcurrent protection of transformers—Traditional and new philosophies for small and large transformers, in *IEEE/PES Transmission & Distribution Conference and Exposition*, 1996, Presented at the training session on "Distribution overcurrent protection philosophies."

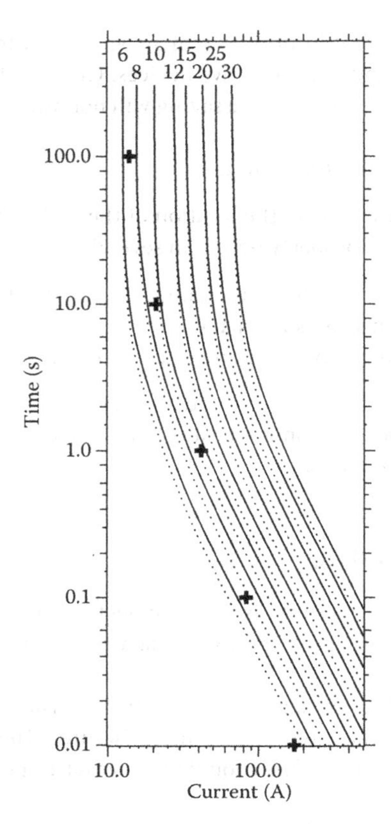

FIGURE 30.11 Transformer inrush and cold-load pickup points for a single-phase, 50 kVA, 7.2 kV transformer. The minimum-melt curves and damage curves (dotted lines) for K-link fuses are also shown.

single-phase transformer which has a full-load current of 6.94 A is shown in Figure 30.11. The cold-load pickup and inrush points are plotted along with K links. The minimum melt time and the damage time (75% of the minimum melt time) are shown. Use the damage curve to coordinate. For this example, a 12 A K link would be selected; the 1 s cold-load pickup point determines the fuse size. Since this point lies between the damage and minimum melt time of the 10 K link, some engineers would pick the 10 K link (not recommended).

Some utilities have major problems from nuisance fuse operations (especially utilities in high-lightning areas). A nuisance operation means that the fuse must be replaced, but the transformer was not permanently damaged. Nuisance fuse operations can be over 1% annually. Some utilities have thousands of nuisance fuse operations per year. A utility in Florida had a region with 57% of total service interruptions due to transformer interruptions, and 63% of the storm-related interruptions required only re-fusing (Plummer et al., 1995). During a storm, multiple transformer fuses can operate on the same circuit. There are differences of opinion as to what is causing the nuisance operations. Some of the possibilities are as follows:

- *Inrush*: Transformer inrush may cause fuse operations even though the inrush points are used in the fuse selection criteria. Reclosing sequences during storms can cause multiple inrush events that can heat up the fuse. In addition, voltage sags can cause inrush (any sudden change in the voltage magnitude or phase angle can cause the transformer to draw inrush).
- *Cold-load pickup*: This is the obvious culprit after an extended interruption (many of the nuisance fuse operations have occurred when there is not an extended interruption).

- *Secondary-side transformer faults*: Secondary-side faults that self-clear can cause some nuisance fuse events.
- *Lightning current*: Lightning current itself can melt small fuses. Arrester placement is key here since the lightning current flows to the low impedance provided by a conducting arrester. If the fuse is upstream of the arrester (which would be the case on a tank-mounted arrester), the lightning surge current flows through the fuse link. If the fuse is downstream, then little current should flow through the fuse.
- *Power-follow current through gapped arresters*: Following operation of a gapped arrester, a few hundred amps of power follow current flows in a gapped silicon carbide arrester until the gap clears (usually just for a half cycle if the gap is in good shape).
- *Transformer saturation from lightning currents*: Lightning can contain multiple strokes and long-duration components that last from 0.1 to 2 s. These currents can saturate distribution transformers. Following saturation, the transformer becomes a low impedance and draws high current from the system through the fuse (Hamel et al., 1990).
- *Animal faults*: Across transformer bushings or arresters.

Several of these causes may add to the total. Nuisance fuse operations have occurred when circuits were out of service. This means that lightning is the cause since any type of inrush would require the system to be energized. Detroit Edison found that 70%–80% of fuse operations were due to lightning (Gabrois et al., 1973). Lightning and inrush events are the most likely cause of nuisance fuse operations. Heavily loaded transformers are more susceptible to nuisance fuse operations because of the preheating of the fuse (a heavily loaded transformer is more susceptible to cold-load pickup as well).

Another method of choosing the transformer fuse size that gives "looser" fusing is the ×2 method (Burke, 1996). Choose a fuse size larger than twice the transformer full load current. A 50 kVA, 7.2 kV single-phase transformer, which has a full-load current of 6.94 A, should have a fuse bigger than 14 A (the next biggest standard size is a 15 A fuse). This applies for any type of fuse (K, T, or other). The factor of two provides a safety margin so that transformer fuses do not operate for inrush or cold-load pickup, and it helps with lightning.

The *fusing ratio* is the ratio of the fuse minimum melt current to the transformer full-load current (some sources also define a fusing ratio as the ratio of fuse-rated current to transformer-rated current which is different from this definition by a factor of two). Tight fusing means the fuse ratio is low. Relatively low fusing ratios have been historically used which has led to the nuisance fuse problems. The tighter fusing given using the Table 30.6 approach results in fusing ratios of 2–4. The looser ×2 method gives a fusing ratio of at least 4 (since the fuse rating is multiplied by two, and the minimum melting current at 300 s is twice the fuse rating). The fusing ratio for the 50 kVA, 7.2 kV transformer with the 12-K fuse is 3.46, and it is 4.32 with the 15-K fuse.

Another strategy that is especially useful in high-lightning areas: Use a standard fuse size for all transformers up to a certain size. This also helps ensure that the wrong fuse is not applied on a given transformer. A standard fuse size of at least 15 T or 20 K results in few nuisance fuse operations (IEEE Std. C62.22-1997). At 12.5 kV, a 20 K fuse should protect a 5 kVA transformer almost as well as it protects a 50 kVA transformer. It may lose some secondary protection relative to a smaller fuse, and a small portion of evolving faults will not be detected as soon; but other than that, there should not be much difference. If fuses get too big, they may start to bump up against tap fuse sizes and limit fuse options for lateral taps.

If looser fusing is used, some argue that overload protection of transformers is lost. Countering that argument, overload protection with fuses is not really possible if the transformer is used for its most economic performance (which means overloading a transformer at peak periods). To avoid nuisance fuse operations from load, we must use a fuse big enough so that thermal overload protection is impossible. It is also argued that most transformer failures start as failures between turns or layers and that a smaller, faster fuse detects this more quickly. Tests have indicated that a smaller fuse might not be much better than a larger fuse at detecting interwinding failures (Lunsford and Tobin, 1997) (pressure-relief valves limit tank pressures very well during this type of failure). All together, the arguments for a smaller fuse are not enough to overcome the concerns with nuisance fuse operations. If overload protection must be used, use a surge resistant fuse (it has a slower characteristic for high-magnitude, short-duration currents).

A few utilities practice group fusing where a lateral fuse provides protection to all of the transformers on the tap. If the transformer failure rate (including bushing faults) is low enough, then this practice will not degrade the overall frequency of interruptions significantly. One of the major disadvantages of this approach is that an internal transformer failure on a tap may be *very* hard to find. This drives up repair time (so the duration reliability numbers suffer but not necessarily the frequency indices). Also, the beneficial feature of being able to switch the transformer with the fused cutout is lost if group fusing is used.

Widely used, completely self-protected transformers (CSPs) have an internal weak-link fuse; an external fuse is not needed (although they may need an external current-limiting fuse to supplement the weak link).

Transformer bushing faults often caused by animals can have different impacts depending on fusing practices. A fault across a primary bushing operates an external transformer fuse. If the transformer is a CSP or group fusing is used, the upstream tap fuse operates (so more customers are affected).

Current-limiting fuses are regularly used on transformers in high fault current areas to provide protection against violent transformer failure. NEMA established tests which were later adopted by ANSI (ANSI C57.12.20-1988) for distribution transformers to be able to withstand internal arcs. Transformers with external fuses are subjected to a test where an internal arcing fault with an arc length of 1 in. (2.54 cm) is maintained for 1/2 to one cycle. It was thought that 1 in. (2.54 cm) was representative of the length that arcs could typically achieve. The current is 8000 A. Under this fault condition, the transformer must not rupture or expel excessive oil. Note that this test does not include all of the possible failure modes and is no guarantee that a transformer will not fail with lower current. For example, a failure with an arc longer than 1 in. has more energy and ruptures the transformer at a lower level of current.

Table 30.7 shows rupture limits for several types of transformers based on tests for the Canadian Electrical Association. If fault current values exceed those given in this table, consider using current-limiting fuses to reduce the chance of violent failures (the CEA report considers the limits provisional and suggests that more tests are needed). At arc energies within this range, the failure probability is on the order of 15%–35%. Note that the 2.5 kA limit for pole-mounted transformers is much less than the ANSI test limit of 8 kA. Based on a series of tests with internal 2 in. (5 cm) arcs, Hamel et al. (2003) recommend considering current-limiting fuses for pole-type transformers when the short-circuit current exceeds 1.7 kA.

For transformers with an internal fuse, completely self-protected (CSP) transformers or padmounted transformers, the arcing test is done to the rating of the fuse which is generally much lower than 8000 A. Table 30.8 shows the maximum fault current ratings based on the ANSI tests. If the available line-to-ground fault current exceeds these values, then consider current-limiting fuses to reduce the possibility of violent failures. Not all utilities use current-limiting fuses in these situations, and in such instances, internal faults have failed transformers violently, blowing the cover.

TABLE 30.7 Transformer Rupture Limits for Internal Faults

Transformer Type		$I \cdot t$, A-s, or C	Current Limit for a One-Cycle Clearing Time, kA
Pole mounted	1ϕ	41	2.5
Pad mounted	1ϕ	150	9
Pad mounted/subway	3ϕ	180	11
Network with switch compartment	3ϕ	90	5.4
Submersible and vault	1ϕ	41	2.5

Source: CEA 288 D 747, *Application Guide for Distribution Fusing*, Canadian Electrical Association, Ottawa, Ontario, Canada, 1998.

TABLE 30.8 Maximum 1/2- to One-Cycle Fault Current Rating on Distribution Transformers Based on the Test in ANSI C57.12.20-1988

Transformer	Maximum Tested Symmetrical Current, A	$\int Idt$ in the ANSI Test, A-s
Overhead transformer	8000	66.7
Under-oil expulsion fuse (based on typical fuse ratings)		
Up to 8.3 kV_{LG}	3500	29.2
Up to 14.4 kV_{LG}	2500	20.8
Up to 25 kV_{LG}	1000	8.3

Source: ANSI C57.12.20-1988, American national standard requirements for overhead-type distribution transformers, 500 kVA and smaller: High-voltage, 67,000 volts and below; low-voltage, 15,000 volts and below, American National Standards Institute, Washington, DC.

If a transformer is applied in a location where the available line-to-ground fault current is higher than shown in Table 30.8, use current-limiting fuses.

30.4 Lateral Tap Fusing and Fuse Coordination

Utilities use two main philosophies to apply tap fuses: fusing based on load and standardized fusing schedules. With fusing based on load, we pick a fuse based on some multiplier of peak load current. The fuse should not operate for cold-load pickup or inrush to prevent nuisance operations. As an example, one utility sizes fuses based on 1.5 times the current from the phase with the highest connected kVA. With standardized fuse sizes, a typical strategy is to apply 100 K links at all taps off of the mains (even if a tap only has one 15-kVA transformer). If using second-level fusing, use 65 K links for these and 40 K fuses for the third level. There is no clear winner; each has advantages and disadvantages:

- *Fusing based on load*: This tends to fuse more tightly. High-impedance faults are somewhat more likely to be detected. Nuisance fuse operations are more likely, especially with utilities that tightly fuse laterals. We are more likely to have load growth cause branch loadings to increase to the point of causing nuisance fuse operations. Fusing based on load helps on circuits that have covered wire because a smaller fuse helps protect against conductor burndown (taps that are more heavily loaded usually have a larger wire, which resists burndown).
- *Standardized fuse sizes*: It is simple: we spend less time coordinating fuses, we do not constantly check loadings, and utilities have less inventory. There is also less chance that the wrong fuse is installed at a location. A disadvantage of this approach is that larger fuses than needed are used at many locations, resulting in higher fault damage at the arc location, longer voltage sags, and more stress on in-line equipment.

Coordinating lateral tap fuses is generally straightforward. The fuse must coordinate with the station recloser or circuit breaker relays. Station ground relays are usually set to coordinate with the largest tap fuse. On the downstream side, a tap fuse should coordinate with the largest transformer fuse. This usually is not a problem.

In addition to sizing a fuse to avoid nuisance operations and coordinating with upstream and downstream protectors, we size fuses to ensure that the fuses provide protection to the line section that they are protecting. The reach of the fuse must exceed the length of the line section. Several methods are used to quantify the reach of a fuse:

- Where the fuse will clear a bolted single line-to-ground fault in 3 s
- Where the bolted single line-to-ground fault current is six times the fuse rating
- Where the fuse will clear a single line-to-ground fault with a 30 Ω resistance in 5 s

In most situations, typical fuse sizes provide sufficient reach by any of these methods. The first two methods are the best; the 30 Ω resistance is overly conservative and difficult to apply.

Reliability needs dictate the number of fuses used. The most common application for line fuses is at tap points. Occasionally, utilities fuse three-phase mains, but a recloser is more commonly used for this purpose. In the southwest United States, in areas with few trees and little lightning, fuses may be rarely used. This is the exception, not the rule. Most utilities fuse most taps off the main line. Some go further and provide several levels of fusing, especially utilities with heavy tree coverage. Returns diminish: Too many fuses leads to situations where fuses do not coordinate, and the extra fusing does not increase reliability significantly. Cutouts themselves contribute to causing faults by providing an easy location where animals, trees, and lightning cause faults, especially if they are poorly installed.

30.5 Station Relay and Recloser Settings

The main feeder circuit breaker relays (or recloser) must be set so that the circuit breaker coordinates with downstream devices, coordinates with upstream devices, and does not have trips from inrush or cold-load pickup. Station relays almost always use phase and ground time–overcurrent relays.

Table 30.9 shows typical settings used by several utilities. Many utilities try to use standardized relay settings at all distribution stations. This has the advantage that relays are less likely to be set wrong, and there is less engineering effort in a coordination study. Some other utilities set each relay based on a coordination study.

Differences exist about the meaning of “peak load.” Some utilities base it on the maximum design emergency load (which is typically something like 600 A). Others use the designed normal load (typically 400 A). Others may base it on some percentage of the total connected distribution transformer kVA.

Instantaneous relay settings vary more than phase relay settings. Several utilities also either disable or do not use an instantaneous relay setting. The instantaneous relay pickup ranges from one to almost 10 times the phase relay pickup.

One reasonable set of base pickup settings is

- *Phase TOC (time–overcurrent) relay*: Use two times the normal designed peak load on the circuit.
- *Ground TOC relay*: Use 0.75 times the normal designed peak load on the circuit.
- *Instantaneous phase and ground relays*: Use two times the TOC relay pickups.

Settings any less than this are prone to false trips from cold-load pickup and inrush.

In addition to avoiding nuisance trips, the relays (or recloser) must provide protection to its line section (to the end of the line or to the next protective device in series). Ensure that the relay has sufficient reach at the minimum operating current of the relay.

TABLE 30.9 Time–Overcurrent and Instantaneous Station Relay Pickup Settings in Amperes on the Primary at Several Utilities

	Phase		Ground		
Utility	TOC	Inst.	TOC	Inst.	Notes
A	720	4000	480	4000	Assumes peak current = 400 A
B	720	1200	360	1200	Full load = 300–400 A
C	600	None	300	530	
D	960	1300	480	600	
E	800	None	340	None	
F	960	2880	240	1920	
G	≥2.25 × current rating	Same as TOC	≥0.6 × current rating	Same as TOC	
	600	600	160	160	Typical settings

For a phase relay, sufficient reach is achieved by ensuring that 75% of the bolted line-to-line fault current at the end of the circuit is greater than the relay's pickup (its minimum operating current). So, if the line-to-line fault current at the end of the circuit is 1000 A, the pickup of the relay should be no more than 750 A. We use the line-to-line fault current because the two types of faults not seen by the ground relay are the three phase and the line to line. Of these, the line-to-line fault has the lower magnitude. The 75% factor provides a safety margin and allows some fault impedance. Another approach is to ensure that the line-to-ground fault current at the end of the circuit is less than the minimum operating current. The line-to-ground fault current is less than the line-to-line fault current, which provides the safety margin.

For a ground relay, ensure that the relay pickup is less than 75% of the line-to-ground fault current at the end of the line or to the next protective device. The ground relay must also coordinate with the largest lateral fuse.

Feeders dedicated to supplying secondary networks, either grid or spot networks, have similar settings as feeders supplying radial loads. Two main differences are related to loading and the ground relay setting. The pickup settings of station circuit breakers may have to account for higher peak loads. Some utilities have phase relay pickups that are similar to radial circuits, from 600 to 800 A, but some have settings above 900 A with one utility having a 1680 A setting (Smith, 1999). Also, if feeders are supplying only network load, and all network transformers are connected delta—grounded wye—the unbalanced current seen by the ground relay is small. Utilities can set a low ground relay setting; some have settings ranging from 40 to 80 A (Smith, 1999). The main limitation on lowering the setting further is that during line-to-ground faults, the unfaulted circuits will backfeed the fault through the zero-sequence capacitance of that circuit. Lower ground-relay settings also help detect turn-to-turn or layer-to-layer faults within the primary windings of network transformers.

30.6 Arc Flash

Arc flash is another situation often requiring overcurrent coordination, much like protecting against conductor burndown or transformer damage. For arc flash, the goal is to greatly reduce the chance of burns to workers if an arc occurs near a worker. We want a protective device to clear the fault before a fault arc could cause incident energy in excess of the rating of the clothing.

The severity of an arc flash event is normally quantified as the incident energy that would reach a worker, normally given in terms of cal/cm^2. Flame-resistant (FR) clothing systems have an arc thermal performance value (ATPV) rating based on ASTM test standards (ASTM F1959, 2006). This rating is the incident energy in cal/cm^2 on the clothing surface that has a 50% probability of causing a second-degree skin burn. The goal of an arc flash analysis is to ensure that workers have an ATPV protection sufficient to handle the incident energy that might be expected in a given work scenario. Out of 14 responses to an Electric Power Research Institute (EPRI) survey, utility minimum ATPV ratings ranged from 4 to 8.7 cal/cm^2 with a median value of 5.4 cal/cm^2.

A number of approaches are available for estimating arc flash. The two most commonly cited methods are the ARCPRO program (Kinectrics, 2002) and the IEEE 1584 method (IEEE Std 1584-2002). The IEEE 1584 method is based on curve-fit regressions to mainly three-phase arc-in-a-box tests and is most suitable for arcs in equipment and other arc-in-a-box scenarios. ARCPRO is based on a single arc in open air, so it is most suitable for overhead, open-air scenarios.

IEEE 1584-2002 was developed by the IEEE Industry Applications Society, a society focusing on industrial and commercial power. IEEE 1584 is the most widely adopted approach to arc flash analysis. The method for estimating arc flash incident energies is based on tests performed at several short-circuit labs. From this test set, regression was used to find equations to best fit the test data. IEEE 1584 assumes a three-phase fault and is mainly geared toward arc-in-a-box evaluations. Above 15 kV, the IEEE 1584 guide and companion spreadsheet default to the Lee method. The Lee method is the oldest arc flash calculation method. It is a simple model of a single, open-air arc. For medium-voltage applications, the Lee method

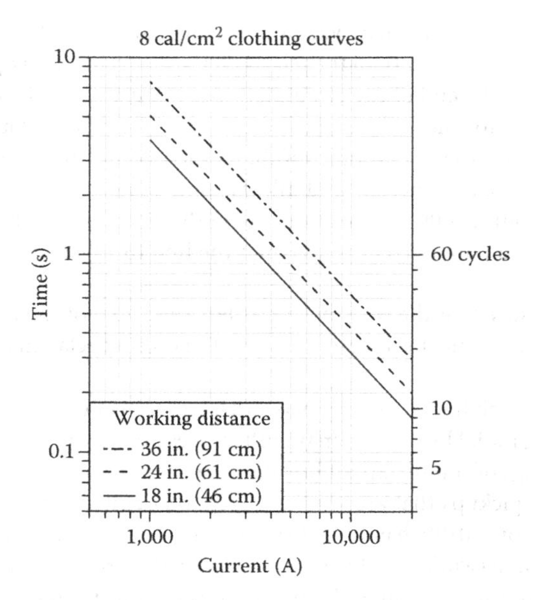

FIGURE 30.12 Arc flash time–current curves for medium-voltage switchgear based on IEEE 1584 and 8 cal/cm² clothing.

produces unrealistically high predictions of incident energies. For 25 and 35 kV scenarios, consider using the 15 kV results from the IEEE spreadsheet as suggested in Short (2011). Figure 30.12 shows time-current curves based on IEEE 1584 that are suitable for medium-voltage switchgear. A protective device should clear before the time indicated on the appropriate curve. For different clothing, the curves can be shifted. For 4 cal/cm² clothing, the protective device must operate twice as fast as with 8 cal/cm² clothing.

ARCPRO is a commercial program for analyzing arc flash incident energies, developed by Kinectrics. The ARCPRO algorithm is based on the work of Bingwu and Chengkang (1991), but it is not completely described in any peer-reviewed paper. The ARCPRO model assumes the following (Cress, 2008): a vertical free burning arc in air; an arc length much greater than arc diameter; a one-arc column, either phase–phase or phase–ground; no electrode region heat transfer; and an optically thin plasma and gas. Cress (2008) reported that ARCPRO was verified with over 300 test points for arc energy and incident energy for currents from 3 to 25 kA, arc durations from 4 to 35 cycles, distances from 8 to 24 in. (20–60 cm), and with gaps from 1 to 12 in. (2.5–30 cm).

The key input parameters for an arc flash study are as follows:

- *Working distance*: Distance from the worker to the fault arc is an important input. Incident energies drop significantly with distance, normally as a power of $1/d^{1.5}$ to $1/d^2$.
- *Arc length*: Some arc flash models like ARCPRO include an arc length. The arc energy increases almost linearly with arc length. Arc lengths and arc voltages are primarily a function of gap spacings, not the system driving voltage. The arc length is different than the shortest gap between energized conductors or between an energized conductor and ground. Because a fault arc can move and elongate, the arc length is normally longer than the gap length. IEEE 1584 includes an arc gap internally in calculations.
- *Fault current*: For medium-voltage arc flash, engineers commonly assume a bolted fault. For low-voltage arc flash (under 1000 V), the arc impedance will reduce the fault current appreciable, so the arcing current is needed. IEEE 1584 suggests the following equation to estimate arcing current:

$$\log_{10} I_a = K + 0.662 \log_{10} I_{bf} + 0.0966V + 0.000526G + 0.5588V \log_{10} I_{bf} - 0.00304G \log_{10} I_{bf}$$

where

I_a is the arc current, kA
I_{bf} is the bolted fault current, kA
$K = -0.153$ for open configurations, -0.097 for box configurations (enclosed equipment)
V is the system voltage, kV
G is the distance between buses, mm (IEEE assumes 32 mm for low-voltage switchgear.)

- *Duration*: The duration is based on the clearing time of the upstream protective device(s) based on its time–current characteristics. For very long or indeterminate clearing times, a worker self-extraction time is sometimes assumed as a maximum duration to consider. IEEE 1584 mentions 2 s for this duration. Even 2 or 3 ft (0.7–1 m) of movement away from the fault or to the side will dramatically reduce incident energies.
- *Fault type*: Whether the fault is a single phase–ground fault, a phase–phase fault, or a three-phase fault will affect the fault current and the duration. For faults in equipment, three-phase faults are commonly assumed as it is likely for the fault to expand to involve all three phases.

The 2007 National Electrical Safety Code (NESC) (IEEE C2-2007) requires an arc flash assessment be performed on systems above 1000 V. They do not provide specifics in general but do offer a table with default assumptions based on an ARCPRO analysis for open-air, single-phase-to-ground faults. The NESC table 410-1 footnotes specify a 15 in. (38 cm) separation distance from the arc to the employee for glove work and arc lengths as follows: 1–15 kV = 2 in. (5 cm), 15.1–25 kV = 4 in. (10 cm), and 25.1–36 kV = 6 in. (15 cm). Figure 30.13 shows time–current curves for 8 cal/cm² clothing based on ARCPRO with the 15 in. (38 cm) working distance along with various arc lengths. Because Short and Eblen (2011) found that the default arc lengths in the NESC are low, longer arc lengths are also given.

As an overcurrent protection problem, two related assumptions are made for arc flash:

1. The incident energy increases linearly with time. If you double the duration, the incident energy doubles. All of the modeling approaches make this same assumption.
2. At lower currents and longer durations, the main impact is due to lower current. Most models have almost a linear relationship between current and incident energy: if the current doubles, the incident energy doubles.

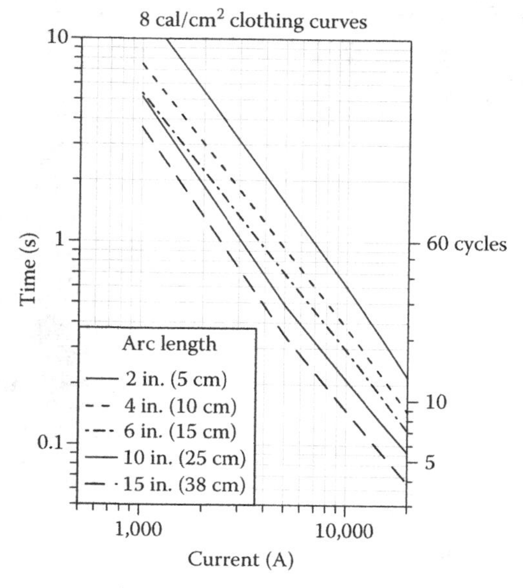

FIGURE 30.13 Arc flash time–current curves for overhead glove work based on ARCPRO with a 15 in. (38 cm) working distance and 8 cal/cm² clothing.

At longer durations, these assumptions are uncertain. As in Figure 30.13, incident energies often become more of a concern where fault currents are lower and durations are longer. Arc flash models have been mostly tested with durations less than 0.5 s. At longer durations, several factors can come into play: the arc can move and/or elongate (increasing energy), the arc may involve additional phases (increasing energy), and the arc may self-extinguish (decreasing energy).

Another consideration is arc impedance. At low voltages, this is important to consider. Above 1000 V, a bolted fault is most appropriate. The 30–40 ohm impedance given by the Rural Electrification Association (REA Bull. 61-2, 1978) is too large; see also Dagenhart (2000) and Burke (2006). The arc in an arc flash scenario involves relatively low arc resistances. A 3 ft (1 m) arc has a voltage of about 1400 V. If the fault current at that point in the line was 1000 A, then the arc resistance is about 1.4 Ω. A 1 ft (0.3 m) arc with the same fault current has a resistance of 0.47 Ω. Most fault arcs have resistances of 0–2 Ω, so that can be used as guidance to find the minimum fault current.

For low-voltage arcs, arc sustainability is an important variable. Below 250 V, it is rare to find conditions where high-current arcs can sustain in an arc flash scenario. At 480 V, arc sustainability is highly dependent on equipment. In the 2012 NESC, a table is provided for clothing to protect workers for different types of equipment. This is partially based on tests documented by Eblen and Short (2010; EPRI 1018693, 2009; EPRI 1020210, 2009). For cases where a coordination study is needed at 480 V (like network protectors), IEEE 1584 is the most appropriate tool.

Arc flash analysis is still in its infancy, and further advances in modeling and protection are expected. Arc flash is equipment specific. In some cases, existing methods do not adequately predict performance. Short and Eblen (2011) show test results for a medium-voltage pad-mounted switch with incident energies three times that predicted by IEEE 1584. They provide an equation to estimate the incident energies. The higher incident energies were caused by the bus configuration which projected arcs out the front of the enclosure (Figure 30.14).

A number of relaying options are available to reduce incident energies, depending on the application. Common approaches include reducing clearing times by enabling a fast trip and disabling reclosing. Other options are available to coordinate relaying times with clothing capabilities and with work practices.

FIGURE 30.14 Example arc flash in a medium-voltage pad-mounted switch.

30.7 Coordinating Devices

Several details often arise when coordinating specific devices. Normally, we want to ensure that the downstream device clears before the upstream device operates over the range of fault currents available at the downstream device. Time–current characteristics of both device normally show us how well two devices coordinate. Because of device differences, some combinations require slightly different approaches. We discuss some of the common combinations in the sections that follow.

30.7.1 Expulsion Fuse–Expulsion Fuse Coordination

When coordinating two fuses, the downstream fuse (referred to as the *protecting* fuse) should operate before the upstream fuse (the *protected* fuse). To achieve this goal, ensure that the total clear time of the protecting fuse is less than the damage time of the protected fuse. The damage time is 75% of the minimum melt time. An example for coordinating a 100 K link with a 65 K link is shown in Figure 30.15. Above a certain current, the two fuses do not coordinate; the protected fuse could suffer damage or melt before the protecting fuse can clear. For high fault currents, coordination is impossible because both fuses can open. The example shows that above 2310 A, the total clear curve of the 65 K is above the damage curve of the 100 K link. Utilities live with this common miscoordination. Table 30.10 lists the maximum coordination currents between K links. In cases where fuses do not coordinate, why have the second fuse? The second fuse still has some value; it adds another sectionalizing point (for a fuse in a cutout), and for a downstream fault, it identifies the fault location to a smaller area. Also, the downstream fuse may operate without damaging the upstream fuse. The amount of damage to the upstream fuse depends on the point of the waveform where the fault occurs (the extra 1/2 + cycle waiting for a current zero causes the extra heating to the protected fuse).

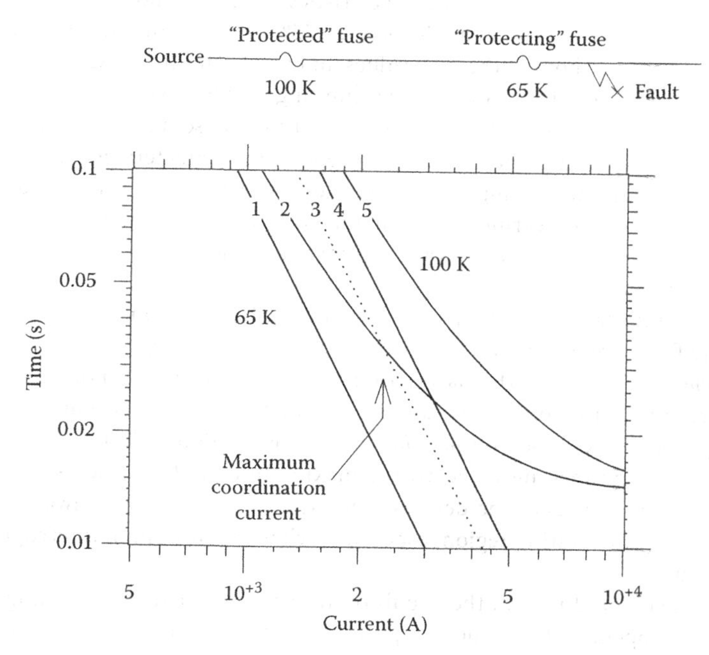

FIGURE 30.15 Example of fuse coordination between a 100-K (the protected fuse) and a 65-K link (the protecting link). (1) 65 K, minimum melt; (2) 65 K, total clearing; (3) 100 K, damage time; (4) 100 K, minimum melt; (5) 100 K, total clearing.

TABLE 30.10 Maximum Fault Currents for Coordination between the Given K Fuse Links

	10 K	12 K	15 K	20 K	25 K	30 K	40 K	50 K	65 K	80 K	100 K	140 K	200 K
6 K	170	310	460	640	840	1060	1410	1800	2230	2930	3670	5890	9190
8 K	20	230	410	610	810	1040	1400	1790	2230	2930	3670	5890	9190
10 K		40	300	550	780	1000	1370	1770	2220	2930	3670	5890	9190
12 K			80	420	690	950	1330	1730	2190	2910	3650	5880	9190
15 K				90	530	840	1250	1670	2120	2870	3640	5870	9190
20 K					100	610	1120	1570	2040	2800	3590	5870	9190
25 K						120	840	1380	1920	2710	3510	5830	9150
30 K							240	1090	1690	2570	3380	5740	9110
40 K								300	1240	2260	3210	5630	9010
50 K									240	1750	2800	5500	8910
65 K										970	2310	5210	8740
80 K											420	4460	8430
100 K												3550	7950
140 K													4210

30.7.2 Current-Limiting Fuse Coordination

Coordinating two current-limiting fuses is similar to coordinating two expulsion links. Plot the time–current characteristics and ensure that the maximum clearing time of the load-side fuse is less than 75% of the minimum-melting time of the source-side fuse over the range of fault currents available at the load-side fuse. The 75% factor accounts for damage to the source-side fuse. Unlike expulsion links, current-limiting fuses can coordinate to very high currents. For coordination at higher currents than are shown on published time–current characteristics (operations faster than 0.01 s), ensure that the maximum clearing I^2t of the load-side fuse is less than 75% of the minimum-melt I^2t of the source-side fuse. Manufacturers provide both of these I^2t values for current-limiting fuses.

Coordinating an expulsion link with a current-limiting fuse follows similar principles. Because the melting and clearing characteristics of current-limiting fuses are so much steeper than those of expulsion links, coordination is sometimes difficult; the operating characteristic curves are more likely to cross over. A load-side current-limiting fuse coordinates over a wide range of fault current. For a source-side current-limiting fuse, the clearing-time limitations of expulsion links (to about 0.8 cycles) prevent coordination at high currents. For currents above this value, either both will operate, or just the current-limiting fuse will operate.

Backup current-limiting fuse coordination requires special attention. To ensure that the CLF does not try to operate for currents below its minimum interrupting rating, the intersection of the expulsion fuse's total clearing curve and the backup fuse's minimum-melting curve must be greater than the maximum interrupting rating of the backup fuse. Normally, we select backup current-limiting fuses for use with expulsion links based on *matched-melt* coordination. Select a backup current-limiting fuse that has a maximum melting I^2t below the maximum clear I^2t of the expulsion element. Also, check the time–current curves of the devices. The expulsion link should always clear for fault currents in the low-current operating region, especially below the minimum interrupting current of the current-limiting fuse.

With matched-melt coordination, the expulsion fuse always operates, including when the backup current-limiting fuse operates. In overhead applications with an expulsion fuse in a cutout, the dropout of the expulsion fuse provides a visible indication when the fuse(s) operate. Also, the backup fuse is unlikely to have full voltage across it.

The maximum melting I^2t of expulsion links is not provided from curves or data. To estimate this, take the minimum melting I^2t calculated from the minimum-melt curve at 0.0125 s, and multiply by 1.2 for tin links or 1.1 for silver links. The multiplier allows for conservatism in minimum-melt curves and for manufacturing tolerances.

Somewhat less conservatively, experience has shown that fuses coordinate well if the maximum melt I^2t of the expulsion link does not exceed twice the minimum melt I^2t of the backup fuse (IEEE Std. C37.48-1997). We can tighten up the backup fuse because, under most practical situations, the backup fuse lets through significantly more I^2t than its minimum-melt value.

Manufacturers of backup current-limiting fuses normally provide coordination recommendations for their fuses, but review of the coordination approach is sometimes appropriate. Backup fuses often use a "K" nomenclature signifying the K link that it coordinates with. For example, a "25 K" backup link coordinates with a K link rated at 25 A or less. Figure 30.16 shows the time–current curves of a 40 K expulsion link and the curves of one manufacturer's 40 K backup current-limiting fuse. This graph extends below the normal cutoff time of 0.01 s to show how the fuses coordinate at high currents. This example shows that the backup fuse does not coordinate using the strict matched-melt criteria (the maximum melting time of the expulsion link is more than the minimum melting time of the backup fuse). The minimum-melt I^2t of the backup fuse is 1.6 times the minimum melt I^2t of the backup fuse, so it meets the relaxed matched-melt criteria since this ratio is less than two.

The *time–current curve crossover coordination* allows a smaller backup current-limiting fuse. As before, the intersection of the expulsion fuse's total clearing curve and the backup fuse's minimum-melting curve must be greater than the maximum interrupting rating of the backup fuse. We do not try to ensure that the backup fuse always melts. We can use a smaller fuse, which reduces the I^2t let through and reduces energy to faults. The backup CLF operates for a wider range of short-circuit currents.

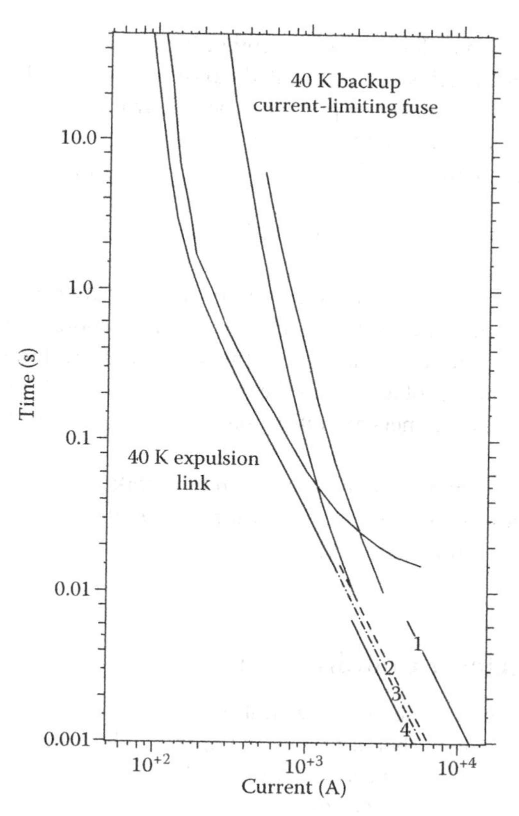

FIGURE 30.16 Coordination between a 40-K expulsion link and a 40-K backup current-limiting fuse using the relaxed matched-melt criteria. (1) Backup total clearing I^2t, (2) expulsion minimum melting time, (3) expulsion maximum melting time, (4) backup minimum-melt I^2t.

With a smaller fuse, the backup fuse can operate before the expulsion link melts for high fault currents. Utilities often use time–current curve crossover coordination for under-oil backup current-limiting fuses. In addition to lowering energy to faults, crossover coordination extends the range of current-limiting fuse protection to larger transformers.

For transformer protection, overload and secondary faults are also considerations for backup current-limiting fuses. Secondary faults at the terminals of the transformer should not damage or melt the backup CLF. One way to do this is to ensure that at the total clearing time of the expulsion link with the bolted secondary fault, the backup fuse's minimum melting current is at least 125% of the secondary fault current (Hi-Tech Fuses, 2002). Also, overload should not damage or melt the backup CLF.

30.7.3 Recloser–Expulsion Fuse Coordination

Normally, we want the recloser's fast curve (A curve) to clear before downstream fuses operate. This saves the fuse for temporary faults (we discuss fuse saving in more detail later in this chapter). Select the delayed curve (B, C, ...) to be above the clearing time of downstream fuses. A permanent fault downstream of a fuse should blow the fuse, not lockout the recloser.

To open the recloser before the fuse blows, Cooper (1990) recommends adjusting the A curve by multiplying the time by a factor of 1.25 for one fast operation, a factor of 1.35 for two fast operations with a reclosing time greater than or equal to 1 s, and a factor of 1.8 for two fast operations with a reclosing time from 25 to 30 cycles. For applications with two or more delayed operations, the fast curve coordinates for fault currents up to the point where the adjusted A curve crosses the expulsion fuse's minimum-melting curve.

On hydraulically controlled reclosers, the trip-coil rating determines the recloser's "pickup." Beyond that, hydraulically controlled reclosers have limited curve selections and no adjustments. Figure 30.17 shows average clearing curves for a single-phase Cooper 4E hydraulically controlled recloser overlayed on top of two K fuse links. For this example, only a limited range of fuses coordinate for low and high fault currents. Fuses larger than a 65 K have significant overlap in the low-current area, leaving more chance that the recloser could lock out for a fault on a lateral tap. The slower delayed curves, such as the C curve shown, reduce the chance of miscoordination for lower fault currents. For the fast-trip A curve, the 40 K link only coordinates with the fast curve for fault currents up to 360 A; smaller links are worse. Since K links are significantly steeper than these recloser curves, we must expect limited coordination for certain combinations. In this instance, T links coordinate over a wider current range because their time–current characteristics match the slope of the recloser characteristics more accurately. Miscoordination is more problematic in the low-current region. If the recloser locks out for faults downstream of a fuse, more customers are interrupted, and crews have a harder time finding the fault (more area to patrol).

Reclosers with electronic controls and relayed circuit breakers offer more flexibility. We can tailor tripping characteristics to coordinate over a wider range of currents. Three-phase reclosers have a ground-trip element that can increase the sensitivity of the recloser and also coordinate better with downstream fuses.

30.7.4 Recloser–Recloser Coordination

For coordinating two reclosers, the curve separation we need depends on the type of recloser. For hydraulically controlled reclosers that are series coil operated, both operate if there is less than a two-cycle separation; both *may* operate for a separation of 2–12 cycles, and both coordinate properly if there are more than 12 cycles of separation. For hydraulically controlled reclosers that use high-voltage solenoid closing (larger reclosers), we need 8 cycles of separation for coordination (if it is less than 2 cycles, both devices operate). This data is for Cooper reclosers (Cooper Power Systems, 1990).

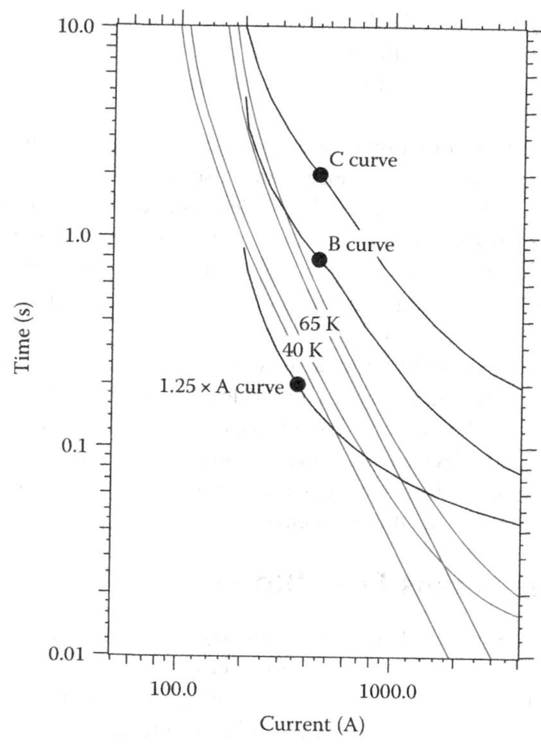

FIGURE 30.17 Example of coordination between K links and a Cooper 4E single-phase hydraulic recloser with a 100 A trip coil (A, B, and C curves shown).

30.7.5 Coordinating Instantaneous Elements

Coordinating instantaneous relay elements or recloser fast curves is difficult. By the nature of an instantaneous element, two in series will both operate if the short-circuit current is above the pickup of both relays.

The most common way to coordinate two instantaneous elements is to raise the pickup of the upstream element. Find a setting where the instantaneous relay will not operate for faults downstream of the second protective device. The upstream relay cannot operate if its pickup is above the available fault current at the location of the downstream element. For this strategy, the instantaneous pickup on the element must be higher than its time–overcurrent pickup. This rules out hydraulic reclosers, which have the same pickup for the fast (A) curve and the delayed curves (B and C), but is not a concern with electronic reclosers because they have the same flexibility as relayed circuit breakers.

Rather than using an instantaneous relay element, we can perform the "fast trip" function with a time–overcurrent relay with a fast characteristic. Now, we might be able to coordinate the fast curve of a line recloser with the substation circuit breaker or recloser.

As another way to coordinate two instantaneous elements, use a time delay on the upstream instantaneous element. Choose enough time delay, 6–10 cycles, to allow the downstream device to clear before the station device operates.

Even with coordinated fast curves (either using a delay or using a fast TOC curve), nuisance momentary interruptions occur for faults cleared by a downstream line recloser. Consider a station recloser R1 and a downstream line recloser R2 each with one fast curve (A) and two delayed curves (B). If a permanent fault occurs downstream of R2, R2 will first operate on its A curve. If the fast curves of R1 and R2 are coordinated, R1 will not operate. After a delay, R2 recloses. The fault is still there, so R2 operates on its delayed curve (its B curve). Now, R1 *does* operate because it is on its A curve which operates before R2's B curve. After R1 recloses, R2 should then clear the fault on its B curve, which should operate before R1's B curve. The fault is still cleared properly, but customers upstream of R2 have extra momentary interruptions.

A more advanced form of coordination called *sequence coordination* removes this problem. Sequence coordination is available on electronic reclosers and also on digital relays controlling circuit breakers. With sequence coordination, the station device detects and counts faults—but does not open—for a fault cleared by a downstream protector on the fast trip. If the fault current occurs again (usually because the fault is permanent), the station device switches to the time–overcurrent element because it counted the first as an operation. Using this form of coordination eliminates the momentary interruption for the entire feeder for permanent faults downstream of a feeder recloser. On a relay or recloser that has sequence coordination, if the device senses current above some minimum trip setting and the current does not last long enough to trip based on the device's fast curve, the device advances its control-sequence counter as if the unit had operated on its fast curve. So when the downstream device moves to its delayed curve, the upstream device with sequence coordination also is operating on its delayed curve. With sequence coordination, for the fast curves, the response curve of the upstream device must still be slower than the clearing curve of the downstream device.

30.8 Fuse Saving versus Fuse Blowing

Fuse saving is a protection scheme where a circuit breaker or recloser is used to operate before a lateral tap fuse. A fuse does not have reclosing capability; a circuit breaker (or recloser) does. Fuse saving is usually implemented with an instantaneous relay on a breaker (or the fast curve on a recloser). The instantaneous trip is disabled after the first fault, so after the breaker recloses, if the fault is still there, the system is time coordinated, so the fuse blows. Because most faults are temporary, fuse saving prevents a number of lateral fuse operations.

The main disadvantage of fuse saving is that all customers on the circuit see a momentary interruption for lateral faults. Because of this, many utilities are switching to a fuse blowing scheme. The instantaneous relay trip is disabled, and the fuse is always allowed to blow. The fuse blowing scheme is also called trip saving or breaker saving. Figure 30.18 shows a comparison of the sequence of events of each mode of operation. Fuse saving is primarily directed at reducing sustained interruptions, and fuse blowing is primarily aimed at reducing the number of momentary interruptions.

30.8.1 Industry Usage

Until the late 1980s, fuse saving was almost universally used. As power quality concerns grew, some utilities switched to a fuse blowing mode. An IEEE survey on distribution protection practices that is done periodically has shown a decrease in the use of fuse saving as shown in Table 30.11.

Another survey done by Power Technologies, Inc., in 1996 showed a mixture of practices at utilities as shown in Table 30.12. A few used fuse blowing because they indicated that fuse saving was not successful. Many of the "mixed practices" utilities decided on a case-by-case basis. Many of these normally used fuse saving but switched to fuse blowing if too many power quality complaints were received.

30.8.2 Effects on Momentary and Sustained Interruptions

The change in the number of momentary interruptions can be estimated simply by using the ratio of the length of the mains to the total length of the circuit including all laterals. For example, if a circuit has 5 mi of mains and 10 mi of laterals, the number of momentaries after switching to fuse blowing would be 1/3 of the number of momentaries with fuse saving ($5/(5 + 10) = 1/3$). This assumes that the mains and laterals have the same fault rate; if the fault rate on laterals is higher (which it often is because of less tree trimming, etc.), the number of momentaries is even less. Note how dramatically we can reduce

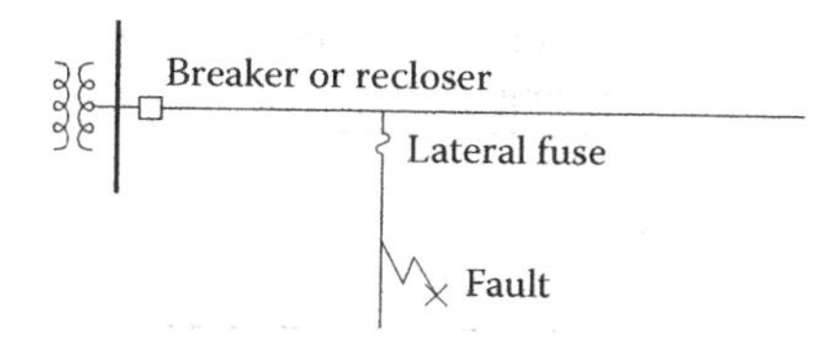

Fuse saving

Temporary fault

1. The circuit breaker operates on the instantaneous relay trip (before the fuse operates).
2. The breaker recloses.
3. The fault is gone, so no other action is necessary.

Permanent fault

1. The circuit breaker operates on the instantaneous relay trip (before the fuse operates).
2. The breaker recloses.
3. The fault is still there.
4. The instantaneous relay is disabled, so the fuse operates.
5. Crews must be sent out to fix the fault and replace the fuse.

Fuse blowing

Temporary fault

1. The fuse operates.
2. Crews must be sent out to replace the fuse.

Permanent fault

1. The fuse operates.
2. Crews must be sent out to fix the fault and replace the fuse.

FIGURE 30.18 Comparison of the sequence of events for fuse saving and fuse blowing for a fault on a lateral.

TABLE 30.11 IEEE Survey Results on the Percentage of Utilities That Use Fuse Saving

Survey Year	Percent of Utilities Using Fuse Saving
1988	91
1994	71
2000	66

Sources: IEEE Working Group on Distribution Protection, *IEEE Trans. Power Deliv.*, 3(2), 514, April 1988; IEEE Working Group on Distribution Protection, *IEEE Trans. Power Deliv.*, 10(1), 176, January 1995; Report to the IEEE Working Group on System Performance, 2002.

momentaries by using fuse blowing. No other methods can so easily eliminate 30%–70% of momentaries. The effect on reliability of going to a fuse blowing scheme is more difficult to estimate. Fuse blowing increases the number of fuse operations by 40%–500% (Dugan et al., 1996; Short, 1999; Short and Ammon, 1997). This will increase the average frequency of sustained interruptions by 10%–60%. Note that there are many variables that can change the ratios. One example is given in Figure 30.19.

TABLE 30.12 1996 Survey on the Usage of Fuse Saving

Use fuse saving	40%
Use a mixture	33%
Use fuse blowing	27%

Source: Short, T.A., Fuse saving and its effect on power quality, *EEI Distribution Committee Meeting*, 1999.

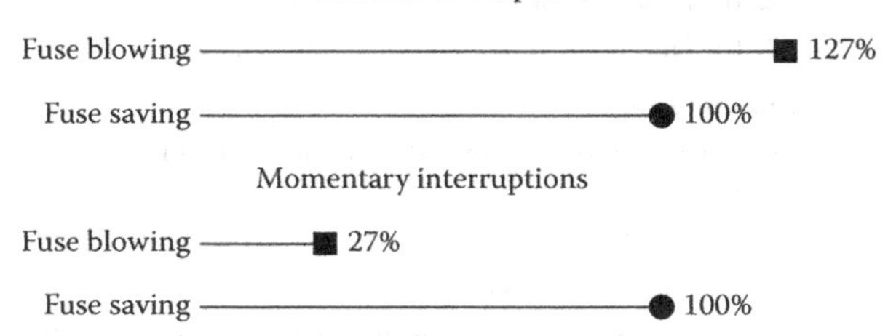

FIGURE 30.19 Comparison of fuse saving and fuse blowing on a hypothetical circuit. Mains: 10 miles (16.1 km), fault rate = 0.5/mile/year (0.8/km/year), 75% temporary. Taps: 10 miles (16.1 km) total, 20 laterals, fault rate = 2/mile/year (3.2/km/year), 75% temporary. It also assumes that fuse saving is 100% successful.

Note that the effect on sustained interruptions is not equally distributed. Customers on the mains see no difference in the number of permanent interruptions. Customers on long laterals may have many more sustained interruptions with a fuse-blowing scheme.

30.8.3 Coordination Limits of Fuse Saving

One of the main reasons that utilities have decided not to use fuse saving is that it is difficult to make it work. Fuses clear quickly relative to circuit breakers, so where fault currents are high, the fuse blows before the breaker trips. This results in a fuse operation and a momentary interruption for all customers on the circuit. K links, the most common lateral fuses, are fast fuses. Most distribution circuit breakers take five-cycles to clear. For fuse saving to work, the breaker must open before the fuse blows, so the fuse needs to survive for the time it takes the instantaneous relay to operate (about one cycle) plus the five-cycles for the breaker. As an illustration, Figure 30.20 shows the limit of coordination of a five-cycle breaker and a 100 K fuse. Fuse saving only coordinates for faults below 1354 A. Smaller fuses have lower current limits. Note that the breaker time is coordinated with the damage time of the fuse.

Tables 30.13 and 30.14 show the limits of coordination of several common lateral fuses for a standard circuit breaker (five-cycle) and for a fast breaker/relay combination (three-cycle circuit breaker and one-cycle relay). Also shown are translations of these fault currents into distances from the substation at 12.47 kV (assuming an 8 kA fault level at the substation). Note that only the larger fuses shown (greater than 100 A) will coordinate for significant portions of the feeder. Smaller fuses used as second- and third-level fuses do not coordinate over the length of most feeders. The situation is even worse at higher voltages. At 24.94 kV, the distances in Tables 30.13 and 30.14 are doubled, so fuse saving is more difficult to achieve at higher voltages. Reclosers are faster than standard five-cycle breakers—the four-cycle total operating time in Table 30.14 is representative of many reclosers.

If smaller K links are used such as 100 K and 65 K fuses (the most common lateral fuses), then fuse saving is not going to work very well. In that case, why use it? There is no sense in having a momentary every time a fuse blows (which is what will happen since the circuit breaker is not fast enough to save the fuse).

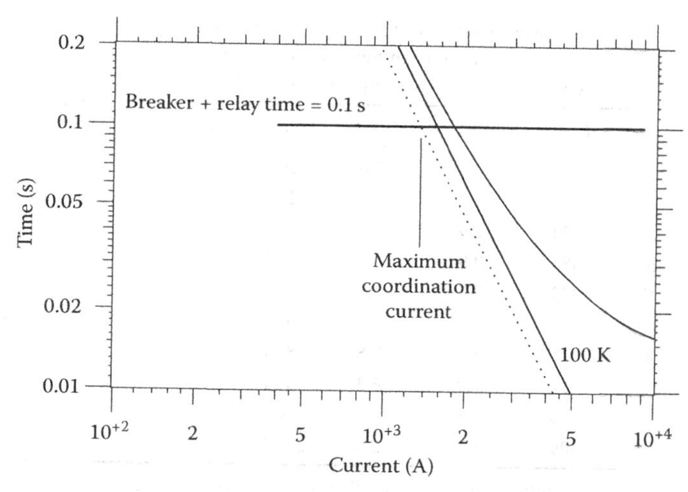

FIGURE 30.20 Coordination of a 100 K lateral fuse with a five-cycle circuit breaker.

TABLE 30.13 Maximum Fault Currents and Critical Distances for Fuse Saving Coordination for Several Common Fuse Links for a Five-Cycle Circuit Breaker and a One-Cycle Relay Time

Fuse	I_c, A	d_c, mi	d_c, km	Fuse	I_c, A	d_c, mi	d_c, km
20 K	254	26.5	42.6	20 T	433	15.5	25.0
25 K	323	20.8	33.5	25 T	552	12.2	19.6
30 K	398	16.9	27.2	30 T	699	9.6	15.5
40 K	520	12.9	20.8	40 T	896	7.5	12.1
50 K	665	10.1	16.3	50 T	1125	6.0	9.7
65 K	816	8.3	13.3	65 T	1428	4.8	7.7
80 K	1078	6.3	10.1	80 T	1790	3.8	6.2
100 K	1354	5.0	8.1	100 T	2277	3.1	4.9
140 K	2162	3.2	5.2	140 T	3447	2.1	3.4
200 K	3401	2.1	3.5	200 T	5436	1.5	2.4

Source: Reprinted from Electric Power Research Institute, 1001665, *Power Quality Improvement Methodology for Wire Companies*, Palo Alto, CA. With permission. Copyright 2003.

Note: I_c, maximum current where fuse saving works; d_c, distance from the substation where fuse saving starts to work for 12.47 kV, 500 kcmil overhead line.

30.8.4 Long-Duration Faults and Damage with Fuse Blowing

Fuse blowing has drawbacks: faults on the mains can last a long time. With fuse saving, main-line faults normally clear in 5 to 7 cycles (0.1 s) on the first shot with the instantaneous element. With fuse blowing, this same fault may last for 0.5–1 s. Much more damage at the fault location occurs during this extra time. Some of the problems that have been identified are as follows:

- *Conductor burndowns*: At the fault, the heat from the fault current arc burns the conductor enough to break it, dropping it to the ground.
- *Damage of inline equipment*: The most common problem has been with inline hot-line clamps. If the connection is not good, the high-current fault arc across the contact can burn the connection apart.
- *Station transformers*: Extra duty on substation transformers.

TABLE 30.14 Maximum Fault Currents and Critical Distances for Fuse Saving Coordination for Several Common Fuse Links for a Three-Cycle Circuit Breaker and a One-Cycle Relay Time

Fuse	I_c, A	d_c, mi	d_c, km	Fuse	I_c, A	d_c, mi	d_c, km
20 K	332	20.3	32.6	20 T	565	11.9	19.2
25 K	424	15.9	25.5	25 T	723	9.3	15.0
30 K	522	12.9	20.8	30 T	920	7.4	11.8
40 K	682	9.9	15.9	40 T	1175	5.8	9.3
50 K	875	7.7	12.4	50 T	1479	4.6	7.4
65 K	1070	6.3	10.2	65 T	1878	3.7	5.9
80 K	1407	4.8	7.8	80 T	2346	3.0	4.8
100 K	1763	3.9	6.3	100 T	2975	2.4	3.9
140 K	2823	2.5	4.1	140 T	4522	1.7	2.7
200 K	4409	1.7	2.8	200 T	7122	1.3	2.0

Source: Reprinted from Electric Power Research Institute, 1001665, *Power Quality Improvement Methodology for Wire Companies*, Palo Alto, CA. With permission. Copyright 2003.

Note: I_c, maximum current where fuse saving works; d_c, distance from the substation where fuse saving starts to work for 12.47 kV, 500 kcmil overhead line.

- *Evolving faults*: Ground faults are more likely to become two- or three-phase faults.
- *Underbuilt*: Faults on underbuilt distribution are more likely to cause faults on the transmission circuit above due to rising arc gases.

A fault current arc will expand after it is initiated. It has been found that the growth of the arc is generally in the vertical direction, and the growth is primarily a function of time and not of current or voltage (Drouet and Nadeau, 1979). The growth of the arc means that a 0.1 s fault on the instantaneous trip (with fuse saving) is less likely to involve other phases or other circuits than a 0.2–1 s fault on the time-delay trip (with fuse blowing).

30.8.5 Long-Duration Voltage Sags with Fuse Blowing

With fuse blowing, voltage sags last longer, especially for faults on the three-phase mains, which have to be cleared by phase or ground time–overcurrent elements. An example is shown in Figure 30.21 where

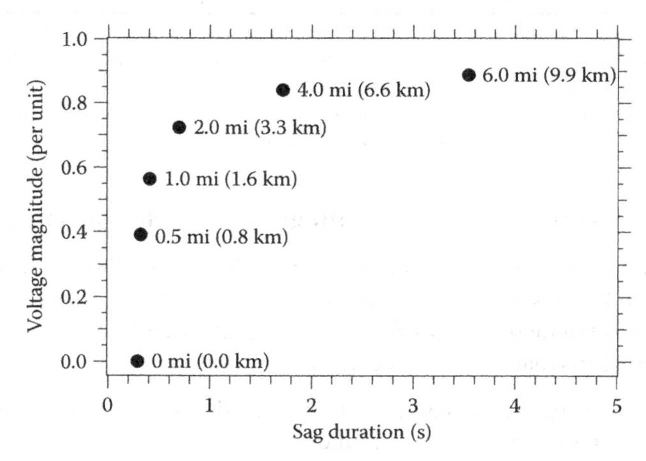

FIGURE 30.21 Magnitudes and durations of substation bus voltage sags for ground faults applied at the given distance with a fuse-blowing scheme. For the same circuit with fuse saving, all of the faults would clear in 0.1 s. *Assumptions:* 12.47 kV, 500 kcmil, all-aluminum conductors. Ground relay: CO-11, TD = 5, pickup = 300 A.

voltage sag magnitudes and durations are shown for faults at various distances from a substation using fuse blowing. For the same circuit with fuse saving, all of the faults would have cleared in 0.1 s. For a fault at the substation, the duration triples. For a fault one mile (1.6 km) from the substation, the duration quadruples. The situation is worse for phase-to-phase faults and three-phase faults because they must be cleared by the phase relays which are generally slower.

30.8.6 Optimal Implementation of Fuse Saving

In order to get a fuse-saving scheme to work, it is necessary to get the substation protective device to open before fuses operate. We can achieve this in several ways:

1. *Slow down the fuse*: Use big, slow fuses (such as a 140 or 200 T) near the substation to ensure proper coordination.
2. *Faster breakers or reclosers*: If three-cycle circuit breakers are used instead of the normal five-cycle breakers, fuse saving coordination is more likely. Some reclosers are even faster than three-cycle breakers.
3. *Limit fault currents*:
   a. Open station bus ties: An open bus tie will reduce the fault current on each feeder and make fuse saving easier. This is the normal operating mode for most utilities.
   b. Use a transformer neutral reactor: A neutral reactor reduces the fault current for single-phase faults (all faults on single-phase taps).
   c. Use line reactors: This reduces the fault current for all types of faults. This has been an uncommon practice. An added advantage, reactors reduce the impact of voltage sags for faults on adjacent feeders.
   d. Specify higher impedance transformers.

We can employ other strategies to limit the impact of momentary interruptions:

- *More downstream reclosers*: Extra downstream devices will reduce the number of momentaries for customers near the substation. It is important to coordinate reclosers with the upstream device (including sequence coordination).
- *Single-phase reclosers.*
- *Immediate reclose.*
- *Switch to a fuse blowing scheme on poor feeders*: For a feeder with many momentaries, disable the instantaneous relay for a time period. Identify poorly performing parts of the circuit during this time. The blown branch fuses provide a convenient fault location method. Once the poor performing sections are identified and improved, switch the circuit back to fuse saving.

30.8.7 Optimal Implementation of Fuse Blowing

Several strategies can optimize a fuse blowing scheme:

- *Fast fuses (or current-limiting fuses)*: If smaller or faster fuses are used, faults clear faster, so voltage sag durations are shorter. Current-limiting fuses also limit the magnitude and duration of the sag. Be careful not to fuse too small, or fuses will operate unnecessarily due to loading, inrush, and cold-load pickup. Note that if smaller fuses are used, it is difficult to switch back to a fuse-saving scheme.
- *Covered wire or small wire*: Watch burndowns on circuits with covered wire or small wire that is protected by the station circuit breaker or recloser. If either of these cases exists, use a modified fuse-blowing scheme with a time-delayed instantaneous element (see the next section).
- *Use single-phase reclosers on longer laterals*: A good way of maintaining some of the reliability of a fuse-saving scheme is to use single-phase reclosers instead of fuses on longer taps. Then, temporary faults on these laterals do not cause permanent interruptions to those customers.

- *More fuses*: Add more second- and third-level fuses to segment the circuit more.
- *Track lateral operations*: Temporary faults on fused laterals cause sustained interruptions. In order to minimize the impacts on lateral customers, track interruptions by lateral. Identify poorly performing laterals, patrol poor sections, then add tree trimming, animal guards, etc.

30.9 Other Protection Schemes

30.9.1 Time Delay on the Instantaneous Element (Fuse Blowing)

An alternative implementation of a fuse-blowing scheme is to use a time delay on the instantaneous trip (rather than removing the instantaneous trip; a definite-time overcurrent relay also could do the same function) (Engelman, 1990). Faults do not last as long as they would if the relay went to a time–overcurrent element; there is less chance of wire burndowns, and voltage sags are of shorter duration for faults on the mains. A common delay is 0.1 s.

An example implementation is shown in Figure 30.22 where a 0.1 s delay is added to the instantaneous. A 100 K fuse link is also shown. For the 100 K link, the scheme is actually a mixture of fuse saving and fuse blowing. For fault currents above roughly 1700 A, it is a fuse-blowing scheme (the fuse clears before the instantaneous relay operates). For currents below 1000 A, the scheme is a fuse-saving scheme (the circuit breaker trips before the fuse is damaged). Between 1000 and 1700 A, one or both devices operate.

Another option sometimes used with this scheme is a high-set instantaneous. The high-set instantaneous has no time delay and is set to clear faults close to the station. This removes the most damaging faults quickly (because they are the most likely to cause damage and cause the most severe voltage sags).

Using a time delay is a better fuse-saving scheme than just removing the instantaneous relay. The disadvantage, and the reason that it is not implemented as much, is that it is usually more difficult and costly to implement. For electromechanical relays, another timer relay must be added, and the relay scheme must be engineered. Many digital relays ease the implementation since they have this time delay option available.

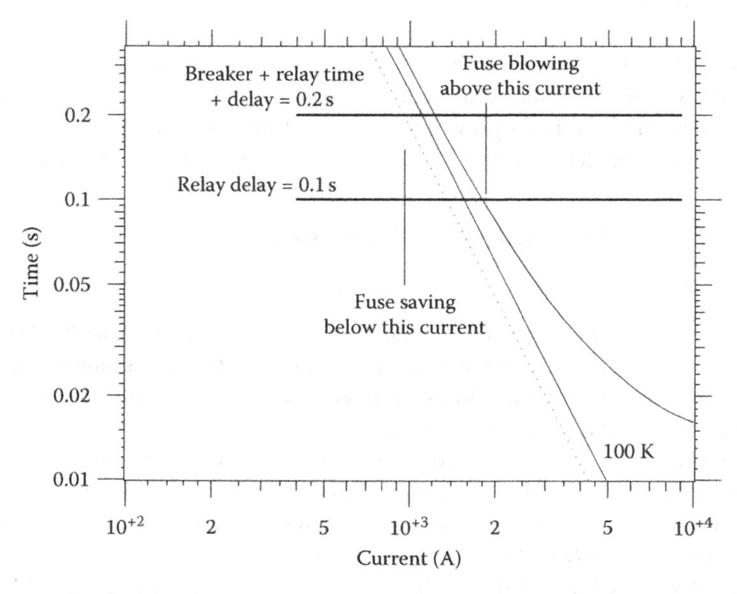

FIGURE 30.22 Example of a delayed instantaneous element used for fuse blowing.

30.9.2 High–Low Combination Scheme

Another option is to use fuse blowing at the substation and fuse saving at downstream reclosers (Burke, 1996).

- *Substation fuse blowing:* Fault currents are high near the substation, so it is difficult to get fuse saving to work here.
- *Recloser fuse saving:* Fault currents are lower downstream, and reclosers are faster, so fuse saving should work well here.

The high–low scheme is easy to implement. The station instantaneous trip is eliminated. Reclosers are operated with a fast trip (the A curve). Most are already in this mode, so no changes are necessary here.

30.9.3 SCADA Control of the Protection Scheme

Another option is to use SCADA to change back and forth between fuse saving and fuse blowing, getting some of the benefits of both schemes. Fuse blowing is the normal operating mode, but operators could switch to a fuse-saving scheme during storms. This avoids clear-sky momentaries while at the same time improving storm restoration. Several factors make fuse saving better during storms:

- Faults are more likely to be temporary during storms (lightning, wind).
- Customers are more forgiving about momentaries during storms.
- Interruptions due to fuse operations last longer during storms (because crews have many repairs to perform). If fuses are blown due to temporary faults, this increases the number of repair locations. Saving fuses reduces the number of interruptions crews will have to address.

In order for SCADA control of fuse saving to work best, we must design the system for fuse saving to work: larger, slower fuses for laterals close to the substation, faster circuit breakers (or use substation reclosers), or possibly even using grounding reactors in the substation to limit fault currents. Likewise, we must design for fuse blowing, so avoid using tree wire (or go to delayed instantaneous relaying rather than removing the instantaneous trip).

Control is more readily available in the substation because the SCADA infrastructure may already be in place. If so, the cost of the SCADA system has already been justified, and this added functionality could be piggybacked on the existing system if there are free channels available. It is feasible to use automation technology to implement remote control of feeder reclosers, but the cost of the communication equipment may not justify having this functionality.

For SCADA control, microprocessor-controlled relays are not needed. A SCADA channel can be used to control a blocking relay on the instantaneous elements of the feeder relays. Alternatively, the SCADA channel could control the delay on the instantaneous relay element (no delay: fuse saving, with delay: fuse blowing). One SCADA channel could control the fuse saving/blowing status of all of the distribution feeders in a station. Alternatively, we could control each feeder independently.

30.9.4 Adaptive Control by Phases

Various protection schemes are classified as adaptive. An adaptive approach to a fuse blowing mode is to adjust the scheme depending on how many phases are faulted:

- *Two- or three-phase fault*: Use the instantaneous; the fault is assumed to be on the three-phase mains. Tripping quickly reduces the duration of voltage sags for faults on the mains.
- *Single-phase fault*: Use fuse blowing (time delay curves or delayed instantaneous relay).

Adaptive control requires microprocessor-based relays. This is not a common scheme, and the expense and complexity are difficult to justify unless the chosen relay comes with this functionality.

30.10 Reclosing Practices

Automatic reclosing is a universally accepted practice on most overhead distribution feeders. On overhead circuits, 50%–80% of faults are temporary, so if a circuit breaker or recloser clears a fault and it *recloses*, most of the time the fault is gone, and customers do not lose power for an extended period of time.

On underground circuits, since virtually all faults are permanent, we do not reclose. A circuit might be considered underground if something like 60%–80% of the circuit is underground. Utility practices vary considerably relative to the exact percentage (IEEE Working Group on Distribution Protection, 1995). A significant number of utilities treat a circuit as underground if as little as 20% is underground while some others put the threshold over 80%.

The first reclose usually happens with a very short delay, either an immediate reclose which means a 1/3–1/2 s dead time (discussed later) or with a 1–5 s delay. Subsequent reclose attempts follow longer delays. The nomenclature is usually stated as 0–15–30 meaning there are three reclose attempts: the first reclose indicated by the "0" is made after no intentional delay (this is an immediate reclose), the second attempt is made following a 15 s dead time, and the final attempt is made after a 30 s dead time. If the fault is still present, the circuit opens and locks open. We also find this specified using circuit breaker terminology as O-0 s-CO-15 s-CO-30 s-CO where "C" means close and "O" means open. Other common cycles that utilities use are 0–30–60–90 and 5–45.

With reclosers and reclosing relays on circuit breakers, the reclosing sequence is reset after an interval that is normally adjustable. This interval is generally set somewhere in the range of 10–2 min. Only a few utilities have reported excessive operations without lockout (IEEE Working Group on Distribution Protection, 1995).

30.10.1 Reclose Attempts and Dead Times

Three reclose attempts is most common as shown in Figure 30.23. More reclose attempts give the fault more chance to clear or burn free. Returns diminish; the chance that the third or fourth reclose attempt is successful is usually small. Additional reclose attempts have the following negative impacts on the system:

- *Additional damage at the fault location*: With each reclose into a fault, arcing does additional damage at the fault location. Faults in equipment do more damage. Cable faults are harder to splice, wire burndowns are more likely, and oil-filled equipment is more likely to rupture. Arcs can start fires. Faults (and the damage the arcs cause) can propagate from one phase to other phases.

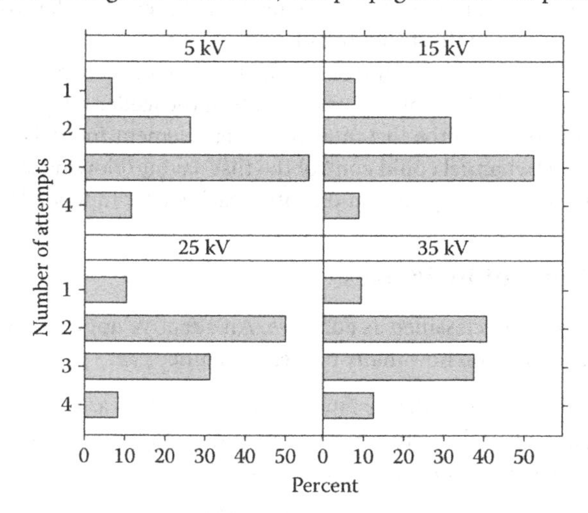

FIGURE 30.23 IEEE survey results on the number of reclose attempts for each voltage class. (Data from IEEE Working Group on Distribution Protection, *IEEE Trans. Power Deliv.*, 10(1), 176, January 1995.)

- *Voltage sags*: With each reclose into a fault, customers on adjacent circuits are hit with another voltage sag. It can be argued that the magnitude and duration of the sag should be about the same, so depending on the type of device, if the customer equipment survived the first sag, it will probably ride through subsequent sags of the same severity. If additional phases become involved in the fault, the voltage sag is more severe.
- *Through-fault damage to transformers*: Each fault subjects transformers to mechanical and thermal stresses. Virginia Power changed their reclosing practices because of excessive transformer failures on their 34.5 kV station transformers due to through faults (Johnston et al., 1978).
- *Through-fault damage to other equipment*: Cables, wires, and especially connectors suffer the thermal and mechanical stresses of the fault.
- *Interrupt ratings of breakers*: Circuit breakers must be derated if the reclose cycle involves more than one reclose attempt within 15 s. This may be a consideration if fault currents are high and breakers are near their ratings. Reclosers do not have to be derated for a complete four-sequence operation. Extra reclose attempts increase the number of operations, which means more frequent breakers and reclosers maintenance.
- *Ratcheting of overcurrent relays*: An induction relay disc turns in response to fault current. After the fault is over, it takes time for the disk to spin back to the neutral position. If this reset is not completed, and another fault occurs, the disk starts spinning from its existing condition, making the relay operate faster than it should. The most common problem area is miscoordination of a substation feeder relay with a downstream feeder recloser. If a fault occurs downstream of the recloser, the induction relay will spin due to the current (but not operate if it is properly coordinated). Multiple recloses by the recloser could ratchet the station relay enough to falsely trip the relay. The normal solution is to take the ratcheting into account when coordinating the relay and recloser, but in some cases modification of the reclosing cycle of the recloser is an option. Another option is to use digital relays, which do not ratchet in this manner.

Given these concerns, the trend has been to decrease the number of reclose attempts. We try to balance the loss in reliability against the problems caused by extra reclose attempts. A major question is how often are the extra reclose attempts successful. Table 30.15 shows the success rate for one utility in a high-lightning area. Table 30.16 shows a second utility with similar reclosing practices but quite different reclose success (more lockouts and lower success rates for the first two reclose attempts). Reclose success rates change based on the types of faults most commonly seen in a region. Another data point with a broader distribution of utilities is obtained in the EPRI distribution power quality study. Table 30.17 shows the number of momentary interruptions (reclose attempts) that do not lead to sustained interruptions (lockouts). The key point is that it is relatively uncommon (but not rare) for the third or fourth reclose attempt to be successful.

We may block reclosing in some cases. It is common to block all reclose attempts when workers are doing maintenance on a circuit to provide an extra level of protection (an instantaneous relay element is also commonly enabled in this situation). Another situation is for very high-current faults. A high-set instantaneous relay covering just the first few hundred feet of circuit detects faults on the substation exit cables.

TABLE 30.15 Reclose Success Rates for a Utility in a High-Lightning Area

Reclosure	Success Rate	Cumulative Success
1st shot (immediate)	83.25%	83.25%
2nd shot (15–45 s)	10.05%	93.30%
3rd shot (120 s)	1.42%	94.72%
Locked out	5.28%	

Source: Westinghouse Electric Corporation, *Applied Protective Relaying*, Monroeville, PA, 1982.

TABLE 30.16 Reclose Success Rates for One 34.5kV Utility

Reclosure	Success Rate	Cumulative Success
1st shot (immediate)	25.3%	25.3%
2nd shot (15 s)	42.1%	67.4%
3rd shot (80 s)	11.6%	79.0%
Locked out	21.0%	

Source: From Johnston, L. et al., *IEEE Trans. Power App. Syst.*, PAS-97(5), 1876, 1978. With permission. Copyright 1978 IEEE.

TABLE 30.17 Number of Interruptions per One Minute Aggregate Period That Do Not Lead to Sustained Interruptions

Number	Percentage
1	87%
2	9%
3	2%
4 or more	2%

Source: EPRI TR-106294-V2, *An Assessment of Distribution System Power Quality: Volume 2: Statistical Summary Report*, Electric Power Research Institute, Palo Alto, CA, 1996.

If it operates, reclosing is disabled. This practice is done to reduce the damage for a failure of one of the station exit cables.

The duration of the open interval—the dead time between reclose attempts—is also a consideration. For a smaller number of reclose attempts, use longer delays to give tree branches and other material more time to clear.

Operator practices must also be considered as part of the reclosing scheme. Not uncommonly, an operator manually recloses the circuit breaker after a feeder lockout (especially during a storm). This sends the breaker or recloser through its whole reclosing cycle along with all of the bad effects (like more equipment damage and more voltage sags) with very little chance of success.

Some engineers and field personnel believe that the purpose of the extra reclose attempts is to *burn* the fault clear. This is dangerous. Faults regularly burn clear on low-voltage systems (<480 V), rarely at distribution primary voltages. Faults can burn clear on primary systems. The most common example is that tree branches or animals can be burned loose. The problem with this concept is that, just as easily, the fault burns the primary conductor, which falls to the ground causing a high-impedance fault. Fires and equipment damage are also more likely with the "burn clear" philosophy.

To reduce the impacts of subsequent reclose attempts, we could switch back to an instantaneous operation after the first time–overcurrent relay operation. If the fault does not clear after the first time–overcurrent relay operation, it means the fault is not downstream of a fuse (or a recloser). The reason to use a time overcurrent relay is to coordinate with the fuse. Since the fuse is out of the picture, why not use a faster trip for subsequent reclose attempts? While not commonly done, we could implement this with digital relays. The setting of the "subsequent reclose" instantaneous relay element should be different than the first-shot instantaneous. Set the pickup at the pickup of the time–overcurrent relay. Because of inrush on subsequent attempts, we may use a fast time–overcurrent curve or an instantaneous element with a short delay (something like five-cycles).

As an example, if a utility uses a 0–15–30–90 s reclosing cycle, the system is subjected to five faults if the system goes through its complete cycle. With the instantaneous operation enabled on the first attempt and disabled on subsequent attempts, we have a very high total duration of the fault current. For a CO-11 ground relay with a time dial of 3, a 2 kA fault clears in roughly 1 s. For the reclosing cycle to lock out, the system has a total fault time of 4.1 s (one 0.1 s fault followed by four 1 s faults). If the instantaneous operation is enabled for reclose attempts 2 through 4, the total fault duration is 1.4 s (one 0.1 s fault followed by a 1 s fault and three 0.1 s faults). This greatly reduces the damage done by certain faults.

30.10.2 Immediate Reclose

An immediate reclose (also called an instantaneous or fast reclose) means having no intentional time delay (or a very short time delay) on the first reclose attempt on circuit breakers and reclosers. Many residential devices such as digital clocks, VCRs, and microwaves can ride through a 1/2 s interruption but not a 5 s interruption, so a fast reclose helps reduce residential complaints.

From a power quality point of view, a faster reclose is better. Some customers may not notice anything more than a quick blink of the lights. Many residential devices such as the digital clocks on alarm clocks, microwaves, and VCRs can ride through a 1/2 s interruption where they usually cannot ride through a 5 s interruption (a first reclose delay used by several utilities).

30.10.2.1 Effect on Sensitive Residential Devices

The most common power quality recorder in the world is the digital clock. Many complaints are due to the "blinking clocks." Using an immediate reclose reduces complaints. Florida Power has reported that a reclosing time of 18–20 cycles nearly eliminates complaints (Dugan et al., 1996). Another utility that has successfully used the immediate reclose is Long Island Lighting Company (now Keyspan) (Short and Ammon, 1997). According to an IEEE survey, a time to first reclose of less than 1 s is the most common practice although the fast reclose practice tends to decline with increasing voltage (see Figure 30.24).

Clocks have a wide range of voltage sensitivity, but most digital clocks will not lose memory for a complete interruption that is less than 0.5 s. So, an immediate reclose helps residential customers ride through momentary interruptions without resetting many devices. Given the wide variation, some customers are sensitive to a 0.5 s interruption. Note that the immediate reclose helps with digital clock-type devices whether it be on radio alarm clocks, VCRs, or microwaves. Fast reclosing does not help with

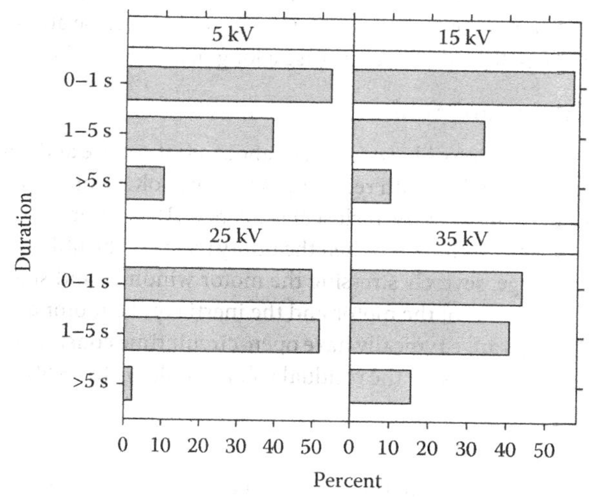

FIGURE 30.24 IEEE survey results of the intervals used before the first reclose attempt for each voltage class. (Data from IEEE Working Group on Distribution Protection, *IEEE Trans. Power Deliv.*, 10(1), 176, January 1995.)

most computers or other computer-based equipment, limiting the power quality improvement of using the immediate reclose to residential customers (no help for commercial or industrial customers).

30.10.2.2 Delay Necessary to Avoid Retriggering Faults

Sometimes a delayed reclose is necessary if there is not enough time to clear the fault. A fault arc needs time to cool, or the reclose could retrigger the arc. Whether the arc strikes again is a function of voltage and structure spacings. A 34.5kV utility (Vepco) added a delay to the first reclose because the probability of success of the first reclose was much less than normal for distribution circuits (Johnston et al., 1978). The success rate for the first attempt after an instantaneous reclose was 25% which is much less than the 70%–80% experienced by most utilities. Another item that added to the low success rate of Vepco's 34.5kV system is that they used a lot of armless design, and the combination of higher voltage and tighter spacings requires a longer time delay for the arc to clear.

With the following equation, we can find the minimum deionization time of an arc based on the line-to-line voltage (Westinghouse Electric Corporation, 1982):

$$t = 10.5 + \frac{V}{34.5}$$

where

t is the minimum deionization time, 60 Hz cycles
V is the rated line-to-line voltage, kV

The deionization time increases only moderately with voltage. Even for a 34.5kV system, the deionization time is 11.5 cycles. This equation is a simplification (separation distances are not included) but does show that arcs rapidly deionize. Many high-voltage transmission lines successfully use a fast reclose. The reclose time for distribution circuit breakers and reclosers varies by design. A typical time is 0.4–0.6 s for an immediate reclose (meaning no intentional delay). The fastest devices (newer vacuum or SF_6 devices) may reclose in as little as 11 cycles. This may prove to be too fast for some applications, so consider adding a small delay of 0.1–0.4 s (especially at 25 or 35kV).

On distribution circuits, other things affect the time to clear a fault besides the deionization of the arc stream. If a temporary fault is caused by a tree limb or animal, time may be needed for the "debris" to fall off the conductors or insulators. Because of this, with an immediate reclose use at least two reclose attempts before lockout. For example, use a 0–15–30 s cycle (three reclose attempts), or if you wish to use two reclose attempts, use a 0–30 or 0–45 s cycle (use a long delay before the last reclose attempt).

30.10.2.3 Reclose Impacts on Motors

Industrial customers with large motors have concerns about a fast reclose and damage to motors and their driven equipment. The major problem with reclosing is that the voltage on a motor will not drop instantly to zero when the utility circuit breaker (or recloser) is opened. The motor has residual voltage, where the magnitude and frequency decay with time. When the utility recloses, the utility voltage can be out of phase with the motor residual voltage, severely stressing the motor windings and shaft and its driven load. The decay time is a function of the size of the motor and the inertia of the motor and its load.

Motors in the 200–2000 hp range typically have open-circuit time constants of 0.5–2 s (Bottrell, 1993). The time constant is the time it takes for the residual voltage to decay to 36.8% of its initial value. Reclose impacts are worse with

- Larger motors
- Capacitor banks—excitation from the capacitor banks can greatly increase the motor decay time
- Synchronous motors and generators—much larger time constants makes synchronous machines more vulnerable to damage than induction machines

On the vast majority of distribution circuits, reclosing impacts will not be a concern because of the following:

- Motors on contactors will drop out. Also, larger motors and synchronous motors normally have an undervoltage relay to trip when voltage is lost.
- Most utility feeders do not have individual motor loads larger than 500 hp.
- Even with feeders with large industrial customers, the non-motor load will be large enough to pull the voltage down to a safe level within the time it takes to do a normal immediate reclose (0.4–0.6 s).

Because of this, we can safely implement an immediate reclose on almost all distribution circuits. One exception is a feeder with an industrial customer that is a majority of the feeder load, and the industrial customer has several large induction or (especially) synchronous motors. Another exception is a feeder with a large rotating distributed generator. In both of these cases, delay the first reclose or, alternatively, use line-side voltage supervision (if voltage is detected downstream of the breaker, reclosing is blocked to prevent an out-of-phase reclosing situation).

30.11 Single-Phase Protective Devices

Many distribution protective devices are single phase or are available in single-phase versions including reclosers, fuses, and sectionalizers. Single-phase protective devices are used widely on distribution systems; taps are almost universally fused. On long single-phase taps, single-phase reclosers are sometimes used. Most utilities also use fuses for three-phase taps. The utilities that do not fuse three-phase taps most often cite the problem of single-phasing motors of three-phase customers. Some utilities use single-phase reclosers that protect three-phase circuits (even in the substation).

Single-phase protective devices on single-phase laterals are widely used, and the benefits are universally accepted. The fuse provides an inexpensive way of isolating faulted circuit sections. The fuse also aids in finding the fault.

Using single-phase interrupters helps on three-phase circuits—only one phase is interrupted for line-to-ground faults. We can easily estimate the effect on individual customers using the number of phases that are faulted on average as shown in Table 30.18. Overall, using single-phase protective devices cuts the average number of interruptions in half. This assumes that all customers are single phase and that the customers are evenly split between phases.

Service to three-phase customers downstream of single-phase interrupters generally improves, too. Three-phase customers have many single-phase loads, and the loads on the unfaulted phases are unaffected by the fault. Three-phase devices may also ride through an event caused by a single-phase fault (although motors may heat up because of the voltage unbalance as discussed in the next section). Single-phase protective devices do have some drawbacks. The main concerns are

- Ferroresonance
- Single-phasing of motors
- Backfeeds

TABLE 30.18 Effect on Interruptions When Using Single-Phase Protective Devices on Three-Phase Circuits

Fault Type	Percent of Faults	Portion Affected	Weighted Effect
Single phase	70%	33%	23%
Two phase	20%	67%	13%
Three phase	10%	100%	10%
		Total	47%

Ferroresonance usually occurs during manual switching of single-pole switching devices (where the load is usually an unloaded transformer). It is less common for ferroresonance to occur downstream of a single-phase protective device that is operating due to a fault. The reason for this is that if there is a fault on the opened phase, the fault prevents an overvoltage on the opened phase. Also, any load on the opened section helps prevent ferroresonant overvoltages. Because ferroresonance will be uncommon with single-phase protective devices, it is usually not a major factor in protective device selection. Still, caution is warranted on small three-phase transformers that may be switched unloaded (especially at 24.94 or 34.5 kV).

With single-phase protective devices, backfeeds can create hazards. During a line-to-ground fault where a single-phase device opens, backfeed through a three-phase load can cause voltage on the load side of an opened protective device. Backfeeds can happen with most types of three-phase distribution transformer connections (even with a grounded-wye–grounded-wye connection). The important points to note are as follows:

- The backfeed voltage is enough to be a safety hazard to workers or the public (e.g., in a wire down situation).
- The available backfeed is a stiff enough source to maintain an arc of significant length. The arc can continue causing damage at the fault location during a backfeed condition. It may also be a low-level sparking and sputtering fault.

Based on these points, single-phasing can cause problems from backfeeding. Whether this constrains use of single-phase protective devices is debatable. Most utilities do use single-phase protective devices, usually with fuses, on three-phase circuits.

Under single-phasing, motors can overheat and fail. Motors have relatively low impedance to negative-sequence voltage; therefore, a small negative-sequence component of the voltage produces a relatively large negative-sequence current. Consequently the effect magnifies; a small negative-sequence voltage appears as a significantly larger percentage of unbalanced current than the percentage of unbalanced voltage.

Loss of one or two phases is a large unbalance. For one phase open, the phase-to-phase voltages become 0.57, 1.0, 0.57 for a wye–wye transformer and 0.88, 0.88, 0.33 for a delta–wye transformer. In either case, the negative-sequence voltage is 0.66 per unit. With such high unbalance, a motor overheats quickly. The negative-sequence impedance of a motor is roughly 15%, so for a 66% negative-sequence voltage, the motor draws a negative-sequence current of 440%.

Most utility service agreements with customers state that it is the customer's responsibility to protect their equipment against single phasing. The best way to protect motors is with a phase-loss relay. Nevertheless, some utilities take measures to reduce the possibility of single-phasing customers' motors, and one way to do that is to limit the use of single-phase protective devices. Other utilities are more aggressive in their use of single-phase protective equipment and leave it up to customers to protect their equipment.

30.11.1 Single-Phase Reclosers with Three-Phase Lockout

Many single-phase reclosers and recloser controls come with a controller option for a single-phase trip and three-phase lockout. Three-phase reclosers that can operate each phase independently are also available. For single-phase faults, only the faulted phase opens. For temporary faults, the recloser successfully clears the fault and closes back in, so there will only be a momentary interruption on the faulted phase. If the fault is still present after the final reclose attempt (a permanent fault), the recloser trips all three phases and will not attempt additional reclosing operations.

Problems of single-phasing motors, backfeeds, and ferroresonance disappear. Single-phasing motors and ferroresonance cause heating, and heating usually takes many minutes for damage to occur. Short-duration single-phasing occurring during a typical reclose cycle does not cause enough heat to do damage. If the fault is permanent, all three phases trip and lock out, so there is no long-term single phasing. A three-phase lockout also reduces the chance of backfeed to a downed wire for a prolonged period.

Single-phase reclosers are available that have high enough continuous and interrupting ratings that utilities can use them in almost all feeder applications and many substation applications.

Another consideration with single-phase reclosers vs. three-phase devices is that a ground relay is often not available on single-phase reclosers. A ground relay provides extra sensitivity for line-to-ground faults. Not having the ground relay is a tradeoff to using single-phase devices. Even if a ground relay is available on a unit with single-phase tripping, if the ground relay operates, it trips all three phases (which defeats the purpose of single-phase tripping).

References

Amundson, R. H., High voltage fuse protection theory & considerations, *IEEE Tutorial Course on Application and Coordination of Reclosers, Sectionalizers, and Fuses*, 1980. Publication 80 EHO157-8-PWR.

ANSI C57.12.20-1988, American national standard requirements for overhead-type distribution transformers, 500 kVA and smaller: High-voltage, 67,000 volts and below; low-voltage, 15,000 volts and below.

ANSI C37.42-1989, American national standard specifications for distribution cutouts and fuse links.

ANSI C37.06-1997, AC high-voltage circuit breakers rated on a symmetrical current basis—Preferred ratings and related required capabilities.

ANSI/IEEE C37.60-1981, IEEE standard requirements for overhead, pad mounted, dry vault, and submersible automatic circuit reclosers and fault interrupters for AC systems.

ASTM F1959, *Standard Test Method for Determining the Arc Rating of Materials for Clothing*, ASTM International, West Conshohocken, PA, 2006.

Benmouyal, G. and Zocholl, S. E., Time–current coordination concepts, in *Western Protective Relay Conference*, Spokane, WA, 1994.

Bingwu, G. and Chengkang, W., The gasdynamic and electromagnetic factors affecting the position of arc roots in a tubular arc heater, *Acta Mechanica Sinica*, 7(3), 199–208, 1991.

Bottrell, G. W., Hazards and benefits of utility reclosing, in *Proceedings of Power Quality*, 1993.

Burke, J. J., Philosophies of distribution system overcurrent protection, in *IEEE/PES Transmission and Distribution Conference*, Los Angeles, CA, 1996.

Burke, J. J., High impedance faults (40 ohms is a fallacy), *IEEE Rural Electric Power Conference*, Albuquerque, NM, April 9–11, 2006.

CEA 288 D 747, *Application Guide for Distribution Fusing*, Canadian Electrical Association, Ottawa, Ontario, Canada, 1998.

Cook, C. J. and Niemira, J. K., Overcurrent protection of transformers—Traditional and new philosophies for small and large transformers, in *IEEE/PES Transmission & Distribution Conference and Exposition*, 1996. Presented at the training session on "Distribution overcurrent protection philosophies."

Cooper Power Systems, *Comparison of Recloser and Breaker Standards*, Reference data R280-90-5, February 1994.

Cooper Power Systems, *Electrical Distribution—System Protection*, 3rd edn., 1990.

Cress, S. L., Arc modeling and hazard assessment, *EPRI Distribution Arc Flash Workshop*, Chicago, IL, April 2008.

Dagenhart, J., The 40-ohm ground-fault phenomenon, *IEEE Transactions on Industry Applications*, 36(1), 30–32, 2000.

Distribution Line Protection Practices Industry Survey Results IEEE Power System Relaying Committee Report, December 2002. http://www.pes-psrc.org/Reports/DISTRIBUTION%20LINE%20PROTECTION%20PRACTICES%20report%20final%20dec20.pdf

Drouet, M. G. and Nadeau, F., Pressure waves due to arcing faults in a substation, *IEEE Transactions on Power Apparatus and Systems*, PAS-98(5), 1632–1635, 1979.

Dugan, R. C., Ray, L. A., Sabin, D. D., Baker, G., Gilker, C., and Sundaram, A., Impact of fast tripping of utility breakers on industrial load interruptions, *IEEE Industry Applications Magazine*, 2(3), 55–64, May/June 1996.

Eblen, M. L. and Short, T. A., Arc flash testing of typical 480-V utility equipment, *IEEE IAS Electrical Safety Workshop*, Memphis, TN, February 1–5, 2010. Paper ESW2010-05.

Engelman, N. G., Relaying changes improve distribution power quality, *Transmission and Distribution*, pp. 72–76, May 1990.

EPRI 1001665, *Power Quality Improvement Methodology for Wire Companies*, Electric Power Research Institute, Palo Alto, CA, 2003.

EPRI 1018693, *Distribution Arc Flash: Analysis Methods and Arc Characteristics*, Electric Power Research Institute, Palo Alto, CA, 2009. http://my.epri.com/portal/server.pt?Abstract_id=000000000001018693

EPRI 1020210, *Distribution Arc Flash: 480-V Padmounted Transformers and Network Protectors*, Electric Power Research Institute, Palo Alto, CA, 2009. http://my.epri.com/portal/server.pt?Abstract_id=000000000001020210

EPRI TR-106294-V2, *An Assessment of Distribution System Power Quality: Volume 2: Statistical Summary Report*, Electric Power Research Institute, Palo Alto, CA, 1996.

Gabrois, G. L., Huber, W. J., and Stoelting, H. O., Blowing of distribution transformer fuses by lightning, in *IEEE/PES Summer Meeting*, Vancouver, British Columbia, Canada, 1973. Paper C73-421-5.

Hamel, A., Dastous, J. B., and Foata, M., Estimating overpressures in pole-type distribution transformers. Part I: Tank withstand evaluation, *IEEE Transactions on Power Delivery*, 18(1), 113–119, 2003.

Hamel, A., St. Jean, G., and Paquette, M., Nuisance fuse operation on MV transformers during storms, *IEEE Transactions on Power Delivery*, 5(4), 1866–1874, October 1990.

Hi-Tech Fuses, Bulletin FS-10: Application of trans-guard EXT and OS fuses, in *Hi-Tech Fuses*, Hickory, NC, 2002.

IEEE C2-2007, *National Electrical Safety Code.*

IEEE Std 1584-2002, *IEEE Guide for Performing Arc-Flash Hazard Calculations.*

IEEE Std. C37.40-1993, IEEE standard service conditions and definitions for high-voltage fuses, distribution enclosed single-pole air switches, fuse disconnecting switches, and accessories.

IEEE Std. C37.112-1996, IEEE standard inverse-time characteristic equations for overcurrent relays.

IEEE Std. C37.48-1997, IEEE guide for the application, operation, and maintenance of high-voltage fuses, distribution enclosed single-pole air switches, fuse disconnecting switches, and accessories.

IEEE Std. C62.22-1997, IEEE guide for the application of metal-oxide surge arresters for alternating-current systems.

IEEE Std. C37.04-1999, IEEE standard rating structure for AC high-voltage circuit breakers.

IEEE Std. C37.010-1999, IEEE application guide for AC high-voltage circuit breakers rated on a symmetrical current basis.

IEEE Working Group on Distribution Protection, Distribution line protection practices—Industry survey results, *IEEE Transactions on Power Delivery*, 10(1), 176–186, January 1995.

IEEE Working Group on Distribution Protection, Distribution line protection practices—Industry survey results, *IEEE Transactions on Power Delivery*, 3(2), 514–524, April 1988.

Johnston, L., Tweed, N. B., Ward, D. J., and Burke, J. J., An analysis of Vepco's 34.5 kV distribution feeder faults as related to through fault failures of substation transformers, *IEEE Transactions on Power Apparatus and Systems*, PAS-97(5), 1876–1884, 1978.

Kinectrics, *User's Guide for ARCPRO*, version 2.0, Toronto, Ontario, Canada, 2002.

Lunsford, J. M. and Tobin, T. J., Detection of and protection for internal low-current winding faults in overhead distribution transformers, *IEEE Transactions on Power Delivery*, 12(3), 1241–1249, July 1997.

Plummer, C. W., Goedde, G. L., Pettit, E. L. J., Godbee, J. S., and Hennessey, M. G., Reduction in distribution transformer failure rates and nuisance outages using improved lightning protection concepts, *IEEE Transactions on Power Delivery*, 10(2), 768–777, April 1995.

REA Bull. 61-2, *Guide for Making a Sectionalizing Study on Rural Electric Systems*, Rural Electrification Administration, U.S. Department of Agriculture, Washington, DC, 1978.

Short, T. A., Fuse saving and its effect on power quality, in *EEI Distribution Committee Meeting*, pp. 87–100, March 29–31, Phoenix, AZ, 1999.

Short, T. A., *Electric Power Distribution Handbook*, CRC Press, Boca Raton, FL, 2004.

Short, T. A., Arc-flash analysis approaches for medium-voltage distribution, *IEEE Transactions on Industry Applications*, 47(4), 1902–1909, 2011.

Short, T. A. and Ammon, R. A., Instantaneous trip relay: Examining its role, *Transmission and Distribution World*, 49(2), 60–63, February 1997.

Short, T. A. and Eblen, M. L., Medium-voltage arc flash in open air and padmounted equipment, *IEEE Rural Electric Power Conference (REPC)*, Chattanooga, TN, April 10–13, 2011.

Smith, D. R., Network primary feeder grounding and protection, in Draft 3 of Chapter 10 of the *IEEE Network Tutorial*, 1999.

Smithley, R. S., Normal relay settings handle cold load, *Electrical World*, pp. 52–54, June 15, 1959.

Westinghouse Electric Corporation, *Applied Protective Relaying*, 1982.

Electric Power Utilization

Andrew P. Hanson

Andrew P. Hanson is a senior manager with The Structure Group. He has 20 years of experience in power system engineering and operations and consulting, having worked at Tampa Electric, Siemens, and ABB and having led the development of an office for a small privately held engineering firm. His expertise is focused on power delivery system operations and planning, having led the development of processes, forecasts, plans, and distribution automation implementations for a number of utilities. Dr. Hanson received his BEE from the Georgia Institute of Technology and his MEE and PhD in electrical engineering from Auburn University.

31

Metering of Electric Power and Energy

John V. Grubbs
Alabama Power Company

Electrical metering deals with two basic quantities: *energy* and *power.* Energy is equivalent to work. Power is the rate of doing work. Power applied (or consumed) for any length of time is energy. In mathematical terms, power integrated over time is energy. The basic electrical unit of energy is the watthour. The basic unit of power is the watt. The watthour meter measures energy (in watthours), while the wattmeter measures the rate of energy, power (in watthours per hour or simply watts). For a constant power level, power multiplied by time is energy. For example, a watthour meter connected for 2 h in a circuit using 500 W (500 Wh/h) will register 1000 Wh.

31.1 The Electromechanical Meter

The electromechanical watthour meter is basically a very specialized electric motor, consisting of

- A *stator* and a *rotor* that together produce torque
- A *brake* that creates a counter torque
- A *register* to count and display the revolutions of the rotor

31.1.1 Single Stator Electromechanical Meter

A two-wire single stator meter is the simplest electromechanical meter. The single stator consists of two electromagnets. One electromagnet is the potential coil connected between the two circuit conductors. The other electromagnet is the current coil connected in series with the load current. Figure 31.1 shows the major components of a single stator meter.

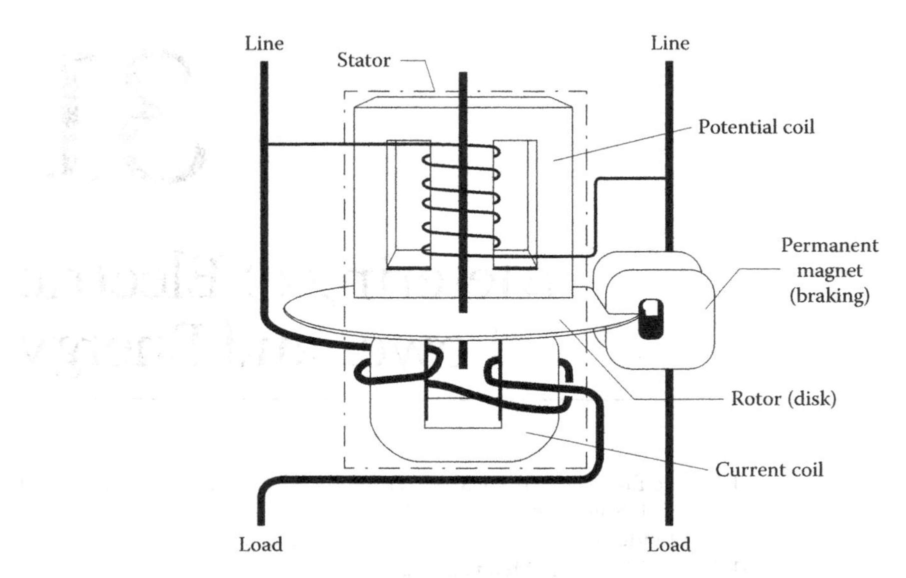

FIGURE 31.1 Main components of electromechanical meter.

The electromagnetic fields of the current coil and the potential coil interact to generate torque on the rotor of the meter. This torque is proportional to the product of the source voltage, the line current, and the cosine of the phase angle between the two. Thus, the torque is also proportional to the power in the metered circuit.

The device described so far is incomplete. In measuring a steady power in a circuit, this meter would generate constant *torque* causing steady acceleration of the rotor. The rotor would spin faster and faster until the torque could no longer overcome friction and other forces acting on the rotor. This ultimate speed would not represent the level of power present in the metered circuit.

To address these problems, designers add a permanent magnet whose magnetic field acts on the rotor. This field interacts with the rotor to cause a *counter torque* proportional to the speed of the rotor. Careful design and adjustment of the magnet strength yields a meter that rotates at a *speed* proportional to power. This speed can be kept relatively slow. The product of the rotor speed and time is revolutions of the rotor. The revolutions are proportional to energy consumed in the metered circuit. One revolution of the rotor represents a fixed number of watthours. The revolutions are easily converted via mechanical gearing or other methods into a display of watthours or, more commonly, kilowatthours.

31.2 Blondel's Theorem

Blondel's theorem of polyphase metering describes the measurement of power in a polyphase system made up of an arbitrary number of conductors. The theorem provides the basis for correctly metering power in polyphase circuits. In simple terms, Blondel's theorem states that the total power in a system of (N) conductors can be properly measured by using (N) wattmeters or watt-measuring elements. The elements are placed such that one current coil is in each of the conductors and one potential coil is connected between each of the conductors and some common point. If this common point is chosen to be one of the (N) conductors, there will be zero voltage across one of the measuring element potential coils. This element will register zero power. *Therefore, the total power is correctly measured by the remaining (N – 1) elements.*

In application, this means that to accurately measure the power in a four-wire three-phase circuit (N = 4), the meter must contain (N – 1) or three measuring elements. Likewise, for a three-wire three-phase circuit (N = 3), the meter must contain two measuring elements. There are meter designs available that, for commercial reasons, employ less than the minimum number of elements (N – 1) for a given circuit configuration. These designs depend on *balanced* phase voltages for proper operation. Their accuracy suffers as voltages become unbalanced.

31.3 The Electronic Meter

Since the 1980s, meters available for common use have evolved from (1) electromechanical mechanisms driving mechanical, geared registers to (2) the same electromechanical devices driving electronic registers to (3) totally electronic (or solid state) designs. All three types remain in wide use, but the type that is growing in use is the solid state meter.

The addition of the electronic register to an electromechanical meter provides a digital display of energy and demand. It supports enhanced capabilities and eliminates some of the mechanical complexity inherent in the geared mechanical registers.

Electronic meters contain no moving mechanical parts—rotors, shafts, gears, bearings. They are built instead around large-scale integrated circuits, other solid state components, and digital logic. Such meters are much more closely related to computers than to electromechanical meters.

The operation of an electronic meter is very different than that described in earlier sections for an electromechanical meter. Electronic circuitry samples the voltage and current waveforms during each electrical cycle and converts these samples to digital quantities. Other circuitry then manipulates these values to determine numerous electrical parameters, such as kW, kWh, kvar, kvarh, kQ, kQh, power factor, kVA, rms current, rms voltage.

Various electronic meter designs also offer some or all of the following capabilities:

- *Time of use (TOU)*. The meter keeps up with energy and demand in multiple daily periods. (See Section 31.4.2.)
- *Bi-directional*. The meter measures (as separate quantities) energy delivered to and received from a customer. This feature is used for a customer that is capable of generating electricity and feeding it back into the utility system.
- *Loss compensation*. The meter can be programmed to automatically calculate watt and var losses in transformers and electrical conductors based on defined or tested loss characteristics of the transformers and conductors. It can internally add or subtract these calculated values from its measured energy and demand. This feature permits metering to be installed at the most economical location. For instance, we can install metering on the secondary (e.g., 4 kV) side of a customer substation, even when the contractual service point is on the primary (e.g., 110 kV) side. The 4 kV metering installation is much less expensive than a corresponding one at 110 kV. Under this situation, the meter compensates its secondary-side energy and demand readings to simulate primary-side readings.
- *Interval data recording*. The meter contains solid state memory in which it can record up to several months of interval-by-interval data. (See Section 31.4.3.)
- *Remote communications*. Built-in communications capabilities permit the meter to be interrogated remotely via telephone, radio, or other communications media.
- *Diagnostics*. The meter checks for the proper voltages, currents, and phase angles on the meter conductors. (See Section on 31.3.3.)
- *Power quality*. The meter can measure and report on momentary voltage or current variations and on harmonic conditions.

Note that many of these features are available only in the more advanced (and expensive) models of electronic meters.

As an example of the benefits offered by electronic meters, consider the following two methods of metering a large customer who is capable of generating and feeding electricity back to the utility. In this example, the metering package must perform these functions:

Measure kWh delivered to the customer
Measure kWh received from the customer
Measure kvarh delivered
Measure kvarh received
Measure kW delivered
Measure kW received
Compensate received quantities for transformer losses
Record the measured quantities for each demand interval
Method A. (2) kW/kWh electromechanical meters with pulse generators (one for delivered, one for received)
(2) kWh electromechanical meters with pulse generators (to measure kvarh)
(2) Phase shifting transformers (used along with the kWh meters to measure kvarh)
(2) Transformer loss compensators
(1) Pulse data recorder
Method B. (1) Electronic meter

Obviously, the electronic installation is much simpler. In addition, it is less expensive to purchase and install and is easier to maintain.

Benefits common to most solid state designs are high accuracy and stability. Another less obvious advantage is in the area of error detection. When an electromechanical meter develops a serious problem, it may produce readings in error by any arbitrary amount. An error of 10%, 20%, or even 30% can go undetected for years, resulting in very large over- or under-billings. However, when an electronic meter develops a problem, it is more likely to produce an obviously bad reading (e.g., all zeroes; all 9s; a demand 100 times larger than normal; or a blank display). This greatly increases the likelihood that the error will be noticed and reported soon after it occurs. The sooner such a problem is recognized and corrected, the less inconvenience and disruption it causes to the utility and to the customer.

31.3.1 Multifunction Meter

Multifunction or *extended function* refers to a meter that can measure reactive or apparent power (e.g., kvar or kVA) in addition to real power (kW). By virtue of their designs, many electronic meters inherently measure the quantities and relationships that define reactive and apparent power. It is a relatively simple step for designers to add meter intelligence to calculate and display these values.

31.3.2 Voltage Ranging and Multiform Meter

Electronic meter designs have introduced many new features to the watthour metering world. Two features, typically found together, offer additional flexibility, simplified application, and opportunities for reduced meter inventories for utilities.

- *Voltage ranging*—Many electronic meters incorporate circuitry that can sense the voltage level of the meter input signals and adjust automatically to meter correctly over a wide range of voltages. For example, a meter with this capability can be installed on either a 120 or 277 V service.
- *Multiform*—Meter form refers to the specific combination of voltage and current signals, how they are applied to the terminals of the meter, and how the meter uses these signals to measure power and energy. For example, a Form 15 m would be used for self-contained application on a 120/240 V four-wire delta service, while a Form 16 m would be used on a self-contained 120/208 V four-wire wye service. A *multiform* 15/16 m can work interchangeably on either of these services.

31.3.3 Site Diagnostic Meter

Newer meter designs incorporate the ability to measure, display, and evaluate the voltage and current magnitudes and phase relationships of the circuits to which they are attached. This capability offers important advantages:

- At the time of installation or reinstallation, the meter analyzes the voltage and current signals and determines if they represent a recognizable service type.
- Also at installation or reinstallation, the meter performs an initial check for wiring errors such as crossed connections or reversed polarities. If it finds an error, it displays an error message so that corrections can be made.
- Throughout its life, the meter continuously evaluates voltage and current conditions. It can detect a problem that develops weeks, months, or years after installation, such as tampering or deteriorated CT or VT wiring.
- Field personnel can switch the meter display into diagnostic mode. It will display voltage and current magnitudes and phase angles for each phase. This provides a quick and very accurate way to obtain information on service characteristics.

If a diagnostic meter detects any error that might affect the accuracy of its measurements, it will lock its display in error mode. The meter continues to make energy and demand measurements in the background. However, these readings cannot be retrieved from the meter until the error is cleared. This ensures the error will be reported the next time someone tries to read the meter.

31.4 Special Metering

31.4.1 Demand Metering

31.4.1.1 What is Demand?

Electrical energy is commonly measured in units of kilowatthours. Electrical power is expressed as kilowatthours per hour or, more commonly, kilowatts.

Demand is defined as power averaged over some specified period. Figure 31.2 shows a sample power curve representing instantaneous power. In the time interval shown, the integrated area under the power curve represents the energy consumed during the interval. This energy, divided by the length of the interval (in hours) yields "demand." In other words, the demand for the interval is that value of power that, if held constant over the interval, would result in an energy consumption equal to that energy the customer actually used.

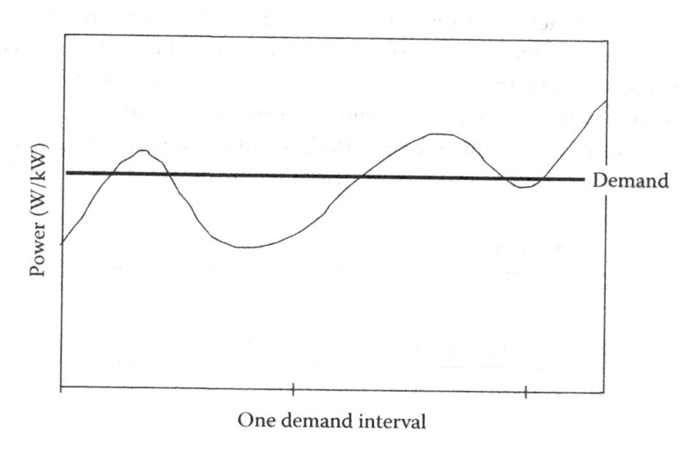

FIGURE 31.2 Instantaneous power vs. demand.

Demand is most frequently expressed in terms of real power (kilowatts). However, demand may also apply to reactive power (kilovars), apparent power (kilovolt-amperes), or other suitable units. Billing for demand is commonly based on a customer's maximum demand reached during the billing period.

31.4.1.2 Why is Demand Metered?

Electrical conductors and transformers needed to serve a customer are selected based on the expected maximum demand for the customer. The equipment must be capable of handling the maximum levels of voltages and currents needed by the customer. A customer with a higher maximum demand requires a greater investment by the utility in equipment. Billing based on energy usage alone does not necessarily relate directly to the cost of equipment needed to serve a customer. Thus, energy billing alone may not equitably distribute to each customer an appropriate share of the utility's costs of doing business.

For example, consider two commercial customers with very simple electricity needs. Customer A has a demand of 25 kW and operates at this level 24 h per day. Customer B has a maximum demand of 100 kW but operates at this level only 4 h per day. For the remaining 20 h of the day, "B" operates at a 10 kW power level.

$$\text{“A” uses } 25\,\text{kW} \times 24\,\text{h} = 600\,\text{kWh per day}$$

$$\text{“B” uses } (100\,\text{kW} \times 4\,\text{h}) + (10\,\text{kW} \times 20\,\text{h}) = 600\,\text{kWh per day}$$

Assuming identical billing rates, each customer would incur the same energy costs. However, the utility's equipment investment will be larger for Customer B than for Customer A. By implementing a charge for demand as well as energy, the utility would bill Customer A for a maximum demand of 25 kW and Customer B for 100 kW. "B" would incur a larger total monthly bill, and each customer's bill would more closely represent the utility's cost to serve.

31.4.1.3 Integrating Demand Meters

By far the most common type of demand meter is the integrating demand meter. It performs two basic functions. First, it measures the *average* power during each *demand interval*. (Common demand interval lengths are 15, 30, or 60 min.) See Table 31.1. The meter makes these measurements interval-by-interval throughout each day. Second, it retains the maximum of these interval measurements.

The demand calculation function of an electronic meter is very simple. The meter measures the energy consumed during a demand interval, then multiplies by the number of demand intervals per hour. In effect, it calculates the energy that would be used if the rate of usage continued for 1 h. The following table illustrates the correspondence between energy and demand for common demand interval lengths.

After each measurement, the meter compares the new demand value to the stored *maximum demand*. If the new value is greater than that stored, the meter replaces the stored value with the new one. Otherwise, it keeps the previously stored value and discards the new value. The meter repeats this process for each interval. At the end of the billing period, the utility records the maximum demand, then resets the stored *maximum demand* to zero. The meter then starts over for the new billing period.

TABLE 31.1 Energy/Demand Comparisons

Demand Interval (min)	Intervals per Hour	Energy During Demand Interval (kWh)	Resulting Demand (kW)
60	1	100	100
30	2	50	100
15	4	25	100

31.4.2 Time of Use Metering

A time of use meter measures and stores energy (and perhaps demand) for multiple periods in a day. For example, a service rate might define one price for energy used between the hours of 12 noon and 6 P.M. and another rate for that used outside this period. The TOU meter will identify the hours from 12 noon until 6 P.M. as "Rate 1." All other hours would be "Rate 2." The meter will maintain separate measurements of Rate 1 energy (and demand) and Rate 2 energy (and demand) for the entire billing period. Actual TOU service rates can be much more complex than this example, including features such as

- More than two periods per day,
- Different periods for weekends and holidays, and
- Different periods for different seasons of the year.

A TOU meter depends on an internal clock/calendar for proper operation. It includes battery backup to maintain its clock time during power outages.

31.4.3 Interval Data Metering

The standard method of gathering billing data from a meter is quite simple. The utility reads the meter at the beginning of the billing period and again at the end of the billing period. From these readings, it determines the energy and maximum demand for that period. This information is adequate to determine the bills for the great majority of customers. However, with the development of more complex service rates and the need to study customer usage patterns, the utility sometimes wants more detail about how a customer uses electricity. One option would be to read the meter daily. That would allow the utility to develop a day-by-day pattern of the customer's usage. However, having someone visit the meter site every day would quickly become very expensive. What if the meter could record usage data every day? The utility would have more detailed usage data, but would only have to visit the meter when it needed the data, for instance, once per month. And if the meter is smart enough to do that, why not have it record data even more often, for instance every hour?

In very simple terms, this is what *interval data metering* does. The interval meter includes sufficient circuitry and intelligence to record usage multiple times per hour. The length of the recording interval is programmable, often over a range from 1 to 60 min. The meter includes sufficient solid state memory to accumulate these interval readings for a minimum of 30 days at 15-min intervals. Obviously, more frequent recording times reduce the days of storage available.

A simple kWh/kW recording meter typically records one set of data representing kWh. This provides the detailed usage patterns that allow the utility to analyze and evaluate customer "load profiles" based on daily, weekly, monthly, or annual bases. An extended function meter is commonly programmed to record two channels of data, e.g., kWh and kvarh. This provides the additional capability of analyzing customers' power factor patterns over the same periods. Though the meter records information in energy units (kWh or kvarh), it is a simple matter to convert this data to equivalent demand (kW or kvar). Since demand represents energy per unit time, simply divide the energy value for one recorder interval by the length of the interval (in hours). If the meter records 16.4 kWh in a 30-min period, the equivalent demand for that period is 16.4 kWh/(0.5 h) = 32.8 kW.

A sample 15-min interval load shape for a 24-h period is shown in the graph in Figure 31.3. The minimum demand for that period was 10.5 kW, occurring during the interval ending at 04:30. The maximum demand was 28.7 kW, occurring during the interval ending at 15:15, or 3:15 P.M.

31.4.4 Pulse Metering

Metering pulses are signals generated in a meter for use outside the meter. Each pulse represents a discrete quantity of the metered value, such as kWh, kVAh, or kvarh. The device receiving the pulses determines

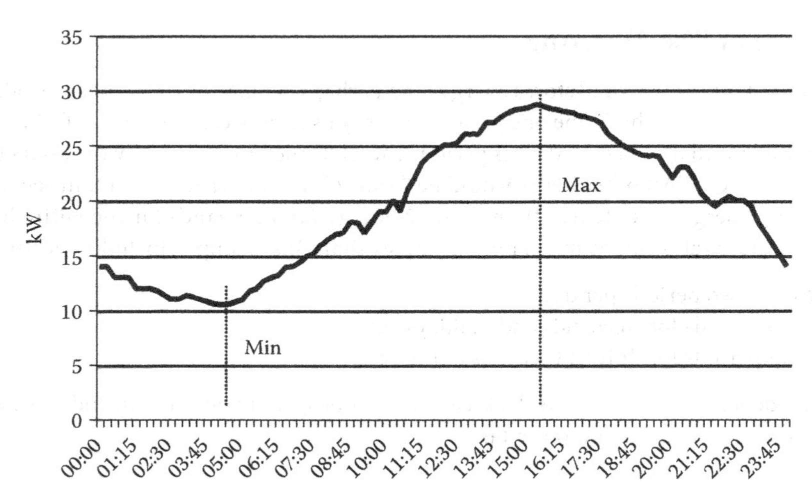

FIGURE 31.3 Graph of interval data.

the energy or demand at the meter by counting the number of pulses occurring in some time interval. A pulse is indicated by the transition (e.g., open to closed) of the circuit at the meter end. Pulses are commonly transmitted on small conductor wire circuits. Common uses of pulses include providing signals to

- Customer's demand indicator
- Customer's energy management system
- A *totalizer* (see Section 31.4.5)
- A metering data recorder
- A telemetering device that converts the pulses to other signal forms (e.g., telephone line tones or optical signals) for transmission over long distances

Pulse metering is installed when customer service requirements, equipment configurations, or other special requirements exceed the capability of conventional metering. Pulse metering is also used to transmit metered data to a remote location.

31.4.4.1 Recording Pulses

A meter pulse represents a quantity of energy, not power. For example, a pulse is properly expressed in terms of watthours (or kWh) rather than watts (or kW). A pulse recorder will associate time with pulses as it records them. If set up for a 15-min recording interval, the recorder counts pulses for 15 min, then records that number of pulses. It then counts pulses for the next 15 min, records that number, and so on, interval after interval, day after day. It is a simple matter to determine the number of pulses recorded in a chosen length of time. Since the number of pulses recorded represents a certain amount of energy, simply divide this energy by the corresponding length of time (in hours) to determine average power for that period.

Example: For a metering installation, we are given that each pulse represents 2400 Wh or 2.4 kWh. In a 15-min period, we record 210 pulses. What is the corresponding energy (kWh) and demand (kW) during this 15-min interval?

$$\text{Total energy in interval} = 2.4\,\text{kWh per pulse} \times 210\,\text{pulses} = 504\,\text{kWh}$$

$$\text{Demand} = \text{Energy/Time} = 504\,\text{kWh}/0.25\,\text{h} = 2016\,\text{kW}$$

Often, a customer asks for the demand value of a pulse, rather than the energy value. The demand value is dependent on demand interval length. The demand pulse value is equal to the energy pulse value divided by the interval length in hours.

For the previous example, the kW pulse value would be:

$$2.4\,\text{kWh per pulse}/0.25\,\text{h} = 9.6\,\text{kW per pulse}$$

and the resulting demand calculation is:

$$\text{Demand} = 9.6\,\text{kW per pulse} \times 210\,\text{pulses} = 2016\,\text{kW}$$

Remember, however, that a pulse demand value is meaningful only for a specific demand interval. In the example above, counting pulses for any period other than 15 min and then applying the kW pulse value will yield incorrect results for demand.

31.4.4.2 Pulse Circuits

Pulse circuits commonly take two forms (Figure 31.4):

- *Form A*, a two-wire circuit where a switch toggles between closed and open. Each transition of the circuit (to open or to closed) represents one pulse.
- *Form C*, a three-wire circuit where the switch flip-flops. Each transition (from closed on one side to closed on the other) represents one pulse.

Use care in interpreting pulse values for these circuits. The value will normally be expressed per *transition*. With Form C circuits, a transition is a change from closed on the first side to closed on the second side. Most receiving equipment interprets this properly. However, with Form A circuits, the transition is defined as a change from open to closed or from closed to open. An initially open Form A circuit that closes, then opens has undergone two (2) transitions. If the receiving equipment counts only circuit closures, it will record only half of the actual transitions. This is not a problem if the applicable pulse value of the Form A circuit is *doubled* from the rated pulse weight per transition. For example, if the value of a Form A meter pulse is 3.2 kWh per transition, the value needed for a piece of equipment that only counted circuit closures would be 3.2 × 2 = 6.4 kWh per pulse.

31.4.5 Totalized Metering

Totalized metering refers to the practice of combining data to make multiple service points look as if they were measured by a single meter. This is done by combining two or more sets of data from separate meters to generate data equivalent to what would be produced by a single "virtual meter" that measured the total load. This combination can be accomplished by:

- Adding recorded interval data from multiple meters, usually on a computer
- Adding (usually on-site) meter pulses from multiple meters by a special piece of metering equipment known as a totalizer
- Paralleling the secondaries of current transformers (CTs) located in multiple circuits and feeding the combined current into a conventional meter (this works only when the service voltages and ratios of the CTs are identical)
- Using a multi-circuit meter, which accepts the voltage and current inputs from multiple services

FIGURE 31.4 Pulse circuits.

TABLE 31.2 Example of Totalized Meter Data

Interval	Meter A	Meter B	Meter C	Totalized (A + B + C)
1	800	600	700	2100
2	780	650	740	2170
3	750	700	500	1950
4	780	680	720	2180

Totalized demand is the sum of the *coincident* demands and is usually less than the sum of the individual peak demands registered by the individual meters. Totalized energy equals the sum of the energies measured by the individual meters.

Table 31.2 illustrates the effects of totalizing a customer served by three delivery (and metering) points. It presents the customer's demands over a period of four demand intervals and illustrates the difference in the maximum totalized demand compared to the sum of the individual meter maximum demands.

The peak kW demand for each meter point is shown in bold. The sum of these demands is 2240 kW. However, when summed interval-by-interval, the peak of the sums is 2180 kW. This is the *totalized demand*. The difference in the two demands, 60 kW, represents a cost savings to the customer. It should be clear why many customers with multiple service points desire to have their demands totalized.

31.5 Instrument Transformers

Instrument transformers is the general name for members of the family of CTs and voltage transformers (VTs) used in metering. They are high-accuracy transformers that convert load currents or voltages to other (usually smaller) values by some fixed ratio. Voltage transformers are also often called potential transformers (PTs). The terms are used interchangeably in this section. CTs and VTs are most commonly used in services where the current and/or voltage levels are too large to be applied directly to the meter.

A current transformer is rated in terms of its nameplate primary current as a ratio to 5 amps secondary current (e.g., 400:5). The CT is not necessarily limited to this nameplate current. Its maximum capacity is found by multiplying its nameplate rating by its *rating factor*. This yields the total current the CT can carry while maintaining its rated accuracy and avoiding thermal overload. For example, a 200:5 CT with a rating factor of 3.0 can be used and will maintain its rated accuracy up to 600 amps. Rating factors for most CTs are based on open-air outdoor conditions. When a CT is installed indoors or inside a cabinet, its rating factor is reduced.

A VTs is rated in terms of its nameplate primary voltage as a ratio to either 115 or 120 V secondary voltage (e.g., 7,200:120 or 115,000:115). These ratios are sometimes listed as an equivalent ratio to 1 (e.g., 60:1 or 1000:1).

Symbols for a CT and a PT connected in a two-wire circuit are shown in Figure 31.5.

31.5.1 Measuring kVA

In many cases, a combination watthour demand meter will provide the billing determinants for small- to medium-sized customers served under rates that require only real power (kW) and energy (kWh).

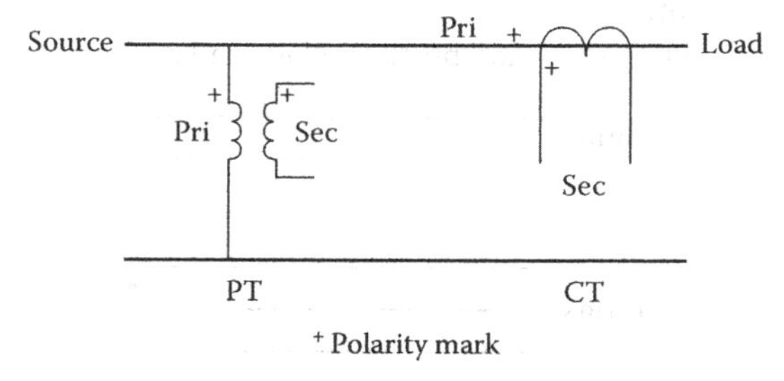

FIGURE 31.5 Instrument transformer symbols.

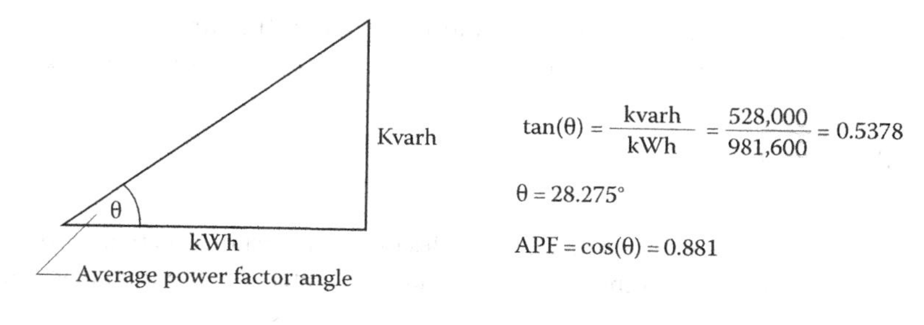

$$\text{kVA demand} = \frac{\text{kW demand}}{\text{APF}} = \frac{1412}{0.881}$$

$$= 1603\ \text{kVA}$$

FIGURE 31.6 Calculation of kVA demand using the average power factor method.

Rates for larger customers often require an *extended function* meter to provide the additional reactive or apparent power capability needed to measure or determine kVA demand. There are two common methods for determining kVA demand for billing.

1. *Actual kVA*. This method directly measures actual kVA, a simple matter for electronic meters.
2. *Average Power Factor (APF) kVA*. This method approaches the measurement of kVA in a more round-about fashion. It was developed when most metering was done with mechanical meters that could directly measure only real energy and power (kWh and kW). With a little help, they could measure kvarh. Those few meters that could measure actual kVA were very complex and demanded frequent maintenance. The APF method of calculating kVA addressed these limitations. It requires three (3) pieces of meter information:
   - Total real energy (kWh)
   - Maximum real demand (kW)
   - Total reactive energy (kvarh)

These can be measured with two standard mechanical meters. The first meter measures kWh and kW. With the help of a special transformer to shift the voltage signals 90° in phase, the second mechanical meter can be made to measure kvarh.

Average power factor kVA is determined by calculating the customer's "APF" over the billing period using the total kWh and kvarh for the period. This APF is then applied to the maximum kW reading to yield APF kVA. An example of this calculation process follows.

Customer: XYZ Corporation

Billing determinants obtained from the meter:

kWh	981,600
kvarh	528,000
kW	1,412

The calculations are shown in Figure 31.6.

31.6 Defining Terms

Class—The class designation of a watthour meter represents the maximum current at which the meter can be operated continuously with acceptable accuracy and without excessive temperature rise. Examples of common watthour meter classes are:

Self-contained—Class 200, 320, or 400
Transformer rated—Class 10 or 20

Test amperes (TA)—The test amperes rating of a watthour meter is the current that is used as a base for adjusting and determining percent registration (accuracy). Typical test current ratings and their relations to meter class are:

Class 10 and 20—TA 2.5
Class 200—TA 30

Self-contained meter—A self-contained meter is one designed and installed so that power flows from the utility system *through* the meter to the customer's load. The meter sees the total load current and full service voltage.

Transformer rated meter—A transformer rated meter is one designed to accept *reduced* levels of current and/or voltage that are directly proportional to the service current and voltage. The primary windings of CTs and/or VTs are placed in the customer's service and see the total load current and full service voltage. The transformer rated meter connects into the secondary windings of these transformers.

Meter element—A meter element is the basic energy and power measurement circuit for one set of meter input signals. It consists of a current measurement device and a voltage measurement device for one phase of the meter inputs. Usually, a meter will have one less element than the number of wires in the circuit being metered. That is, a four-wire wye or delta circuit will be metered by a three-element meter; a three-wire delta circuit will be metered by a two-element meter, although there are numerous exceptions.

CT PT ratio—A number or factor obtained by multiplying the current transformer ratio by the PT ratio. Example: If a meter is connected to 7200:120 V PTs (60:1) and 600:5 CTs (120:1), the CT PT ratio is $60 \times 120 = 7200$. A metering installation may have CTs but no PT in which case the CT PT ratio is just the CT ratio.

Meter multiplier—Also called the dial constant or kilowatthour constant, this is the multiplier used to convert meter kWh readings to actual kWh. The meter multiplier is the CT PT ratio. For a self-contained meter, this constant is 1.

Further Information

Further information and more detail on many of the topics related to metering can be found in the *Handbook for Electricity Metering*, published by Edison Electric Institute. This authoritative book provides extensive explanations of many aspects of metering, from fundamentals of how meters and instrument transformers operate, to meter testing, wiring, and installation.

32

Basic Electric Power Utilization: Loads, Load Characterization and Load Modeling

Andrew P. Hanson
The Structure Group

Utilization is the "end result" of the generation, transmission, and distribution of electric power. The energy carried by the transmission and distribution system is turned into useful work, light, heat, or a combination of these items at the utilization point. Understanding and characterizing the utilization of electric power is critical for proper planning and operation of power systems. Improper characterization of utilization can result of over or under building of power system facilities and stressing of system equipment beyond design capabilities. This chapter describes some of the basic concepts used to characterize and model loads in electric power systems.

The term *load* refers to a device or collection of devices that draw energy from the power system. Individual loads (devices) range from small light bulbs to large induction motors to arc furnaces. The term *load* is often somewhat arbitrarily applied, at times being used to describe a specific device, and other times referring to an entire facility and even being used to describe the lumped power requirements of power system components and connected utilization devices downstream of a specific point in large-scale system studies.

32.1 Basic Load Characterization

A number of terms are used to characterize the magnitude and intensity of loads. Several such terms are defined and uses outlined below.

Energy—Energy use (over a specified period of time) is a key identifying parameter for power system loads. Energy use is often recorded for various portions of the power system (e.g., homes, businesses, feeders, substations, districts). Utilities report aggregate system energy use over a variety of time frames (daily, weekly, monthly, and annually). System energy use is tied directly to sales and thus is often used as a measure of the utility or system performance from one period to another.

Demand—Loads require specific amounts of energy over short periods of time. Demand is a measure of this energy and is expressed in terms of power (kilowatts or Megawatts). Instantaneous demand is the peak instantaneous power use of a device, facility, or system. Demand, as commonly referred to in utility discussions, is an integrated demand value, most often integrated over 10, 15, or 30 min. Integrated demand values are determined by dividing the energy used by the time interval of measurement or the demand interval.

$$\text{Demand} = \frac{\text{Energy Use Over Demand Interval}}{\text{Demand Interval}} \tag{32.1}$$

Integrated demand values can be much lower than peak instantaneous demand values for a load or facility.

Demand factor—Demand factor is a ratio of the maximum demand to the total connected load of a system or the part of the system under consideration. Demand factor is often used to express the expected diversity of individual loads within a facility prior to construction. Use of demand factors allows facility power system equipment to be sized appropriately for the expected loads (*National Electric Code*, 1996).

$$\text{Demand Factor} = \frac{\text{Maximum Demand}}{\text{Total Connected Load}} \tag{32.2}$$

Load factor—Load factor is similar to demand factor and is calculated from the energy use, the demand, and the period of time associated with the measurement.

$$\text{Load Factor} = \frac{\text{Energy Use}}{\text{Demand} \times \text{Time}} \tag{32.3}$$

A high load factor is typically desirable, indicating that a load or group of loads operates near its peak most of the time, allowing the greatest benefit to be derived from any facilities installed to serve the load.

32.2 Composite Loads and Composite Load Characterization

It is impractical to model each individual load connected to a power system to the level of detail at which power is delivered to each individual utilization device. Loads are normally lumped together to represent all of the "downstream" power system components and individual connected loads. This grouping occurs as a result of metering all downstream power use from a certain point in the power system, or as a result of model simplification in which effects of the downstream power system and connected loads are represented by a single load in system analysis.

32.2.1 Coincidence and Diversity

Although individual loads vary unpredictably from hour to hour and minute to minute, an averaging effect occurs as many loads are examined in aggregate. This effect begins at individual facilities (home, commercial establishment, or industrial establishment) where all devices are seldom if ever in operation at the same instant. Progressing from an individual facility to the distribution and transmission systems, the effect is compounded, resulting in somewhat predictable load characteristics.

Diversity is a measure of the dispersion of the individual loads of a system under observation over time. Diversity is generally low in individual commercial and industrial installations. However, at a feeder level, diversity is a significant factor, allowing more economical choices for equipment since the feeder needs to supply power to the aggregate peak load of the connected customers, not the sum of the customer individual (noncoincident) peak loads.

Groups of customers of the same class (i.e., residential, commercial, industrial) tend to have an aggregate peak load per customer that decreases as the number of customers increases. This tendency is termed *coincidence* and has significant impact on the planning and construction of power systems (Willis, 1997). For example, load diversity would allow a feeder or substation to serve a number of customers whose individual (noncoincident) peak demands may exceed the feeder or substation rating by a factor of two or more.

$$\text{Coincidence Factor} = \frac{\text{Aggregate Demand for a Group of Customers}}{\text{Sum of Individual Customer Demands}} \tag{32.4}$$

Note that there is a minor but significant difference between coincidence (and its representation as a coincidence factor) and the demand factor discussed above. The coincidence factor is based on the *observed* peak demand for individuals and groups, whereas the demand factor is based on the *connected* load.

32.2.2 Load Curves and Load Duration

Load curves and load duration curves graphically convey very detailed information about the characteristics of loads over time. Load curves typically display the load of a customer class, feeder, or other portion of a power system over a 24 h period. Load duration curves display the cumulative amount of time that load levels are experienced over a period of time.

Load curves represent the demand of a load or groups of load over a period of time, typically 24 h. The curves provide "typical" load levels for a customer class on an hour-by-hour or minute-by-minute basis. The curves themselves represent the demand of a certain class of customers or portion of the system. The area under the curve represents the corresponding energy use over the time period under consideration. Load curves provide easily interpreted information regarding the peak load duration as well as the variation between minimum and maximum load levels. Load curves provide key information for daily load forecasts allowing planners and operators to ensure system capacity is available to meet customer needs. Three sample load curves (for residential, commercial, and industrial customer classes) are shown in Figures 32.1 through 32.3.

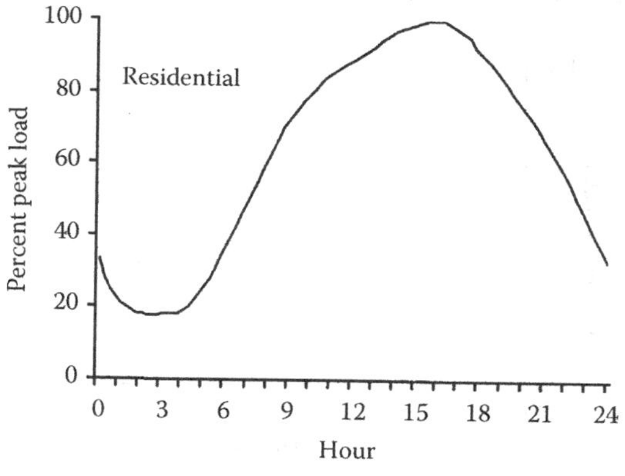

FIGURE 32.1 Residential load curve.

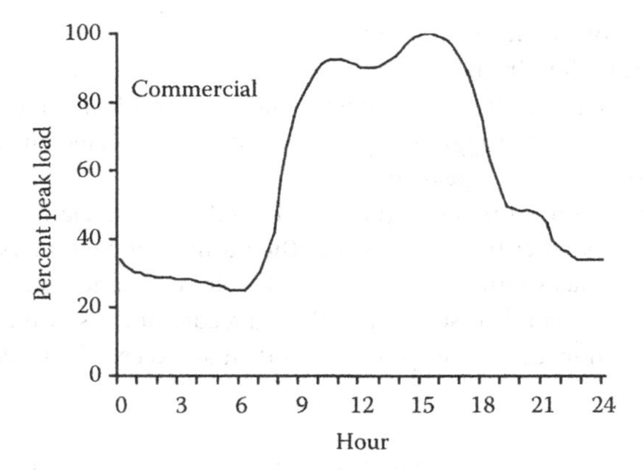

FIGURE 32.2 Commercial load curve.

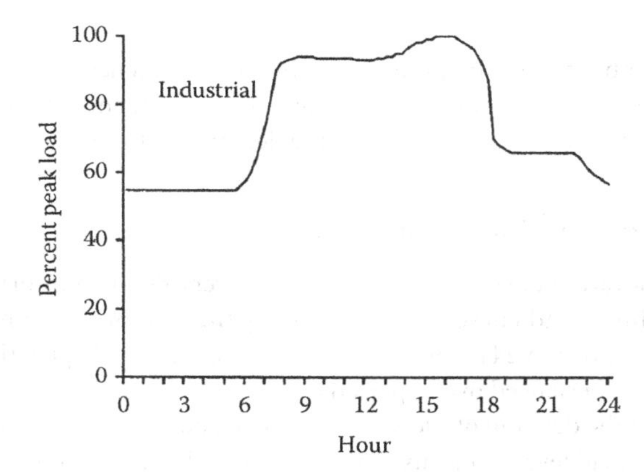

FIGURE 32.3 Industrial load curve.

Load curves can also be developed on a feeder or substation basis, as a composite representation of the load profile of a portion of the system.

Load duration curves quickly convey the duration of the peak period for a portion of a power system over a given period of time. Load duration curves plot the cumulative amount of time that load levels are seen over a specified time period. The information conveyed graphically in a load duration curve, although more detailed, is analogous to the information provided by the load factor discussed above. A sample load duration curve is shown in Figure 32.4.

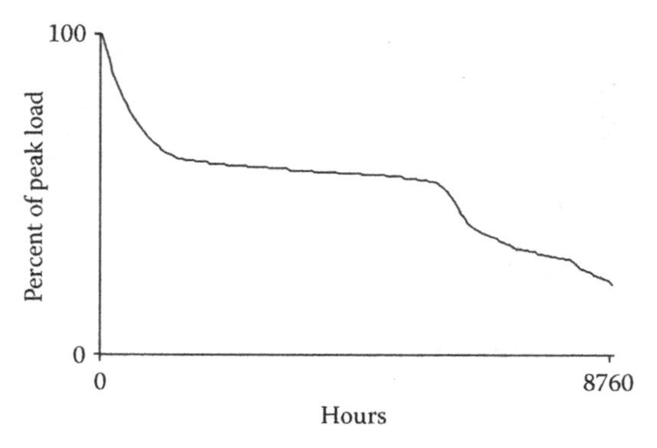

FIGURE 32.4 Annual load duration curve.

Load duration curves are often characterized by very sharp ascents to the peak load value. The shape of the remainder of the curves vary based on utilization patterns, size, and content of the system for which the load duration curve is plotted.

32.3 Composite Load Modeling

Load models can generally be divided into a variety of categories for modeling purposes. The appropriate load model depends largely on the application. For example, for switching transient analyses, simple load models as combinations of time-invariant circuit elements (resistors, inductors, capacitors) and/or voltage sources are usually sufficient. Power flow analyses are performed for a specific operating point at a specific frequency, allowing loads to be modeled primarily as constant impedance or constant power. However, midterm and extended term transient stability analyses require that load voltage and frequency dependencies be modeled, requiring more complex aggregate load models. Two load models are discussed below.

Composite loads exhibit dependencies on frequency and voltage. Both linear (Elgerd, 1982; Gross, 1986) and exponential models (Arrillaga and Arnold, 1990) are used for addressing these dependencies. *Linear voltage and frequency dependence model*—The linear model provides excellent representation of load variations as frequency and voltages vary by small amounts about a nominal point.

$$P = P_{nominal} + \frac{\partial P}{\partial |\bar{V}|}\Delta|\bar{V}| + \frac{\partial P}{\partial f}\Delta f \tag{32.5}$$

$$Q = Q_{nominal} + \frac{\partial Q}{\partial |\bar{V}|}\Delta|\bar{V}| + \frac{\partial Q}{\partial f}\Delta f \tag{32.6}$$

where

$P_{nominal}$, $Q_{nominal}$ are the real and reactive power under nominal conditions
$\partial P/\partial|\bar{V}|, \partial P/\partial f, \partial Q/\partial|\bar{V}|, \partial Q/\partial f$ are the rates of change of real and reactive power with respect to voltage magnitude and frequency
$\Delta|\bar{V}|$, Δf are the deviations in voltage magnitude and frequency from nominal values

The values for the partial derivatives with respect to voltage and frequency can be determined through analysis of metered load data recorded during system disturbances or in the case of very simple loads, through calculations based on the equivalent circuit models of individual components.

Exponential voltage and frequency dependence model—The exponential model provides load characteristics useful in midterm and extended term stability simulations in which the changes in system frequency and voltage are explicitly modeled in each time step.

$$P = P_{nominal}|\bar{V}|^{pv} f^{pf} \tag{32.7}$$

$$Q = Q_{nominal}|\bar{V}|^{qv} f^{qf} \tag{32.8}$$

where

$P_{nominal}$, $Q_{nominal}$ are the real and reactive power of the load under nominal conditions
$|\bar{V}|$ is the voltage magnitude in per unit
f is the frequency in per unit
pv, pf, qv, and qf are the exponential modeling parameters for the voltage and frequency dependence of the real and reactive power portions of the load, respectively

32.4 Other Load-Related Issues

32.4.1 Cold Load Pickup

Following periods of extended service interruption, the advantages provided by load diversity are often lost. The term *cold load pickup* refers to the energization of the loads associated with a circuit or substation following an extended interruption during which much of the diversity normally encountered in power systems is lost.

For example, if a feeder suffers an outage, interrupting all customers on the feeder during a particularly cold day, the homes and businesses will cool to levels below the individual thermostat settings. This situation eliminates the diversity normally experienced, where only a fraction of the heating will be required to operate at any given time. Once power is restored, the heating at all customer locations served by the feeder will attempt to operate to bring the building temperatures back to levels near the thermostat settings. The load experienced by the feeder following reenergization can be far in excess of the design loading due to lack of load diversity.

Cold load pickup can result in a number of adverse power system reactions. Individual service transformers can become overloaded under cold load pickup conditions, resulting in loss of life and possible failure due to overheating. Feeder load levels can exceed protective device ratings/settings, resulting in customer interruptions following initial service restoration. Additionally, the heavily loaded system conditions can result in conductors sagging below their designed minimum clearance levels, creating safety concerns.

32.4.2 Harmonics and Other Nonsinusoidal Loads

Electronic loads that draw current from the power system in a nonsinusoidal manner represent a significant portion of the load connected to modern power systems. These loads cause distortions of the generally sinusoidal characteristics traditionally observed. Harmonic loads include power electronic based devices (rectifiers, motor drives, switched mode power supplies, etc.) and arc furnaces. More details on power electronics and their effects on power system operation can be found in the power electronics section of this handbook.

References

Arrillaga, J. and Arnold, C.P., *Computer Analysis of Power Systems*, John Wiley & Sons, West Sussex, U.K., 1990.

Elgerd, O.I., *Electric Energy Systems Theory: An Introduction*, 2nd ed., McGraw Hill Publishing Company, New York, 1982.

Gross, C.A., *Power System Analysis*, 2nd ed., John Wiley & Sons, New York, 1986.

National Electric Code, NFPA 70, Article 100, Quincy, MA, 1996.

Willis, H.L., *Power Distribution Planning Reference Book*, Marcel-Dekker, Inc., New York, 1997.

Further Information

The references provide a brief treatment of loads and their characteristics. More detailed load characteristics for specific industries can be found in specific industry trade publications. For example, specific characteristics of loads encountered in the steel industry can be found in Fruehan, R.J., Ed., *The Making, Shaping and Treating of Steel*, 11th ed., AISE Steel Foundation, Pittsburgh, PA, 1998.

The quarterly journals *IEEE Transactions on Power Systems* and *IEEE Transactions on Power Delivery* contain numerous papers on load modeling, as well as short and long term load forecasting. Papers in these journals also track recent developments in these areas.

Information on load modeling for long term load forecasting for power system planning can be found the following references respectively:

Willis, H.L., *Spatial Electric Load Forecasting*, Marcel-Dekker, Inc., New York, 1996.

Stoll, H.G., *Least Cost Electric Utility Planning*, John Wiley & Sons, New York, 1989.

33

Electric Power Utilization: Motors

Charles A. Gross
Auburn University

A major application of electric energy is in its conversion to mechanical energy. Electromagnetic, or "EM" devices designed for this purpose are commonly called "motors." Actually the machine is the central component of an integrated system consisting of the source, controller, motor, and load. For specialized applications, the system may be, and frequently is, designed as a integrated whole. Many household appliances (e.g., a vacuum cleaner) have in one unit, the controller, the motor, and the load. However, there remain a large number of important stand-alone applications that require the selection of a proper motor and associated control, for a particular load. It is this general issue that is the subject of this chapter.

The reader is cautioned that there is no "magic bullet" to deal with all motor-load applications. Like many engineering problems, there is an artistic, as well as a scientific dimension to its solution. Likewise, each individual application has its own peculiar characteristics, and requires significant experience to manage. Nevertheless, a systematic formulation of the issues can be useful to a beginner in this area of design, and even for experienced engineers faced with a new or unusual application.

33.1 Some General Perspectives

Consider the general situation in Figure 33.1a. The flow of energy through the system is from left to right, or from electrical source to mechanical load. Also, note the positive definitions of currents, voltages, speed, and torques. These definitions are collectively called the "motor convention," and are logically used when motor applications are under study. Likewise, when generator applications are considered, the sign conventions of Figure 33.1b (called generation convention) will be adopted. This means that variables will be positive under "normal" conditions (motors operating in the motor mode, generators

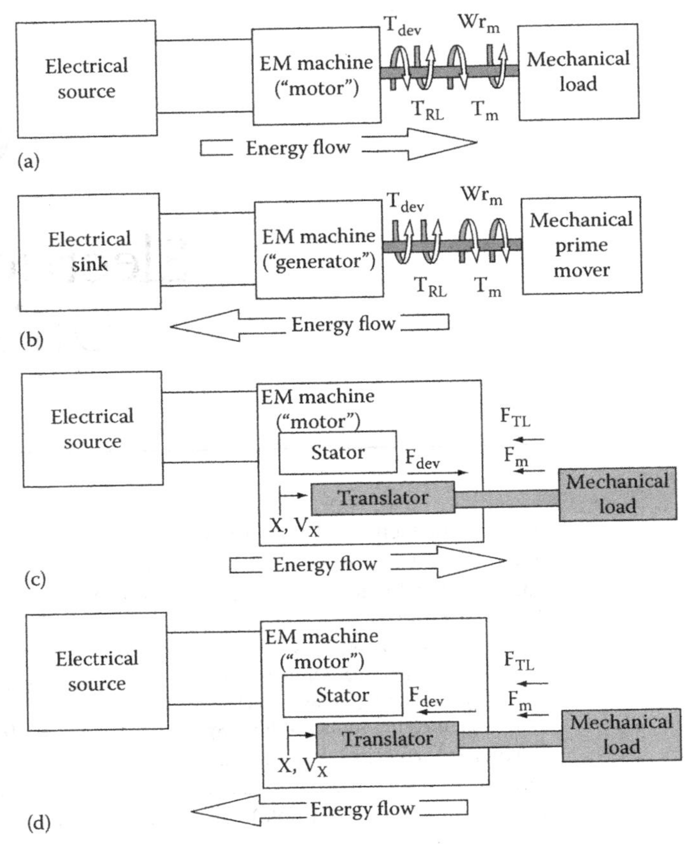

FIGURE 33.1 Motor and generator sign conventions for EM machines. (a) The EM rotational machine; motor convention. (b) The EM rotational machine; generator convention. (c) The EM translational machine; motor convention. (d) The EM translational machine; generator convention.

in the generator mode), and negative under some abnormal conditions (motors running "backwards," for example). Using motor convention:

$$T_{dev} - (T_m + T_{RL}) = T_{dev} - T'_m = J\left(\frac{d\omega_{rm}}{dt}\right) \tag{33.1}$$

where

T_{dev} is the EM torque, produced by the motor, Nm
T_m is the torque absorbed by the mechanical load, including the load losses and that used for useful mechanical work, Nm
T_{RL} is the rotational loss torque, internal to the motor, Nm
$T'_m = T_m + T_{RL}$ is equivalent load torque, Nm
J is the mass polar moment of inertia of all rotating parts, kg-m^2
ω_{rm} is the angular velocity of rotating parts, rad/s

Observe that whenever $T_{dev} > T'_m$, the system accelerates; if $T_{dev} < T'_m$, the system decelerates. The system will inherently seek out the equilibrium condition of $T_{dev} = T'_m$, which will determine the running speed. In general, the steady state running speed for any motor-load system occurs at the intersection of the motor and load torque-speed characteristics, i.e., where $T_{dev} = T'_m$. If $T_{dev} > T'_m$, the system

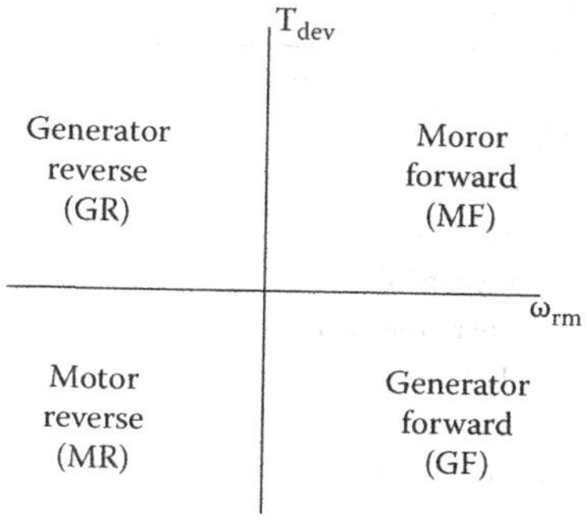

T_{dev}

Generator reverse (GR)

Moror forward (MF)

ω_{rm}

Motor reverse (MR)

Generator forward (GF)

FIGURE 33.2 Operating modes.

is accelerating; for $T_{dev} < T'_m$, the system decelerates. Thus, torque-speed characteristics for motors and loads are necessary for the design of a speed (or position) control system.

The corresponding system powers are:

$P_{dev} = T_{dev}\ \omega_{rm}$ is the EM power, converted by the motor into mechanical form, W
$P_m = T_m\ \omega_{rm}$ is the power absorbed by the mechanical load, including the load losses and that used for useful mechanical work, W
$P_{RL} = T_{RL}\ \omega_{rm}$ is the rotational power loss, internal to the motor, W

33.2 Operating Modes

Equation 33.1 implies that torque and speed are positive. Consider positive speed as "forward," meaning rotation in the "normal" direction, which should be obvious in a specific application. "Reverse" is defined to mean rotation in the direction opposite to "forward," and corresponds to $\omega_{rm} < 0$. Positive EM torque is in the positive speed direction. Using motor convention, first quadrant operation means that (1) speed is positive ("forward") and (2) T_{dev} is positive (also forward), and transferring energy from motor to load ("motoring"). There are four possible operating modes specific to the four quadrants of Figure 33.2. In any application, a primary consideration is to determine which of these operating modes will be required.

33.3 Motor, Enclosure, and Controller Types

The general types of enclosures, motors, and controllers are summarized in Tables 33.1 through 33.3.

33.4 System Design

The design of a proper motor-enclosure-controller system for a particular application is a significant engineering problem requiring engineering expertise and experience. The following issues must be faced and resolved.

33.4.1 Load Requirements

1. The steady-state duty cycle with torque-speed (position) requirements at each load step.
2. What operating modes are required.
3. Dynamic performance requirements, including starting and stopping, and maximum and minimum accelerations.
4. The relevant torque-speed (position) characteristics.
5. All load inertias (J).

TABLE 33.1 General Enclosure Types

Types
Open
Drip-proof
Splash-proof
Semi-guarded
Weather protected
Type I
Type II
Totally enclosed
Nonventilated
Fan-cooled
Explosion-proof
Dust-ignition-proof
Water-proof
Pipe-ventilated
Water-cooled
Water-air-cooled
Air-to-air-cooled
Air-over-cooled

See NEMA Standard MG 1.1.25–1.1.27 for definitions.

6. Coupling options (direct drive, belt-drive, gearing).
7. Reliability of service. How critical is a system failure?
8. Future modifications.

33.4.2 Environmental Requirements

1. Ambient atmospheric conditions (pressure, temperature, humidity, content)
2. Indoor, outdoor application
3. Wet, dry location
4. Ventilation
5. Acceptable acoustical noise levels
6. Electrical/mechanical hazards to personnel
7. Accessibility for inspection and maintenance

33.4.3 Electrical Source Options

1. DC–AC
2. If AC, single- and/or three-phase
3. Voltage level
4. Frequency
5. Capacity (kVA)
6. Protection options
7. Power quality specifications

33.4.4 Preliminary System Design

Based on the information compiled in the steps above, select an appropriate enclosure, motor type, and controller. In general, the enclosure entries, reading from top to bottom in Table 33.1, are from

TABLE 33.2 General Motor Types

Type
DC motors (commutator devices)
Permanent magnet field
Wound field
Series
Shunt
Compound
AC motors
Single-phase
Cage rotor
Split phase
Resistance-start
Capacitor start
Single capacitor (start-run)
Capacitor start/capacitor run
Shaded pole
Wound rotor
Repulsion
Repulsion start/induction run
Universal
Synchronous
Hysteresis
Three-phase
Synchronous
Permanent magnet field
Wound field
Induction
Cage rotor
NEMA Design A,B,C,D,F
Wound rotor

See NEMA Standard MG 1.1.1–1.1.21 for definitions.

simplest (and cheapest) to most complex (and expensive). Select the simplest enclosure that meets all the environmental constraints. Next, select a motor and controller combination from Tables 33.2 and 33.3. This requires personal experience and/or consulting with engineers with experience relevant to the application.

In general, DC motors are expensive and require more maintenance, but have excellent speed and position control options. Single-phase AC motors are limited to about 5 kW, but may be desirable in locations where three-phase service is not available and control specifications are not critical.

Three-phase AC synchronous motors are not amenable to frequent starting and stopping, but are ideal for medium and high power applications which run at essentially fixed speeds. Three-phase AC cage rotor induction motors are versatile and economical, and will be the preferred choice for most applications, particularly in the medium power range. Three-phase AC wound rotor induction motors are expensive, and only appropriate for some unusual applications.

The controller must be compatible with the motor selected; the best choice is the most economical that meets all load specifications. If the engineer's experience with the application under study is lacking, two or more systems should be selected.

TABLE 33.3 General Motor Controllers

Type
DC motor controllers
Electromechanical
Armature starting resistance; rheostat field control
Power electronic drive
Phase converters: 1, 2, 4 quadrant drives
Chopper control: 1, 2, 4 quadrant drives
AC motor controllers
Single-phase
Electromechanical
Across-the-line: protection only
Step-reduced voltage
Power electronic drive
Armature control: 1, 2, 4 quadrant drives
Three-phase induction
Cage rotor
Electromechanical
Across-the-line: protection only
Step-reduced voltage
Power electronic drive (ASDs)
Variable voltage source inverter
Variable current source inverter
Chopper voltage source inverter
PWM voltage source inverter
Vector control
Wound rotor
Variable rotor resistance
Power electronic rotor power recovery
Three-phase synchronous
Same as cage rotor induction
Brushless DC control

33.4.5 System Ratings

Based on the steps above, select appropriate power, voltage, and frequency ratings. For cyclic loads, the power rating may tentatively be selected based on the "rms horsepower" method (calculating the rms power requirements over the load cycle).

33.4.6 System Data Acquisition

Request data from at least two vendors on all systems selected in the steps above, including:

- Circuit diagrams
- Performance test data
- Equivalent circuit values, including inertia constants
- Cost data
- Warranties and guarantees

33.4.7 Engineering Studies

Perform the following studies using data from the system data acquisition step above.

1. Steady state performance. Verify that each candidate system meets all steady state load requirements.
2. Dynamic performance. Verify that each system meets all dynamic load requirements.
3. Load cycle efficiency. Determine the energy efficiency over the load cycle.
4. Provide a cost estimate for each system, including capital investment, maintenance, and annual operating costs.
5. Perform a power quality assessment.

Based on these studies, select a final system design.

33.4.8 Final System Design

Request a competitive bid on the final design from appropriate vendors. Select a vendor based on cost, expectation of continuing technical support, reputation, warranties, and past customer experience.

33.4.9 Field Testing

Whenever practical, customer and vendor engineers should design and perform field tests on the installed system, demonstrating that it meets or exceeds all specifications. If multiple units are involved, one proto-unit should be installed, tested, and commissioned before delivery is made on the balance of the order.

Further Information

The design of a properly engineered motor-controller system for a particular application requires access to several technical resources, including standards, the technical literature, manufacturers' publications, textbooks, and handbooks. The following section provides a list of references and resource material that the author recommends for work in this area. In many cases, more recent versions of publications listed are available and should be used.

Organizations

American National Standards Institute (ANSI), 1430 Broadway, New York, 10018.

Institute of Electrical and Electronics Engineers (IEEE), 445 Hoes Lane, Piscataway, NJ 08855.

International Organization for Standardization (ISO) 1, rue de Varembe, 1211 Geneva 20, Switzerland.

American Society for Testing and Materials (ASTM), 1916 Race Street, Philadelphia, PA 19103.

National Electrical Manufacturers Association (NEMA), 2101 L Street, NW, Washington, DC 20037.

National Fire Protection Association (NFPA), Batterymarch Park, Quincy, MA 02269.

The Rubber Manufacturers Association, Inc., 1400 K Street, NW, Suite 300, Washington, DC 20005.

Mechanical Power Transmission Association, 1717 Howard Street, Evanston, IL 60201.

Standards ANSI/IEEE Std 304-1982, Test Procedure for Evaluation and Classification of Insulation Systems for DC Machines.

ANSI/IEEE Std 115-1983, Test Procedures for Synchronous Machines.

ANSI/IEEE Std 100-1984, IEEE Standard Dictionary of Electrical and Electronics Terms.

ANSI/IEEE Std 114-1984, Test Procedure for Single-Phase Induction Motors.

ANSI/IEEE Std 117-1985, Standard Test Procedure for Evaluation of Systems of Insulating Materials for Random-Wound AC Electric Machinery.

ANSI/NFPA 70-1998, National Electrical Code.

IEEE Std 1-1969, General Principles for Temperature Limits in the Rating of Electric Equipment.

IEEE Std 85-1980, Test Procedure for Airborne Sound Measurements on Rotating Electric Machinery.

IEEE Std 112-1984, Standard Test Procedure for Potyphase Induction Motors and Generators.

IEEE Std 113-1985, Guide on Test Procedures for DC Machines.

ISO R-1000, SI Units and Recommendations for the Use of their Multiples and of Certain Other Units.

NEMA MG 2-1983, Safety Standard for Construction and Guide for Selection, Installation and Use of Electric Motors and Generators.

NEMA MG 13-1984, Frame Assignments for Alternating-Current Integral-horsepower Induction Motors.

NEMA MG 3-1984, Sound Level Prediction for Installed Rotating Electrical Machines.

NEMA MG 1-1987, Motors and Generators.

Books (An Abridged Sample)

Acarnley, P.P., *Stepping Motors*, 2nd ed., Peter Peregrinus, Ltd., London, U.K., 1984.

Anderson, L.R., *Electric Machines and Transformers*, Reston Publishing, Reston, VA, 1981.

Bergseth, F.R. and Venkata, S.S., *Introduction to Electric Energy Devices*, Prentice-Hall, Englewood Cliffs, NJ, 1987.

Bose, B.K., *Power Electronics and AC Drives*, Prentice-Hall, Englewood Cliffs, NJ, 1985.

Brown, D. and Hamilton 111, E.P., *Electromechanical Energy Conversion*, Macmillan, New York, 1984.

Chapman, S.J., *Electric Machinery Fundamentals*, McGraw-Hill, New York, 1985.

DC Motors, Speed Controls, Servo Systems—An Engineering Handbook, 5th ed., Electro-Craft Corporation, Hopkins, MN, 1980.

Del Toro, V., *Electric Machinery and Power Systems*, Prentice-Hall, Englewood Cliffs, NJ, 1986.

Electro-Craft Corporation, *DC Motors, Speed Controls, Servo Systems*, 3rd ed., Pergamon Press, Ltd., Oxford, U.K., 1977.

Fitzgerald, A.E., Kingsley, Jr., C., and Umans, S.D., *Electric Machinery*, 5th ed., McGraw-Hill, New York, 1990.

Gonen, T., *Engineering Economy for Engineering Managers*, Wiley, New York, 1990.

Kenjo, T. and Nagamori, S., *Permanent-Magnet and Brush-Less DC Motors*, Clarendon Press, Oxford, U.K., 1985.

Krause, P.C. and Wasynezk, O., *Electromechanical Machines and Devices*, McGraw-Hill, New York, 1989.

Krein, P., *Elements of Power Electronics*, Oxford Press, New York, 1998.

Moha, N., Undeland, T.M., and Robbins, W.P., *Power Electronics: Converters, Application, and Design*, 2nd ed., John Wiley & Sons, New York, 1995.

Nasar, S.A., ed., *Handbook of Electric Machines*, McGraw-Hill, New York, 1987.

Nasar, S.A. and Boldea, I., *Linear Motion Electric Machines*, John Wiley & Sons, New York, 1976.

Patrick, D.R. and Fardo, S.W., *Rotating Electrical Machines and Power Systems*, Prentice-Hall, Englewood Cliffs, NJ, 1985.

Ramshaw, R. and Van Heeswijk, R.G., *Energy Conversion: Electric Motors and Generators*, Saunders College Publishing, Orlando, FL, 1990.

Rashid, M.H., *Power Electronics: Circuits, Devices, and Applications*, 2nd ed., Prentice-Hall, Englewood Cliffs, NJ, 1993.

Sarma, M.S., *Electric Machines: Steady-State Theory and Dynamic Performance*, Brown Publishers, Dubuque, IA, 1985.

Smeatson, R.W., ed., *Motor Application and Maintenance Handbook*, McGraw-Hill, New York, 1969.

Stein, R. and Hunt, W.T., *Electric Power System Components: Transformers and Rotating Machines*, Van Nostrand Reinhold Co., New York, 1979.

Veinott, C.G. and Martin, J.E., *Fractional- and Subfractional-Horsepower Electric Motors*, 4th ed., McGraw-Hill, New York, 1986.

Wenick, E.H., ed., *Electric Motor Handbook*, McGraw-Hill, London, U.K., 1978.

34

Linear Electric Motors

Jacek F. Gieras
University of Technology and Life Sciences

Linear electric motors belong to the group of special electrical machines that convert electrical energy directly into mechanical energy of translatory motion. Linear electric motors can drive a linear-motion load without intermediate gears, screws, or crank shafts. Linear electric motors can be classified as follows:

- DC motors
- Induction motors
- Synchronous motors, including reluctance and stepping motors
- Oscillating motors
- Hybrid motors

The application of DC linear motor is marginal. The most popular are permanent magnet (PM) linear synchronous motors (LSMs) and linear induction motors (LIMs), which are manufactured commercially in several countries and are finding many applications.

A linear motor can be obtained by cutting a rotary motor along its radius from the center axis of the shaft to the external surface of the stator core and rolling it out flat (Figure 34.1).

34.1 Linear Synchronous Motors

34.1.1 Basic Geometries and Constructions

An LSM is a linear motor in which the mechanical motion is in synchronism with the magnetic field, i.e., the mechanical speed is the same as the speed of the traveling magnetic field. The thrust (propulsion force) can be generated as an action of

- Traveling magnetic field produced by a polyphase winding and an array of magnetic poles N, S, …, N, S or a variable reluctance ferromagnetic rail (LSMs with AC armature windings)
- Magnetic field produced by electronically switched DC windings and an array of magnetic poles N, S, …, N, S or variable reluctance ferromagnetic rail (linear stepping or switched reluctance motors)

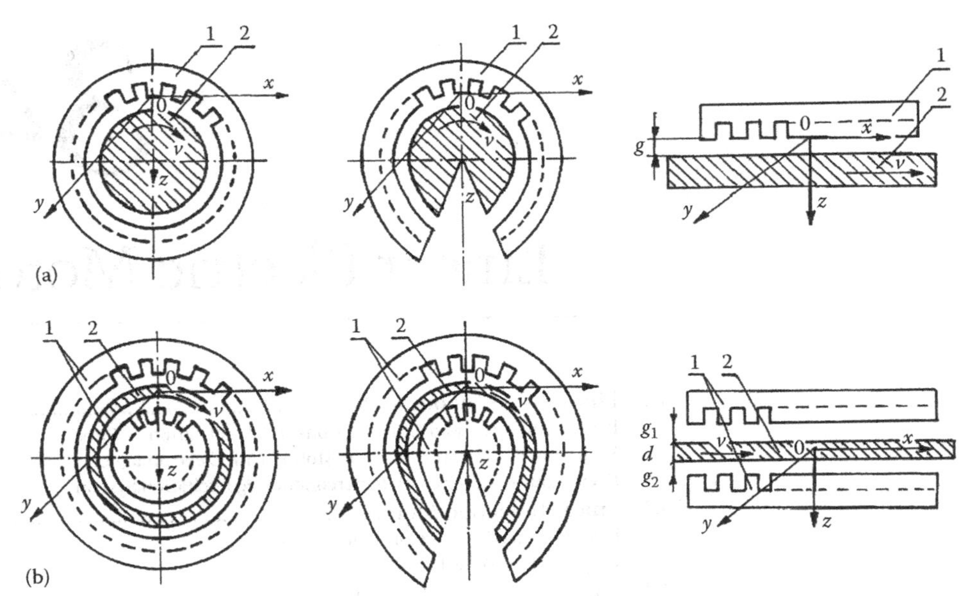

FIGURE 34.1 Evolution of a rotary induction motor: (a) solid-rotor induction motor into a flat, single-sided LIM and (b) hollow-rotor induction motor into a flat double-sided LIM. 1, primary; 2, secondary.

The part producing the traveling magnetic field is called the *armature* or *forcer*. The part that provides the DC magnetic flux or variable reluctance is called the *field excitation system* (if the excitation system exists) or *salient-pole rail, reaction rail*, or *variable reluctance platen*. The terms *primary* and *secondary* should rather be avoided, as they are only justified for LIMs [6] or transformers. The operation of an LSM does not depend on which part is movable and which one is stationary.

Traditionally, AC polyphase synchronous motors are motors with DC electromagnetic excitation, the propulsion force of which has two components: (1) due to the traveling magnetic field and DC current magnetic flux (synchronous component) and (2) due to the traveling magnetic field and variable reluctance in d- and q-axes (reluctance component). Replacement of DC electromagnets with PMs is common, except for LSMs for magnetically levitated vehicles. PM brushless LSMs can be divided into two groups:

- PM LSMs in which the input current waveforms are sinusoidal and produce a traveling magnetic field
- PM DC linear brushless motors (LBMs) with position feedback, in which the input rectangular or trapezoidal current waveforms are precisely synchronized with the speed and position of the moving part

Construction of magnetic and electric circuits of LSMs belonging to both groups is the same. LSMs can be designed as flat motors (Figure 34.2) or tubular motors (Figure 34.3). In DC brushless motors, the information about the position of the moving part is usually provided by an absolute position sensor. This control scheme corresponds to an *electronic commutation*, functionally equivalent to the mechanical commutation in DC commutator motors. Therefore, motors with square (trapezoidal) current waveforms are called *DC brushless motors*.

Instead of DC or PM excitation, the difference between the d- and q-axes reluctances and the traveling magnetic field can generate the reluctance component of the thrust. Such a motor is called the AC *variable reluctance* LSM. Different reluctances in the d- and q-axes can be created by making salient ferromagnetic poles using ferromagnetic and nonferromagnetic materials or using anisotropic ferromagnetic materials. The operation of LBMs can be regarded as a special case of the operation of LSMs.

FIGURE 34.2 Flat three-phase PM linear motors. (Photo courtesy of Kollmorgen, Radford, VA.)

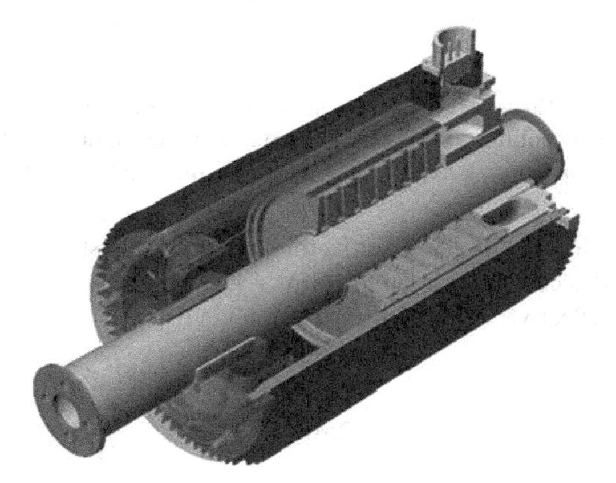

FIGURE 34.3 Tubular PM LSM. Moving rod (reaction rail) contains circular PMs (Photo courtesy of California Linear Drives, Carlsbad, CA.)

In the case of LSMs operating on the principle of the traveling magnetic field, the speed v of the moving part is equal to synchronous speed v_s, i.e.,

$$v = v_s = 2f\tau = \frac{\omega}{\pi}\tau \tag{34.1}$$

The *synchronous speed* v_s of the traveling magnetic field depends only on the input frequency f (angular input frequency $\omega = 2\pi f$) and pole pitch τ. It does not depend on the number of poles $2p$.

As for any other linear-motion electrical machine, the useful force (thrust) F_x is directly proportional to the output power P_{out} and inversely proportional to the speed $v = v_s$, i.e.,

$$F_x = \frac{P_{out}}{v_s} \tag{34.2}$$

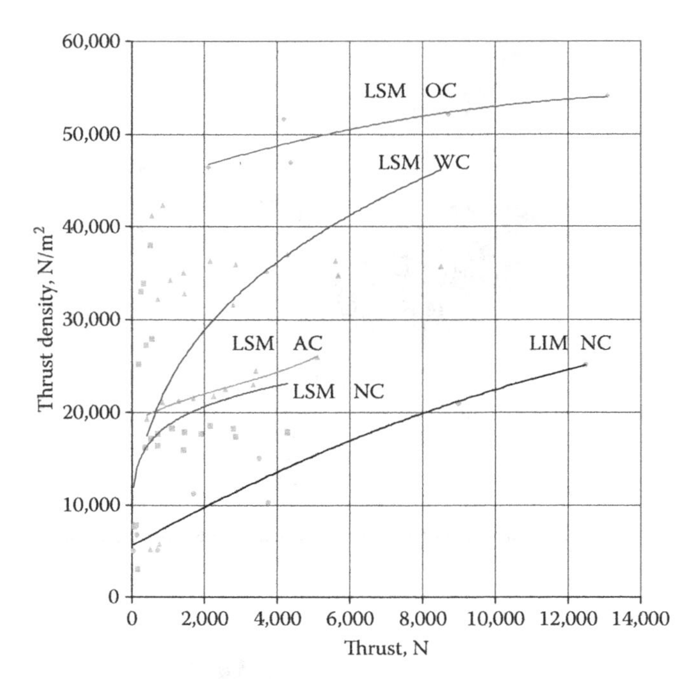

FIGURE 34.4 Comparison of thrust density for single-sided LIMs and LSMs. AC, air cooling; NC, natural cooling; OC, oil cooling; WC, water cooling. (From Gieras, J.F., Status of linear motors in the United States, *International Symposium on Linear Drives for Industry Applications (LDIA'03)*, Birmingham, U.K., pp. 169–176, 2003.)

Direct electromechanical drives with LSMs for factory automation systems can achieve speeds exceeding 600 m/min = 36 km/h and acceleration of up to 360 m/s² [7]. The thrust density, i.e., thrust per active surface $2p\tau L_i$, where L_i is the effective width of the stack.

$$f_x = \frac{F_x}{2p\tau L_i} \quad (\mathrm{N/m^2}) \tag{34.3}$$

of LSMs is higher than that of LIMs (Figure 34.4).

The polyphase (usually three-phase) armature winding can be distributed in slots, made in the form of concentrated-parameter coils or made as a coreless (air-cored) winding layer. PMs are the most popular field excitation systems for short traveling distances (less than 10 m), for example, factory transportation or automation systems. A long PM rail would be expensive. Electromagnetic excitation is used in high-speed passenger transportation systems operating on the principle of magnetic levitation (maglev). The German system, *Transrapid*, uses vehicle-mounted steel core excitation electromagnets and stationary slotted armatures. Japanese MLX001 test train sets use onboard superconducting (SC) air-cored electromagnets and a stationary three-phase air-cored armature winding distributed along the guideway (Yamanashi Maglev Test Line).

34.1.2 Classification

LSMs can be classified according to whether they are

- Flat (planar) or tubular (cylindrical)
- Single sided or double sided
- Slotted or slotless

- Iron cored or air cored
- Transverse flux or longitudinal flux

The topologies mentioned earlier are possible for nearly all types of excitation systems. LSMs operating on the principle of the traveling magnetic field can have the following excitation systems:

- PMs in the reaction rail
- PMs in the armature (passive reaction rail)
- Electromagnetic excitation system (with winding)
- SC excitation system
- Passive reaction rail with saliency and neither PMs nor windings (variable reluctance motors)

LSMs with electronically switched DC armature windings are designed either as linear stepping motors or linear switched reluctance motors.

34.1.2.1 PM Motors with Active Reaction Rail

Figure 34.5a shows a single-sided flat LSM with the armature winding located in slots and surface PMs. Figure 34.5b shows a similar motor with buried-type PMs. In surface arrangement of PMs, the yoke (back iron) of the reaction rail is ferromagnetic, and PMs are magnetized in the normal direction (perpendicular to the active surface). Buried PMs are magnetized in the direction of the traveling magnetic field, and the yoke is nonferromagnetic, for example, made of aluminum (Al). Otherwise, the bottom leakage flux would be greater than the linkage flux, as shown in Figure 34.6. The same effect occurs in buried-type PM rotors of rotary machines in which the shaft must also be nonferromagnetic [2,3,9].

The so-called *Halbach array* of PMs also does not require any ferromagnetic yoke and excites stronger magnetic flux density and closer to the sinusoids than a conventional PM array. The key concept of the Halbach array is that the magnetization vector should rotate as a function of distance along the array (Figure 34.7).

It is recommended that a PM LSM be furnished with a *damper*. A rotary synchronous motor has a cage damper winding embedded in pole shoe slots. When the speed is different from the synchronous speed, electric currents are induced in damper circuits. The action of the armature magnetic field and damper currents allows for asynchronous starting, damps the oscillations, and helps to return to synchronous operation when the speed decreases or increases. Also, a damper circuit reduces the backward-traveling magnetic field. It would be rather difficult to furnish PMs with a cage winding so that the damper of PM LSMs has the form of an Al cover (Figure 34.8a) or solid steel pole shoes (Figure 34.8b). In addition, steel pole shoes or Al cover (shield) can protect brittle PMs against mechanical damage.

The *detent force*, i.e., attractive force between PMs and the armature ferromagnetic teeth, force ripple, and some higher-space harmonics, can be reduced with the aid of skewed assembly of PMs. Skewed PMs can be arranged in one row (Figure 34.9a), two rows (Figure 34.9b), or even more rows.

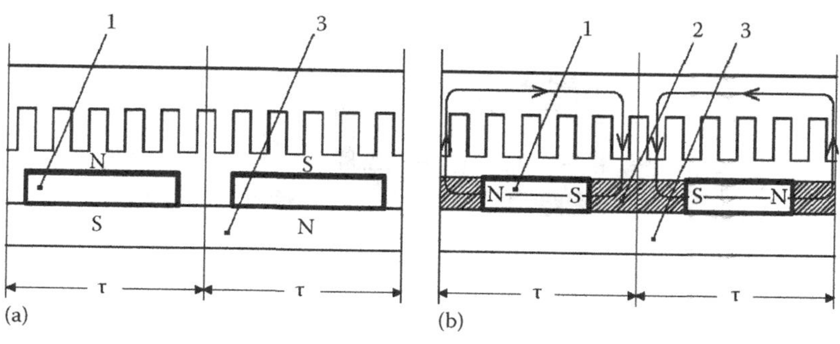

FIGURE 34.5 Single-sided flat PM LSMs with slotted armature core and (a) surface PMs and (b) buried PMs. 1, PM; 2, mild steel pole; 3, yoke.

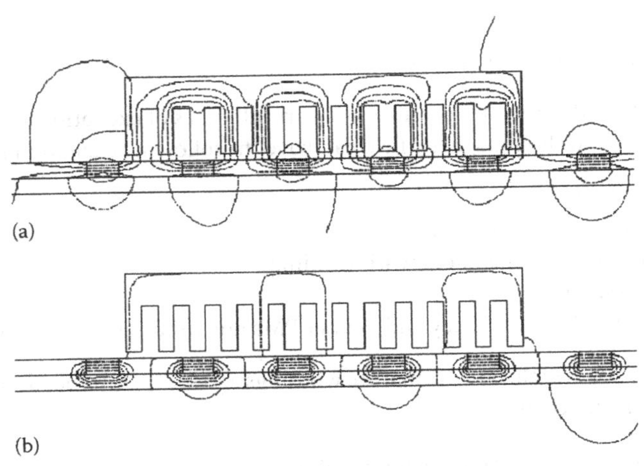

FIGURE 34.6 Magnetic flux distribution in the longitudinal sections of buried-type PM LSMs: (a) nonferromagnetic yoke and (b) ferromagnetic yoke (back iron).

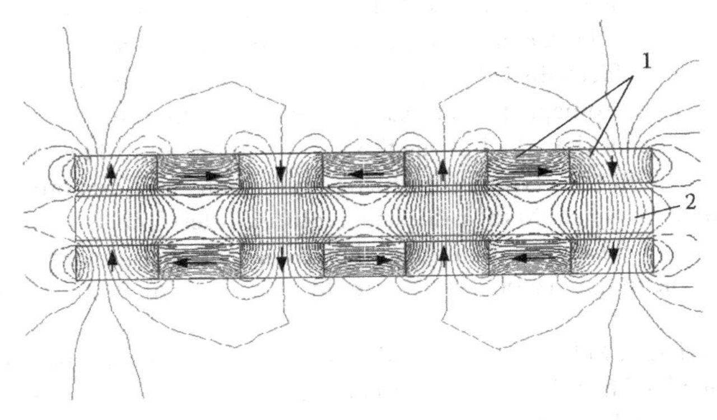

FIGURE 34.7 Double-sided LSM with Halbach array of PMs. 1, PMs; 2, coreless armature winding.

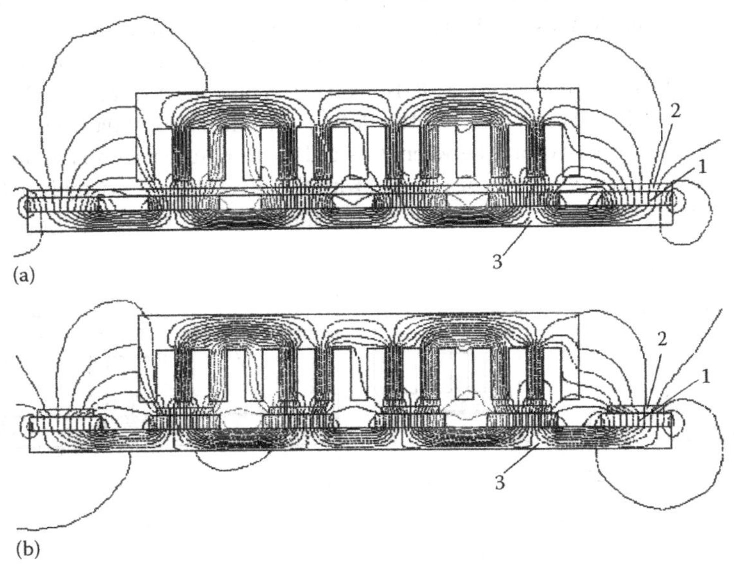

FIGURE 34.8 Dampers of surface-type PM LSMs: (a) Al cover (shield) and (b) solid steel pole shoes. 1, PM; 2, damper; 3, yoke.

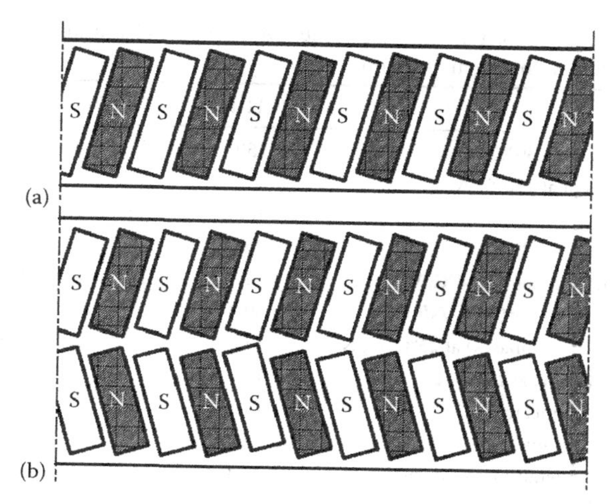

FIGURE 34.9 Skewed PMs in flat LSMs: (a) one row and (b) two rows.

TABLE 34.1 Flat Three-Phase, Single-Sided PM LBMs with Natural Cooling Systems Manufactured by Anorad, Hauppauge, NY

Parameter	LCD-T-1	LCD-T-2-P	LCD-T-3-P	LCD-T-4-P
Continuous thrust at 25°C, N	163	245	327	490
Continuous current at 25°C, A	4.2	6.3	8.5	12.7
Continuous thrust at 125°C, N	139	208	277	416
Continuous current at 125°C, A	3.6	5.4	7.2	10.8
Peak thrust (0.25 s), N	303	455	606	909
Peak current (0.25 s), A	9.2	13.8	18.4	27.6
Peak force (1.0 s), N	248	373	497	745
Peak current (1.0 s), A	7.3	11.0	14.7	22.0
Continuous power losses at 125°C, W	58	87	115	173
Armature constant, k_E, Vs/m	12.9			
Thrust constant (three phases), k_F, N/A	38.6			
Resistance per phase at 25°C, Ω	3.2	2.2	1.6	1.1
Inductance, mH	14.3	9.5	7.1	4.8
PM pole pitch, mm	23.45			
Maximum winding temperature, °C	125			
Armature assembly mass, kg	1.8	2.4	3.6	4.8
PM assembly mass, kg/m	6.4			
Normal attractive force, N	1036	1555	2073	3109

Specification data of flat, single-sided PM LBMs manufactured by Anorad are shown in Table 34.1 [1], and motors manufactured by Kollmorgen are shown in Table 34.2 [11]. The temperature 25°C, 125°C, or 130°C for the thrust, current, resistance, and power loss is the temperature of the armature winding.

The electromotive force (EMF) constant k_E in Tables 34.1 and 34.2 for sinusoidal operation is defined according to the equation expressing the EMF (induced voltage) excited by PMs without the armature reaction, i.e.,

$$E_f = k_E v_s \tag{34.4}$$

where v_s is the synchronous speed according to Equation 34.1. Thus, the armature constant k_E multiplied by the synchronous speed v_s gives the EMF E_f.

TABLE 34.2 Flat Three-Phase, Single-Sided PM LBMs with Natural Cooling Systems Manufactured by Kollmorgen, Radford, VA

Parameter	IC11-030	IC11-050	IC11-100	IC11-200
Continuous thrust at 130°C, N	150	275	600	1260
Continuous current at 130°C, A	4.0	4.4	4.8	5.0
Peak thrust, N	300	500	1000	2000
Peak current, A	7.9	7.9	7.9	7.9
Continuous power losses at 130°C, W	64	106	210	418
Armature constant, at 25°C, k_E, Vs/m	30.9	51.4	102.8	205.7
Thrust constant (three phases) at 25°C, k_F, N/A	37.8	62.9	125.9	251.9
Resistance, line to line, at 25°C, Ω	1.9	2.6	4.4	8.0
Inductance, line to line, mH	17.3	27.8	54.1	106.6
Electrical time constant, ms	8.9	10.5	12.3	13.4
Thermal resistance winding to external structure, °C/W	1.64	0.99	0.50	0.25
Maximum winding temperature, °C	130			
Armature assembly mass, kg	2.0	3.2	6.2	12.2
PM assembly mass, kg/m	5.5	7.6	12.8	26.9
Normal attractive force, N	1440	2430	4900	9850

The thrust constant k_F in Tables 34.1 and 34.2 is defined according to the simplified equation for the developed electromagnetic thrust, i.e.,

$$F_{dx} = k_F I_a \cos \Psi \tag{34.5}$$

for a sinusoidally excited LSM with equal reluctances in the d- and q-axes and for the angle between the armature current I_a and the q-axis $\Psi = 0°$ ($\cos \Psi = 1$). Thus, the thrust constant k_F times the armature current I_a gives the electromagnetic thrust developed by the LSM.

Double-sided, flat PM LSMs consist of two external armature systems and one internal excitation system (Figure 34.10a), or one internal armature system and two external excitation systems (Figure 34.10b). In the second case, a linear Gramme's armature winding can be used.

In *slotless motors*, the primary winding is uniformly distributed on a smooth armature core or does not have any armature core. Slotless PM LSMs are detent force–free motors, provide lower torque ripple, and, at high input frequency, can achieve higher efficiency than slotted LSMs. On the other hand, larger nonferromagnetic air gap requires more PM material, and the thrust density (thrust per mass or volume) is lower than that of slotted motors (Table 34.3). The input current is higher as synchronous reactances

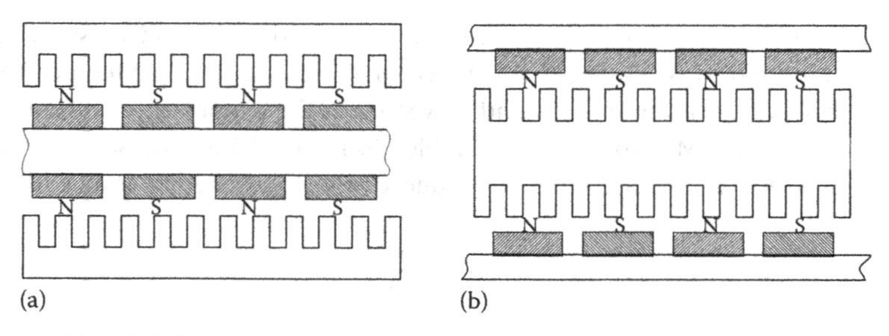

FIGURE 34.10 Double-sided flat PM LSMs with (a) two external armature systems and (b) one internal armature system.

TABLE 34.3 Slotted versus Slotless LSMs

Quantity	Slotted LSM	Slotless LSM
Higher thrust density	X	
Higher efficiency in the lower speed range	X	
Higher efficiency in the higher speed range		x
Lower input current	X	
Less PM material	X	
Lower winding cost		x
Lower thrust pulsations		x
Lower acoustic noise		x

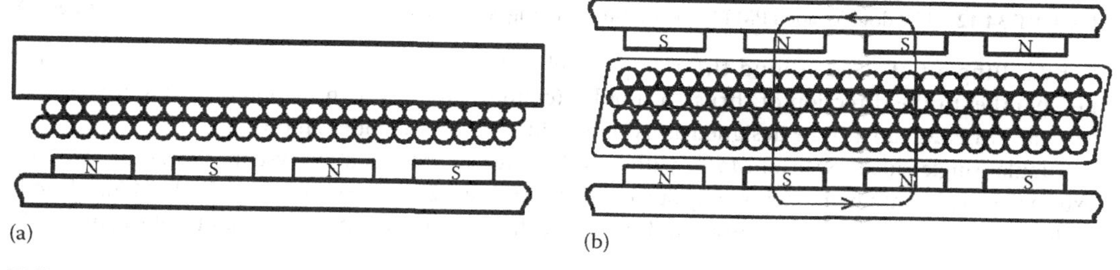

FIGURE 34.11 Flat slotless PM LSMs: (a) single-sided with armature core and (b) double-sided with inner air-cored armature winding.

in the d- and q-axes can decrease to a low undesired value due to the absence of teeth. Figure 34.11a shows a single-sided flat slotless motor with armature core, and Figure 34.11b shows a double-sided slotless motor with inner air-cored armature winding (moving coil motor).

Table 34.4 contains performance specifications of double-sided PM LBMs with inner three-phase air-cored armature winding manufactured by Trilogy Systems Corporation, Webster, TX (Figure 34.12).

TABLE 34.4 Flat Double-Sided PM LBMs with Inner Three-Phase Air-Cored Series-Coil Armature Winding Manufactured by Trilogy Systems Corporation, Webster, TX

Parameter	310-2	310-4	310-6
Continuous thrust, N	111.2	209.1	314.9
Continuous power for sinusoidal operation, W	87	152	230
Peak thrust, N	356	712	1068
Peak power, W	900	1800	2700
Peak/continuous current, A	10.0/2.8	10.0/2.6	10.0/2.6
Thrust constant k_F for sinusoidal operation, N/A	40.0	80.0	120.0
Thrust constant k_F for trapezoidal operation with Hall sensors, N/A	35.1	72.5	109.5
Resistance per phase, Ω	8.6	17.2	25.8
Inductance ±0.5 mH	6.0	12.0	18.0
Heat dissipation constant for natural cooling, W/°C	1.10	2.01	3.01
Heat dissipation constant for forced air cooling, W/°C	1.30	2.40	3.55
Heat dissipation constant for liquid cooling, W/°C	1.54	2.85	4.21
Number of poles	2	4	6
Coil length, mm	142.2	264.2	386.1
Coil mass, kg	0.55	1.03	1.53
Mass of PM excitation systems, kg/m	12.67 or 8.38		

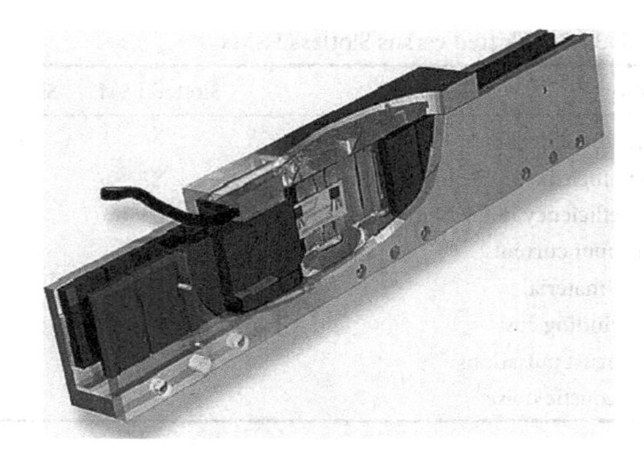

FIGURE 34.12 Flat double-sided PM LSM with inner moving coil. (Photo courtesy of Trilogy Systems, Webster, TX.)

By rolling a flat LSM around the axis parallel to the direction of the traveling magnetic field, i.e., parallel to the direction of thrust, a tubular (cylindrical) LSM can be obtained (Figure 34.13). A tubular PM LSM can also be designed as a double-sided motor or slotless motor.

Tubular single-sided LSMs LinMoT®* with movable internal PM excitation system (slider) and stationary external armature are manufactured by Sulzer Electronics AG, Zurich, Switzerland (Table 34.5). All active motor parts, bearings, position sensors, and electronics have been integrated into a rigid metal cylinder [14].

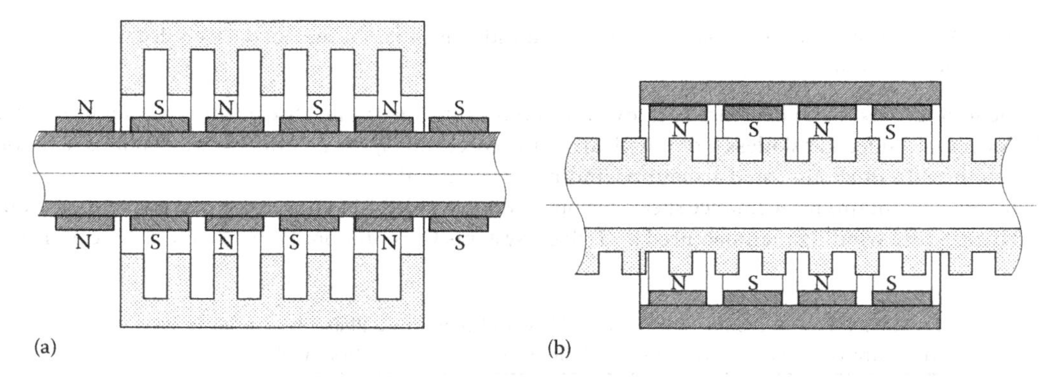

FIGURE 34.13 Single-sided slotted tubular PM LSMs: (a) with external armature system and (b) with external excitation system.

TABLE 34.5 Data of Tubular LSMs LinMot® Manufactured by Sulzer Electronics AG, Zürich, Switzerland

Parameter	P01 23 × 80	P01 23 × 160	P01 37 × 120	P01 37 × 240
Number of phases	2			
PMs	NdFeB			
Maximum stroke, m	0.210	0.340	1.400	1.460
Maximum force, N	33	60	122	204
Maximum acceleration, m/s²	280	350	247	268
Maximum speed, m/s	2.4	4.2	4.0	3.1
Stator (armature) length, m	0.177	0.257	0.227	0.347
Stator outer diameter, mm	23	23	37	37
Stator mass, kg	0.265	0.450	0.740	1.385
Slider diameter, mm	12	12	20	20
Maximum temperature of the armature winding, °C	90			

* LinMot® is a registered trademark of Sulzer Electronics AG, Zürich, Switzerland.

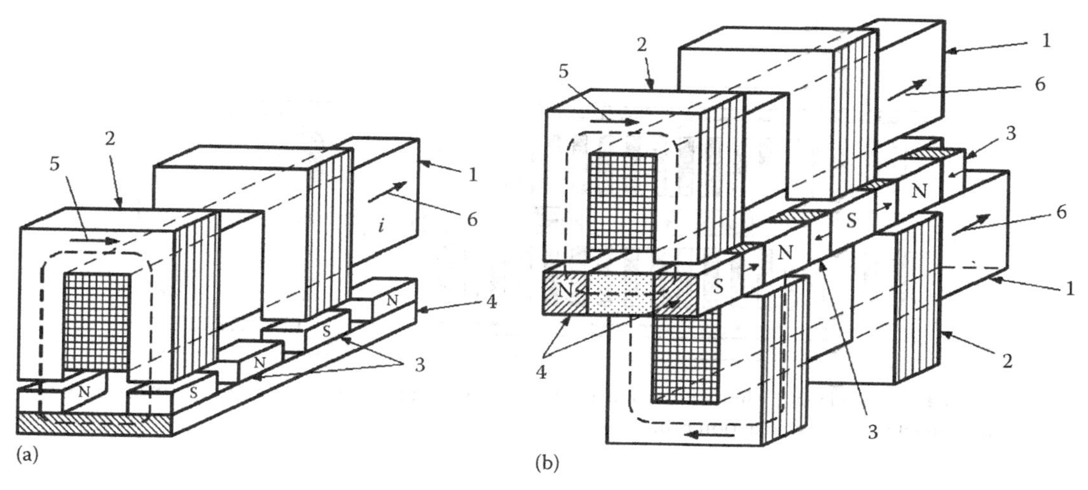

FIGURE 34.14 Transverse flux PM LSM: (a) single-sided and (b) double-sided. 1, armature winding; 2, armature laminated core; 3, PM; 4, back ferromagnetic core; 5, magnetic flux; 6, armature current.

All the aforementioned PM LSMs are motors with *longitudinal magnetic flux*, the lines of which lie in the plane parallel to the direction of the traveling magnetic field. LSMs can also be designed as *transverse magnetic flux* motors, in which the lines of magnetic flux are perpendicular to the direction of the traveling field. Figure 34.14a shows a single-sided transverse flux LSM in which PMs are arranged in two rows. A pair of parallel PMs creates a two pole flux excitation system. A double-sided configuration of transverse flux motor is possible; however, it is complicated and expensive (Figure 34.14b).

34.1.2.2 PM Motors with Passive Reaction Rail

The drawback of PM LSMs is the large amount of PM material that must be used to design the excitation system. Normally, expensive rare-earth PMs are requested. If a small PM LSM uses, say, 10 kg of NdFeB per 1 m of the reaction rail, and 1 kg of good-quality NdFeB costs U.S.$ 130, the cost of the reaction rail without assembly amounts to U.S.$ 1300 per 1 m. This price cannot be acceptable, for example, in passenger transportation systems.

A cheaper solution is to apply the PM excitation system to the short armature that magnetizes the long reaction rail and creates magnetic poles in it. Such a linear motor is called the *homopolar* LSM.

The homopolar LSM as described in [5,17] is a double-sided AC linear motor that consists of two polyphase armature systems connected mechanically and magnetically by a ferromagnetic U-type yoke (Figure 34.15). Each armature consists of a typical slotted linear motor stack with polyphase armature winding and PMs located between the stack and U-type yoke. Since the armature and excitation systems are combined together, the armature stack is oversized as compared with a conventional steel-cored LSM. The PMs can also be replaced by electromagnets [17,19]. The variable reluctance reaction rail is passive. The saliency is created by using ferromagnetic (solid or laminated) cubes separated by a nonferromagnetic material. The reaction rail poles are magnetized by the armature PMs through the air gap. The traveling magnetic field of the polyphase armature winding and salient poles of the reaction rail produce the thrust. Such a homopolar LSM has been proposed for the propulsion of maglev trains of *Swissmetro* [17].

Further simplification of the double-sided configuration can be made to obtain a single-sided PM LSM shown in Figure 34.16.

34.1.3 Flux-Switching PM Linear Motors

Large gantry systems and machining centers require powerful linear motors, preferably with PM-free reaction rail. Siemens 1FN6 PM LSMs with a magnet-free reaction rail belong to the group of the so

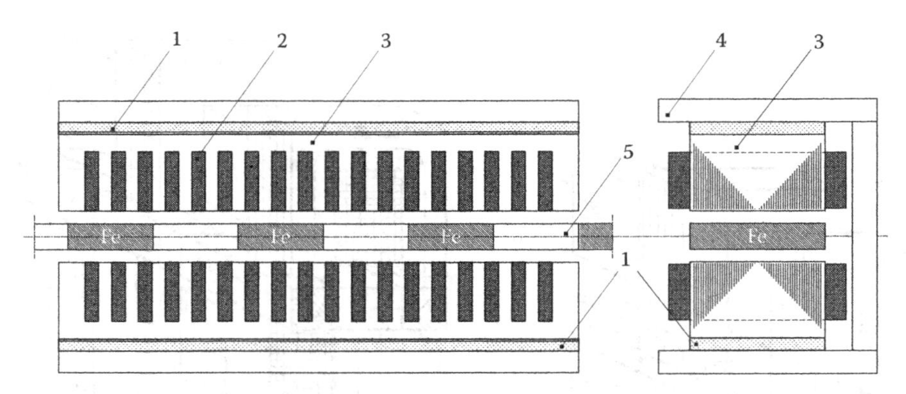

FIGURE 34.15 Double-sided homopolar PM LSM with passive reaction rail. 1, PM; 2, armature winding; 3, armature stack; 4, yoke; 5, reaction rail.

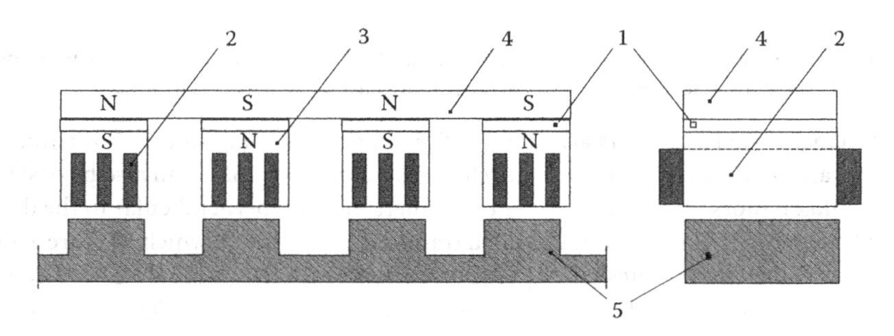

FIGURE 34.16 Single-sided PM LSM with a passive reaction rail. 1, PM; 2, armature winding; 3, armature stack; 4, yoke; 5, ferromagnetic reaction rail.

TABLE 34.6 Specifications of 1FN6 PM LSMs Manufactured by Siemens, Erlangen, Germany

Armature Unit	Rated Thrust N	Max. Thrust N	Max. Speed at Rated Thrust m/min	Max. Speed at Max. Thrust m/min	Rated Current A	Max. Current A
1FN6008-1LC17	235–350	900	263	103	1.7–2.6	9.0
1FN6008-1LC37	235–350	900	541	224	3.5–5.3	18.0
1FN6016-1LC30	470–710	1800	419	176	5.4–8.0	28.0
1FN6016-1LC17	935–1400	3590	263	101	7.0–10.5	36.0
1FN6024-1LC12	705–1060	2690	176	69	3.5–5.3	18.0
1FN6024-1LC20	705–1060	2690	277	114	5.4–8.0	28.0
1FN6024-1LG10	2110–3170	8080	172	62	10.5–16.0	54.0
1FN6024-1LG17	2110–3170	8080	270	102	16.2–24.3	84.0

Source: Synchronous Linear Motor 1FN6, *The Electrical Gear Rack*, Siemens AG Industry Sector, Drive Technologies, Motion Control, Erlangen, Germany, 2008.

called *flux-switching PM machines* [10,16]. The armature system is air cooled, degree of protection IP23, class of insulation F, line voltage from 400 to 480 V, rated thrust from 235 to 2110 N, maximum velocity at rated thrust from 170 to 540 m/min (Table 34.6), overload capacity 3.8 of rated thrust, modular type construction [20]. These LSMs operate with Siemens *Sinamics* or *Simodrive* solid-state converters and external encoders. According to Siemens [20], these new LSMs produce thrust forces and velocities equivalent to competitive classical models for light-duty machine tool, machine accessory, and material handling applications.

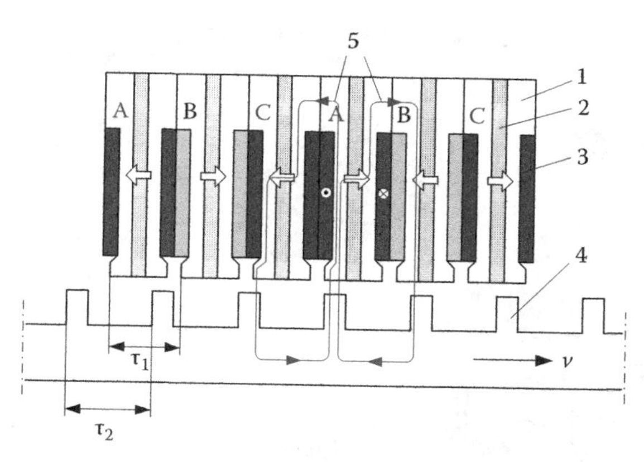

FIGURE 34.17 Construction of flux-switching LSM with PM-free reaction rail. 1, laminated armature core; 2, PM; 3, armature coil (phase C); 4, toothed passive steel reaction rail; 5, linkage magnetic flux (phase A is on).

The magnet-free reaction rail is easy to install and does not require the safety considerations of standard PM reaction rails. Without PMs, there is no problem with ferrous chips and other debris being attracted to these sections. Maintenance becomes a simple matter of installing a wiper or brush on the moving part of the slide.

The flux-switching LSM comprises an armature section that is equipped with coils and PMs as well as a nonmagnetic, toothed reaction rail section (Figure 34.17). The key design innovation is an LSM in which PMs are integrated directly into the lamination of the armature core along with the individual windings for each phase. Both magnitudes and polarities of the linkage flux in the armature winding vary periodically along with the reaction rail movement. The magnetic flux between the armature core and steel reaction rail is controlled by switching the three-phase armature currents according to a designated algorithm. The passive reaction rail consists of milled steel with poles (teeth) and is much simpler to manufacture.

34.1.4 Motors with Electromagnetic Excitation

The electromagnetic excitation system of an LSM is similar to the salient-pole rotor of a rotary synchronous motor. Figure 34.18 shows a flat single-sided LSM with salient ferromagnetic poles and *DC field excitation winding*. The poles and pole shoes can be made of solid steel, laminated steel, or sintered powder. If the electromagnetic excitation system is integrated with the moving part, the DC current can be delivered with the aid of brushes and contact bars, inductive power transfer (IPT) systems, linear transformers, or linear brushless exciters.

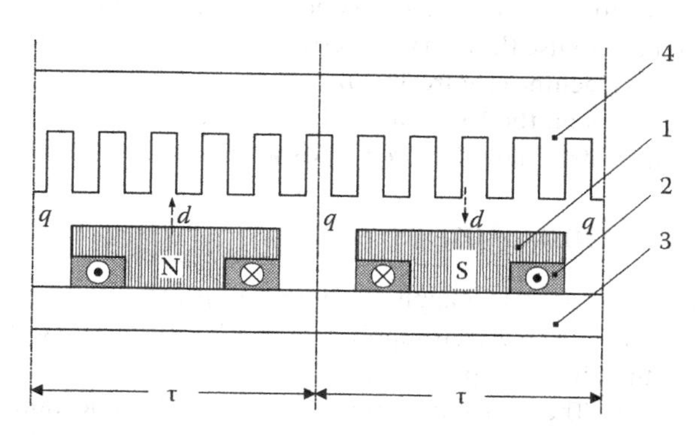

FIGURE 34.18 Electromagnetic excitation system of a flat single-sided iron-cored LSM. 1, salient pole; 2, DC excitation winding; 3, ferromagnetic rail (yoke); 4, armature system.

FIGURE 34.19 Transrapid 09 maglev train driven by LSMs with electromagnetic excitation. (Photo courtesy of Thyssen Transrapid System GmbH, Munich, Germany.)

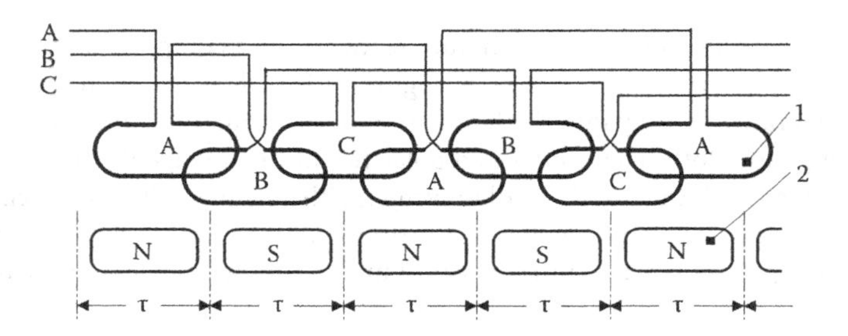

FIGURE 34.20 Three-phase air-cored LSM with SC excitation system. 1, armature coils; 2, SC excitation coils.

LSMs with electromagnetic excitation (wound-field reaction rail) are used in German *Transrapid* maglev train (Figure 34.19).

34.1.5 Motors with Superconducting Excitation System

In large-power LSMs, the electromagnets with ferromagnetic core that produce the excitation flux can be replaced by coreless SC electromagnets. Since the magnetic flux density produced by the SC electromagnet is greater than the saturation magnetic flux density of the best laminated alloys ($B_{sat} \approx 2.4$ T for cobalt alloy), there is no need to use the armature ferromagnetic core. An LSM with SC field excitation system is a totally air-cored machine (Figure 34.20).

Experimental maglev trains on the Yamanashi Maglev Test Line (YMTL) in Yamanashi Prefecture (west from Tokyo), Japan, are driven by air-cored LSMs with SC excitation systems (Figure 34.21).

34.2 Linear Induction Motors

LIMs have found the widest prospects for applications in *transportation systems*, beginning with electrical traction on small passenger or material supply cars (used at airports, exhibitions, electrohighways, elevators) and ending with pallet transportation, wafer transportation, belt conveyors, transportation systems of bulk materials, etc. The second important place for LIM applications is in industry, i.e., *manufacturing processes* (machine tools, hammers, presses, mills, separators, automated manufacturing systems, strip tensioners, textile shuttles, index tables, turntables, disk saws for wood, sliding doors, robots, etc.).

FIGURE 34.21 YMTL in Yamanashi Prefecture (near Tokyo), Japan. YMTL maglev trains use air-cored LSMs with SC excitation system. (Photo courtesy of Central Japan Railway Company and Railway Technical Research Institute, Tokyo, Japan.)

LIMs can also play an important part in *industrial investigations and tests*, for example, high acceleration of model aircraft in aerodynamic tunnels; high acceleration of vessels in laboratory pools; propulsion of mixers, shakers, and vibrators; and adjusting *x–y* tables and instruments. There is also a possibility of using LIMs in *consumer electronics* (sound and vision equipment, knitting machines, curtains) and in offices (transportation of documents, letters, and cash). The *Handbook of Linear Motor Applications* [21] printed in Japan in 1986 contains about 50 examples of applications of LIMs in operation or in the process of implementation.

34.2.1 Basic Geometries and Constructions

A LIM can be obtained by cutting in the same way either a cage rotor induction motor or a wound rotor induction motor. The stator becomes the *primary*, and the rotor becomes the *secondary* [12]. The secondary of a LIM can be simplified by using a solid steel core and replacing the cage (ladder) or slip-ring winding with a high-conductivity nonferromagnetic plate (Al or Cu). The nonferromagnetic plate is a secondary electric circuit with distributed parameters, and the ferromagnetic core is a conductor both for the magnetic flux and the electric current. It does not matter from the principle of operation point of view which part (primary or secondary) is in motion. Thus, the *flat, single-sided LIM* can be obtained from a solid rotor induction motor (Figure 34.1a) and the *flat, double-sided LIM* can be obtained from a hollow-rotor induction motor with wound external and internal stator (Figure 34.1b). In a double-sided LIM, the secondary ferromagnetic core is not necessary, since the magnetic flux excited by one of the primary windings after passing through the air gaps and nonferromagnetic secondary is then closed up by the core of the second primary unit.

Theoretically, a *double-sided LIM* with primary windings located on two cores, in comparison with a single-sided LIM exciting the same MMF, has twice the air gap magnetic flux density. Therefore, the *thrust* of such a motor is four times greater, assuming the same dimensions. If only one primary core is wound, the output parameters of a double-sided LIM are the same as those for a single-sided LIM with laminated secondary back iron. The fundamental advantage of double-sided LIMs is the elimination of the normal attractive force between the primary and the secondary because the secondary is usually nonferromagnetic.

Flat LIMs can have primary cores consisting of an array of cores arranged in parallel at appropriate distances and connected magnetically by additional yokes perpendicular to the direction of the

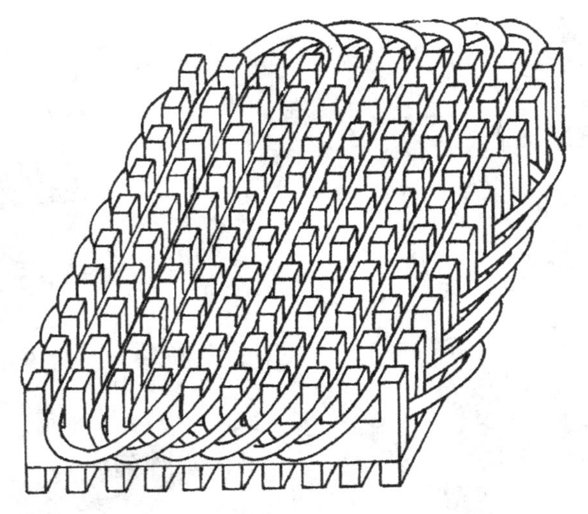

FIGURE 34.22 The primary of a flat LIM with two DOFs.

traveling field. A magnetic circuit designed in such a way makes it possible to apply two windings, in general multiphase windings, with perpendicular conductors, as shown in Figure 34.22. Adjusting the current in each winding, the secondary can be moved in two perpendicular directions and can be positioned at any point of the x–y plane. A *flat LIM with two degrees of freedom* (DOFs) can be designed both as single-sided and double-sided machines.

By rolling a flat, single-sided or double-sided LIM around the axis parallel to the direction of the traveling magnetic field, i.e., parallel to the direction of the thrust, a tubular motor can be obtained (Figure 34.23).

A *tubular (cylindrical) LIM*, similar to a flat LIM, can be designed both as single-sided and double-sided machines and can have a square or rectangular cross section, as in Figure 34.24. There are possible

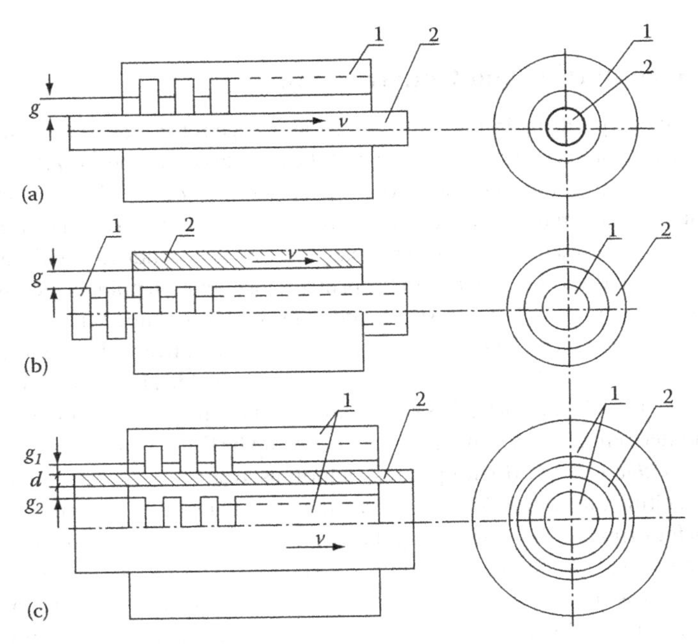

FIGURE 34.23 Tubular LIMs: (a) single-sided with an external short primary, (b) single-sided with an external short secondary, and (c) double-sided with short primary. 1, primary; 2, secondary.

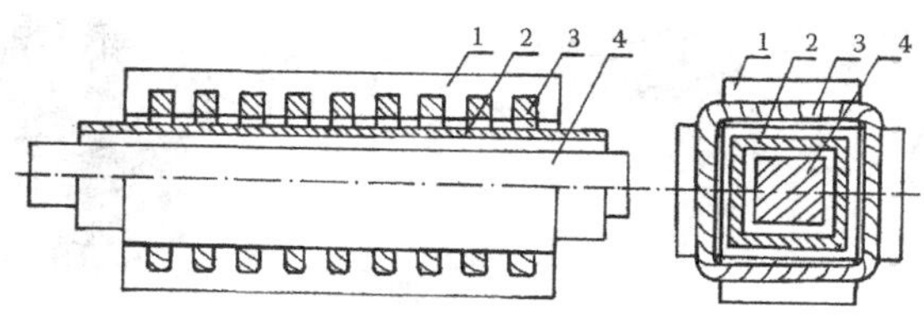

FIGURE 34.24 Tubular, double-sided LIM with a square cross section: 1, primary; 2, secondary; 3, primary coil; 4, internal core.

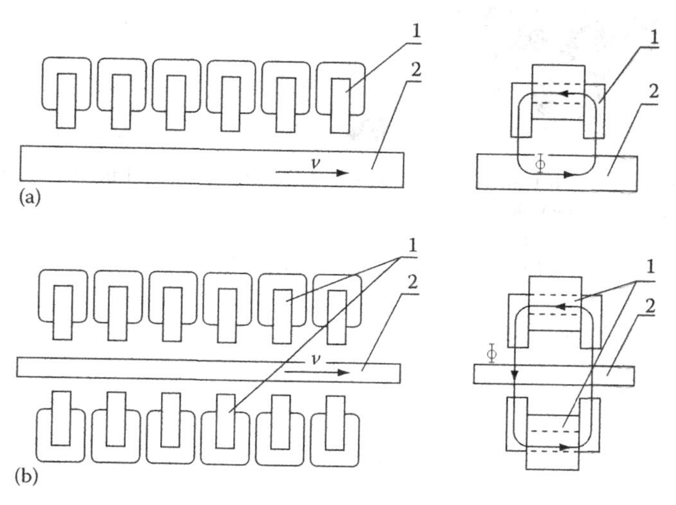

FIGURE 34.25 Flat LIM with transverse magnetic flux and salient poles: (a) single-sided and (b) double-sided. 1, primary; 2, secondary.

configurations other than that in Figure 34.23, regarding the length of the secondary with respect to the length of the primary.

All the aforementioned LIMs are motors with *longitudinal magnetic flux*, i.e., the lines of magnetic flux lie in the plane parallel to the direction of the traveling magnetic field. A LIM can also be designed in such a way as to obtain magnetic flux lines perpendicular to the direction of the traveling field. Such motors are said to have *transverse magnetic flux* (Figure 34.25).

The fundamental advantage of a LIM with transverse magnetic flux in comparison with a LIM with longitudinal magnetic flux is the lower magnetizing current necessary due to the shorter magnetic flux paths. The significant disadvantage is lower thrust. A flat LIM with transverse magnetic flux usually has a primary winding of concentrated coils located on salient poles.

A flat, single-sided LIM with transverse flux can produce not only thrust but also *electrodynamic suspension*. In the design shown in Figure 34.26, the secondary is suspended electrodynamically and propelled and stabilized laterally by the primary magnetic field [8]. The primary magnetic circuit consists of E-shaped laminations assembled in two rows. The short secondary is made of light Al alloy in the form of a boat. The lateral sides of the boat are inclined by 60° with respect to the active surface. Such a shape provides maximum normal repulsive force and maximum lateral stabilization.

According to their geometry, the LIMs can be divided into the following groups:

- With movable primary or movable secondary
- Single sided and double sided
- Flat and tubular
- With short primary and short secondary
- With longitudinal and transverse magnetic flux

FIGURE 34.26 Flat LIM with transverse magnetic flux, salient poles, and nonferromagnetic secondary propelled, suspended, and stabilized electrodynamically. (From Gieras, J.F. et al., Analytical calculation of electrodynamic levitation forces in a special-purpose linear induction motor, *International Electric Machines and Drives Conference (IEMDC'11)*, Niagara Falls, Ontario, Canada, 2011 [on CD-ROM].)

The linear speed of a LIM is

$$v = (1 - s)v_s \tag{34.6}$$

where
s is the slip
v_s is the synchronous speed according to Equation 34.1

Neglecting the *longitudinal end effect* [6], the electromagnetic thrust developed by a LIM can be expressed in a similar way as for a rotary induction motor, i.e.,

$$F_{dx} = \frac{m_1}{v_s}\left(I_2'\right)^2 \frac{R_2'}{s} \tag{34.7}$$

where
m_1 is the number of phases of the primary winding
v_s is the synchronous speed according to Equation 34.1
I_2' is the secondary current referred to the primary system
R_2' is the secondary resistance referred to the primary system
s is the slip

The secondary resistance R_2' must include the so-called *edge effect* [6].

34.2.2 Propulsion of Wheel-on-Rail Vehicles

Modern railway systems and electrical traction should meet the following requirements:

- High level of automatization and computerization
- Propulsion and braking independent of adhesion which in turn is affected, first of all, by climate and weather
- Low level of noise, sometimes below 70 dB (A)

- Ability to cope with high slopes, at least 6%, and sharp bends with radius of curvature less than 20 m
- No pollution to natural environment and landscape
- High reliability

The congestion problems of big cities should be solved by creating collective transport forms that can be implemented without affecting a highly populated city. For example, there are many cities in Italy where the historical center has remained the same since the Renaissance. A heavy railway might be a completely wrong solution and have a notable impact on the city planning. An adequate solution might be a light railway, a *people mover*, with transport capacity of 10,000–20,000 passengers per hour, to replace the traditional transport nets and to integrate the existing railway nets.

All the requirements mentioned earlier can be met by using LIMs as propulsion machines. Replacing electrical rotary motors with linear motors in traction drives (electrical locomotives) generally does not require new tracks but only their adjustment to linear drives. Single-sided LIMs (Figure 34.27a and b) are best, since the normal attractive force of these can strengthen the adhesion of wheels and rails. The air gap is from 10 to 15 mm. Most frequently, the motor car has two LIMs with short primaries (Figure 34.27a) assembled in series. The double-layer secondary consists of a solid back iron and an Al cap and is located between the rails. Cables for computer communication with the vehicle and, quite often, collectors for electrical energy delivery to the vehicle are located along the track. Three-phase LIMs are fed from variable-voltage, variable-frequency (VVVF) voltage-source inverters. A linear propulsion system of wheel-on-rail vehicles allows more flexibility (Figure 34.28), reducing noise on bends and wear of the wheels and rails.

Double-sided LIMs (Figure 34.27c and d) have found very limited application due to the technical difficulty of eliminating faults caused by bends in the secondary with primary cores on either side. It is more difficult to keep a small and uniform air gap in double-sided LIMs than that in single-sided LIMs.

The LIM-*driven wheel-on-rail vehicles* are built for low velocities, i.e., less than 100 km/h, and for short routes, i.e., less than 50 km (urban transit systems and short-distance trains). At present, LIM-driven trains operate in Toronto and Vancouver, BC, Canada; in Detroit, MI (Figure 34.29); Tokyo (Figure 34.30) and Osaka, Japan; and Kuala Lumpur, Malaysia.

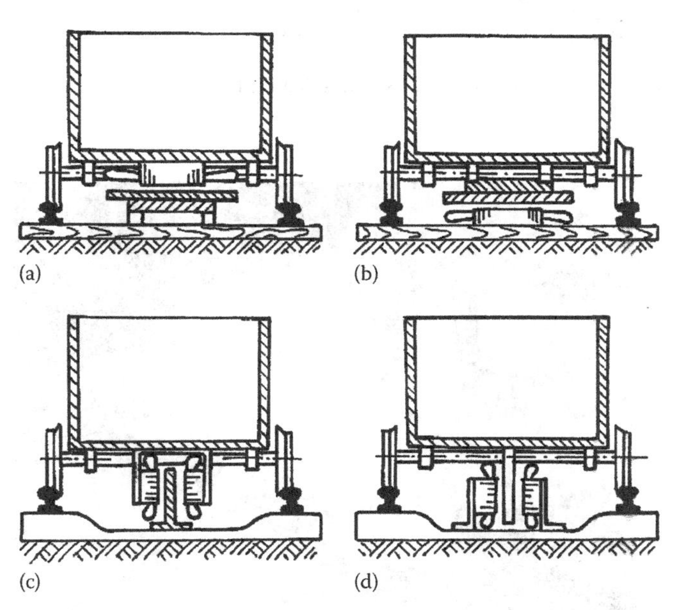

FIGURE 34.27 LIM-driven wheel-on-rail cars: (a) single-sided LIM with short primary mounted on the undercarriage, (b) single-sided LIM with short secondary mounted on the undercarriage, (c) double-sided LIM with short primary mounted on the undercarriage, and (d) double-sided LIM with short secondary mounted on the undercarriage.

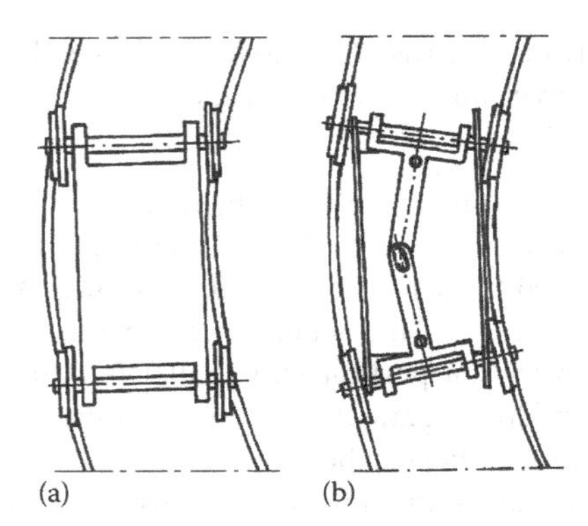

FIGURE 34.28 The undercarriage of a wheel-on-rail vehicle: (a) rotary motor propulsion and (b) LIM propulsion.

FIGURE 34.29 Detroit people mover driven by single-sided LIMs.

FIGURE 34.30 Toei Oedo subway line in Tokyo driven by single-sided LIMs.

TABLE 34.7 Design Data of Single-Sided, Three-Phase LIMs for Propulsion of Vehicles

	LIM					
Quantity	JLMDR	ICTS	KU	CIGGT	GEC	Unit
Pullout thrust at frequency given in the following, F_x	12.5	9.0	3.5	1.7	0.7	kN
Input frequency, f	20.0	40.0	25.0	40.0	60.0	Hz
Rated phase current, V_1	275.0	465.0	130.0	200.0	200.0	A
Number of poles, $2p$	8	6	4	6	4	—
Number of turns per phase, N_1	128	96	128	108	48	—
Equivalent diameter of conductor, d_1	—	8.93	5.28	1.115	8.1	mm
Number of parallel conductors	—	1	—	19	1	—
Effective width of primary core, L_i	0.23	0.216	0.29	0.101	0.1715	m
Pole pitch, τ	0.27	0.2868	0.30	0.25	0.20	m
Length of single end connection, l_e	—	0.3483	—	0.2955	0.3685	m
Coil pitch, w_c	0.225	0.1673	0.25	0.1944	0.1555	m
Air gap, g	15.0	12.6	12.0	15.0	18.2	mm
Number of slots, $z_1(z_1')$	96(106)	72(79)	48(58)	54(61)	36(43)	—
Width of slot, b_{11}	—	15.6	17.0	15.0	13.08	mm
Width of slot opening, b_{14}	—	15.6	—	10.44	13.08	mm
Depth of slot, h_{11}	—	53.0	38.0	34.21	61.47	mm
Height of yoke, h_{1y}	—	43.6	39.3	71.63	50.0	mm
Conductivity of back iron at 20°C, σ_{Fe}	—	4.46	9.52	4.46	5.12	$\times 10^6$ S/m
Conductivity of Al cap at 20°C, σ_{Al}	—	30.0	30.3	32.3	21.5	$\times 10^6$ S/m
Width of back iron, w	0.3	0.24	0.3	0.111	0.1715	m
Thickness of back iron, h_{sec}	19.0	12.5	25.0	25.4	47.4	mm
Thickness of Al cap, d	5.0	4.5	5.0	4.5(2.5)	3.2	mm
Thickness of Al cap behind Fe core, t_{ov}	5.0	17.0	5.0	12.7	3.2	mm
Width of Al cap	0.30	0.32	0.40	0.201	0.2985	m

JLMDR, Japanese Linear Motor Driven Railcar, Japan; ICTS, Intermediate Capacity Transit System, Canada; KU, Kyushu University LIM, Japan; CIGGT, Canadian Inst. of Guided Ground Transport, Canada; GEC, General Electric Company, United States; (z_1') is the number of half filled slots.

The design data of some single-sided traction LIMs are presented in Table 34.7 and double-sided traction LIMs in Table 34.8.

34.3 Variable Reluctance Motors

The simplest construction of a *variable reluctance LSM or linear reluctance motor* (LRM) is that shown in Figure 34.18 with DC excitation winding being removed. However, the thrust of such a motor would be low as the ratio of d-axis permeance to q-axis permeance is low. Better performance can be obtained when using *flux barriers* [18] or steel laminations [15]. To make flux barriers, any nonferromagnetic materials can be used. To obtain high permeance (low reluctance) in the d-axis and low permeance in the q-axis, steel laminations should be oriented in such a way as to create high permeance for the d-axis magnetic flux.

Figure 34.31a shows a variable reluctance platen with flux barriers, and Figure 34.31b shows how to arrange steel laminations to obtain different reluctances in the d- and q-axes. The platen can be composed of segments, the length of which is equal to the pole pitch τ. Each segment consists of semicircular *lamellas* cut out from electrotechnical sheet. A filling, for example, epoxy resin, is used to make the segment rigid and robust. By putting the segments together, a platen of any desired length can be obtained.

TABLE 34.8 Design Data of Double-Sided, Three-Phase LIMs for Propulsion of Vehicles

	LIM		
Quantity	GEC	LIMVR	Unit
Maximum thrust at input frequency, F_x, given in the following	0.85	16.68	kN
Input frequency, f	60.0	173.0	Hz
Input phase current, I_1	200.0	2000.0	A
Number of poles, $2p$	4	10	—
Number of turns per phase, N_1	144	100	—
Diameter of conductor, d_1	0.75	1.10	mm
Effective width of primary core, L_i	0.0869	0.254	m
Pole pitch, τ	0.1795	0.3556	m
Coil pitch, w_c	0.05	0.05	m
Number of slots, $z_1(z_1')$	36(45)	150(160)	—
Width of slot, b_{11}	13.7	16.0	mm
Width of slot opening, b_{14}	13.7	—	mm
Depth of slot, h_1	63.0	—	mm
Height of yoke, h_{1y}	23.9	—	mm
Width of tooth, c_1	20.0	7.7	mm
Resultant air gap, $2g + d$	38.1	38.075	mm
Air gap, g	2×12.7	2×11.1	mm
Thickness of secondary, d	12.7	15.875(hollow)	mm
Effective thickness of Al secondary	12.7	7.2	mm
Width of secondary, w	≥ 0.3046	—	m
Conductivity of Al cap at 20°C, σ_{Al}	≈ 24.34	≈ 24.0	$\times 10^6$ S/m

GEC, General Electric Company, United States; LIMVR, LIM Research Vehicle, Pueblo, CO, United States; (z_1') is the number of half filled slots.

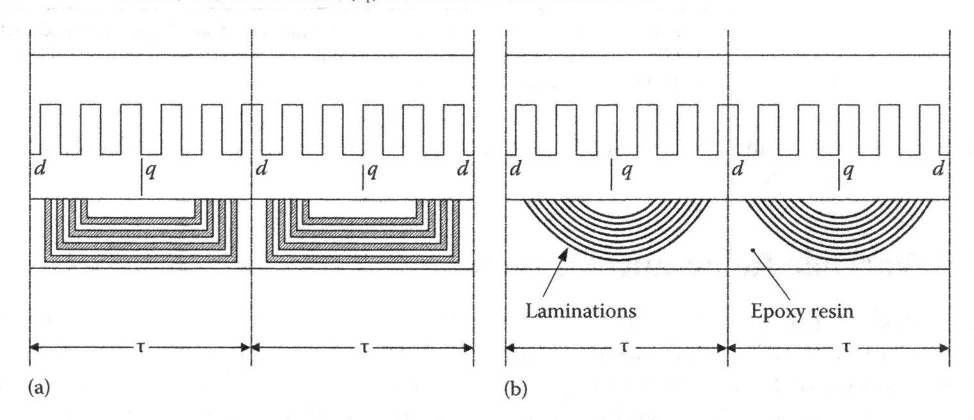

FIGURE 34.31 Variable reluctance LSMs with (a) flux barriers and (b) steel laminations.

34.4 Stepping Motors

A *linear stepping motor* has a concentrated armature winding wound on salient poles and PM excitation rail or variable reluctance platen. The thrust is generated as an action of the armature magnetic flux and PM flux (active platen), or the armature magnetic flux and salient ferromagnetic poles (variable reluctance platen). Stepping motors have no position feedback.

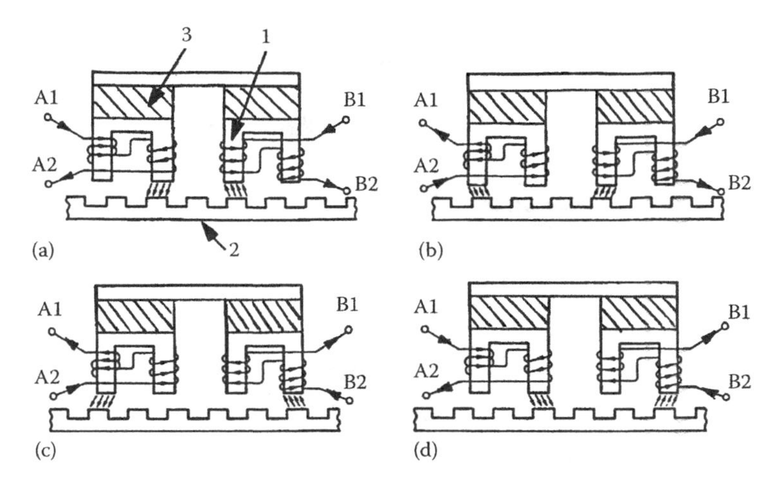

FIGURE 34.32 Principle of operation of an HLSM: (a) initial position, (b) 1/4 tooth pitch displacement of the forcer, (c) 2/4 tooth pitch displacement, and (d) 3/4 tooth pitch displacement. 1, forcer; 2, platen; 3, PM.

So far, only stepping linear motors of hybrid construction (PM, winding and variable reluctance air gap) have found practical applications.

The *hybrid linear stepping motor* (HLSM), as shown in Figure 34.32, consists of two parts: the *forcer* (also called the *slider*) and the variable reluctance platen [4]. Both of them are evenly toothed and made of high-permeability steel. This is an early design of the HLSM, the so-called Sawyer linear motor. The forcer is the moving part with two rare-earth magnets and two concentrated-parameter windings. The tooth pitch of the forcer matches the tooth pitch on the platen. However, the tooth pitches on the forcer poles are spaced 1/4 or 1/2 pitch from one pole to the next. This arrangement allows for the PM flux to be controlled at any level between minimum and maximum by the winding so that the forcer and the platen line up at a maximum permeance position. The HLSM is fed with two-phase currents (90° out of phase), similarly as a rotary stepping motor. The forcer moves 1/4 tooth pitch per each full step.

There is a very small air gap between the two parts that is maintained by strong air flow produced by an air compressor. The average air pressure is about 300–400kPa and depends on how many phases are excited.

Table 34.9 shows specification data of HLSMs (Figure 34.33) manufactured by Tokyo Aircraft Instrument Co., Ltd., Tokyo, Japan [13]. The *holding force* is the amount of external force required to break the forcer away from its rest position at rated current applied to the motor. The *step-to-step accuracy* is a measure of the maximum deviation from the true position in the length of each step.

TABLE 34.9 Data of HLSMs Manufactured by Tokyo Aircraft Instrument Co., Ltd., Tokyo, Japan

Parameter	LP02-20A	LP04-20A	LP04-30A	LP60-20A
Driver	Bipolar chopper			
Voltage, V	24 DC			
Resolution, mm	0.2	0.4	0.4	0.423
Holding force, N	20	20	29.5	20
Step-to-step accuracy, mm	±0.03			
Cumulative accuracy, mm	±0.2			
Maximum start–stop speed, mm/s	60	120	120	127
Maximum speed, mm/s	400	600	500	600
Maximum load mass, kg	3.0	3.0	5.0	3.0
Effective stroke, mm	330	300	360	310
Mass, kg	1.4	1.2	2.8	1.4

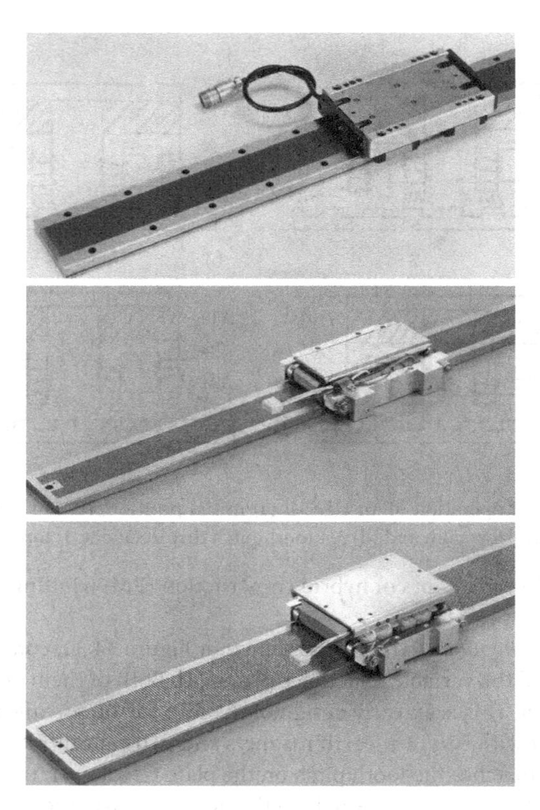

FIGURE 34.33 PM linear stepping motors. (Photo courtesy of Tokyo Aircraft Instrument, Co., Ltd., Tokyo, Japan.)

This value is different for full-step and microstepping drives. The *maximum start–stop speed* is the maximum speed that can be used when starting or stopping the motor without ramping that does not cause the motor to fall out of synchronism or lose steps. The *maximum speed* is the maximum linear speed that can be achieved without the motor stalling or falling out of synchronism. The *maximum load mass* is the maximum allowable mass applied to the forcer against the scale that does not result in mechanical damage. The *full-step resolution* is the position increment obtained when the currents are switched from one winding to the next winding. This is the typical resolution obtained with full-step drives, and it is strictly a function of the motor construction. The *microstepping resolution* is the position increment obtained when the full-step resolution is divided electronically by proportioning the currents in the two windings. This resolution is typically 10–250 times smaller than the full-step resolution [13].

HLSMs are regarded as an excellent solution to positioning systems that require a high accuracy and rapid acceleration. With a microprocessor controlled *microstepping mode*, a smooth operation with standard resolution of a few hundred steps/mm can be obtained. The advantages such as high efficiency, high throughput, mechanical simplicity, high reliability, precise open-loop operation, and low inertia of the system have made these kind of motors more and more attractive in such applications as factory automation, high speed positioning, computer peripherals, facsimile machines, numerically controlled machine tools, automated medical equipment, automated laboratory equipment, and welding robots. This motor is especially suitable for machine tools, printers, plotters, and computer-controlled material handling in which a high positioning accuracy and repeatability are the key problems.

When two or four forcers mounted at 90° and a special grooved platen ("waffle plate") are used, the x–y motion (two DOFs) in a single plane is obtained (Figure 34.34).

Specification data of the x–y HLSMs manufactured by Normag Northern Magnetics, Inc., Santa Clarita, CA, are given in Table 34.10.

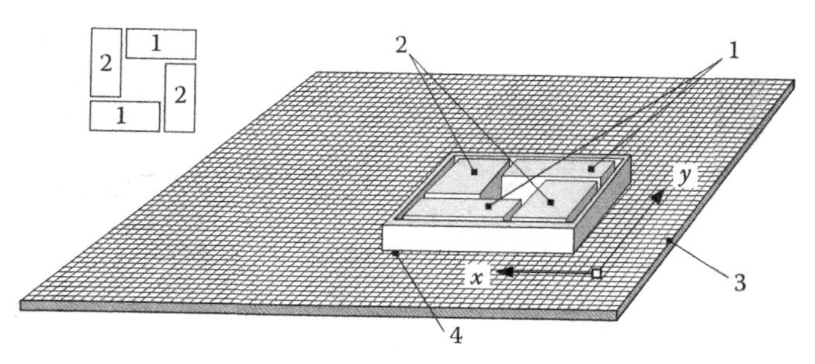

FIGURE 34.34 HLSM with a four-unit forcer to obtain the x–y motion: 1, forcers for the x-direction; 2, forcers for the y-direction; 3, platen; 4, air pressure.

TABLE 34.10 Data of x–y HLSMs Manufactured by Normag Northern Magnetics, Inc., Santa Clarita, CA

Parameter	4XY0602-2-0	4XY2002-2-0	4XY2004-2-0	4XY2504-2-0
Number of forcer units per axis	1	1	2	2
Number of phases	2	2	2(4)	2(4)
Static thrust, N	13.3	40.0	98.0	133.0
Thrust at 1 m/s, N	11.1	31.1	71.2	98.0
Normal attractive force, N	160.0	400.0	1440.0	1800.0
Resistance per phase, Ω	2.9	3.3	1.6	1.9
Inductance per phase, mH	1.5	4.0	2.0	2.3
Input phase current, A	2.0	2.0	4.0	4.0
Air gap, mm	0.02			
Maximum temperature, °C	110			
Mass, kg	3.2	0.72	2.0	1.5
Repeatability, mm	0.00254			
Resolution, mm	0.00254			
Bearing type	Air			

34.5 Switched Reluctance Motors

The topology of a linear switched reluctance motor is similar to that of a stepping motor with variable reluctance platen. In addition, it is equipped with position sensors. The *turn-on* and *turn-off* instant of the input current is synchronized with the position of the moving part. The thrust is very sensitive to the turn-on and turn-off instant.

In the case of a linear stepping or linear switched reluctance motor, the speed v of the moving part is

$$v = v_s = f_{sw}\tau \tag{34.8}$$

where

f_{sw} is the fundamental switching frequency in one armature phase winding
τ is the pole pitch of the reaction rail

For a rotary stepping or switched reluctance motor, $f_{sw} = 2p_r n$, where $2p_r$ is the number of rotor poles and n is rotational speed in rev/s.

A *linear switched reluctance motor* has a doubly salient magnetic circuit with a polyphase winding on the armature. Longitudinal and transverse flux designs are shown in Figure 34.35. A linear switched reluctance motor allows precise speed and position-controlled linear motion at low speeds and is not subject to design constraints (minimum speed limited by minimum feasible pole pitch) of linear AC motors.

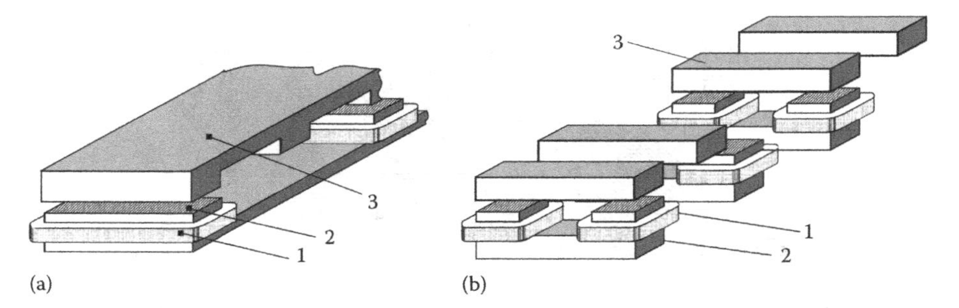

FIGURE 34.35 Linear switched reluctance motor configurations: (a) longitudinal flux design and (b) transverse flux design. 1, armature winding; 2, armature stack; 3, platen.

34.6 Linear Positioning Stages

Linear motors are now playing a key role in *advanced precision linear positioning. Linear precision positioning systems* can be classified into open-loop systems with HLSMs and closed-loop servo systems with LSMs, LBMs, or LIMs.

A PM LSM–driven *positioning stage* is shown in Figure 34.36. A stationary base is made of Al, steel, ceramic, or granite plate. It provides a stable, precise, and flat platform to which all stationary positioning components are attached. The base of the stage is attached to the host system with the aid of mounting screws.

The *moving table* accommodates all moving positioning components. To achieve maximum acceleration, the mass of the moving table should be as small as possible, and usually, Al is used as a lightweight material. A number of mounting holes on the moving table is necessary to fix the payload to the mounting table.

Linear bearing rails provide a precise guidance to the moving table. A minimum of one bearing rail is required. Linear ball bearing or air bearings are attached to each rail.

The armature of a linear motor is fastened to the moving table, and reaction rail (PM excitation system) is built in the base between the rails (Figure 34.36).

A linear encoder is needed to obtain precise control of position of the table, velocity, and acceleration. The read head of the encoder is attached to the moving table.

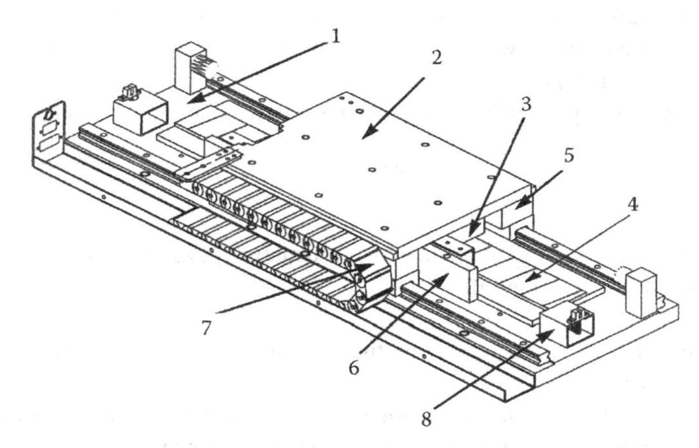

FIGURE 34.36 Linear positioning stage driven by PM LBM. 1, base; 2, moving table; 3, armature of LBM; 4, PMs; 5, linear bearing; 6, encoder; 7, cable carrier; 8, limit switch. (Courtesy of Normag, Santa Clarita, CA.)

Noncontact limit switches fixed to the base provide an overtravel protection and initial homing. A cable carrier accommodates and routes electrical cables between the moving table and stationary connector box fixed to the base.

An HLSM-driven linear precision stage is of similar construction. Instead of PMs between bearing rails, it has a variable reluctance platen. HLSMs usually need air bearings, and, in addition to the electrical cables, an air hose between air bearings and the compressor is required.

Linear positioning stages are used in semiconductor technology, electronic assembly, quality assurance, laser cutting, optical scanning, water jet cutting, gantry systems (x, y, z stages), color printers, plotters, and Cartesian coordinate robotics.

References

1. Anorad Linear Motors, *Information Brochure*, Anorad, Hauppauge, New York, 2007, www.anorad.com
2. Boldea, I. and Nasar, S.A., *Linear Motion Electromagnetic Systems*, John Wiley & Sons, New York, 1985.
3. Boldea, I. and Nasar, S.A., *Linear Electric Actuators and Generators*, Cambridge University Press, New York, 2005.
4. Compumotor Digiplan, *Positioning Control Systems and Drives*, Parker Hannifin Corporation, Rohnert Park, CA, 2011.
5. Everes, W., Henneberger, G., Wunderlich, H., and Selig, A., A linear homopolar motor for a transportation system, *2nd International Symposium on Linear Drives for Industry Applications (LDIA'98)*, Tokyo, Japan, 1998, pp. 46–49.
6. Gieras, J.F., *Linear Induction Drives*, Clarendon Press, Oxford, U.K., 1994.
7. Gieras, J.F., Status of linear motors in the United States, *International Symposium on Linear Drives for Industry Applications (LDIA'03)*, Birmingham, U.K., 2003, pp. 169–176.
8. Gieras, J.F., Gientkowski, Z., Mews, J., and Splawski, P., Analytical calculation of electrodynamic levitation forces in a special-purpose linear induction motor, *International Electric Machines and Drives Conference (IEMDC'11)*, Niagara Falls, Ontario, Canada, 2011 [on CD-ROM].
9. Gieras, J.F., Piech, Z.J., and Tomczuk, B.Z., *Linear Synchronous Motors: Transportation and Automation Systems*, Taylor & Francis (CRC Press), Boca Raton, FL, 2011.
10. Hoang, E., Ahmed, A.H.B., and Lucidarme, J., Switching flux permanent magnet polyphase synchronous machines, *7th European Conference on Power Electronics and Applications (EPE'97)*, Vol. 3, Trondheim, Norway, 1977, pp. 903–908.
11. *Kollmorgen Linear Motors Aim to Cut Cost of Semiconductors and Electronics Manufacture*, Kollmorgen, Radford, VA, 1997.
12. Laithwaite, E.R., A *History of Linear Electric Motors*, Macmillan, London, U.K., 1987.
13. Linear Step Motor, *Information Brochure*, Tokyo Aircraft Instrument Co., Ltd., Tokyo, Japan, 1998.
14. *LinMot Design Manual*, Sulzer Electronics, Ltd, Zürich, Switzerland, 1999.
15. Locci, N. and Marongiu, I., Modelling and testing a new linear reluctance motor, *International Conference on Electrical Machines (ICEM'92)*, Vol. 2, Manchester, U.K., 1992, pp. 706–710.
16. Rauch, S.E. and Johnson, L.J., Design principles of flux-switch alternators, *AIEE Trans.*, Part III, 74(12), 1261–1269, 1955.
17. Rosenmayr, M., Casat, Glavitsch, A., and Stemmler, H., Swissmetro—Power supply for a high-power-propulsion system with short stator linear motors, *15th International Conference on Magnetically Levitated Systems and Linear Drives Maglev'98*, Mount Fuji, Yamanashi, Japan, 1998, pp. 280–286.
18. Sanada, M., Morimoto, S., and Takeda, Y., Reluctance equalization design of multi flux barrier construction for linear synchronous reluctance motors, *2nd International Symposium on Linear Drives for Industrial Applications (LDIA'98)*, Tokyo, Japan, 1998, pp. 259–262.

19. Seok-Myeong, J. and Sang-Sub, J., Design and analysis of the linear homopolar synchronous motor for integrated magnetic propulsion and suspension, *2nd International Symposium on Linear Drives for Industrial Applications (LDIA'98)*, Tokyo, Japan, 1998, pp. 74–77.
20. Synchronous Linear Motor 1FN6, *The Electrical Gear Rack*, Siemens AG Industry Sector, Drive Technologies, Motion Control, Erlangen, Germany, 2008.
21. Yamada, H., *Handbook of Linear Motor Applications* (in Japanese), Kogyo Chosaki Publishing Co., Tokyo, Japan, 1986.

VI

Power Quality

S. Mark Halpin

S. Mark Halpin was born in 1965 in Sandersville, Georgia. He received his PhD from Auburn University, Auburn, Alabama, in 1992. He has worked in both the utility and production industries in many technical capacities. He held the Tennessee Valley Authority Endowed Professorship at Mississippi State University and is presently the Alabama Power Company Distinguished Professor at Auburn University. His areas of technical specialty include power and energy systems, electric utility system planning and operation, and power quality. He has been involved in IEEE and IEC standards related to power and energy systems for over 15 years, where he has made numerous technical, administrative, and leadership contributions. He is also active in the CIGRE and CIRED working groups, charged with the technical development of contributions for IEC standards. He has published over 150 articles, book chapters, and special publications related to power and energy, including technical, economic, and policy concerns.

Dr. Halpin received the IEEE Industry Applications Society Outstanding Young Member Award in 1997 and the IEEE Power Engineering Society Alabama Chapter Outstanding Engineer Award in 2005. He was elected to the grade of IEEE fellow in 2005. In 2006, he received the IEEE Charles Proteus Steinmetz Technical Field Award. He is a former president (2007–2008) of the IEEE Industry Applications Society and a distinguished lecturer for 2010 and 2011.

35

Introduction

S. Mark Halpin
Auburn University

Electric power quality has emerged as a major area of electric power engineering. The predominant reason for this emergence is the increase in sensitivity of end-use equipment. This chapter is devoted to various aspects of power quality as it impacts utility companies and their customers and includes material on (1) grounding, (2) voltage sags, (3) harmonics, (4) voltage flicker, and (5) long-term monitoring. While these five topics do not cover all aspects of power quality, they provide the reader with a broad-based overview that should serve to increase overall understanding of problems related to power quality.

Proper grounding of equipment is essential for safe and proper operation of sensitive electronic equipment. In times past, it was thought by some that equipment grounding as specified in the United States by the National Electric Code was in contrast with methods needed to insure power quality. Since those early times, significant evidence has emerged to support the position that, in the vast majority of instances, grounding according to the National Electric Code is essential to insure proper and trouble-free equipment operation, and also to insure the safety of associated personnel.

Other than poor grounding practices, voltage sags due primarily to system faults are probably the most significant of all power quality problems. Voltage sags due to short circuits are often seen at distances very remote from the fault point, thereby affecting a potentially large number of utility customers. Coupled with the wide-area impact of a fault event is the fact that there is no effective preventive for all power system faults. End-use equipment will, therefore, be exposed to short periods of reduced voltage which may or may not lead to malfunctions.

Like voltage sags, the concerns associated with flicker are also related to voltage variations. Voltage flicker, however, is tied to the likelihood of a human observer to become annoyed by the variations in the output of a lamp when the supply voltage amplitude is varying. In most cases, voltage flicker considers (at least approximately) periodic voltage fluctuations with frequencies less than about 30–35 Hz that are small in size. Human perception, rather than equipment malfunction, is the relevant factor when considering voltage flicker.

For many periodic waveform (either voltage or current) variations, the power of classical Fourier series theory can be applied. The terms in the Fourier series are called harmonics; relevant harmonic terms may have frequencies above or below the fundamental power system frequency. In most cases, nonfundamental frequency equipment currents produce voltages in the power delivery system at those same frequencies. This voltage distortion is present in the supply to other end-use equipment and can lead to improper operation of the equipment.

Harmonics, like most other power quality problems, require significant amounts of measured data in order for the problem to be diagnosed accurately. Monitoring may be short- or long-term and may be relatively cheap or very costly and often represents the majority of the work required to develop power quality solutions.

In summary, the power quality problems associated with grounding, voltage sags, harmonics, and voltage flicker are those most often encountered in practice. It should be recognized that the voltage and current transients associated with common events like lightning strokes and capacitor switching can also negatively impact end-use equipment. Because transients are covered in a separate chapter of this book, they are not considered further in this chapter.

36

Wiring and Grounding for Power Quality

Christopher J. Melhorn
EPRI PEAC Corporation

Perhaps one of the most common problems related to power quality is wiring and grounding. It has been reported that approximately 70%–80% of all power quality related problems can be attributed to faulty connections and/or wiring. This chapter describes wiring and grounding issues as they relate to power quality. It is not intended to replace or supercede the National Electric Code (NEC) or any local codes concerning grounding.

36.1 Definitions and Standards

Defining grounding terminology is outside the scope of this chapter. There are several publications on the topic of grounding that define grounding terminology in various levels of detail. The reader is referred to these publications for the definitions of grounding terminology.

The following is a list of standards and recommended practice pertaining to wiring and grounding issues. See the section on References for complete information.

National Electric Code Handbook, 1996 edition.

IEEE Std. 1100–1999. *IEEE Recommended Practice for Powering and Grounding Electronic Equipment.*

IEEE Std. 142–1991. *IEEE Recommended Practice for Grounding Industrial and Commercial Power Systems.*

Guideline on Electrical Power for ADP Installations, Federal Information Processing Standards (FIPS) Publication 94, September 1983.

Electrical Power Systems Quality

36.1.1 The National Electric Code

NFPAs *National Electrical Code Handbook* pulls together all the extra facts, figures, and explanations readers need to interpret the 1999 NEC. It includes the entire text of the Code, plus expert commentary, real-world examples, diagrams, and illustrations that clarify requirements. Code text appears in blue type and commentary stands out in black. It also includes a user-friendly index that references article numbers to be consistent with the Code.

Several definitions of grounding terms pertinent to discussions in this article have been included for reader convenience. The following definitions were taken from various publications as cited.

36.1.2 From the *IEEE Dictionary—Std. 100*

Grounding: A conducting connection, whether intentional or accidental, by which an electric circuit or equipment is connected to the earth, or to some conducting body of relatively large extent that serves in place of the earth. It is used for establishing and maintaining the potential of the earth (or of the conducting body) or approximately that potential, on conductors connected to it; and for conducting ground current to and from the earth (or the conducting body).

36.1.3 Green Book (IEEE Std. 142) Definitions

Ungrounded system: A system, circuit, or apparatus without an intentional connection to ground, except through potential indicating or measuring devices or other very high impedance devices.

Grounded system: A system of conductors in which at least one conductor or point (usually the middle wire or neutral point of transformer or generator windings) is intentionally grounded, either solidly or through an impedance.

36.1.4 NEC Definitions

Refer to Figure 36.1.

Bonding jumper, main: The connector between the grounded circuit conductor (neutral) and the equipment-grounding conductor at the service entrance.

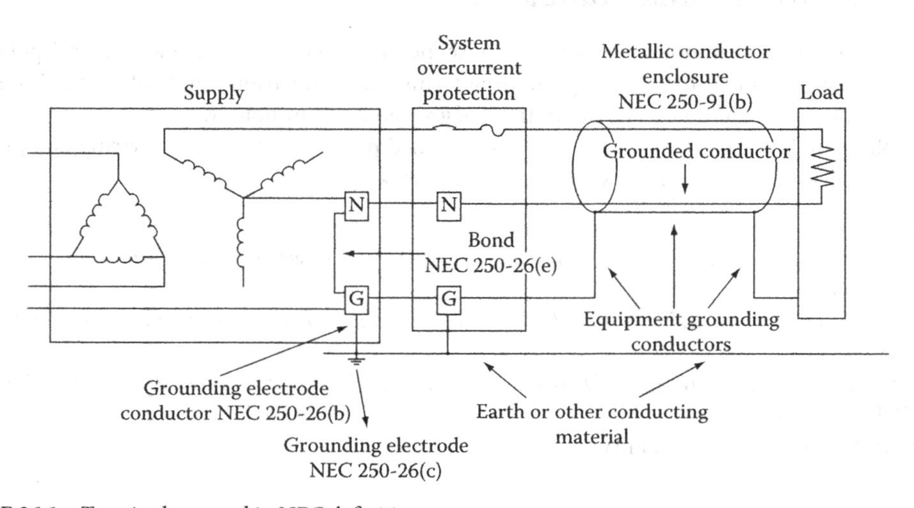

FIGURE 36.1 Terminology used in NEC definitions.

Conduit/Enclosure bond: (bonding definition) The permanent joining of metallic parts to form an electrically conductive path which will assure electrical continuity and the capacity to conduct safely any current likely to be imposed.

Grounded: Connected to earth or to some conducting body that serves in place of the earth.

Grounded conductor: A system or circuit conductor that is intentionally grounded (the grounded conductor is normally referred to as the neutral conductor).

Grounding conductor: A conductor used to connect equipment or the grounded circuit of a wiring system to a grounding electrode or electrodes.

Grounding conductor, equipment: The conductor used to connect the noncurrent-carrying metal parts of equipment, raceways, and other enclosures to the system grounded conductor and/or the grounding electrode conductor at the service equipment or at the source of a separately derived system.

Grounding electrode conductor: The conductor used to connect the grounding electrode to the equipment-grounding conductor and/or to the grounded conductor of the circuit at the service equipment or at the source of a separately derived system.

Grounding electrode: The grounding electrode shall be as near as practicable to and preferably in the same area as the grounding conductor connection to the system. The grounding electrode shall be: (1) the nearest available effectively grounded structural metal member of the structure; or (2) the nearest available effectively grounded metal water pipe; or (3) other electrodes (Sections 250–81 and 250–83) where electrodes specified in (1) and (2) are not available.

Grounding electrode system: Defined in NEC Section 250–81 as including: (a) metal underground water pipe; (b) metal frame of the building; (c) concrete-encased electrode; and (d) ground ring. When these elements are available, they are required to be bonded together to form the grounding electrode system. Where a metal underground water pipe is the only grounding electrode available, it must be supplemented by one of the grounding electrodes specified in Section 250–81 or 250–83.

Separately derived systems: A premises wiring system whose power is derived from generator, transformer, or converter windings and has no direct electrical connection, including a solidly connected grounded circuit conductor, to supply conductors originating in another system.

36.2 Reasons for Grounding

There are three basic reasons for grounding a power system: personal safety, protective device operation, and noise control. All three of these reasons will be addressed.

36.2.1 Personal Safety

The most important reason for grounding a device on a power system is personal safety. The safety ground, as it is sometimes called, is provided to reduce or eliminate the chance of a high touch potential if a fault occurs in a piece of electrical equipment. Touch potential is defined as the voltage potential between any two conducting materials that can be touched simultaneously by an individual or animal.

Figure 36.2 illustrates a dangerous touch potential situation. The "hot" conductor in the piece of equipment has come in contact with the case of the equipment. Under normal conditions, with the safety ground intact, the protective device would operate when this condition occurred. However, in Figure 36.2, the safety ground is missing. This allows the case of the equipment to float above ground since the case of the equipment is not grounded through its base. In other words, the voltage potential between the equipment case and ground is the same as the voltage potential between the hot leg and ground. If the operator would come in contact with the case and ground (the floor), serious injury could result.

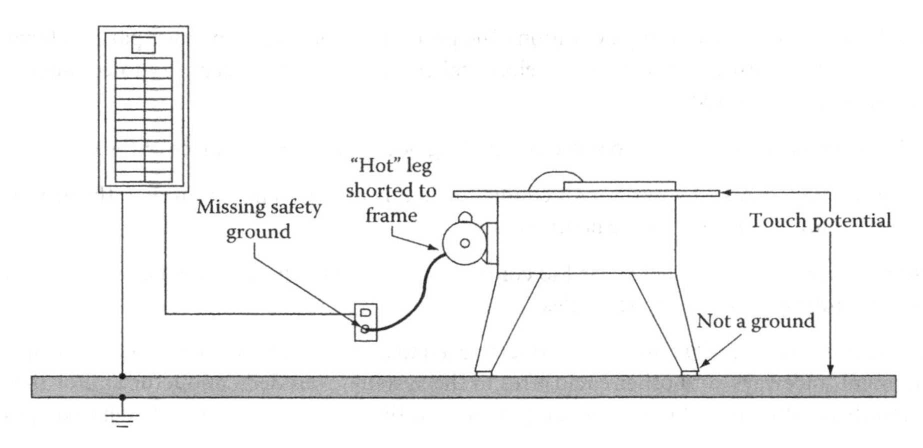

FIGURE 36.2 Illustration of a dangerous touch potential situation.

In recent years, manufacturers of handheld equipment, drills, saws, hair dryers, etc., have developed double insulated equipment. This equipment generally does not have a safety ground. However, there is never any conducting material for the operator to contact and therefore there is no touch potential hazard. If the equipment becomes faulted, the case or housing of the equipment is not energized.

36.2.2 Protective Device Operation

As mentioned in the previous section, there must be a path for fault current to return to the source if protective devices are to operate during fault conditions. The NEC requires that an effective grounding path must be mechanically and electrically continuous (NEC 250–51), have the capacity to carry any fault currents imposed on it without damage (NEC 250–75). The NEC also states that the ground path must have sufficiently low impedance to limit the voltage and facilitate protective device operation. Finally, the earth cannot serve as the equipment-grounding path (NEC-250–91(c)).

The formula to determine the maximum circuit impedance for the grounding path is

$$\text{Ground Path Impedance} = \frac{\text{Maximum Voltage to Ground}}{\text{Overcurrect Protection Rating} \times 5}$$

Table 36.1 gives examples of maximum ground path circuit impedances required for proper protective device operation.

36.2.3 Noise Control

Noise control is the third main reason for grounding. Noise is defined as unwanted voltages and currents on a grounding system. This includes signals from all sources whether it is radiated or conducted. As stated, the primary reason for grounding is safety and is regulated by the NEC and local codes. Any changes to the grounding system to improve performance or eliminate noise control must be in addition to the minimum NEC requirements.

TABLE 36.1 Example Ground Impedance Values

Protective Device Rating (A)	Voltage to Ground 120 V (Ω)	Voltage to Ground 277 V (Ω)
20	1.20	2.77
40	0.60	1.39
50	0.48	1.11
60	0.40	0.92
100	0.24	0.55

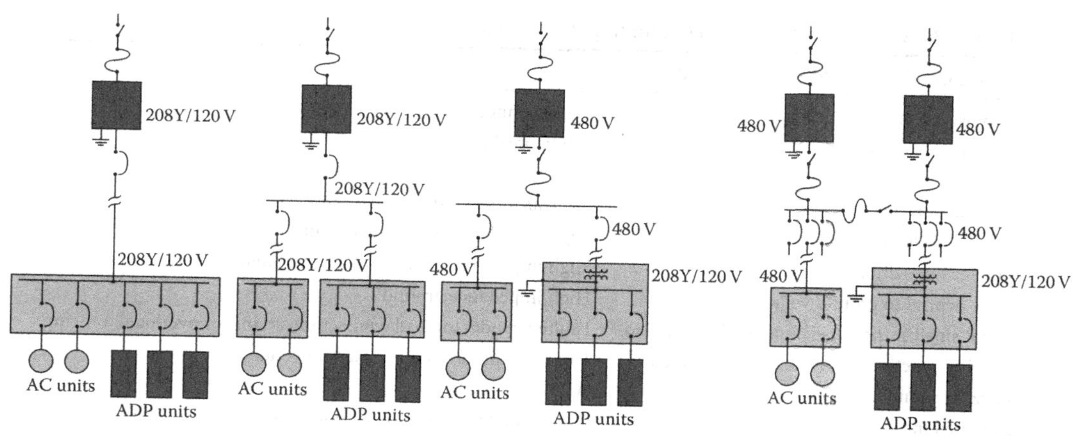

FIGURE 36.3 Separation of loads for noise control.

When potential differences occur between different grounding systems, insulation can be stressed and circulating currents can be created in low voltage cables (e.g., communications cables). In today's electrical environment, buildings that are separated by large physical distances are typically tied together via a communication circuit. An example of this would be a college campus that may cover several square miles. Each building has its own grounding system. If these grounding systems are not tied together, a potential difference on the grounding circuit for the communication cable can occur. The idea behind grounding for noise control is to create an equipotential grounding system, which in turn limits or even eliminates the potential differences between the grounding systems. If the there is an equipotential grounding system and currents are injected into the ground system, the potential of the whole grounding system will rise and fall and potential differences will not occur.

Supplemental conductors, ground reference grids, and ground plates can all be used to improve the performance of the system as it relates to power quality. Optically isolated communications can also improve the performance of the system. By using the opto-isolators, connecting the communications to different ground planes is avoided. All improvements to the grounding system must be done in addition to the requirements for safety.

Separation of loads is another method used to control noise. Figure 36.3 illustrates this point. Figure 36.3 shows four different connection schemes. Each system from left to right improves noise control.

As seen in Figure 36.3, the best case would be the complete separation (system on the far right) of the ADP units from the motor loads and other equipment. Conversely, the worst condition is on the left of Figure 36.3 where the ADP units are served from the same circuit as the motor loads.

36.3 Typical Wiring and Grounding Problems

In this section, typical wiring and grounding problems, as related to power quality, are presented. Possible solutions are given for these problems as well as the possible causes for the problems being observed on the grounding system Dugan et al. (1995), and Holt (1993). (See Table 36.2.)

The following list is just a sample of problems that can occur on the grounding system.

- Isolated grounds
- Ground loops
- Missing safety ground
- Multiple neutral-to-ground bonds
- Additional ground rods
- Insufficient neutral conductors

TABLE 36.2 Typical Wiring and Grounding Problems and Causes

Wiring Condition or Problem Observed	Possible Cause
Impulse, voltage drop out	Loose connections
Impulse, voltage drop out	Faulty breaker
Ground currents	Extra neutral-to-ground bond
Ground currents	Neutral-to-ground reversal
Extreme voltage fluctuations	High impedance in neutral circuit
Voltage fluctuations	High impedance neutral-to-ground bonds
High neutral to ground voltage	High impedance ground
Burnt smell at the panel, junction box, or load	Faulted conductor, bad connection, arcing, or overloaded wiring
Panel or junction box is warm to the touch	Faulty circuit breaker or bad connection
Buzzing sound	Arcing
Scorched insulation	Overloaded wiring, faulted conductor, or bad connection
Scorched panel or junction box	Bad connection, faulted conductor
No voltage at load equipment	Tripped breaker, bad connection, or faulted conductor
Intermittent voltage at the load equipment	Bad connection or arcing

36.3.1 Insulated Grounds

Insulated grounds in themselves are not a grounding problem. However, improperly used insulated grounds can be a problem. Insulated grounds are used to control noise on the grounding system. This is accomplished by using insulated ground receptacles, which are indicated by a "Δ" on the face of the outlet. Insulated ground receptacles are often orange in color. Figure 36.4 illustrates a properly wired insulated ground circuit.

The 1996 NEC has this to say about insulated grounds.

> **NEC 250-74. Connecting Receptacle Grounding Terminal to Box.** An equipment bonding jumper shall be used to connect the grounding terminal of a grounding-type receptacle to a grounded box.
>
> *Exception No. 4. Where required for the reduction of electrical noise (electromagnetic interference) on the grounding circuit, a receptacle in which the grounding terminal is purposely insulated from the receptacle mounting means shall be permitted. The receptacle grounding terminal shall be grounded by an insulated equipment grounding conductor run with the circuit conductors. This grounding conductor shall be permitted to pass through one or more panelboards without connection to the panelboard grounding terminal as permitted in Section 384–20, Exception so as to terminate*

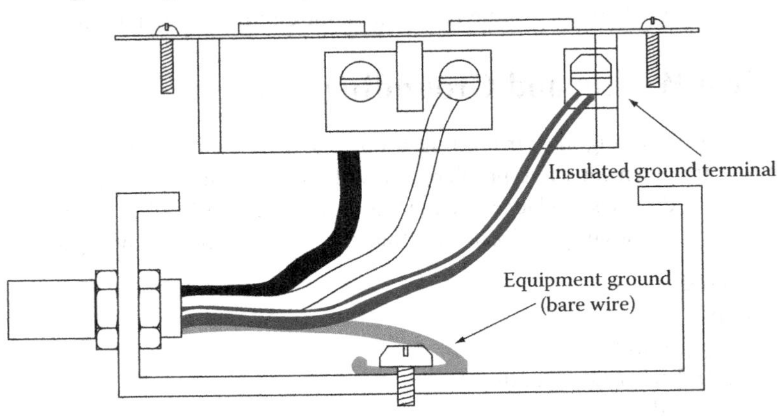

FIGURE 36.4 Properly wired isolated ground circuit.

within the same building or structure directly at an equipment grounding conductor terminal of the applicable derived system or source.

(FPN): Use of an isolated equipment grounding conductor does not relieve the requirement for grounding the raceway system and outlet box.

NEC 517–16. Receptacles with Insulated Grounding Terminals. Receptacles with insulated grounding terminals, as permitted in Section 250–74, Exception No. 4, shall be identified; such identification shall be visible after installation.

(FPN): Caution is important in specifying such a system with receptacles having insulated grounding terminals, since the grounding impedance is controlled only by the grounding conductors and does not benefit functionally from any parallel grounding paths.

The following is a list of pitfalls that should be avoided when installing insulated ground circuits.

- Running an insulated ground circuit to a regular receptacle.
- Sharing the conduit of an insulated ground circuit with another circuit.
- Installing an insulated ground receptacle in a two-gang box with another circuit.
- Not running the insulated ground circuit in a metal cable armor or conduit.
- Do not assume that an insulated ground receptacle has a truly insulated ground.

36.3.2 Ground Loops

Ground loops can occur for several reasons. One is when two or more pieces of equipment share a common circuit like a communication circuit, but have separate grounding systems (Figure 36.5).

To avoid this problem, only one ground should be used for grounding systems in a building. More than one grounding electrode can be used, but they must be tied together (NEC 250–81, 250–83, and 250–84) as illustrated in Figure 36.6.

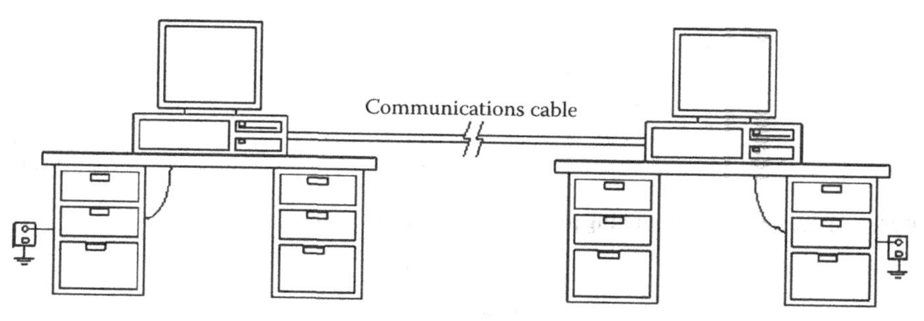

FIGURE 36.5 Circuit with a ground loop.

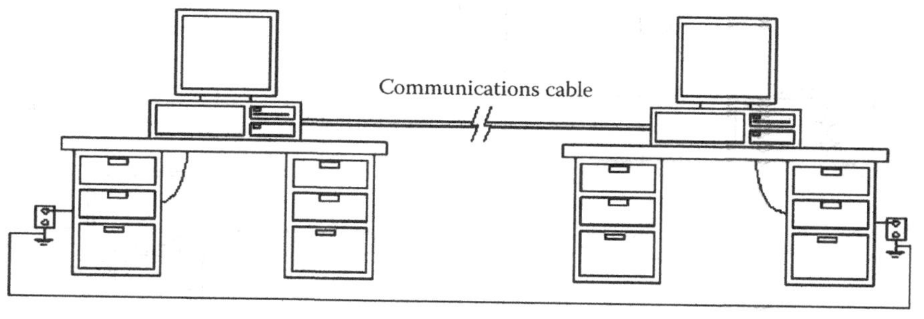

FIGURE 36.6 Grounding electrodes must be bonded together.

36.3.3 Missing Safety Ground

As discussed previously, a missing safety ground poses a serious problem. Missing safety grounds usually occur because the safety ground has been bypassed. This is typical in buildings where the 120-V outlets only have two conductors. Modern equipment is typically equipped with a plug that has three prongs, one of which is a ground prong. When using this equipment on a two-prong outlet, a grounding plug adapter or "cheater plug" can be employed provided there is an equipment ground present in the outlet box. This device allows the use of a three-prong device in a two-prong outlet. When properly connected, the safety ground remains intact. Figure 36.7 illustrates the proper use of the cheater plug.

If an equipment ground is not present in the outlet box, then the grounding plug adapter should not be used. If the equipment grounding conductor is present, the preferred method for solving the missing safety ground problem is to install a new three-prong outlet in the outlet box. This method insures that the grounding conductor will not be bypassed. The NEC discusses equipment grounding conductors in detail in Section 250—Grounding.

36.3.4 Multiple Neutral to Ground Bonds

Another misconception when grounding equipment is that the neutral must be tied to the grounding conductor. Only one neutral-to-ground bond is permitted in a system or sub-system. This typically occurs at the service entrance to a facility unless there is a separately derived system. A separately derived system is defined as a system that receives its power from the windings of a transformer, generator, or some type of converter. Separately derived systems must be grounded in accordance with NEC 250–26.

The neutral should be kept separate from the grounding conductor in all panels and junction boxes that are downline from the service entrance. Extra neutral-to-ground bonds in a power system will cause neutral currents to flow on the ground system. This flow of current on the ground system occurs because of the parallel paths. Figures 36.8 and 36.9 illustrate this effect.

As seen in Figure 36.9, neutral current can find its way onto the ground system due to the extra neutral-to-ground bond in the secondary panel board. Notice that not only will current flow in the

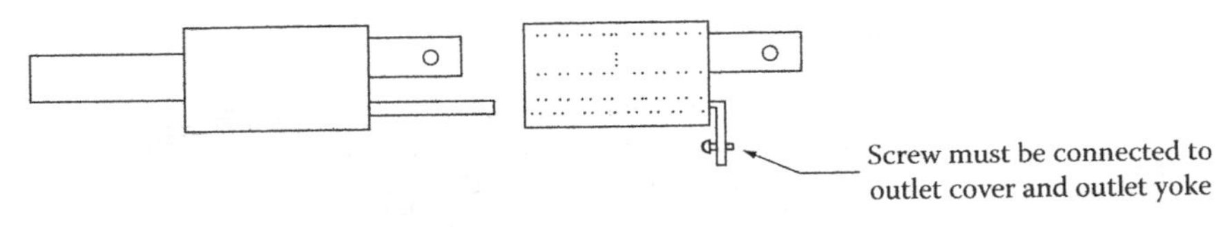

FIGURE 36.7 Proper use of a grounding plug adapter or "cheater plug."

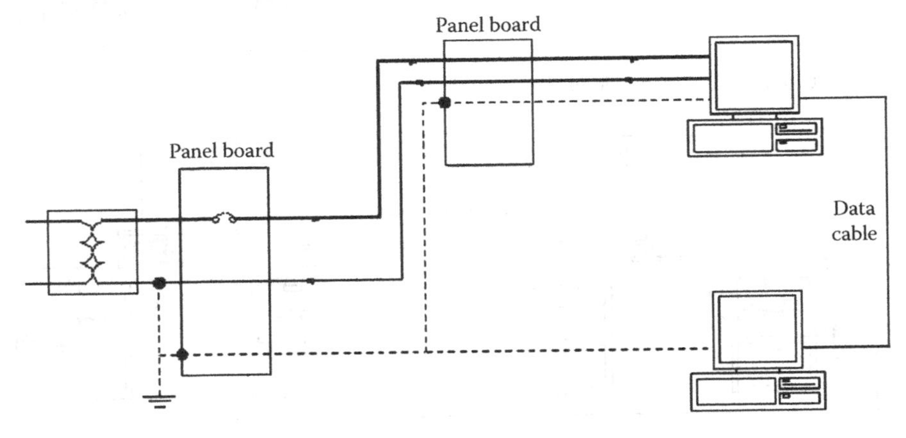

FIGURE 36.8 Neutral current flow with one neutral-to-ground bond.

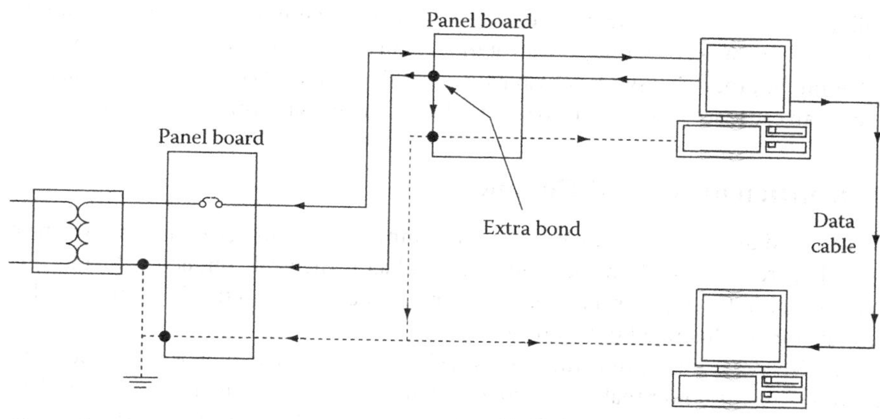

FIGURE 36.9 Neutral current flow with and extra neutral-to-ground bond.

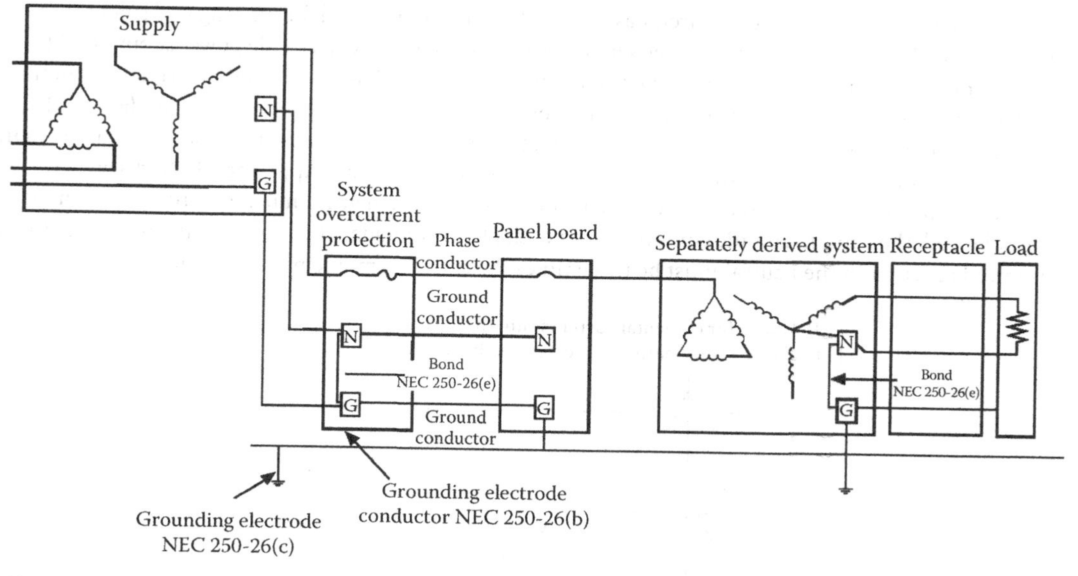

FIGURE 36.10 Example of the use of a separately derived system.

ground wire for the power system, but currents can flow in the shield wire for the communication cable between the two PCs.

If the neutral-to-ground bond needs to be reestablished (high neutral-to-ground voltages), this can be accomplished by creating a separately derived system as defined above. Figure 36.10 illustrates a separately derived system.

36.3.5 Additional Ground Rods

Additional ground rods are another common problem in grounding systems. Ground rods for a facility or building should be part of the grounding system. The ground rods should be connected where all the building grounding electrodes are bonded together. Isolated grounds can be used as described in the NEC's Isolated Ground section, but should not be confused with isolated ground rods, which are not permitted.

The main problem with additional ground rods is that they create secondary paths for transient currents, such as lightning strikes, to flow. When a facility incorporates the use of one ground rod, any currents caused by lightning will enter the building ground system at one point. The ground potential of the entire facility will rise and fall together. However, if there is more than one ground rod for the facility, the transient current enters the facility's grounding system at more than one location and a portion of the transient current will flow on the grounding system causing the ground potential of equipment to rise at different levels. This, in turn, can cause severe transient voltage problems and possible conductor overload conditions.

36.3.6 Insufficient Neutral Conductor

With the increased use of electronic equipment in commercial buildings, there is a growing concern for the increased current imposed on the grounded conductor (neutral conductor) Melhorn (1995, 1996, 1997). With a typical three-phase load that is balanced, there is theoretically no current flowing in the neutral conductor, as illustrated in Figure 36.11.

However, PCs, laser printers, and other pieces of electronic office equipment all use the same basic technology for receiving the power that they need to operate. Figure 36.12 illustrates the typical power supply of a PC. The input power is generally 120 V AC, single phase. The internal electronic parts require various levels of DC voltage (e.g., ±5, 12 V DC) to operate. This DC voltage is obtained by converting the AC voltage through some type of rectifier circuit as shown. The capacitor is used for filtering and smoothing the rectified AC signal. These types of power supplies are referred to as switch mode power supplies (SMPS).

The concern with devices that incorporate the use of SMPS is that they introduce triplen harmonics into the power system. Triplen harmonics are those that are odd multiples of the fundamental frequency component ($h = 3, 9, 15, 21, \ldots$). For a system that has balanced single-phase loads as illustrated in Figure 36.13, fundamental and third harmonic components are present. Applying Kirchoff's current law at node N shows that the fundamental current component in the neutral must be zero. But when loads are balanced, the third harmonic components in each phase coincide. Therefore, the magnitude of third harmonic current in the neutral must be three times the third harmonic phase current.

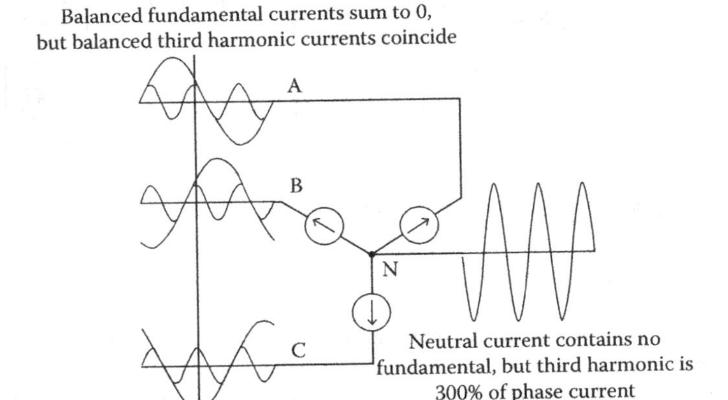

FIGURE 36.11 A balanced three-phase system.

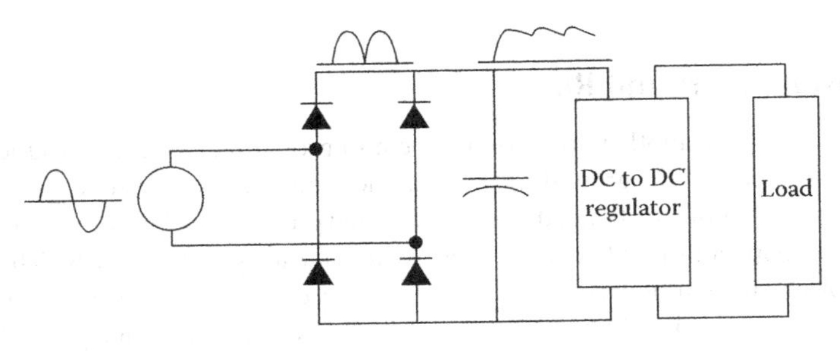

FIGURE 36.12 The basic one-line for a SMPS.

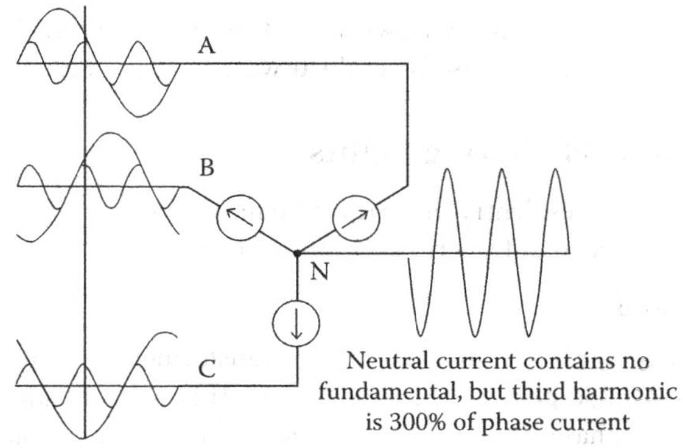

FIGURE 36.13 Balanced single-phase loads.

This becomes a problem in office buildings when multiple single-phase loads are supplied from a three-phase system. Separate neutral wires are run with each circuit, therefore the neutral current will be equivalent to the line current. However, when the multiple neutral currents are returned to the panel or transformer serving the loads, the triplen currents will add in the common neutral for the panel and this can cause overheating and eventually even cause failure of the neutral conductor. If office partitions are used, the same, often undersized neutral conductor is run in the partition with three-phase conductors. Each receptacle is fed from a separate phase in order to balance the load current. However, a single neutral is usually shared by all three phases. This can lead to disastrous results if the partition electrical receptacles are used to supply nonlinear loads rich in triplen harmonics.

Under the worst conditions, the neutral current will never exceed 173% of the phase current. Figure 36.13 illustrates a case where a three-phase panel is used to serve multiple single-phase SMPS PCs.

36.3.7 Summary

As discussed previously, the three main reasons for grounding in electrical systems are

1. Personal safety
2. Proper protective device operation
3. Noise control

By following the guidelines found below, the objectives for grounding can be accomplished.

- All equipment should have a safety ground. A safety ground conductor.
- Avoid load currents on the grounding system.
- Place all equipment in a system on the same equipotential reference.

Table 36.3 summarizes typical wiring and grounding issues.

TABLE 36.3 Summary of Wiring and Grounding Issues

Summary Issues
Good power quality and noise control practices do not conflict with safety requirements.
Wiring and grounding problems cause a majority of equipment interference problems.
Make an effort to put sensitive equipment on dedicated circuits.
The grounded conductor, neutral conductor, should be bonded to the ground at the transformer or main panel, but not at other panel down line except as allowed by separately derived systems.

36.4 Case Study

This section presents a case study involving wiring and grounding issues. The purpose of this case study is to inform the reader on the procedures used to evaluate wiring and grounding problems and present solutions.

36.4.1 Case Study: Flickering Lights

This case study concerns a residential electrical system. The homeowners were experiencing light flicker when loads were energized and deenergized in their homes.

36.4.1.1 Background

Residential systems are served from single-phase transformers employing a spilt secondary winding, often referred to as a single-phase three-wire system. This type of transformer is used to deliver both 120- and 240-V single-phase power to the residential loads. The primary of the transformer is often served from a 12 to 15 kV distribution system by the local utility. Figure 36.14 illustrates the concept of a split-phase system.

When this type of service is operating properly, 120 V can be measured from either leg to the neutral conductor. Due to the polarity of the secondary windings in the transformer, the polarity of each 120 V leg is opposite the other, thus allowing a total of 240 V between the legs as illustrated. The proper operation of this type of system is dependent on the physical connection of the neutral conductor or center tap of the secondary winding. If the neutral connection is removed, 240 V will remain across the two legs, but the line-to-neutral voltage for either phase can be shifted, causing either a low or high voltage from line to neutral.

Most loads in a residential dwelling, i.e., lighting, televisions, microwaves, home electronics, etc., are operated from 120 V. However, there are a few major loads that incorporate the use of the 240 V available. These loads include electric water heaters, electric stoves and ovens, heat pumps, etc.

36.4.1.2 The Problem

In this case, there were problems in the residence that caused the homeowner to question the integrity of the power system serving his home. On occasion, the lights would flicker erratically when the washing machine and dryer were operating at the same time. When large single-phase loads were operated, low power incandescent light bulb intensity would flicker.

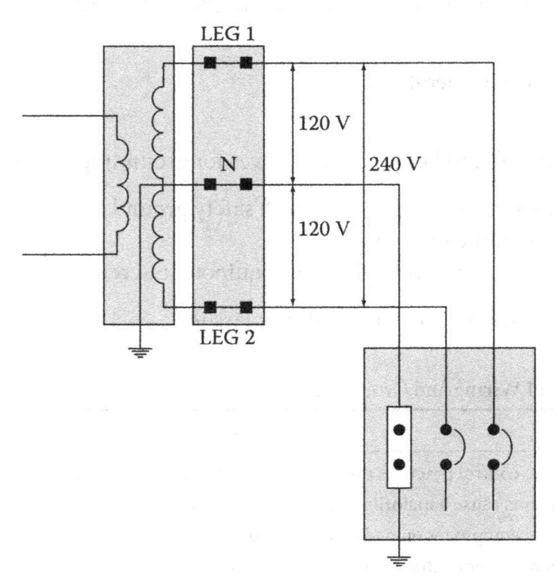

FIGURE 36.14 Split-phase system serving a residential customer.

Measurements were performed at several 120-V outlets throughout the house. When the microwave was operated, the voltage at several of the 120-V outlets would increase from 120 V nominal to 128 V. The voltage would return to normal after the microwave was turned off. The voltage would also increase when a 1500 W space heater was operated. It was determined that the voltage would decrease to approximately 112 V on the leg from which the large load was served. After the measurements confirmed suspicions of high and low voltages during heavy load operation, finding the source of the problem was the next task at hand.

The hunt began at the service entrance to the house. A visual inspection was made of the meter base and socket after the meter was removed by the local utility. It was discovered that one of the neutral connectors was loose. While attempting to tighten this connector, the connector fell off of the meter socket into the bottom of the meter base (see Figure 36.15). Could this loose connector have been the cause of the flickering voltage? Let's examine the effects of the loose neutral connection.

Figures 36.16 and 36.17 will be referred to several times during this discussion. Under normal conditions with a solid neutral connection (Figure 36.16), load current flows through each leg and is returned to the source through the neutral conductor. There is very little impedance in either the hot or the neutral conductor; therefore, no appreciable voltage drop exists.

When the neutral is loose or missing, a significant voltage can develop across the neutral connection in the meter base, as illustrated in Figure 36.17. When a large load is connected across Leg 1 to N and the other leg is lightly loaded (i.e., Leg 1 to N is approximately 10 times the load on Leg 2 to N), the current flowing through

FIGURE 36.15 Actual residential meter base. Notice the missing neutral clamp on load side of meter.

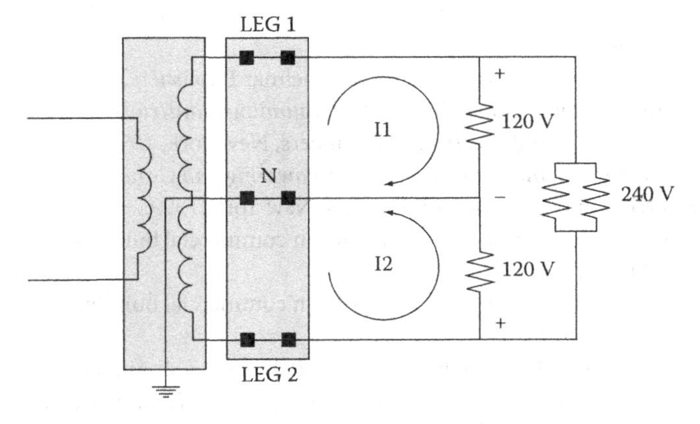

FIGURE 36.16 The effects of a solid neutral connection in the meter base.

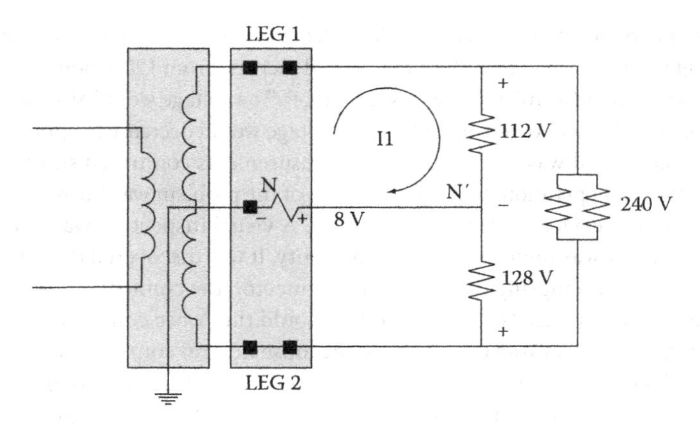

FIGURE 36.17 The effects of a loose neutral connection in the meter base.

the neutral will develop a voltage across the loose connection. This voltage is in phase with the voltage from Leg 1 to N′ (see Figure 36.17) and the total voltage from Leg 1 to N will be 120 V. However, the voltage supplied to any loads connected from Leg 2 to N′ will rise to 128 V, as illustrated in Figure 36.17. The total voltage across the Leg 1 and Leg 2 must remain constant at 240 V. It should be noted that the voltage from Leg 2 to N will be 120 V since the voltage across the loose connection is 180° out of phase with the Leg 2 to N′ voltage.

Therefore, with the missing neutral connection, the voltage from Leg 2 to N′ would rise, causing the light flicker. This explains the rise in voltage when a large load was energized on the system.

36.4.1.3 The Solution

The solution in this case was simple—replace the failed connector.

36.4.1.4 Conclusions

Over time, the neutral connector had become loose. This loose connection caused heating, which in turn caused the threads on the connector to become worn, and the connector failed. After replacing the connector in the meter base, the flickering light phenomena disappeared.

On systems of this type, if a voltage rise occurs when loads are energized, it is a good indication that the neutral connection may be loose or missing.

References

Dugan, R.C. et al., *Electrical Power Systems Quality*, McGraw-Hill, New York, 1995.

Guideline on Electrical Power for ADP Installations, Federal Information Processing Standards (FIPS) Publication 94, September 1983.

Holt, C.M., *Understanding the National Electric Code*, Delmar Publishers, Inc., Albany, NY, 1993.

IEEE Std. 142-1991, *IEEE Recommended Practice for Grounding Industrial and Commercial Power Systems*, The Institute of Electrical and Electronics Engineers, New York, 1991.

IEEE Std. 1100-1999, *IEEE Recommended Practice for Powering and Grounding Electronic Equipment*, The Institute of Electrical and Electronics Engineers, New York, 1999.

Melhorn, C.J., Coping with non-linear computer loads in commercial buildings—Part I, *Emf-Emi Control* 2, 5, September/October, 1995.

Melhorn, C.J., Coping with non-linear computer loads in commercial building—Part II, *Emf-Emi Control* 2, 6, January/February, 1996.

Melhorn, C., Flickering lights—A case of faulty wiring, *PQToday*, 3, 4, August 1997.

National Fire Protection Association, *National Electrical Code Handbook*, National Fire Protection Agency, Quincy, MA, 1996.

37

Harmonics in Power Systems

S. Mark Halpin
Auburn University

Power system harmonics are not a new topic, but the proliferation of high-power electronics used in motor drives and power controllers has necessitated increased research and development in many areas relating to harmonics. For many years, high-voltage direct current (HVDC) stations have been a major focus area for the study of power system harmonics due to their rectifier and inverter stations. Roughly two decades ago, electronic devices that could handle several kW up to several MW became commercially viable and reliable products. This technological advance in electronics led to the widespread use of numerous converter topologies, all of which represent nonlinear elements in the power system.

Even though the power semiconductor converter is largely responsible for the large-scale interest in power system harmonics, other types of equipment also present a nonlinear characteristic to the power system. In broad terms, loads that produce harmonics can be grouped into three main categories covering (1) arcing loads, (2) semiconductor converter loads, and (3) loads with magnetic saturation of iron cores. Arcing loads, like electric arc furnaces and florescent lamps, tend to produce harmonics across a wide range of frequencies with a generally decreasing relationship with frequency. Semiconductor loads, such as adjustable-speed motor drives, tend to produce certain harmonic patterns with relatively predictable amplitudes at known harmonics. Saturated magnetic elements, like overexcited transformers, also tend to produce certain "characteristic" harmonics. Like arcing loads, both semiconductor converters and saturated magnetics produce harmonics that generally decrease with frequency.

Regardless of the load category, the same fundamental theory can be used to study power quality problems associated with harmonics. In most cases, any periodic distorted power system waveform (voltage, current, flux, etc.) can be represented as a series consisting of a DC term and an infinite sum of sinusoidal terms as shown in Equation 37.1 where ω_0 is the fundamental power frequency.

$$f(t) = F_0 + \sum_{i=1}^{\infty} \sqrt{2}F_i \cos(i\omega_0 t + \theta_i) \tag{37.1}$$

A vast amount of theoretical mathematics has been devoted to the evaluation of the terms in the infinite sum in Equation 37.1, but such rigor is beyond the scope of this chapter. For the purposes here, it is reasonable to presume that instrumentation is available that will provide both the magnitude F_i and the phase angle θ_i for each term in the series. Taken together, the magnitude and phase of the ith term completely describe the ith harmonic.

It should be noted that not all loads produce harmonics that are integer multiples of the power frequency. These noninteger multiple harmonics are generally referred to as interharmonics and are commonly produced by arcing loads and cycloconverters. All harmonic terms, both integer and non-integer multiples of the power frequency, are analytically treated in the same manner, usually based on the principle of superposition.

In practice, the infinite sum in Equation 37.1 is reduced to about 50 terms; most measuring instruments do not report harmonics higher than the 50th multiple (2500–3000 Hz for 50–60 Hz systems). The reporting can be in the form of a tabular listing of harmonic magnitudes and angles or in the form of a magnitude and phase spectrum. In each case, the information provided is the same and can be used to reproduce the original waveform by direct substitution into Equation 37.1 with satisfactory accuracy. As an example, Figure 37.1 shows the (primary) current waveform drawn by a small industrial plant. Table 37.1 shows a table of the first 31 harmonic magnitudes and angles. Figure 37.2 shows a bar graph magnitude spectrum for this same waveform. These data are widely available from many commercial instruments; the choice of instrument makes little difference in most cases.

A fundamental presumption when analyzing distorted waveforms using Fourier methods is that the waveform is in steady state. In practice, waveform distortion varies widely and is dependent on both load levels and system conditions. It is typical to assume that a steady-state condition exists at the instant at which the measurement is taken, but the next measurement at the next time could be markedly different. As examples, Figures 37.3 and 37.4 show time plots of fifth harmonic voltage and the total harmonic distortion, respectively, of the same waveform measured on a 115 kV transmission system

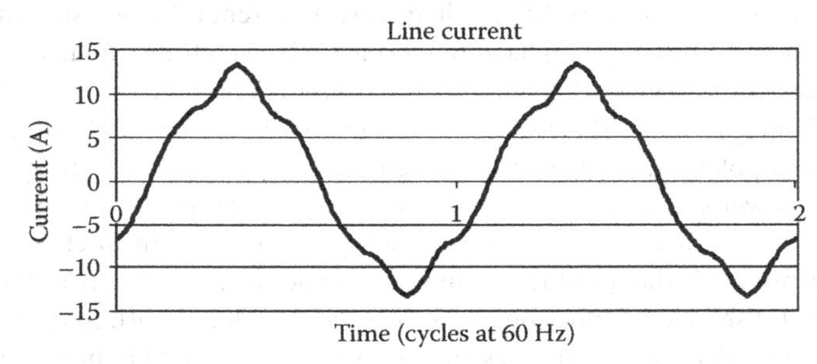

FIGURE 37.1 Current waveform.

TABLE 37.1 Current Harmonic Magnitudes and Phase Angles

Harmonic #	Current (A_{rms})	Phase (Deg)	Harmonic #	Current (A_{rms})	Phase (Deg)
1	8.36	−65	2	0.01	−167
3	0.13	43	4	0.01	95
5	0.76	102	6	0.01	8
7	0.21	−129	8	0	−148
9	0.02	−94	10	0	78
11	0.08	28	12	0	−89
13	0.04	−172	14	0	126
15	0	159	16	0	45
17	0.02	−18	18	0	−117
19	0.01	153	20	0	22
21	0	119	22	0	26
23	0.01	−76	24	0	143
25	0	0	26	0	150
27	0	74	28	0	143
29	0	50	30	0	−13
31	0	−180			

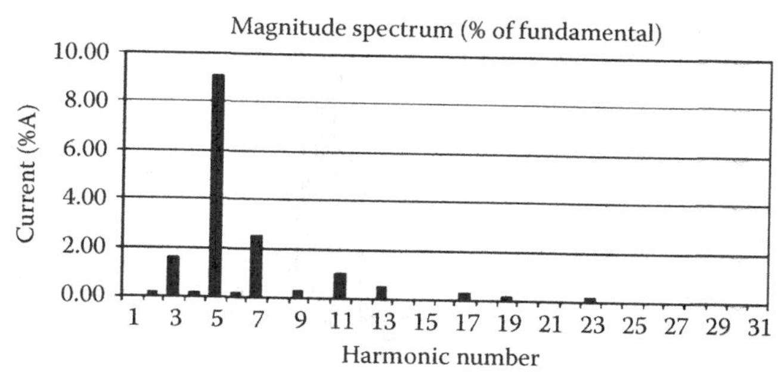

FIGURE 37.2 Harmonic magnitude spectrum.

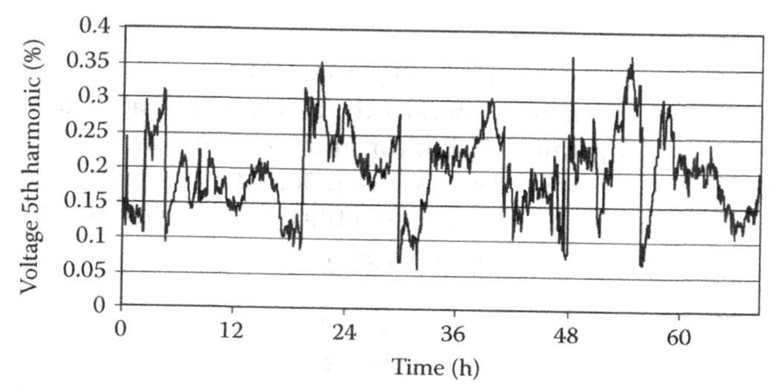

FIGURE 37.3 Example of time-varying nature of harmonics.

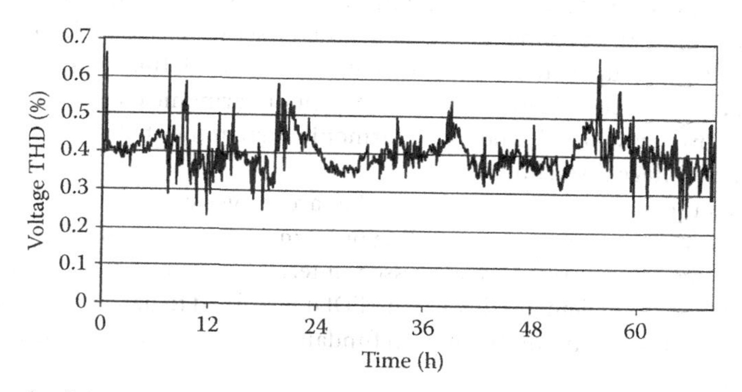

FIGURE 37.4 Example of time-varying nature of voltage THD.

near a 5 MW customer. Note that the THD is fundamentally defined in Equation 37.2, with 50 often used in practice as the upper limit on the infinite summation.

$$\text{THD}(\%) = \frac{\sqrt{\sum_{i=2}^{\infty} F_i^2}}{F_1} * 100\% \tag{37.2}$$

Because harmonic levels are never constant, it is difficult to establish utility-side or manufacturing-side limits for these quantities. In general, a probabilistic representation is used to describe harmonic quantities in terms of percentiles. Often, the 95th and 99th percentiles are used for design or operating limits. Figure 37.5 shows a histogram of the voltage THD in Figure 37.4, and also includes a cumulative probability curve derived from the frequency distribution. Any percentile of interest can be readily calculated from the cumulative probability curve.

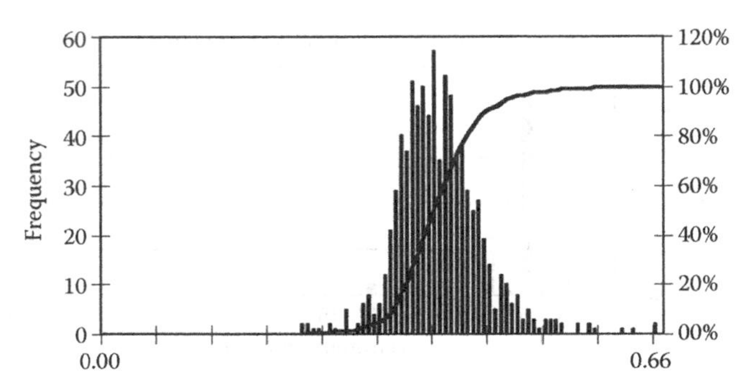

FIGURE 37.5 Probabilistic representation of voltage THD.

Both the Institute of Electrical and Electronics Engineers (IEEE) and the International Electrotechnical Commission (IEC) recognize the need to consider the time-varying nature of harmonics when determining harmonic levels that are permissible. Both organizations publish harmonic limits, but the degree to which the various limits can be applied varies widely. Both IEEE and IEC publish "system-level" harmonic limits that are intended to be applied from the utility point-of-view in order to limit power system harmonics to acceptable levels. The IEC, however, goes further and also publishes harmonic limits for individual pieces of equipment.

The IEEE limits are covered in two documents, IEEE 519-1992 and IEEE 519A (draft). These documents suggest that harmonics in the power system be limited by two different methods. One set of harmonic limits is for the harmonic current that a user can inject into the utility system at the point where other customers are or could be (in the future) served. (Note that this point in the system is often called the point of common coupling, or PCC.) The other set of harmonic limits is for the harmonic voltage that the utility can supply to any customer at the PCC. With this two-part approach, customers insure that they do not inject an "unreasonable" amount of harmonic current into the system, and the utility insures that any "reasonable" amount of harmonic current injected by any and all customers does not lead to excessive voltage distortion.

Table 37.2 shows the harmonic current limits that are suggested for utility customers. The table is broken into various rows and columns depending on harmonic number, short circuit to load ratio, and voltage level. Note that all quantities are expressed in terms of a percentage of the maximum demand current (I_L in the table). Total demand distortion (TDD) is defined to be the rms value of all harmonics, in amperes, divided by the maximum (12 month) fundamental frequency load current, I_L, with this ratio then multiplied by 100%.

The intent of the harmonic current limits is to permit larger customers, who in concept pay a greater share of the cost of power delivery equipment, to inject a greater portion of the harmonic current (in amperes) that the utility can absorb without producing excessive voltage distortion. Furthermore, customers served at transmission level voltage have more restricted injection limits than do customers served at lower voltage because harmonics in the high voltage network have the potential to adversely impact a greater number of other users through voltage distortion.

Table 37.3 gives the IEEE 519-1992 voltage distortion limits. Similar to the current limits, the permissible distortion is decreased at higher voltage levels in an effort to minimize potential problems for the majority of system users. Note that Tables 37.2 and 37.3 are given here for illustrative purposes only; the reader is strongly advised to consider additional material listed at the end of this chapter prior to trying to apply the limits.

The IEC formulates similar limit tables with the same intent: limit harmonic current injections so that voltage distortion problems are not created; the utility will correct voltage distortion problems if they exist and if all customers are within the specified harmonic current limits. Because the numbers suggested by the IEC are similar (but not identical) to those given in Tables 37.2 and 37.3, the IEC tables for system-level harmonic limits given in IEC 1000-3-6 are not repeated here.

TABLE 37.2 IEEE-519 Harmonic Current Limits

I_{SC}/I_L [a]	h < 11	11 ≤ h < 17	17 ≤ h < 23	23 ≤ h < 35	35 ≤ h	TDD
			$V_{supply} \leq 69\,kV$			
<20[b]	4.0	2.0	1.5	0.6	0.3	5.0
20–50	7.0	3.5	2.5	1.0	0.5	8.0
50–100	10.0	4.5	4.0	1.5	0.7	12.0
100–1000	12.0	5.5	5.0	2.0	1.0	15.0
>1000	15.0	7.0	6.0	2.5	1.4	20.0
			$69\,kV < V_{supply} \leq 161\,kV$			
<20[b]	2.0	1.0	0.75	0.3	0.15	2.5
20–50	3.5	1.75	1.25	0.5	0.25	4.0
50–100	5.0	2.25	2.0	1.25	0.35	6.0
100–1000	6.0	2.75	2.5	1.0	0.5	7.5
>1000	7.5	3.5	3.0	1.25	0.7	10.0
			$V_{supply} > 161\,kV$			
<50	2.0	1.0	0.75	0.3	0.15	2.5
≥50	3.5	1.75	1.25	0.5	0.25	4.0

Note: Even harmonics are limited to 25% of the odd harmonic limits above. Current distortions that result in a DC offset, e.g., half wave converters, are not allowed.

[a] I_{SC} = maximum short-circuit current at PCC; I_L = maximum demand load current (fundamental frequency component) at PCC.

[b] All power generation equipment is limited to these values of current distortion, regardless of actual I_{SC}/I_L.

TABLE 37.3 IEEE 519-1992 Voltage Harmonic Limits

Bus Voltage at PCC (V_{L-L})	Individual Harmonic Voltage Distortion (%)	Total Voltage Distortion—THD_{V_n} (%)
$V_n \leq 69\,kV$	3.0	5.0
$69\,kV < V_n \leq 161\,kV$	1.5	2.5
$V_n > 161\,kV$	1.0	1.5

Note: High-voltage systems can have up to 2.0% THD where the cause is an HVDC terminal that will attenuate by the time it is tapped for a user.

While the IEEE harmonic limits are designed for application at the three-phase PCC, the IEC goes further and provides limits appropriate for single-phase and three-phase individual equipment types. The most notable feature of these equipment limits is the "mA per W" manner in which they are proposed. For a wide variety of harmonic-producing loads, the steady-state (normal operation) harmonic currents are limited by prescribing a certain harmonic current, in mA, for each watt of power rating. The IEC also provides a specific waveshape for some load types that represents the most distorted current waveform allowed. Equipment covered by such limits include personal computers (power supplies) and single-phase battery charging equipment.

Even though limits exist, problems related to harmonics often arise from single, large "point source" harmonic loads as well as from numerous distributed smaller loads. In these situations, it is necessary to conduct a measurement, modeling, and analysis campaign that is designed to gather data and develop a solution. As previously mentioned, there are many commercially available instruments that can provide harmonic measurement information both at a single "snapshot" in time as well as continuous monitoring over time. How this information is used to develop problem solutions, however, can be a very complex issue.

Computer-assisted harmonic studies generally require significantly more input data than load flow or short circuit studies. Because high frequencies (up to 2–3 kHz) are under consideration, it is important to have mathematically correct equipment models and the data to use in them. Assuming that this data is available, there are a variety of commercially available software tools for actually performing the studies.

Most harmonic studies are performed in the frequency domain using sinusoidal steady-state techniques. (Note that other techniques, including full time-domain simulation, are sometimes used for specific problems.) A power system equivalent circuit is prepared for each frequency to be analyzed (recall that the Fourier series representation of a waveform is based on harmonic terms of known frequencies), and then basic circuit analysis techniques are used to determine voltages and currents of interest at that frequency. Most harmonic producing loads are modeled using a current source at each frequency that the load produces (arc furnaces are sometimes modeled using voltage sources), and network currents and voltages are determined based on these load currents. Recognize that at each frequency, voltage and current solutions are obtained from an equivalent circuit that is valid at that frequency only; the principle of superposition is used to "reconstruct" the Fourier series for any desired quantity in the network from the solutions of multiple equivalent circuits. Depending on the software tool used, the results can be presented in tabular form, spectral form, or as a waveform as shown in Table 37.1 and Figures 37.1 and 37.2, respectively. An example voltage magnitude spectrum obtained from a harmonic study of a distribution primary circuit is shown in Figure 37.6.

Regardless of the presentation format of the results, it is possible to use this type of frequency-domain harmonic analysis procedure to predict the impact of harmonic producing loads at any location in any power system. However, it is often impractical to consider a complete model of a large system, especially when unbalanced conditions must be considered. Of particular importance, however, are the locations of capacitor banks.

When electrically in parallel with network inductive reactance, capacitor banks produce a parallel resonance condition that tends to amplify voltage harmonics for a given current harmonic injection. When electrically in series with network inductive reactance, capacitor banks produce a series resonance condition that tends to amplify current harmonics for a given voltage distortion. In either case, harmonic levels far in excess of what are expected can be produced. Fortunately, a relatively simple calculation procedure called a frequency scan, can be used to indicate potential resonance problems. Figure 37.7 shows an example of a frequency scan conducted on the positive sequence network model of a distribution circuit. Note that the distribution primary included the standard feeder optimization capacitors.

A frequency scan result is actually a plot of impedance vs. frequency. Two types of results are available: driving point and transfer impedance scans. The driving point frequency scan shown in Figure 37.7 indicates how much voltage would be produced at a given bus and frequency for a 1 A current injection

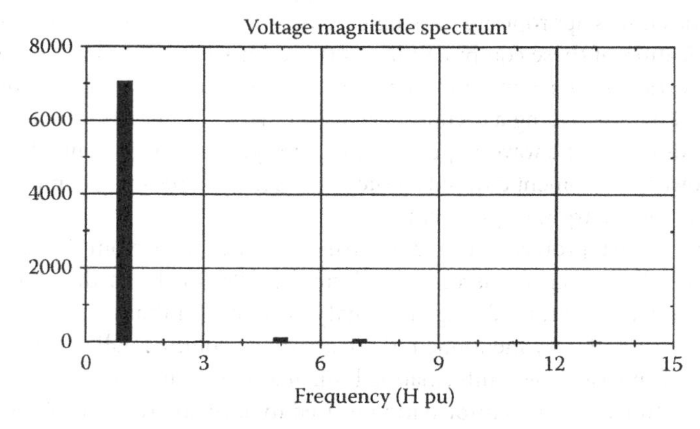

FIGURE 37.6 Sample magnitude spectrum results from a harmonic study.

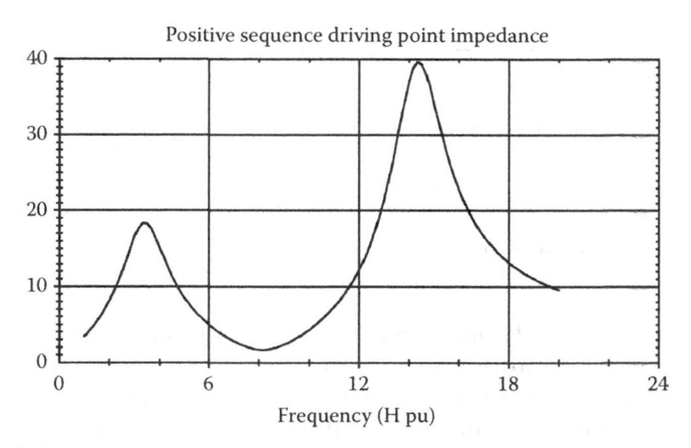

FIGURE 37.7 Sample frequency scan results.

at that same location and frequency. Where necessary, the principle of linearity can be used to scale the 1 A injection to the level actually injected by specific equipment. In other words, the driving point impedance predicts how a customer's harmonic producing load could impact the voltage at that load's terminals. Local maximums, or peaks, in the scan plot indicate parallel resonance conditions. Local minimums, or valleys, in the scan plot indicate series resonance.

A transfer impedance scan predicts how a customer's harmonic producing load at one location can impact voltage distortions at other (possibly very remote) locations. In general, to assess the ability of a relatively small current injection to produce a significant voltage distortion (due to resonance) at remote locations (due to transfer impedance) is the primary goal of every harmonic study.

Should a harmonic study indicate a potential problem (violation of limits, for example), two categories of solutions are available: (1) reduce the harmonics at their point of origin (before they enter the system), or (2) apply filtering to reduce undesirable harmonics. Many methods for reducing harmonics at their origin are available; for example, using various transformer connections to cancel certain harmonics has been extremely effective in practice. In most cases, however, reducing or eliminating harmonics at their origin is effective only in the design or expansion stage of a new facility. For existing facilities, harmonic filters often provide the least-cost solution.

Harmonic filters can be subdivided into two types: active and passive. Active filters are only now becoming commercially viable products for high-power applications and operate as follows. For a load that injects certain harmonic currents into the supply system, a DC to AC inverter can be controlled such that the inverter supplies the harmonic current for the load, while allowing the power system to supply the power frequency current for the load. Figure 37.8 shows a diagram of such an active filter application.

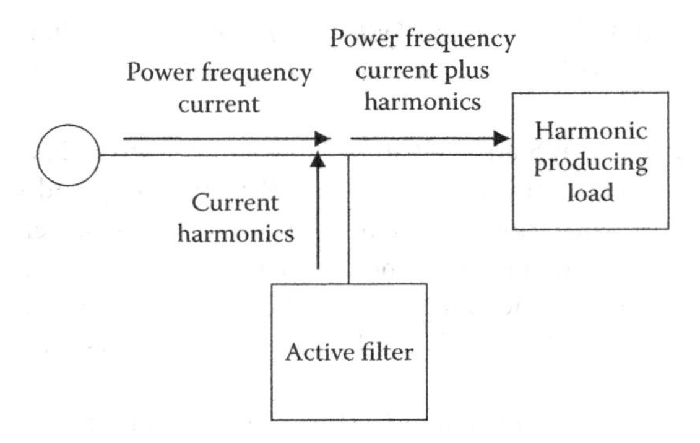

FIGURE 37.8 Active filter concept diagram.

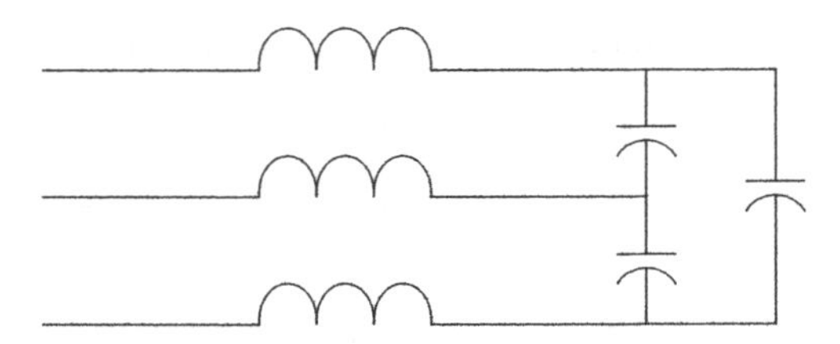

FIGURE 37.9 Typical passive filter design.

For high power applications or for applications where power factor correction capacitors already exist, it is typically more cost effective to use passive filtering. Passive filtering is based on the series resonance principle (recall that a low impedance at a specific frequency is a series-resonant characteristic) and can be easily implemented. Figure 37.9 shows a typical three-phase harmonic filter (many other designs are also used) that is commonly used to filter fifth or seventh harmonics.

It should be noted that passive filtering cannot always make use of existing capacitor banks. In filter applications, the capacitors will typically be exposed continuously to voltages greater than their ratings (which were determined based on their original application). 600 V capacitors, for example, may be required for 480 V filter applications. Even with the potential cost of new capacitors, passive filtering still appears to offer the most cost effective solution to the harmonic problem at this time.

In conclusion, power system harmonics have been carefully considered for many years and have received a significant increase in research and development activity as a direct result of the proliferation of high-power semiconductors. Fortunately, harmonic measurement equipment is readily available, and the underlying theory used to evaluate harmonics analytically (with computer assistance) is well understood. Limits for harmonic voltages and currents have been suggested by multiple standards-making bodies, but care must be used because the suggested limits are not necessarily equivalent.

Regardless of which limit numbers are appropriate for a given application, multiple options are available to help meet the levels required. As with all power quality problems, however, accurate study on the "front end" usually will reveal possible problems in the design stage, and a lower-cost solution can be implemented before problems arise.

The material presented here is not intended to be all-inclusive. The suggested reading provides further documents, including both IEEE and IEC standards, recommended practices, and technical papers and reports that provide the knowledge base required to apply the standards properly.

Further Information

Arrillaga, J., Bradley, D., and Bodger, P., *Power System Harmonics*, John Wiley, New York, 1985.

Dugan, R.C., McGranaghan, M.F., and Beaty, H.W., *Electrical Power Systems Quality*, McGraw-Hill, New York, 1996.

Heydt, G.T., *Electric Power Quality*, Stars in a Circle Publications, West LaFayette, IN, 1991.

IEC 61000-4-7, Electromagnetic compatibility (EMC)—Part 4: Testing and measurement techniques—Section 7: General guide on harmonics and interharmonics measurements and instrumentation, for power supply systems and equipment connected thereto, Ed. 1.0 b: 1991.

IEC 61000-3-6 TR3, Electromagnetic compatibility (EMC)—Part 3: Limits—Section 6: Assessment of emission limits for distorting loads in MV and HV power systems—Basic EMC Publication, Ed. 1.0 b: 1996.

IEC 61000-3-2, Electromagnetic compatibility (EMC)—Part 3-2: Limits—Limits for harmonic current emissions (equipment input current < = 16 A per phase), Ed. 1.2 b: 1998.

IEEE Harmonics Modeling and Simulation Task Force, IEEE Special Publication #98-TP-125-0: *IEEE Tutorial on Harmonics Modeling and Simulation*, IEEE Press, New York, 1998.

IEEE Standard 519-1992, *Recommended Practices and Requirements for Harmonic Control in Electrical Power Systems*, IEEE Press, New York, April 1993.

Mohan, N., Undeland, T.M., and Robbins, W.P., *Power Electronics: Converters, Applications, and Design*, John Wiley, New York, 1989.

P519A Task Force of the Harmonics Working Group and SCC20-Power Quality, *Guide for Applying Harmonic Limits on Power Systems* (draft), IEEE, New York, May 1996.

UIE, *Guide to Quality of Electrical Supply for Industrial Installations, Part 3: Harmonics*, 1998.

38

Voltage Sags

Math H.J. Bollen
Swedish Transmission Research Institute

Voltage sags are short duration reductions in rms voltage, mainly caused by short circuits and starting of large motors. The interest in voltage sags is due to the problems they cause on several types of equipment. Adjustable-speed drives, process-control equipment, and computers are especially notorious for their sensitivity (Conrad et al., 1991; McGranaghan et al., 1993). Some pieces of equipment trip when the rms voltage drops below 90% for longer than one or two cycles. Such a piece of equipment will trip tens of times a year. If this is the process-control equipment of a paper mill, one can imagine that the costs due to voltage sags can be enormous. A voltage sag is not as damaging to industry as a (long or short) interruption, but as there are far more voltage sags than interruptions, the total damage due to sags is still larger. Another important aspect of voltage sags is that they are hard to mitigate. Short interruptions and many long interruptions can be prevented via simple, although expensive measures in the local distribution network. Voltage sags at equipment terminals can be due to short-circuit faults hundreds of kilometers away in the transmission system. It will be clear that there is no simple method to prevent them.

38.1 Voltage Sag Characteristics

An example of a voltage sag is shown in Figure 38.1.* The voltage amplitude drops to a value of about 20% of its pre-event value for about two and a half cycles, after which the voltage recovers again. The event shown in Figure 38.1 can be characterized as a voltage sag down to 20% (of the pre-event voltage) for 2.5 cycles (of the fundamental frequency). This event can be characterized as a voltage sag with a magnitude of 20% and a duration of 2.5 cycles.

* The datafile containing these measurements was obtained from a Website with test data set up for IEEE project group P1159.2: http://grouper.ieee.org/groups/1159/2/index.html.

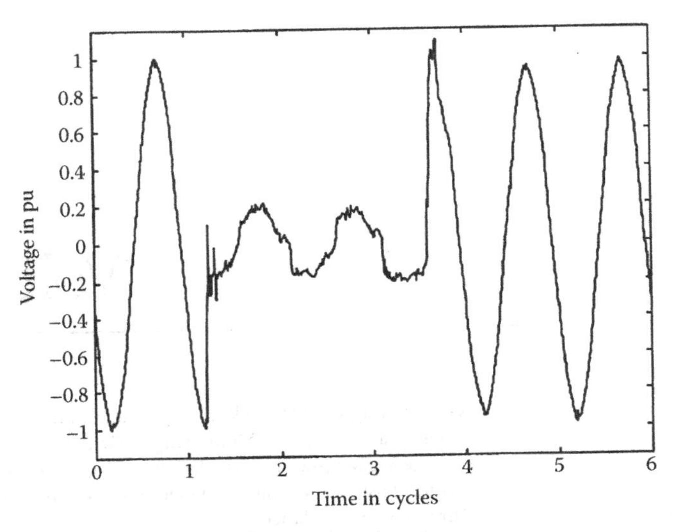

FIGURE 38.1 A voltage sag—voltage in one phase in time domain.

38.1.1 Voltage Sag Magnitude: Monitoring

The magnitude of a voltage sag is determined from the rms voltage. The rms voltage for the sag in Figure 38.1 is shown in Figure 38.2. The rms voltage has been calculated over a one-cycle sliding window:

$$V_{rms}(k) = \sqrt{\frac{1}{N} \sum_{i=k-N+1}^{i=k} v(i)^2} \tag{38.1}$$

with N the number of samples per cycle, and $v(i)$ the sampled voltage in time domain. The rms voltage as shown in Figure 38.2 does not immediately drop to a lower value, but takes one cycle for the transition. This is due to the finite length of the window used to calculate the rms value. We also see that the rms value during the sag is not completely constant and that the voltage does not immediately recover after the fault.

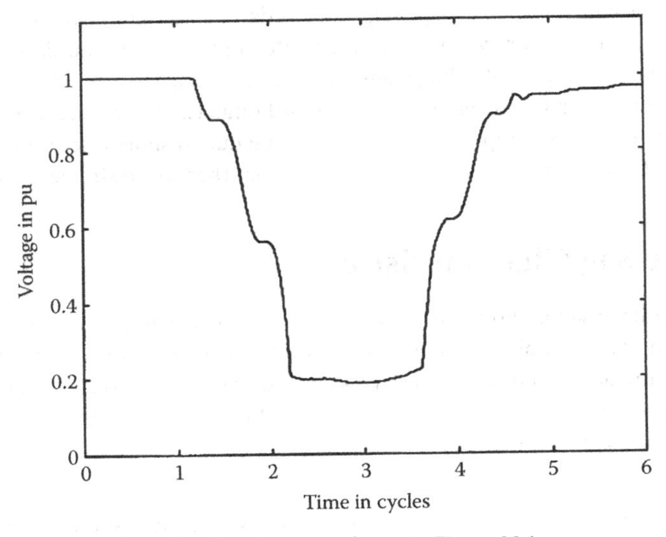

FIGURE 38.2 One-cycle rms voltage for the voltage sag shown in Figure 38.1.

There are various ways of obtaining the sag magnitude from the rms voltages. Most power quality monitors take the lowest value obtained during the event. As sags normally have a constant rms value during the deep part of the sag, using the lowest value is an acceptable approximation.

The sag is characterized through the remaining voltage during the event. This is then given as a percentage of the nominal voltage. Thus, a 70% sag in a 230-V system means that the voltage dropped to 161 V. The confusion with this terminology is clear. One could be tricked into thinking that a 70% sag refers to a drop of 70%, thus a remaining voltage of 30%. The recommendation is therefore to use the phrase "a sag down to 70%." Characterizing the sag through the actual drop in rms voltage can solve this ambiguity, but this will introduce new ambiguities like the choice of the reference voltage.

38.1.2 Origin of Voltage Sags

Consider the distribution network shown in Figure 38.3, where the numbers (1 through 5) indicate fault positions and the letters (A through D) loads. A fault in the transmission network, fault position 1, will cause a serious sag for both substations bordering the faulted line. This sag is transferred down to all customers fed from these two substations. As there is normally no generation connected at lower voltage levels, there is nothing to keep up the voltage. The result is that all customers (A, B, C, and D) experience a deep sag. The sag experienced by A is likely to be somewhat less deep, as the generators connected to that substation will keep up the voltage. A fault at position 2 will not cause much voltage drop for customer A. The impedance of the transformers between the transmission and the subtransmission system are large enough to considerably limit the voltage drop at high-voltage side of the transformer. The sag experienced by customer A is further mitigated by the generators feeding into its local transmission substation. The fault at position 2 will, however, cause a deep sag at both subtransmission substations and thus for all customers fed from here (B, C, and D). A fault at position 3 will cause a short or long interruption for customer D when the protection clears the fault. Customer C will only experience a deep sag. Customer B will experience a shallow sag due to the fault at position 3, again due to the transformer impedance. Customer A will probably not notice anything from this fault. Fault 4 causes a deep sag for customer C and a shallow one for customer D. For fault 5, the result is the other way around: a deep sag for customer D and a shallow one for customer C. Customers A and B will not experience any significant drop in voltage due to faults 4 and 5.

38.1.3 Voltage Sag Magnitude: Calculation

To quantify sag magnitude in radial systems, the voltage divider model, shown in Figure 38.4, can be used, where Z_S is the source impedance at the point-of-common coupling; and Z_F is the impedance

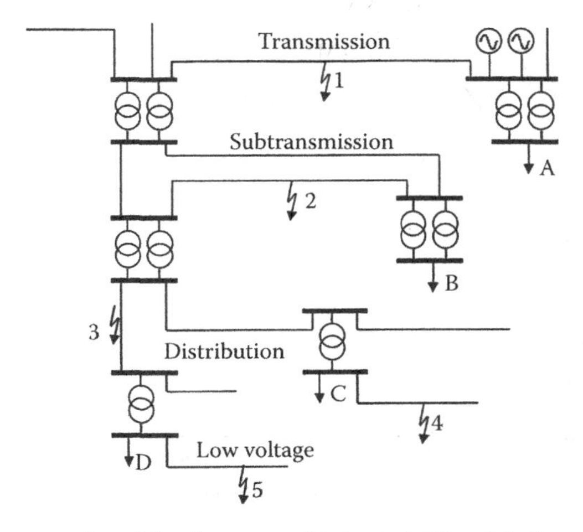

FIGURE 38.3 Distribution network with load positions (A through D) and fault positions (1 through 5).

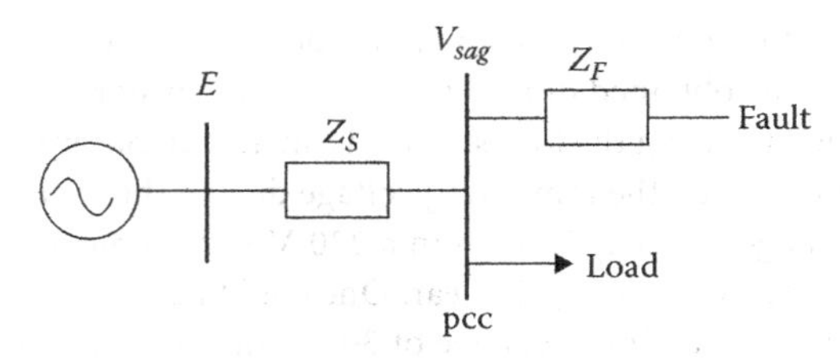

FIGURE 38.4 Voltage divider model for a voltage sag.

between the point-of-common coupling and the fault. The point-of-common coupling (pcc) is the point from which both the fault and the load are fed. In other words, it is the place where the load current branches off from the fault current. In the voltage divider model, the load current before, as well as during the fault is neglected. The voltage at the pcc is found from

$$V_{sag} = \frac{Z_F}{Z_S + Z_F} \tag{38.2}$$

where it is assumed that the pre-event voltage is exactly 1 pu, thus $E = 1$. The same expression can be derived for constant-impedance load, where E is the pre-event voltage at the pcc. We see from Equation 38.2 that the sag becomes deeper for faults electrically closer to the customer (when Z_F becomes smaller), and for weaker systems (when Z_S becomes larger).

Equation 38.2 can be used to calculate the sag magnitude as a function of the distance to the fault. Therefore, we write $Z_F = zd$, with z the impedance of the feeder per unit length and d the distance between the fault and the pcc, leading to

$$V_{sag} = \frac{zd}{Z_S + zd} \tag{38.3}$$

This expression has been used to calculate the sag magnitude as a function of the distance to the fault for a typical 11 kV overhead line, resulting in Figure 38.5. For the calculations, a 150-mm² overhead line was

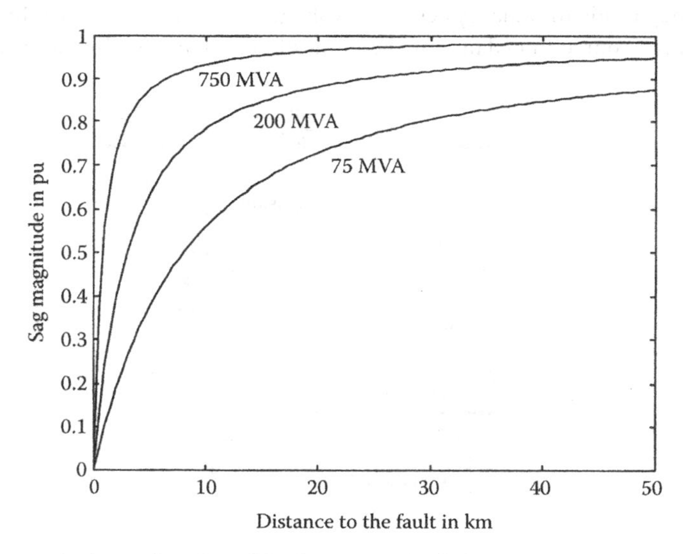

FIGURE 38.5 Sag magnitude as a function of the distance to the fault.

used and fault levels of 750, 200, and 75 MVA. The fault level is used to calculate the source impedance at the pcc and the feeder impedance is used to calculate the impedance between the pcc and the fault. It is assumed that the source impedance is purely reactive, thus $Z_S = j\,0.161\,\Omega$ for the 750 MVA source. The impedance of the 150 mm^2 overhead line is $z = 0.117 + j\,0.315\,\Omega$/km.

38.1.4 Propagation of Voltage Sags

It is also possible to calculate the sag magnitude directly from fault levels at the pcc and at the fault position. Let S_{FLT} be the fault level at the fault position and S_{PCC} at the point-of-common coupling. The voltage at the pcc can be written as

$$V_{sag} = 1 - \frac{S_{FLT}}{S_{PCC}} \tag{38.4}$$

This equation can be used to calculate the magnitude of sags due to faults at voltage levels other than the point-of-common coupling. Consider typical fault levels as shown in Table 38.1. This data has been used to obtain Table 38.2, showing the effect of a short circuit fault at a lower voltage level than the pcc. We can see that sags are significantly "damped" when they propagate upwards in the power system. In a sags study, we typically only have to take faults one voltage level down from the pcc into account. And even those are seldom of serious concern. Note, however, that faults at a lower voltage level may be associated with a longer fault-clearing time and thus a longer sag duration. This especially holds for faults on distribution feeders, where fault-clearing times in excess of 1 s are possible.

38.1.5 Critical Distance

Equation 38.3 gives the voltage as a function of distance to the fault. From this equation we can obtain the distance at which a fault will lead to a sag of a certain magnitude V. If we assume equal X/R ratio of source and feeder, we get the following equation:

$$d_{crit} = \frac{Z_S}{z} \times \frac{V}{1-V} \tag{38.5}$$

TABLE 38.1 Typical Fault Levels at Different Voltage Levels

Voltage Level	Fault Level (MVA)
400 V	20
11 kV	200
33 kV	900
132 kV	3,000
400 kV	17,000

TABLE 38.2 Propagation of Voltage Sags to Higher Voltage Levels

	Point-of-Common Coupling at:				
Fault at:	400 V	11 kV	33 kV	132 kV	400 kV
400 V	—	90%	98%	99%	100%
11 kV	—	—	78%	93%	99%
33 kV	—	—	—	70%	95%
132 kV	—	—	—	—	82%

TABLE 38.3 Critical Distance for Faults at Different Voltage Levels

Nominal Voltage	Short-Circuit Level (MVA)	Feeder Impedance (mΩ/km)	Critical Distance
400 V	20	230	35 m
11 kV	200	310	2 km
33 kV	900	340	4 km
132 kV	3,000	450	13 km
400 kV	10,000	290	55 km

We refer to this distance as the critical distance. Suppose that a piece of equipment trips when the voltage drops below a certain level (the critical voltage). The definition of critical distance is such that each fault within the critical distance will cause the equipment to trip. This concept can be used to estimate the expected number of equipment trips due to voltage sags (Bollen, 1998). The critical distance has been calculated for different voltage levels, using typical fault levels and feeder impedances. The data used and the results obtained are summarized in Table 38.3 for the critical voltage of 50%. Note how the critical distance increases for higher voltage levels. A customer will be exposed to much more kilometers of transmission lines than of distribution feeder. This effect is understood by writing Equation 38.5 as a function of the short-circuit current I_{flt} at the pcc:

$$d_{crit} = \frac{V_{nom}}{zI_{flt}} \times \frac{V}{1-V} \tag{38.6}$$

with V_{nom} the nominal voltage. As both z and I_{flt} are of similar magnitude for different voltage levels, one can conclude from Equation 38.6 that the critical distance increases proportionally with the voltage level.

38.1.6 Voltage Sag Duration

It was shown before, the drop in voltage during a sag is due to a short circuit being present in the system. The moment the short circuit fault is cleared by the protection, the voltage starts to return to its original value. The duration of a sag is thus determined by the fault-clearing time. However, the actual duration of a sag is normally longer than the fault-clearing time.

Measurement of sag duration is less trivial than it might appear. From a recording the sag duration may be obvious, but to come up with an automatic way for a power quality monitor to obtain the sag duration is no longer straightforward. The commonly used definition of sag duration is the number of cycles during which the rms voltage is below a given threshold. This threshold will be somewhat different for each monitor but typical values are around 90% of the nominal voltage. A power quality monitor will typically calculate the rms value once every cycle.

The main problem is that the so-called post-fault sag will affect the sag duration. When the fault is cleared, the voltage does not recover immediately. This is mainly due to the reenergizing and reacceleration of induction motor load (Bollen, 1995). This post-fault sag can last several seconds, much longer than the actual sag. Therefore, the sag duration as defined before, is no longer equal to the fault-clearing time. More seriously, different power quality monitors will give different values for the sag duration. As the rms voltage recovers slowly, a small difference in threshold setting may already lead to a serious difference in recorded sag duration (Bollen, 1999).

Generally speaking, faults in transmission systems are cleared faster than faults in distribution systems. In transmission systems, the critical fault-clearing time is rather small. Thus, fast protection and fast circuit breakers are essential. Also, transmission and subtransmission systems are normally operated as a grid, requiring distance protection or differential protection, both of which allow for fast clearing of the fault. The principal form of protection in distribution systems is overcurrent protection. This requires a certain amount of time-grading, which increases the fault-clearing time. An exception is

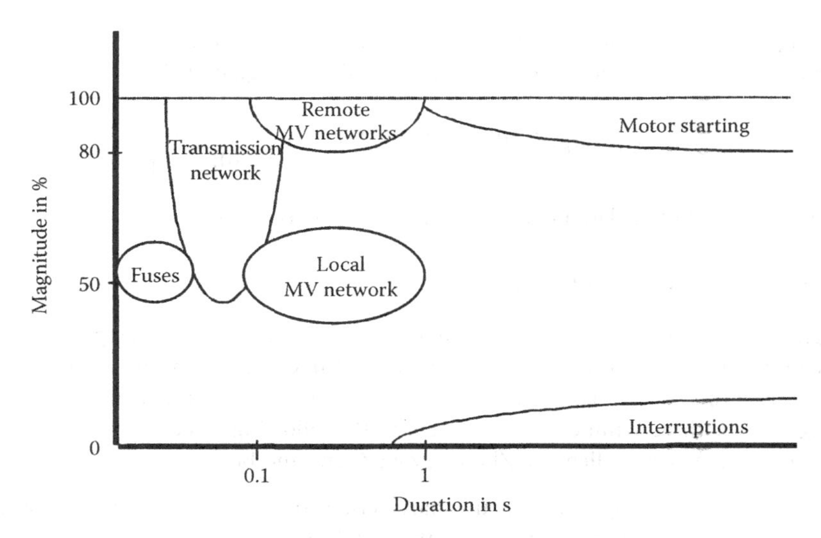

FIGURE 38.6 Sags of different origin in a magnitude-duration plot.

formed by systems in which current-limiting fuses are used. These have the ability to clear a fault within one half-cycle. In overhead distribution systems, the instantaneous trip of the recloser will lead to a short sag duration, but the clearing of a permanent fault will give a sag of much longer duration.

The so-called magnitude-duration plot is a common tool used to show the quality of supply at a certain location or the average quality of supply of a number of locations. Voltage sags due to faults can be shown in such a plot, as well as sags due to motor starting, and even long and short interruptions. Different underlying causes lead to events in different parts of the magnitude-duration plot, as shown in Figure 38.6.

38.1.7 Phase-Angle Jumps

A short circuit in a power system not only causes a drop in voltage magnitude, but also a change in the phase angle of the voltage. This sudden change in phase angle is called a "phase-angle jump." The phase-angle jump is visible in a time-domain plot of the sag as a shift in voltage zero-crossing between the pre-event and the during-event voltage. With reference to Figure 38.4 and Equation 38.2, the phase-angle jump is the argument of V_{sag}, thus the difference in argument between Z_F and $Z_S + Z_F$. If source and feeder impedance have equal X/R ratio, there will be no phase-angle jump in the voltage at the pcc. This is the case for faults in transmission systems, but normally not for faults in distribution systems. The latter may have phase-angle jumps up to a few tens of degrees (Bollen, 1999; Bollen et al., 1996).

Figure 38.4 shows a single-phase circuit, which is a valid model for three-phase faults in a three-phase system. For nonsymmetrical faults, the analysis becomes much more complicated. A consequence of nonsymmetrical faults (single-phase, phase-to-phase, two-phase-to-ground) is that single-phase load experiences a phase-angle jump even for equal X/R ratio of feeder and source impedance (Bollen, 1997, 1999).

To obtain the phase-angle jump from the measured voltage waveshape, the phase angle of the voltage during the event must be compared with the phase angle of the voltage before the event. The phase angle of the voltage can be obtained from the voltage zero-crossings or from the argument of the fundamental component of the voltage. The fundamental component can be obtained by using a discrete Fourier transform algorithm. Let $V_1(t)$ be the fundamental component obtained from a window $(t-T,t)$, with T one cycle of the power frequency, and let $t = 0$ correspond to the moment of sag initiation. In case there is no chance in voltage magnitude or phase angle, the fundamental component as a function of time is found from

$$V_1(t) = V_1(0)e^{j\omega t} \tag{38.7}$$

The phase-angle jump, as a function of time, is the difference in phase angle between the actual fundamental component and the "synchronous voltage" according to Equation 38.7:

$$\phi(t) = \arg\{V_1(t)\} - \arg\{V_1(0)e^{j\omega t}\} = \arg\left\{\frac{V_1(t)}{V_1(0)}e^{-j\omega t}\right\} \tag{38.8}$$

Note that the argument of the latter expression is always between −180° and +180°.

38.1.8 Three-Phase Unbalance

For three-phase equipment, three voltages need to be considered when analyzing a voltage sag event at the equipment terminals. For this, a characterization of three-phase unbalanced voltage sags is introduced. The basis of this characterization is the theory of symmetrical components. Instead of the three-phase voltages or the three symmetrical components, the following three (complex) values are used to characterize the voltage sag (Bollen and Zhang, 1999; Zhang and Bollen, 1998):

- The "characteristic voltage" is the main characteristic of the event. It indicates the severity of the sag, and can be treated in the same way as the remaining voltage for a sag experienced by a single-phase event.
- The "PN factor" is a correction factor for the effect of the load on the voltages during the event. The PN factor is normally close to unity and can then be neglected. Exceptions are systems with a large amount of dynamic load, and sags due to two-phase-to-ground faults.
- The "zero-sequence voltage," which is normally not transferred to the equipment terminals, rarely affects equipment behavior. The zero-sequence voltage can be neglected in most studies.

Neglecting the zero-sequence voltage, it can be shown that there are two types of three-phase unbalanced sags, denoted as types C and D. Type A is a balanced sag due to a three-phase fault. Type B is the sag due to a single-phase fault, which turns into type D after removal of the zero-sequence voltage. The three complex voltages for a type C sag are written as follows:

$$\begin{aligned} V_a &= F \\ V_b &= -\frac{1}{2}F - \frac{1}{2}jV\sqrt{3} \\ V_c &= -\frac{1}{2}F + \frac{1}{2}jV\sqrt{3} \end{aligned} \tag{38.9}$$

where V is the characteristic voltage and F the PN factor. The (characteristic) sag magnitude is defined as the absolute value of the characteristic voltage; the (characteristic) phase-angle jump is the argument of the characteristic voltage. For a sag of type D, the expressions for the three voltage phasors are as follows:

$$\begin{aligned} V_a &= V \\ V_b &= -\frac{1}{2}V - \frac{1}{2}jF\sqrt{3} \\ V_c &= -\frac{1}{2}V + \frac{1}{2}jF\sqrt{3} \end{aligned} \tag{38.10}$$

Sag type D is due to a phase-to-phase fault, or due to a single-phase fault behind a Δy-transformer, or a phase-to-phase fault behind two Δy-transformers, etc. Sag type C is due to a single-phase fault, or due to a phase-to-phase fault behind a Δy-transformer, etc. When using characteristic voltage for a three-phase unbalanced sag, the same single-phase scheme as in Figure 38.4 can be used to study the transfer of voltage sags in the system (Bollen, 1997, 1999).

38.2 Equipment Voltage Tolerance

38.2.1 Voltage Tolerance Requirement

Generally speaking, electrical equipment prefers a constant rms voltage. That is what the equipment has been designed for and that is where it will operate best. The other extreme is zero voltage for a longer period of time. In that case the equipment will simply stop operating completely. For each piece of equipment there is a maximum interruption duration, after which it will continue to operate correctly. A rather simple test will give this duration. The same test can be done for a voltage of 10% (of nominal), for a voltage of 20%, etc. If the voltage becomes high enough, the equipment will be able to operate on it indefinitely. Connecting the points obtained by performing these tests results in the so-called "voltage-tolerance curve" (Key, 1979). An example of a voltage-tolerance curve is shown in Figure 38.7: the requirements for IT-equipment as recommended by the Information Technology Industry Council (ITIC, 1999). Strictly speaking, one can claim that this is not a voltage-tolerance curve as described above, but a requirement for the voltage tolerance. One could refer to this as a voltage-tolerance requirement and to the result of equipment tests as a voltage-tolerance performance. We see in Figure 38.7 that IT equipment has to withstand a voltage sag down to zero for 1.1 cycle, down to 70% for 30 cycles, and that the equipment should be able to operate normally for any voltage of 90% or higher.

38.2.2 Voltage Tolerance Performance

Voltage-tolerance (performance) curves for personal computers are shown in Figure 38.8. The curves are the result of equipment tests performed in the United States (EPRI, 1994) and in Japan (Sekine et al., 1992). The shape of all the curves in Figure 38.8 is close to rectangular. This is typical for many types of equipment, so that the voltage tolerance may be given by only two values, maximum duration and minimum voltage, instead of by a full curve. From the tests summarized in Figure 38.8 it is found that the voltage tolerance of personal computers varies over a wide range: 30–170 ms, 50%–70% being the range containing half of the models. The extreme values found are 8 ms, 88% and 210 ms, 30%.

Voltage-tolerance tests have also been performed on process-control equipment: PLCs, monitoring relays, motor contactors. This equipment is even more sensitive to voltage sags than personal computers. The majority of devices tested tripped between one and three cycles. A small minority was able to tolerate sags up to 15 cycles in duration. The minimum voltage varies over a wider range: from 50% to 80% for most devices, with exceptions of 20% and 30%. Unfortunately, the latter two both tripped in three cycles (Bollen, 1999).

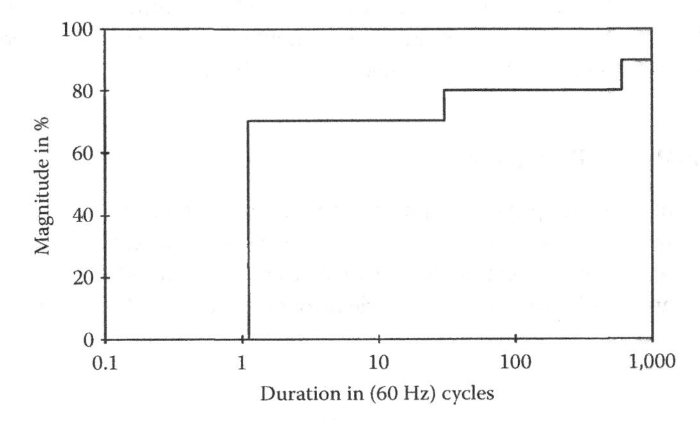

FIGURE 38.7 Voltage-tolerance requirement for IT equipment.

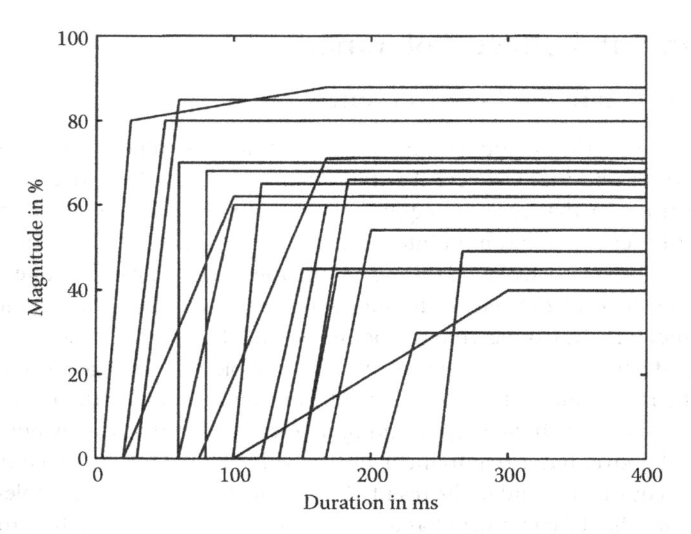

FIGURE 38.8 Voltage-tolerance performance for personal computers.

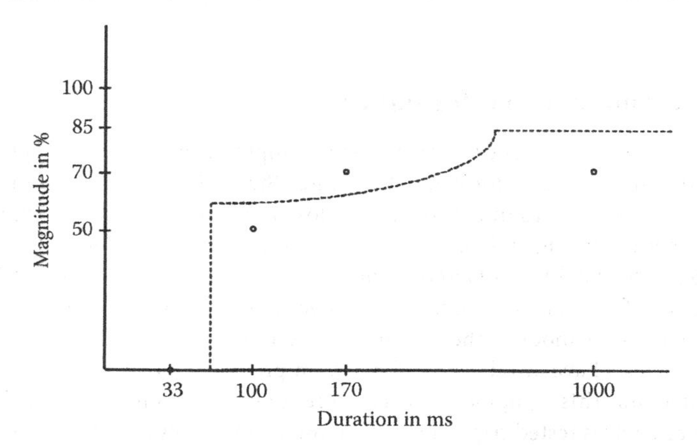

FIGURE 38.9 Average voltage-tolerance curve for adjustable-speed drives.

From performance testing of adjustable-speed drives, an "average voltage-tolerance curve" has been obtained. This curve is shown in Figure 38.9. The sags for which the drive was tested are indicated as circles. It has further been assumed that the drives can operate indefinitely on 85% voltage. Voltage tolerance is defined here as "automatic speed recovery, without reaching zero speed." For sensitive production processes, more strict requirements will hold (Bollen, 1999).

38.2.3 Single-Phase Rectifiers

The sensitivity of most single-phase equipment can be understood from the equivalent scheme in Figure 38.10. The power supply to a computer, process-control equipment, consumer electronics, etc. consists of a single-phase (four-pulse) rectifier together with a capacitor and a DC/DC converter. During normal operation the capacitor is charged twice a cycle through the diodes. The result is a DC voltage ripple:

$$\varepsilon = \frac{PT}{2V_0^2 C} \tag{38.11}$$

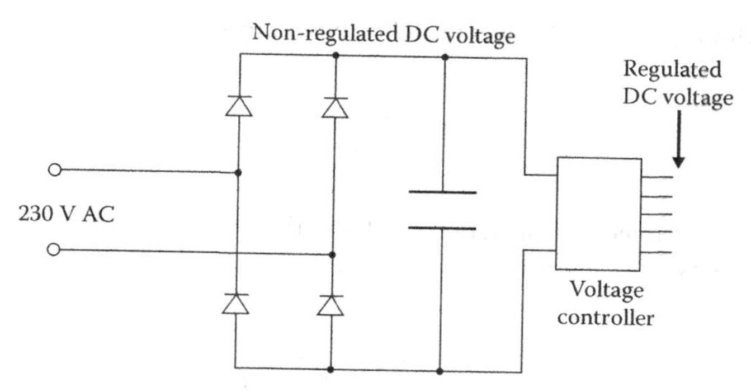

FIGURE 38.10 Typical power supply to sensitive single-phase equipment.

with P the DC bus active-power load, T one cycle of the power frequency, V_0 the maximum DC bus voltage, and C the size of the capacitor.

During a voltage sag or interruption, the capacitor continues to discharge until the DC bus voltage has dropped below the peak of the supply voltage. A new steady state is reached, but at a lower DC bus voltage and with a larger ripple. The resulting DC bus voltage for a sag down to 50% is shown in Figure 38.11, together with the absolute value of the supply voltage. If the new steady state is below the minimum operating voltage of the DC/DC converter, or below a certain protection setting, the equipment will trip. During the decaying DC bus voltage, the capacitor voltage $V(t)$ can be obtained from the law of conservation of energy:

$$\frac{1}{2}CV^2 = \frac{1}{2}CV_0^2 - Pt \tag{38.12}$$

where a constant DC bus load P has been assumed. From Equation 38.12 the voltage as a function of time is obtained:

$$V(t) = \sqrt{V_0^2 - \frac{2P}{C}t} \tag{38.13}$$

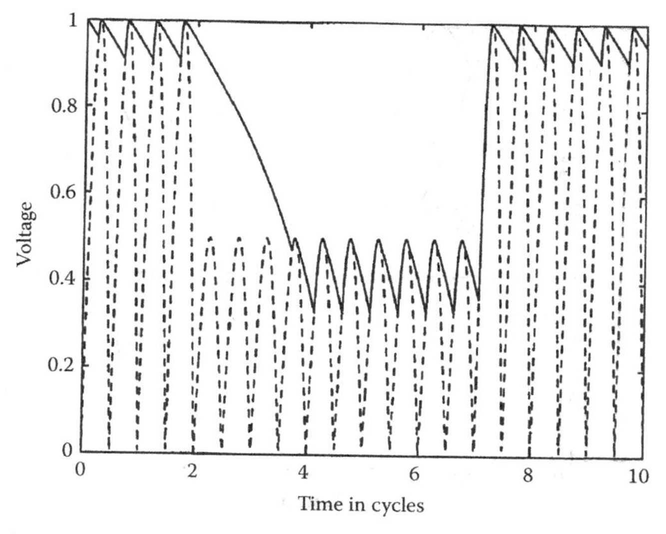

FIGURE 38.11 Absolute value of AC voltage (dashed) and DC bus voltage (solid line) for a sag down to 50%.

Combining this with Equation 38.11 gives the following expression:

$$V(t) = V_0\sqrt{1 - 4\varepsilon\frac{t}{T}} \tag{38.14}$$

The larger the DC ripple in normal operation, the faster the drop in DC bus voltage during a sag. From Equation 38.14 the maximum duration of zero voltage t_{max} is calculated for a minimum operating voltage V_{min}, resulting in

$$t_{max} = \frac{1 - (V_{min}/V_0)^2}{4\varepsilon} T \tag{38.15}$$

38.2.4 Three-Phase Rectifiers

The performance of equipment fed through three-phase rectifiers becomes somewhat more complicated. The main equipment belonging to this category is formed by AC and DC adjustable-speed drives. One of the complications is that the operation of the equipment is affected by the three voltages, which are not necessarily the same during the voltage sag. For non-controlled (six pulse) diode rectifiers, a similar model can be used as for single-phase rectifiers. The operation of three-phase controlled rectifiers can become very complicated and application-specific (Bollen, 1996). Therefore, only noncontrolled rectifiers will be discussed here. For voltage sags due to three-phase faults, the DC bus voltage behind the (three-phase) rectifier will decay until a new steady state is reached at a lower voltage level, with a larger ripple. To calculate the DC bus voltage as a function of time, and the time-to-trip, the same equation as for the single-phase rectifier can be used.

For unbalanced voltage sags, a distinction needs to be made between the two types (C and D), as introduced in Section 38.1.8. Figure 38.12 shows AC and DC side voltages for a sag of type C with $V = 0.5$ pu and $F = 1$. For this sag, the voltage drops in two phases where the third phase stays at its presage value. Three capacitor sizes are used (Bollen and Zhang, 1999); a "large" capacitance is defined as a value that leads to an initial decay of the DC voltage equal to 10%, which is 433 F/kW for a 620 V drive. In the same way, "small" capacitance corresponds to 75% per cycle initial decay, and 57.8 F/kW for

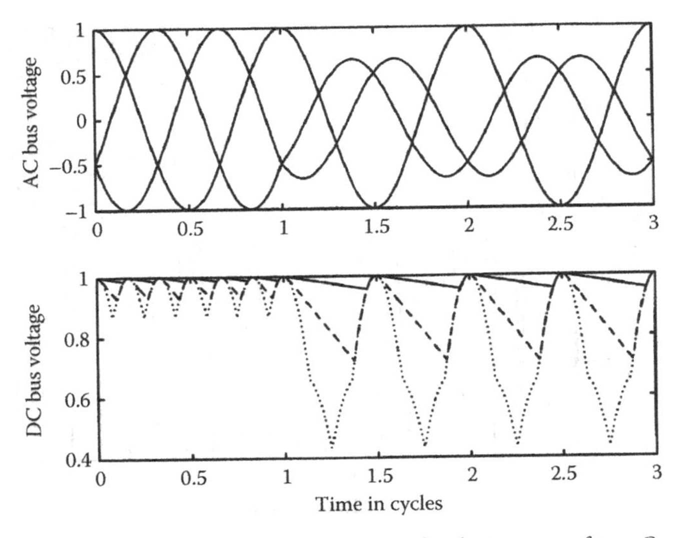

FIGURE 38.12 AC and DC side voltages for a three-phase rectifier during a sag of type C.

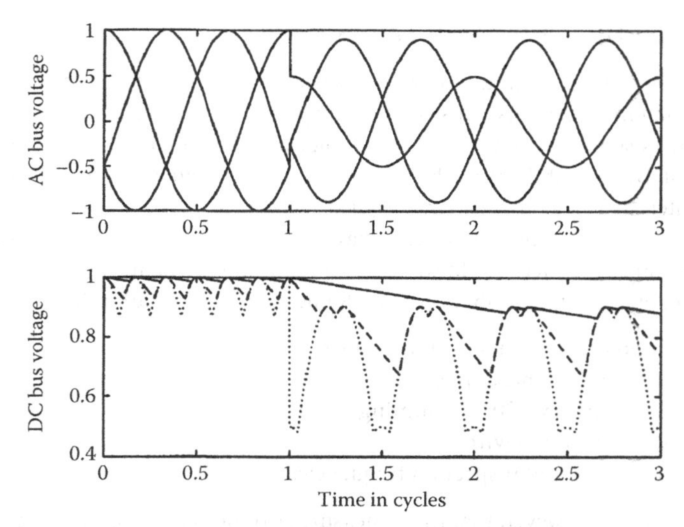

FIGURE 38.13 AC and DC side voltages for a three-phase rectifier during a sag of type D.

a 620V drive. It turns out that even for the small capacitance, the DC bus voltage remains above 70%. For the large capacitance value, the DC bus voltage is hardly affected by the voltage sag. It is easy to understand that this is also the case for type C sags with an even lower characteristic magnitude V (Bollen, 1999; Bollen and Zhang, 1999).

Figure 38.13 shows the equivalent results for a sag of type D, again with $V = 0.5$ and $F = 1$. As all three AC voltages show a drop in voltage magnitude, the DC bus voltage will drop even for a large capacitor. But the effect is still much less than for a three-phase (balanced) sag.

The effect of a lower PN factor ($F < 1$) is that even the highest voltage shows a drop for a type C sag, so that the DC bus voltage will always show a small drop. Also for a type D sag, a lower PN factor will lead to an additional drop in DC bus voltage (Bollen and Zhang, 1999).

38.3 Mitigation of Voltage Sags

38.3.1 From Fault to Trip

To understand the various ways of mitigation, the mechanism leading to an equipment trip needs to be understood. The equipment trip is what makes the event a problem; if there are no equipment trips, there is no voltage sag problem. The underlying event of the equipment trip is a short-circuit fault. At the fault position, the voltage drops to zero, or to a very low value. This zero voltage is changed into an event of a certain magnitude and duration at the interface between the equipment and the power system. The short-circuit fault will always cause a voltage sag for some customers. If the fault takes place in a radial part of the system, the protection intervention clearing the fault will also lead to an interruption. If there is sufficient redundancy present, the short circuit will only lead to a voltage sag. If the resulting event exceeds a certain severity, it will cause an equipment trip.

Based on this reasoning, it is possible to distinguish between the following mitigation methods:

- Reducing the number of short-circuit faults.
- Reducing the fault-clearing time.
- Changing the system such that short-circuit faults result in less severe events at the equipment terminals or at the customer interface.
- Connecting mitigation equipment between the sensitive equipment and the supply.
- Improving the immunity of the equipment.

38.3.2 Reducing the Number of Faults

Reducing the number of short-circuit faults in a system not only reduces the sag frequency, but also the frequency of long interruptions. This is thus a very effective way of improving the quality of supply and many customers suggest this as the obvious solution when a voltage sag or interruption problem occurs. Unfortunately, most of the time the solution is not that obvious. A short circuit not only leads to a voltage sag or interruption at the customer interface, but may also cause damage to utility equipment and plant. Therefore, most utilities will already have reduced the fault frequency as far as economically feasible. In individual cases, there could still be room for improvement, e.g., when the majority of trips are due to faults on one or two distribution lines. Some examples of fault mitigation are

- Replace overhead lines by underground cables.
- Use special wires for overhead lines.
- Implement a strict policy of tree trimming.
- Install additional shielding wires.
- Increase maintenance and inspection frequencies.

One has to keep in mind, however, that these measures can be very expensive, especially for transmission systems, and that their costs have to be weighted against the consequences of the equipment trips.

38.3.3 Reducing the Fault-Clearing Time

Reducing the fault-clearing time does not reduce the number of events, but only their severity. It does not do anything to reduce to number of interruptions, but can significantly limit the sag duration. The ultimate reduction of fault-clearing time is achieved by using current-limiting fuses, able to clear a fault within one half-cycle. The recently introduced static circuit breaker has the same characteristics: fault-clearing time within one half-cycle. Additionally, several types of fault-current limiters have been proposed that do not actually clear the fault, but significantly reduce the fault current magnitude within one or two cycles. One important restriction of all these devices is that they can only be used for low- and medium-voltage systems. The maximum operating voltage is a few tens of kilovolts.

But the fault-clearing time is not only the time needed to open the breaker, but also the time needed for the protection to make a decision. To achieve a serious reduction in fault-clearing time, it is necessary to reduce any grading margins, thereby possibly allowing for a certain loss of selectivity.

38.3.4 Changing the Power System

By implementing changes in the supply system, the severity of the event can be reduced. Here again, the costs may become very high, especially for transmission and subtransmission voltage levels. In industrial systems, such improvements more often outweigh the costs, especially when already included in the design stage. Some examples of mitigation methods especially directed toward voltage sags are

- Install a generator near the sensitive load. The generators will keep up the voltage during a remote sag. The reduction in voltage drop is equal to the percentage contribution of the generator station to the fault current. In case a combined-heat-and-power station is planned, it is worth it to consider the position of its electrical connection to the supply.
- Split buses or substations in the supply path to limit the number of feeders in the exposed area.
- Install current-limiting coils at strategic places in the system to increase the "electrical distance" to the fault. The drawback of this method is that this may make the event worse for other customers.
- Feed the bus with the sensitive equipment from two or more substations. A voltage sag in one substation will be mitigated by the infeed from the other substations. The more independent the substations are, the more the mitigation effect. The best mitigation effect is by feeding from two different transmission substations. Introducing the second infeed increases the number of sags, but reduces their severity.

38.3.5 Installing Mitigation Equipment

The most commonly applied method of mitigation is the installation of additional equipment at the system-equipment interface. Also recent developments point toward a continued interest in this way of mitigation. The popularity of mitigation equipment is explained by it being the only place where the customer has control over the situation. Both changes in the supply as well as improvement of the equipment are often completely outside of the control of the end user. Some examples of mitigation equipment are

- *Uninterruptable power supply* (*UPS*). This is the most commonly used device to protect low-power equipment (computers, etc.) against voltage sags and interruptions. During the sag or interruption, the power supply is taken over by an internal battery. The battery can supply the load for, typically, between 15 and 30 min.
- *Static transfer switch*. A static transfer switch switches the load from the supply with the sag to another supply within a few milliseconds. This limits the duration of a sag to less than one half-cycle, assuming that a suitable alternate supply is available.
- *Dynamic voltage restorer* (*DVR*). This device uses modern power electronic components to insert a series voltage source between the supply and the load. The voltage source compensates for the voltage drop due to the sag. Some devices use internal energy storage to make up for the drop in active power supplied by the system. They can only mitigate sags up to a maximum duration. Other devices take the same amount of active power from the supply by increasing the current. These can only mitigate sags down to a minimum magnitude. The same holds for devices boosting the voltage through a transformer with static tap changer.
- *Motor-generator sets*. Motor-generator sets are the classical solution for sag and interruption mitigation with large equipment. They are obviously not suitable for an office environment but the noise and the maintenance requirements are often no problem in an industrial environment. Some manufacturers combine the motor-generator set with a backup generator; others combine it with power-electronic converters to obtain a longer ride-through time.

38.3.6 Improving Equipment Voltage Tolerance

Improvement of equipment voltage tolerance is probably the most effective solution against equipment trips due to voltage sags. But as a short-time solution, it is often not suitable. In many cases, a customer only finds out about equipment performance after it has been installed. Even most adjustable-speed drives have become off-the-shelf equipment where the customer has no influence on the specifications. Only large industrial equipment is custom-made for a certain application, which enables the incorporation of voltage-tolerance requirements in the specification.

Apart from improving large equipment (drives, process-control computers), a thorough inspection of the immunity of all contactors, relays, sensors, etc. can significantly improve the voltage tolerance of the process.

38.3.7 Different Events and Mitigation Methods

Figure 38.6 showed the magnitude and duration of voltage sags and interruptions resulting from various system events. For different events, different mitigation strategies apply.

Sags due to short-circuit faults in the transmission and subtransmission system are characterized by a short duration, typically up to 100 ms. These sags are very hard to mitigate at the source and improvements in the system are seldom feasible. The only way of mitigating these events is by improvement of the equipment or, where this turns out to be unfeasible, installing mitigation equipment. For low-power equipment, a UPS is a straightforward solution; for high-power equipment and for complete installations, several competing tools are emerging.

The duration of sags due to distribution system faults depends on the type of protection used—ranging from less than a cycle for current-limiting fuses up to several seconds for overcurrent relays in underground or industrial distribution systems. The long sag duration also enables equipment to trip due to faults on distribution feeders fed from other HV/MV substations. For deep long-duration sags, equipment improvement becomes more difficult and system improvement easier. The latter could well become the preferred solution, although a critical assessment of the various options is certainly needed.

Sags due to faults in remote distribution systems and sags due to motor starting should not lead to equipment tripping for sags down to 85%. If there are problems, the equipment needs to be improved. If equipment trips occur for long-duration sags in the 70%–80% magnitude range, changes in the system have to be considered as an option.

For interruptions, especially the longer ones, equipment improvement is no longer feasible. System improvements or a UPS in combination with an emergency generator are possible solutions here.

References

Bollen, M.H.J., The influence of motor reacceleration on voltage sags. *IEEE Trans. Ind. Appl.*, 31(4): 667–674, July 1995.

Bollen, M.H.J., Characterization of voltage sags experienced by three-phase adjustable-speed drives. *IEEE Trans. Power Del.*, 12(4): 1666–1671, October 1997.

Bollen, M.H.J., Method of critical distances for stochastic assessment of voltage sags. *IEE Proc. Gener. Transm. Distrib.*, 145(1): 70–76, January 1998.

Bollen, M.H.J., *Solving Power Quality Problems, Voltage Sags and Interruptions*. IEEE Press, New York, 1999.

Bollen, M.H.J., Wang, P., and Jenkins, N., Analysis and consequences of the phase jump associated with a voltage sag, in *Power Systems Computation Conference*, Dresden, Germany, August 1996.

Bollen, M.H.J. and Zhang, L.D., Analysis of voltage tolerance of adjustable-speed drives for three-phase balanced and unbalanced sags, in *IEEE Industrial and Commercial Power Systems Technical Conference*, Sparks, Nevada, May 1999.

Conrad, L., Little, K., and Grigg, C., Predicting and preventing problems associated with remote fault-clearing voltage dips. *IEEE Trans. Ind. Appl.*, 27(1): 167–172, January 1991.

Information Technology Industry Council, Interteq, http://www.itic.com, 1999.

Key, T.S., Diagnosing power-quality related computer problems. *IEEE Trans. Ind. Appl.*, 15(4): 381–393, July 1979.

McGranaghan, M.F., Mueller, D.R., and Samotej, M.J., Voltage sags in industrial power systems. *IEEE Trans. Ind. Appl.*, 29(2): 397–403, March 1993.

PQTN Brief 7: *Undervoltage Ride-through Performance of Off-the-Shelf Personal Computers*. EPRI Power Electronics Application Centre, Knoxville, TN, 1994.

Sekine, Y., Yamamoto, T., Mori, S., Saito, N., and Kurokawa, H., Present state of momentary voltage dip interferences and the countermeasures in Japan, in *International Conference on Large Electric Networks (CIGRE), 34th Session*, Paris, France, September 1992.

Zhang, L.D. and Bollen, M.H.J., A method for characterizing unbalanced voltage dips (sags) with symmetrical components. *IEEE Power Eng. Rev.*, 18(7): 50–52, July 1998.

Further Information

Bollen, M.H.J., Fast assessment methods for voltage sags in distribution systems. *IEEE Trans. Ind. Appl.*, 31(6): 1414–1423, November 1996.

Bollen, M.H.J., Tayjasajant, T., and Yalcinkaya, G., Assessment of the number of voltage sags experienced by a large industrial customer. *IEEE Trans. Ind. Appl.*, 33(6): 1465–1471, November 1997.

Collins, E.R. and Morgan, R.L., A three-phase sag generator for testing industrial equipment. *IEEE Trans. Power Del.*, 11(1): 526–532, January 1996.

Conrad, L.E. and Bollen, M.H.J., Voltage sag coordination for reliable plant operation. *IEEE Trans. Ind. Appl.*, 33(6): 1459–1464, November 1997.

Diliberti, T.W., Wagner, V.E., Staniak, J.P., Sheppard, S.L., and Orfloff, T.L., Power quality requirements of a large industrial user: A case study, in *IEEE Industrial and Commercial Power Systems Technical Conference*, Detroit, MI, pp. 1–4, May 1990.

Dorr, D.S., Point of utilization power quality study results. *IEEE Trans. Ind. Appl.*, 31(4): 658–666, July 1995.

Dorr, D.S., Hughes, M.B., Gruzs, T.M., Jurewicz, R.E., and McClaine, J.L., Interpreting recent power quality surveys to define the electrical environment. *IEEE Trans. Ind. Appl.*, 33(6): 1480–1487, November 1997.

Dorr, D.S., Mansoor, A., Morinec, A.G., and Worley, J.C., Effects of power line voltage variations on different types of 400-W high-pressure sodium ballasts. *IEEE Trans. Ind. Appl.*, 33(2): 472–476, March 1997.

Dugan, R.C., McGranaghan, M.F., and Beaty, H.W., *Electric Power Systems Quality*. McGraw-Hill, New York, 1996.

European standard EN-50160, *Voltage Characteristics of Electricity Supplied by Public Distribution Systems*. CENELEC, Brussels, Belgium, 1994.

Gunther, E.W. and Mehta, H., A survey of distribution system power quality—Preliminary results. *IEEE Trans. Power Del.*, 10(1): 322–329, January 1995.

Holtz, J., Lotzhat, W., and Stadfeld, S., Controlled AC drives with ride-through capacity at power interruption. *IEEE Trans. Ind. Appl.*, 30(5): 1275–1283, September 1994.

IEEE Std. 1100–1992, *IEEE Recommended Practice for Powering and Grounding Sensitive Electronic Equipment*.

IEEE Std. 1159–1995, *IEEE Recommended Practice for Monitoring Electric Power Quality*, IEEE, New York, 1995.

IEEE Std. 493–1997, *IEEE Recommended Practice for the Design of Reliable Industrial and Commercial Power Systems* (The Gold Book).

IEEE Std. 1346–1998, *IEEE Recommended Practice for Evaluating Electric Power System Compatibility with Electronic Process Equipment*.

IEC 61000-4-11, *Electromagnetic Compatibility (EMC) Voltage Dips, Short Interruptions and Voltage Variations Immunity Tests*.

Kojovic, L.J. and Hassler, S., Application of current limiting fuses in distribution systems for improved power quality and protection. *IEEE Trans. Power Del.*, 12(2): 791–800, April 1997.

Koval, D.O., Bocancea, R.A., and Hughes, M.B., Canadian national power quality survey: Frequency of industrial and commercial voltage sags. *IEEE Trans. Ind. Appl.*, 35(5): 904–910, September 1998.

Koval, D.O. and Leonard, J.J., Rural power profiles. *IEEE Trans. Ind. Appl.*, 30(2): 469–475, March–April 1994.

Lamoree, J., Mueller, D., Vinett, P., Jones, W., and Samotyj, M., Voltage sag analysis case studies. *IEEE Trans. Ind. Appl.*, 30(4): 1083–1089, July 1994.

Mansoor, A., Collins, E.R., and Morgan, R.L., Effects of unsymmetrical voltage sags on adjustable-speed drives, in *7th IEEE International Conference on Harmonics and Quality of Power (ICHPQ)*, Las Vegas, NV, pp. 467–472, October 1996.

Middlekauff, S.W. and Collins, E.R., System and customer impact: Considerations for series custom power devices. *IEEE Trans. Power Del.*, 13(1): 278–282, January 1998.

Pilay, P., (Ed.), Motor drive/power systems interactions, IEEE Industry Applications Society Tutorial Course, October 1997.

Pumar, C., Amantegui, J., Torrealday, J.R., and Ugarte, C., A comparison between DC and AC drives as regards their behavior in the presence of voltage dips: New techniques for reducing the susceptibility of AC drives, in *International Conference on Electricity Distribution (CIRED)*, Birmingham, U.K., pp. 9/1–9/5, June 2–5, 1997.

Rioual, P., Pouliquen, H., and Louis, J.-P., Regulation of a PWM rectifier in the unbalanced network state using a generalized model. *IEEE Trans. Power Electron.*, 11(3): 495–502, May 1996.

Schwartzenberg, J.W. and DeDoncker, R.W., 15 kV medium voltage transfer switch, in *IEEE Industry Applications Society Annual Meeting*, Orlando, FL, pp. 2515–2520, October 1995.

Smith, R.K., Slade, P.G., Sarkozi, M., Stacey, E.J., Bonk, J.J., and Mehta, H., Solid state distribution current limiter and circuit breaker: Application requirements and control strategies. *IEEE Trans. Power Del.*, 8(3): 1155–1164, July 1993.

Strangas, E.G., Wagner, V.E., and Unruh, T.D., Variable speed drives evaluation test, in *IEEE Industry Applications Society Annual Meeting*, San Diego, CA, pp. 2239–2243, October 1996.

van Zyl, A., Enslin, J.H.R., and Spée, R., Converter-based solution to power quality problems on radial distribution lines. *IEEE Trans. Ind. Appl.*, 32(6): 1323–1330, November 1996.

Wang, P., Jenkins, N., and Bollen, M.H.J., Experimental investigation of voltage sag mitigation by an advanced static VAr compensator. *IEEE Trans. Power Del.*, 13(4): 1461–1467, October 1998.

Woodley, N.H., Morgan, L., and Sundaram, A., Experience with an inverter-based dynamic voltage restorer. *IEEE Trans. Power Del.*, 14(3): 1181–1186, July 1999.

Woodley, N., Sarkozi, M., Lopez, F., Tahiliani, V., and Malkin, P., Solid-state 13-kV distribution class circuit breaker: Planning, development and demonstration, in *IEEE Conference on Trends in Distribution Switchgear*, London, U.K., pp. 163–167, November 1994.

Yalçinkaya, G., Bollen, M.H.J., and Crossley, P.A., Characterisation of voltage sags in industrial distribution systems. *IEEE Trans. Ind. Appl.*, 34(4): 682–688, July 1998.

Zhang, L.D. and Bollen, M.H.J., A method for characterisation of three-phase unbalanced dips (sags) from recorded voltage waveshapes, in *International Telecommunications Energy Conference (INTELEC)*, Copenhagen, Denmark, June 1999.

39

Voltage Fluctuations and Lamp Flicker in Power Systems

S. Mark Halpin
Auburn University

Voltage flicker is a problem that has existed in the power industry for many years. Many types of end-use equipment can create voltage flicker, and many types of solution methods are available. Fortunately, the problem is not overly complex, and it can often be analyzed using fairly simple methods. In many cases, however, solutions can be expensive. Perhaps the most difficult aspect of the voltage flicker problem has been the development of a widely accepted definition of just what "flicker" is and how it can be quantified in terms of measurable quantities.

To electric utility engineers, voltage flicker is considered in terms of magnitude and rate of change of voltage fluctuations. To the utility customer, however, flicker is considered in terms of "my lights are flickering." The necessary presence of a human observer to "see" the change in lamp (intensity) output in response to a change in supply voltage is the most complex factor for which to account. Significant research, dating back to the early twentieth century, has been devoted to establishing an accurate correlation between voltage changes and observer perceptions. This correlation is essential so that a readily measurable quantity, supply voltage, can be used to predict a human response.

The early work regarding voltage flicker considered voltage flicker to be a single-frequency modulation of the power frequency voltage. Both sinusoidal and square wave modulations were considered as shown mathematically in Equations 39.1 and 39.2, with most work concentrating on square wave modulation.

$$v(t) = \sqrt{2}V_{rms}\cos(\omega t)\{1.0 + V\cos(\omega_m t)\} \quad (39.1)$$

$$v(t) = \sqrt{2}V_{rms}\cos(\omega t)\{1.0 + V\text{square}(\omega_m t)\} \quad (39.2)$$

Based on Equations 39.1 and 39.2, the voltage flicker magnitude can be expressed as a percentage of the root-mean-square (rms) voltage, where the term "V" in the two equations represents the percentage. While both the magnitude of the fluctuations ("V") and the "shape" of the modulating waveform are obviously important, the frequency of the modulation is also extremely relevant and is explicitly represented as ω_m. For sinusoidal flicker (given by Equation 39.1), the total waveform appears as shown in Figure 39.1 with the modulating waveform shown explicitly. A similar waveform can be easily created for square-wave modulation.

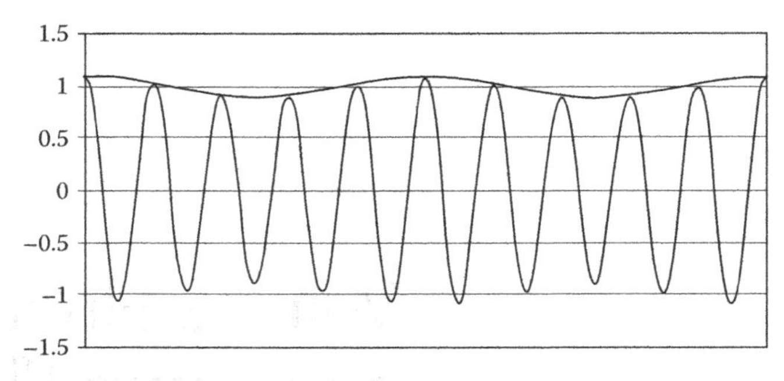

FIGURE 39.1 Sinusoidal voltage flicker.

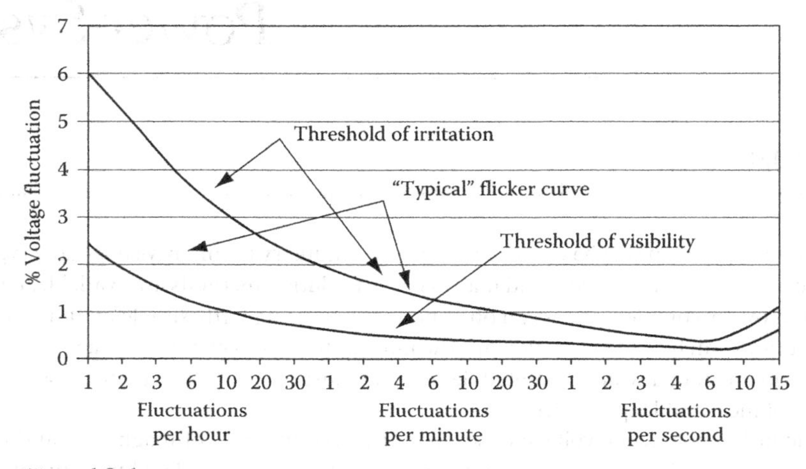

FIGURE 39.2 Typical flicker curves.

To correlate the voltage change percentage, V, at a certain frequency, ω_m, with human perceptions, early research led to the widespread use of what is known as a flicker curve to predict possible observer complaints. Flicker curves are still in widespread use, particularly in the United States. A typical flicker curve is shown in Figure 39.2 and is based on tests conducted by the General Electric Company. It is important to realize that these curves are developed based on square wave modulation. Voltage changes from one level to another are considered to be "instantaneous" in nature, which may or may not be an accurate representation of actual equipment-produced voltage fluctuations.

The curve of Figure 39.2 requires some explanation in order to understand its application. The "threshold of visibility" corresponds to certain fluctuation magnitude and frequency pairs that represent the borderline above which an observer can just perceive lamp (intensity) output variations in a 120 V, 60 Hz, 60 W incandescent bulb. The "threshold of irritation" corresponds to certain fluctuation magnitude and frequency pairs that represent the borderline above which the majority of observers would be irritated by lamp (intensity) output variations for the same lamp type. Two conclusions are immediately apparent from these two curves: (1) even small percentage changes in supply voltage can be noticed by persons observing lamp output, and (2) the frequency of the voltage fluctuations is an important consideration, with the frequency range from 6 to 10 Hz being the most sensitive.

Most utility companies do not permit excessive voltage fluctuations on their system, regardless of the frequency. For this reason, a "typical" utility flicker curve will follow either the "threshold of irritation" or the "threshold of visibility" curve as long as the chosen curve lies below some established value (2% in Figure 39.2). By requiring that voltage fluctuations not exceed the "borderline of visibility" curve, the utility is insuring conservative criteria that should minimize potential problems due to voltage fluctuations.

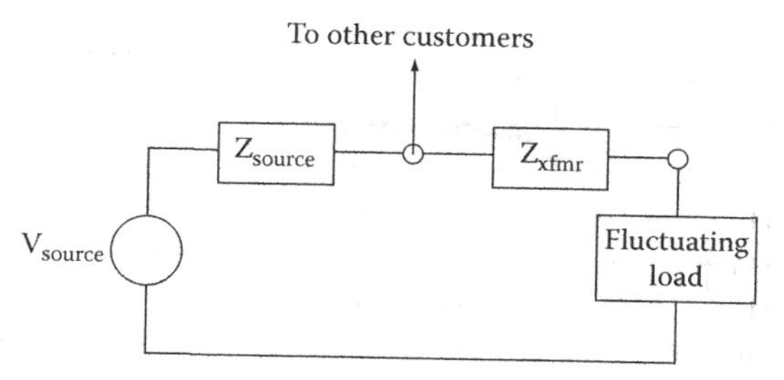

FIGURE 39.3 Example circuit for flicker calculations.

For many years, the generic flicker curve has served the utility industry well. Fluctuating motor loads like car shredders, wood chippers, and many others can be fairly well characterized in terms of a duty cycle and a maximum torque. From this information, engineers can predict the magnitude and frequency of voltage changes anywhere in the supplying transmission and distribution system. Voltage fluctuations associated with motor starting events are also easily translated into a point (or points) on the flicker curve, and many utilities have based their motor starting criteria on this method for many years. Other loads, most notably arcing loads, cannot be represented as a single flicker magnitude and frequency term. For these types of loads, utility engineers typically presume either worst-case or most-likely variations for analytical evaluations.

Regardless of the type of load, the typical calculation procedure involves either basic load flow or simple voltage division calculations. Figure 39.3 shows an example positive sequence circuit with all data assumed in per-unit on consistent bases.

For fluctuating loads that are best represented by a constant power model (arc furnaces and load torque variations on a running motor), basic load flow techniques can be used to determine the full-load and no-load (or "normal condition") voltages at the "critical" or "point of common coupling" bus where other customers might be served. For fluctuating loads that are best represented by a constant impedance model (motor starting), basic circuit analysis techniques readily provide the full-load and no-load ("normal condition") voltages at the critical bus. Regardless of the modeling and calculation procedures used, equations similar to Equation 39.3 can be used to determine the percentage voltage change for use in conjunction with a flicker curve. Of course, accurate information regarding the frequency of the assumed fluctuation is absolutely necessary. Note that Equation 39.3 represents an over-simplification and should therefore not be used in cases where the fluctuations are frequent enough to impact the average rms value (measured over several seconds up to a minute). Other more elaborate formulas are available for these situations.

$$\%\text{Voltage change} = \left(1.0 - \frac{V_{\text{full load}}}{V_{\text{normal}}}\right) * 100\% \tag{39.3}$$

From a utility engineer's viewpoint, the decision to either serve or deny service to a fluctuating load is often based on the result of Equation 39.3 (or a more complex version of Equation 39.3) including information about the frequency at which the calculated change occurs. From this simplified discussion, several questions arise:

1. How are fluctuating loads taken into account when the nature of the fluctuations is not constant in magnitude?
2. How are fluctuating loads taken into account when the nature of the fluctuations is not constant in frequency?
3. How are static compensators and other high response speed mitigation devices included in the calculations?

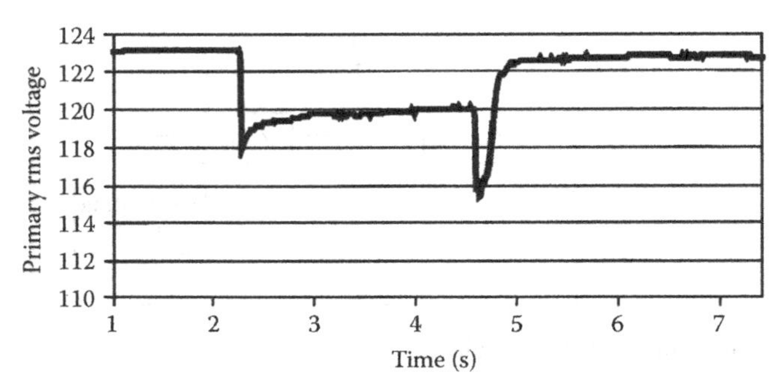

FIGURE 39.4 Poorly timed motor starter voltage fluctuation.

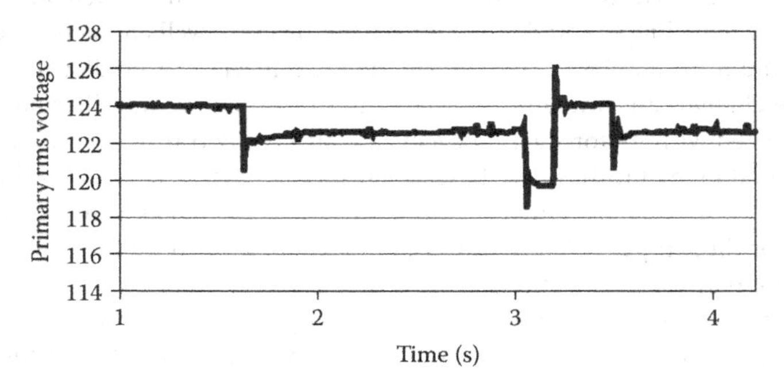

FIGURE 39.5 Adaptive-var compensator effects.

As examples, consider the rms voltage plots (on 120V bases) shown in Figures 39.4 and 39.5. Figure 39.4 shows an rms plot associated with a poorly timed two-step reduced-voltage motor starter. Figure 39.5 shows a motor starting event when the motor is compensated by an adaptive-var compensator. Questions 1–3 are clearly difficult to answer for these plots, so it would be very difficult to apply the basic flicker curve.

In many cases of practical interest, "rules of thumb" are often used to answer approximately these and other related questions so that the simple flicker curve can be used effectively. However, these assumptions and approaches must be conservative in nature and may result in costly equipment modifications prior to connection of certain fluctuating loads. In modern environment, it is imperative that end-users operate at the least total cost. It is equally important that end-use fluctuating loads not create problems for other users. Due to the conservative and approximate nature of the flicker curve methodology, there is often significant room for negotiation, and the matter is often not settled considering only engineering results.

For roughly three decades, certain engineering groups have recognized the limitations of the flicker curve methods and have developed alternative approaches based on an instrument called a flicker meter. This work, driven strongly in Europe by the International Union for Electroheat (UIE) and the International Electrotechnical Commission (IEC), appears to offer solutions to many of the problems with the flicker curve methodology. Many years of industrial experience have been obtained with the flicker meter approach, and its output has been well-correlated with complaints of utility customers. At this time, the Institute of Electrical and Electronics Engineers (IEEE) is working toward adopting the flicker meter methodology for use in North America.

The flicker meter is a continuous time measuring system that takes voltage as an input and produces three output indices that are related to customer perception. These outputs are: (1) instantaneous flicker sensation, P_{inst}, (2) short-term flicker severity, P_{st}, and (3) long-term flicker severity, P_{lt}. A block diagram of an analog flicker meter is shown in Figure 39.6.

The flicker meter takes into account both the physical aspects of engineering (how does the lamp [intensity] output vary with voltage?) and the physiological aspects of human observers (how fast can

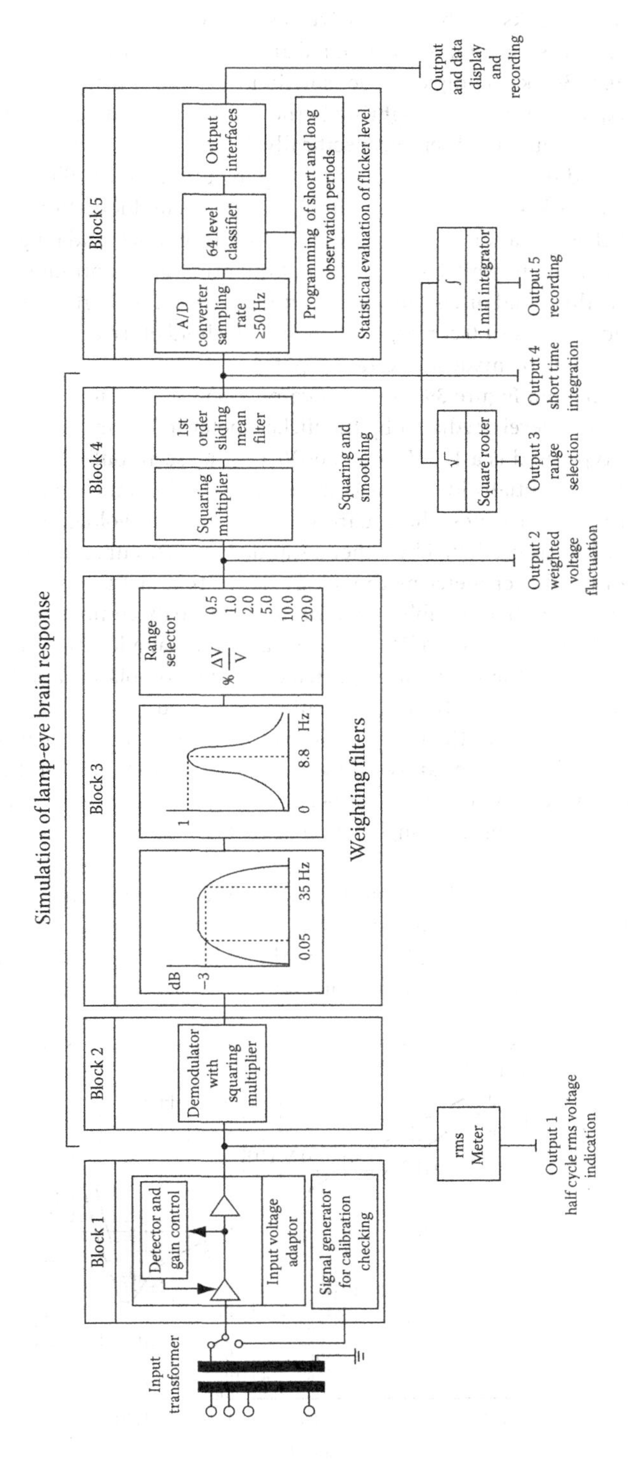

FIGURE 39.6 Flicker meter block diagram.

the human eye respond to light changes?). Each of the five basic blocks in Figure 39.6 contribute to one or both of these aspects. While a detailed discussion of the flicker meter is beyond the scope of this chapter, the function of the blocks can be summarized as follows.

Blocks 1 and 2 act to process the input voltage signal and to partially isolate only the modulating term in Equations 39.1 or 39.2. Block 3 completes the isolation of the modulating signal through complex filtering and applies frequency-sensitive weighting to the "pure" modulating signal. Block 4 models the physiological response of the human observer, specifically the short-term memory tendency of the brain to correlate the voltage modulating signal with a human perception ability. Block 5 performs statistical analysis on the output of Block 4 to capture the cumulative effects of fluctuations over time.

The instantaneous flicker sensation is the output of Block 4. The short- and long-term severity indices are the outputs of Block 5. P_{inst} is available as an output quantity on a continuous basis, and a value of 1.0 corresponds with the threshold of visibility curve in Figure 39.2. A single P_{st} value is available as an output every 10 min, and a value of 1.0 corresponds to the threshold of irritation curve in Figure 39.2. Of course, a comparison can only be made for certain inputs.

For square wave modulation, Figure 39.7 shows a comparison of the "irritation level" given by IEEE Std. 141 (Red Book) and that level predicted by the flicker meter to be "irritating" (P_{st} = 1.0). For these comparisons, the lamp type used is a 120 V, 60 Hz, 60 W incandescent bulb. Note that the flicker curve taken from IEEE Std. 141 is essentially identical to the "borderline of irritation" curve given in Figure 39.2.

As Figure 39.7 clearly demonstrates, the square wave modulation voltage fluctuations that lead to irritation are nearly identical as predicted by either a standard flicker curve or a flicker meter.

The real advantage of the flicker meter methodology lies in that fact that the continuous time measurement system can easily predict possible irritation for arbitrarily complex modulation waveforms. As an example, Figure 39.8 shows a plot of P_{st} over a 3-day period at a location serving a small electric arc furnace. (**Note:** In this case, there were no reported customer complaints and P_{st} was well below the irritation threshold value of 1.0 during the entire monitoring period.)

Due to the very random nature of the fluctuations associated with an arc furnace, the flicker curve methodology cannot be used directly as an accurate predictor of irritation levels because it is appropriate only for the "sudden" voltage fluctuations associated with square wave modulation. The trade-off required for more accurate flicker prediction, however, is that the inherent simplicity of the basic flicker curve is lost.

For the basic flicker curve, simple calculations based on circuit and equipment models in Figure 39.3 can be used. Data for these models is readily available, and time-tested assumptions are widely known for cases when exact data are not available. Because the flicker meter is a continuous-time system, continuous-time voltage input data is required for its use. For existing fluctuating loads, it is reasonable

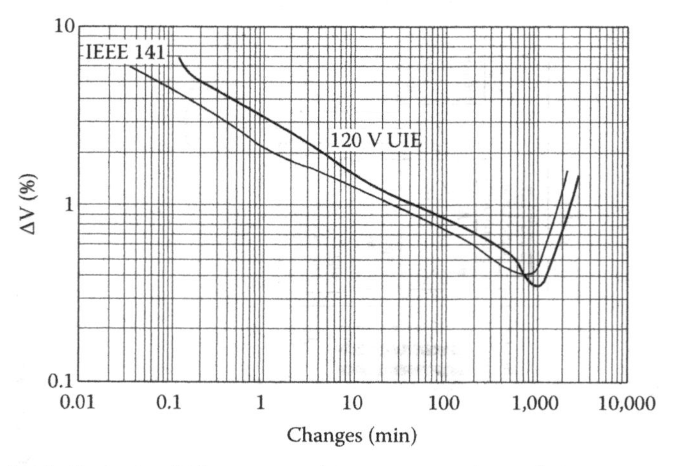

FIGURE 39.7 Threshold of irritation flicker curve and P_{st} = 1.0 curve from a flicker meter.

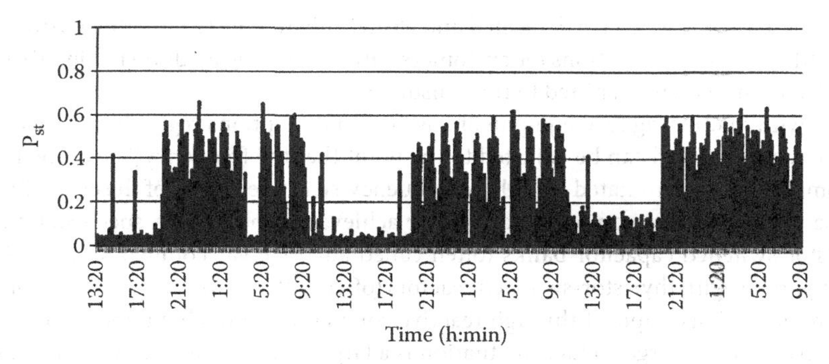

FIGURE 39.8 Short-term flicker severity example plot.

to presume that a flicker meter can be connected and used to predict whether or not the fluctuations are irritating. However, it is necessary to be able to predict potential flicker problems prior to the connection of a fluctuating load well before it is possible to measure anything.

There are three possible solutions to the apparent "prediction" dilemma associated with the flicker meter approach. The most basic approach is to locate an existing fluctuating load that is similar to the one under consideration and simply measure the flicker produced by the existing load. Of course, the engineer is responsible for making sure that the existing installation is nearly identical to the one proposed. While the fluctuating load equipment itself might be identical, supply system characteristics will almost never be the same.

Because the short-term flicker severity output of the flicker meter, P_{st}, is linearly dependent on voltage fluctuation magnitude over a wide range, it is possible to linearly scale the P_{st} measurements from one location to predict those at another location where the supply impedance is different. (In most cases, voltage fluctuations are directly related to the supply impedance; a system with 10% higher supply impedance would expect 10% greater voltage fluctuation for the same load change.) In evaluations where it is not possible to measure another existing fluctuating load, other approaches must be used.

If detailed system and load data are known, a time-domain simulation can be used to generate a continuous-time series of voltage data points. These points could then be used as inputs to a simulated flicker meter to predict the short-term flicker severity, P_{st}. This approach, however, is usually too intensive and time-consuming to be appropriate for most applications. For these situations, "shape factors" have been proposed that predict a P_{st} value for various types of fluctuations.

Shape factors are simple curves that can be used to predict, without simulation or measurement, the P_{st} that would be measured if the load were connected. Different curves exist for different "shapes" of voltage variation. Curves exist for simple square and triangular variations, as well as for more complex variations such as motor starting. To use a shape factor, an engineer must have some knowledge of (1) the magnitude of the fluctuation, (2) the shape of the fluctuation, including the time spent at each voltage level if the shape is complex, (3) rise time and fall times between voltage levels, and (4) the rate at which the shape repeats. In some cases, this level of data is not available, and assumptions are often made (on the conservative side). It is interesting to note that the extreme of the conservative choices is a rectangular fluctuation at a known frequency; which is exactly the data required to use the basic flicker curve of Figure 39.2.

Using either the flicker curve for simple evaluations or the flicker meter methodology for more complex evaluations, it is possible to predict if a given fluctuating load will produce complaints from other customers. In the event that complaints are predicted, modifications must be made prior to granting service. The possible modifications can be made either on the utility side or on the customer (load) side (or both), or some type of compensation equipment can be installed.

In most cases, the most effective, but not least cost, ways to reduce or eliminate flicker complaints are to either (1) reduce the supply system impedance of the whole path from source to fluctuating load or

(2) serve the fluctuating load from a dedicated and electrically remote (from other customers) circuit. In most cases, utility revenue projections for customers with fluctuating loads do not justify such expenses, and the burden of mitigation is shifted to the consumer.

Customers with fluctuating load equipment have two main options regarding voltage flicker mitigation. In some cases, the load can be adjusted to the point that the frequency(ies) of the fluctuations are such that complaints are eliminated (recall the frequency-sensitive nature of the entire flicker problem). In other cases, direct voltage compensation can be achieved through high-speed static compensators. Either thyristor-switched capacitor banks (often called adaptive var compensators or AVCs) or fixed capacitors in parallel with thyristor-switched reactors (often called static var compensators or SVCs) can be used to provide voltage support through reactive compensation in about one cycle. For loads where the main contributor to a large voltage fluctuation is a large reactive power change, reactive compensators can significantly reduce or eliminate the potential for flicker complaints. In cases where voltage fluctuations are due to large real power changes, reactive compensation offers only small improvements and can, in some cases, make the problem worse.

In conclusion, it is almost always necessary to measure/predict flicker levels under a variety of possible conditions, both with and without mitigation equipment and procedures in effect. In very simple cases, a basic flicker curve will provide acceptable results. In more complex cases, however, an intensive measurement, modeling, and simulation effort may be required in order to minimize potential flicker complaints.

While this chapter has addressed the basic issues associated with voltage flicker complaints, prediction, and measurement, it is not intended to be all-inclusive. A number of relevant publications, papers, reports, and standards are given for further reading, and the reader should certainly consider these documents carefully in addition to what is provided here.

Further Information

Bergeron, R., Power quality measurement protocol: CEA guide to performing power quality surveys, CEA Report 220 D 771, May 1996.

IEC 1000-3-3, *Electromagnetic Compatibility (EMC) Part 3: Limits—Part 3: Limitation of Voltage Fluctuations and Flicker in Low-Voltage Supply Systems for Equipment with Rated Current ≤ 16 A*, 1994.

IEC 1000-3-5, *Electromagnetic Compatibility (EMC) Part 3: Limits—Part 5: Limitation of Voltage Fluctuations and Flicker in Low-Voltage Supply Systems for Equipment with Rated Current > 16 A*, 1994.

IEC 1000-3-7, *Electromagnetic Compatibility (EMC) Part 3: Limits—Part 7: Assessment of Emission Limits for Fluctuating Loads in MV and HV Power Systems*, 1996.

IEC 1000-3-11, *Electromagnetic Compatibility (EMC) Part 3: Limits—Part 11: Limitation of Voltage Changes, Voltage Fluctuations, and Flicker in Public Low Voltage Supply Systems with Rated Current ≤ 75 A and Subject to Conditional Connection*, 1996.

IEC 61000-4-15, *Flickermeter-Functional and Design Specifications*, 1997.

IEC Publication 868, *Flickermeter-Functional and Design Specifications*, 1986.

IEEE Standard 141-1993, *Recommended Practice for Power Distribution in Industrial Plants*, IEEE, New York, 1993.

Sakulin, M. and Key, T.S., UIE/IEC flicker standard for use in North America: Measuring techniques and practical applications. In *Proceedings of PQA'97*, Columbus, OH, March 1997.

Seebald, R.C., Buch, J.F., and Ward, D.J., Flicker limitations of electric utilities. *IEEE Trans. Power Appar. Syst.*, PAS-104(9): 1385–1390, September 1985.

UIE WG on Disturbances, *Flicker Measurement and Evaluations: 2nd Revised Edition*, 1992.

UIE WG on Disturbances, *Connection of Fluctuating Loads*, 1998.

UIE WG on Disturbances, *Guide to Quality of Electrical Supply for Industrial Installations, Part 5: Flicker and Voltage Fluctuations*, 1999.

Xenis, C.P. and Perine, W., Slide rule yields lamp flicker data. *Electrical World*, October 1937.

40

Power Quality Monitoring

Patrick Coleman
Alabama Power Company

Many power quality problems are caused by inadequate wiring or improper grounding. These problems can be detected by simple examination of the wiring and grounding systems. Another large population of power quality problems can be solved by spotchecks of voltage, current, or harmonics using hand held meters. Some problems, however, are intermittent and require longer-term monitoring for solution.

Long-term power quality monitoring is largely a problem of data management. If an rms value of voltage and current is recorded each electrical cycle, for a three-phase system, about 6 GB of data will be produced each day. Some equipment is disrupted by changes in the voltage waveshape that may not affect the rms value of the waveform. Recording the voltage and current waveforms will result in about 132 GB of data per day. While modern data storage technologies may make it feasible to record every electrical cycle, the task of detecting power quality problems within this mass of data is daunting indeed.

Most commercially available power quality monitoring equipment attempts to reduce the recorded data to manageable levels. Each manufacturer has a generally proprietary data reduction algorithm. It is critical that the user understand the algorithm used in order to properly interpret the results.

40.1 Selecting a Monitoring Point

Power quality monitoring is usually done to either solve an existing power quality problem, or to determine the electrical environment prior to installing new sensitive equipment. For new equipment, it is easy to argue that the monitoring equipment should be installed at the point nearest the point of connection of the new equipment. For power quality problems affecting existing equipment, there is frequently pressure to determine if the problem is being caused by some external source, i.e., the utility. This leads to the installation of monitoring equipment at the service point to try to detect the source of the problem. This is usually not the optimum location for monitoring equipment. Most studies suggest that 80% of power quality problems originate within the facility. A monitor installed on the equipment being affected will detect problems originating within the facility, as well as problems originating on the utility. Each type of event has distinguishing characteristics to assist the engineer in correctly identifying the source of the disturbance.

40.2 What to Monitor

At minimum, the input voltage to the affected equipment should be monitored. If the equipment is single phase, the monitored voltage should include at least the line-to-neutral voltage and the neutral-to-ground voltages. If possible, the line-to-ground voltage should also be monitored. For three-phase equipment, the voltages may either be monitored line to neutral, or line to line. Line-to-neutral voltages are easier to understand, but most three-phase equipment operates on line-to-line voltages. Usually, it is preferable to monitor the voltage line to line for three-phase equipment.

If the monitoring equipment has voltage thresholds which can be adjusted, the thresholds should be set to match the sensitive equipment voltage requirements. If the requirements are not known, a good starting point is usually the nominal equipment voltage plus or minus 10%.

In most sensitive equipment, the connection to the source is a rectifier, and the critical voltages are DC. In some cases, it may be necessary to monitor the critical DC voltages. Some commercial power quality monitors are capable of monitoring AC and DC simultaneously, while others are AC only.

It is frequently useful to monitor current as well as voltage. For example, if the problem is being caused by voltage sags, the reaction of the current during the sag can help determine the source of the sag. If the current doubles when the voltage sags 10%, then the cause of the sag is on the load side of the current monitor point. If the current increases or decreases 10%–20% during a 10% voltage sag, then the cause of the sag is on the source side of the current monitoring point.

Sensitive equipment can also be affected by other environmental factors such as temperature, humidity, static, harmonics, magnetic fields, radio frequency interference (RFI), and operator error or sabotage. Some commercial monitors can record some of these factors, but it may be necessary to install more than one monitor to cover every possible source of disturbance.

It can also be useful to record power quantity data while searching for power quality problems. For example, the author found a shortcut to the source of a disturbance affecting a wide area by using the power quantity data. The recordings revealed an increase in demand of 2500 kW immediately after the disturbance. Asking a few questions quickly led to a nearby plant with a 2500 kW switched load that was found to be malfunctioning.

40.3 Selecting a Monitor

Commercially available monitors fall into two basic categories: line disturbance analyzers and voltage recorders. The line between the categories is becoming blurred as new models are developed. Voltage recorders are primarily designed to record voltage and current stripchart data, but some models are able to capture waveforms under certain circumstances. Line disturbance analyzers are designed to capture voltage events that may affect sensitive equipment. Generally, line disturbance analyzers are not good voltage recorders, but newer models are better than previous designs at recording voltage stripcharts.

In order to select the best monitor for the job, it is necessary to have an idea of the type of disturbance to be recorded, and an idea of the operating characteristics of the available disturbance analyzers. For example, a common power quality problem is nuisance tripping of variable speed drives. Variable speed drives may trip due to the waveform disturbance created by power factor correction capacitor switching, or due to high or low steady state voltage, or, in some cases, due to excessive voltage imbalance. If the drive trips due to high voltage or waveform disturbances, the drive diagnostics will usually indicate an overvoltage code as the cause of the trip. If the voltage is not balanced, the drive will draw significantly unbalanced currents. The current imbalance may reach a level that causes the drive to trip for input overcurrent. Selecting a monitor for variable speed drive tripping can be a challenge. Most line disturbance analyzers can easily capture the waveshape disturbance of capacitor switching, but they are not good voltage recorders, and may not do a good job of reporting high steady state voltage. Many line disturbance analyzers cannot capture voltage unbalance at all, nor will they respond to current events unless there is a corresponding voltage event. Most voltage and current recorders can easily capture the high steady state voltage that leads to a drive trip, but they may not capture the capacitor switching waveshape disturbance. Many voltage recorders can capture voltage imbalance, current imbalance, and some of them will trigger a capture of voltage and current during a current event, such as the drive tripping off.

To select the best monitor for the job, it is necessary to understand the characteristics of the available monitors. The following sections will discuss the various types of data that may be needed for a power quality investigation, and the characteristics of some commercially available monitors.

40.3.1 Voltage

The most commonly recorded parameter in power quality investigations is the rms voltage delivered to the equipment. Manufacturers of recording equipment use a variety of techniques to reduce the volume of the data recorded. The most common method of data reduction is to record Min/Max/Average data over some interval. Figure 40.1 shows a strip chart of rms voltages recorded on a cycle-by-cycle basis. Figure 40.2 shows a Min/Max/Average chart for the same time period. A common recording period is 1 week. Typical recorders will use a recording interval of 2–5 min. Each recording interval will produce three numbers: the rms voltage of the highest 1 cycle, the lowest 1 cycle, and the average of every cycle during the interval. This is a simple, easily understood recording method, and it is easily implemented by the manufacturer. There are several drawbacks to this method. If there are several events during a recording interval, only the event with the largest deviation is recorded. Unless the recorder records

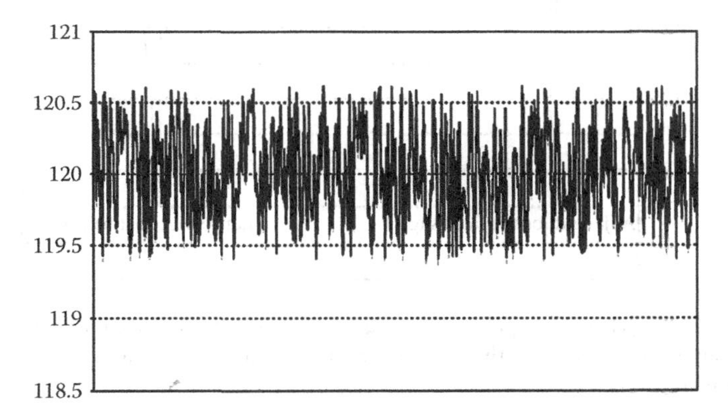

FIGURE 40.1 RMS voltage stripchart, taken cycle by cycle.

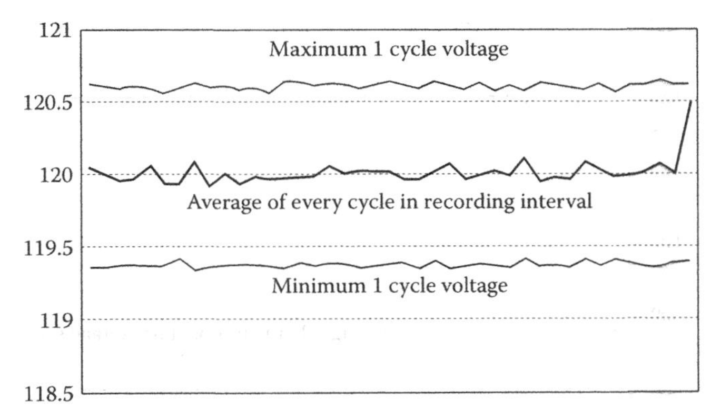

FIGURE 40.2 Min/Max/Average stripchart, showing the minimum single cycle voltage, the maximum single cycle voltage, and the average of every cycle in a recording interval. Compare to the Figure 40.1 stripchart data.

the event in some other manner, there is no time-stamp associated with the events, and no duration available. The most critical deficiency is the lack of a voltage profile during the event. The voltage profile provides significant clues to the source of the event. For example, if the event is a voltage sag, the minimum voltage may be the same for an event caused by a distant fault on the utility system, and for a nearby large motor start. For the distant fault, however, the voltage will sag nearly instantaneously, stay at a fairly constant level for 3–10 cycles, and almost instantly recover to full voltage, or possibly a slightly higher voltage if the faulted section of the utility system is separated. For a nearby motor start, the voltage will drop nearly instantaneously, and almost immediately begin a gradual recovery over 30–180 cycles to a voltage somewhat lower than before. Figure 40.3 shows a cycle-by-cycle recording of a simulated adjacent feeder fault, followed by a simulation of a voltage sag caused by a large motor start. Figure 40.4 shows a Min/Max/Average recording of the same two events. The events look quite similar when captured by the Min/Max/Average recorder, while the cycle-by-cycle recorder reveals the difference in the voltage recovery profile.

Some line disturbance analyzers allow the user to set thresholds for voltage events. If the voltage exceeds these thresholds, a short duration stripchart is captured showing the voltage profile during the event. This short duration stripchart is in addition to the long duration recordings, meaning that the engineer must look at several different charts to find the needed information.

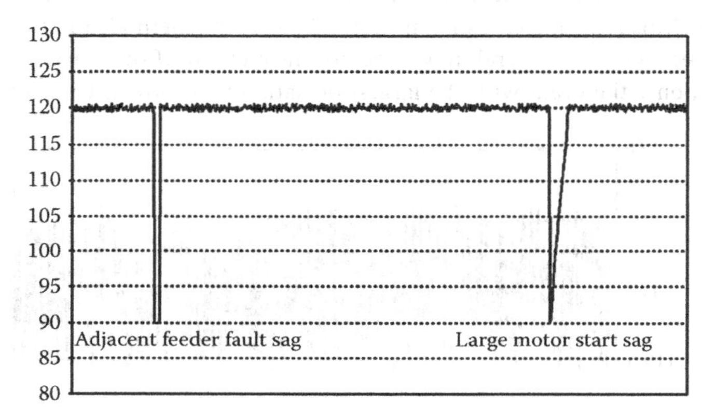

FIGURE 40.3 Cycle-by-cycle rms stripchart showing two voltage sags. The sag on the left is due to an adjacent feeder fault on the supply substation, and the sag on the right is due to a large motor start. Note the difference in the voltage profile during recovery.

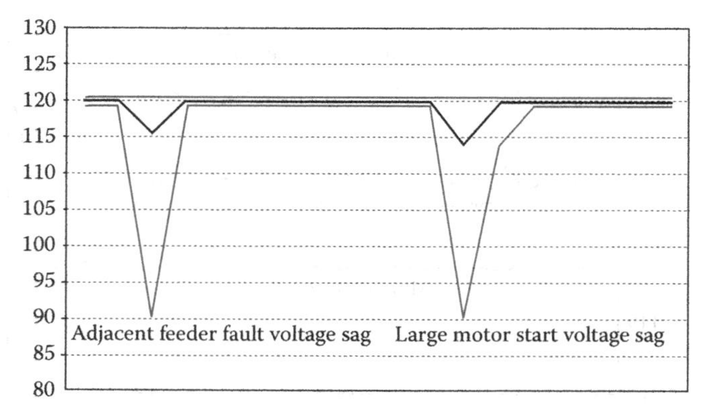

FIGURE 40.4 Min/Max/Average stripchart of the same voltage sags as Figure 40.3. Note that both sags look almost identical. Without the recovery detail found in Figure 40.3, it is difficult to determine a cause for the voltage sags.

Some voltage recorders have user-programmable thresholds, and record deviations at a higher resolution than voltages that fall within the thresholds. These deviations are incorporated into the stripchart, so the user need only open the stripchart to determine, at a glance, if there are any significant events. If there are events to be examined, the engineer can immediately "zoom in" on the portion of the stripchart with the event.

Some voltage recorders do not have user-settable thresholds, but rather choose to capture events based either on fixed default thresholds or on some type of significant change. For some users, fixed thresholds are an advantage, while others are uncomfortable with the lack of control over the meter function. In units with fixed thresholds, if the environment is normally somewhat disturbed, such as on a welder circuit at a motor control center, the meter memory may fill up with insignificant events and the monitor may not be able to record a significant event when it occurs. For this reason, monitors with fixed thresholds should not be used in electrically noisy environments.

40.3.2 Voltage Waveform Disturbances

Some equipment can be disturbed by changes in the voltage waveform. These waveform changes may not significantly affect the rms voltage, yet may still cause equipment to malfunction. An rms-only recorder may not detect the cause of the malfunction. Most line disturbance analyzers have some mechanism to detect and record changes in voltage waveforms. Some machines compare portions of successive waveforms, and capture the waveform if there is a significant deviation in any portion of the waveform. Others capture waveforms if there is a significant change in the rms value of successive waveforms. Another method is to capture waveforms if there is a significant change in the voltage total harmonic distortion (THD) between successive cycles.

The most common voltage waveform change that may cause equipment malfunction is the disturbance created by power factor correction capacitor switching. When capacitors are energized, a disturbance is created that lasts about 1 cycle, but does not result in a significant change in the rms voltage. Figure 40.5 shows a typical power factor correction capacitor switching event.

40.3.3 Current Recordings

Most modern recorders are capable of simultaneous voltage and current recordings. Current recordings can be useful in identifying the cause of power quality disturbances. For example, if a 20% voltage sag (to 80% of full voltage) is accompanied by a small change in current (plus or minus about 30%), the cause of the voltage sag is usually upstream (toward the utility source) of the monitoring point. If the sag is accompanied by a large increase in current (about 100%), the cause of the sag is downstream (toward the load) of the monitoring point. Figure 40.6 shows the rms voltage and current captured during a motor start downstream of the monitor. Notice the large current increase during starting and the corresponding small decrease in voltage.

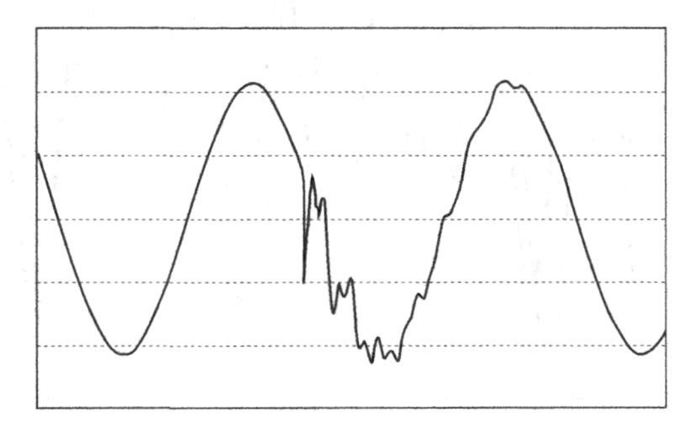

FIGURE 40.5 Typical voltage waveform disturbance caused by power factor correction capacitor energization.

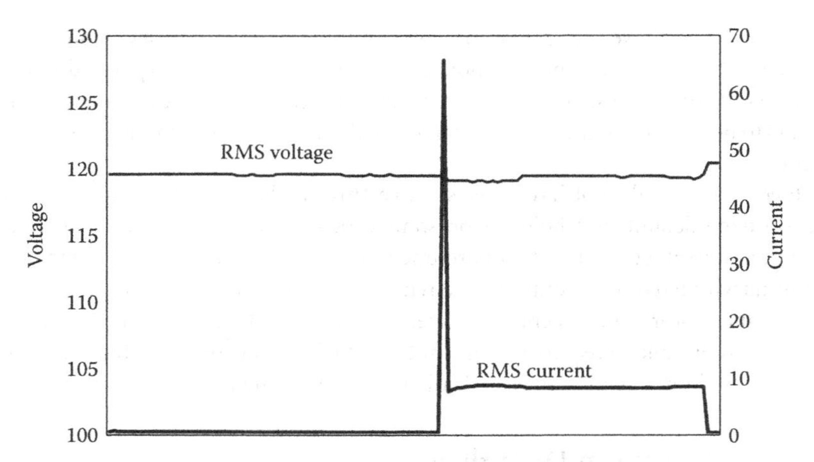

FIGURE 40.6 RMS stripcharts of voltage and current during a large current increase due to a motor start downstream of the monitor point.

Some monitors allow the user to select current thresholds that will cause the monitor to capture both voltage and current when the current exceeds the threshold. This can be useful for detecting over- and under-currents that may not result in a voltage disturbance. For example, if a small, unattended machine is tripping off unexpectedly, it would be useful to have a snapshot of the voltage and current just prior to the trip. A threshold can be set to trigger a snapshot when the current goes to zero. This snapshot can be used to determine if the input voltage or current was the cause of the machine trip.

40.3.4 Current Waveshape Disturbances

Very few monitors are capable of capturing changes in current waveshape. It is usually not necessary to capture changes in current waveshape, but in some special cases this can be useful data. For example, inrush current waveforms can provide more useful information than inrush current rms data. Figure 40.7 shows a significant change in the current waveform when the current changes from zero to nearly 100 A peak. The shape of the waveform, and the phase shift with respect to the voltage waveform, confirm that this current increase was due to an induction motor start. Figure 40.7 shows the first few cycles of the event shown in Figure 40.6.

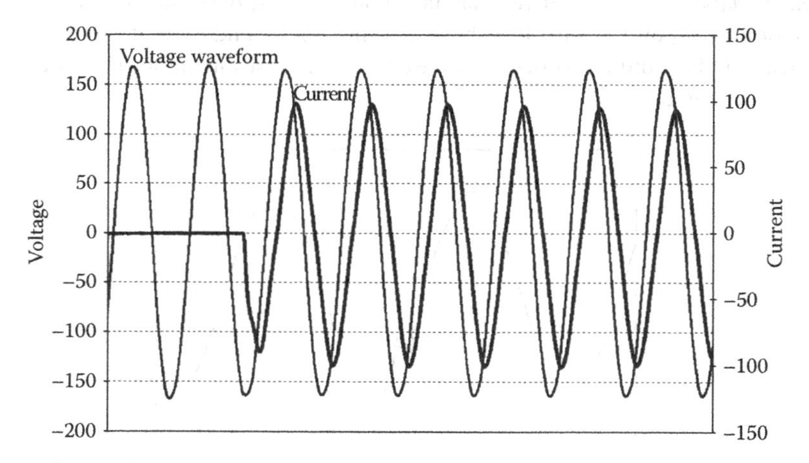

FIGURE 40.7 Voltage and current waveforms for the first few cycles of the current increase illustrated in Figure 40.6.

40.3.5 Harmonics

Harmonic distortion is a growing area of concern. Many commercially available monitors are capable of capturing harmonic snapshots. Some monitors have the ability to capture harmonic stripchart data. In this area, it is critical that the monitor produce accurate data. Some commercially available monitors have deficiencies in measuring harmonics. Monitors generally capture a sample of the voltage and current waveforms, and perform a Fast Fourier Transform to produce a harmonic spectrum. According to the Nyquist Sampling Theorem, the input waveform must be sampled at least twice the highest frequency that is present in the waveform. Some manufacturers interpret this to mean the highest frequency of interest, and adjust their sample rates accordingly. If the input signal contains a frequency that is above the maximum frequency that can be correctly sampled, the high frequency signal may be "aliased," that is, it may be incorrectly identified as a lower frequency harmonic. This may lead the engineer to search for a solution to a harmonic problem that does not exist. The aliasing problem can be alleviated by sampling at higher sample rates, and by filtering out frequencies above the highest frequency of interest. The sample rate is usually found in the manufacturer's literature, but the presence of an antialiasing filter is not usually mentioned in the literature.

40.3.6 Flicker

Some users define flicker as the voltage sag that occurs when a large motor starts. Other users regard flicker as the frequent, small changes in voltage that occur due to the operation of arc furnaces, welders, chippers, shredders, and other varying loads. Nearly any monitor is capable of adequately capturing voltage sags due to occasional motor starts. The second definition of flicker is more difficult to monitor. In the absence of standards, several manufacturers have developed proprietary "flicker" meters. In recent years, an effort has been made to standardize the definition of "flicker," and to standardize the performance of flicker meters. At the time of this writing, several monitor manufacturers are attempting to incorporate the standardized flicker function into their existing products.

40.3.7 High Frequency Noise

Sensitive electronic equipment can be susceptible to higher frequency signals imposed on the voltage waveform. These signals may be induced on the conductors by sources such as radio transmitters or arcing devices such as fluorescent lamps, or they may be conductively coupled by sources such as power line carrier energy management systems. A few manufacturers include detection circuitry for high frequency signals imposed on the voltage waveform.

40.3.8 Other Quantities

It may be necessary to find a way to monitor other quantities that may affect sensitive equipment. Examples of other quantities are temperature, humidity, vibration, static electricity, magnetic fields, fluid flow, and air flow. In some cases, it may also become necessary to monitor for vandalism or sabotage. Most power quality monitors cannot record these quantities, but other devices exist that can be used in conjunction with power quality monitors to find a solution to the problem.

40.4 Summary

Most power quality problems can be solved with simple hand-tools and attention to detail. Some problems, however, are not so easily identified, and it may be necessary to monitor to correctly identify the problem. Successful monitoring involves several steps. First, determine if it is really necessary to monitor. Second, decide on a location for the monitor. Generally, the monitor should be installed close to the affected equipment. Third, decide what quantities need to be monitored, such as voltage, current, harmonics, and power data. Try to determine the types of events that can disturb the equipment, and select a meter that is capable of detecting those types of events. Fourth, decide on a monitoring period. Usually, a good first choice is at least 1 business cycle, or at least 1 day, and more commonly, 1 week. It may be necessary to monitor until the problem recurs. Some monitors can record indefinitely by discarding older data to make space for new data. These monitors can be installed and left until the problem recurs. When the problem recurs, the monitoring should be stopped before the event data is discarded.

After the monitoring period ends, the most difficult task begins—interpreting the data. Modern power quality monitors produce reams of data during a disturbance. Data interpretation is largely a matter of experience, and Ohm's law. There are many examples of disturbance data in books such as *The BMI Handbook of Power Signatures, Second Edition*, and the *Dranetz Field Handbook for Power Quality Analysis*.

Index

A

D

E

F

G

H

I

K

L

M

N

O

R

S

T

U

V

W

Y

Z

9781439856284